Heidemarie Halbritter • Silvia Ulrich
Friðgeir Grímsson • Martina Weber
Reinhard Zetter • Michael Hesse
Ralf Buchner • Matthias Svojtka
Andrea Frosch-Radivo

Illustrated Pollen Terminology

Second Edition

Heidemarie Halbritter
Division of Structural and Functional Botany
Department of Botany and Biodiversity Research
University of Vienna
Vienna
Austria

Friðgeir Grímsson
Department of Palaeontology
University of Vienna
Vienna
Austria

Reinhard Zetter
Department of Palaeontology
University of Vienna
Vienna
Austria

Ralf Buchner
Division of Structural and Functional Botany
Department of Botany and Biodiversity Research
University of Vienna
Vienna
Austria

Andrea Frosch-Radivo
Division of Structural and Functional Botany
Department of Botany and Biodiversity Research
University of Vienna
Vienna
Austria

Silvia Ulrich
Division of Structural and Functional Botany
Department of Botany and Biodiversity Research
University of Vienna
Vienna
Austria

Martina Weber
Division of Structural and Functional Botany
Department of Botany and Biodiversity Research
University of Vienna
Vienna
Austria

Michael Hesse
Division of Structural and Functional Botany
Department of Botany and Biodiversity Research
University of Vienna
Vienna
Austria

Matthias Svojtka
Division of Structural and Functional Botany
Department of Botany and Biodiversity Research
University of Vienna
Vienna
Austria

https://doi.org/10.1007/978-3-319-71365-6

Preface to the First Edition

There are more things in heaven and earth, than are dreamt of in our philosophy.

Shakespeare, Hamlet to Horatio

The principal aim in compiling this book was to provide the reader with first-hand information about the structure and outlook of the extremely manifold pollen in seed plants. This book should not be seen as a mere collection of striking and/or informative light and electron micrographs. Each of the micrographs is intended to convey a specific message related to properties and functions of the pollen grains shown. The authors hope that the book will be useful for experienced researchers as well as for beginners in palynology, but also for medicine, biochemistry, or even for lawyers and artists as an aid and guide for the evaluation and interpretation of pollen features.

Vienna, Austria

Michael Hesse
Heidemarie Halbritter
Reinhard Zetter
Martina Weber
Ralf Buchner
Andrea Frosch-Radivo
Silvia Ulrich

Preface to the Second Edition

The first edition of this book *Pollen Terminology: An illustrated handbook* was extremely successful and well received by the worldwide palynological community. As with the first edition, our main intention when compiling the second edition was to provide both scientists and the public with an easily understandable and primarily illustrative access to the hidden beautiful world of pollen and the fascinating subject of palynology. This new edition titled *Illustrated Pollen Terminology* allowed us to improve many aspects of our book and to illustrate in more detail the important concepts and various preparation techniques applied in (paleo)palynology. It is our hope that this edition will become the guidance tool for all students of palynology, as well as reference work and illustrated encyclopedia for the more advanced scientists.

Vienna, Austria

Heidemarie Halbritter
Silvia Ulrich
Friðgeir Grímsson
Martina Weber
Reinhard Zetter
Michael Hesse
Ralf Buchner
Matthias Svojtka
Andrea Frosch-Radivo

Contents

Part I General Chapters

Abbreviations

DMP	2,2-Dimethoxypropane
$KMnO_4$	Potassium permanganate
LM	Light microscope
PA	Periodic acid
PA+TCH+SP	Thiéry test
PA+TCH+SP (short)	Modified Thiéry test
TCH+SP	Lipid test
Pb	Lead citrate
SEM	Scanning electron microscope
SP	Silver proteinate
TCH	Thiocarbohydrazide
TEM	Transmission electron microscope
U	Uranyl acetate

Prefixes

a-	Prefix meaning absent
bi-	Prefix for two
brevi-	Prefix meaning short
di-	Prefix meaning two
eu-	Prefix meaning true
hetero-	Prefix meaning different
hexa-	Prefix meaning six
homo-	Prefix meaning equal
in-	Prefix meaning absent
infra-	Prefix meaning beneath
inter-	Prefix for in between
intra-	Prefix for within
iso-	Prefix meaning identical
meso-	Prefix meaning middle
micro-	Prefix for small; features between 1 and 0.5 µm
mono-	Prefix meaning one
nano-	Prefix for very small; features between 0.5 and 0.1 µm
panto-	Prefix for global
penta-	Prefix meaning five

peri-	See panto-
poly-	Prefix for many
prae-	Prefix for before
semi-	Prefix for half
stephano-	Prefix meaning equatorially situated
sub-	Prefix for less than
supra-	Prefix for above
syn-	Prefix for together
tetra-	Prefix meaning four
tri-	Prefix meaning three

Introduction

Illustrated Pollen Terminology is a collection of palynological terms and well-illustrated with light and electron microscope images. The focus of this book is on the pollen of seed plants, predominantly angiosperms; therefore, it rarely explains features unique to spores or gymnosperms. A strict rationalization of terms on the basis of practical criteria has been attempted for this book. Where necessary, definitions have been reworded, newly circumscribed, or brought into focus. In addition, consistent application of EM techniques and the nowadays better understanding of pollen features have made redefinition of some terms necessary.

Since 1994, the *Glossary of Pollen and Spore Terminology*, by W. Punt, S. Blackmore, S. Nilsson, and A. L. Thomas, was the standard reference in palynology (Punt et al. 1994). In 1999 the online version by Peter Hoen appeared, with several additions. A new version published in 2007 provided informative schematic drawings containing the essentials of each term, mostly using LM observations (Punt et al. 2007). Although extremely useful for overview purposes, drawings cannot show the full range of features. This can only be achieved with various LM, SEM, and TEM micrographs, which demonstrate the stunning diversity of features as seen in this book.

This book is divided into four parts. The first part comprises the "General Chapters". The first chapter "Palynology: History and Systematic Aspects" provides a comprehensive overview of the history of palynological research and the origin and development of categories and classification systems as well as the systematic value of pollen. The following chapter "Pollen Development" explains the formation and development of a pollen grain (microsporogenesis and microgametogenesis). The third chapter "Pollen Morphology and Ultrastructure" gives a thorough overview on all aspects of pollen features, both structure and sculpture, that need to be considered when studying pollen grains. There are many features observed with microscopes that can be misleading or misinterpreted, most of them have been summarized in the consecutive chapter "Misinterpretations in Palynology." In the fifth chapter "How to Describe and Illustrate Pollen Grains" examples are given as to how to properly present palynological data. The chapter thereafter "Methods in Palynology" includes detailed and illustrated protocols of most methods and techniques used when studying recent and fossil pollen grains with LM, SEM, and TEM. The second part "Illustrated Pollen Terms" is the main part of this book, and comprises 6 chapters. All terms in this part are defined and comprehensively illustrated, and if necessary features are highlighted for easy recognition. The world's most comprehensive database on recent pollen, *PalDat* (http://www.paldat.org/), is the main source of pictures. Each term is illustrated with LM or EM pictures in order to point out the character range of a term (or, more precisely, to show the full range of a single character). In the third part, all terms are listed along with their definition in the "Glossary of Palynological Terms". Numbers following terms refer to the respective page(s) where the terms are discussed. Numbers in bold relate to illustrations in the chapters of the "Illustrated Pollen Terms." The fourth part "Annex" comprises the "Picture Copyrights" and the "Index" listing all plant names occouring in this book.

General Chapters

Contents

Palynology: History and Systematic Aspects

H. Halbritter et al., *Illustrated Pollen Terminology*, https://doi.org/10.1007/978-3-319-71365-6_1

Palynology is the science of palynomorphs, a general term for all entities found in palynological preparations (e.g., pollen, spores, cysts, diatoms). A dominating object of the palynomorph spectrum is the pollen grain. The term palynology was coined by Hyde and Williams (1955; Fig. 1). It is a combination of the Greek verb paluno (παλύνω, "I strew or sprinkle"), palunein (παλύνειν, "to strew or sprinkle"), the Greek noun pale (παλη, in the sense of "dust, fine meal," and very close to the Latin word pollen, meaning "fine flour, dust"), and the Greek noun logos (λογος, "word, speech").

The History of Palynology

Assyrians are said to have known the principles of pollination (they practiced hand pollination of date palms), but it is unclear if they recognized the nature of pollen itself. The invention of the first microscopes and especially the compound microscope in the late sixteenth century represents the starting point of a new fascinating era. Some of the most important findings and scientists within the long tradition of light microscopy are mentioned here. For a more comprehensive overview, see Wodehouse (1935) and Ducker and Knox (1985).

Following the invention of the simple microscope by J. Janssen and Z. Janssen in 1590, the first compound microscope was developed by Hooke (1665). This was an important contribution to the study of pollen morphology. Malpighi in his "Anatomia Plantarum" was the first to describe pollen grains as having germination furrows while Grew noted in his famous work "The Anatomy of Plants" the constancy of pollen characters within the same species (Fig. 2; Grew 1682; Malpighi 1901). They are both considered the founders of pollen morphology. Camerarius described several pollination experiments and communicated the results in his letters about plant sexuality to Valentini (Camerarius 1694). He stated that male "seed dust" is necessary for seed development. Von Linné (also known before his ennoblement as Carl Nilsson Linnœus) first used the term pollen (in 1750). In the 18th and the early nineteenth centuries, there was considerable progress in pollen research and the understanding of pollination. In 1749, Gleditsch demonstrated in a spectacular experiment (Experimentum Berolinense) the central role of pollen in double fertilization. He organized the transport of an inflorescence from a male fan palm in Leipzig to a hitherto "sterile" female fan palm growing in a greenhouse in Berlin. After pollination, the female flowers produced fertile seeds for the first time in the palms lifetime (Gleditsch 1751, 1765). Placing male inflorescences within groups of female date palms for pollination was according to Theophrast already

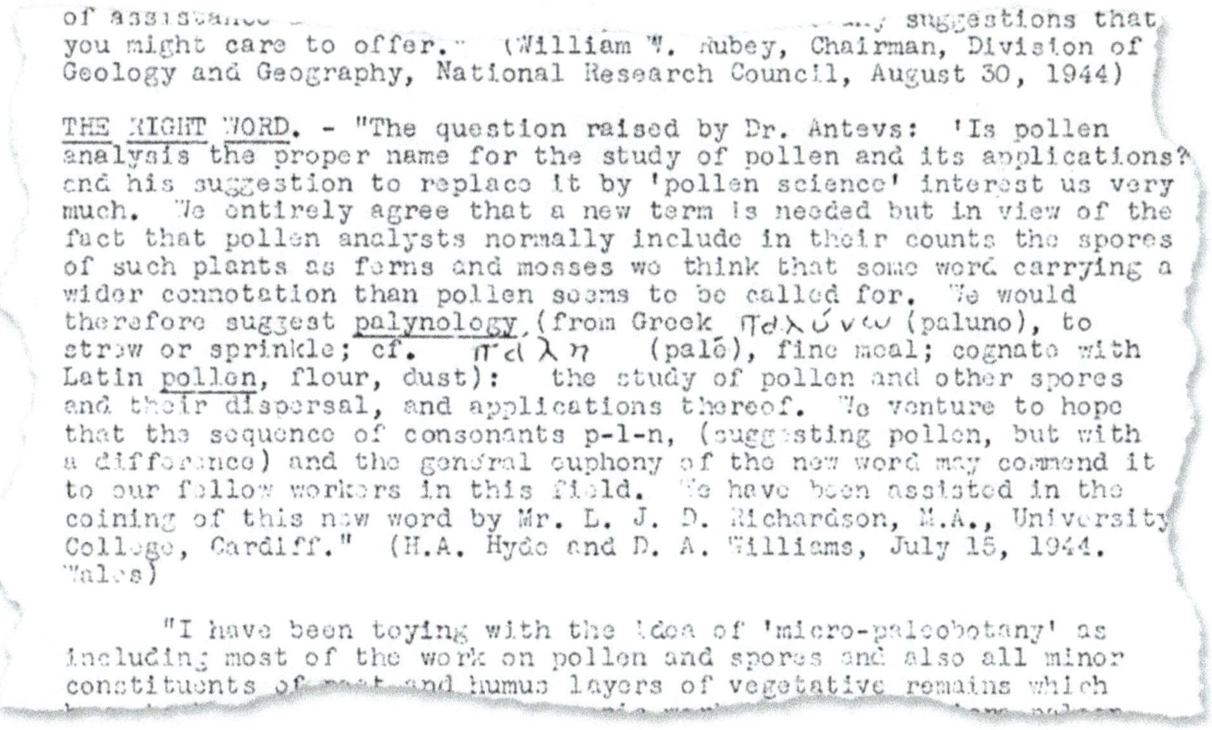
of assistance ... suggestions that you might care to offer." (William W. Rubey, Chairman, Division of Geology and Geography, National Research Council, August 30, 1944)

THE RIGHT WORD. - "The question raised by Dr. Antevs: 'Is pollen analysis the proper name for the study of pollen and its applications?' and his suggestion to replace it by 'pollen science' interest us very much. We entirely agree that a new term is needed but in view of the fact that pollen analysts normally include in their counts the spores of such plants as ferns and mosses we think that some word carrying a wider connotation than pollen seems to be called for. We would therefore suggest palynology (from Greek παλύνω (paluno), to strew or sprinkle; cf. παλη (palē), fine meal; cognate with Latin pollen, flour, dust): the study of pollen and other spores and their dispersal, and applications thereof. We venture to hope that the sequence of consonants p-l-n, (suggesting pollen, but with a difference) and the general euphony of the new word may commend it to our fellow workers in this field. We have been assisted in the coining of this new word by Mr. L. J. D. Richardson, M.A., University College, Cardiff." (H.A. Hyde and D. A. Williams, July 15, 1944. Wales)

"I have been toying with the idea of 'micro-paleobotany' as including most of the work on pollen and spores and also all minor constituents of peat and humus layers of vegetative remains which

Fig. 1 The right word. Excerpt from Hyde and Williams (1955). Pollen Analysis Circular no. 8, p. 6

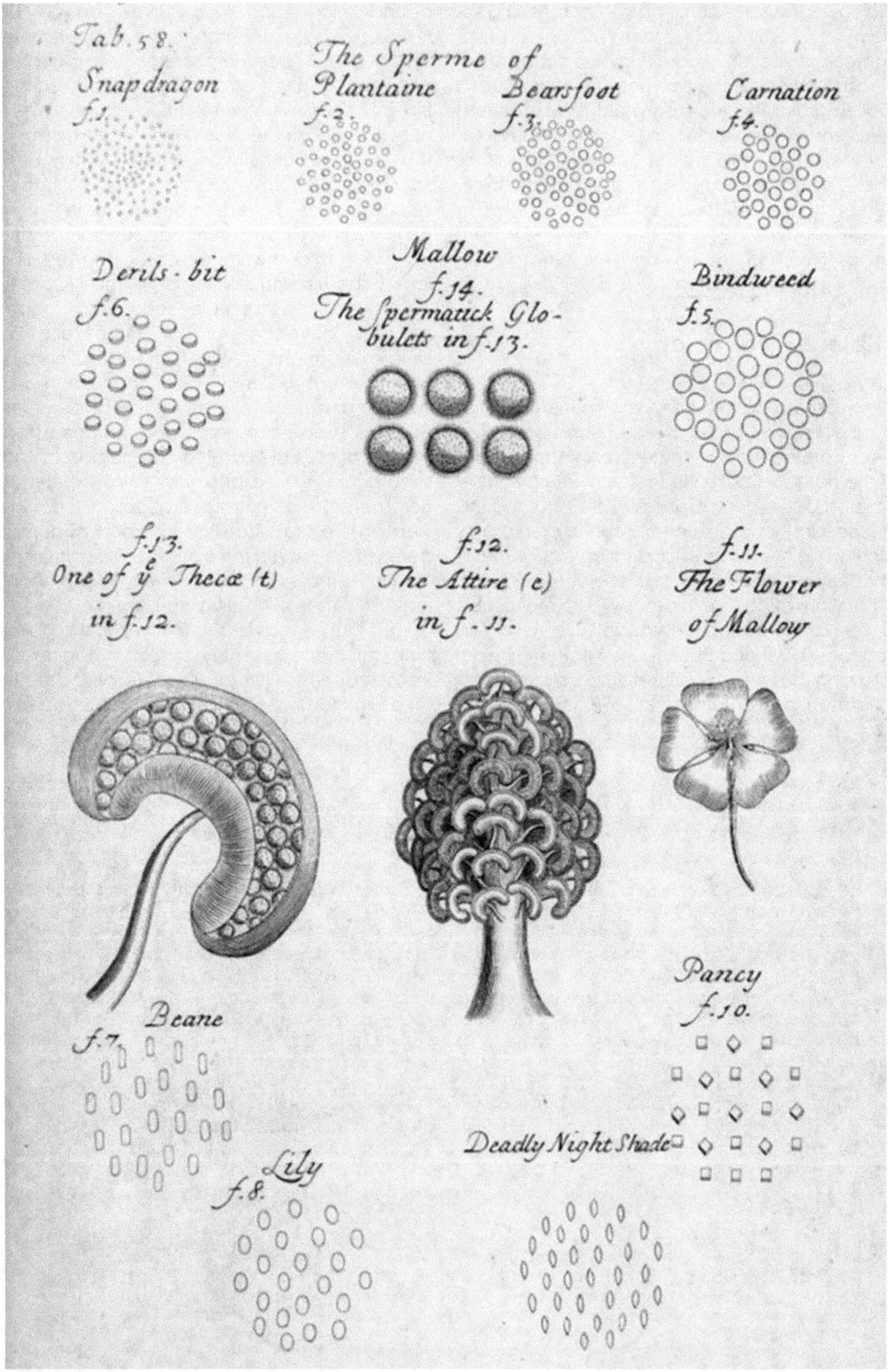

Fig. 2 First drawings of pollen. Grew (1682) "The anatomy of plants"

practiced by the Assyrians and Egyptians. Kölreuter, together with Sprengel, the founder of research on flower biology, perceived the importance of insects in flower pollination and discovered for the first time that pollen plays an important role in determining the characters of the offspring (Kölreuter 1761-1766; Sprengel 1793). Kölreuter (1806, 1811) also discovered that the pollen wall is consisting of two distinct layers and made the first attempt to classify pollen based on their morphology. Sprengel recognized pores and furrows in the pollen wall and demonstrated the effects of cross pollination, dichogamy, and distinguished between entomo- and anemophily (Candolle and Sprengel 1821). Moreover, he also realized that every plant species has a characteristic pollen type (Sprengel 1804).

During the first half of the nineteenth century, some fundamental insights into pollen morphology and physiology were achieved. Purkinje made the first attempt for a palynological terminology by classifying pollen based on their morphology (Purkinje 1830). Wodehouse (1935) pointed out that "Purkinje's system of nomenclature deserved much more attention than was ever given to it by subsequent investigators. A system of this kind, had it been put into use, would have saved much confusion." Brown gave the first description of the origin and role of the pollen tube (Brown 1828, 1833). He credited Bauer as the first observer of the pollen tube's nature, of the double wall in *Asclepias* pollen, and for his minute drawings of *Asclepias* pollen. His brother Bauer, a great botanical artist, was the first to recognize compound pollen in *Acacia* and orchids. Cavolini described and illustrated the filiform pollen of sea grasses *Zostera* and *Cymodocea* (Cavolini 1792).

Göppert and Ehrenberg were the first to describe and depict fossil pollen grains and spores (Göppert 1837, 1848; Ehrenberg 1838). In 1834 von Mohl wrote his fundamental work entitled "Über den Bau und die Formen von Pollenkörner/On the structure and diversity of pollen grains," which was a major contribution to the knowledge of pollen structure and descriptive classification. von Mohl and Fritzsche recognized the principal layers of the pollen wall and published new surveys on pollen morphology (von Mohl 1835; Fritzsche 1837). The term pollenin goes back to von Grotthuss (1814), John (1814), Stolze (1816), and Fritzsche (1834). The terms "exine," "intine," and "Zwischenkörper" were established by Fritzsche and published in his book "Über den Pollen" (Figs. 3 and 4; Fritzsche 1837). He also demonstrated that apertures are predetermined in most angiosperm pollen while others are inaperturate. Zetzsche first coined the term "sporopollenin" to describe the resistant chemical substance present in the outer wall of both pollen grains and spores (Zetzsche and Huggler 1928; Zetzsche and Vicari 1931; Zetzsche et al. 1931). Campbell reported pollen of seagrasses (*Naias* and *Zannichellia*) to be thin walled, without exospore (exine), and two-celled. Moreover, Campbell described the mitotic division of the generative cell into two (sperm) cells (Campbell 1897). Hofmeister and Strasburger provided ground-breaking insights into the development and internal structure of pollen and fertilization (involves the fusion of a single sperm nucleus and the egg nucleus) and investigated the bi- and tricellular pollen condition of many angiosperms (Strasburger 1884; Hofmeister 1849). The role of the second sperm nucleus in the pollen tube remained unexplained until double fertilization was discovered by Guignard (1891, 1899); Nawaschin (1898). Nägeli studied the ontogeny of pollen grains within anthers and was the first to recognize the callose wall (Nägeli 1842). Schacht described differences in exine patterning, exine thickness, and apertures covered by an operculum. He also used cytochemical staining techniques to detect pollen reserves. He was also the first to cut sections of embedded pollen with razor blades for anatomical studies (Schacht 1856/59). Strasburger described the basic concepts of pollen wall development already in 1889, but major break-through in pollen wall ontogeny was achieved much later by Heslop-Harrison (1975). The first successful classification of orchidaceous plants based on pollen features was made by Lindley (1836). Later, Fischer recognized the potential of pollen morphology in aiding the phylogenetic position of angiosperms (Fischer 1890).

Paleopalynology was established at the end of the nineteenth century, when P. Reinsch published the first photomicrographs of fossil pollen and spores from Russian coals (Reinsch 1884). He also described methods for the extraction of palynomorphs from coal samples with concentrated potassium hydroxide (KOH) and hydrofluoric acid (HF). Von Post published the first pollen diagram (profile) using exclusively arboreal pollen (von Post 1916). Already before and especially after the Second World War, Schopf as well as Potonié published their impressive publications devoted to fossil spores and pollen (e.g., Potonié 1956; Schopf 1957, 1964). Schopf established the systematic study of palynomorphs, while Potonié was one of the first who recognized the stratigraphic value of paleopalynology, applying his "turmal classification" system (Potonié 1934; see also "The Treme System and the NPC-Classification" below). The rise of stratigraphic palynology started shortly before 1950 and played a prominent role in petroleum explorations during the second half of the twentieth century (Manten 1966).

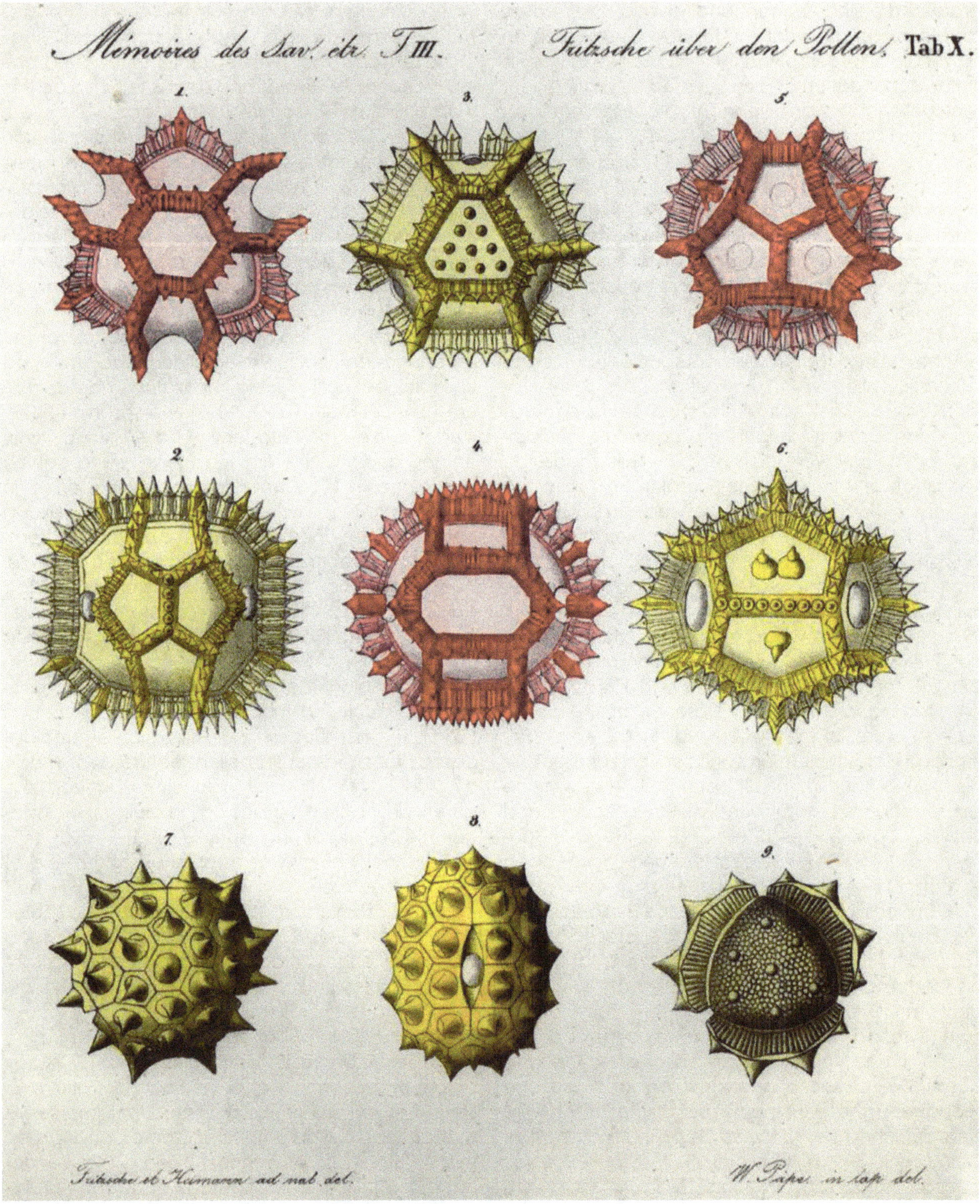

Fig. 3 **Detailed drawings of pollen**. Fritzsche (1837) "Über den Pollen"

The key role of palynology in stratigraphy depends upon the fact that the natural biopolymer sporopollenin in the spore/pollen walls is extremely resistant; thus, pollen/spores are often abundantly preserved in sedimentary rocks.

The twentieth century up to ca 1960 was dominated by the skillful use of the LM, with many new findings; for example, the LO-analysis, a method for analyzing patterns of exine organization by light microscopy, focusing at different levels distinct

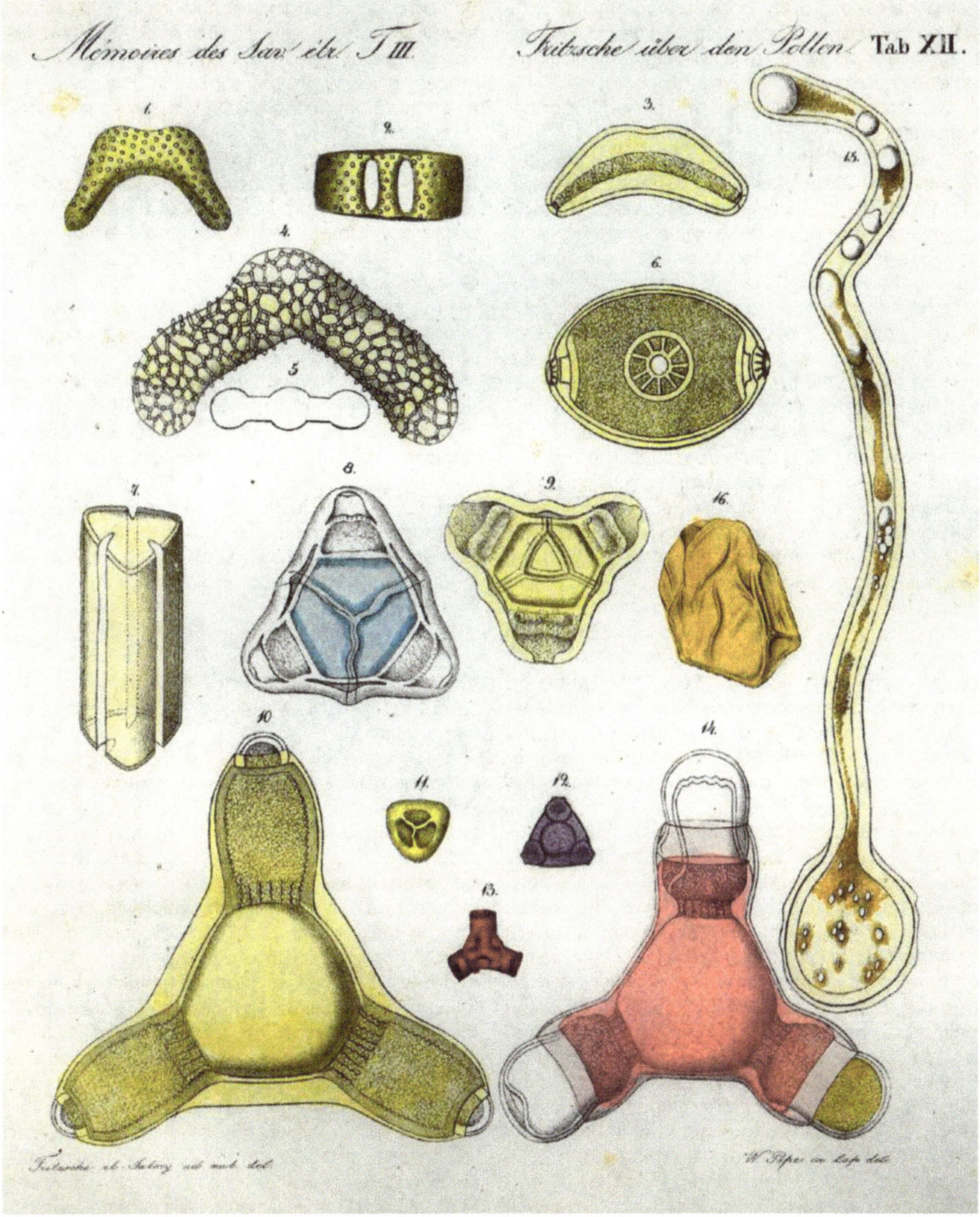

Fig. 4 First detailed drawings of pollen. Fritzsche (1837) "Über den Pollen"

features appear bright (*L* = Lux) or dark (*O* = Obscuritas). Textbooks by Wodehouse, Erdtman, or Fægri and Iversen summarized the knowledge of palynology from that time, but are still in good use (Erdtman 1943, 1952, 1957, 1969; Fægri and Iversen 1950, 1989; Wodehouse 1935). During this time palynology also became more diverse and applied in numerous fields among others: aeropalynology, biostratigraphy, copropalynology, cryopalynology, forensic palynology, iatropalynology, melissopalynology, paleopalynology, archeology, paleoclimatology, and palynotaxonomy.

Electron Microscopy with its two most important instrument types, the Transmission Electron Microscope (TEM) and the Scanning Electron Microscope (SEM), facilitated major breakthroughs in palynology. The TEM revealed new and stunning insights into pollen wall development and stratification. This prompted authors to publish new descriptions and create new terms. As pointed out by Knox: "The terminology applied to the pollen wall is daunting, especially as it has been developed from early light microscopy work, and then transposed to the images seen in the transmission and scanning electron microscopes" (Knox 1984, p. 204).

One of the first reports on the ultrastructure of recent pollen using TEM were published by H. Fernandez-Moran and A. O. Dahl (1952), and by K. Mühlethaler (1953). The first reports on the ultrastructure of fossil pollen were published by Ehrlich and Hall (1959; Pettitt and Chaloner (1964). During the 1950s and early 1960s considerable progress in TEM preparation methods (from fixation to microtome sectioning and staining) took place. EM-based information on ornamentation details of pollen grains was rare up to the mid-1960s. Only TEM-based casts or replica methods were available, all of them with limited resolution and depth of focus (e.g., the single-stage carbon replica technique; Mühlethaler 1955; Bradley 1958; Rowley and Flynn 1966). The time-consuming and laborious TEM replica procedures were an obstacle to extensive surveys of pollen morphology and later replaced by SEM (Harley and Ferguson 1990). The introduction of SEM in palynology in the mid of the 1960s was a key innovation in the study of the fine relief (sculpture) of pollen and spore surfaces. Advantages of SEM include the relatively simple and rapid preparation methods and the supreme depth of focus. SEM was considered from the very first moment as the quantum leap in EM (Hay and Sandberg 1967). The first SEM micrographs of pollen grains were published by Thornhill et al. (1965) and Erdtman and Dunbar (1966). Since then palynologists have been provided with a plethora of beautiful micrographs. Like Blackmore noted "The scanning electron microscope has provided a greater impetus to palynology than any other technical development during the history of the subject." Blackmore (1992). The LM with basic and advanced equipment, such as the fluorescent super-resolution microscopy, is overcoming the Abbe limit of LM resolution (especially STED microscopy, Hell 2009). The super-resolution LM and the two main types of EM form an expedient combination of imaging techniques. The LM remains the "workhorse method" (Traverse 2007; see the compendia by Reille 1992, 1995, 1998), but is limited regarding various morphological and structural features. Therefore, the role of SEM as an essential part in illustrating exine sculpture and ornamentation cannot be overrated (Harley and Ferguson 1990). The TEM still plays an important role, for example, in elucidating the complex steps of exine formation and development (e.g., Blackmore et al. 2007, 2010; Gabarayeva and Grigorjeva 2010; Gabarayeva et al. 2010).

The first and especially the second half of the twentieth century saw palynology at its peak, combining light microscopy with electron microscopy techniques. In addition to the above-mentioned scientists, other great palynologists have also promoted our science toward its present multifaceted appearance. These include among others: B. Albert, H.-J. Beug, G. El Ghazali, F. Firbas, M. Harley, J. Jansonius, W. Klaus, G. O. W. Kremp, B. Lugardon, S. Nadot, A. Maurizio, J. Muller, S. Nilsson, J. R. Rowley, J. J. Skvarla, H. Straka, G. Thanikaimoni, R. H. Tschudy, M. van Campo, T. van der Hammen, and A. Le Thomas.

Categories, Classification Systems and Systematic Value of Pollen Features

For the scientist, categories are essential for classifying natural characters in their diversity, defining their range and placing them in a systematic order. In addition to the theoretical concept, categorization always depends on the manner in which a feature is perceived: i.e. on the **visibility** of a feature, and/or their specific value. Categorization also greatly depends on the technical equipment and method(s) used, as well as on the **subjective interpretation** of character(s) (see "Methods in Palynology"). Thus, categorization of features is difficult to standardize. An example is the category **pollen size**: there is not just a natural size variation within a single anther/flower/taxon, dimensions may also vary depending on the preparation method(s) used, and the observer's evaluation. Moreover, sometimes the size of a pollen grain is just at the boundary between two adjacent pollen size categories (for size categories: see "Pollen Morphology and Ultrastructure").

When describing and categorizing pollen, two basic groupings are known from the literature: pollen type and pollen class. **Pollen type** is a general term categorizing pollen grains by a distinct combination of characters and is used in connection with systematics, affiliating the pollen type with a distinct

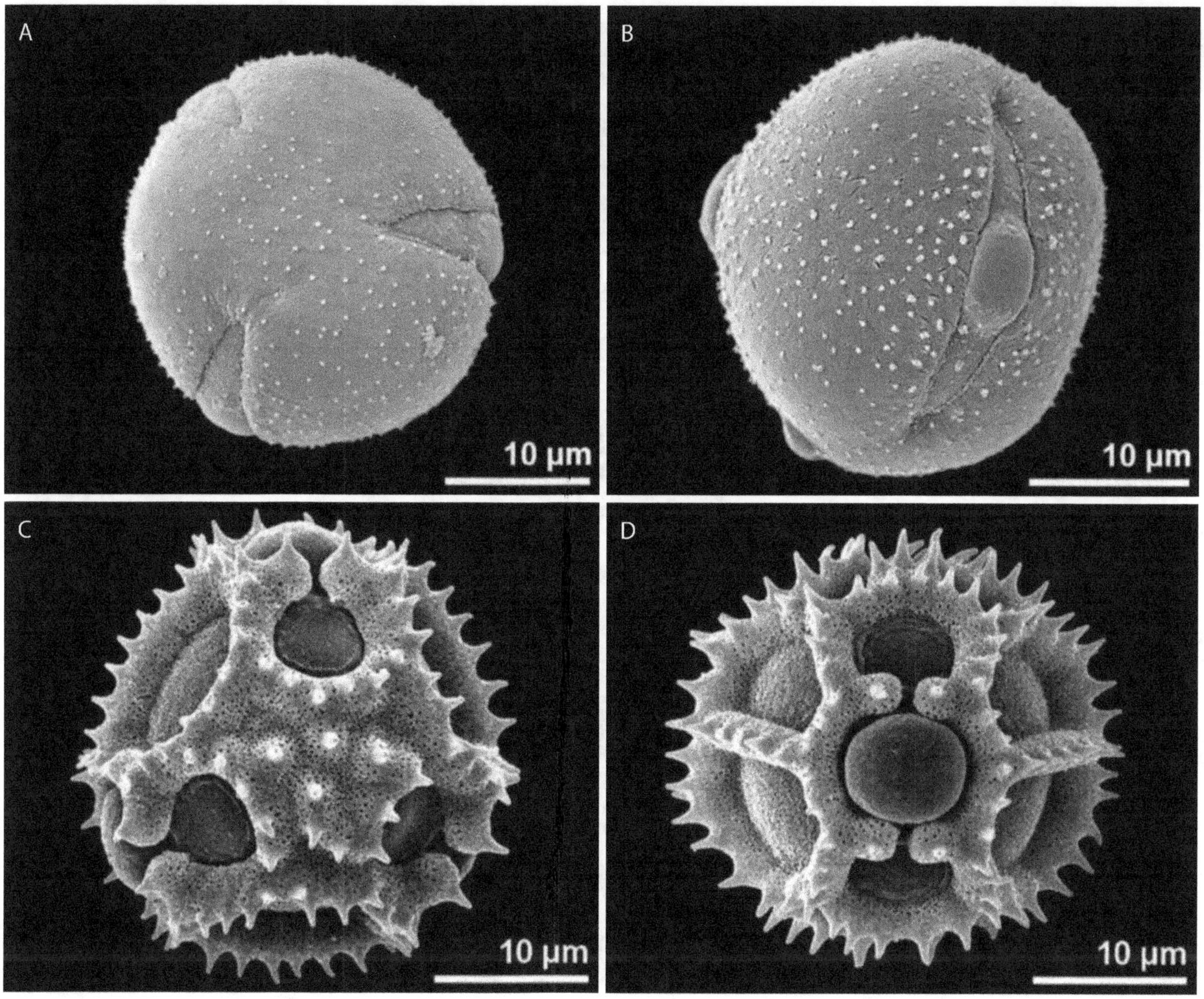

Fig. 5 **Pollen type vs pollen class. A-B.** *Polygonum aviculare*, Polygonaceae; all *Polygonum* pollen sharing the combined features observed here belong to the *Polygonum* aviculare type. This pollen can also be included in the pollen class "tricolporate". **C-D.** *Leontodon saxatilis*, Asteraceae; all Asteraceae pollen sharing the combined features observed here (lophate, tricolporate, echinate) belong to the *Leontodon* type, characteristic for the "Liguliflorae" group within the Asteraceae. This pollen can also be included in the pollen class "tricolporate"

taxon/a (e.g., *Polygonum aviculare* type/*Leontodon* type, Fig. 5). The term "pollen type" is sometimes (colloquially) misused: for example, *Croton* type, which is a distinct feature of ornamentation and is correctly termed *Croton* pattern.

Pollen class is an artificial grouping of pollen grains that share a single or more, distinctive characters (see "Illustrated Pollen Terms"). Pollen classes can refer to pollen units (e.g., polyads, tetrads), to shape (e.g., saccate, polygonal, heteropolar, arcus), to aperture type and location (e.g., inaperturate, sulcate, ulcerate, colpate, colporate, porate, synaperturate, spiraperturate), or to an extremely distinctive ornamentation character (e.g., lophate, clypeate). These classes can be useful in identification keys as they have a good diagnostic, although mostly no systematic, value. In general, a pollen grain may belong to more than one pollen class; in such cases, the more significant feature should be ranked first (e.g., *Pistia*: plicate-inaperturate, *Hemigraphis*: plicate-colporate, *Typha*: tetrads-ulcerate, *Rhododendron*: tetrads-colporate).

Many terms in palynology were coined at a time when only LM observations were available. Mainly for historical reasons, inconsequent nomenclatural applications, enumerations of synonyms, and even differing definitions have been found for one and the same term. During the twentieth century, questions of terminology became more and more problematic. The main reasons were the increasing numbers of publications in palynology, dealing with sometimes insufficiently described or "uncommon" pollen features, and simultaneously the advent of manifold applied fields of palynology. For various

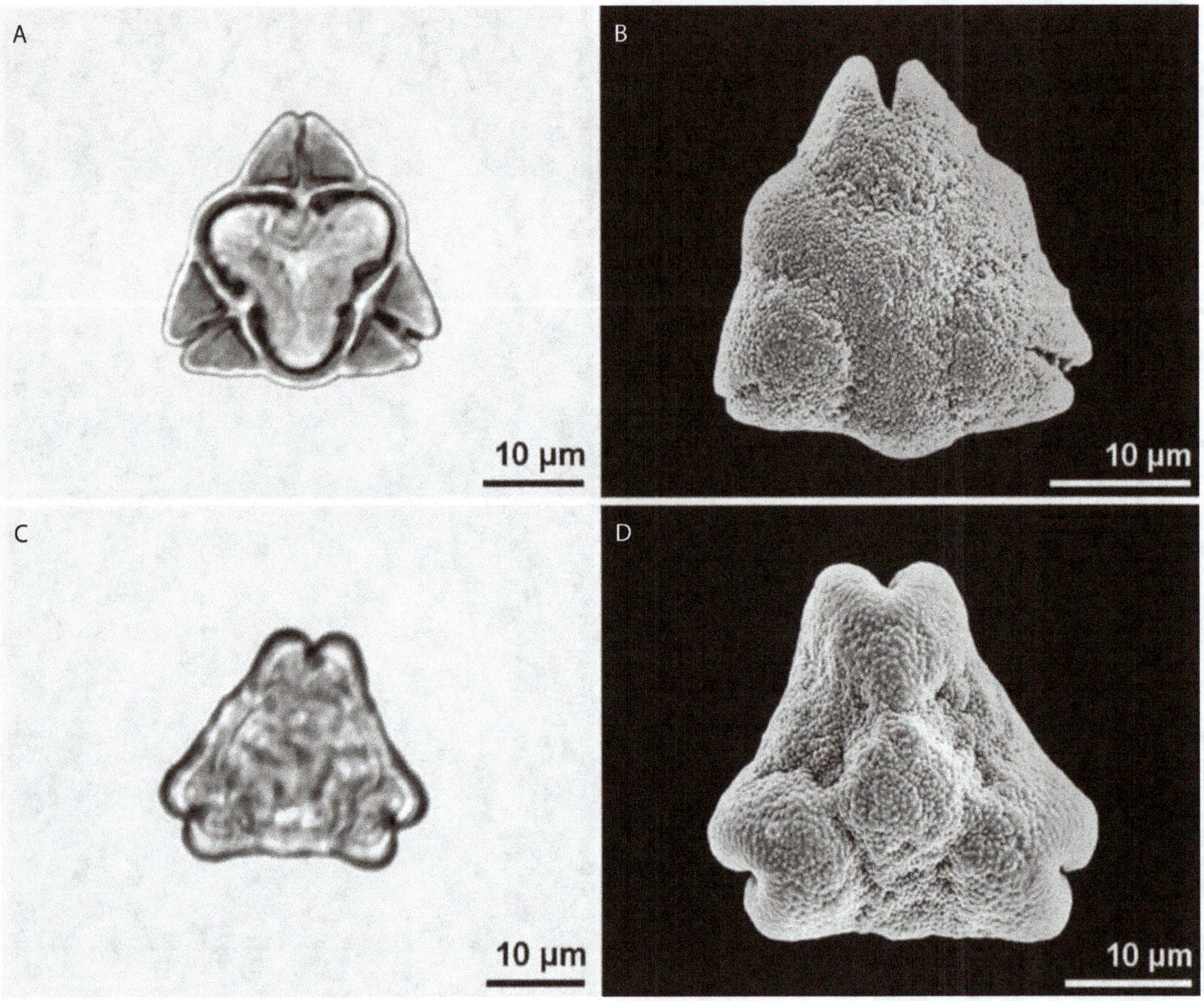

Fig. 6 Nomenclature in Paleopalynology. A. *Oculopollis* sp., fossil, Upper Cretaceous, Hungary, polar view. **B.** *Oculopollis* sp., fossil, Upper Cretaceous, Hungary, polar view. **C.** *Trudopollis* sp., fossil, Upper Cretaceous, Hungary, polar view. **D.** *Trudopollis* sp., fossil, Upper Cretaceous, Hungary, polar view

reasons, nearly all authors used their own terminology. Nonetheless, in the 1950s attempts were made to restrict the wording and to state the definitions of terms more precisely (Erdtman 1947; Erdtman and Vishnu-Mittre 1956). A limited list of pollen morphological terms and definitions was published as early as 1950 by Iversen and Troels-Smith. Later, Kremp (1968), in his famous encyclopedia, provided a monumental enumeration of all known terms. Reitsma (1970) took the first resolute step to overcome the problem of synonyms in palynological terminology, though unfortunately not taking into account the variation range of palynological features. Fægri and Iversen (1989, 4th ed.) restricted their glossary of terms exclusively used in their book. Moore et al. (1991, 2nd ed.) provided a glossary of selected terms used in their pollen and spore keys. Standardization came with the glossary by Punt et al. (1994, 2007). The main advance of their concise and comprehensive terminology is the consistent use of drawings and the critical comments on terms.

A complex category issue in (Paleo-) Palynology is the nomenclature question. In Paleopalynology, for morphotaxa often form-generic names are used. The nomenclature of form-genera is either artificial when the relationship is not known at all (e.g., *Oculopollis* and *Trudopollis* from the Normapolles group, Fig. 6), or semi-biological, when reference to an extant taxon is suspected but not proven (e.g., *Liliacidites*). However, if reference to extant taxa is certain, then a biological nomenclature is possible (e.g., *Quercus* sp.).

The Turmal System

A quite different classification and nomenclature is **Potonié's turmal system**. This is an artificial, informal, neutral suprageneric classification scheme for fossil (especially Carboniferous or Permian) pollen and spores. It is subdivided into a hierarchy of progressively finer units (ranks): anteturma, turma, subturma, infraturma, subinfraturma, and corresponds mostly to morphological features (for details see Traverse 2007).

The Treme System and the NPC-Classification

The "**-treme system**" of aperture configuration as an alternative or an addition to the traditional nomenclature was introduced by Erdtman and Straka (1961). The suffix -treme is derived from trema (pl. tremata) and is synonymous with aperture. In combination with prefixes such as cata-, ana-, zono-, and panto-, the position of germination sites in relation to pollen polarity can be designated. Catatreme indicates the proximal, anatreme the distal, zonotreme the equatorial, and pantotreme the global position of apertures. Other prefixes such as mono-, di-, tri-, tetra- indicate the number of apertures irrespective of their position.

The **NPC-classification** by Erdtman and Straka (1961), resting upon the -treme system, is a morphological system for classifying pollen and spores. This system is based on the aperture features: their number (N), position (P), and character (C). Their NPC-system for spore/pollen classification was used as a diagnostic tool in systematics. As an example, the three apertures (N_3) of pollen grains, having a zonotreme position (P_4) and being colporate (C_5) have the NPC-formula 345 (Fig. 7; Erdtman and Straka 1961). Taxa with the same general NPC-formula are grouped together, those showing a different formula, separately. This system does not work in, e.g., heteroaperturate or inaperturate (formula 000) pollen, or pollen tetrads. Unfortunately, the NPC-system ignores other pollen characters including shape and ornamentation that are indispensable for a complete description.

Systematic Value of Pollen Features

One of the main research interests in palynology focuses on the taxon-specific patterns of the pollen wall, how they developed and evolved. Moreover, pollen can provide phylogenetic evidence important to plant systematics (Hesse and Blackmore 2013). The reconstruction of phylogenies has continuously developed. The advances in modern phylogenetic approaches are resulting in constant changes in plant systematics, even whole genomes are being used together with multiple DNA analyses for a better insight into relationships (Stuessy and Funk 2013). Critically evaluated pollen features may be a useful tool for systematics with a significant diagnostic value, supporting or contradicting the results of molecular studies ("The palynological compass" sensu Blackmore 2000; Hesse and Blackmore 2013). Palynological features are very valuable, especially in delimiting taxa (Ulrich et al. 2012). Regarding multiple-gene tree studies with conflicting results, pollen data combined with other morphological evidence (e.g., floral characters) have more recently become an important indicator of which tree may be the best representative (Stuessy and Funk 2013; Ulrich et al. 2012, 2013). Furthermore, pollen morphological studies proved to be indispensable for the understanding of evolutionary

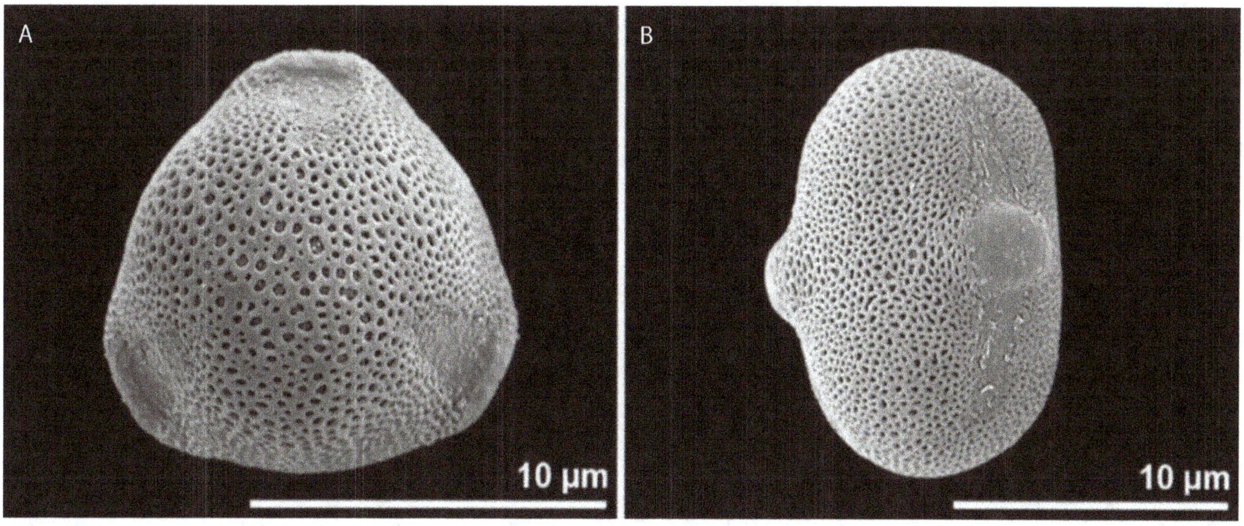

Fig. 7 NPC-classification of pollen. A-B. *Androsace chamaejasme*, Primulaceae, a tricolporate pollen with the formula $N_3P_4C_5$

processes and systematics. For taxonomic studies, pollen features that have value for the lower and higher taxonomic levels should be obtained by a combined study using LM, SEM, and TEM (Stuessy 1979).

Alternation of generations is a unique feature of plants that occurs in green algae, mosses, ferns, gymnosperms, and angiosperms. Pollen grains develop in anthers as the result of meiosis and mitoses (two in angiosperms, three to five in gymnosperms) and represent an extra generation, the highly reduced male gametophyte. Therefore, pollen grains are not simply small parts of a plant like leaves or seeds; they are the complete (hidden) haploid counterpart to the more dominant plant, which represents the diploid generation (Kessler and Harley 2004). During dispersal, pollen grains are completely separated from the parent plant and perfectly adapted for their role — the transfer of male genetic material — and are able to resist hostile environmental stresses on their way to the female flower parts. Usually, pollen does not suffer to the same extent from the various and harsh selective pressures to which the diploid plant is subjected. Because selective pressures (e.g., temperature, precipitation) upon pollen characters are predominantly absent or low, compared to those on the diploid plant, pollen features may remain constant for millions of years, meaning pollen features can be conservative and of taxonomic value (Wodehouse 1928, 1935; Hao et al. 2001; Grímsson et al. 2014, 2016, 2017a, b). Therefore, identical and rare conditions in fossil vs recent pollen probably belong to only one group and were not invented independently in distant groups (e.g., fossil *Spinizonocolpites* pollen and recent *Nypa* pollen, Arecaceae; Zetter and Hofmann 2001; Gee 2001). Selective pressures might concern especially the pollen aperture number, but also the pollen sculpture and the mode of pollination ecology (Furness and Rudall 2004). Pollen features are, if used for a systematic purpose, at least as important as any other morphological character of the diploid generation. For this reason pollen morphology claims a crucial role in, e.g., systematics and palynostratigraphy, for example in elucidating the early history of angiosperms. Angiosperm pollen from the Early Cretaceous are usually sulcate (typical for basal angiosperms) with a columellate infratectum (which is restricted to angiosperms). The first appearance of dispersed tricolpate pollen, typical for eudicots, is not known before the latest Barremian, is rare in the Aptian of Southern Laurasia and Northern Gondwana, but is ubiquitous in the Albian of both provinces. Tricolporate pollen appears first in the late Albian, and triporate pollen in the middle Cenomanian (Doyle and Endress 2010; Friis et al. 2011; Doyle 2012). For a detailed overview of structural pollen diversification and of the stratigraphic appearance of major angiosperm pollen types during the Cretaceous, see Friis et al. (2011) and Mendes et al. (2014).

Palynological data may be helpful at all levels of systematics, especially in angiosperms (c.f. Stuessy 2009). When pollen of a taxon (representing family/ies or genus/era) is characteristic and similar among species they are termed **stenopalynous** (Fig. 8), and occur, for example, in Poaceae, Lamiaceae, Asclepiadaceae, Brassicaceae, Asteroideae, and Cichorioideae. On the contrary, **eurypalynous** (Fig. 9) taxa are heterogeneous and pollen can vary among others in size, aperture, and in exine stratification. Examples for eurypalynous groups are Acanthaceae (Sarawichit 2012) and Araceae (Harley and Baker 2001; Ulrich et al. 2017).

At the highest taxonomic level (e.g., angiosperms vs gymnosperms, dicots vs monocots), a columellate exine condition occurs exclusively in angiosperms. A lamellate endexine is typical for gymnosperms, whereas the angiosperm endexine is usually not lamellate, except in immature stages (*Orobanche hederae*). But in very few cases there is a continuously lamellate endexine present, like in *Ambrosia* (Furness and Rudall 1999a, b, Weber and Ulrich 2010). In inaperturate pollen of Araceae the endexine is exceptionally thick and spongy, which may be a functional benefit and of systematic value. A strong phylogenetic signal comes from the aperture arrangement: the "tricolpate" condition is a synapomorphy for eudicots, tricolporate pollen occurs only in core eudicots, while sulcate pollen is a plesiomorphic condition in basal angiosperms (Nadot et al. 2006). Palynologists have long wondered about the two fundamental evolutionary shifts occurring at the base of the eudicot clade, both in aperture position (from distal to equatorial) and number (from one to three or more). Most probably, these changes in pollen morphology have a systematic and simultaneously a functional background. The shift from a distal, single aperture to equatorially or globally situated apertures, increases the number of possible germination sites (Furness and Rudall 2004). Pollen morphology does not support sharp delimitation between dicots and monocots, as dicotyledonous pollen characters also occur in some monocots and conversely. In early-diverging angiosperms the formation of pollen features appears to be more plastic than in dicots (especially in eudicots). Manifold combinations of pollen features are typical for basal angiosperms and even for the most basal eudicots, the Ranunculales. All of them are more or less eurypalynous. In contrast, late-divergent eudicots are often stenopalynous and

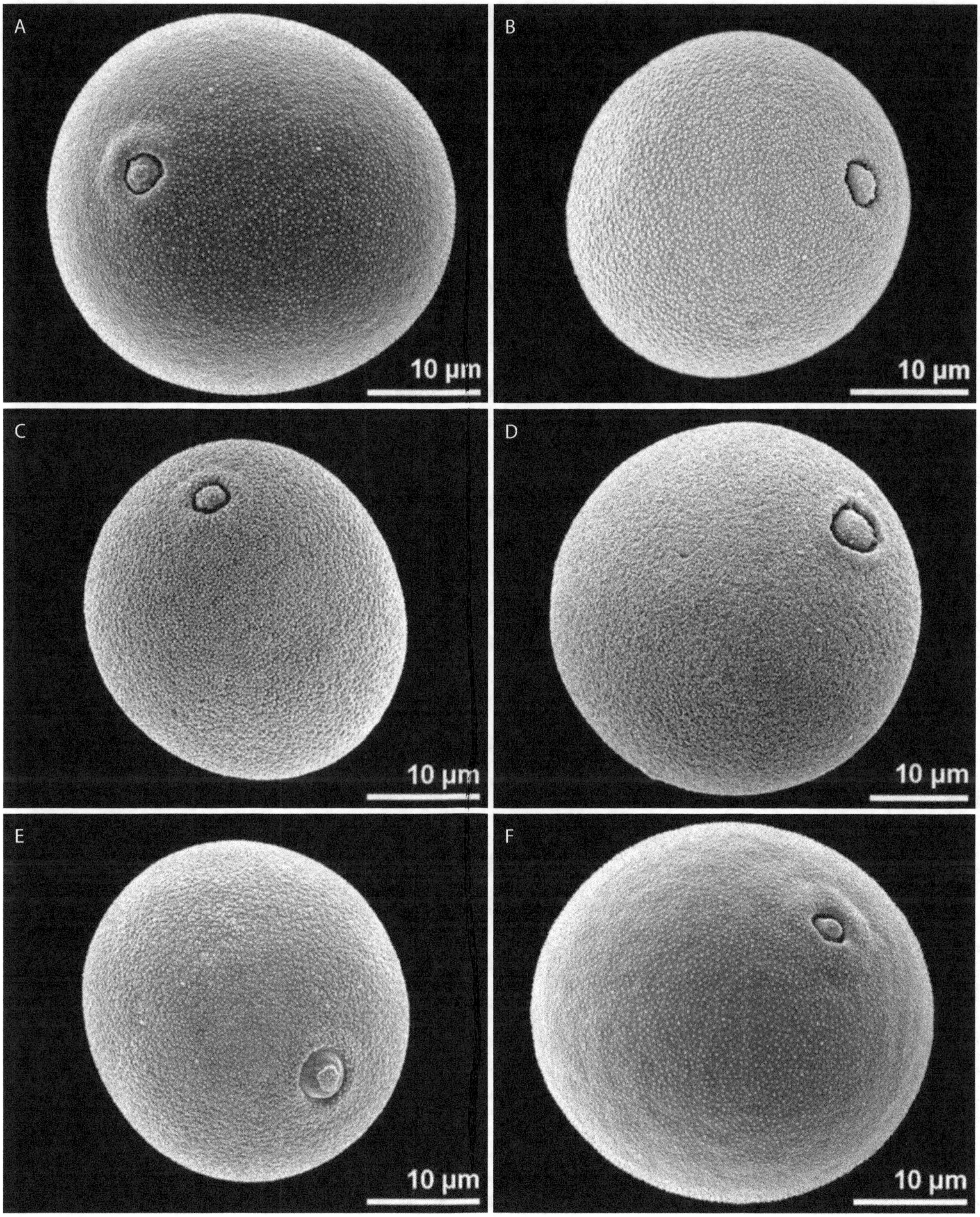

Fig. 8 Stenopalynous taxa (family level). Pollen of different Poaceae all look very similar, for example in *Alopecurus* (**A**), *Cutandia* (**B**), *Dactylis* (**C**), *Fargesia* (**D**), *Poa* (**E**), *Sesleria* (**F**) the pollen is spherical, ulcerate with nano-sized sculpture elements

Fig. 9 Eurypalynous taxa (family level). A-F. Pollen of different Araceae genera look very different. **A.** *Ambrosina* pollen is plicate and inaperturate. **B.** *Dracunculus* pollen is verrucate and inaperturate. **C.** *Pinellia* pollen is echinate and inaperturate. **D.** *Cyrtosperma* pollen is reticulate and ulcerate. **E.** *Anthurium* pollen is reticulate-microechinate and diporate. **F.** *Monstera* pollen is psilate, with ring-like aperture

appear somewhat "poor" regarding the diversity of pollen features (Hesse et al. 2000). In general, the richness and variation of morphological features in pollen decreases in eudicots (Furness and Rudall 1999a). In Alismatales, many pollen features are adaptive and related to their aquatic/semiaquatic habitat, e.g., thin-walled, inaperturate pollen have evolved iteratively, even filamentous pollen is not rare (Furness and Banks 2010).

Fine example for adaptive and simultaneously systematic values is the ring-like aperture found especially in monocots, while only few occur in dicots. A ring-like aperture was probably the best way to a target-oriented harmomegathic movement, to contract or expand a large area adapted for pollen tube formation. This type of aperture might be relict of early angiosperms, before the advent of the "eudicot-tricolpates".

Examples for diagnostic features at lower taxonomic levels (family) are saccate pollen, typical for Pinaceae and Podocarpaceae. A small papilla is characteristic for Taxodioideae pollen (see "Illustrated Pollen Terms"). Another example for a strong phylogenetic signal comes from an aroid subfamily, the aperigoniate Aroideae (Araceae). They are characterized by several synapomorphies: inaperturate pollen, often with an outermost non-sporopollenin layer (exine absent) and a thick spongy endexine. The absence of callose in pollen development is the reason for this uncommon wall structure, that differs from all other currently known angiosperms (Anger and Weber 2006; Hesse 2006a, b).

At the lowest taxonomic level (genus, species) a combination of distinct morphological and structural features usually refers to a particular genus or species. Even very inconspicuous features can represent an example of systematic value, like the *Pinus* subgenus *Strobus* (Haploxylon) type and the *Pinus* subgenus *Pinus* (Diploxylon) type (see "Pollen Morphology and Ultrastructure"). Another example is the large genus *Amorphophallus* (Araceae), showing high diversity in ornamentation (e.g., Ulrich et al. 2017). As a result of the harmomegathic effect, the shape of pollen may change, which is enabled by the elasticity of the exine and infoldings of the apertures. The aperture type and arrangement may lead to characteristic infoldings. Therefore, the shape of pollen in dry state can be typical for a family or genus (e.g., Halbritter and Hesse 2004). For example, tricolporate pollen of the genus *Chaenarrhinum* (Plantaginaceae) is heteropolar. The heteropolarity is only apparent in dry condition. Also, tricolpate pollen of Lamiaceae is highly characteristic in dry condition: it is prolate, extremely flattened, and with apertures arranged in a very distinct manner (Fig. 10; see also "harmomegathic effect" in "Pollen Morphology and Ultrastructure").

Future Perspective

Nowadays, palynology serves as an indispensable tool for various applied sciences such as systematics (Doyle and Endress 2010; Dransfield et al. 2008), melissopalynology (Jones and Bryant 1996), and forensics (e.g., Mildenhall et al. 2006; Bryant 2013; Weber and Ulrich 2016), but should also stand alone as a basic field in science. In general, compared to the sporophyte the male gametophyte in seed plants is poorly investigated. From ca. 260.000 to 422.000 plant species (e.g., Thorne 2002; Govaerts 2003; Scotland and Wortley 2003; *The Plant List* currently accepts 350.699 species) only about 10% have been studied with respect to pollen grain morphology, and regarding pollen ultrastructure it is even much less. Therefore, it is important to continue classical and more advanced palynological studies.

Despite the long tradition of palynology and its application in many fields, it should be considered why it is important and where it is heading in the near future. In the twenty-first century, no matter what role palynology will play, being a basic field of science or more probably a bundle of applied fields, a vital issue will be the increase of our knowledge of pollen grains and in this context the enhancement of pollen terminology. Online pollen databases (efficient in data storage, data transmitting and dissemination) will get more and more important for the exchange of pollen and spore information (for example, *PalDat;* Weber and Ulrich 2017). Journals are nowadays published simultaneously in print as well as in electronic format, both have manifold advantages and disadvantages. Nevertheless, illustrated monographs, like this one, will retain their role of detailed information and long-living documentation.

Fig. 10 Characteristic shape of pollen in dry condition. A-B. *Lamium maculatum,* Lamiaceae, pollen in hydrated and dry condition. **C-D.** *Microrrhinum minus,* Plantaginaceae, pollen in hydrated and dry condition. **E-F.** *Scutellaria baicalensis,* Lamiaceae, pollen in hydrated and dry condition

References

Anger E, Weber M (2006) Pollen wall formation in *Arum alpinum*. Ann Bot 97: 239–244

Blackmore S (1992) Scanning electron microscopy in palynology. In: Nilsson S, Praglowski J (eds) Erdtman's Handbook of Palynology. 2nd edition, Munksgaard, Copenhagen, p. 403–431

Blackmore S (2000) The palynological compass: the contribution of palynology to systematics. In: Nordenstam B, El-Ghazaly G, Kassas M (eds) Plant Systematics for the 21st Century. Portland Press, London, p. 161–177

Blackmore S, Wortley A, Skvarla JJ, Rowley JR (2007) Pollen wall development in flowering plants. New Phytol 174: 483–498

Blackmore S, Wortley AH, Skvarla JJ, Gabarayeva NI, Rowley JR (2010) Developmental origins of structural diversity in pollen walls of Compositae. Plant Syst Evol 284: 17–32

Bradley DE (1958) The study of pollen grain surfaces in the Electron Microscope. New Phytol 57: 226–229

Brown R (1828) A brief account of microscopical observations made in the months of June, July, and August, 1827, on the particles contained in the pollen of plants; and on the general existence of active molecules in organic and inorganic bodies. Richard Taylor, London

Brown R (1833) On the organs and mode of fecundation in Orchideae and Asclepiadeae. In: The miscellaneous botanical works by Robert Brown. The Ray Society, London (1866)

Bryant VM (2013) Pollen and spore use in forensics. In: Jamieson J, Moenssens A (eds) Wiley Encyclopedia of Forensic Science, 2nd edition, John Wiley & Sons Ltd, Chichester, U.K.

Camerarius R J (1694) Ueber das Geschlecht der Pflanzen (De sexu plantarum epistola). Uebersetzt und herausgegeben von M. Mobius. Ostwald's Klassiker der exakten Wissenschaften 105

Campbell DH (1897) A morphological study of *Naias* and *Zannichellia*. Proc Calif Acad, Bot 3(1): 1–70

Candolle AP, Sprengel K (1821) Elements of the philosophy of plants. Edinburgh, printed for William Blackwood

Cavolini F (1792) *Zosterae oceanicae* Linnei ΑΝΗΣΙΣ. Contemplatus est Philippus Caulinus Neapolitanus. Annis 1787 et 1791. Neapoli

Doyle JA (2012) Molecular and fossil evidence on the origin of Angiosperms. Ann Rev Earth Planet Sci 40: 301–326

Doyle JA, Endress PK (2010) Integrating Early Cretaceous fossils into the phylogeny of living angiosperms: Magnoliidae and eudicots. J Syst Evol 48: 1–35

Dransfield J, Uhl NW, Asmussen CB, Baker WJ, Harley MM, Lewis CE (2008) Genera Palmarum. The Evolution and Classification of Palms. Kew Publishing, Kew

Ducker S, Knox B (1985) Pollen and pollination: a historical review. Taxon 34: 401–419

Ehrenberg CG (1838) Über die Bildung der Kreidefelsen und des Kreidemergels durch unsichtbare Organismen. Abh Kgl Akademie Wiss Berlin 1838: 59–147

Ehrlich HG, Hall JW (1959) The ultrastructure of eocene pollen. Grana Palynol 2: 32–35

Erdtman G (1943) An introduction to pollen analysis. Chronica Botanica, Waltham, Mass

Erdtman G (1947) Suggestions for the classification of fossil and recent pollen grains and spores. Svensk Bot Tidskr 41: 104–114

Erdtman G (1952) Pollen Morphology and Plant Taxonomy. Angiosperms. Almqvist & Wiksell, Stockholm

Erdtman G (1957) Pollen and Spore Morphology. Plant Taxonomy. Gymnospermae, Pteridophyta, Bryophyta. Almqvist & Wiksell, Stockholm

Erdtman G (1969) Handbook of Palynology – An Introduction to the Study of Polllen Grains and Spores. Munksgaard, Copenhagen

Erdtman G, Dunbar A (1966) Notes on electron micrographs illustrating the pollen morphology in *Armeria maritima* and *Armeria sibirica*. Grana Palynol 6: 338–354

Erdtman G, Straka H (1961) Cormophyte spore classification. Geol Fören Förenhandl 83: 65–78

Erdtman G, Vishnu-Mittre (1956) On terminology in pollen and spore morphology. The Palaeobotanist 5: 109–111

Fægri K, Iversen J (1950) Textbook of modern pollen analysis. Munksgaard, Copenhagen

Fægri K, Iversen J (1989) Textbook of Pollen analysis. 4th edition, John Wiley & Sons, Chichester

Fernandez-Moran H, Dahl AO (1952) Electron microscopy of ultrathin frozen sections of pollen grains. Science 116: 465–467

Fischer H (1890) Beiträge zur vergleichenden Morphologie der Pollenkörner. Thesis, Breslau

Friis EM, Crane PR, Pedersen KR (2011) Early Flowers and Angiosperm Evolution. Cambridge University Press, Cambridge

Fritzsche J (1834) Ueber den Pollen der Pflanzen und das Pollenin. Ann Phys 108: 481–492

Fritzsche J (1837) Über den Pollen. Mém Sav Étrang Acad Sci Pétersbourg 3: 649–672

Furness CA, Banks H (2010) Pollen evolution in the early–divergent monocot order Alismatales. Int J Plant Sci 171: 713–739

Furness CA, Rudall PJ (1999a) Microsporogenesis in Monocotyledons. Ann Bot 84: 475–499

Furness CA, Rudall PJ (1999b) Inaperturate pollen in monocotyledons. Int J Pl Sci 160: 395–414

Furness CA, Rudall PJ (2004) Pollen aperture evolution – a crucial factor for eudicot success? Trends Plant Sci 9: 154–158

Gabarayeva NI, Grigorjeva VV (2010) Sporoderm ontogeny in *Chamaedorea microspadix* (Arecaceae): self–assembly as the underlying cause of development. Grana 49: 91–114

Gabarayeva NI, Grigorjeva VV, Rowley JR (2010) A new look at sporoderm ontogeny in *Persea americana* and the hidden side of development. Ann Bot 105: 939–955

Gee CT (2001) The mangrove palm Nypa in the geologic past of the New World. Wetl Ecol Manag 9: 181–194

Gleditsch JG (1751) Essai d'une Fécondation artificielle, fait sur l'espèce de Palmier qu'on nomme, Palma dactylifera folio flabelliformi. – Histoire de l'Académie Royale des Sciences et Belles Lettres de Berlin année 1749: 103–108

Gleditsch JG (1765) Kurze Nachricht von einer künstlichen wohlgelungenen Befruchtung eines Palmbaumes im Königlichen Kräutergarten zu Berlin. – Vermischte Physikalisch–Botanisch–Ökonomische Abh 1: 94–104

Göppert HR (1837) De floribus in statu fossili, commentatio botanica. Thesis, Breslau

Göppert HR (1848) Über das Vorkommen von Pollen im fossilen Zustande. Neues Jahrbuch für Mineralogie, Geognosie, Geologie und Petrefaktenkunde 11: 338–340

Govaerts R (2003) How many species of seed plants are there? – a response. Taxon 52: 583–584

Grew N (1682) The Anatomy of plants, with an idea of a philosophical history of plants, and several other lectures, read before the Royal Society. W. Rawlins, London

Grímsson F, Grimm GW, Zetter R, Denk T (2016) Cretaceous and Paleogene Fagaceae from North America and Greenland: evidence for a Late Cretaceous split between *Fagus* and the remaining Fagaceae. Acta Palaeobotanica 56: 247–305

Grímsson F, Grimm GW, Zetter R (2017a) Tiny pollen grains: first evidence of Saururaceae from the Late Cretaceous of western North America. Peer J 5:e3434 https://doi.org/10.7717/peerj.3434

Grímsson F, Kapli P, Hofmann C, Zetter R, Grimm GW (2017b) Eocene Loranthaceae pollen pushes back divergence ages for major splits in the family. PeerJ 5:e3373 https://doi.org/10.7717/peerj.3373

Grímsson F, Zetter R, Halbritter H, Grimm GW (2014) *Aponogeton* pollen from the Cretaceous and Paleogene of North America and West Greenland: Implications for the origin and palaeobiogeography of the genus. Rev Palaeobot Palynol 200: 161–187

Guignard L (1891) Nouvelles études sur la fécondation. Ann Sc Nat, Bot ser 7, 14: 163–296

Guignard L (1899) Sur les antherozoides et la double copulation sexuelle chez les végétaux angiosperms. Rev Gen Bot 11: 129–135

Halbritter H, Hesse M (2004) Principal modes of infoldings in tricolp(or)ate Angiosperm pollen. Grana 43: 1–14

Hao G, Chye M-L, Saunders RMK (2001) A phylogenetic analysis of the Schisandraceae based on morphology and nuclear ribosomal ITS sequences. Bot J Linn Soc 135: 401–411

Harley MM, Baker WJ (2001) Pollen aperture morphology in Arecaceae: application within phylogenetic analyses, and a summary of the fossil record of palm-like pollen. Grana 40: 45–77

Harley MM, Ferguson IK (1990) The role of the SEM in pollen morphology and plant systematics. In: Claugher D (ed) Scanning Electron Microscopy in Taxonomy and Functional Morphology. Syst Ass Special Volume 41: 45–68. Clarendon Press, Oxford

Hay WW, Sandberg PA (1967) The Scanning Electron Microscope, a major break-through for micropaleontology. Micropaleontology 13: 407–418

Hell SW (2009) Microscopy and its focal switch. Nature Methods 6: 24–32

Heslop-Harrison J (1975) The physiology of the pollen grain surface. Proc R Soc, London B 190: 275–299

Hesse M (2006a) Reason and consequences of the lack of a sporopollenin ektexine in Aroideae (Araceae). Flora 201: 421–428

Hesse M (2006b) Conventional and novel modes of exine patterning in members of the Araceae – the consequence of ecological paradigm shifts? Protoplasma 228: 145–149

Hesse M, Blackmore S (2013) Editorial: Preface to the Special Focus manuscripts. Plant Syst Evol 299: 1011–1012

Hesse M, Weber M, Halbritter H (2000) A comparative study of the polyplicate pollen types in Arales, Laurales, Zingiberales and Gnetales. In: Harley MM, Morton CM, Blackmore S (eds) Pollen and spores: morphology and biology. Royal Botanic Gardens, Kew, p. 227–239

Hofmeister W (1849) Die Entstehung des Embryo der Phanerogamen. Friedrich Hofmeister, Leipzig

Hooke R (1665) Micrographia, or, Some physiological descriptions of minute bodies made by magnifying glasses, with observations and inquiries thereupon. Printed by Jo. Martyn and Ja. Allestry, London

Hyde HA (1955) Oncus, a new term in pollen morphology. New Phytol 54: 255

Iversen J, Troels-Smith J (1950) Pollenmorfologiske definitioner og typer. Pollenmorphologische Definitionen und Typen. Danm Geol Unders, ser 4, 3: 1–54

John JF (1814) Ueber den Befruchtungsstaub, nebst einer Analyse des Tulpenpollens. J Chem Phys 12: 244–252

Jones GD, Bryant VM Jr. (1996) Melissopalynology. In: Jansonius J, McGregor DC (eds) Palynology: principles and applications. American Association of Stratigraphic Palynologists Foundation, vol 3, AASP Foundation, Dallas, p. 933–938

Kesseler R, Harley MM (2004) Pollen. The hidden sexuality of flowers. Papadakis Publisher, London

Knox RB (1984) The pollen grain. In: Johri BM (ed) Embryology of Angiosperms. Springer, Berlin

Kölreuter JG (1761-1766) Vorläufige Nachricht von einigen das Geschlecht der Pflanzen betreffenden Versuchen und Beobachtungen. 4 Vol., Gleditsch, Leipzig

Kölreuter JG (1806) De antherarum pulvere. Nova acta Academiae Scientiarum Imperialis Petropolitanae 15: 359–398

Kölreuter JG (1811) Dissertationis de antherarum pulvere continuato. Mem Acad Sci Petersbourg 3: 159–199

Kremp GOW (1968) Morphologic Encyclopedia of Palynology. 2nd edition, Arizona Press, Tucson

Lindley J (1836) A natural system of botany; or, A systematic view of the organization, natural affinities, and geographical distribution of the whole vegetable kingdom: together with the uses of the most important species in medicine, the arts, and rural or domestic economy (2nd edition), Longman, London

Linnaeus C (1750) Sponsalia plantarum. J. G. Wahlbom, Stockholm. Facs. edition, Rediviva No. 19, Stockholm 1971

Malpighi M (1901) Die Anatomie der Pflanzen. I und II Theil, London 1675 und 1679. Bearbeitet von M. Möbius. Ostwald's Klassiker der exakten Wissenschaften Nr. 120, pp. 163

Manten AA (1966) Half a century of modern palynology. Earth-Sci Rev 2: 277–316

Mendes MM, Dinis J, Pais J, Friis EM (2014) Vegetational composition of the Early Cretaceous Chicalhão flora (Lusitanian Basin, western Portugal) based on palynological and mesofossil assemblages. Rev Palaeobot Palynol 200: 65–81

Mildenhall DC, Wiltshire PEJ, Bryant VM (2006) Forensic palynology: Why do it and how it works. Forensic Sci Int 163: 163–172

Moore PD, Webb JA, Collinson ME (1991) Pollen analysis. 2nd edition. Blackwell Scientific Publication, Oxford

Mühlethaler K (1953) Untersuchungen über die Struktur der Pollenmembran. Mikroskopie 8: 103–110

Mühlethaler K (1955) Die Struktur einiger Pollenmembranen. Planta 46: 1–13

Nadot S, Forchioni A, Penet L, Sannier J, Ressayre A (2006) Links between early pollen development and aperture pattern in monocots. Protoplasma 228: 55–64

Nägeli K (1842) Zur Entwicklungsgeschichte des Pollens bei den Phanerogamen. Orell, Füssli & Comp., Zürich

Nawaschin S (1898) Resultate einer Revision der Befruchtungsvorgänge bei *Lilium martagon* und *Fritillaria tenella*. Bull Acad Imp Sci St. Petersbourg, ser. 5, 9: 377–382

Pettitt JM, Chaloner WG (1964) The ultrastructure of the Mesozoic pollen Classopollis. Pollen Spores 6: 611–620

Potonié R (1934) I. Zur Morphologie der fossilen Pollen und Sporen. Arb Inst Paläobotanik Petrographie Brennsteine 4: 5–24

Potonié R (1956) Synopsis der Gattungen der Sporae dispersae, I. Teil: Sporites. Beih Geol Jahrb 23: 1–103

Punt W, Blackmore S, Nilsson S, Le Thomas A (1994) Glossary of Pollen and Spore Terminology. LPP Foundation, Laboratory of Palaeobotany and Palynology, University of Utrecht, Utrecht. LPP Contributions Series 1

Punt W, Hoen PP, Blackmore S, Nilsson S, Le Thomas A (2007) Glossary of pollen and spore terminology. Rev Palaeobot Palynol 143: 1–81

Purkinje JE (1830) De Cellulis antherarum fibrosis nec non de granorum pollinarium formis: Commentatio phytotomica. Grueson, Breslau

Reille M (1992) Pollen et Spores d'Europe et d'Afrique du Nord. Laboratoire de Botanique Historique et Palynologie, Marseille

Reille M (1995) Pollen et Spores d'Europe et d'Afrique du Nord, Supplement 1. Laboratoire de Botanique Historique et Palynologie, Marseille

Reille M (1998) Pollen et Spores d'Europe et d'Afrique du Nord, Supplement 2. Laboratoire de Botanique Historique et Palynologie, Marseille

Reinsch P (1884) Micro-Palaeophytologia formationis carboniferae. Krische, Erlangen

Reitsma TJ (1970) Suggestions towards unification of descriptive terminology of angiosperm pollen grains. Rev Palaeobot Palynol 10: 39–60

Rowley JR, Flynn JJ (1966) Single-stage carbon replicas of microspores. Stain Technol 41: 287–290

Sarawichit P (2012) Pollen and orbicular walls of selected species of *Justicieae* (Acanthaceae) and their systematic significance. Thesis, University of Vienna

Schacht H (1856/59) Lehrbuch der Anatomie und Physiologie der Gewächse. 2 Vol., Müller, Berlin

Schopf JM (1957) Spores and related plant microfossils – Paleozoic. In: Ladd HS (ed) Treatise on marine ecology and paleoecology 2, Paleoecology: Geological Scociety of America Memoir 67/2, p. 703–707

Schopf JM (1964) Practical problems and principles in study of plant microfossils. In: Cross, AT (ed) Palynology in oil exploration – A symposium: Society of Economic Paleontologists and Mineralogists Special Publication 11, p. 29–57

Scotland RW, Wortley AH (2003) How many species of seed plants are there? Taxon 52: 101–104

Sprengel CK (1793) Das entdeckte Geheimnis der Natur im Bau und in der Befruchtung der Blumen. Vieweg, Berlin

Sprengel K (1804) Anleitung zur Kenntniß der Gewächse. In Briefen. Erste Sammlung, Kimmel, Halle

Stolze GH (1816) Der Pollen der Pflanzen in chemischer Hinsicht; nebst einer Analyse des Pollens der Haselnusstaude (*Corylus avellana* Linn.). Jahrb Pharm 17: 159–187

Strasburger E (1884) Neue Untersuchungen über den Befruchtungsvorgang bei den Phanerogamen als Grundlage für eine Theorie der Zeugung. Fischer, Jena

Stuessy TF (1979) Ultrastructural data for the practicing plant systematist. Am Zool 19: 621–635

Stuessy TF (2009) Plant Taxonomy: The Systematic Evaluation of Comparative Data. 2nd edition, Columbia University Press, New York

Stuessy TF, Funk VA (2013) New trends in plant systematics – Introduction. Taxon 62: 873–875

Thorne RF (2002) How many species of seed plants are there? Taxon 51: 511–522

Thornhill JW, Matta RK, Wood WH (1965) Examining three-dimensional microstructures with the scanning electron microscope. Grana Palynol 6: 3–6

Traverse A (2007) Paleopalynology. 2nd ed, Springer, Dordrecht

Ulrich S, Hesse M, Bröderbauer D, Bogner J, Weber M, Halbritter H (2013) *Calla palustris* (Araceae): New insights with special regard to its controversial systematic position and to closely related genera. Taxon 62: 701–712

Ulrich S, Hesse M, Bröderbauer D, Wong SY, Boyce PC (2012) *Schismatoglottis* and *Apoballis* (Araceae: Schismatoglottideae): A new example for the significance of pollen morphology in Araceae systematics. Taxon 61: 281–292

Ulrich S, Hesse M, Weber M, Halbritter H (2017) *Amorphophallus*: New insights into pollen morphology and the chemical nature of the pollen wall. Grana 56: 1–36

Von Grotthuss T (1814) Analysis des Tulpensamenstaubs. J Chem Phys 11: 281–380

Von Mohl H (1835) Sur la structure et les formes des grains de pollen. Ann Sci nat 2. Ser., 3: 148–180, 220–236, 304–346

Von Post L (1916) Einige südschwedische Quellmoore. Bull Geol Inst Univ Uppsala 15: 219–278

Weber M, Ulrich S (2010) The endexine: a frequently overlooked pollen wall layer and a simple method for detection. Grana 49: 83–90

Weber M, Ulrich S (2016) Forensic Palynology: How pollen in dry grass can link to a crime scene. In: Kars H, van den Eijkel L (eds) Soil in Criminal and Environmental Forensics: Proceedings of the Soil Forensics Special, 6th European Academy of Forensic Science Conference. The Hague, Springer, p. 15–23

Weber M, Ulrich S (2017) *PalDat* 3.0 – second revision of the database, including a free online publication tool. Grana 56: 257–262

Wodehouse RP (1928) The phylogenetic value of pollen grain characters. Ann Bot 42: 891–934

Wodehouse RP (1935) Pollen grains. Their structure, identification and significance in science and medicine. McGraw-Hill, New York

Zetter R, Hofmann C (2001) New aspects of the palynoflora of the lowermost Eocene (Krappfeld Area, Carinthia). In: Piller WE, Rasser MW (eds) Paleogene of the Eastern Alps. Österreichische Akademie der Wissenschaften, Schriftenreihe der Erdwissenschaftlichen Kommissionen 14, p. 473–507

Zetzsche F, Huggler K (1928) Untersuchungen über die Membran der Sporen und Pollen. I. 1. *Lycopodium clavatum* L. Ann Chem 461: 89–108

Zetzsche F, Kalt P, Leichti J, Ziegler E (1931) Zur Konstitution des Lycopodiumsporonins, des Tasmanins und des Lange-Sporonins. J Prakt Chem 148: 67–84

Zetzsche F, Vicari H (1931) Untersuchungen über die Membran der Sporen und Pollen II, *Lycopodium clavatum* L., Untersuchungen über die Membran der Sporen und Pollen III. *Picea orientalis*, *Pinus silvestris* L., *Corylus avellana* L. Helv Chim Acta 14: 58–67

Pollen Development

H. Halbritter et al., *Illustrated Pollen Terminology*, https://doi.org/10.1007/978-3-319-71365-6_2

Microsporogenesis and Microgametogenesis

Pollen is source and transport unit for the male gametes (or their progenitor cell). The unicellular pollen grain represents the microspore of seed plants, the multicellular pollen grain the male gametophytic generation. The development of a pollen grain includes **microsporogenesis** and **microgametogenesis** (Figs. 1 and 2, Gomez et al. 2015; Keijzer and Willemse 1988). Microsporogenesis starts with the differentiation of microspore mother cells (MMC) respectively **pollen mother cells** (PMC). These diploid cells become enclosed by a thick **callose** wall and undergo meiosis, forming a tetrad of four haploid **microspores**, each encased in another callose wall insulating them from each other and from the surrounding diploid tapetal cells (Figs. 1 C-E, and 2). Cytokinesis following meiotic nuclear divisions is accompanied by the formation of cleavage planes determined by the configuration and orientation of the meiotic spindle axes. In the case of **successive cytokinesis**, planes are formed after the first and second meiotic divisions leading to the formation of various microspore tetrad types (see "Pollen Morphology and Ultrastructure"). During **simultaneous cytokinesis** the cleavage planes are formed simultaneously after the second meiotic division and microspores become arranged in a **tetrahedral tetrad** (Furness and Rudall 1999, 2001).

Pollen wall formation starts while the microspores are arranged in tetrads, encased by callose. The first step starts with the deposition of **primexine**, a fibrillar polysaccharidic material, on the surface of the microspores. The primexine forms a template where sporopollenin precursors and subsequently **sporopollenin** are deposited, building the final pollen wall (Fig. 1 E). Apertures are formed where the endoplasmic reticulum has prevented the deposition of primexine.

During pollen formation and maturation the **tapetum** plays an important role, usually forming a single layer of cells circumscribing the loculus. Tapetal cells are specialized and have a short lifespan. They finally lose their cellular organization and are reabsorbed. Two types of tapetum are known: the **secretory** (or glandular or parietal) and the **amoeboid** (or periplasmodial). In the secretory type (e.g., in Apiaceae) the tapetal cells remain stationary until they finish their physiological functions. In the amoeboid type (e.g., in Araceae) cells lose their individuality at an early developmental stage by degeneration of the cell walls (Furness and Rudall 1999, 2001). The protoplasts then fuse and intrude into the locule where they enclose the pollen grains (Fig. 3). The tapetum plays an important role during several stages of pollen development (Pacini 1997). Its main function is the nourishment of the microspores, but it also synthesizes enzymes (e.g., callase), exine precursors, pollen coatings, forms Ubisch bodies (orbicules) and viscin threads (both equivalents to the ektexine). The most striking material produced by the tapetum is **pollenkitt** (and **tryphine** in Brassicaceae), a sticky, heterogeneous material composed of neutral lipids, flavonoids, carotenoids, proteins and polysaccharides. Pollenkitt serves numerous functions: keeping pollen grains together during transport, protecting pollen (from water loss, ultraviolet radiation, hydrolysis and exocellular enzymes), and maintaining sporophytic proteins inside exine cavities.

Microgametogenesis (Fig. 1 G-K) in angiosperms includes first and second pollen mitosis, leading to the formation of the male gametes, the **sperm cells** (Mccormick 1993; Cresti et al. 1992). Microgametogenesis starts with formation of a central vacuole within the uninucleate microspore, pushing the nucleus towards the pollen wall. As long as the nucleus is in a central position within the cytoplasm, the cell is called a **microspore** (Fig. 1 F). With the dislocation of the microspore nucleus the cell becomes the young **pollen grain** (Fig. 1 G).

The **first pollen mitosis** is followed by an asymmetric cell division, leading to the formation of a smaller generative cell and a larger **vegetative cell** with a **vegetative nucleus** (Figs. 4 and 5). Subsequently, the generative cell detaches from the pollen wall and is finally located within the cytoplasm of the vegetative cell (Fig. 1 I). The **generative cell**, sparse in organelles, becomes **spindle-shaped** and the shape of the generative nucleus changes correspondingly (Figs. 6 and 7).

The **second pollen mitosis** includes a symmetric cell division, and divides the generative cell into two **sperm cells** (Figs. 8 and 9), the final stage of gametophytic development (Fig. 1 J-K). Angiosperm pollen is either **two-celled** (75%) or **three-celled** (25% of investigated taxa) at the time of anthesis (Brewbaker 1967; Edlund et al. 2004; Williams et al. 2014). In the latter case the second pollen mitosis takes place in the **pollen tube** (Fig. 1 K), after **germination** of the pollen grain on a stigma or on a corresponding structure (Figs. 10 and 11, Edlund et al. 2004, Mascarenhas 1993). In some families, genera with three- as well as two-celled/nuclear pollen grains occur (e.g., Araceae, Brewbaker 1967).

Microgametogenesis in gymnosperms includes several mitotic divisions. Normally, pollen grains of conifers, cycads and allies are multicelled at anthesis, and comprise prothallial cell(s), a large tube cell and a small antheridial cell. The tube cell becomes a pollen tube; the antheridial cell undergoes division into the stalk cell and the spermatogenous cell, the latter finally dividing into the male gametes (sperm cells or spermatozoids).

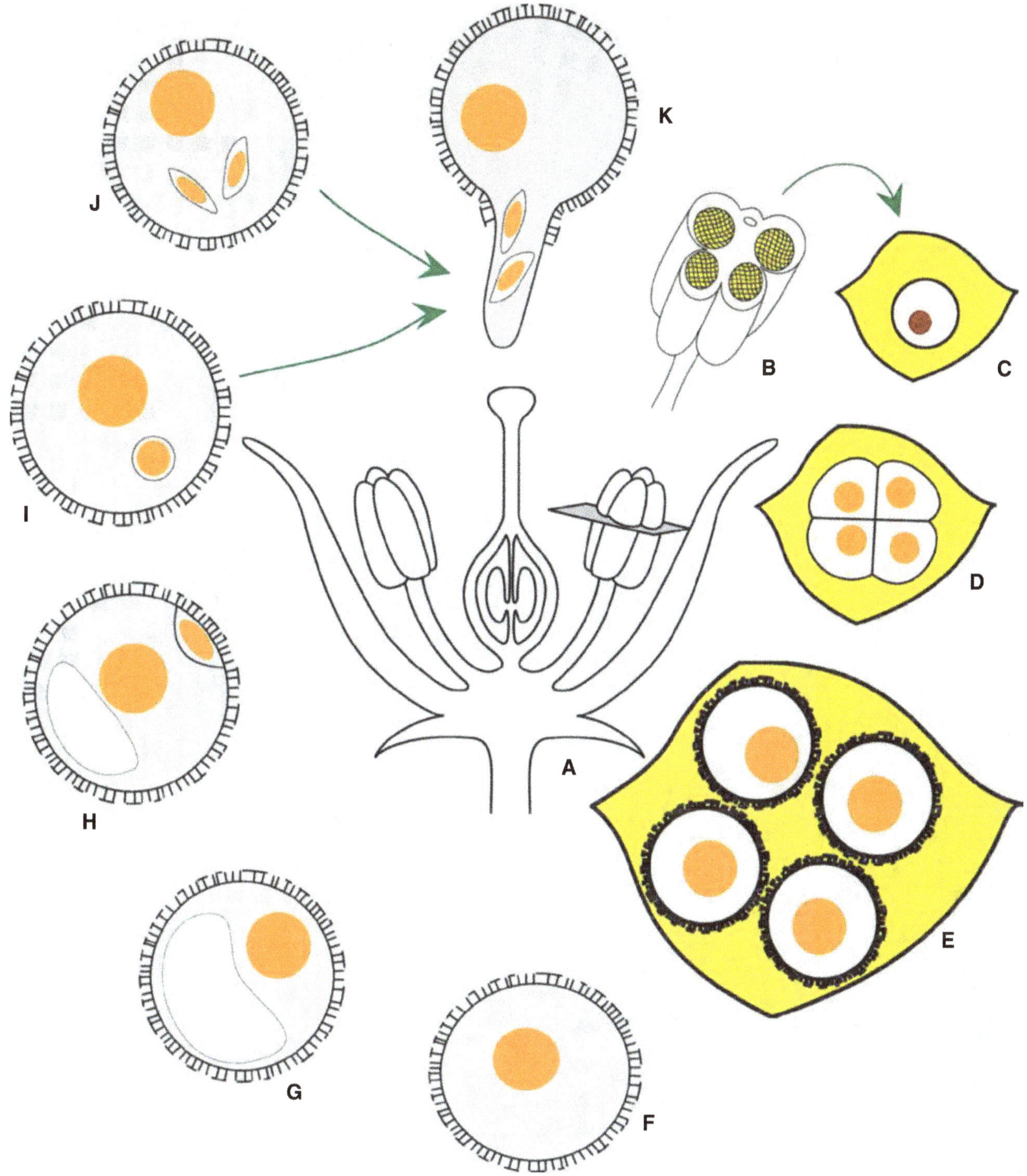

Fig. 1 Pollen development in angiosperms. A. Schematic illustration of an angiosperm flower. **B**. Cross section of anther. **C**. Pollen mother cell (PMC) encased in callose (diploid nucleus dark red). **D**. Tetrad of four haploid microspores encased in callose (haploid nucleus orange). **E**. Pollen wall formation and separation of microspores. **F**. A single free microspore with central haploid nucleus. **G**. Beginning of gametogenesis, formation of a central vacuole (white). **H**. First pollen mitosis, lens-shaped generative cell with generative nucleus attached to pollen wall. **I**. Two-celled pollen grain, generative cell detached from pollen wall. **J**. Three-celled pollen grain after second pollen mitosis, note two sperm cells with sperm nuclei. **K**. Germination can occur from either a two-celled pollen grain, followed by the formation of sperm cells, or from a three-celled pollen grain (pathways indicated by green arrows)

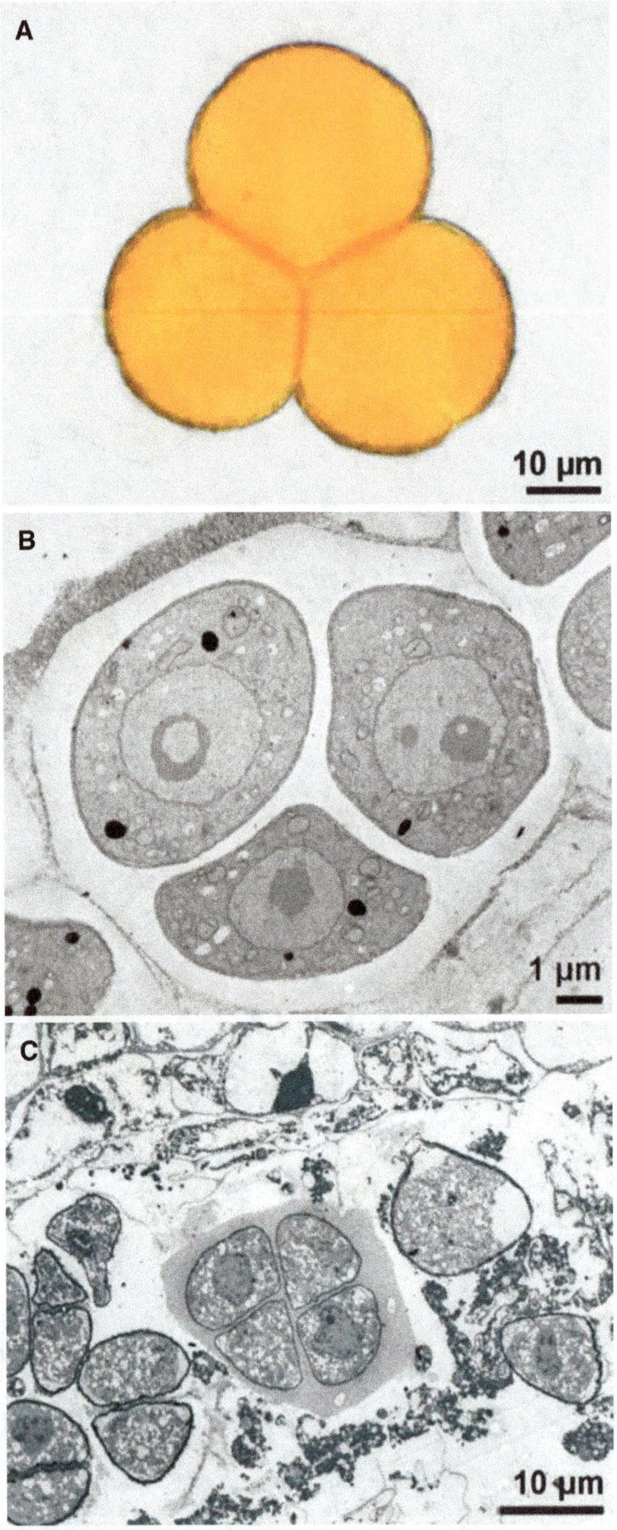

Fig. 2 Microsporogenesis. A. *Scrophularia nodosa*, Scrophulariaceae, tetrad tetrahedral, iodine. **B.** *Spirea* sp., Rosaceae, tetrad tetrahedral, Thiéry test. **C.** *Orobanche hederae*, Orobanchaceae, tetrad planar, potassium iodine

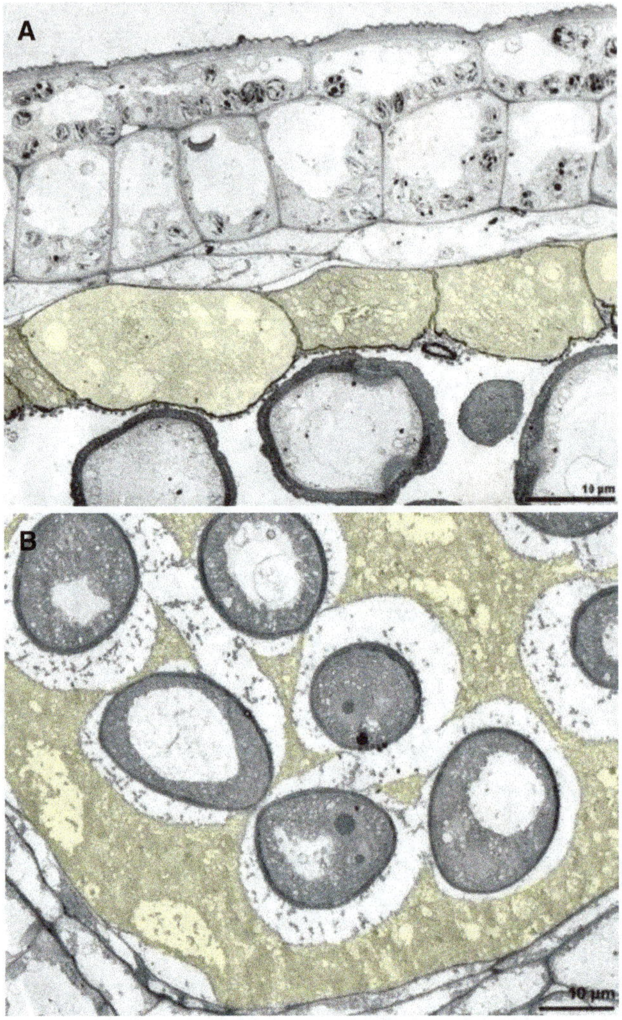

Fig. 3 Tapetum types. A. *Hacquetia epipactis*, Apiaceae, secretory tapetum in young anther, Thiéry test. **B.** *Zantedeschia aethiopica*, Araceae, amoeboid tapetum, U+Pb

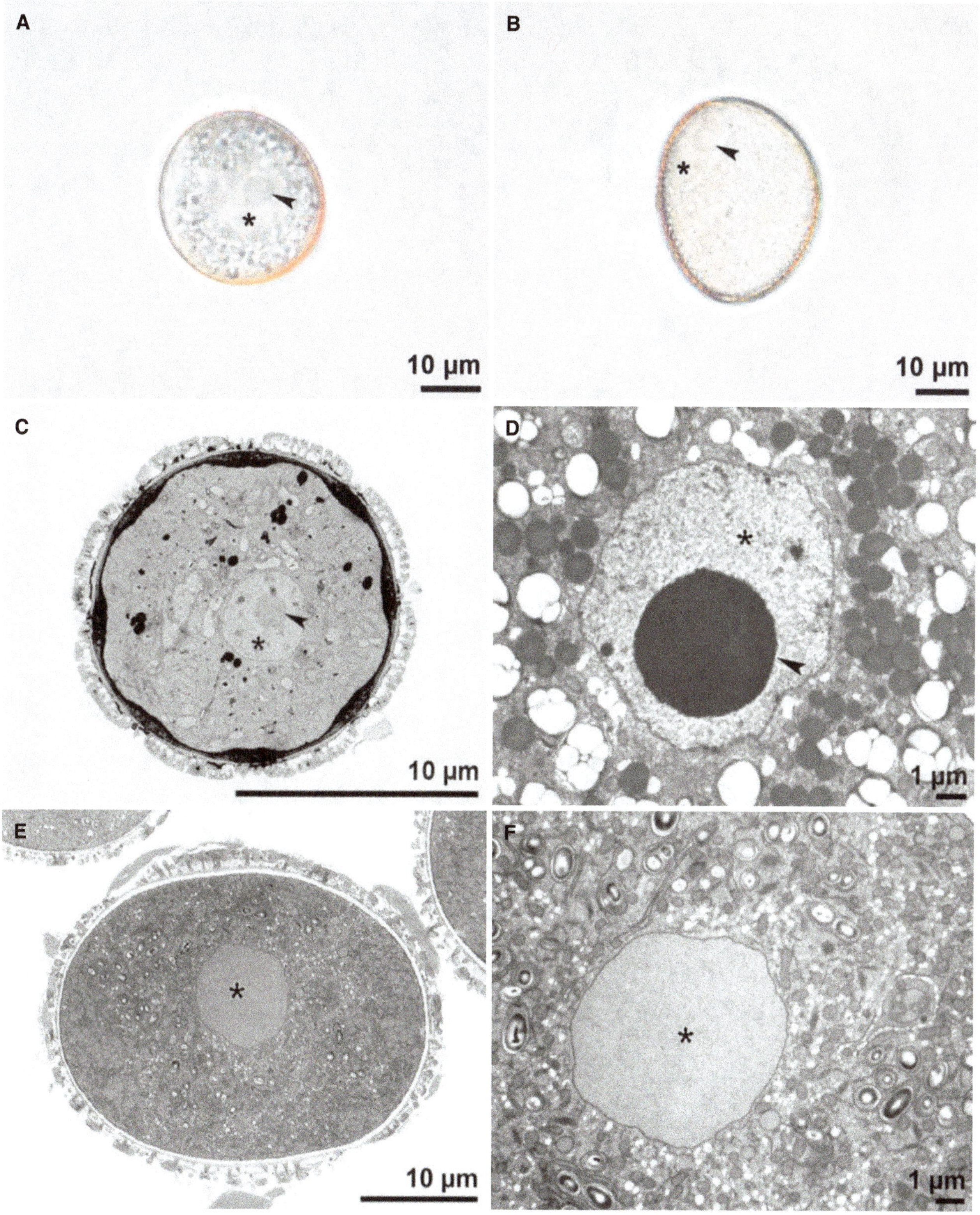

Fig. 4 Variability of the vegetative nucleus in LM and TEM (cross sections). A-B. *Dracontium asperum*, Araceae, pollen hydrated in water, vegetative cell with vegetative nucleus (asterisk) and nucleolus (arrowhead). **C.** *Galium odoratum*, Rubiaceae, vegetative cytoplasm and nucleus (asterisk) with nucleolus (arrowhead), U+Pb. **D.** *Salvia nemorosa*, Lamiaceae, vegetative nucleus (asterisk) with nucleolus (arrowhead) surrounded by cytoplasm, U+Pb. **E.** *Thymus glabrescens*, Lamiaceae, vegetative cytoplasm and nucleus (asterisk), modified Thiéry test. **F.** *Thymus glabrescens*, Lamiaceae, vegetative nucleus (asterisk) surrounded by cytoplasm, modified Thiéry test

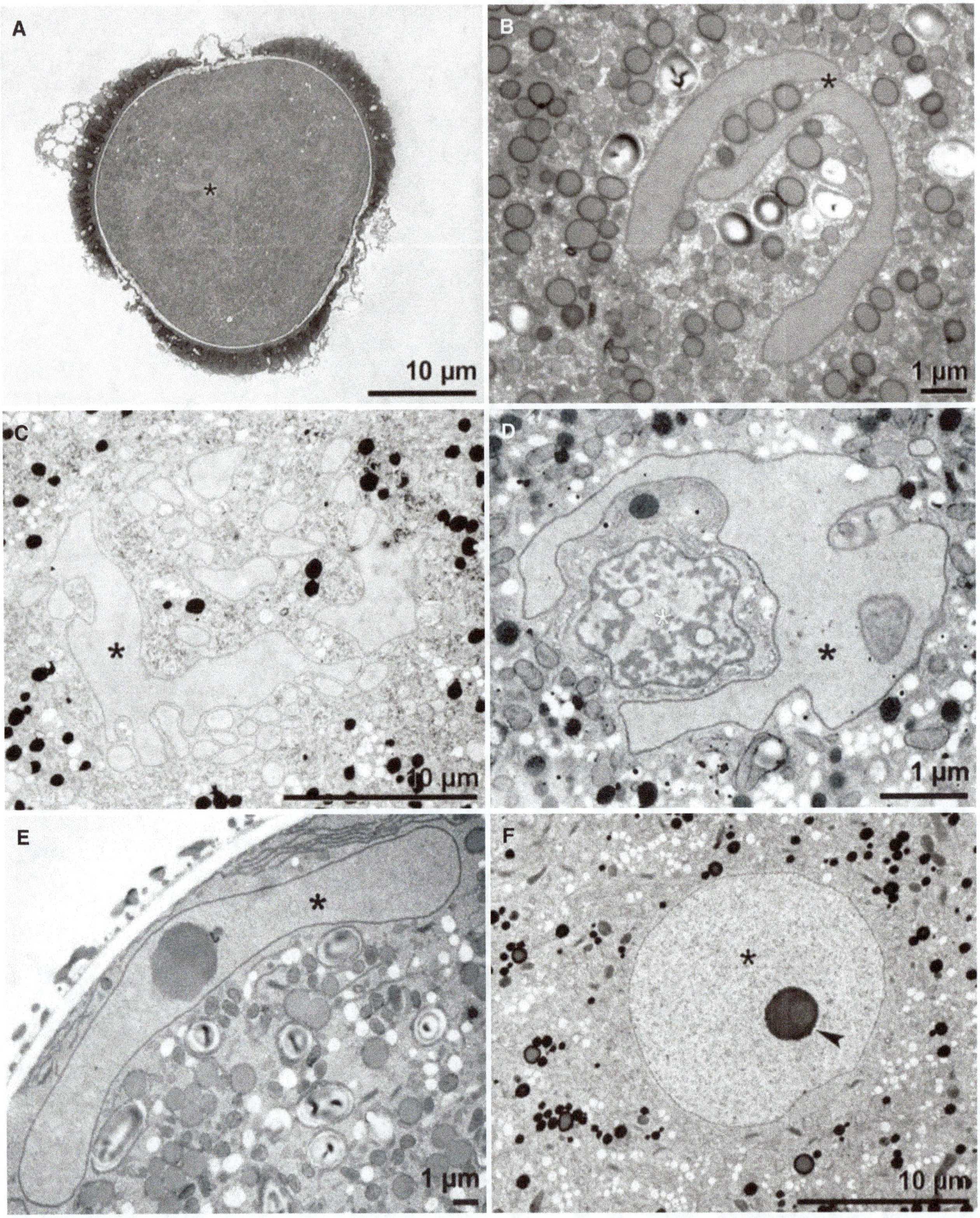

Fig. 5 Variability of the vegetative nucleus in TEM (cross sections). A. *Brassica napus*, Brassicaceae, vegetative cytoplasm and nucleus (asterisk), modified Thiéry test. **B.** *Salvia verticillata*, Lamiaceae, vegetative cytoplasm and nucleus (asterisk), modified Thiéry test. **C.** *Iris pumila*, Iridaceae, vegetative cytoplasm and nucleus (asterisk), modified Thiéry test. **D.** *Consolida regalis*, Ranunculaceae, vegetative nucleus (black asterisk) and generative cell (white asterisk), modified Thiéry test. **E.** *Acinos alpinus*, Lamiaceae, vegetative nucleus (asterisk) with nucleolus (arrowhead) surrounded by cytoplasm, modified Thiéry test. **F.** *Stachys palustris*, Lamiaceae, vegetative nucleus (asterisk) with nucleolus (arrowhead) surrounded by cytoplasm, U+Pb

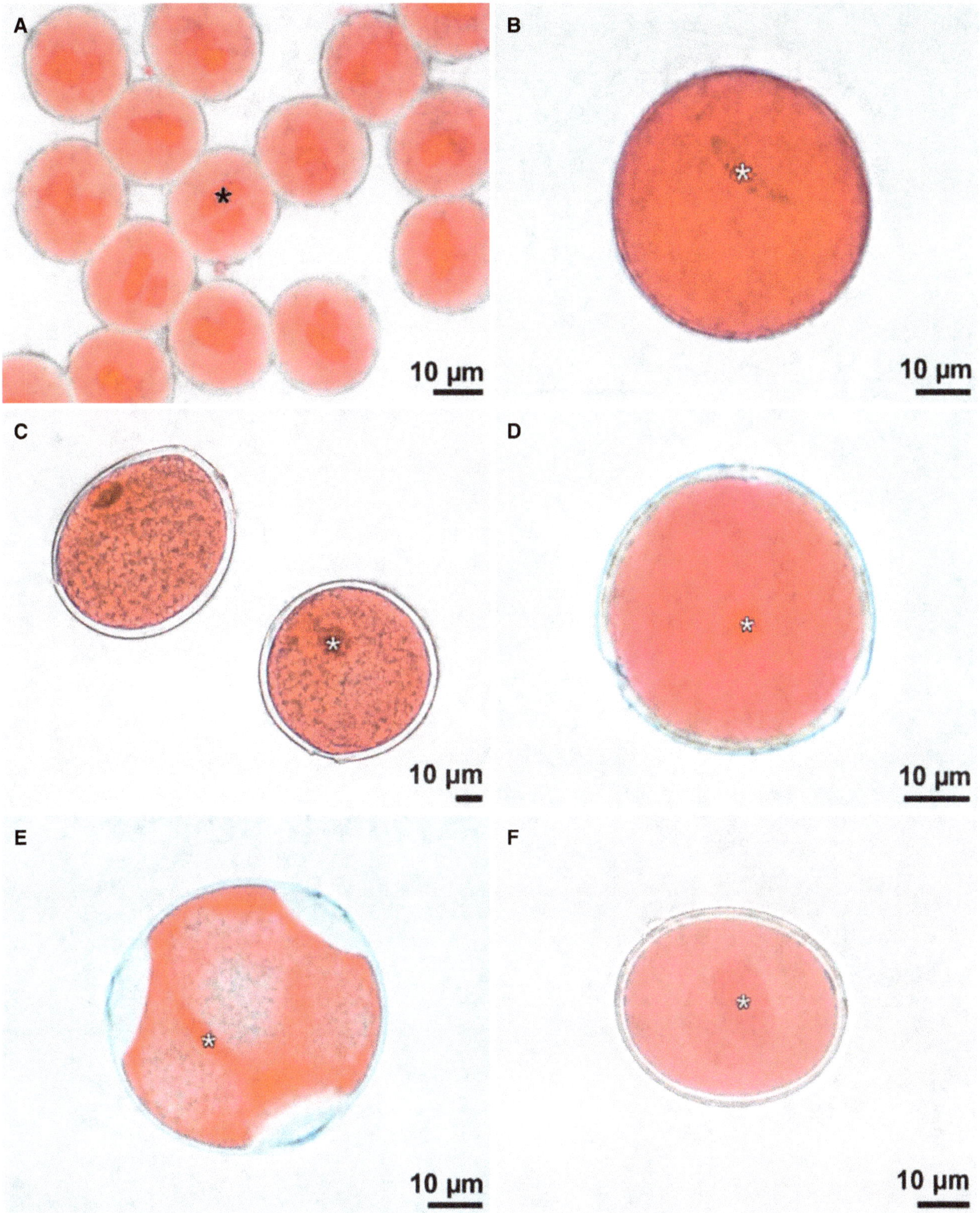

Fig. 6 Generative cell and nucleus stained with acetocarmine in LM. A. *Melampyrum nemorosum*, Orobanchaceae, spindle-shaped generative cell/nucleus (asterisk) and vegetative nucleus. **B.** *Betonica officinalis*, Lamiaceae, spindle-shaped generative cell/nucleus (asterisk). **C.** *Anchomanes welwitschii*, Araceae generative cell/nucleus (asterisk). **D.** *Quercus robur*, Fagaceae, generative cell/nucleus (asterisk). **E.** *Carpinus betulus*, Betulaceae, spindle-shaped generative cell/nucleus (asterisk). **F.** *Asterostigma lividum*, Araceae, generative cell/nucleus (asterisk) and vegetative nucleus

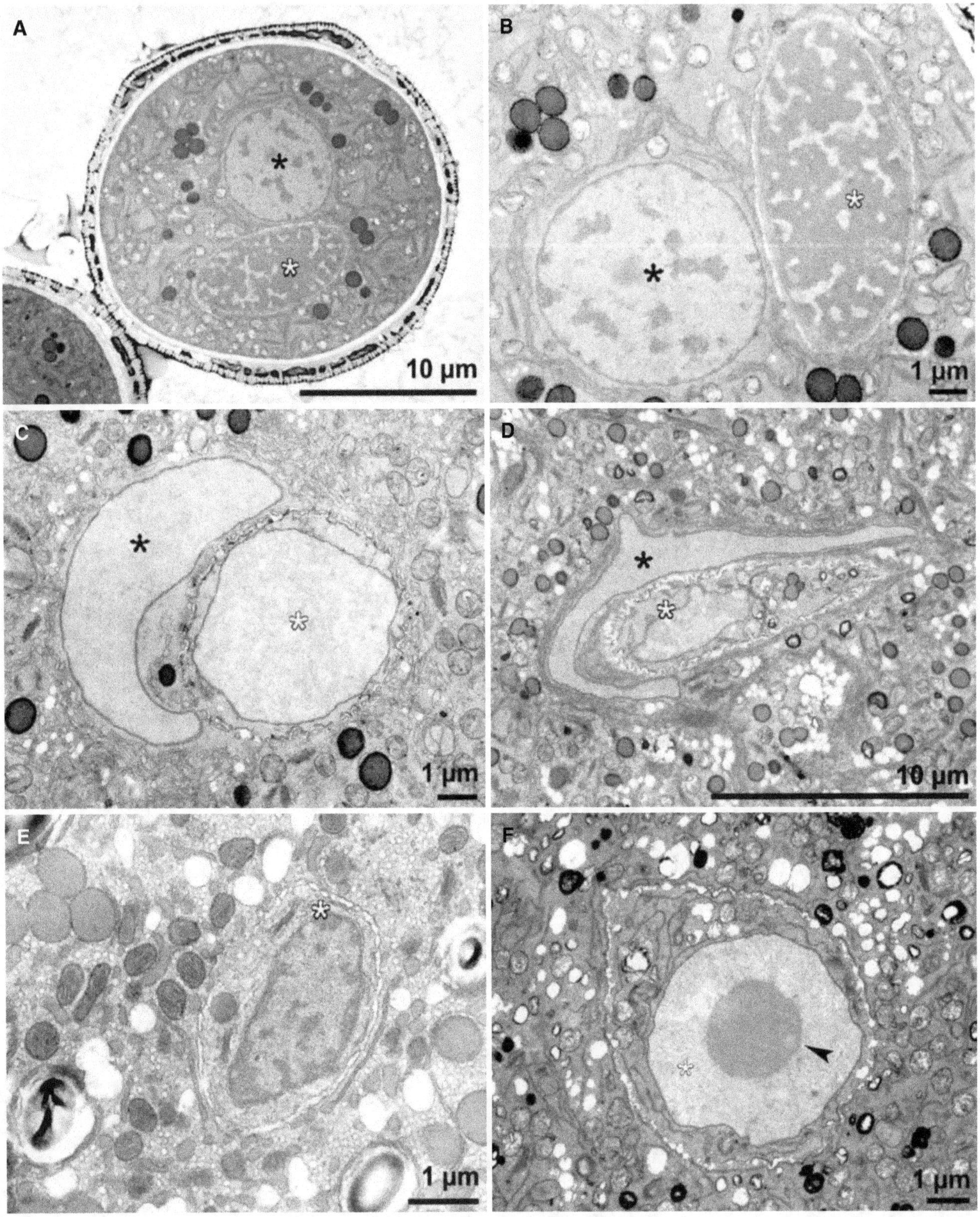

Fig. 7 Variability of the generative cell and nucleus in TEM (cross sections). A. *Melampyrum nemorosum*, Orobanchaceae, pollen in overview, vegetative nucleus (black asterisk), generative cell/nucleus (white asterisk), modified Thiéry test. **B.** *Melampyrum nemorosum*, Orobanchaceae, vegetative nucleus (black asterisk), generative cell/nucleus (white asterisk), modified Thiéry test. **C.** *Betonica officinalis*, Lamiaceae, vegetative nucleus (black asterisk) and generative cell/nucleus (white asterisk) surrounded by cytoplasm, modified Thiéry test. **D.** *Ajuga reptans*, Lamiaceae, vegetative nucleus (black asterisk) and generative cell/nucleus (white asterisk) surrounded by cytoplasm, modified Thiéry test. **E.** *Acinos alpinus*, Lamiaceae, generative cell/nucleus (white asterisk) surrounded by cytoplasm, modified Thiéry test. **F.** *Stachys palustris*, Lamiaceae, generative cell/nucleus (white asterisk) with nucleolus (arrowhead) surrounded by cytoplasm, modified Thiéry test

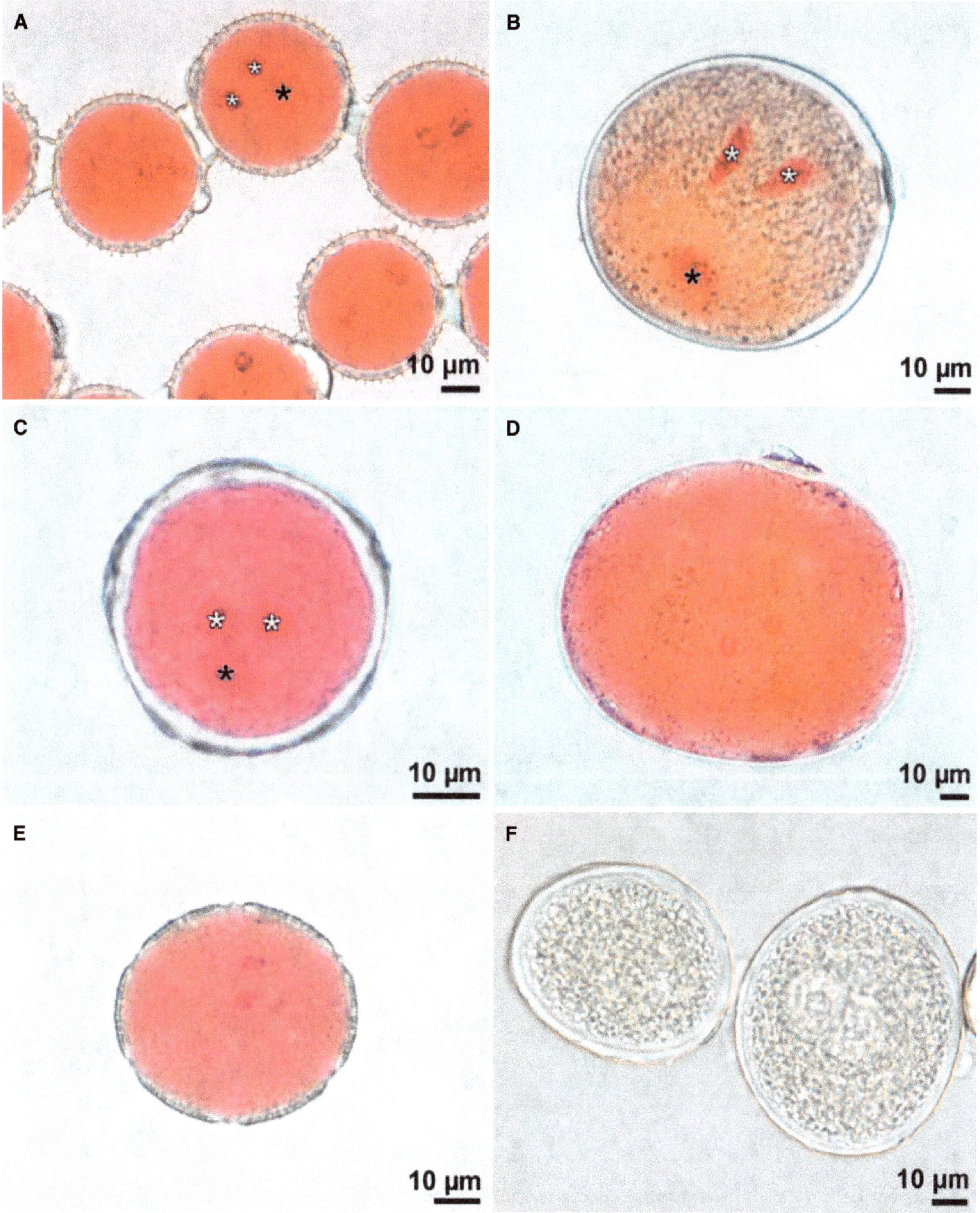

Fig. 8 Sperm cells of different species in LM. A. *Filarum manserichense*, Araceae, stained pollen showing two sperm cells (white asterisks) and vegetative nucleus (black asterisk), acetocarmine. **B.** *Triticum aestivum*, Poaceae, stained pollen showing two sperm cells (white asterisks) and vegetative nucleus (black asterisk), acetocarmine. **C.** *Ulmus minor*, Ulmaceae, stained pollen showing two sperm cells (white asterisks) and vegetative nucleus (black asterisk), acetocarmine. **D.** *Zea mays*, Poaceae, stained pollen showing two sperm cells, acetocarmine. **E.** *Thymus odoratissimus*, Lamiaceae, stained pollen showing two sperm cells, acetocarmine. **F.** *Amorphophallus taurostigma*, Araceae, pollen showing two sperm cells with nuclei, glycerine

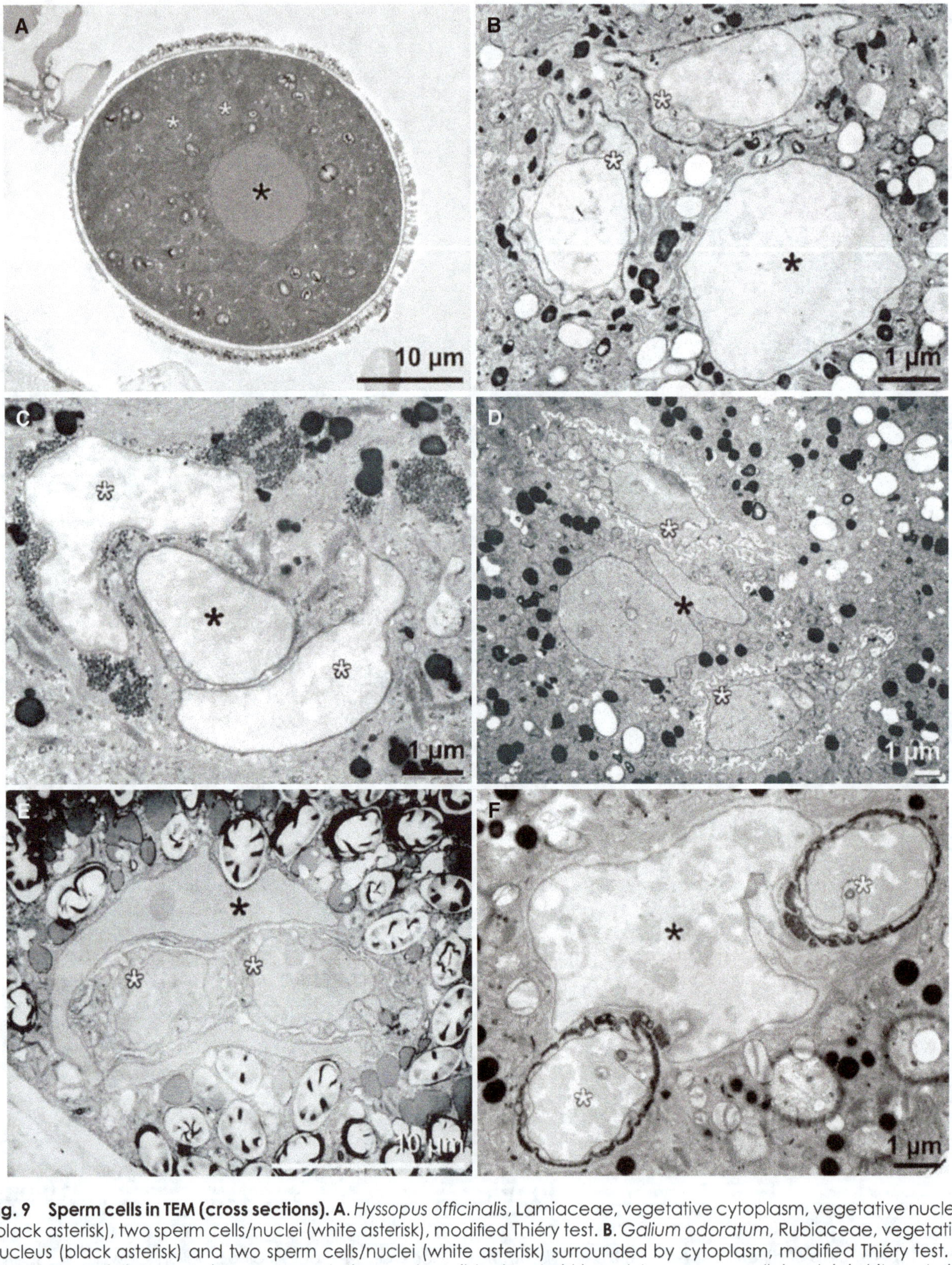

Fig. 9 Sperm cells in TEM (cross sections). A. *Hyssopus officinalis*, Lamiaceae, vegetative cytoplasm, vegetative nucleus (black asterisk), two sperm cells/nuclei (white asterisk), modified Thiéry test. **B.** *Galium odoratum*, Rubiaceae, vegetative nucleus (black asterisk) and two sperm cells/nuclei (white asterisk) surrounded by cytoplasm, modified Thiéry test. **C.** *Smyrnium perfoliatum*, Apiaceae, vegetative nucleus (black asterisk) and two sperm cells/nuclei (white asterisk) surrounded by cytoplasm, Thiéry test. **D.** *Jasminum nudiflorum*, Oleaceae, vegetative nucleus (black asterisk) and two sperm cells/nuclei (white asterisk) surrounded by cytoplasm, Lipid-test. **E.** *Zantedeschia aetiopica*, Araceae, vegetative nucleus (black asterisk) and two sperm cells/nuclei (white asterisk) surrounded by cytoplasm; sperm cells still in contact with each other and enclosed by the vegetative nucleus, modified Thiéry test. **F.** *Melampyrum pratense*, Orobanchaceae, vegetative nucleus (black asterisk) and two sperm cells/nuclei (white asterisk) surrounded by cytoplasm, modified Thiéry test

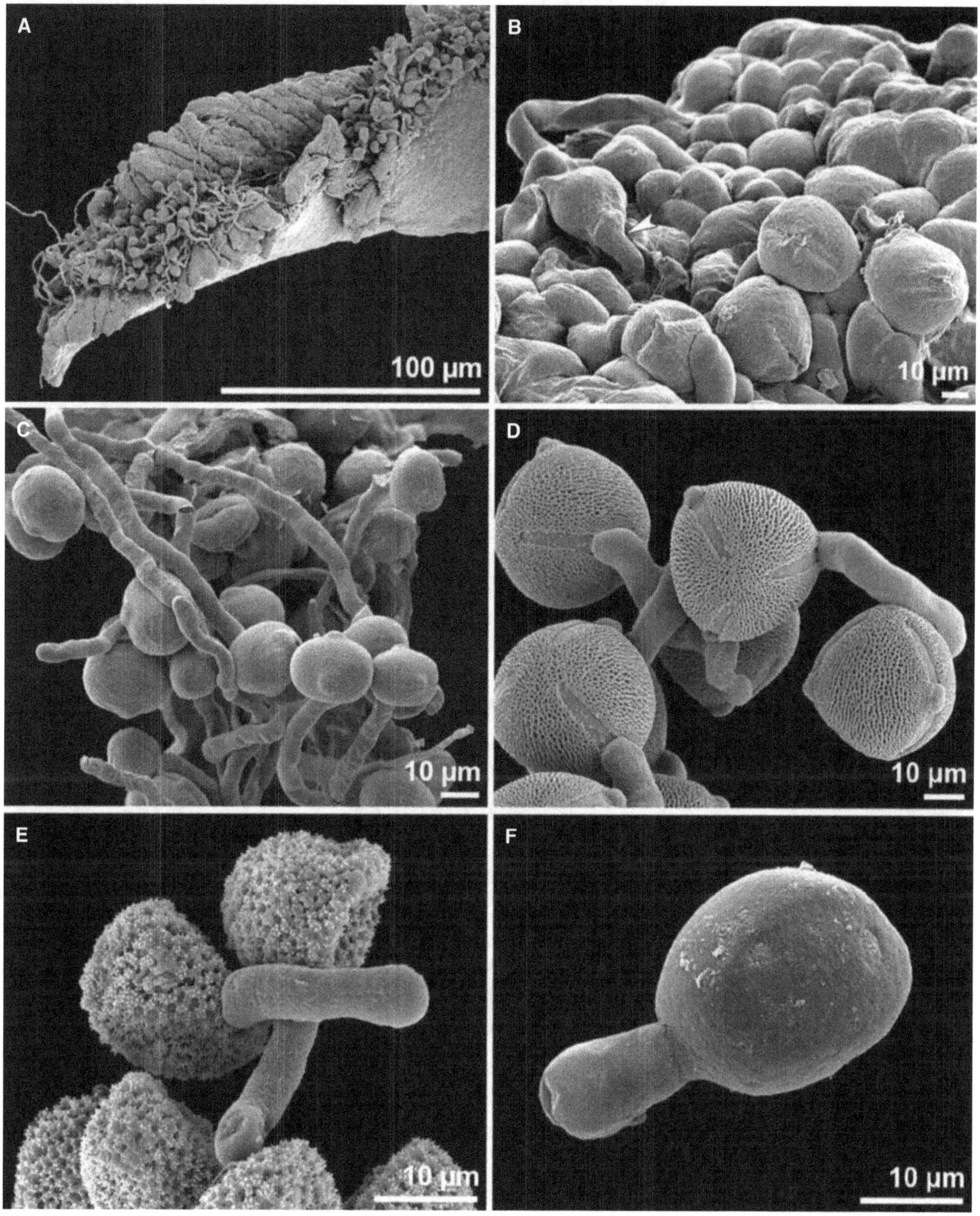

Fig. 10 Pollen germination and pollen tubes in SEM. A. *Cryptanthus bromelioides*, Bromeliaceae, sulcate pollen germinating on stigma. **B.** *Prunus* sp., Rosaceae, tricolporate pollen, note germinating pollen on stigma (left side). **C.** *Oxytropis jacquinii*, Fabaceae, tricolporate pollen. **D.** *Tuberaria guttata*, Cistaceae, tricolporate pollen. **E.** *Anthurium gracile*, Araceae, inaperturate pollen. **F.** *Vanilla pompona*, Orchidaceae, porate pollen

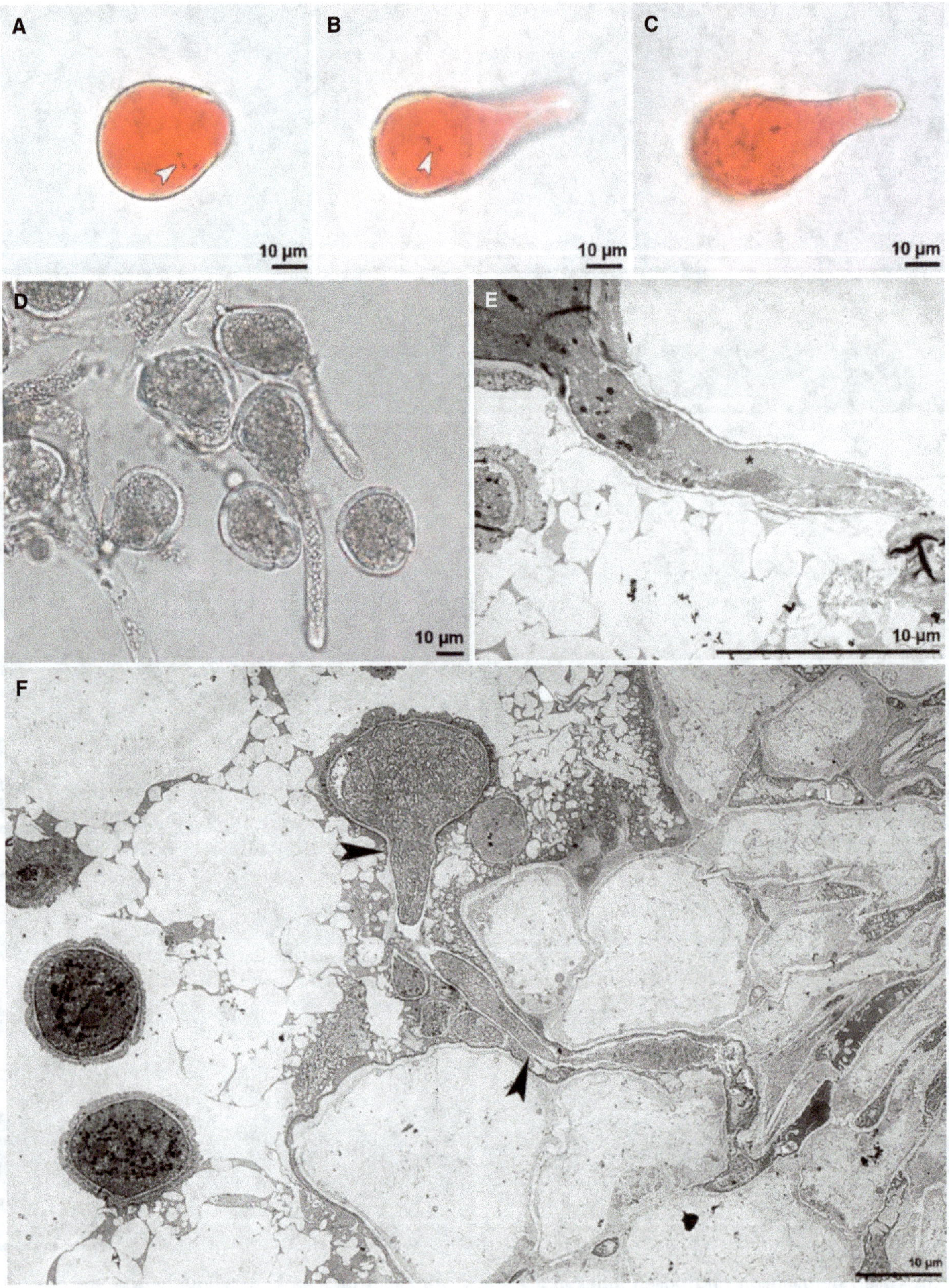

Fig. 11 Pollen germination and pollen tubes in LM and TEM. A-C. *Arum cylindraceum*, Araceae, three-celled, inaperturate pollen, germination can occur anywhere on the pollen surface, staining with acetocarmine, note the two sperm nuclei (arrowhead) staining dark red with acetocarmine, pictures. **B-C** showing optical section and upper focus. **D.** *Colocasia antiquorum*, Araceae, pollen grains germinating in water. **E-F.** *Smyrnium perfoliatum*, Apiaceae, TEM sections of germinating pollen (arrowheads), Thiéry test; detail of pollen tube with sperm nucleus (**E**, asterisk)

References

Brewbaker JL (1967) The distribution and phylogenetic significance of binucleate and trinucleate pollen grains in the angiosperms. Am J Bot 54: 1069–1083

Cresti M, Blackmore S, Van Went JL (1992) Atlas of sexual reproduction in flowering plants. Springer, Berlin, Heidelberg

Edlund AF, Swanson R, Preuss D (2004) Pollen and Stigma Structure and Function: The Role of Diversity in Pollination. Plant Cell 16: 84–97

Furness CA, Rudall PJ (2001) Pollen and anther characters in monocot systematics. Grana, 40: 17–25

Furness CA, Rudall PJ (1999) Microsporogenesis in Monocotyledons. Ann Bot 84: 475–499

Gomez JF, Talle B, Wilson ZA (2015) Anther and pollen development: A conserved developmental pathway. J Integr Plant Biol 57: 876–891

Keijzer CJ, Willemse MTM (1988) Tissue interactions in the developing locule of *Gasteria verrucosa* during microsporogenesis. Acta Bot Neerl 37: 493–508

Mascarenhas JP (1993) Molecular mechanisms of pollen tube growth and differentiation. Plant Cell 5: 1303–1314

McCormick S (1993) Male gametophyte development. Plant Cell 5: 1265–1275

Pacini E (1997) Tapetum character states: analytical keys for tapetum types and activities. Can J Bot 75: 1448–1459

Williams JH, Taylor ML, O'Meara BC (2014) Related evolution of tricellular (and bicellular) pollen. Am J Bot 101: 559–571

Pollen Morphology and Ultrastructure

H. Halbritter et al., *Illustrated Pollen Terminology*, https://doi.org/10.1007/978-3-319-71365-6_3

The study of pollen should encompass all structural and ornamental aspects of the grain. Pollen morphology is studied using LM and SEM and is important to visualize the general features of a pollen grain, including, e.g., symmetry, shape, size, aperture number and location, as well as ornamentation. TEM investigations are used to highlight the stratification and the uniqueness of pollen wall layers as well as cytoplasmic features. The following sections explain the most important structural and sculptural pollen features a palynologist should observe.

Polarity and Shape

Mature pollen is shed in **dispersal units**. When the post-meiotic products become separated the dispersal unit is a single pollen grain, a **monad**. Post-meiotic products also become partly separated or remain permanently united, resulting in **dyads** (a rare combination), **tetrads** or **polyads**. **Pollinaria** are dispersal units of two pollinia including a sterile, interconnecting appendage (see "Glossary of Palynological Terms").

Pollen shape and aperture location relate directly to pollen **polarity.** The polarity is determined by the spatial orientation of the microspore in the meiotic tetrad and can be examined in the **tetrad stage** (Fig. 1). The **polar axis** of each microspore/pollen runs from the **proximal pole,** orientated towards the tetrad center, to the **distal pole** of the microspore/pollen (Fig. 2). The **equatorial plane** is located at the microspore's center, perpendicular to the polar axis (Fig. 2). Therefore, the **equatorial plane** divides the microspore/pollen into a proximal and a distal half, comparable to the northern and southern hemisphere of our planet Earth.

The polarity gives rise to the polar and to the equatorial view. In dicots there is usually one polar

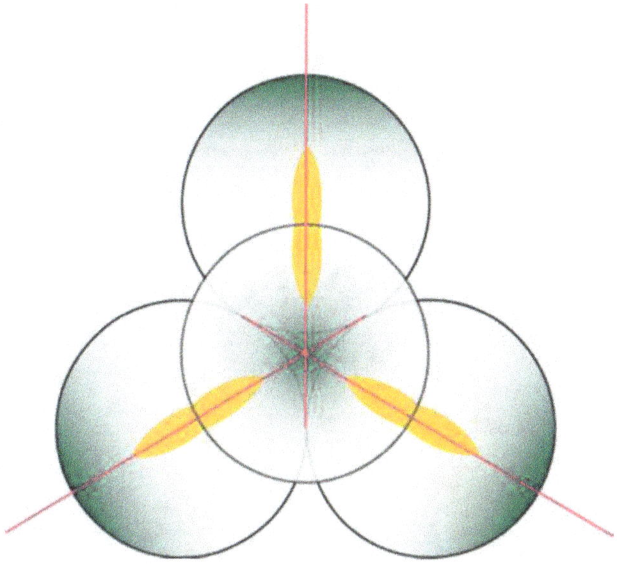

Fig. 1 Tetrad stage. Orientation of microspores/pollen in the tetrad; distal poles shaded green

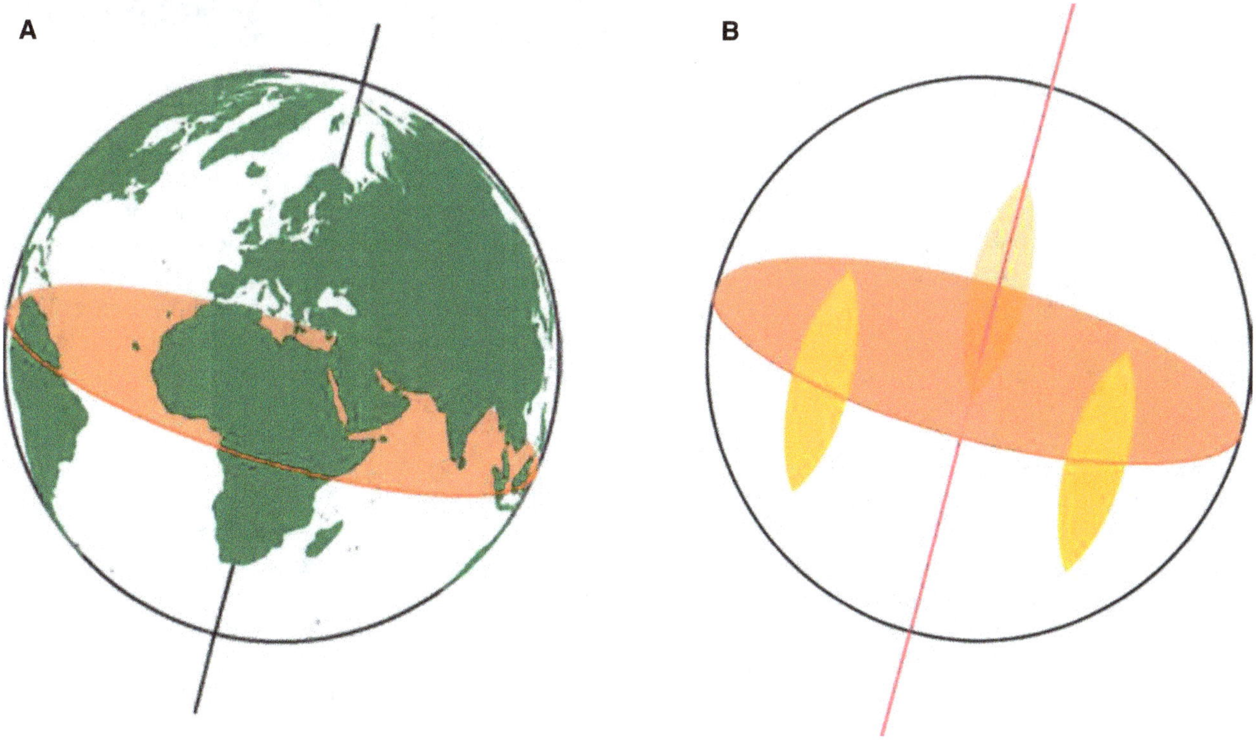

Fig. 2 Polar axis and equatorial plane. A-B. Polar axis and equatorial plane

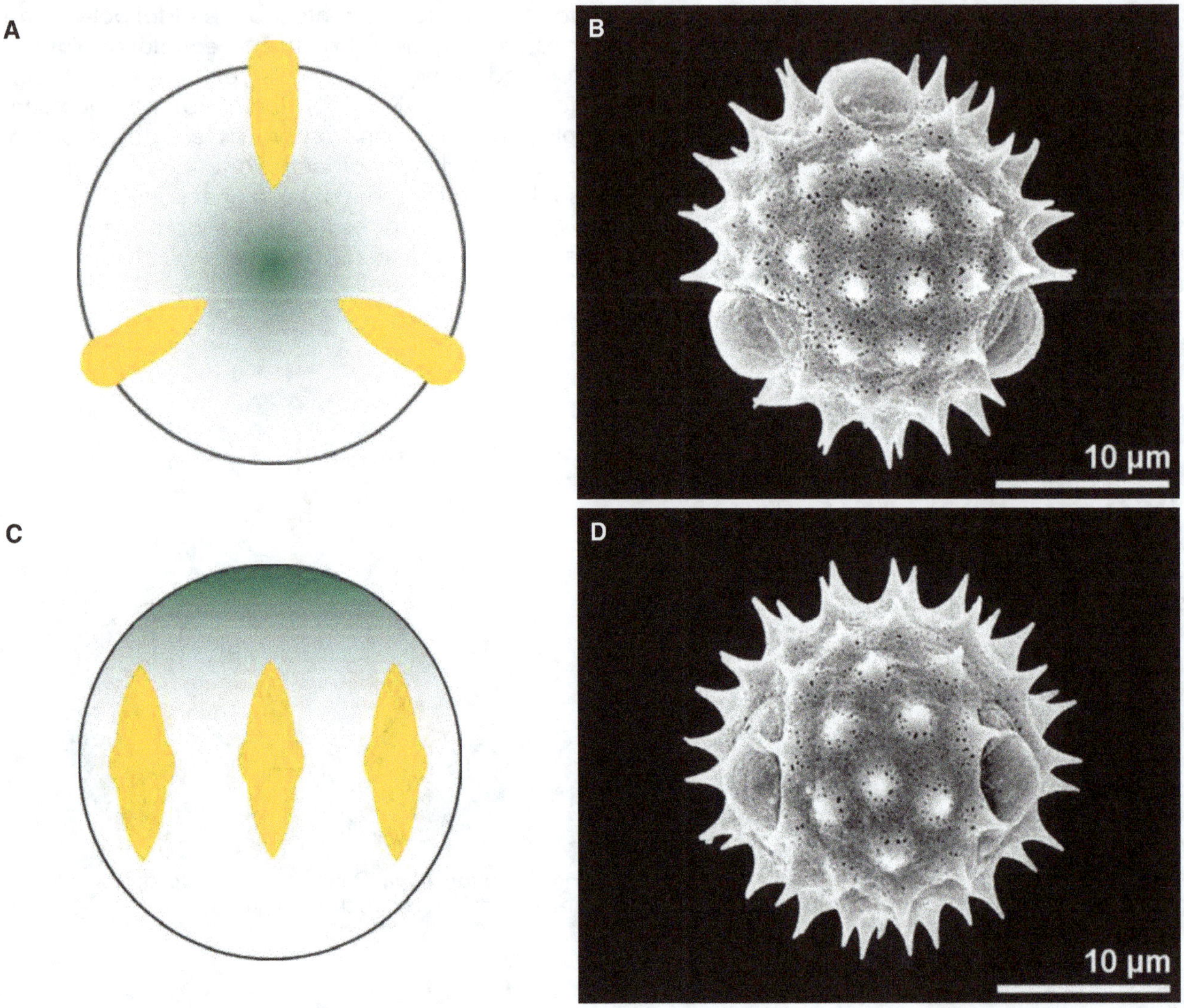

Fig. 3 Polarity of pollen in dicots. A-B. *Bellis perennis*, Asteraceae, polar view. **C-D.** *Bellis perennis*, Asteraceae, equatorial view

and one equatorial view (Fig. 3). In monocots, due to the mostly distal position of apertures, there are four views: distal polar, proximal polar, and two different equatorial views (Fig. 4).

Isopolar pollen has identical proximal and distal poles, thus the equatorial plane is a symmetry plane. In **heteropolar** pollen the proximal and distal halves differ (Fig. 5).

The various arrangements of the four microspores within **tetrads** depend on the simultaneous or successive type of cytokinesis and on the type of intersporal wall formation. The spatial arrangement of microspores after **simultaneous** cytokinesis is a **tetrahedral** (or rarely decussate) **tetrad** (Fig. 6A). This tetrad types may have systematic relevance, e.g., all species within the genus *Rhododendron* are characterized by tetrahedral tetrads. The spatial arrangement of microspores after **successive** cytokinesis leads to different morphological tetrad types, which can be differentiated into **planar** (tetragonal, linear, T-shaped) and/or **non-planar** (decussate or tetrahedral) tetrads (Fig. 6B). These morphotypes have no systematic relevance, as tetrads may vary within a genus/species, e.g., in *Typha* tetrads may be tetragonal, T-shaped and/or linear (Furness and Rudall 2001; Copenhaver 2005; see also "Illustrated Pollen Terms").

P/E ratio (Fig. 7) refers to the length of the polar axis (P) between the two poles compared to the equatorial diameter (E). In **isodiametric** pollen the polar axis is ± equal to the equatorial diameter. In **prolate** pollen the polar axis is longer than the equatorial diameter. In **oblate** pollen the polar axis is shorter than the equatorial diameter. **Pollen shape** refers to the 3-dimensional form of a pollen grain in relation to the P/E ratio. A pollen grain can, for

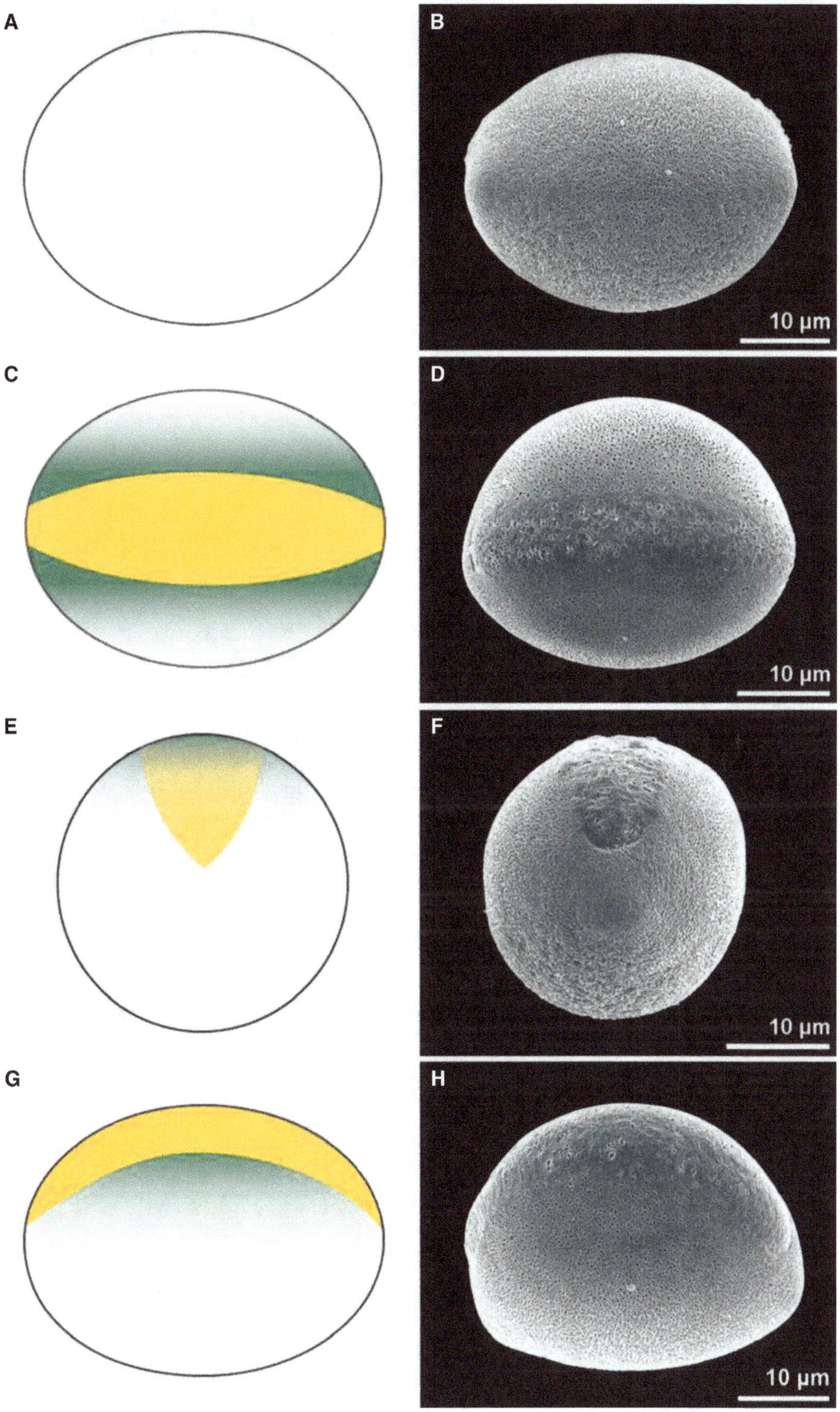

Fig. 4 Polarity of pollen in monocots (*Allium paradoxum*, Alliaceae). A-B. Proximal polar view. **C-D.** Distal polar view. **E-F.** Equatorial view (short axis). **G-H.** Equatorial view (long axis)

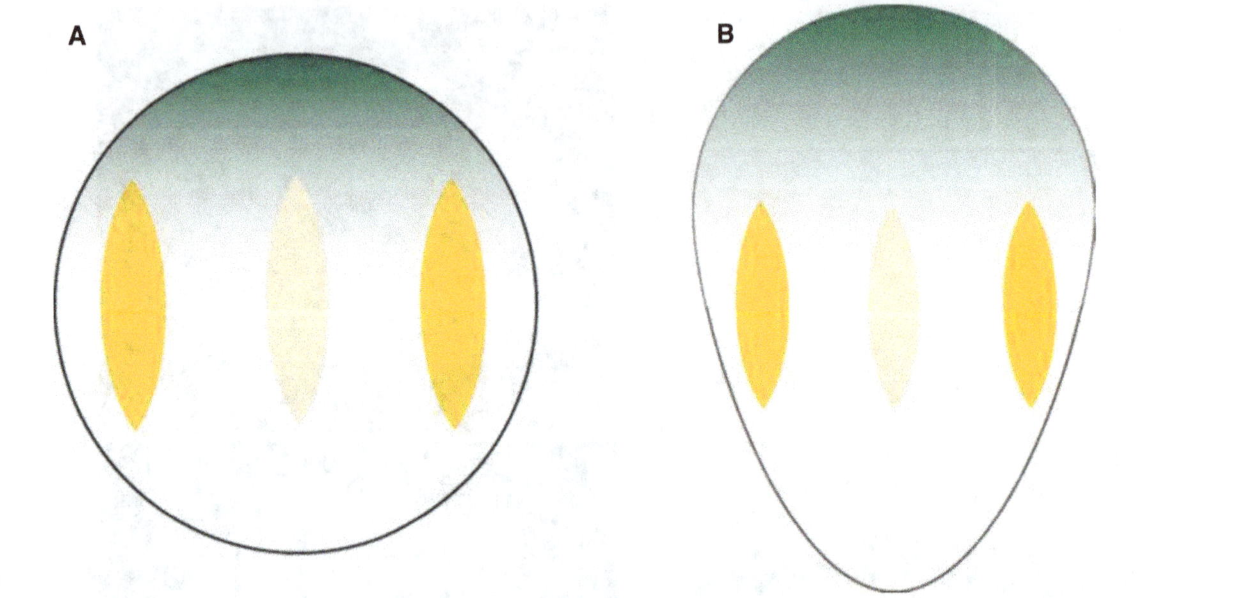

Fig. 5 Pollen symmetry. A. Isopolar pollen. **B.** Heteropolar pollen

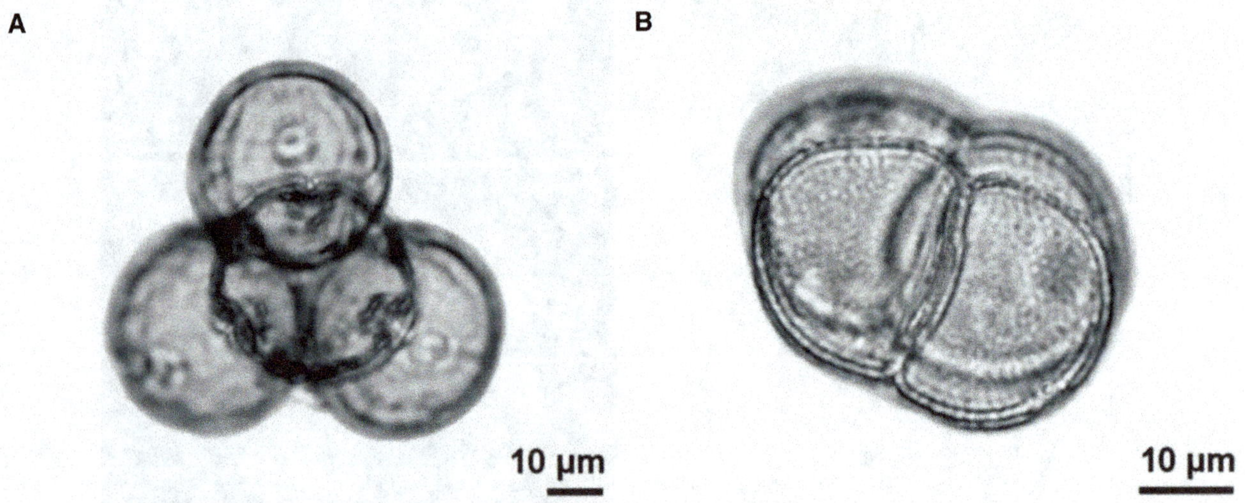

Fig. 6 Pollen arrangements in tetrads. A. Tetrahedral tetrad, *Fagus* sp., Fagaceae, fossil, Quaternary, Austria; apical view. **B.** Planar tetrad, *Typha latifolia*, Typhaceae, fossil, Quaternary, Austria

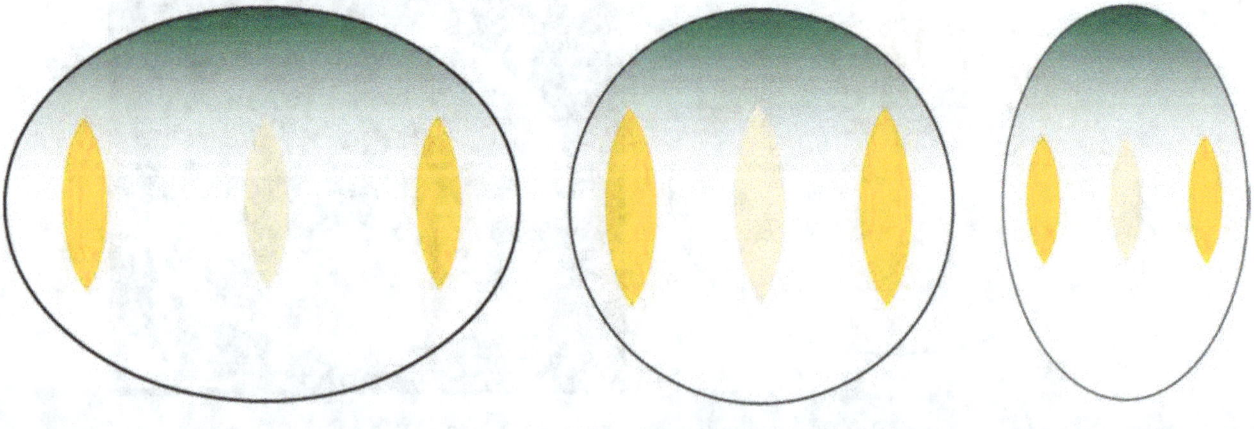

Fig. 7 P/E ratio of pollen. Schematic drawings of oblate (left), isodiametric (middle), and prolate (right) pollen

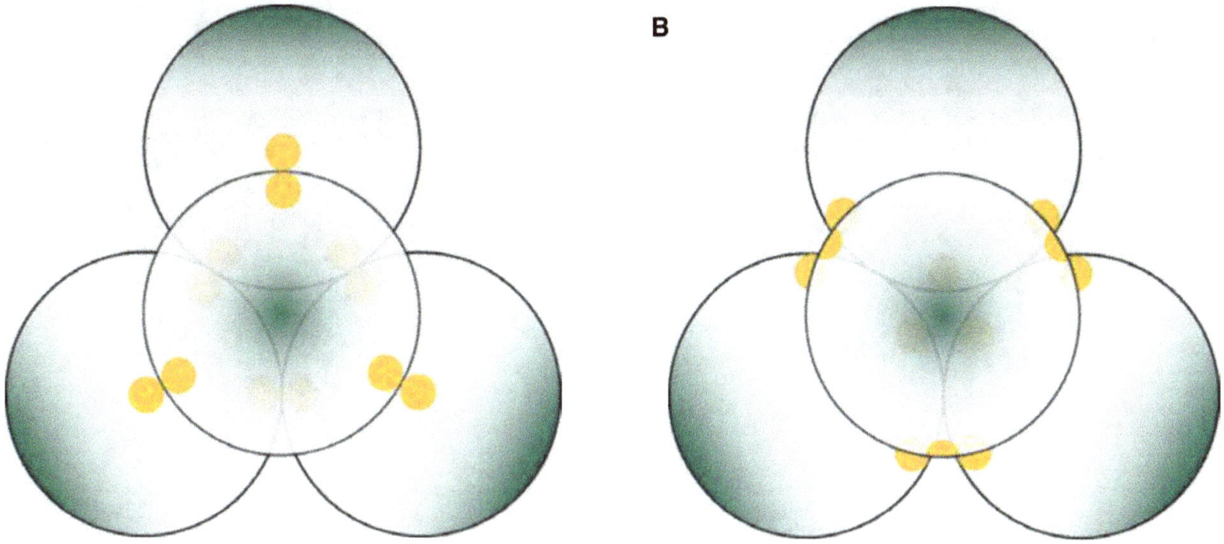

Fig. 8 Aperture arrangement. A. Fischer's law, apertures in pairs. **B.** Garside's law, apertures in a group of three

example, be spheroid-, cup-, boat-, cube-, tetrahedral-, triangular dipyramid-, hexafoil dipyramid-, triangular prism-, pentagonal prism-, or hexagonal prism shaped (see "Illustrated Pollen Terms").

In pollen grains with three apertures, two types of aperture arrangement occur after simultaneous cytokinesis (Fig. 8). **Fischer's law** refers to the most frequent arrangement where a pair of apertures occurs at six points in a tetrad (e.g., Ericaceae, permanent tetrads). **Garside's law** refers to the unusual arrangement of apertures where a group of three apertures occur at four points in the tetrad (probably restricted to Proteaceae, no permanent tetrads; Blackmore and Barnes 1995) (Fig. 8).

Apertures

An **aperture** is a region of the pollen wall that differs significantly from its surroundings in morphology and/or anatomy. The aperture is presumed to function as the site of germination and to play a role in harmomegathy. Pollen grains lacking apertures are called **inaperturate** (Furness 2007). The aperture definition fits both angiosperm and gymnosperm pollen, but in gymnosperms the type of aperture (e.g., leptoma; germination area) usually differs from that in angiosperms.

The polarity of the pollen grain determines the aperture terminology. A circular aperture is termed a **porus** if situated equatorially or globally; if situated distally, it is called an **ulcus.** An elongated aperture is termed a **colpus** if situated equatorially or globally; if situated distally, it is termed a **sulcus**. A combination of porus and colpus is termed a **colporus**; colpori are situated equatorially or globally. In **heteroaperturate** pollen two different types of apertures (single and/or combined) are present in a combination of colpi with colpori or pori. A circular or elliptic aperture with indistinct margins is termed a **poroid**. Additional rare combinations of ekto- and endoapertures, mostly observed in LM, include pororate and colporoidate (Fig. 9). Pollen grains that have compound apertures composed of circular ektopori and endopori are termed **pororate**. Compound apertures composed of a colpus (ektoaperture) with an indistinct endoaperture are termed **colporoidate**. When the colpus has a clear bulge in the equatorial region of a pollen grain it is termed **geniculum** (Fig. 9D).

The number of equatorial apertures (pori, colpi, colpori) is indicated by the prefixes di-, tri-, tetra-, penta- or hexa-. Writing numbers instead of prefixes is in common use, e.g., 4-porate or tetraporate, 6-colpate or hexacolpate. In this book we prefer the use of prefixes. For pollen grains with more than three apertures, positioned at the equator, the term **stephanoaperturate (stephanoporate, stephanocolpate, stephanocolporate)** is used together with the aperture number (e.g., stephano(4)porate or 4-porate, stephanoporate). Pollen grains with globally distributed apertures are termed **pantoaperturate.**

Apertures are normally covered by an exinous layer, the **aperture membrane**. The aperture membrane can be **ornamented**, e.g., covered with various exine elements, or it is **psilate** (smooth). The aperture can also be covered by an **operculum**, a distinctly delimited exine structure, covering the aperture like a lid (Halbritter and Hesse 1995; Furness and Rudall 2003).

Number, type, and position of apertures are genetically determined and usually the same within

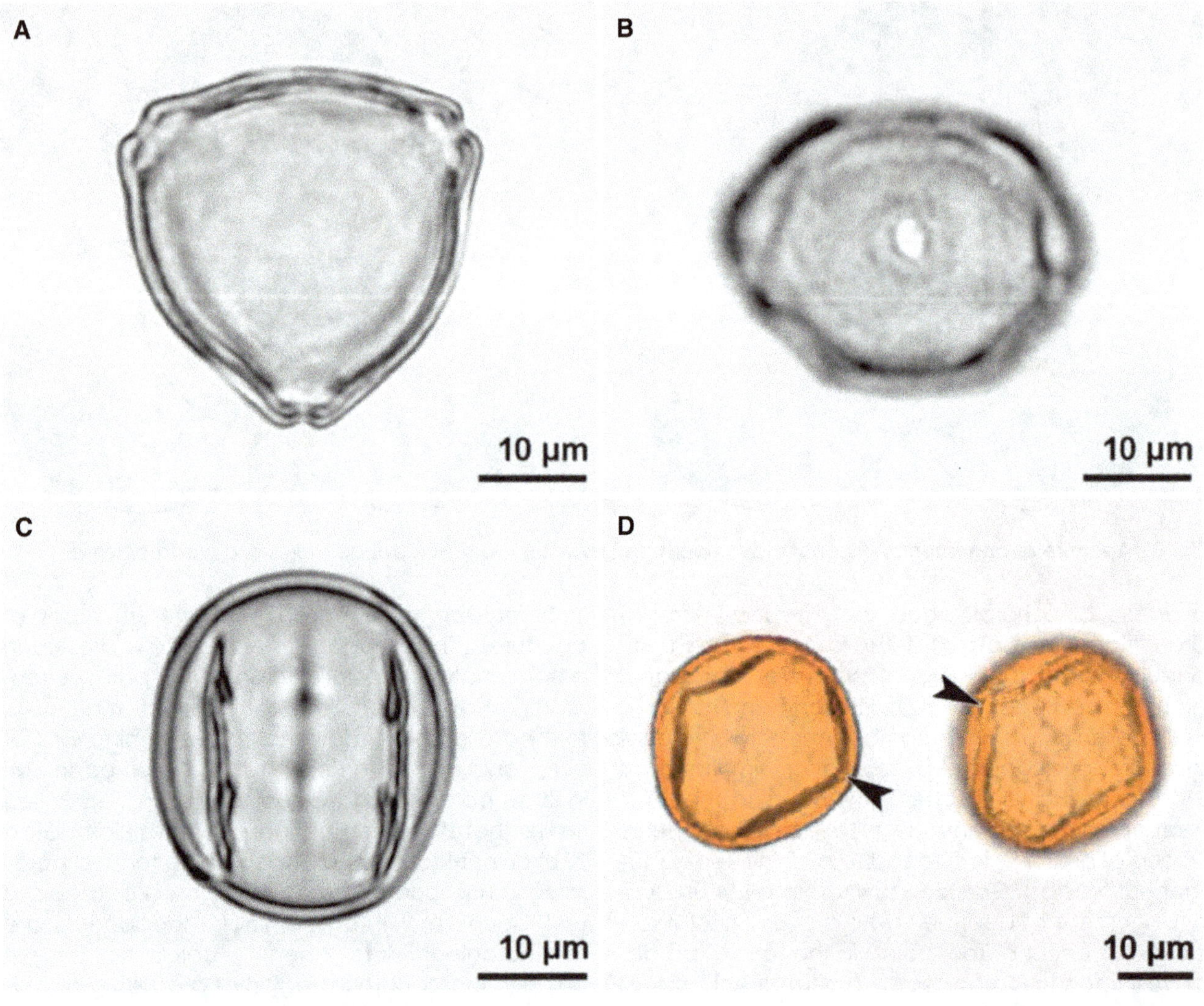

Fig. 9 Special aperture features observed in LM. A-B. *Corylus* sp., Betulaceae, fossil, Quaternary, Austria, pororate pollen in polar and equatorial view. **C.** *Eucommia* sp., Eucommiaceae, fossil, middle Miocene, Austria, colporoidate pollen in equatorial view. **D.** *Quercus petrea*, Fagaceae, pollen with geniculum (arrowhead), optical section (left) and upper focus (right)

a species, but may also vary (e.g., *Alnus* is usually 5-porate, but number of pori can vary from 3 to 6).

A **pseudocolpus** occurs in heteroaperturate pollen and is presumed to be non-functional. Pseudocolpi mostly alternate with colpori (e.g., in Boraginaceae, Lythraceae) or are flanking each colporus (in Acanthaceae). For examples, see "Illustrated Pollen Tems." Pseudocolpi are believed to play a role in **harmomegathy**, but their effect has been poorly studied.

Pre-(prae-)pollen (Fig. 10) is characterized by proximal and sometimes additional distal apertures, and by presumed proximal germination. Pre-pollen are microspores of certain extinct basal seed plants occurring from the Late Devonian until the Cretaceous. Proximal germination is typical for spores.

Spores germinate at the **tetrad mark** (Fig. 11), the so-called **laesura** (for an extensive overview, see

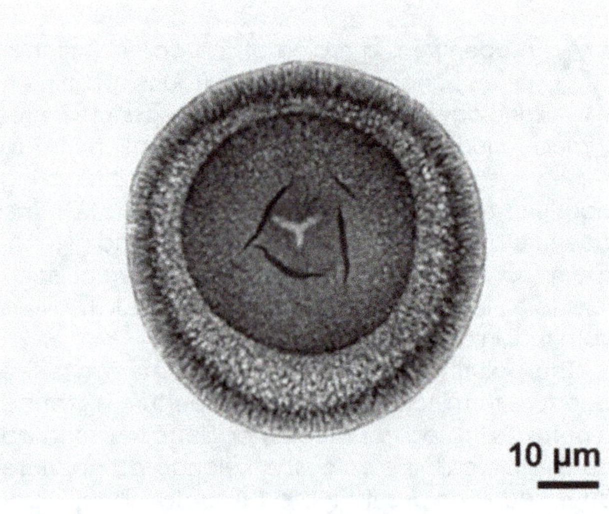

Fig. 10 Pre-pollen. *Nuskoissporites* sp., fossil, Permian, Austria, polar view

Tryon and Lugardon 1991). The tetrad mark is situated at the proximale pole (proximal germination).

Proximal germination is a rare exception in seed plants (Fig. 12), e.g., *Beschorneria yuccoides* (Agavaceae) and *Annona muricata* (Annonaceae). In the two cases, this proximally situated aperture (germination area) is functionally replacing the dysfunctional sulcus (Hesse et al. 2009). In *Beschorneria*, pollen grains forming the tetrads are loosely interconnected and separate frequently. In this special case, the sulcus (distal) is not functional, whereas the proximal face, with a highly reduced exine, functions as germination site. In *Annona*, the microspores rotate within the tetrad during development and the original distally placed sulcus becomes proximally positioned (Tsou and Fu 2002).

The aperture usually acts as the (exclusive) **germination** site. In inaperturate angiosperm pollen the pollen tube can protrude at any given site. In taxoid gymnosperm pollen the exine ruptures during hydration at a specialized region, the leptoma, and is subsequently shed (Fig. 13A-B). The protoplast (enclosed by the intine) is released and a pollen tube can be formed anywhere (resembling functionally an inaperturate pollen grain). Furthermore some angiosperm taxa shed the exine before pollen tube formation, e.g., in some Annonaceae, Araceae. Within the Araceae, a shed pollen wall has been observed in several taxa, e.g., *Amorphophallus*, *Taccarum* (Ulrich et al. 2017). The outer pollen wall (composed of polysaccharide) splits immediately in water and sheds soon afterwards. Subsequently, the naked protoplast is

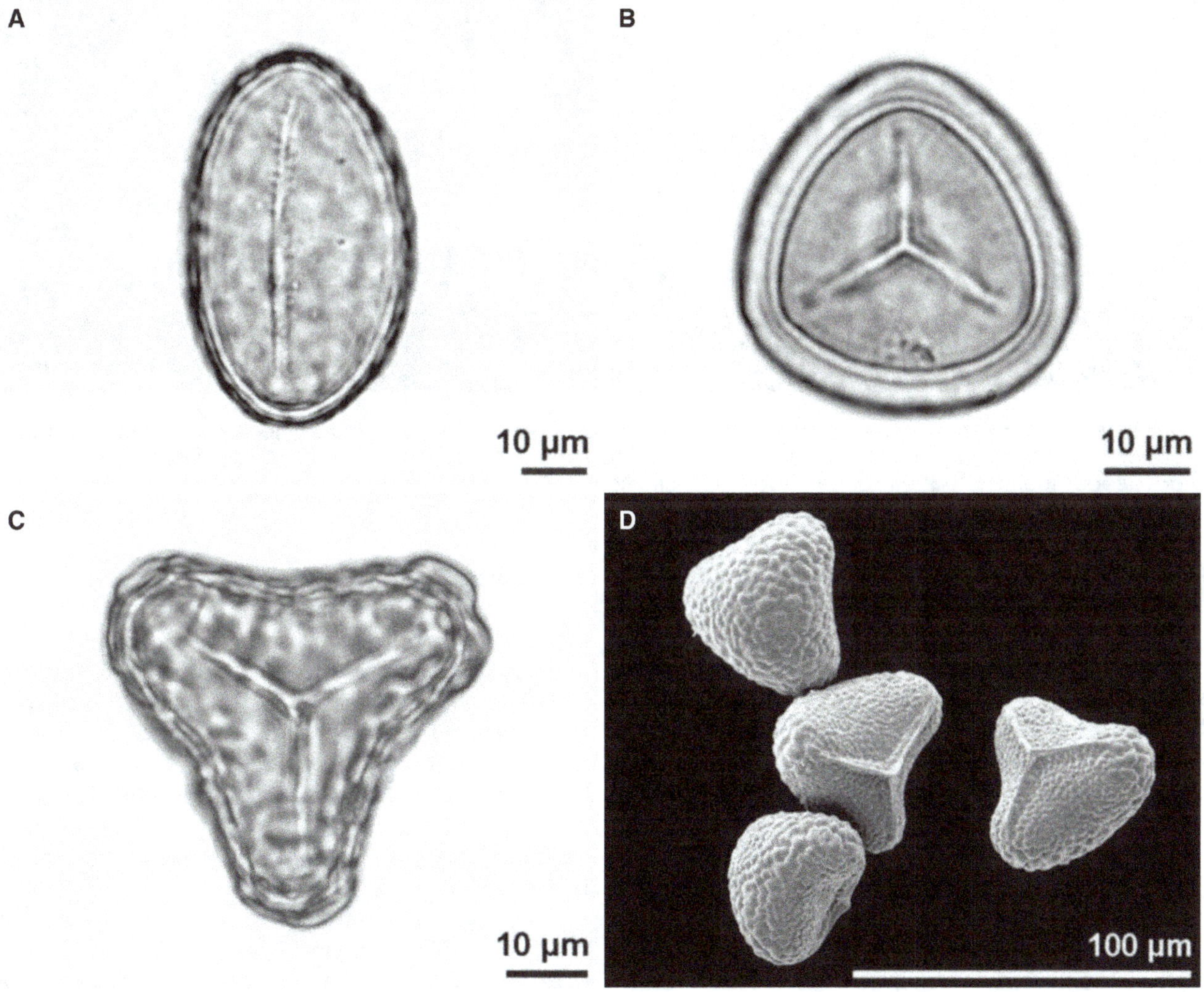

Fig. 11 Tetrad mark in spores. A. *Polypodium* sp., Polypodiaceae, fossil, monolete tetrad mark, middle Miocene, Austria, polar view. **B.** *Sphagnum* sp. Sphagnaceae, fossil, trilete tetrad mark, middle Miocene, Austria, polar view. **C.** Pteridaceae indet., fossil, middle Miocene, Austria, trilete tetrad mark, polar view. **D.** *Cryptogamma crispa*, Pteridaceae, trilete tetrad mark

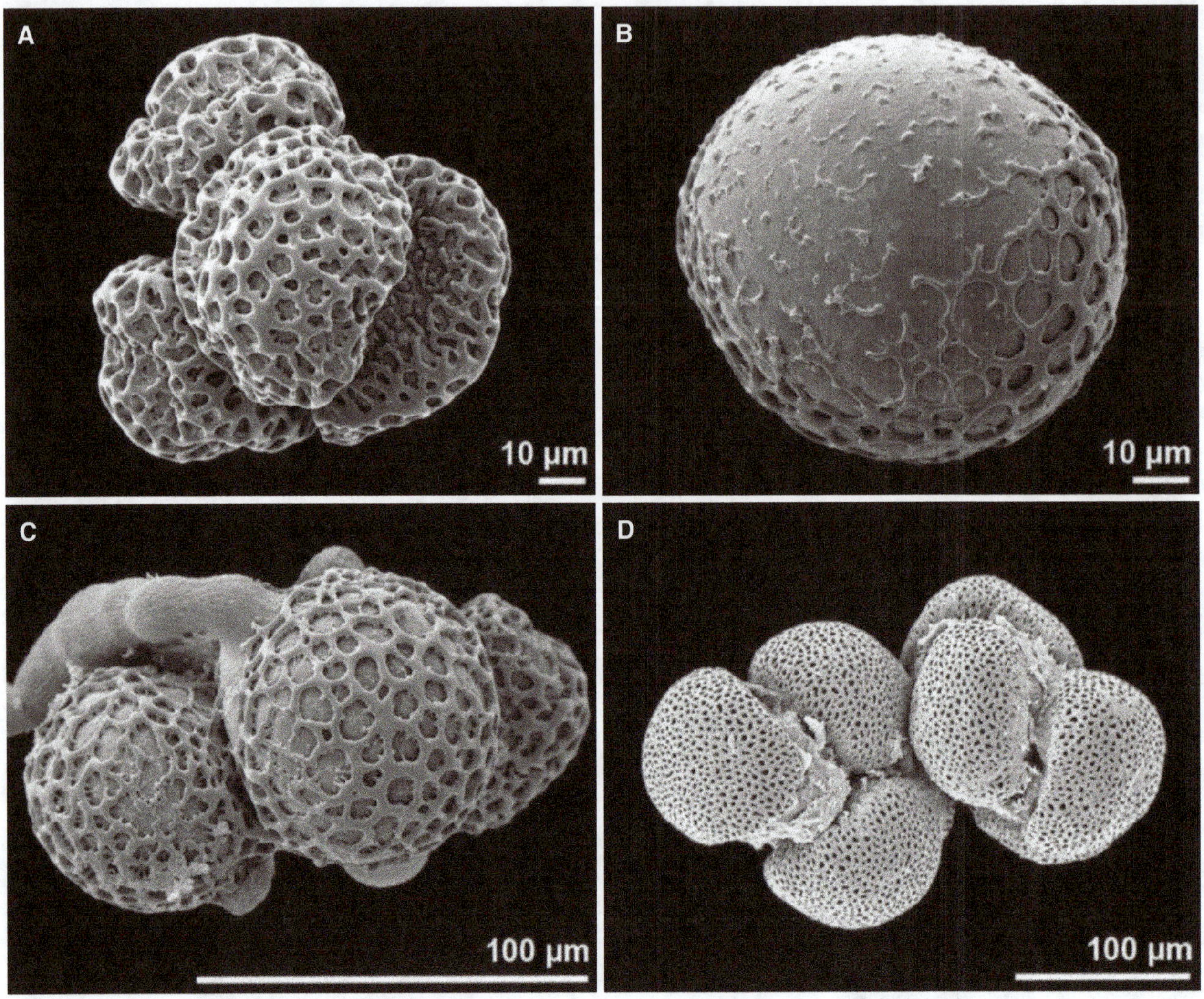

Fig. 12 Proximal germination. A. *Beschorneria yuccoides* (Agavaceae), dry tetrad. **B.** *Beschorneria yuccoides* (Agavaceae), monad, proximal polar view, proximal face functions as germination site. **C.** *Beschorneria yuccoides* (Agavaceae), germinated tetrad, note proximal germination. **D.** *Annona muricata* (Annonaceae), mature tetrads, sulcus hidden in proximal position

floating in water and germinates about 1 hour after shedding (Fig. 13C-D).

During germination, usually a single pollen tube is formed. In some cases, instant pollen **tube-like structures** are simultaneously developed at all apertures (Fig. 14). The formation of these pollen tube-like structures, in relation with moisture, is interpreted as a pre-germinative process that takes place during dehiscence (Blackmore and Cannon 1983).

Pollen Wall

The internal construction of the pollen wall is termed **structure**. Ornamenting elements on the pollen surface (ornamentation) are summarized under the term **sculpture** or sculpturing. However, it is not always possible to distinguish between structure and sculpture (e.g., free-standing columellae).

Structure

In general, the **pollen wall (sporoderm)** of seed plants is formed by two main layers: the outer **exine** and the inner **intine** (Fig. 15). The exine consists mainly of **sporopollenin**, which is an acetolysis- and decay-resistant biopolymer. The intine is mainly composed of cellulose and pectin. Commonly, the pollen wall in aperture regions is characterized by the reduction of exine structures or by a deviant exine, and a thick, often bilayered intine.

Two layers within the exine are distinguished: an inner endexine and an outer ektexine. In **tectate** pollen the ektexine usually consists of a basal **foot layer**, an **infratectum** (e.g., columellae) and a **tectum**, the **endexine** is a mainly unstructured layer (Fig. 16A–C). There are many deviations from this principal construction: layers may be thickened, variably structured or lacking. When the pollen

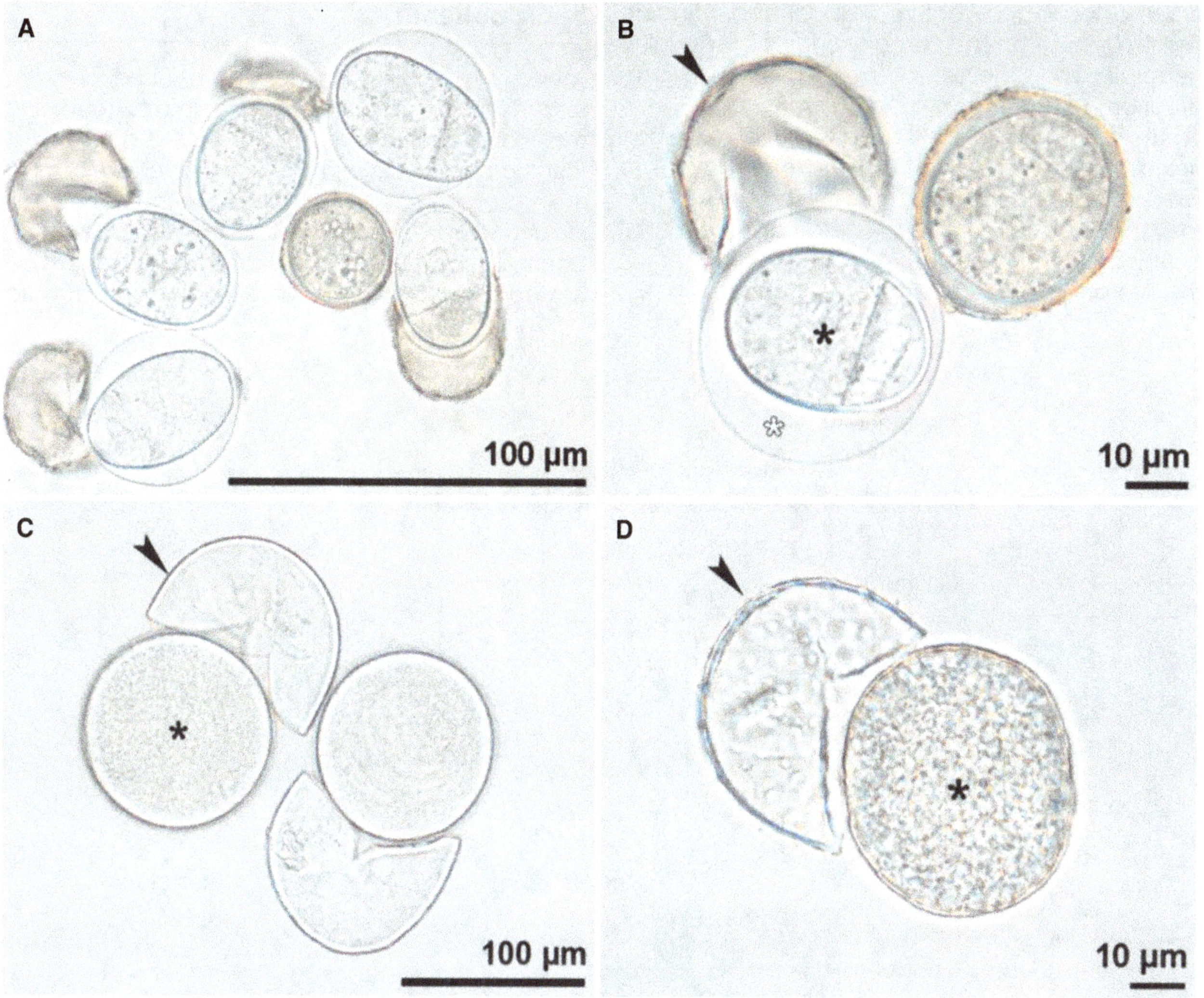

Fig. 13 Exine/pollen wall shedding. A. *Cephalotaxus* sp., Cephalotaxaceae, fresh pollen in water. **B.** exine (*arrowhead*) shedding prior to pollen tube formation, released protoplast (black asterisk) enclosed by a thick swelled intine (white asterisk). **C.** *Taccarum weddellianum*, Araceae, pollen wall shedding, released protoplast (black asterisk) and shed outer pollen wall (arrowhead). **D.** *Amorphophallus mangelsdorffii*, Araceae, pollen wall shedding, released protoplast (black asterisk) and shed outer pollen wall (arrowhead)

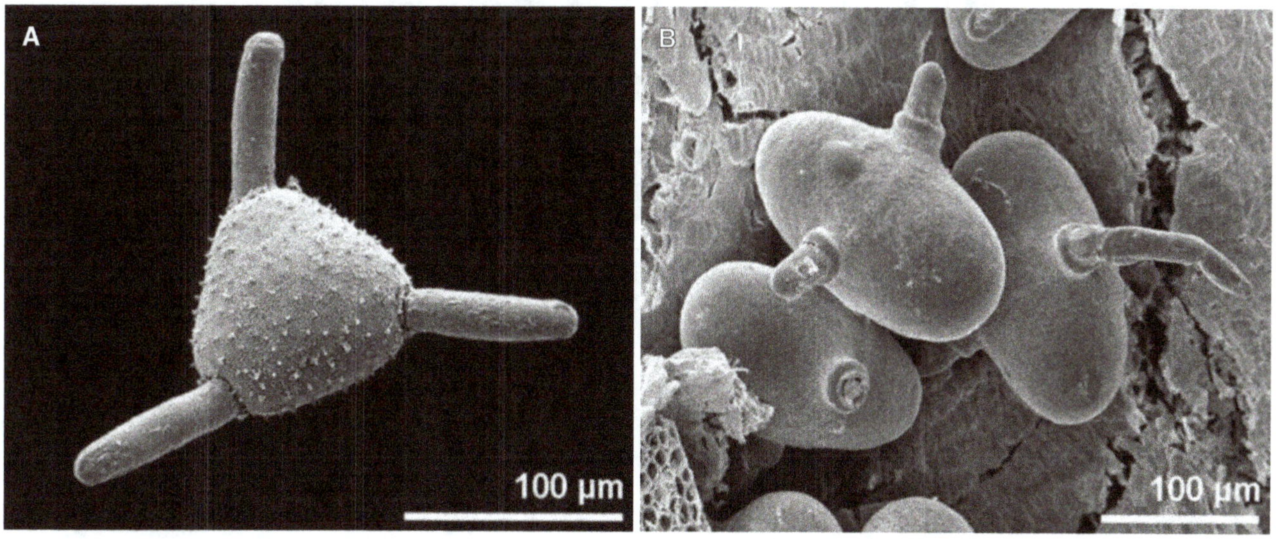

Fig. 14 Instant pollen tubes. A. *Scabiosa caucasica*, Dipsacaceae. **B.** *Morina longifolia*, Morinaceae

grain is lacking a tectum it is termed **atectate** (Fig. 16D–F). In apertural regions the pollen wall is generally characterized by a different exine construction.

The terms **sexine** for the outer, structured, and **nexine** for the inner, unstructured exine layer are widely used in light microscopy, but do not fully correspond to ekt- and endexine, respectively. When a cavity between the sexine and nexine is present in the interapertural area, this is termed **cavea** (Fig. 17).

Sporopollenin

John (1814) and Braconnot (1829) introduced the terms "pollenin" and "sporonin" for the resistant exine material of pollen and spores. Zetzsche et al. (1931) then combined the terms into "sporopollenin," that is the major component of the exine found in most pollen and spores, except in filiform seagrass pollen (e.g., Dobritsa et al. 2009; Jardine et al. 2015). Sporopollenin is a complex biopolymer and extremely resistant to

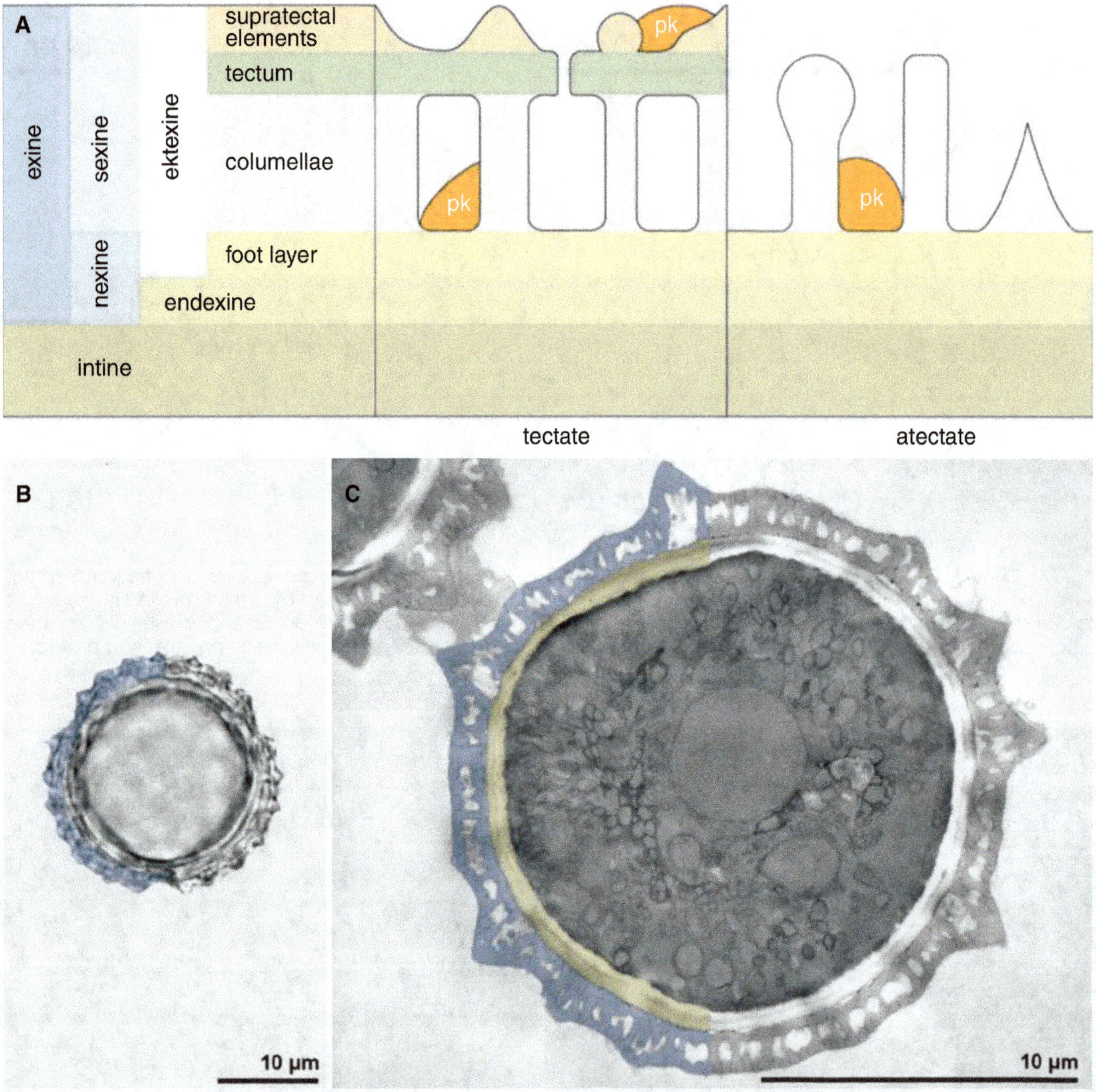

Fig. 15 Pollen wall stratification. A. Schematic cross section of pollen wall, pk: pollenkitt. **B.** *Ambrosia artemisiifolia*, Asteraceae, optical view showing both intine and exine; acetolyzed. **C.** *Ambrosia artemisiifolia*, Asteraceae, cross section showing both intine (yellow) and exine (blue); modified Thíery-test

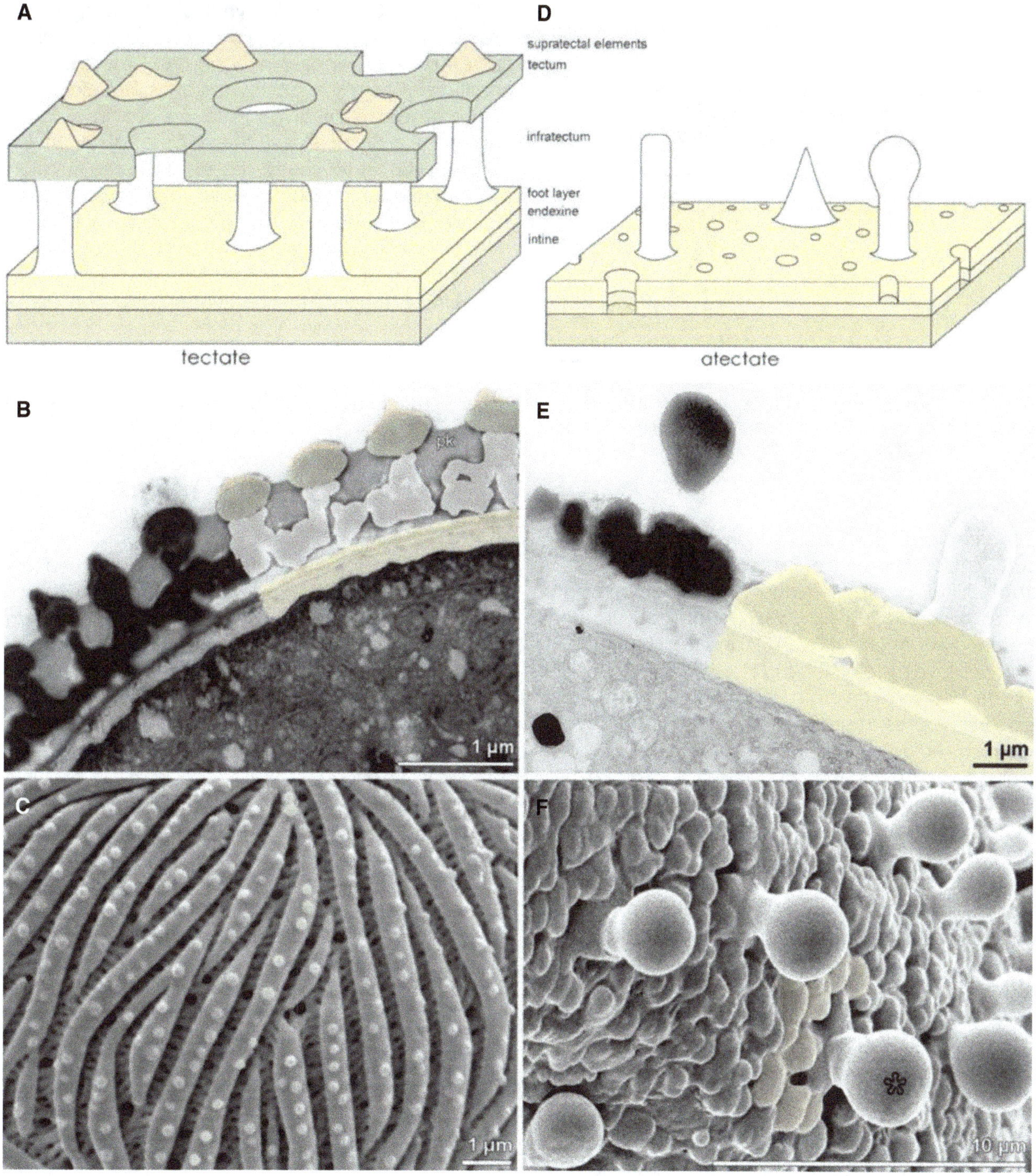

Fig. 16 Tectate vs atectate pollen wall. A-C. Tectate pollen wall. **A.** 3D-model. **B.** *Saxifraga scardica*, Saxifragaceae, cross section showing pollen wall stratification: tectum with internal tectum and supratectal elements, columellate infratectum, very thin footlayer, thin compact-continuous endexine, monolayered intine (colors refer to picture **A**), *pk* pollenkitt. **C.** *Saxifraga scardica*, Saxifragaceae, exine surface in SEM, sculpture striate with nanoechinate suprasculpture (colored). **D-F.** Atectate pollen wall. **D.** 3D-model. **E.** *Iris pumila*, Iridaceae, cross section showing pollen wall stratification: tectum and infratectum lacking, compact-continuous footlayer, monolayered intine (colors refer to picture **D**). **F.** *Iris pumila*, Iridaceae, exine surface (foot layer) in SEM, sculpture verrucate (colored) and clavate (asterisk)

decay as well as to chemical and mechanical damage (e.g., Steemans et al. 2010). However, in the environment, both biotic and abiotic factors are involved in pollen decomposition. Biotic factors are, for instance, the intrusion of bacteria and fungi (e.g., Elsik 1971; Havinga 1971, 1984; Skvarla et al. 1997; Phuphumirat et al. 2011). Abiotic factors include the pH-value of the substrate (e.g., Bryant and Hall 1993),

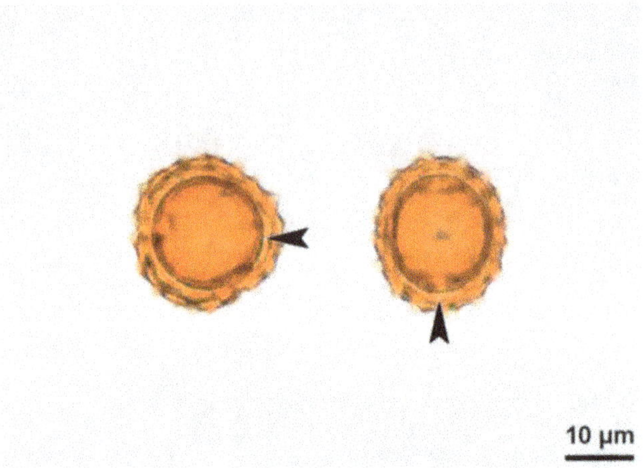

Fig. 17 Cavea. *Xanthium spinosum*, Asteraceae, acetolyzed pollen in polar (left) and equatorial (right) view showing exine cavity (cavea) between sexine and nexine (arrowheads)

oxidation/reduction (e.g., Twiddle and Bunting 2010), autoxidation by UV-light and oxygen (e.g., Jardine et al. 2015), destruction due to mechanic impact, water or fire (Cushing 1967; Bryant et al. 1994; Phuphumirat et al. 2011, 2015), and rapid changes in moisture levels (Halbritter and Hesse 2004).

The preservation status and the amount of pollen and spores in sediments depends on several factors, including rapid anaerobic burial and embedding in mud or peat, absence of any microbial destruction or sapropel, and the exclusion of oxygen (Klaus 1960, 1987; Playford and Dettmann 1996; Traverse 1988, 2007).

Recent studies on the **composition of sporopollenin** suggest that it may have two different types of chemical structures, oxygenated aromatic compounds and aliphatic compounds (e.g., Wiermann et al. 2001; Dobritsa et al. 2009; Gabarayeva and Grigorjeva 2010; Gabarayeva et al. 2010; Steemans et al. 2010; Colpitts et al. 2011). Although its exact structure remains unknown, sporopollenin is believed to compose oxidative polymers of carotenoids, polyunsaturated fatty acids, and conjugated phenols (Diego-Taboada et al. 2014). Some authors are using the plural form "sporopollenins," because there is evidence for several types of sporopollenin in ferns, gymnosperms, and angiosperms (Hemsley et al. 1993; de Leeuw et al. 2006). According to Diego-Taboada et al. (2014) sporopollenin in plants share a common aliphatic core, but depending on the taxon, contain different aromatic side chains. The **chemical constitutional formula** of sporopollenin is also unknown. The **empirical formula** of sporopollenin has highly variable amounts of H- and O-numbers. A generalized formula is $C_{90}H_{142}O_{36}$ (Traverse 1988; Riding and Kyffin-Hughes 2004).

The precise location of **synthesis** of sporopollenin precursors in tapetal cells and the mechanisms of secretion of sporopollenin monomers before polymerization in the microspore walls are still unclear, just as the processes involved in sporopollenin production at the cellular level (Lallemand et al. 2013). Liu and Fan (2013) reviewed the molecular regulation of sporopollenin biosynthesis, which probably includes a framework of catalytic enzyme reactions. As shown in the study by Colpitts et al. (2011), genes responsible for sporopollenin biosynthesis in *Arabidopsis* lead to the conclusion, that the pathway of sporopollenin biosynthesis seems well conserved in land plants since nearly 500 mya.

The question if sporopollenin is of sporophytic or gametophytic origin is still controversial. Probably both sources are involved. Most authors agree that sporopollenin is predominantly produced by the tapetum (Pacini and Franchi 1991; Blackmore et al. 2000; Wallace et al. 2015; Ariizumi and Toryama 2011; Quilichini et al. 2014).

The investigation of fossil pollen and spores revealed that **fossilized sporopollenin** appears chemically very different to sporopollenin found in modern plants (Fraser et al. 2011). During fossilization (coalification) and by diagenetic processes the chemical composition of sporopollenin is modified. Especially at high temperatures, above 200 °C, sporopollenin undergoes a series of chemical changes (Yule et al. 2000; Fraser et al. 2014).

Sporopollenin biochemistry appears to have remained relatively stable since at least the Middle Pennsylvanian (approx. 310 mya). Fraser et al. (2012, 2014) postulated that the structure of sporopollenin has remained constant since plants invaded land during the Middle Ordovician (470-458 mya). A recent comprehensive review on sporopollenin and other biopolymers (de Leeuw et al. 2006) suggests that there may have been multiple forms and configurations of sporopollenin over geological time.

The sporopollenin wall is regarded as a synapomorphy in land plants and allowed land dispersal during the Silurian, perhaps already during the Middle Ordovician (Rubinstein et al. 2010; Wellman 2010).

Chemically Related Biomacromolecules

Sporopollenin is not unique in pollen/spore walls. Cell walls of some algae and dinoflagellates may contain chemically related biomacromolecules, named **algaenan** and **dinosporin** (Versteegh et al. 2012; Bogus et al. 2012). Like sporopollenin these resistant biomacromolecules may also fossilize. They have been reported in, e.g., *Chlorella* (He et al. 2016), *Spirogyra* (Simons et al. 1983), and *Coleochaete* (Ueno 2009). Furthermore, "sporopollenin-like" biomacromolecules have been found in megaspores and "massulae" of water ferns (Salviniales) (van Bergen et al. 1993), as well as in fruiting bodies of cellular slime molds (Maeda 1984).

The Angiosperm Pollen Wall

In angiosperms the **ektexine** consists in general of **tectum**, **infratectum**, and **foot layer**. The outer layer, the more-or-less continuous tectum, can be covered by **supratectal elements**. The infratectum beneath is **columellate** or **granular** (a second layer of columellae may form an internal tectum). However, as, e.g., Doyle (2005) has pointed out intermediate conditions are common. Even the alveolate infratectum, that by definition is restricted to gymnosperms, can also be found in some angiosperms (see "Illustrated Pollen Terms"). The foot layer may be either continuous, discontinuous or absent. The **endexine** can be described as continuous or discontinuous, spongy or compact, overall present, in apertures only, or even completely absent. Some typical deviations of the wall thickness are termed: **arcus**, **annulus**, **tenuitas** (see "Illustrated Pollen Terms") and **costa** (a thickening of the nexine/endexine bordering an endoaperture; Fig. 18).

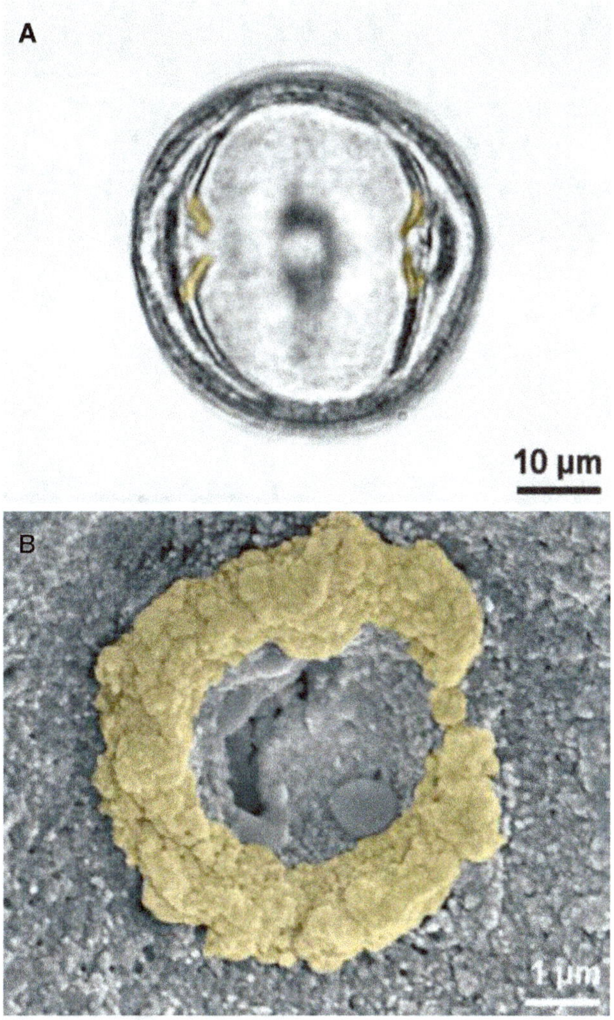

Fig. 18 Costa. A. *Nyssa* sp. Nyssaceae, fossil, middle Miocene, Austria, equatorial view (costa highlighted). **B.** *Austobuxus nitidus*, Picrodendraceae, view on the thickening around the endoaperture on the inner side of the wall

The Gymnosperm Pollen Wall

The gymnosperms comprise cycads, *Ginkgo*, conifers and Gnetales. The basic stratification (ektexine, endexine, and intine) of the gymnosperm pollen wall is identical to that of angiosperms. Still, the gymnosperm pollen wall differs from that of an angiosperm by having (1) a lamellate endexine in mature pollen, and (2) an infratectum that is never columellate (Van Campo and Lugardon 1973). The infratectum is either **alveolate** or **granular**.

A special terminology applies to saccate pollen, i.e. in Pinaceae and Podocarpaceae (Fig. 19). **Saccus** is an exinous expansion forming an air sac, with an alveolate infratectum. **Corpus** is the central body of a saccate pollen grain. **Cappa** is the thick walled proximal face of the corpus. **Leptoma** in conifer pollen refers to a thinning of the pollen wall on the distal face, presumed to function as germination area. Most frequently, two sacci are present (e.g., *Abies, Pinus, Picea;* Pinaceae), in some taxa even three (*Dacrycarpus, Microstrobus;* Podocarpaceae), or only a single one (*Tsuga;* Pinaceae).

The function and evolutionary significance of saccate pollen have been subject of much confusion. The sacci of Pinaceae and Podocarpaceae are reported to play an aerodynamic role, thus being of adaptive significance for wind pollination (Schwendemann et al. 2007; Grega et al. 2013). In fact, their functional role is to float in a liquid pollination droplet towards the ovule ("flotation hypothesis" by Leslie 2010). The flotation system is interpreted as ancestral in conifers. The absence of sacci in, e.g., Cupressaceae and Taxaceae might reflect the loss of "drop mechanism," correlated with the change of pollination mode (shift to upwards orientation of the ovules) (Doyle 2010).

In *Pinus,* pollen can be grouped into two morphotypes (Fig. 20) of systematic value (Grímsson and Zetter 2011). The *Pinus* subgenus *Strobus* (**haploxylon) type** is characterized by pollen grains with broadly attached half-spherical air sacs—in LM the leptoma shows dotted thickenings (seen as dark spots). The *Pinus* subgenus *Pinus* (**diploxylon) type** is characterized by pollen grains with narrowly attached, spherical air sacs often with nodula on nexine area—the leptoma does not show any thickenings.

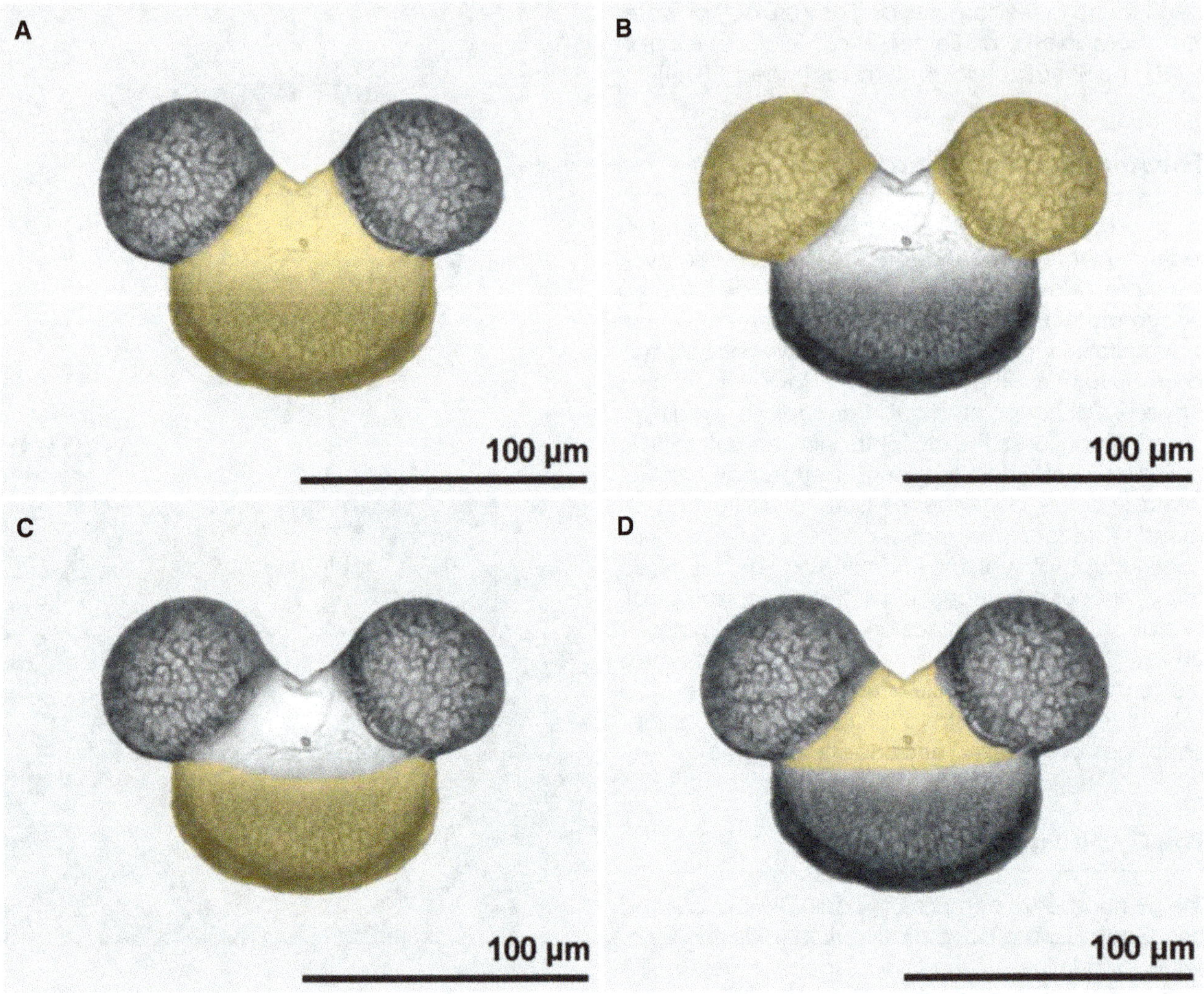

Fig. 19 Pollen terminology in saccate gymnosperm pollen. A-D. *Abies* sp., Pinaceae, bisaccate pollen, fossil, Quaternary, Austria, equatorial view. **A.** Corpus highlighted. **B.** Sacci highlighted. **C.** Cappa highlighted. **D.** Leptoma highlighted

Sculpture: Ornamentation

The terms ornamentation and sculpture applies to surface features of a pollen. The term sculpture is restricted by some authors to surface features in tectate pollen grains (e.g., Praglowski 1975; Punt et al. 2007). **Sculpture elements** (areola, clava, echinus, foveola, fossula, granulum, gemma, plicae, reticulum, rugulae, striae, verruca) can be extremely variable in both size and shape. Based on size many sculpture/ornamentation elements smaller than 1 µm can be described with the prefix micro- (1–0.5 µm) or nano- (0.5–0.1 µm). Also, the boundary between two ornamentation types can be diffuse. For example, "gemmae" and "clavae" are very variable and sometimes hard to differentiate. Combinations of different sculpture/ornamentation elements are common, such as the combination reticulate and foveolate, or echinate and perforate. With a combined sculpture, the pollen ornamentation should then be described in a defined order, with the most eye-catching feature mentioned first, followed by the others. For example, *Aristolochia* pollen is verrucate-perforate, as the verrucae are more prominent than the small perforations (Fig. 21). In the Caryophyllaceae, there are numerous, more-or-less regularly arranged microechini and perforations. In some taxa the microechini are more prominent (microechinate-perforate), in others the perforations (perforate-microechinate) (Fig. 22). In case none of the features are eye-catching, the dominant feature might be a subjective decision of the palynologist e.g., in taxa, where two features are on a par (microechinate and perforate). A more complex example is *Sanchezia nobilis* (Acanthaceae, Fig. 23): is it plicate and reticulate? Should the rod-like elements

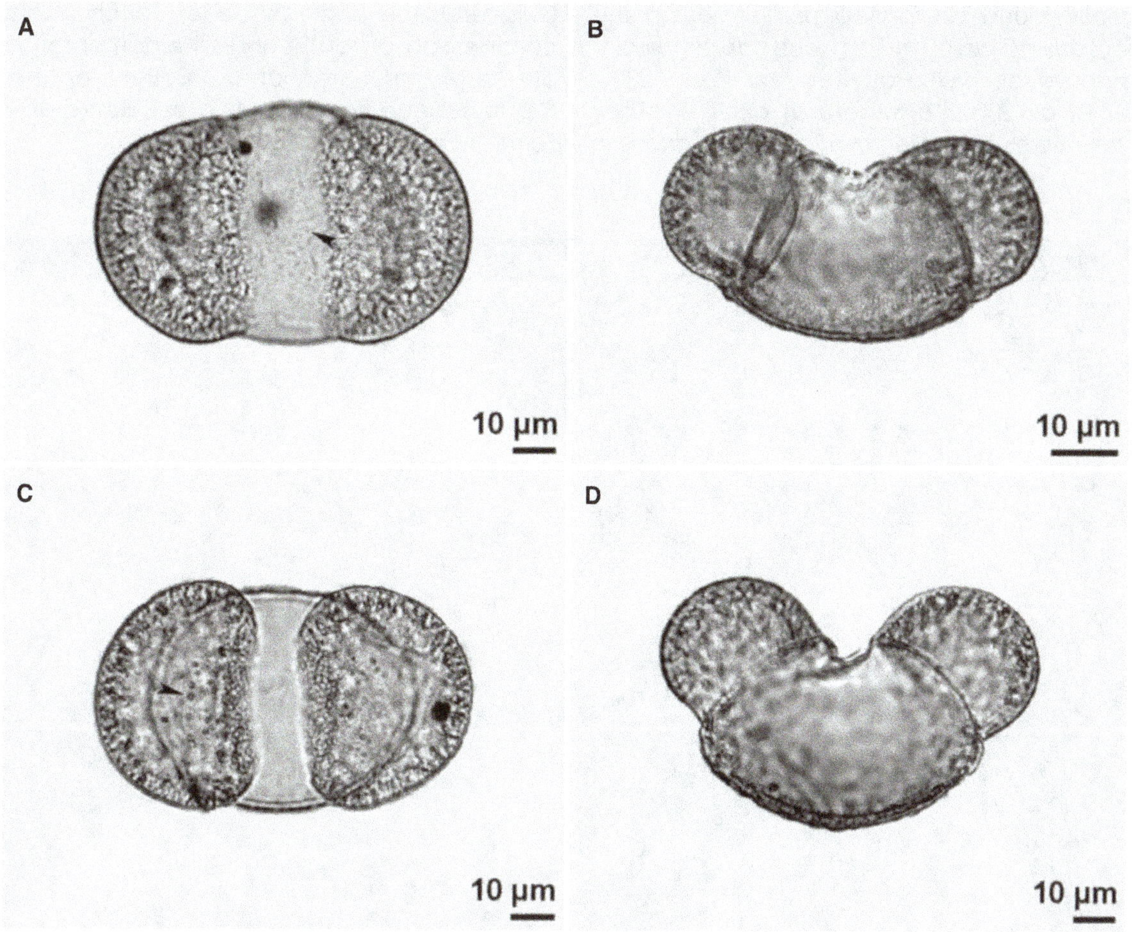

Fig. 20 Pollen types in saccate Pinus pollen (fossil, middle Miocene, Austria). A. *Pinus* subgenus *Strobus* (haploxylon), polar view, thickenings (arrowhead). **B.** *Pinus* subgenus *Strobus* (haploxylon), equatorial view. **C.** *Pinus* subgenus *Pinus* (diploxylon), polar view, nodula (arrowhead). **D.** *Pinus* subgenus *Pinus* (diploxylon), equatorial view

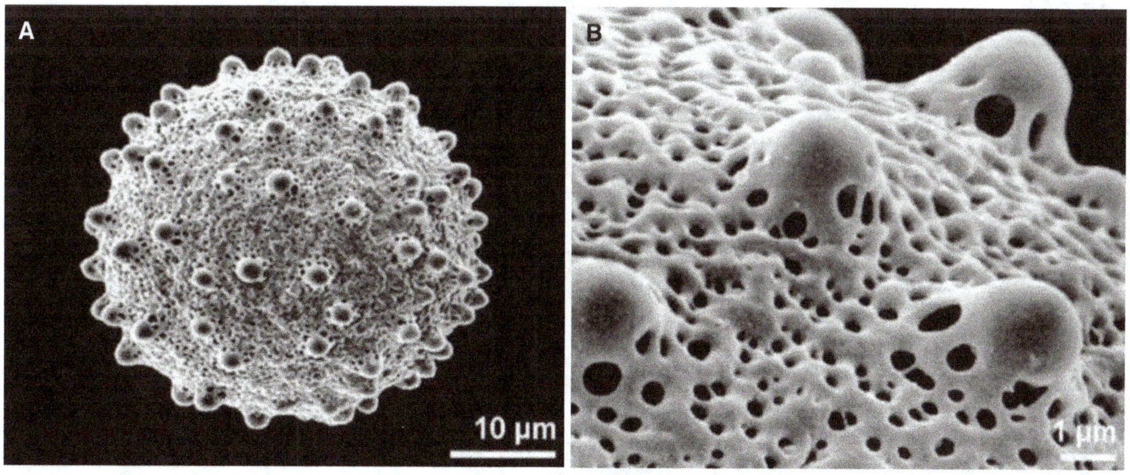

Fig. 21 Combined sculpture elements. A-B. *Aristolochia arborea* (Aristolochiaceae), verrucate, perforate

be termed clavae or free-standing columellae? Is the aperture a porus or a colporus? *PalDat* (www.paldat.org) might provide the answers?

Sculpture/ornamentation elements are often deviating and can be distributed regularly or irregularly over the pollen surface, restricted or absent from distinct areas (polar vs equatorial, interapertural vs aperture area; Fig. 24).

Ubisch bodies (orbicules) are sporopollenin elements produced by the tapetum. Ubisch bodies are usually found as isolated particles lining the mature locular wall, or between pollen grains (Huysmans et al. 1998; Halbritter and Hesse 2005; Vinckier et al. 2005; Verstraete et al. 2014). They often resemble the pollen wall ornamentation. In Cupressaceae and Taxaceae, Ubisch bodies are considered part of the pollen ornamentation and are especially frequent on the leptoma of Cupressaceae (for examples, see "Illustrated Pollen Terms").

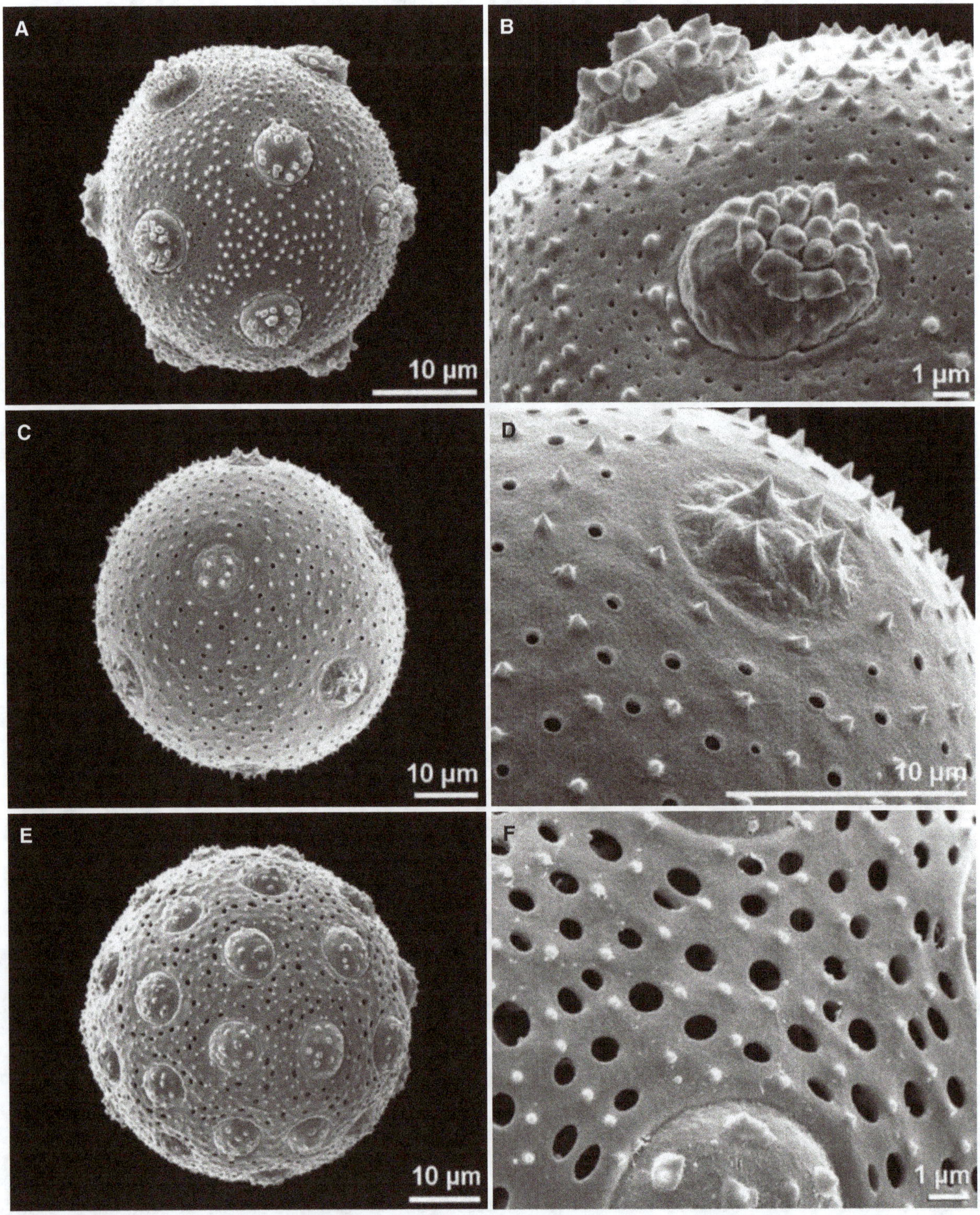

Fig. 22 Combined sculpture elements. A-B. *Stellaria media*, Caryophyllaceae, microechinate and perforate. **C-D.** *Saponaria officinalis*, Caryophyllaceae, microechinate and perforate. **E-F.** *Silene succulenta*, Caryophyllaceae, perforate and nanoechinate

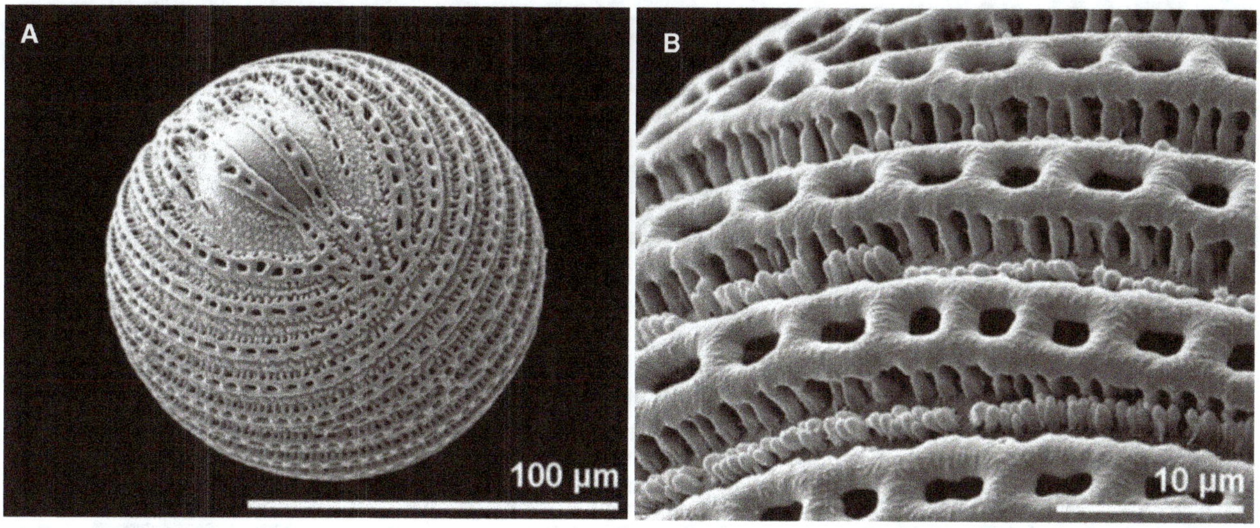

Fig. 23 Interpretation of sculpture elements. A-B. *Sanchezia nobilis*, Acanthaceae, oblique equatorial view and surface detail

Ornamentation in LM vs. SEM

An accurate description of pollen ornamentation depends on the optical magnification used and particularly on the point resolution. Even the SEM at low resolution may not be sufficient to distinguish pollen grains unequivocally (see "Methods in Palynology"). Depending on the type of microscope used for pollen analysis, some pollen features may remain hidden. For LM studies, the term **scabrate** is used, describing minute sculpture elements of undefined shape and size close to the resolution limit of the LM. For example, *Juglans* pollen is scabrate in LM as well as under low magnification SEM, but is nanoechinate at high resolution SEM (Fig. 25A-B).

The descriptive terms may differ whether LM or SEM is used and should be described for both. For example, *Ulmus* pollen seen in LM is described as **verrucate**. Using low SEM magnification the ornamentation is **rugulate to verrucate** (Fig. 25C-E). High SEM magnification shows additional **granula** (≤0.1 µm).

Another example for different interpretations in LM vs SEM is the term **psilate**. Many pollen grains that appear psilate in LM show a distinct ornamentation using high SEM magnification. For example, pollen of *Allium ursinum* is psilate in LM, but is striate and perforate in SEM (Fig. 25F-G).

Terms with nano- or micro- can only be observed in SEM (see "Methods in Palynology"). For example, the term **granulate** should only be used when describing pollen ornamentation under SEM. When minute sculptural elements are observed under high resolution SEM, it is possible to distinguish real "granula" (sculpture element of different/indefinable shape, ≤ than 0.1 µm) from other nano- and/or micro-sculpture elements. For example, the allegedly granulate ornamentation of many Poaceae is in fact nanoechinate, the pointed ends of the echini are seen best in profile and not from top view (see "Illustrated Pollen Terms").

Role of Pollen Ornamentation in Pollination

Depending on the pollination mode the outer pollen wall may be either highly ornamented, often with plenty of pollen coatings (mainly pollenkitt; Pacini and Hesse 2005), or with a more or less psilate pollen surface. The pollen wall of zoophilous plants, as well as autogamous plants, is usually highly ornamented and the thick exine consists of high amounts of sporopollenin (Fægri and Iversen 1989). Pollen of anemophilous plants are known to have less ornamentation and less sporopollenin (Friedman and Barrett 2009). Usually psilate pollen in temperate and boreal zones is indicative for anemophily (Fægri and Iversen 1989), whereas in the tropics it is also indicative for zoophily (Furness and Rudall 1999). For example, in Aroideae (e.g., *Montrichardia*, *Dieffenbachia*, *Philodendron*, *Gearum*) psilate pollen usually equipped with pollenkitt is adapted for entomophily (Weber and Halbritter 2007).

Functional Value of Exine Reduction

Layers of the basic pollen wall type may vary and be partly or totally reduced (for examples, see

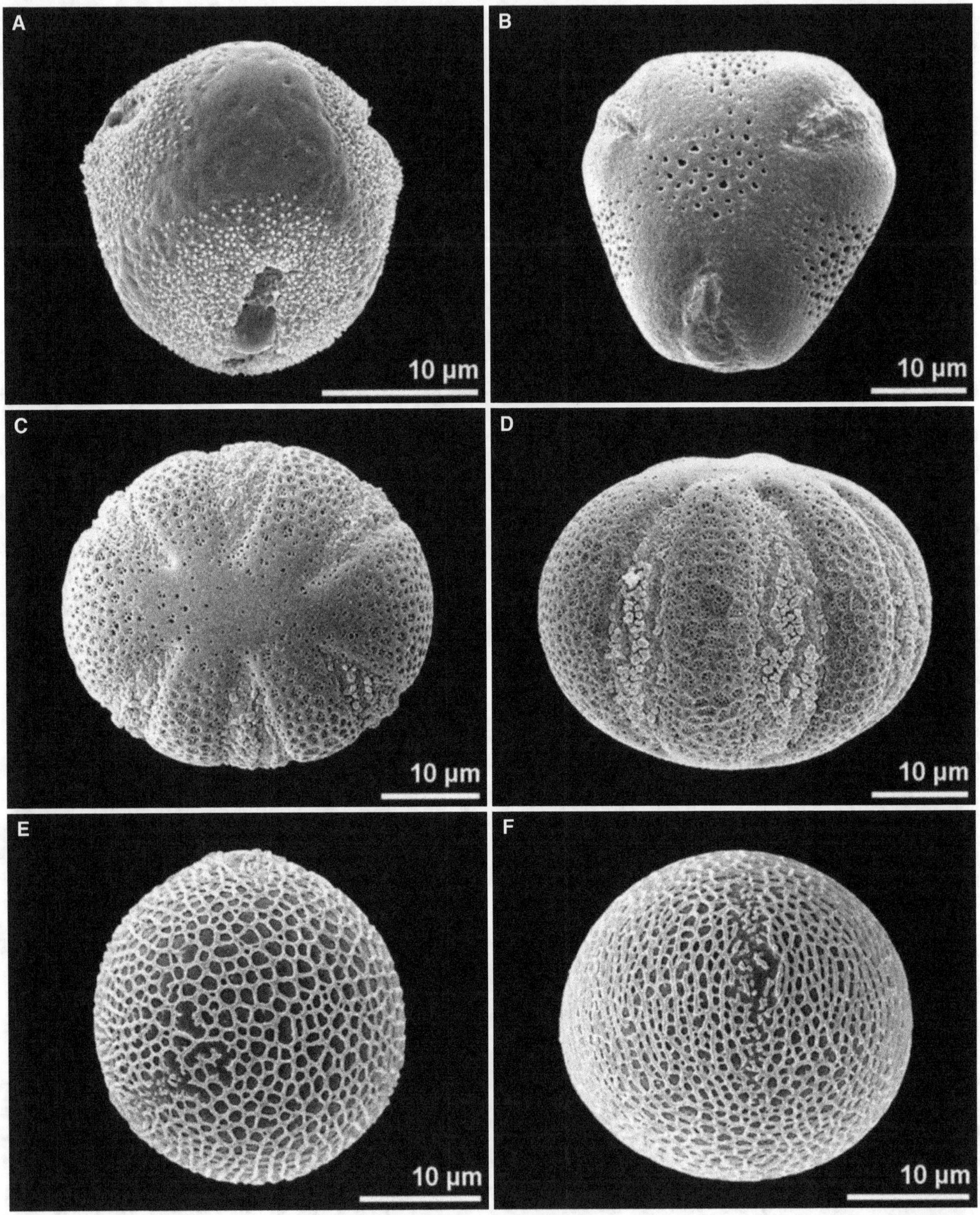

Fig. 24 Pollen surface variation. A. *Fallopia convolvulus*, Polygonaceae, polar view, polar area psilate to perforate and regions around apertures microechinate. **B.** *Sideritis montana*, Lamiaceae, polar view, polar and interapertural areas perforate to foveolate and regions around apertures psilate. **C-D.** *Salvia austriaca*, Lamiaceae, pollen bireticulate, except psilate polar areas (polar and equatorial view). **E-F.** *Solandra longiflora*, Solanaceae, polar area reticulate, equatorial region striato-reticulate (polar and equatorial view)

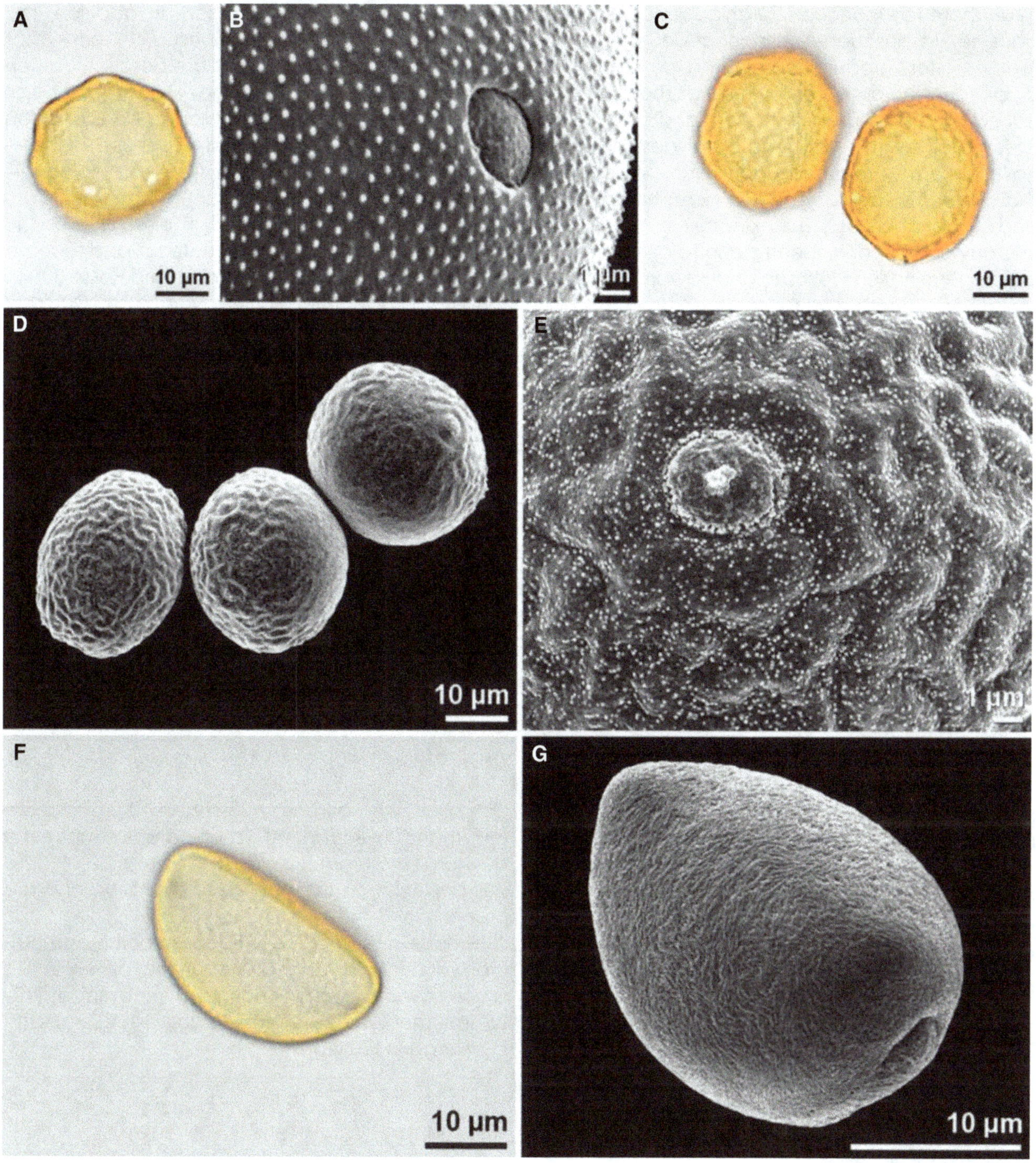

Fig. 25 Ornamentation in LM vs SEM. **A-B**. *Juglans* sp., Juglandaceae. **A**. Scabrate to psilate, LM. **B**. Nanoechinate, SEM. **C-E.** *Ulmus laevis*, Ulmaceae. **C**. Rugulate, LM. **D**. Rugulate to verrucate, low magnification, SEM. **E**. Verrucate, granulate, high magnification SEM. **F-G.** *Allium ursinum*, Amaryllidaceae. **F**. Psilate to scabrate, LM. **G**. Rugulate-perforate, low magnification, SEM

"Pollen Wall" in "Illustrated Pollen Terms"). The sporopollenin ektexine is lacking e.g., in some genera of Monimiaceae and Lauraceae (Walker 1976), in the aquatic Ceratophyllaceae (Takahashi 1995), in many genera of Aroideae, and in the inaperturate filiform pollen of seagrasses, *Posidonia*. An absent exine is an adaptation to hydrophily and correlated with, e.g., aquatic habits, anemophily, and pollinia (Furness 2007). Interestingly, exine reduction has evolved iteratively in angiosperms, especially throughout the monocots. Orchidaceae, Asclepiadaceae, Mimosaceae, Annonaceae, and other families often produce compound pollen, where usually only the outermost pollen wall show the typical ektexine structure with tectum and columellae. Pollen grains within calymmate polyads or tetrads have extremely reduced and fragile pollen walls, that probably facilitates pollen germination (Knox and McConchie 1986). The extreme exine reduction in many orchid pollinia seems also to correlate with pollen germination (Johnson and Edwards 2000).

Harmomegathy: The Harmomegathic Effect

Pollen grains are able to absorb and release water (+ various liquids); thus, each pollen grain exists in two morphologically different conditions, **dry** and **hydrated** (Fig. 26). Harmomegathic mechanisms, e.g., infolding of the pollen wall (Rowley and Skvarla 2000), accommodate the change of the osmotic pressure in the cytoplasm during hydration or dehydration. These mechanisms are denoted as harmomegathic effect, also known as Wodehouse effect. The main purpose of the harmomegathic effect is to protect the male gametophyte against desiccation during pollen presentation and dispersal, and is often related to pollination biology.

In mature anthers, pollen is turgescent before shedding. After anther dehiscence and during pollen presentation, water loss takes place and the pollen grain becomes typically infolded. Various pollen wall features are involved in the harmomegathic effect:

- **Position, number, and type of apertures:** the most important features
- **Thinned or thickened regions within the pollen wall:** in particular, internal belts or endoapertures. If the ektexine is considerably reduced, its role is taken over by other wall strata, namely, by a thick endexine or intine. On the other hand, if the exine is extremely rigid, then the harmomegathic effect is only marginal
- **Ornamentation type**
- **Pollen size:** small, thin-walled pollen grains which are usually less infolded
- **Pollen coatings:** if abundant, pollen coatings have an insulating influence that reduces the harmomegathic effect

The combination of these features is influencing the mode of infolding. Terms used for common morphotypes of dry pollen include: apertures sunken, boat-shaped, cup-shaped, interapertural area infolded, irregularly infolded, not infolded. In addition, the pollen shape can be described with terms that might be helpful for an adequate description such as barrel-like, disk-like, or kidney-like. The mode of infolding and/or shape of pollen in dry condition may be typical for a family and/or genus and therefore of systematic relevance (see "Palynology — History and Systematic Aspects").

The harmomegathic effect is also observed in pollen taken from herbarium material, and to some degree in fossil material (Halbritter and Hesse 2004). This effect is to some degree reversible: rehydrated pollen at the stigma, or under laboratory conditions (various liquids), is again turgescent and largely recalls the shape before shedding. A second dehydration does not necessarily result in the typical dry shape but, if pollen walls are sufficiently stable, the harmomegathic effect can be induced several times in the same way. In pollen with thin walls, the susceptible internal structure may become damaged, and the harmomegathic effect may result in different and randomly shaped pollen. Infoldings of the pollen wall after acetolysis treatment are mostly not comparable with those observed in dry condition.

Size

Pollen **size** varies from less than 10 µm to more than 100 µm (Fig. 27). To indicate pollen size the largest diameter is used (Hesse et al. 2009). The size depends on the degree of hydration and the preparation method (Reitsma 1969, see also "Methods in Palynology"). Because of this and natural variation, a range categorizing pollen size is recommended: very small (<10 µm), small (10–25 µm), medium (26–50 µm), large (51–100 µm), and very large (>100 µm).

Heterostyly and Pollen Dimorphism

In **heterostylous** (long-styled and short-styled) species two different pollen types occur, where pollen size and number of apertures or the ornamentation may differ. In *Linum flavum* (Linaceae) pollen of the short-styled morph is baculate, and the long-styled morph clavate (Fig. 28). In *Primula veris* (Primulaceae) the pollen of the short-styled morph is larger and has more apertures than pollen of the long-styled morph (Fig. 29A). In the tristylous species *Lythrum salicaria*

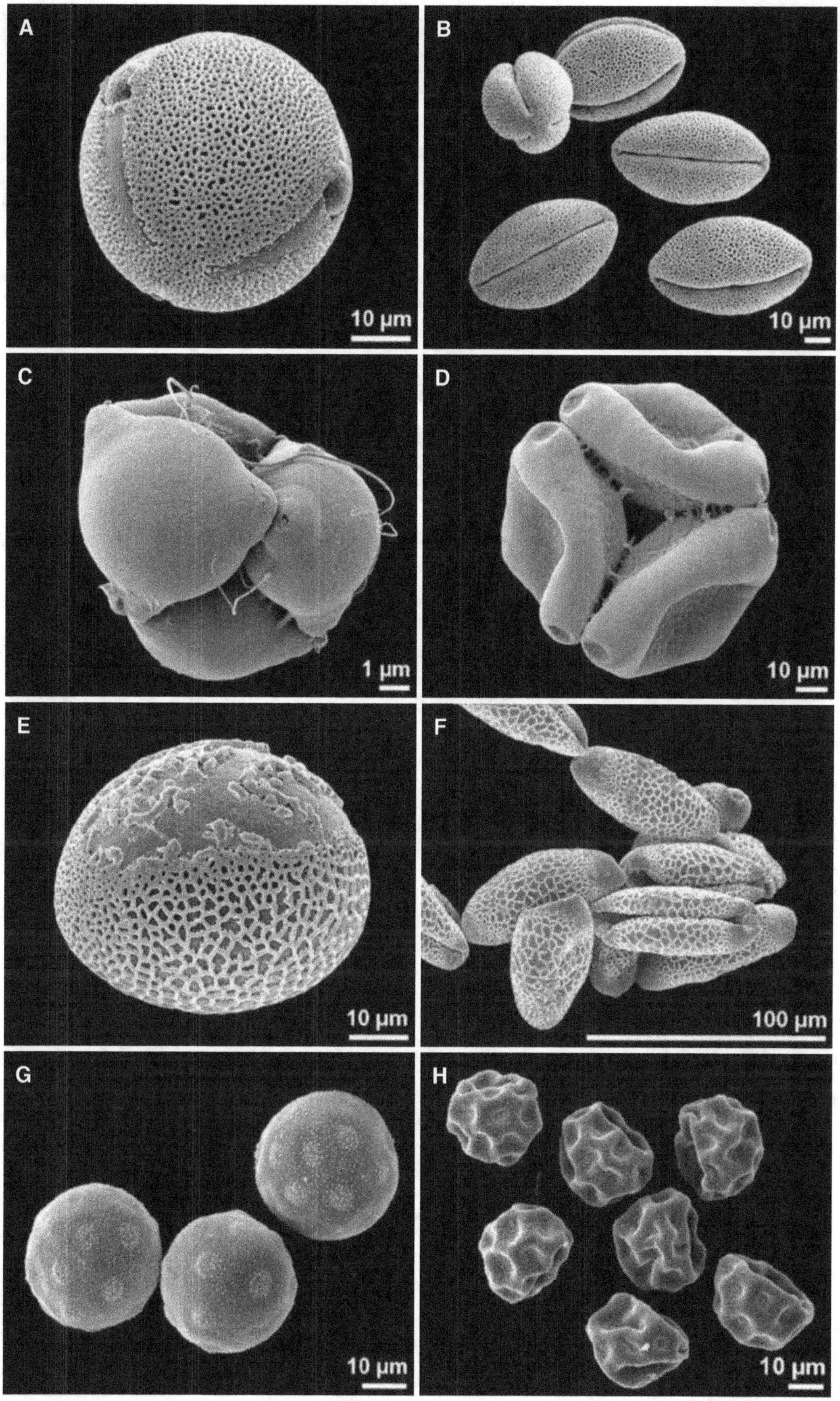

Fig. 26 Harmomegathic effect—hydrated vs dry pollen. A-B. *Cistus creticus*, Cistaceae. **A**. Spheroidal, outline circular. **B**. Prolate, outline lobate, apertures infolded. **C-D.** *Epilobium palustre*, Onagraceae, tetrad. **C**. Oblate, outline triangular. **D**. Interapertural area sunken. **E-F.** *Vriesea pabstii*, Bromeliaceae. **E**. Oblate, outline elliptic. **F**. Boat-shaped. **G-H.** *Alisma lanceolatum*, Alismataceae. **G**. Spheroidal, outline circular. **H**. Irregularly infolded

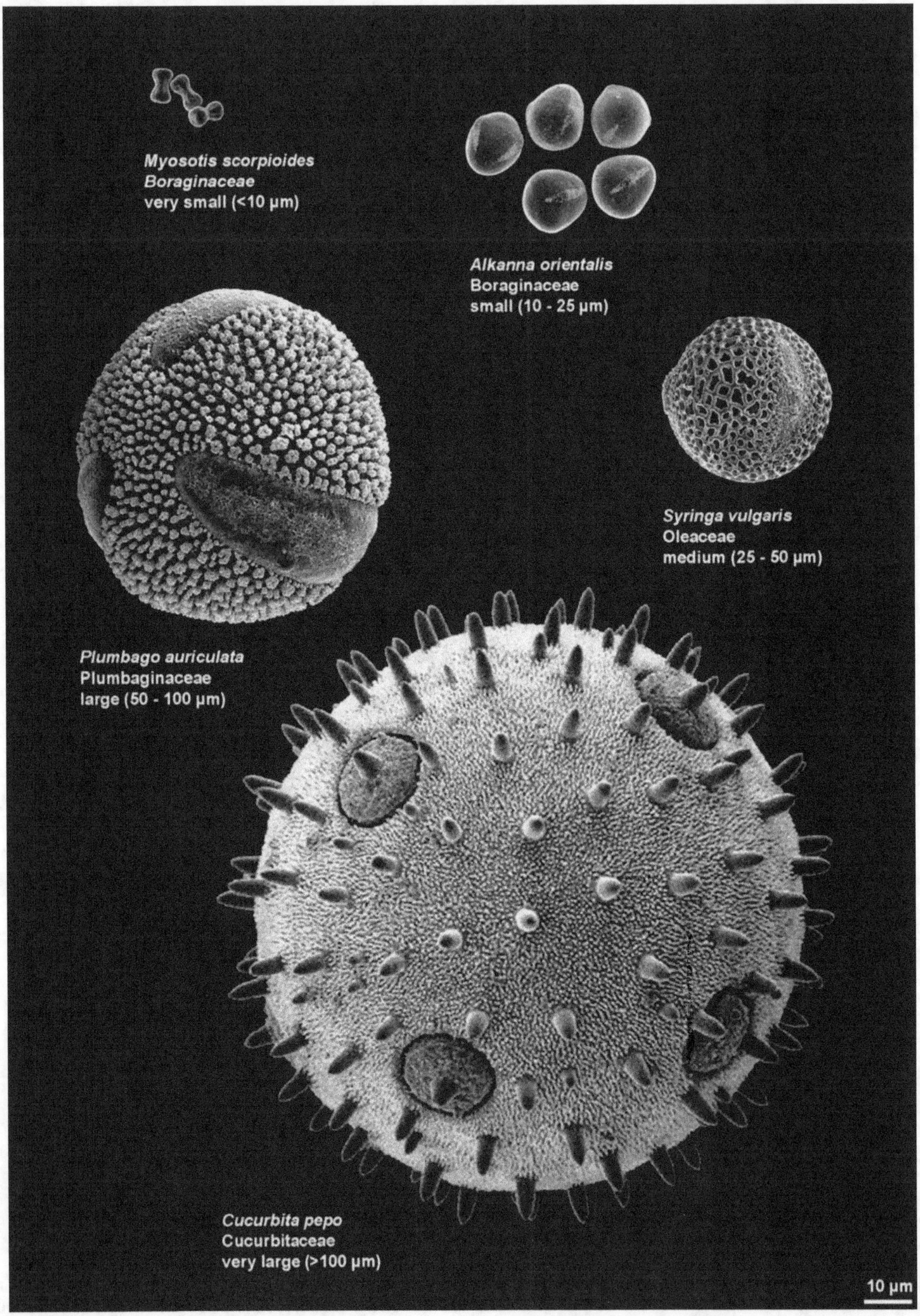

Fig. 27 Pollen size categories.

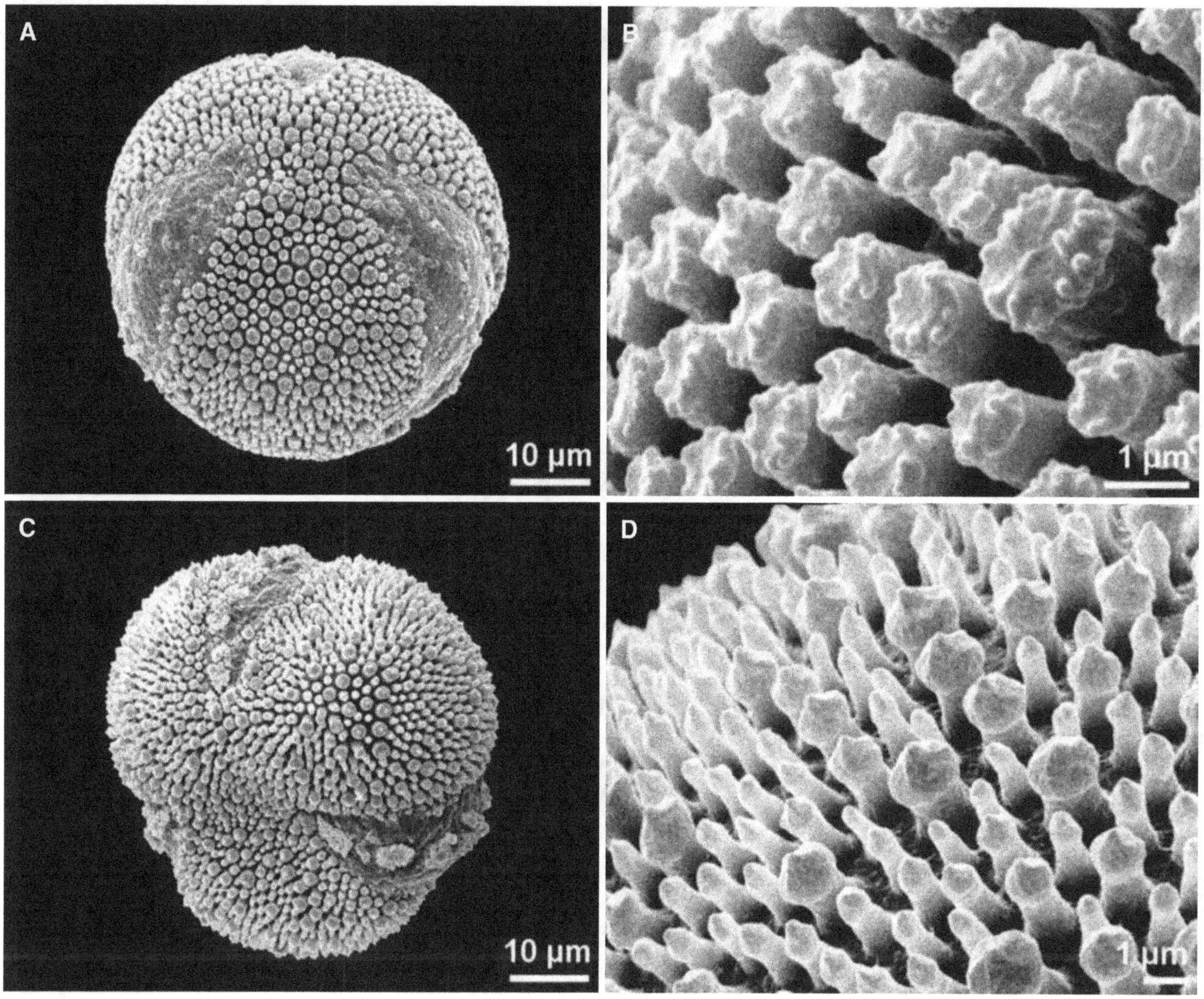

Fig. 28 Pollen dimorphism — different ornamentation. A-D. *Linum flavum*, Linaceae, (**A-B**) short-styled morph, baculate, (**B-C**) long-styled morph, clavate

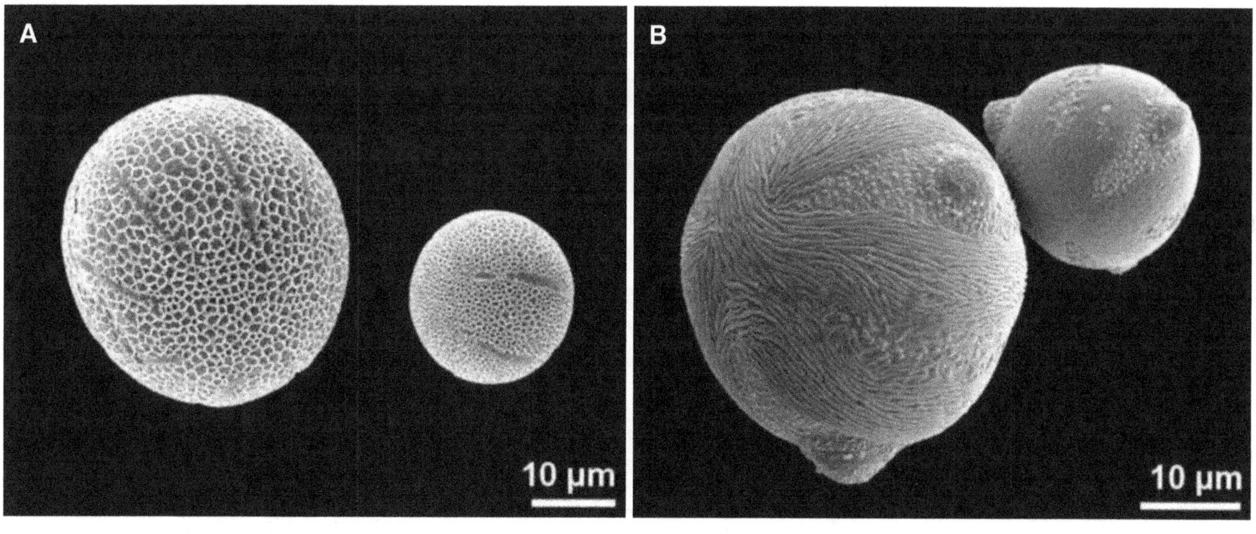

Fig. 29 Pollen dimorphism — different size. A. *Primula veris*, Primulaceae, short-styled morph (left), long-styled morph (right). **B.** *Lythrum salicaria* (Lythraceae), medium-styled morph, dimorphic pollen

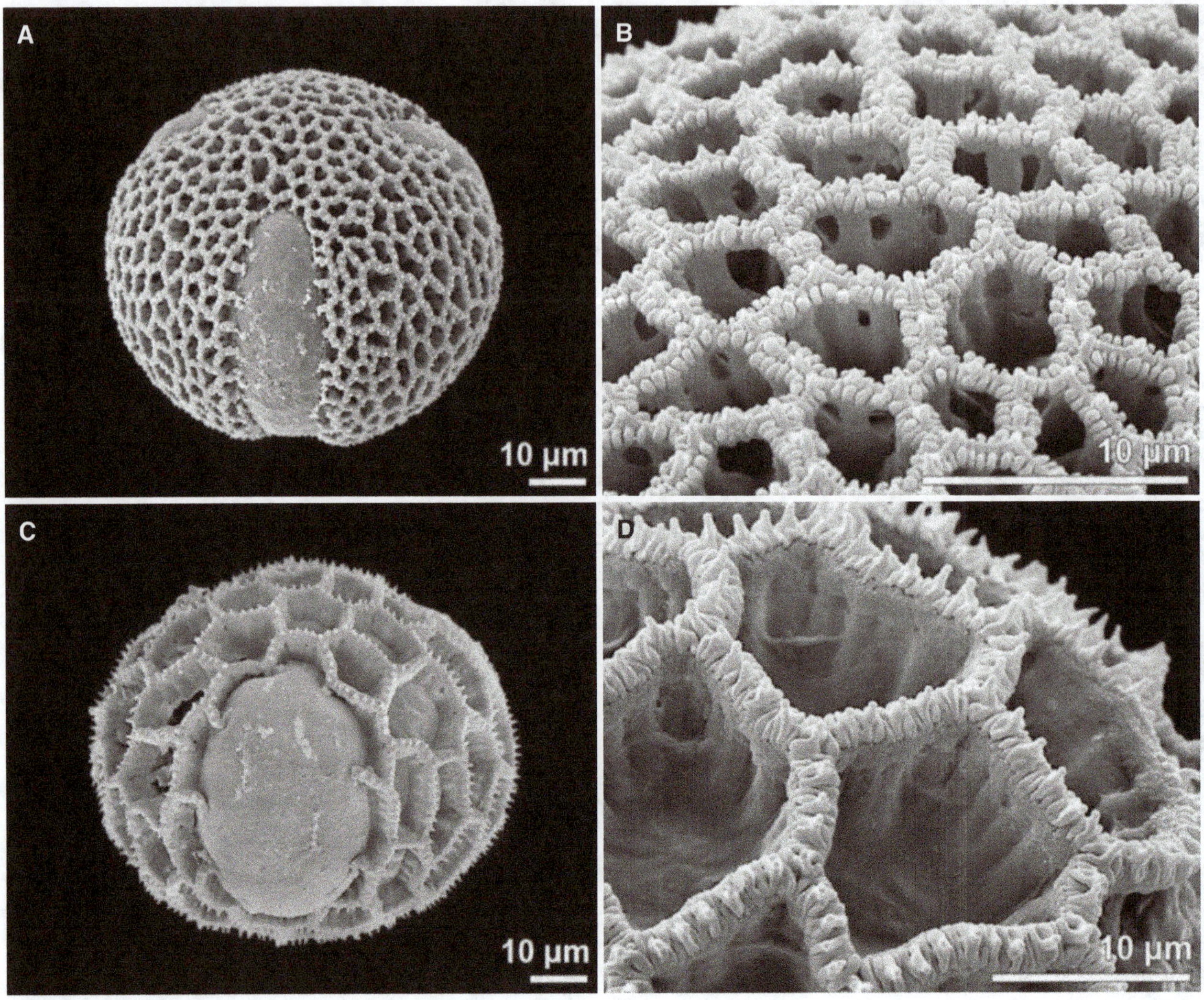

Fig. 30 Pollen dimorphism — different ornamentation. A-D. *Armeria alpina*, Plumbaginaceae. **A-B.** Morph 1, reticulate. **C-D.** Morph 2, reticulate

(Lythraceae) pollen is dimorphic, with different size, ornamentation, and even color of the pollen grains (blue and yellow). Pollen of the long-styled morph is small sized (about 20 µm), short styled morph is medium sized (about 35 µm) and medium-styled morph is small to medium sized (within a single anther) (Fig. 29B). In some Plumbaginaceae, for example in distylous species of *Armeria*, pollen dimorphism (reticulate, different size of lumina and suprasculpture elements) is correlated with dimorphic stigmatic papillae, but style and stamen lengths are monomorphic (Ganders 1979; Fig. 30).

Aberrant Pollen Grains

Aberrant pollen grains are often ignored but occur regularly in small percentages in nearly all anthers and may vary from one individual to another (Pozhidaev 2000a, b; Banks et al. 2007). These aberrant pollen grains can differ from the typical pollen type of the species in shape and dimension, in number and arrangement of apertures, and in ornamentation type (Fig. 31). Reasons for the production of deviating pollen forms are genetically (polyploidy),

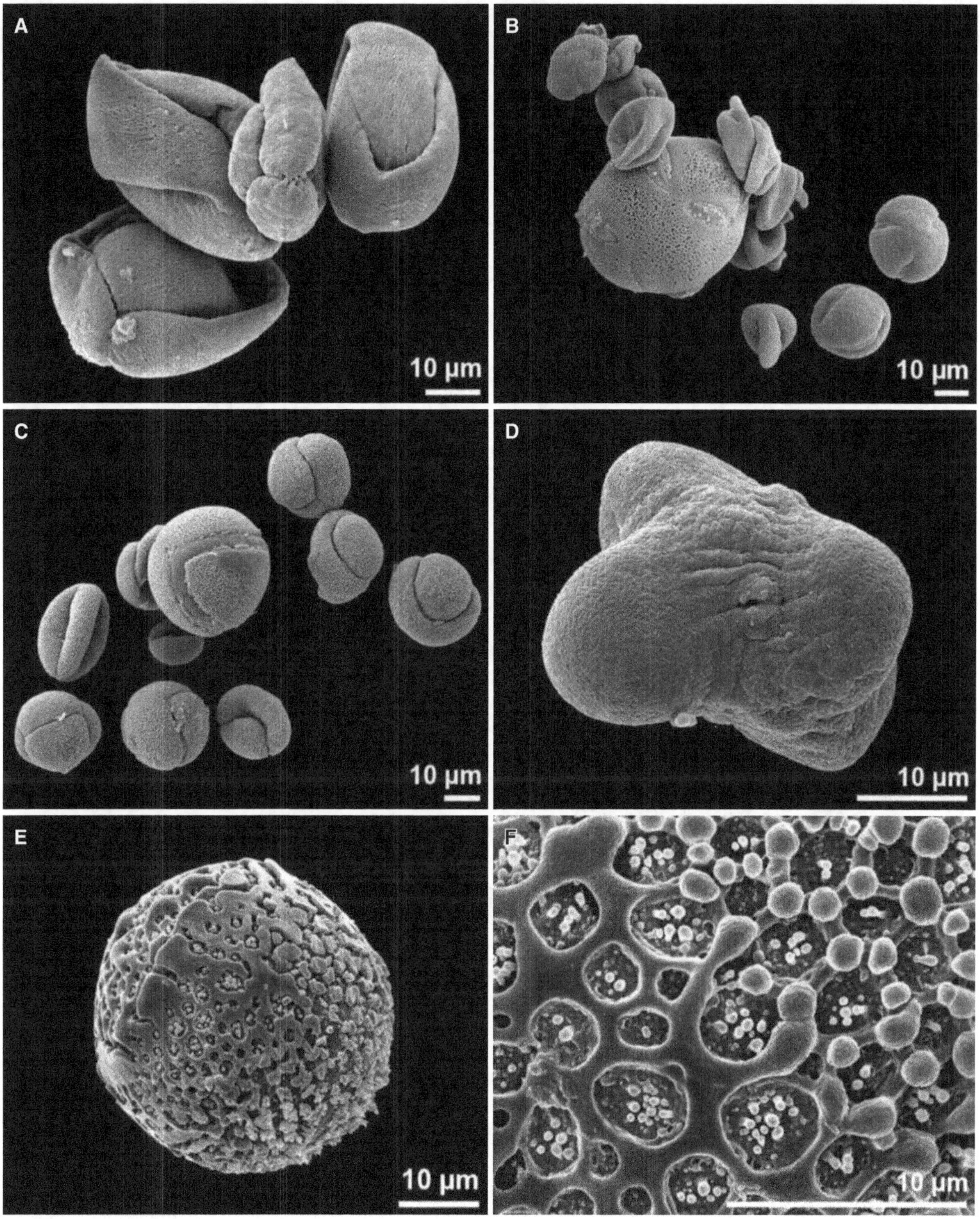

Fig. 31 Aberrant pollen grains. A. *Malus sieboldii*, Rosaceae, irregular aperture arrangement (usually tricolporate). **B.** *Oxalis* sp., Oxalidaceae, many aborted pollen grains, giant pollen (usually tricolpate). **C.** *Scaevola* sp., Goodeniaceae, pollen varies in size and aperture arrangement. **D.** *Scandix pecten-veneris*, Apiaceae, "double" pollen grain (usually tricolporate). **E.** *Codiaeum*-hybrid, Euphorbiaceae, ornamentation intermediate between parent plant species e.g., croton pattern and reticulate with free-standing columellae. **F.** *Codiaeum*-hybrid, Euphorbiaceae, surface detail

chemically, or environmentally induced. Such deviating, malformed pollen is frequently found in cultivated plants, ornamental plants, agricultural crops, annual plants, plants with asexual reproduction (autogamic plants, apomicts), and hybrids. Some species of apomicts, agricultural crops or cultivated plants (e.g., *Malus sieboldii*) produce only malformed pollen.

References

Ariizumi T, Toryama K (2011) Genetic regulation of sporopollenin synthesis and pollen exine development. Annu Rev Plant Biol 62: 437–460

Banks H, Stafford P, Crane PR (2007) Aperture variation in the pollen of *Nelumbo* (Nelumbonaceae). Grana 46: 157–163

Blackmore S, Barnes SH (1995) Garside's rule and the microspore tetrads of *Grevillea rosmarinifolia* A. Cunningham and *Dryandra polycephala* Bentham (Proteaceae). Rev Palaeobot Palynol 85: 111–121

Blackmore S, Cannon SM (1983) Palynology and systematics of Morinaceae. Rev Palaeobot Palynol 40: 207–226

Blackmore S, Takahashi M, Uehara K (2000) A preliminary phylogenetic analysis of sporogenesis in pteridophytes. In: Harley MM, Morton CM, Blackmore S (eds) Pollen and spores: morphology and biology. Royal Botanic Gardens, Kew, p. 109–124

Bogus K, Harding IC, King A, Charles AJ, Zonneveld KAF, Versteegh GJM (2012) The composition and diversity of dinosporin in species of the *Apectodinium complex* (Dinoflagellata). Rev Palaeobot Palynol 183: 21–31

Braconnot H (1829) Recherches chimiques sur le pollen du *Typha latifolia*, Lin., famille de typhacées. Ann Chim Phys 42: 91–105

Bryant VM, Hall SA (1993) Archaeological palynology in the United States: A critique. Am Antiquity 58: 277–286

Bryant VM, Holloway RG, Jones JG, Carlson DL (1994) Pollen preservation in alkaline soils of the American southwest. In: Traverse A (ed) Sedimentation of organic particles. Cambridge University Press, Cambridge, New York, Melbourne, Madrid, Cape Town, Singapore, Sao Paolo, p. 47–58

Colpitts CC, Kim SS, Posehn SE, Jepson C, Kim SY, Wiedemann G, Reski R, Wee AGH, Douglas CJ, Suh D-Y (2011) PpASCL, a moss ortholog of anther-specific chalcone synthase-like enzymes, is a hydroxyalkylpyrone synthase involved in an evolutionarily conserved sporopollenin biosynthesis pathway. New Phytol 192: 855–868

Copenhaver GP (2005) A compendium of plant species producing pollen tetrads. J North Carolina Acad Sci 12: 17–35

Cushing EJ (1967) Evidence for differential pollen preservation in late Quaternary sediments in Minnesota. Rev Palaeobot Palynol 4: 87–101

De Leeuw JW, Versteegh GJM, Van Bergen PF (2006) Biomacromolecules of algae and plants and their fossil analogues. Plant Ecology 182: 209–233

Diego-Taboada A, Beckett ST, Atkin SL, Mackenzie G (2014) Hollow pollen shells to enhance drug delivery. Pharmaceutics 6: 80–96

Dobritsa AA, Shrestha J, Morant M, Pinot F, Matsuno M, Swanson R, Lindberg Møller B, Preuss D (2009) CYP704B1 is a Long-Chain Fatty Acid v-Hydroxylase essential for sporopollenin synthesis in pollen of Arabidopsis. Plant Physiol 151: 574–589

Doyle J (2005) Early evolution of angiosperm pollen as inferred from molecular and morphological phylogenetic analyses. Grana 44: 227–251

Doyle JA (2010) Function and evolution of saccate pollen. New Phytol 188: 6–9

Elsik WC (1971) Microbial degradation of sporopollenin. In: Brooks J, Grant PR, Muir MD, Van Gijzel P, Shaw G (eds) Sporopollenin. Academic Press, London New York, p. 480–511

Fægri K, Iversen J (1989) Textbook of Pollen analysis. 4th edition, John Wiley & Sons, Chichester

Fraser WT, Scott AC, Forbes AES, Glasspool IJ, Plotnick RE, Kenig F, Lomax BH (2012) Evolutionary stasis of sporopollenin biochemistry revealed by unaltered Pennsylvanian spores. New Phytol 196: 397–401

Fraser WT, Sephton MA, Watson JS, Self S, Lomax BH, James DI, Wellman CH, Callaghan TV, Beerling DJ (2011) UV-B absorbing pigments in spores: biochemical responses to shade in a high-latitude birch forest and implications for sporopollenin-based proxies of past environmental change. Polar Res 30, 8312, https://doi.org/10.3402/polar.v30o0.8312

Fraser WT, Watson JS, Sephton MA, Lomax BH, Harrington G, Gosling WD, Self S (2014) Changes in spore chemistry and appearance with increasing maturity. Rev Palaeobot Palynol 201: 41–46

Friedman J, Barrett SCH (2009) Wind of change: new insights on the ecology and evolution of pollination and mating in wind-pollinated plants. Ann Bot 103: 1515–1527

Furness CA (2007) Why does some pollen lack apertures? A review of inaperturate pollen in eudicots. Bot J Linn Soc 155: 29–48

Furness CA, Rudall PJ (1999) Microsporogenesis in Monocotyledons. Ann Bot 84: 475–499

Furness CA, Rudall PJ (2001) Pollen and anther characters in monocot systematics. Grana 40: 17–25

Furness CA, Rudall PJ (2003) Apertures with lids: distribution and significance of operculate pollen in Monocotyledons. Int J Plant Sci 164: 835–854

Gabarayeva NI, Grigorjeva VV (2010) Sporoderm ontogeny in *Chamaedorea microspadix* (Arecaceae): self-assembly as the underlying cause of development. Grana 49: 91–114

Gabarayeva NI, Grigorjeva VV, Rowley JR (2010) A new look at sporoderm ontogeny in *Persea americana* and the hidden side of development. Ann Bot 105: 939–955

Ganders FR (1979) The biology of heterostyly. NZ J Bot 17(4): 607–635

Grega L, Anderson S, Cheetham M, Clemente M, Colletti A, Moy W, Talarico D, Thatcher SL, Osborn JM (2013) Aerodynamic characteristics of saccate pollen grains. Int J Plant Sci 174: 499–510

Grímsson F, Zetter R (2011) Combined LM and SEM study of the Middle Miocene (Sarmatian) palynoflora from the Lavanttal Basin, Austria: Part II. Pinophyta (Cupressaceae, Pinaceae and Sciadopityaceae). Grana 50: 262–310

Halbritter H, Hesse M (1995) The convergent evolution of exine shields in Angiosperm pollen. Grana 34: 108–119

Halbritter H, Hesse M (2004) Principal modes of infoldings in tricolp(or)ate Angiosperm pollen. Grana 43: 1–14

Halbritter H, Hesse M (2005) Specific ornamentation of orbicular walls and pollen grains, as exemplified by Acanthaceae. Grana 44: 308–313

Havinga AJ (1971) An experimental investigation into the decay of pollen and spores in various soil types. In: Brooks J, Grant PR, Muir MD, Van Gijzel P (eds) Sporopollenin. Academic Press, London, New York, p. 446–479

Havinga AJ (1984) A 20-year experimental investigation into the differential corrosion susceptibility of pollen and spores in various soil types. Pollen Spores 26: 541–558

He X, Dai J, Wu Q (2016) Identification of Sporopollenin as the Outer Layer of Cell Wall in Microalga *Chlorella protothecoides*. Front Microbiol 7: 1047

Hesse M, Halbritter H, Zetter R, Weber M, Buchner R, Frosch-Radivo A, Ulrich S (2009) Pollen Terminology. An illustrated Handbook. Springer, Vienna

Hemsley AR, Barrie PJ, Chaloner WG, Scott AC (1993) The composition of sporopollenin and its use in living and fossil plant systematics. Grana 32, Suppl 1: 2–11

Huysmans S, El-Ghazaly G, Smets E (1998) Orbicules in angiosperms: morphology, function, distribution, and relation with tapetum types. Bot Rev 64: 240–272

Jardine PE, Fraser WT, Lomax BH, Gosling WD (2015) The impact of oxidation on spore and pollen chemistry. J Micropalaeontol 24: 139–149

John JF (1814) Ueber den Befruchtungsstaub, nebst einer Analyse des Tulpenpollens. J Chem Phys 12: 244–252

Johnson ST, Edwards TJ (2000) The structure and function of orchid pollinaria. Plant Syst Evol 222: 243–269

Klaus W (1960) Sporen der karnischen Stufe der ostalpinen Trias. In: Oberhauser R, Kristan-Tollmann E, Kollmann K, Klaus W (eds) Beiträge zur Mikropaläontologie der alpinen Trias. Jahrb Geol Bundesanstalt, Sonderband 5: 107–184

Klaus W (1987) Einführung in die Paläobotanik. Fossile Pflanzenwelt und Rohstoffbildung, Band I. Grundlagen – Kohlebildung – Arbeitsmethoden/ Palynologie. Deuticke, Wien

Knox RB, McConchie CA (1986) Structure and function of compound pollen. In: Blackmore S, Ferguson IK (eds) Pollen and Spores, Form and Function. Linnean Society of London, London, p. 265–282

Lallemand B, Erhardt M, Heitz T, Legrand M (2013) Sporopollenin biosynthetic enzymes interact and constitute a metabolon localized to the endoplasmic reticulum of tapetum cells. Plant Physiol 162: 616–625

Leslie AB (2010) Flotation preferentially selects saccate pollen during conifer pollination. New Phytol 188: 273–279

Liu L, Fan X (2013) Tapetum: regulation and role in sporopollenin biosynthesis in *Arabidopsis*. Plant Mol Biol 83: 165–175

Maeda Y (1984) The presence and location of sporopollenin in fruiting bodies of the cellular slime moulds. J Cell Sci 66: 297–308

Pacini E, Franchi GG (1991) Role of the tapetum in pollen and spore dispersal. Plant Syst Evol, Suppl. 7: 1–11

Pacini E, Hesse M (2005) Pollenkitt – its composition, forms and functions. Flora 200: 399–415

PalDat – a palynological database (2000 onwards, www.paldat.org)

Phuphumirat W, Gleason FH, Phongpaichit S, Mildenhall DC (2011) The infection of pollen by zoosporic fungi in tropical soils and its impact on pollen preservation: a preliminary study. Nova Hedwigia 92: 233–244

Phuphumirat W, Zetter R, Hofmann C-C, Ferguson DK (2015) Pollen degradation in mangrove sediments: A short-term experiment. Rev Palaeobot Palynol 221: 106–116

Playford G, Dettmann ME (1996) Spores. In: Jansonius J, McGregor DC (eds) Palynology: principles and applications. American Association of Stratigraphic Palynologists Foundation, vol. 1, AASP Foundation, Dallas, p. 227–260

Pozhidaev AE (2000a) Pollen variety and aperture patterning. In: Harley MM, Morton CM, Blackmore S (eds) Pollen and Spores: Morphology and Biology. Royal Botanic Gardens, Kew, p. 205–225

Pozhidaev AE (2000b) Hypothetical way of pollen aperture patterning. 2: Formation of polycolpate patterns and pseudoaperture geometry. Rev Palaeobot Palynol 109: 235–254

Praglowski J (1975) Importance de la mise au point des terms "structure" de l'exine. Bull Soc Bot France, Coll Palynologie 122: 75–78

Punt W, Hoen PP, Blackmore S, Nilsson S, Le Thomas A (2007) Glossary of pollen and spore terminology. Rev Palaeobot Palynol 143: 1–81

Quilichini TD, Douglas CJ, Samuels AL (2014) New views of tapetum ultrastructure and pollen exine development in *Arabidopsis thaliana*. Ann Bot 114: 1189–120

Reitsma TJ (1969) Size modification of recent pollen grains under different treatments. Rev Palaeobot Palynol 9: 175–202

Riding JB, Kyffin-Hughes JE (2004) A review of the laboratory preparation of palynomorphs with description of an effective non-acid technique. Rev Bras Paleontolog 7: 13–44

Rowley JR, Skvarla JJ (2000) The elasticity of the exine. Grana 37: 1–7

Rubinstein CV, Gerrienne P, de la Puente GS, Astini RA, Steemans P (2010) Early Middle Ordovician evidence for land plants in Argentina (eastern Gondwana). New Phytol 188: 365–369

Schwendemann AB, Wang G, Mertz ML, McWilliams RT, Thatcher SL, Osborn JM (2007) Aerodynamics of saccate pollen and its implications for wind pollination. Am J Bot 94: 1371–1381

Simons J, Van Beem AP, De Vries PJR (1983) Structure and chemical composition of the spore wall in *Spirogyra* (Zygnemataceae, Chlorophyceae). Acta Bot Neerl 31: 359–370

Skvarla JJ, Rowley JR, Chissoe WF (1997) Exine resistance to fungal infestations in Strelitziaceae. Taiwania 42: 17–27

Steemans P, Lepot K, Marshall CP, Le Herisseé A, Javaux EJ (2010) FTIP characterisation of the chemical composition of Silurian miospores (cryptospores and trilete spores) from Gotland, Sweden. Rev Palaeobot Palynol 162: 577–590

Takahashi M (1995) Development of structure-less pollen wall in *Ceratophyllum demersum* L. (Ceratophyllaceae). J Plant Res 108: 205–208

Traverse A (1988) Paleopalynology. Unwin Hyman, Boston

Traverse A (2007) Paleopalynology. 2nd ed, Springer, Dordrecht

Tryon AF, Lugardon B (1991) Spores of the Pteridophyta: Surface, wall structure and diversity based on electron microscopy studies. Springer, New York

Tsou C-H, Fu Y-L (2002) Tetrad pollen formation in *Annona* (Annonaceae): Proexine formation and binding mechanism. Am J Bot 89: 734–747

Twiddle CL, Bunting MJ (2010) Experimental investigations into the preservation of pollen grains: A pilot study of four pollen types. Rev Palaeobot Palynol 162: 621–630

Ueno R (2009) Visualization of sporopollenin–containing pathogenic green micro–alga Prototheca wickerhamii by fluorescent in situ hybridization (FISH). Can J Micro 55: 465–472

Ulrich S, Hesse M, Weber M, Halbritter H (2017) Amorphophallus: New insights into pollen morphology and the chemical nature of the pollen wall. Grana 56: 1–36

Van Bergen PF, Collinson ME, de Leeuw JW (1993) Chemical composition and ultrastructure of fossil and extant salvinialean microspore massulae and megaspores. Grana 32, Suppl 1: 18–30

Van Campo M, Lugardon B (1973) Structure grenue infratectal de l'ectexine des pollens de quelques Gymnospermes et Angiospermes. Pollen Spores 15: 171–189

Versteegh GJM, Blokker P, Bogus KA, Harding IC, Lewis J, Oltmanns S, Rochon A, Zonneveld KAF (2012) Infra red spectroscopy, flash pyrolysis, thermally assisted hydrolysis and methylation (THM) in the presence of tetramethylammonium hydroxide (TMAH) of cultured and sediment–derived *Lingulodinium polyedrum* (Dinoflagellata) cyst walls. Org Geochem 43: 92–102

Verstraete B, Moon H–K, Smets E, Huysmans S (2014) Orbicules in flowering plants: A phylogenetic perspective on their form and function. Bot Rev 80: 107–134

Vinckier S, Cadot P, Smets E (2005) The manifold characters of orbicules: structural diversity, systematic significance, and vectors for allergens. Grana 44: 300–307

Walker JW (1976) Evolutionary significance of the exine in the pollen of primitive angiosperms. In: Ferguson IK, Muller J (eds) The evolutionary significance of the exine. Academic Press, London, p. 251–308

Wallace S, Chater CC, Kamisugi Y, Cuming AC, Wellman CH, Beerling DJ, Fleming AJ (2015) Conservation of Male Sterility 2 function during spore and pollen wall development supports an evolutionarily early recruitment of a core component in the sporopollenin biosynthetic pathway. New Phytol 205: 390–401

Weber M, Halbritter H (2007) Exploding pollen in *Montrichardia arborescens* (Araceae). Plant Syst Evol 263: 51–57

Wellman CH (2010) The invasion of the land by plants: when and where? New Phytol 188: 306–309

Wiermann R, Ahlers F, Schmitz–Thom I (2001) Sporopollenin. In: Hofrichter M, Steinbüchel A (eds) Biopolymers 1: Lignin, Humic Substances and Coal, Wiley–VCH Weinheim, p. 209–227

Yule BL, Roberts S, Marshall JEA (2000) The thermal evolution of sporopollenin. Org Geochem 31: 859–870

Zetzsche F, Kalt P, Leichti J, Ziegler E (1931) Zur Konstitution des Lycopodiumsporonins, des Tasmanins und des Lange–Sporonins. J Prakt Chem 148: 67–84

Misinterpretations in Palynology

H. Halbritter et al., *Illustrated Pollen Terminology*, https://doi.org/10.1007/978-3-319-71365-6_4

The description of pollen ornamentation depends on three major parameters (1) the interpretations of the palynologist (which are subjective), (2) the pollen terminology applied, and (3) the magnification, resolution, and methods used.

The application of different preparation and staining methods and a combined analysis with light microscopy, scanning- and transmission electron microscopy are essential for the interpretation of pollen characters. Investigation of recent and fossil pollen material often reveals interesting features that in some cases may be misinterpreted. To demonstrate the wide range of possible misinterpretations, the following examples are given:

Example 1: Tripartite Feature in Gymnosperms — Impression Mark

Mature pollen of conifers, such as *Abies*, *Larix*, and *Pseudotsuga*, often shows proximally a Y-shaped bulge on the proximal polar side, comparable to a tetrad mark, which is called an **impression mark** (Fig. 1; Harley 1999). The mark results from the close proximity of the four pollen grains at the post-meiotic tetrad phase and is retained afterwards and is not a germination feature. Impression marks are also found in palm pollen. Note: the term tetrad mark is restricted to spores, where it is a germination feature.

Example 2: Tripartite Feature in Angiosperms — Triangular Tenuitas

Superficially similar features in angiosperms are not comparable to those observed in gymnosperms. In recent and fossil Sapindaceae a three-armed feature (more precisely a triangle) is found. *Cardiospermum* has a narrow **triangular tenuitas** (thinning) at the proximal pole, whereas other recent and subfossil Sapindaceae show such a feature at both poles (Fig. 2).

Example 3: Tripartite Feature in Angiosperms — Synaperture

Triangular pollen as found in Myrtaceae, some Primulaceae (*Primula farinosa* or *P. denticulata*) and Loranthaceae is characterized by a tripartite feature in both polar areas (Fig. 3). These are in fact

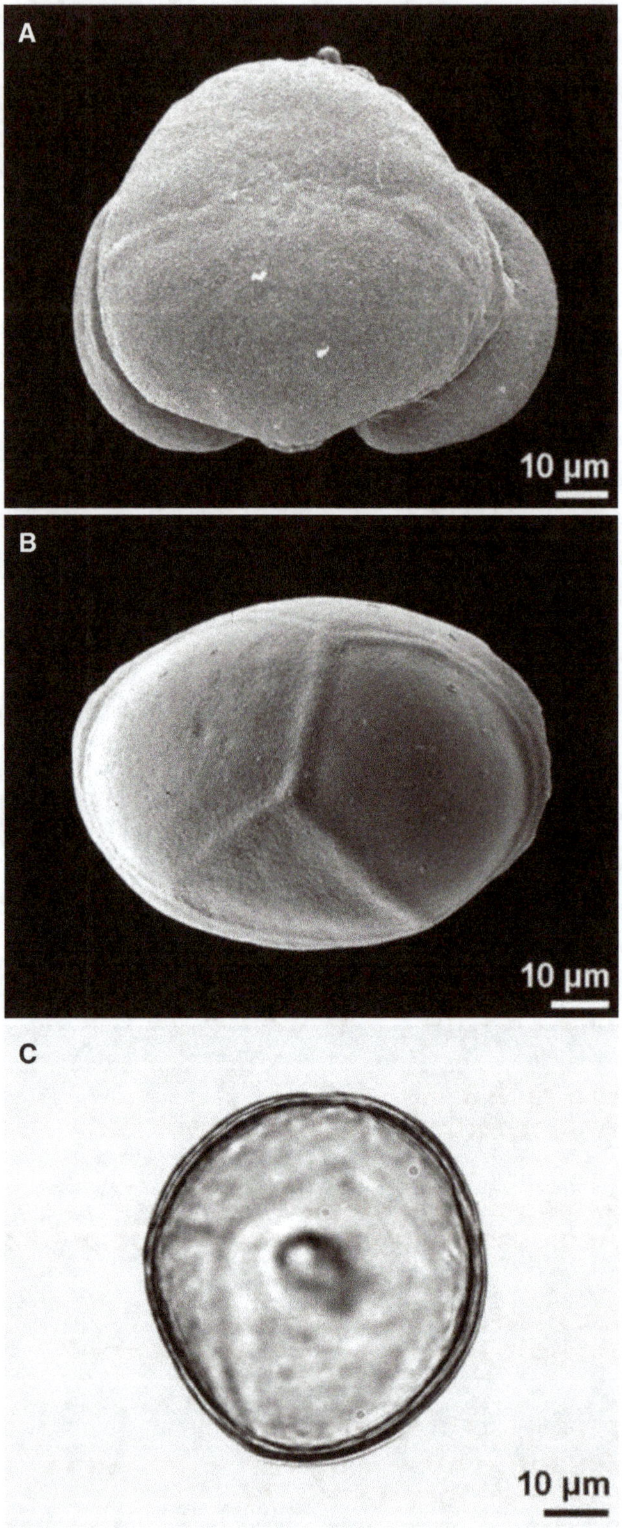

Fig. 1 Impression mark. A. *Abies cephalonica*, Pinaceae, proximal polar view, indistinct impression mark. **B-C.** *Larix* sp., Pinaceae, fossil, middle Miocene, Austria, proximal polar view, Y-shaped impression mark in SEM (**B**) and LM (**C**)

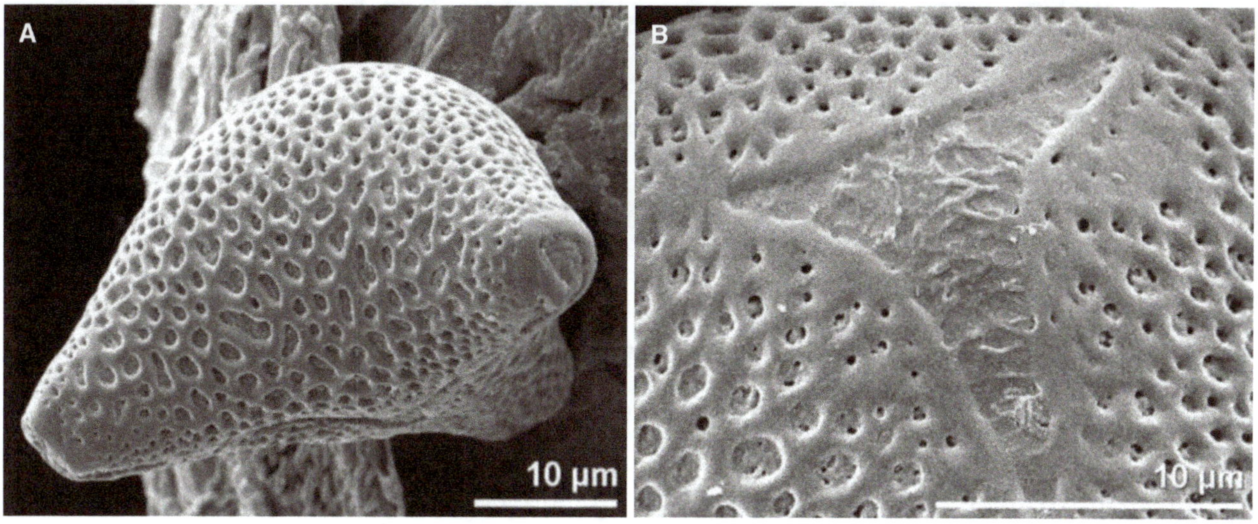

Fig. 2 Triangular tenuitas. A-B. *Cardiospermum corindum*, Sapindaceae, tricolporate, equatorial view (**A**), proximal pole with triangular thinning area (**B**)

Fig. 3 Synaperturate pollen. A-B. *Melaleuca armillaris*, Myrthaceae, syncolporate, polar view (**A**), close-up of polar area (**B**). **C.** *Primula denticulata*, Primulaceae, syncolpate, polar view. **D.** *Primula farinosa*, Primulaceae, syncolpate, dry pollen

three colpi, extending towards and merging at the poles. The pollen is therefore synaperturate (syncolpate, syncolporate). In for example, *Primula* the colpi dissect in the polar area, leaving a triangular field at both poles.

Example 4: Tripartite Feature in Angiosperms — Trichotomosulcus

Another tripartite feature is the **trichotomosulcus** (Harley 2004), a three-armed sulcus occurring exclusively distally, as, e.g., in *Dianella*. Trichotomosulcate pollen has been discussed in relation to the evolution of the tricolpate dicot condition, but so far without success (Fig. 4).

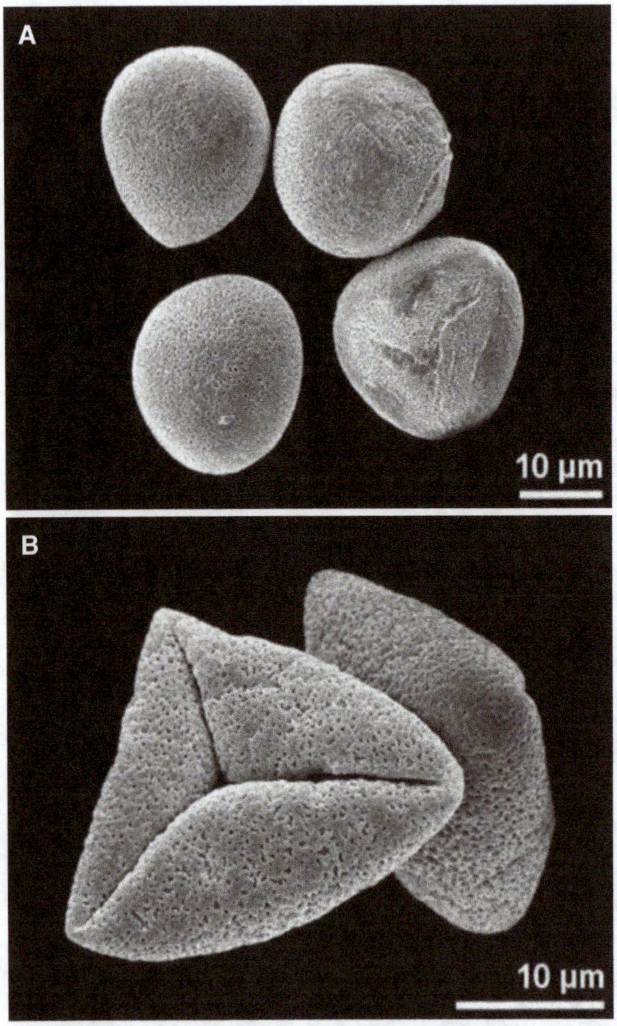

Fig. 4 Trichotomosulcus. A-B. *Dianella tasmanica*, Phormiaceae, trichotomosulcus (**A**), dry pollen, aperture infolded (**B**)

Example 5: Tripartite Feature in Angiosperms — Sulci vs. Colpi vs. Tenuitas

The angiosperm-like pollen of the fossil genus *Eucommiidites* is "trisulcate": a broad distal sulcus and two narrower additional "sulci" (at angles of c. 120° seen from the main sulcus; Fig. 5). This feature was erroneously interpreted as tricolpate pollen (with colpi equatorially situated).

A similar arrangement of a distal sulcus and two small additional sulci on the proximal face was described, for example, in some species of *Tulipa* (Liliaceae) and *Tinantia* (formerly *Commelinantia*, Commelinaceae), but these cases were never interpreted as equivalent to a tricolpate condition (Harley 2004) (Fig. 6). The two small additional sulci may also be interpreted as tenuitates. In some cases the three "sulci" are of similar size. The aperture condition is very similar to a tricolpate one.

Example 6: Tripartite Feature in Angiosperms — Triradiate Aperture

Another three-armed feature is the triradiate aperture in *Thesium alpinum* (Santalaceae) pollen. The heteropolar pollen is 3-aperturate, with apertures placed in the three tapered edges of a tetrahedron (Feuer 1977). Each aperture has a very inconspicuous triradiate outline, which is situated equatorially. Two of the arms point towards the neighboring tetrahedron edge and are rather short; the third, elongated arm is directed towards the rounded pole (Fig. 7).

Example 7: Apertures in Angiosperms — Planaperturate

Sometimes apertures are inconspicuous and not discernible at first sight. In pollen of *Pachira aquatica* (Malvaceae) three large, more-or-less hemispherical areas are seen equatorially, which may at first sight be interpreted as pores. However, a detailed observation reveals **planaperturate** pollen grains with three short colpi (Fig. 8).

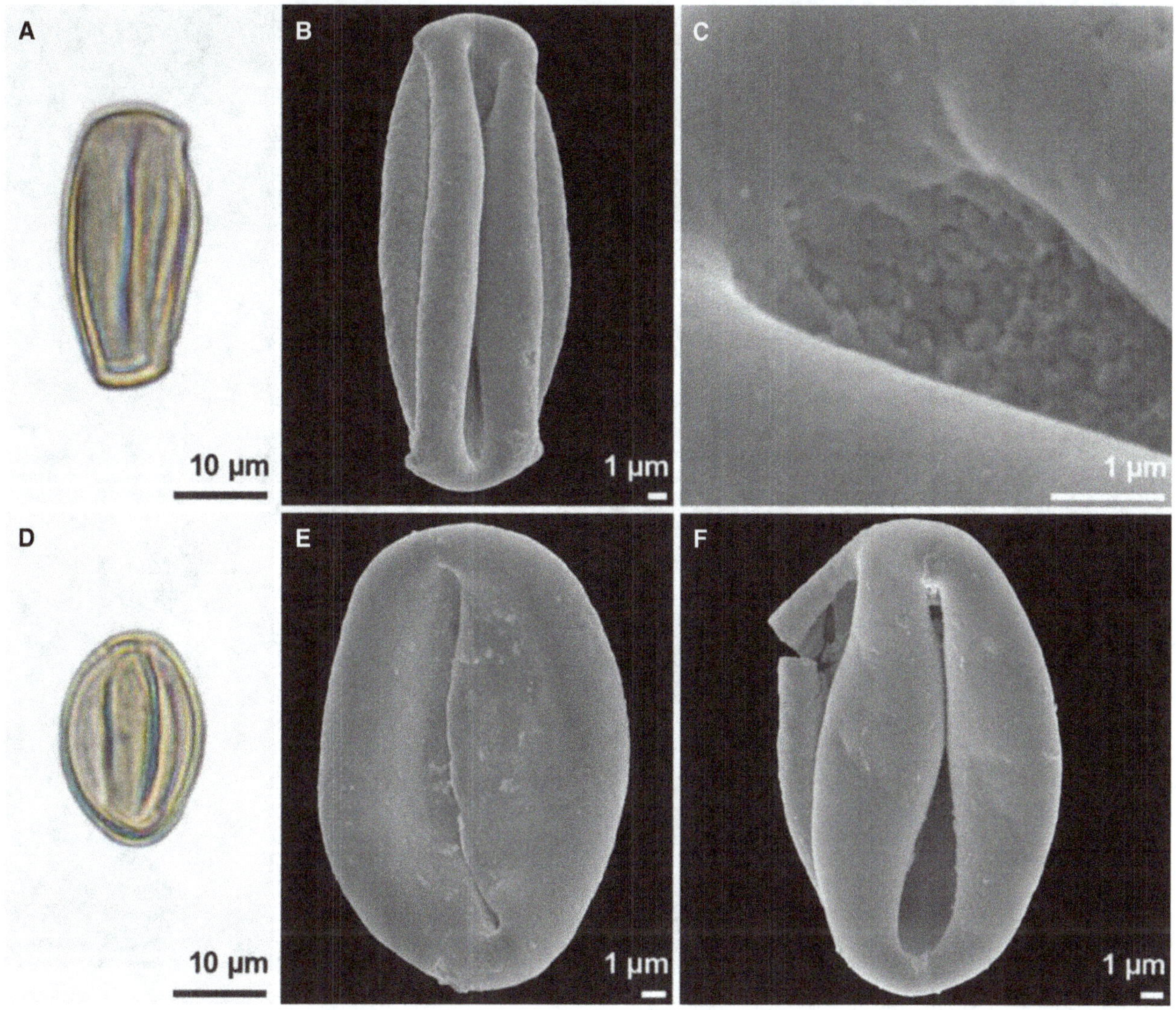

Fig. 5 Trisulcate pollen. A-C. *Eucommiidites* sp., fossil pollen, Lower Cretaceous of U.S.A., main sulcus with membrane seen in center of pollen grain, flanked by additional narrow sulci on each side (at angles of c. 120°, **A-B**), close-up showing sulcus membrane of main sulcus (**C**). **D-F.** *Eucommiidites* sp., fossil pollen, Lower Cretaceous of U.S.A., Narrow lateral sulcus (**E**), same grain turned showing the main broad sulcus and one narrow lateral sulcus (**F**)

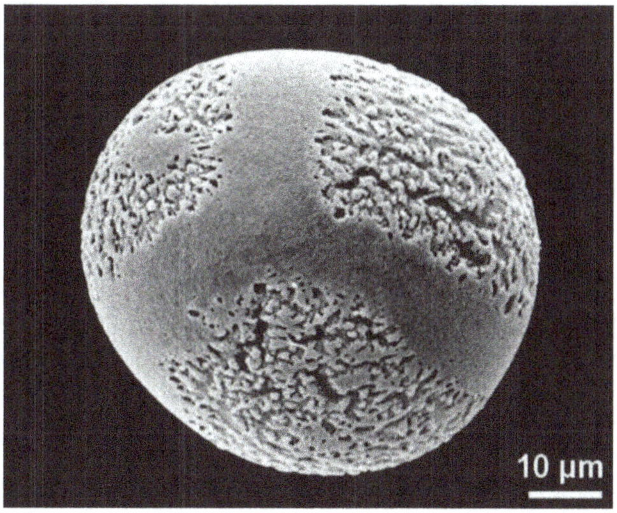

Fig. 6 Trisulcate pollen. *Tulipa kaufmanniana*, Liliaceae, trisulcate or sulcate with two tenuitates, equatorial view

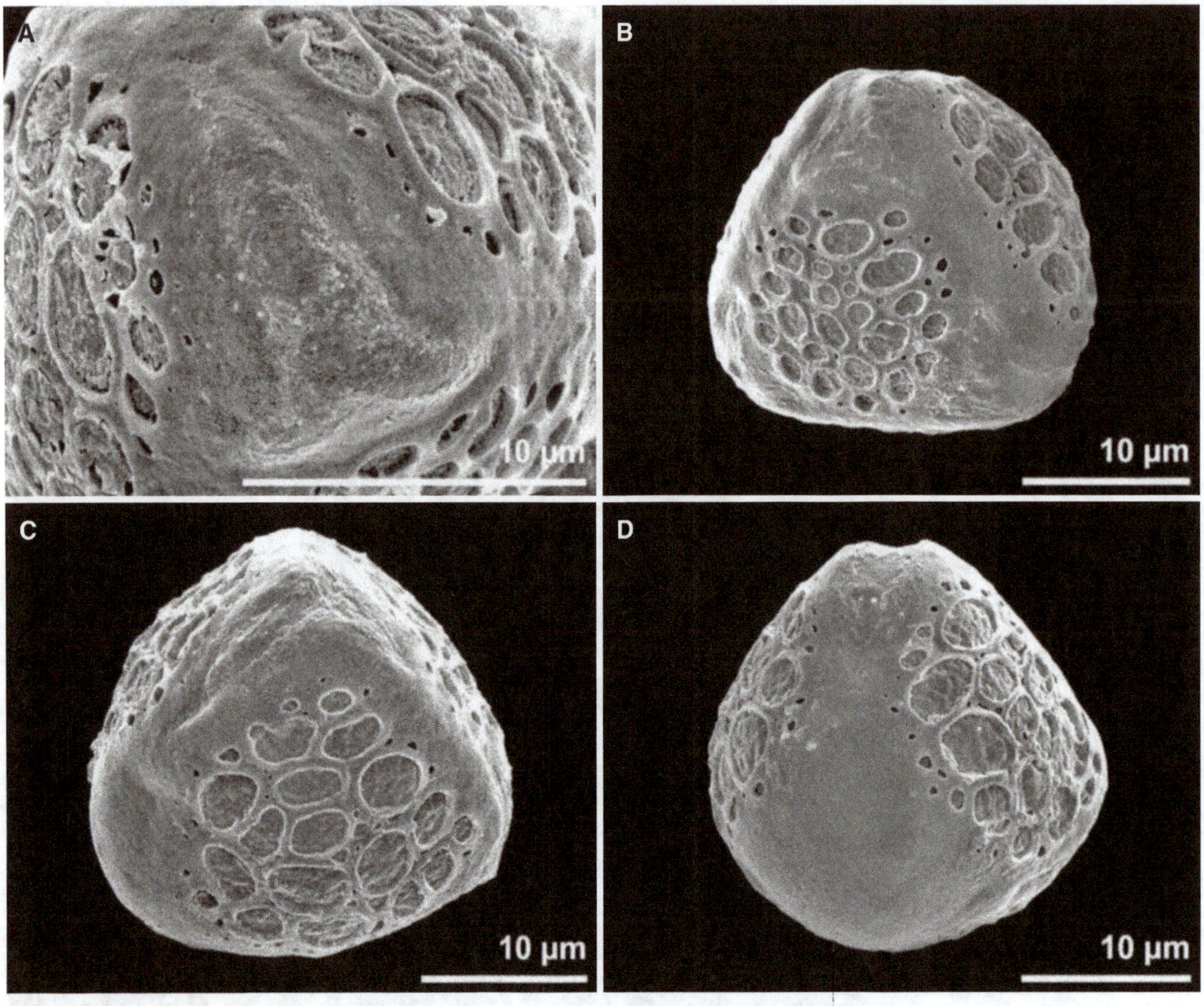

Fig. 7 Triradiate aperture. A-D. *Thesium alpinum*, Santalaceae. **A.** Tricolpate, heteropolar, triradiate colpus. **B.** Polar view (flattened pole). **C.** Equatorial view. **D.** Polar view (rounded pole)

Example 8: Apertures in Angiosperms — Inconspicuous Pori

In *Calliandra emarginata* (Mimosaceae) the monads forming a polyad are separated by narrow groove-like depressions. At low magnification the presence and localization of the apertures remain indistinct; high SEM magnification reveals that the apertures are very inconspicuous pores, situated equatorially, usually at the conjunction of three or four monads (Fig. 9 A, B).

Also, the aperture condition may be overlooked due to other eye-catching features. The clypeate pollen of *Phyllanthus* x *elongatus* (Euphorbiaceae) seems to be inaperturate. Only close-ups reveal the inconspicuous few pores between the exine shields (Fig. 9 C, D).

Example 9: Apertures in Angiosperms — Inconspicuous Colpi

The disc-like pollen of *Oryctanthus* sp. (Loranthaceae) shows at both poles conspicuous circular depressions that are not apertures (Feuer and Kuijt 1985; Grímsson et al. 2018). The pollen is according to Grímsson et al. (2018) demi(3)colpate, with

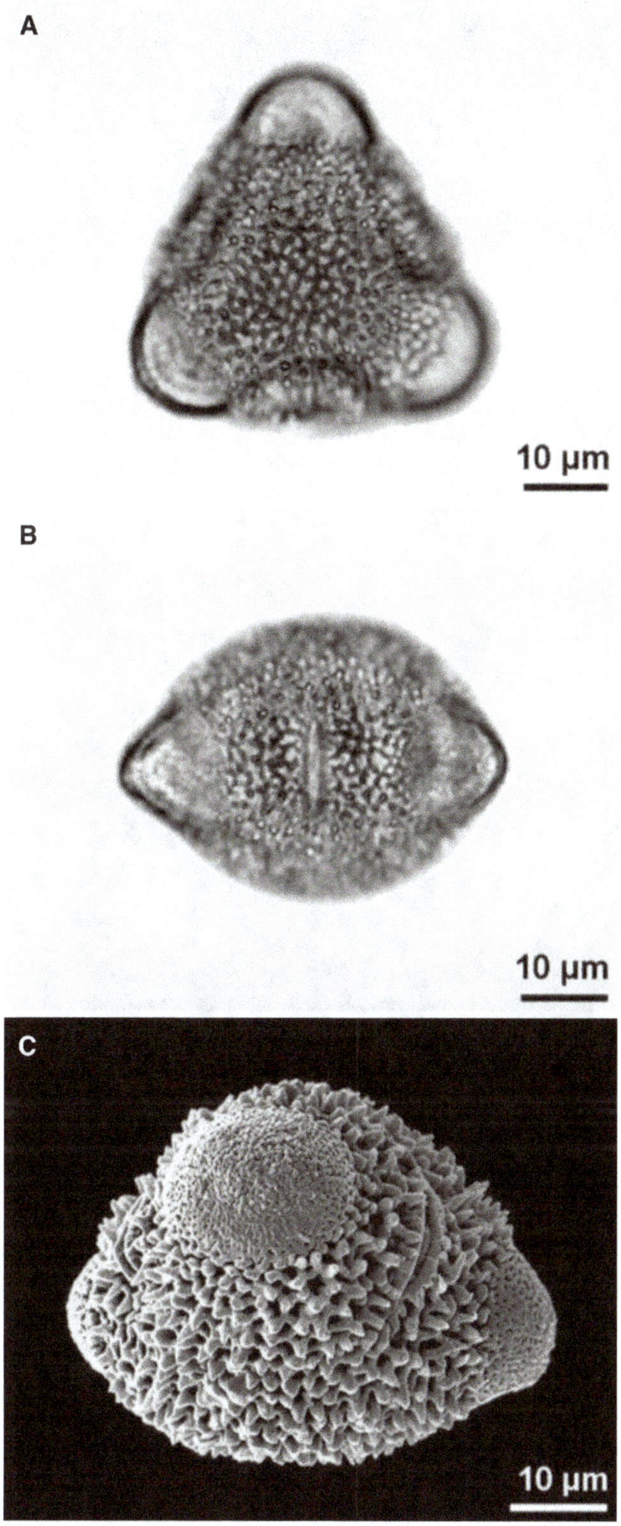

Fig. 8 Planaperturate pollen. A-C. *Pachira aquatica*, Malvaceae, polar view (**A**), equatorial view (**B**), oblique equatorial view (**C**)

inconspicuous slit-like colpi positioned between the polar depressions (Fig. 10). Another example are some Asteraceae pollen, where the colpi are often inconspicuous or not visible in SEM, but obvious in LM.

Example 10: Apertures in Angiosperms — Hidden Apertures

Recent and fossil triaperturate (colpate or porate) pollen of *Trapa* (Trapaceae) is distinguished by unique meridional exine ridges (crests) covering the apertures (Zetter and Ferguson 2001) (Fig. 11).

Example 11: Apertures in Angiosperms — Ring-like Apertures vs. Colpate-Operculate

The apertures in *Passiflora* cf. *incarnata* may be interpreted as three ring-like apertures or may be interpreted as pori (or colpi) each with an operculum. In other species of *Passiflora* e.g., *P. citrina* and *P. suberosa*, the apertures are both narrower and stephanocolpate (Fig. 12).

Example 12: Apertures in Angiosperms — Tenuitas vs. Poroid

Tenuitas is a general term for a pollen wall thinning (Kremp 1968; Harley 2004; Punt et al. 2007). It is normally found additional to apertures, e.g., in *Myosotis* (Fig. 13). A circular tenuitas can be mistaken for a **poroid**, which is a circular or elliptic aperture with an indistinct margin (see also "Illustrated Pollen Terms").

Example 13: Apertures in Angiosperms — Infoldings vs. Apertures

When pollen is infolded it can be hard to distinguish the apertures. Pollen of *Sparganium erectum* (Sparganiaceae) is in dry stage infolded, boat-shaped, and would be considered as sulcate. In fact, *Sparganium* pollen is ulcerate, the ulcus is seen clearly in the hydrated, spherical pollen stage (Fig. 14).

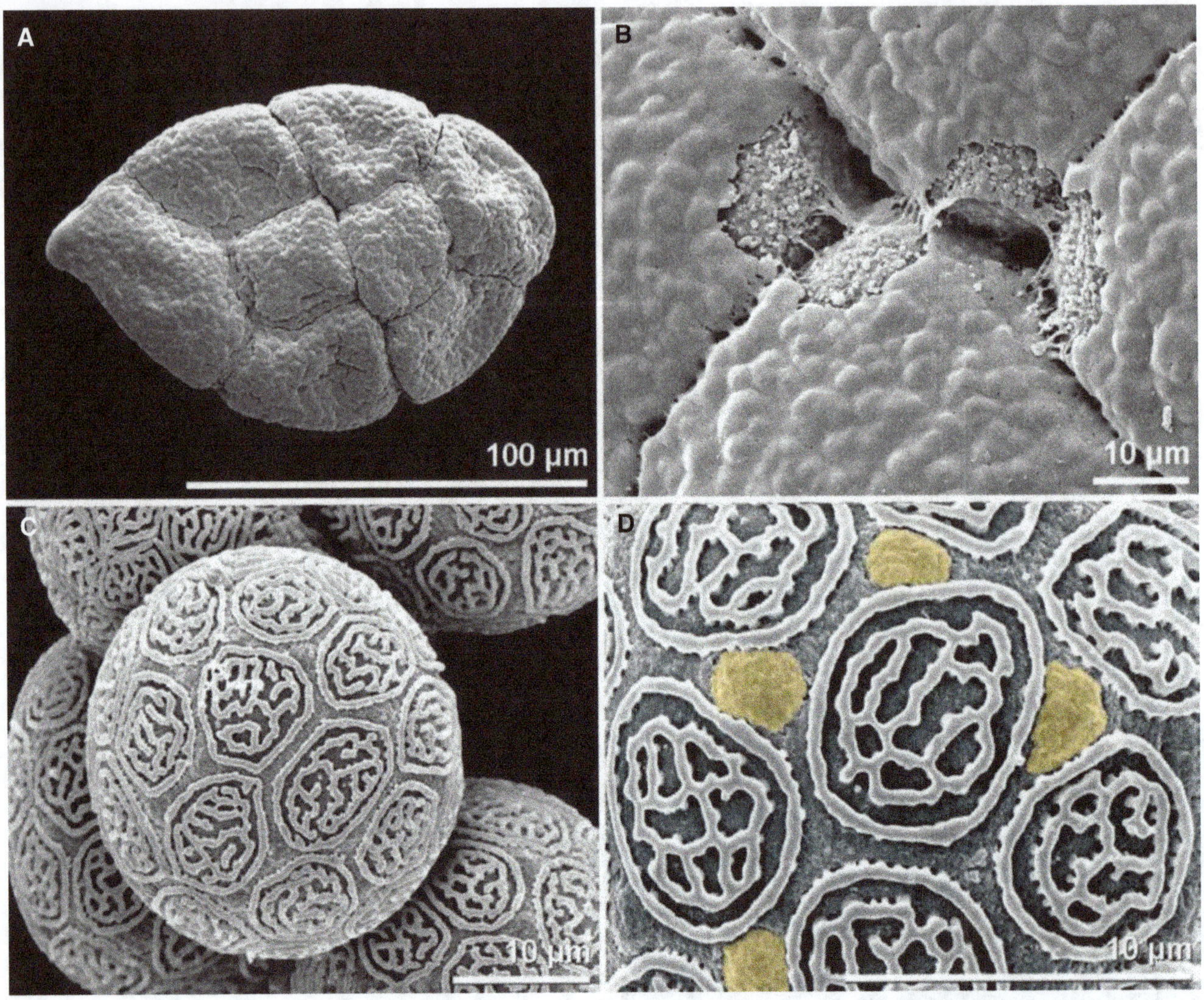

Fig. 9 Apertures in *Calliandra* and *Phyllanthus*. A-B. *Calliandra tergemina*, Fabaceae, polyad, dry state (**A**). Apertures (pori) at the junction of four monads (**B**). **C-D.** *Phyllanthus x elongatus*, Euphorbiaceae, clypeate, seemingly inaperturate (**C**). Inconspicuous pores (colored) between the exine shields (**D**)

Example 14: Apertures in Angiosperms — Ulcerate-Operculate vs. Ring-like Aperture

Nymphaea alba (Nymphaeaceae) pollen has asymmetrical halves divided by a ring-like aperture (Fig. 15). The features of the smaller distal half may be misinterpreted as a large ulcus with a conspicuous operculum. Ultrastructural studies and germination experiments support the interpretation of a ring-like aperture (Gabarayeva and Rowley 1994; Hesse and Zetter 2005).

Example 15: Apertures in Angiosperms — Disulcate vs. Dicolpate

The term disulcate defines two elongated apertures situated usually distally (but not directly at the distal pole), running parallel to or even in the equator (Fig. 16). If the apertures are running meridionally, pollen would be dicolpate (Halbritter and Hesse 1993). To distinguish if the pollen is disulcate or dicolpate it is important to study the pollen in tetrad arrangement to clarify the polarity and position of apertures (see Fig. 3 in "Methods in Palynology").

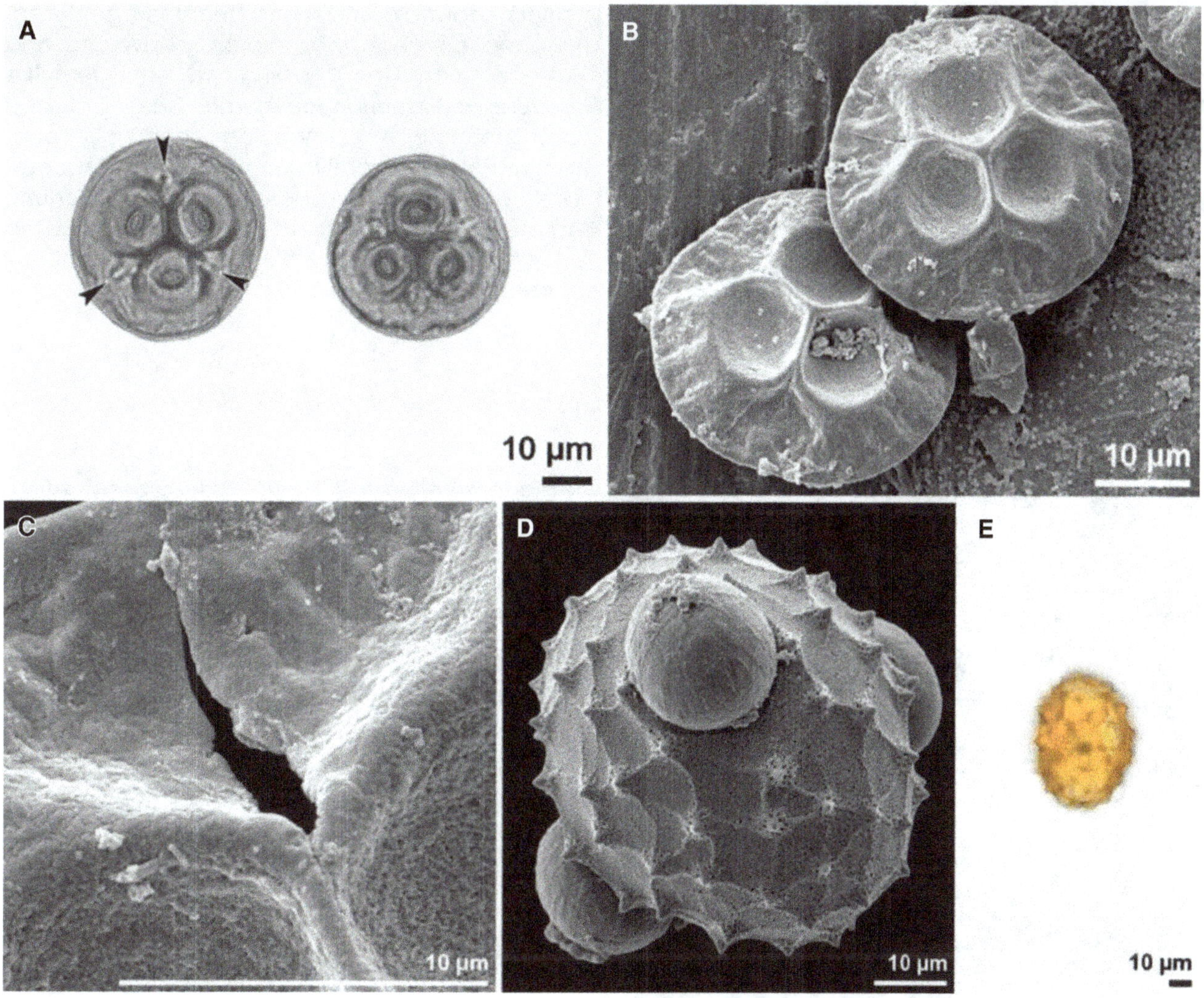

Fig. 10 Apertures in *Oryctanthus*. A-C. *Oryctanthus alveolatus*, Loranthaceae, acetolyzed pollen, arrowheads point to colpi, LM (**A**). two grains in polar view, SEM (**B**). close-up showing colpus (**C**). **D-E.** *Carthamus lanatus*, Asteraceae, hydrated pollen, pollen in SEM seem porate (**D**). Acetolyzed pollen, colporus (highlighted) only visible in LM (**E**)

Examples for taxa with disulcate pollen are the monocots *Tofieldia calyculata* with one sulcus distally, the other proximally, *Uvularia grandiflora*, *Eichhornia crassipes* (Hesse et al. 2009), some *Dioscorea* species (Schols et al. 2005), *Pontederia cordata* (Halbritter 2016), *Calla palustris* (Ulrich et al. 2013), and the magnoliid *Calycanthus floridus* (Huynh 1976).

Example 16: Apertures in Angiosperms — Zon-, Zono-, Zoni-, Zona- vs. Ring-like Aperture and Stephanoaperturate Pollen

Terms combining the basic prefix zon- together with its linguistic derivatives are a source of endless confusion, misunderstanding and superfluous inflation of terms. The prefix include **zon-** (in zonorate, for a ring-like endoaperture, the os, at the equator), the outdated, rarely used **zoni-** (however, with two quite different terminological applications), but especially **zona-** (indicating exclusively a ring-like feature situated anywhere) and **zono-** (indicating any feature located strictly equatorially).

Terms for ring-like (aperture) features include zona-aperturate, zona-sulculus (addressing the polarity by anazona-sulculus and catazona-sulculus), zona-sulcus, zonate, zono-aperturate, and also related names (e.g., "fully zonate condition" sensu Grayum 1992). Even the misleading and contradictory **zon<u>o</u>**-sulcus (a sulcus cannot be situated equatorially) is used instead of the correct, but phonetically confusable, **zon<u>a</u>**-sulcus. Even the

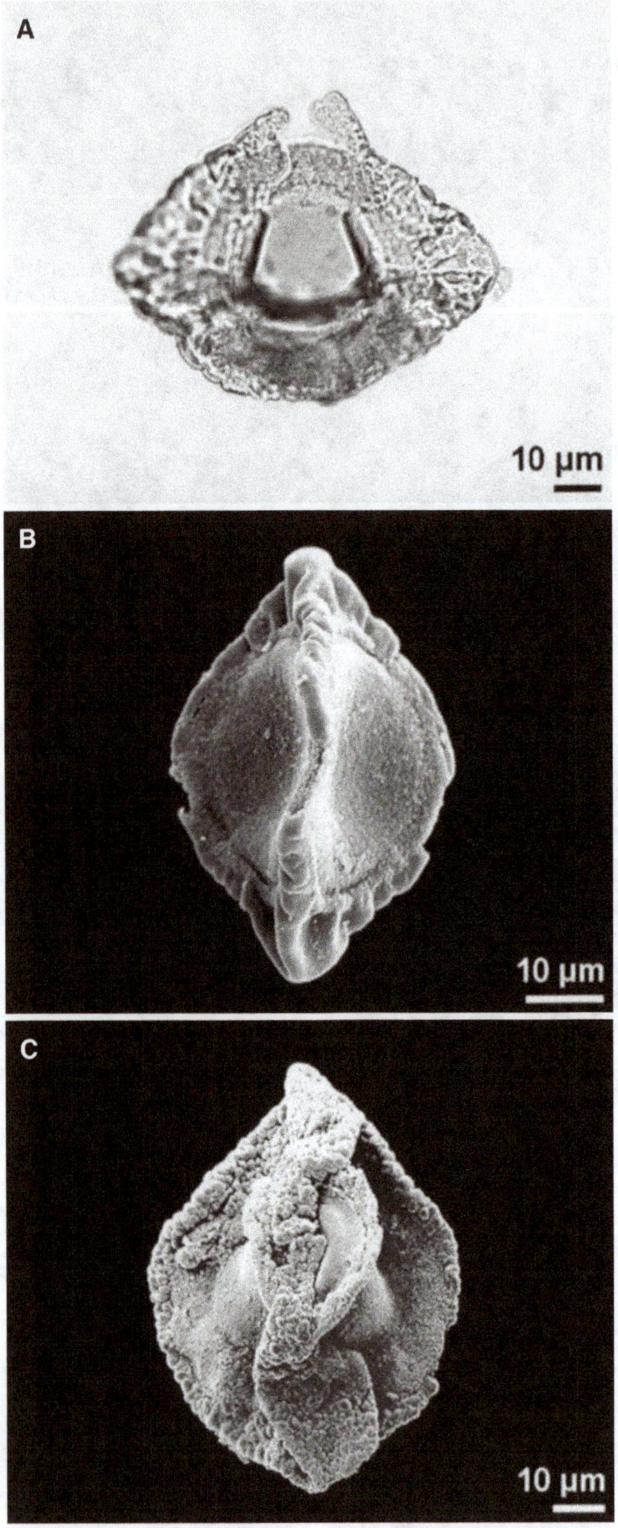

Fig. 11 Apertures in *Trapa*. A-C. *Trapa* sp., Trapaceae, fossil, late Miocene, Austria, equatorial view, crest broken, LM (**A**). Equatorial view, crest partly broken, colpus visible in SEM (**B**). Equatorial view, SEM (**C**)

trained palynologist may become confused. Therefore, all these terms should be avoided and we recommend the following two terms: **ring-like aperture and stephanoaperturate** (see "Illustrated Pollen Terms"). Any encircling aperture ("zona-aperturate"), irrespective of meridional or equatorial location, is simply called a **ring-like aperture**. Any case with more than three apertures at the equator ("zono-aperturate") is called **stephanoaperturate**.

Example 17: Magnification Effect — Retipilate vs. Reticulum Cristatum

The term retipilate (reticuloid) describes a reticulum formed by pila instead of muri (Erdtman 1952). Combined investigations based on LM and SEM have revealed that the given examples *Callitriche* (Punt et al. 2007) and *Cuscuta lupuliformis* (Erdtman 1952) do not fit the definition of retipilate. In fact, the reticulum consists of muri with prominent suprasculpture elements and are without isolated pilae. Such ornamentation is termed reticulum cristatum (a special type of reticulum; muri with prominent suprasculpture elements; Fig. 17, see also "Illustrated Pollen Terms"). So far no example for retipilate sensu Erdtman (1952) is currently known.

Example 18: Dispersal Units — Massula vs. Polyad

For a pollen dispersal unit of more than four pollen grains two terms are in use, **massula** and **polyad** (Fig. 18). The application of both terms is confusing and inconsistent in the literature. Often, the various authors employ the terms more or less interchangeably and do not provide a sharp delimitation (Walker 1971; Wagenitz 2003; Punt et al. 2007; Traverse 2007). These terms, however, are not exchangeable for historical and practical reasons (see extensive review by Teppner 2007).

The term massula was coined by Richard (1817) for parts of a pollinium in some Orchidaceae and should be used for the subunits of orchid sectile pollinia/pollinaria. Massulae within one and the same pollinium are variable and different in shape, size, and numbers of pollen grains. Unfortunately, the term massula has also been used to designate compound pollen in various other families, e.g.

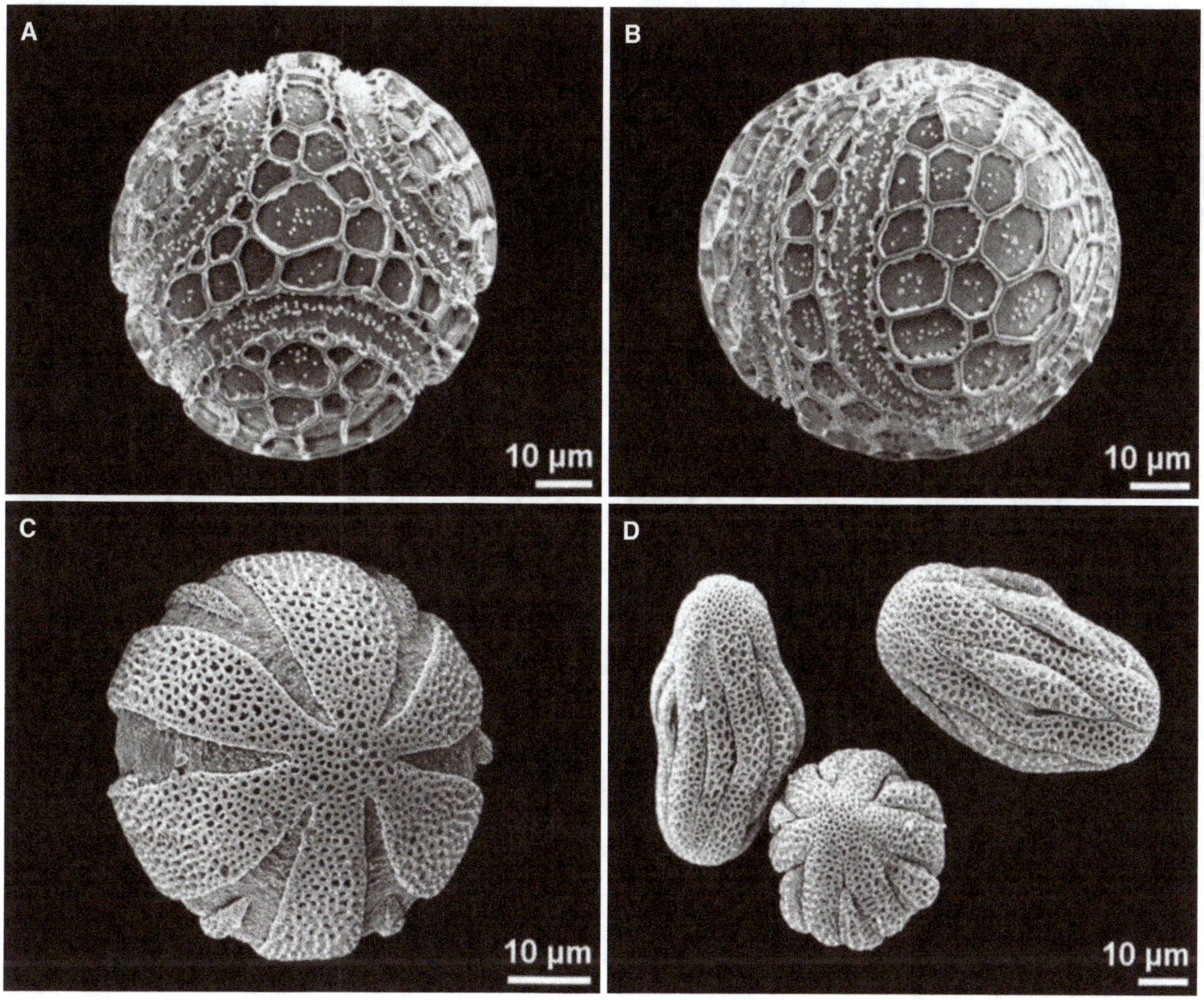

Fig. 12 Apertures in *Passiflora*. A-B. *Passiflora* cf. *incarnata*, Passifloraceae; colpate, operculate aperture, polar view (**A**), equatorial view (**B**). **C.** *Passiflora citrina*, Passifloraceae, stephanocolpate, operculate, polar view. **D.** *Passiflora suberosa*, Passifloraceae, stephanocolpate, operculate, dry pollen

Fabaceae-Mimosoideae, producing dispersal units of more than four pollen grains (e.g., Wettstein 1907; Wagenitz 2003; Punt et al. 2007). For these the term polyad — coined by Iversen and Troels-Smith (1950) — should be used, denoting a symmetric dispersal unit of more than four regularly arranged and permanently united pollen grains. Polyads, currently known to occur in Fabaceae (Mimosoideae), Gentianaceae, Hippocrateaceae, Celastraceae and Annonaceae, contain a specific number of pollen grains (a multiple of four: 8, 12, 16, 24, 32, 48, 64) and show a species-specific shape.

Example 19: Preparation Effect — Psilate vs. Ornamented

Ornamentation sometimes depends on the **preparation method**. A striking example is pollen of many Aroideae (Araceae), that are ornamented (e.g., echinate, striate, verrucate) in fresh or dry condition, but become psilate following acetolysis (Fig. 19). The outer pollen wall layer and ornamentation elements are composed of polysaccharide (lack sporopollenin) and are therefore destroyed during acetolysis (Weber et al. 1999; Ulrich et al. 2017).

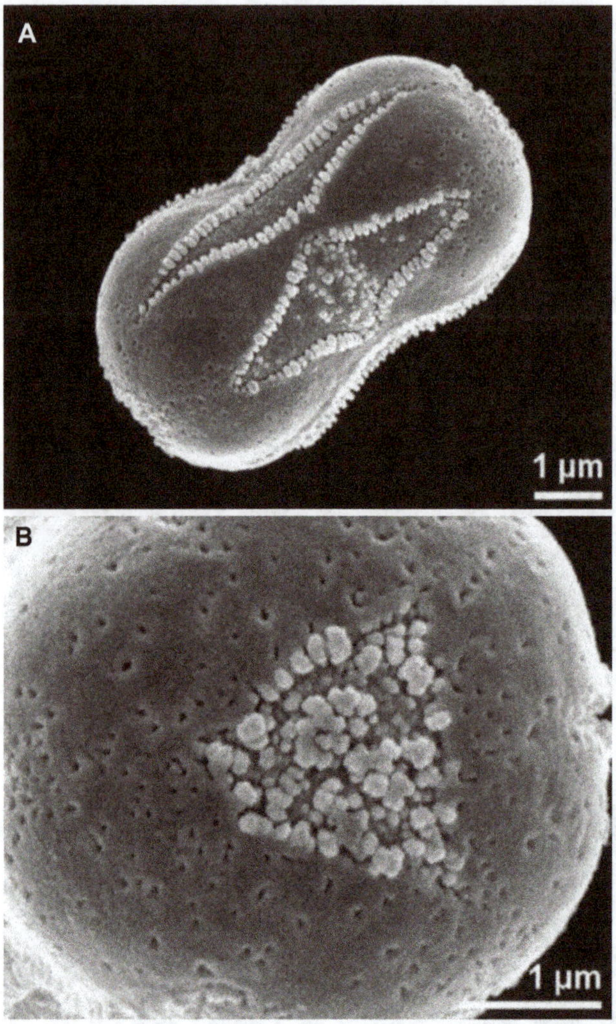

Fig. 13 Tenuitas vs. poroid. A-B. *Myosotis palustris*, Boraginaceae, equatorial view, heteroaperturate, alternating colpori and pseudocolpi (**A**), polar view, polar area with triangular tenuitas (**B**)

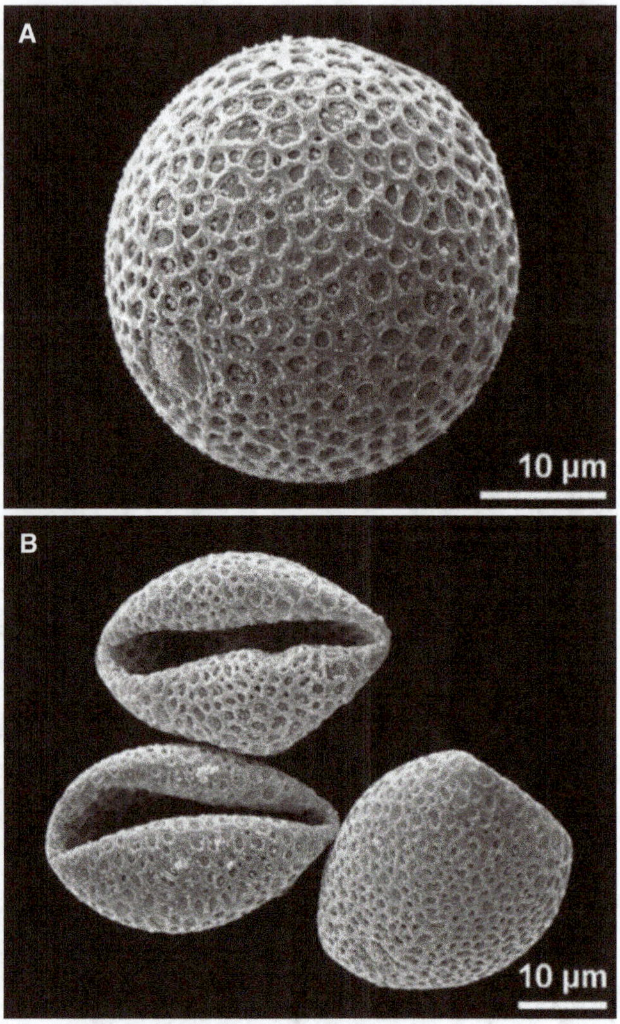

Fig. 14 Apertures in *Sparganium*. A-B. *Sparganium erectum*, Sparganiaceae, ulcerate, equatorial view hydrated pollen (**A**), boat-shaped, dry pollen (**B**)

Example 20: Preparation Effect — Areolate-Fossulate vs. Verrucate

The dehydration process with 2,2-dimethoxypropane (DMP) and critical point drying (CPD) for SEM investigations can affect the ornamentation. An example for different interpretations in relation to a varying degree of hydration is *Trichosanthes anguina* (Cucurbitaceae), where the ornamentation can reflect different degrees of hydration. The ornamentation can be described as areolate and fossulate in partially hydrated condition or verrucate and perforate in fully hydrated condition (Fig. 20).

Example 21: Preparation Effect — Striate vs. Striato-reticulate

The ornamentation of *Amorphophallus longituberosus* pollen in dry condition or hydrated in water is striate, but after critical point drying it becomes striate to reticulate. The striate to reticulate ornamentation of *Amorphophallus longituberosus* is a result of an expanding thin surface layer (Fig. 21 D). During rehydration, the expansion of the thin layer itself forms a reticulum (Fig. 21 C), which finally ruptures partly or completely (Ulrich et al. 2017).

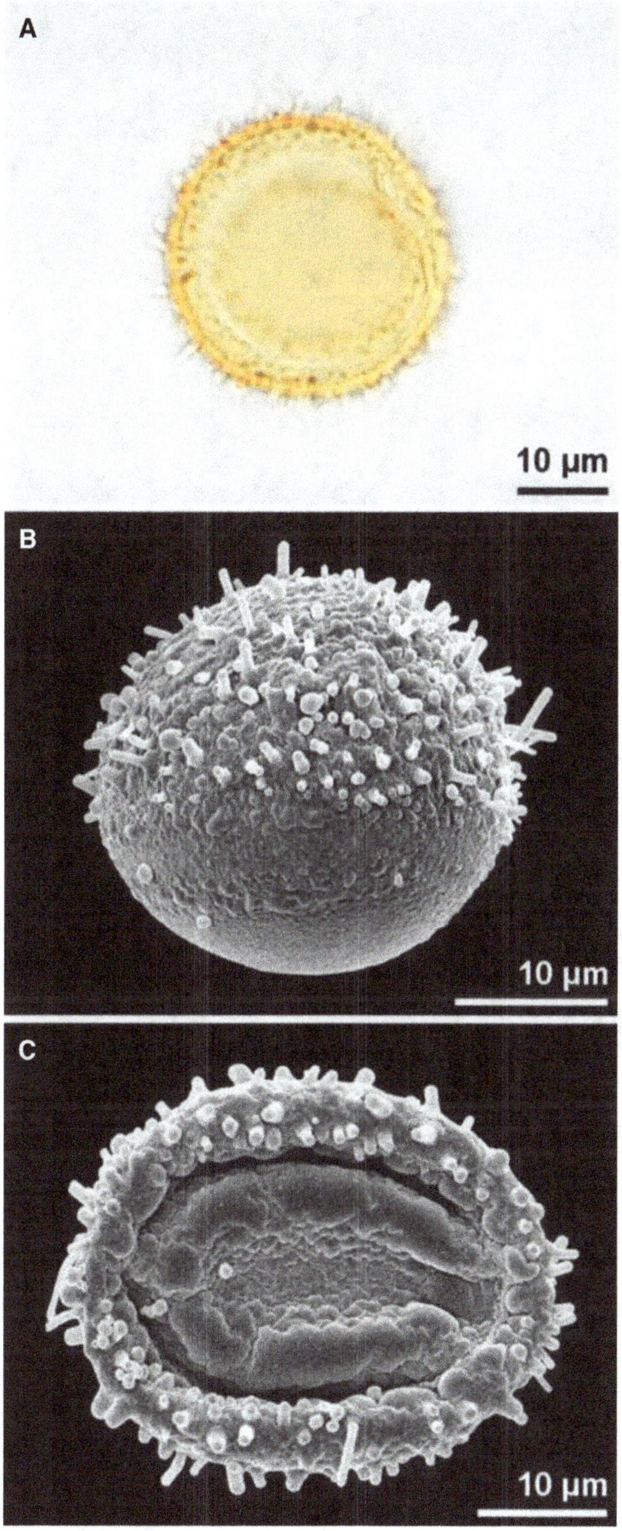

Fig. 15 Apertures in Nymphaea. A-C. *Nymphaea* sp., Nymphaeaceae; ring-like aperture, polar view (**A**), Ring-like aperture, equatorial view (**B**), dry pollen, cup-shaped (**C**)

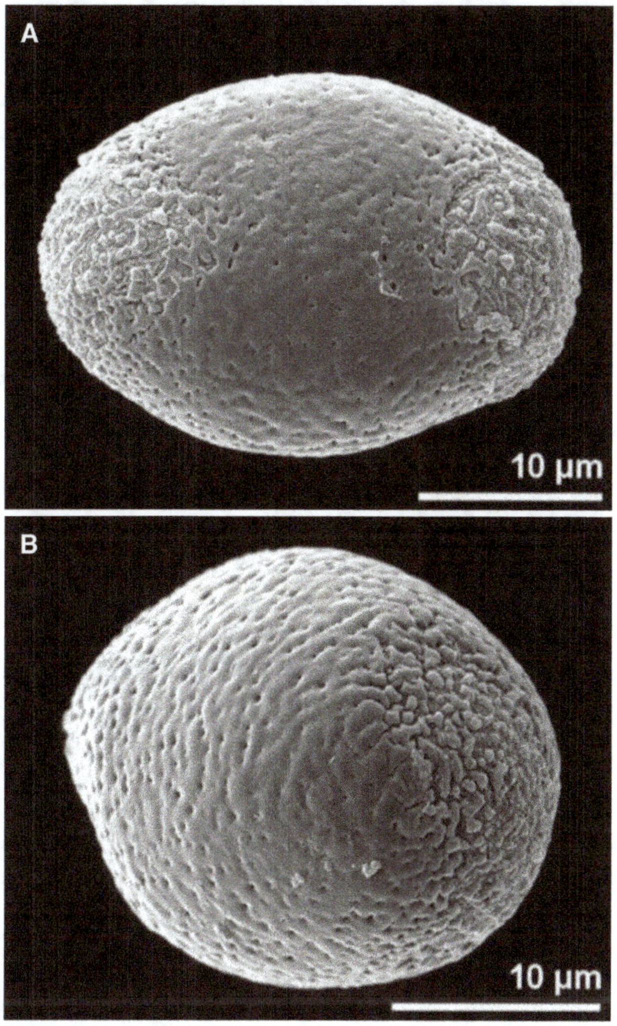

Fig. 16 Disulcate. A-B. *Calla palustris*, Araceae, polar and equatorial view

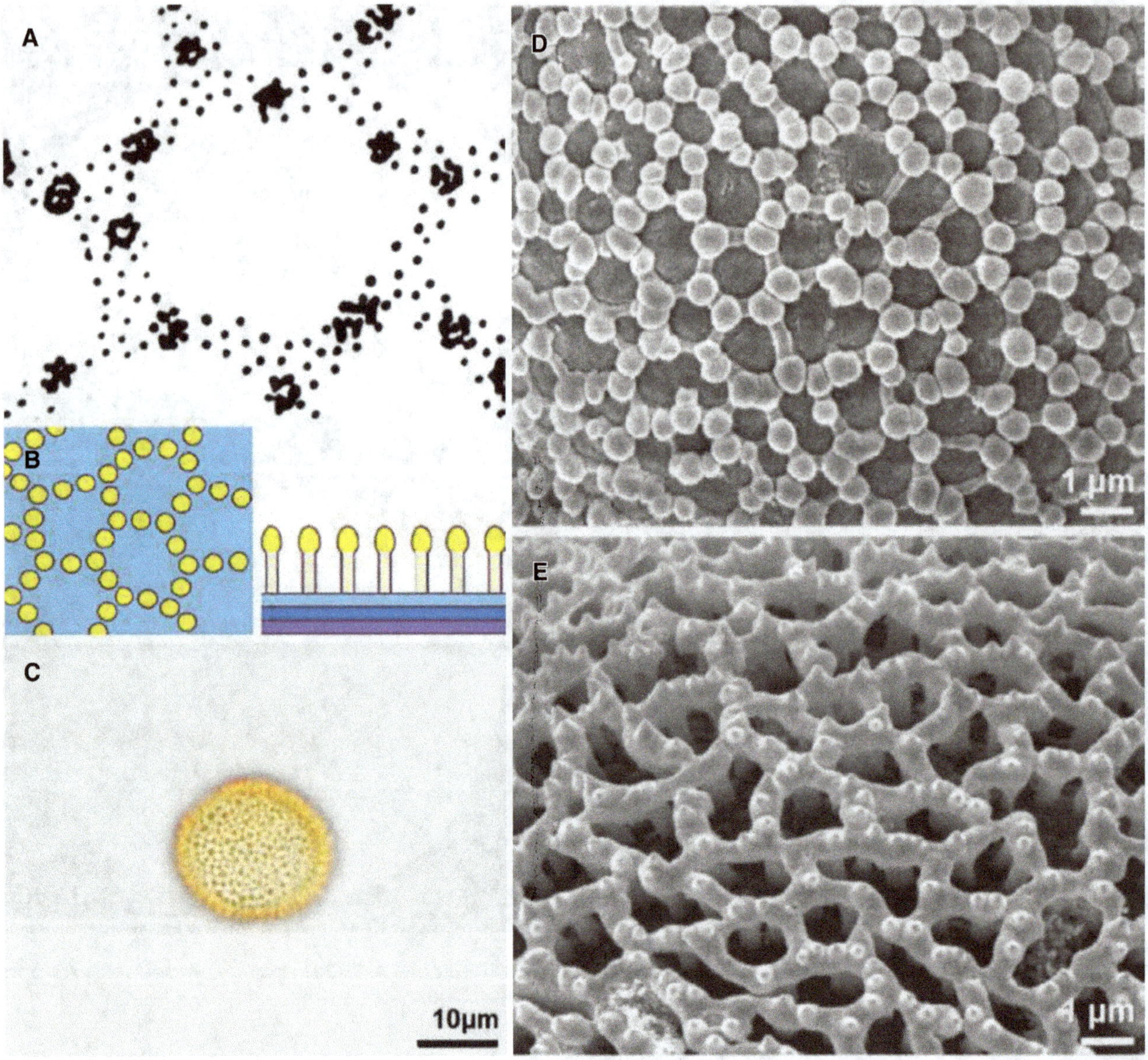

Fig. 17 Retipilate vs. reticulum cristatum. A. Drawing from Erdtman (1952). **B.** Drawings from Punt et al. (2007). **C.** *Callitriche palustris*, Plantaginaceae, acetolyzed pollen in LM. **D.** *Callitriche polymorpha*, Plantaginaceae, reticulum cristatum with small gemmae (suprasculpture) on thin muri. **E.** *Cuscuta lupuliformis*, Convolvulaceae, reticulum cristatum with nanoechini (suprasculpture)

Example 22: Staining Methods — Absence or Presence of Endexine

The staining behavior of the endexine is very heterogeneous, even within the same plant family or the same genus (Weber and Ulrich 2010). Therefore, the endexine is often reported as absent even though the layer is actually present. In most studies on pollen ultrastructure, sections are stained with uranyl acetate and lead citrate only. To truly distinguish the presence of endexine one should/must apply potassium permanganate which stains the endexine electron dense (Fig. 22, see also "Methods in Palynology").

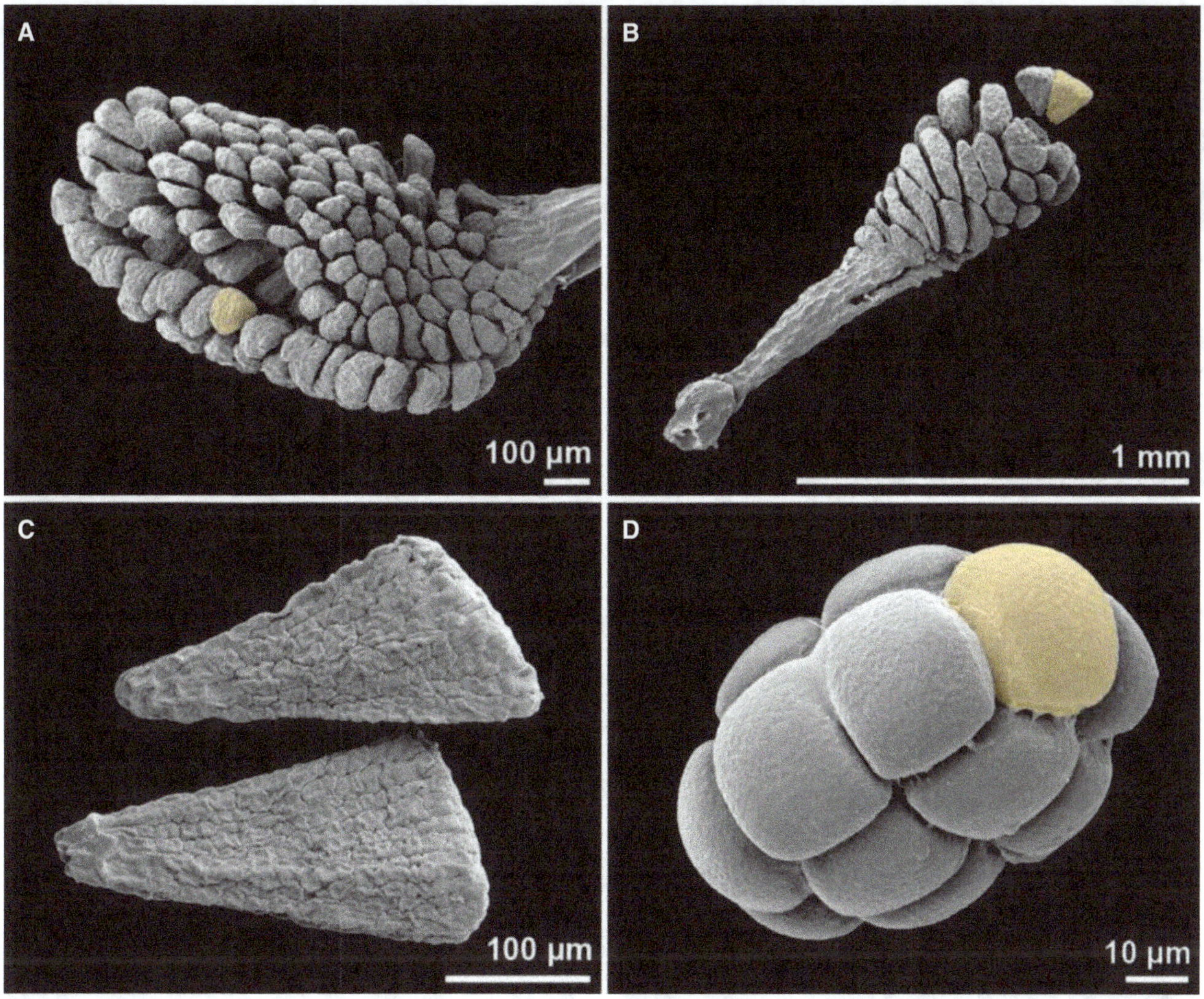

Fig. 18 Massula vs. polyad. A. *Habenaria* sp., Orchidaceae, pollinium composed of numerous massulae (massula highlighted). **B.** *Orchis ustulata*, Orchidaceae, pollinium composed of numerous massulae, two massulae partly segregated (massula highlighted). **C.** *Ludisia discolor*, Orchidaceae, 2 segregated massulae. **D.** *Albizia julibrissin*, Fabaceae, polyad (monad highlighted)

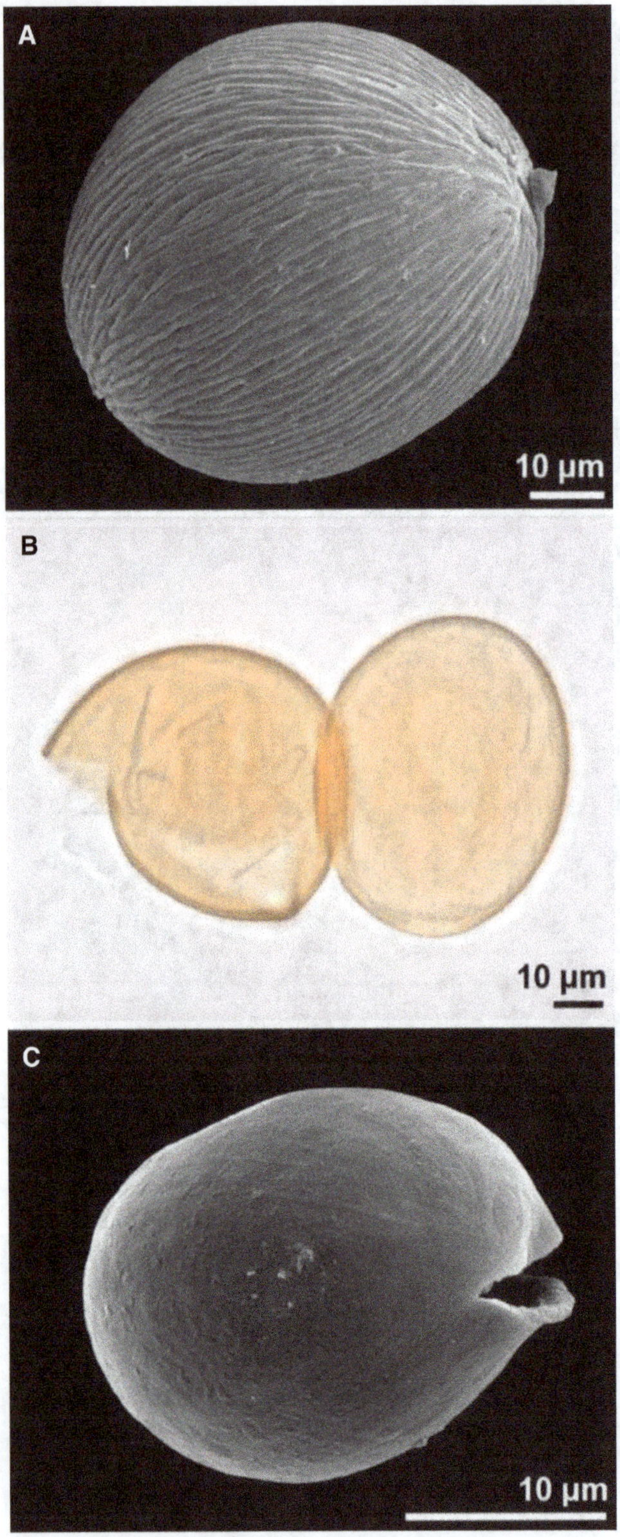

Fig. 19 Preparation effect — psilate vs. ornamented. A-C. *Amorphophallus krausei*, Araceae, pollen striate in hydrated condition (**A**), psilate after acetolysis, LM (**B**) and SEM (**C**)

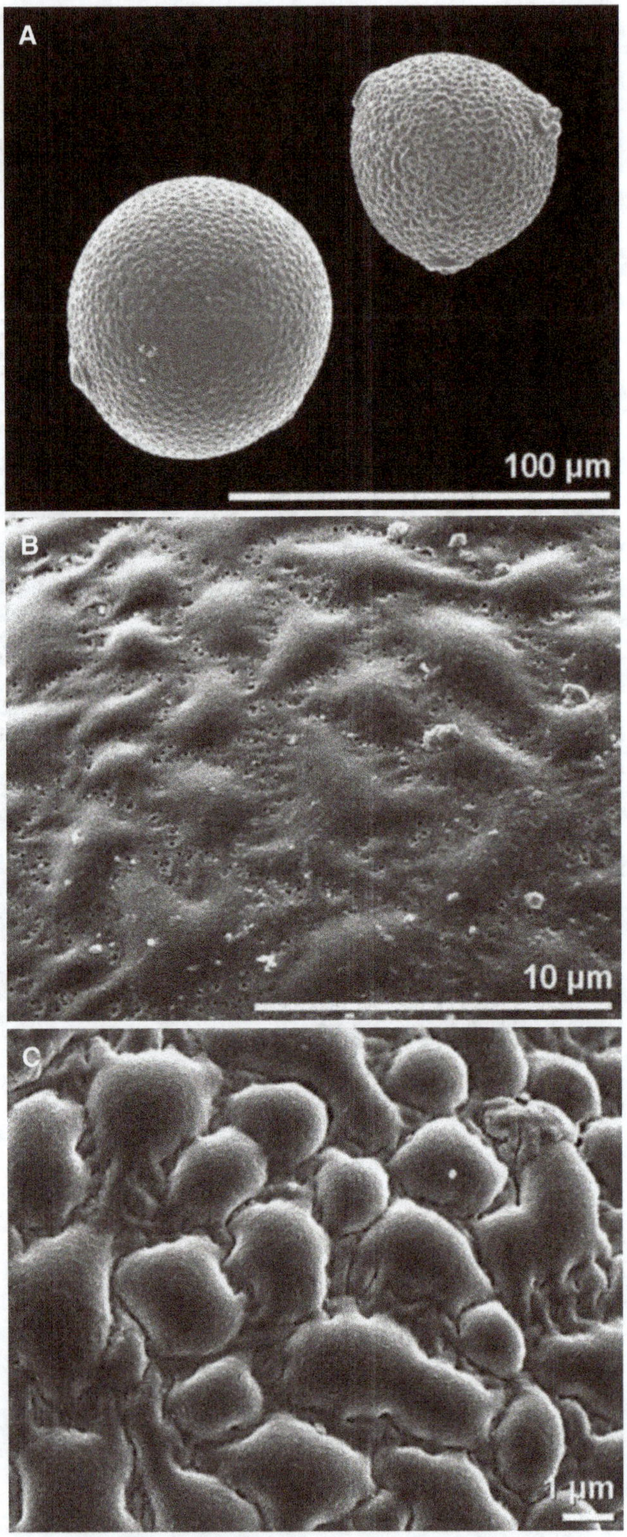

Fig. 20 Preparation effect on ornamentation. A-C. *Trichosanthes anguina*, Cucurbitaceae. **A.** Pollen at different state of hydration: fully hydrated (left), less hydrated (right). **B.** Hydrated pollen, surface detail, verrucate, perforate. **C.** Less hydrated, surface detail, areolate-fossulate

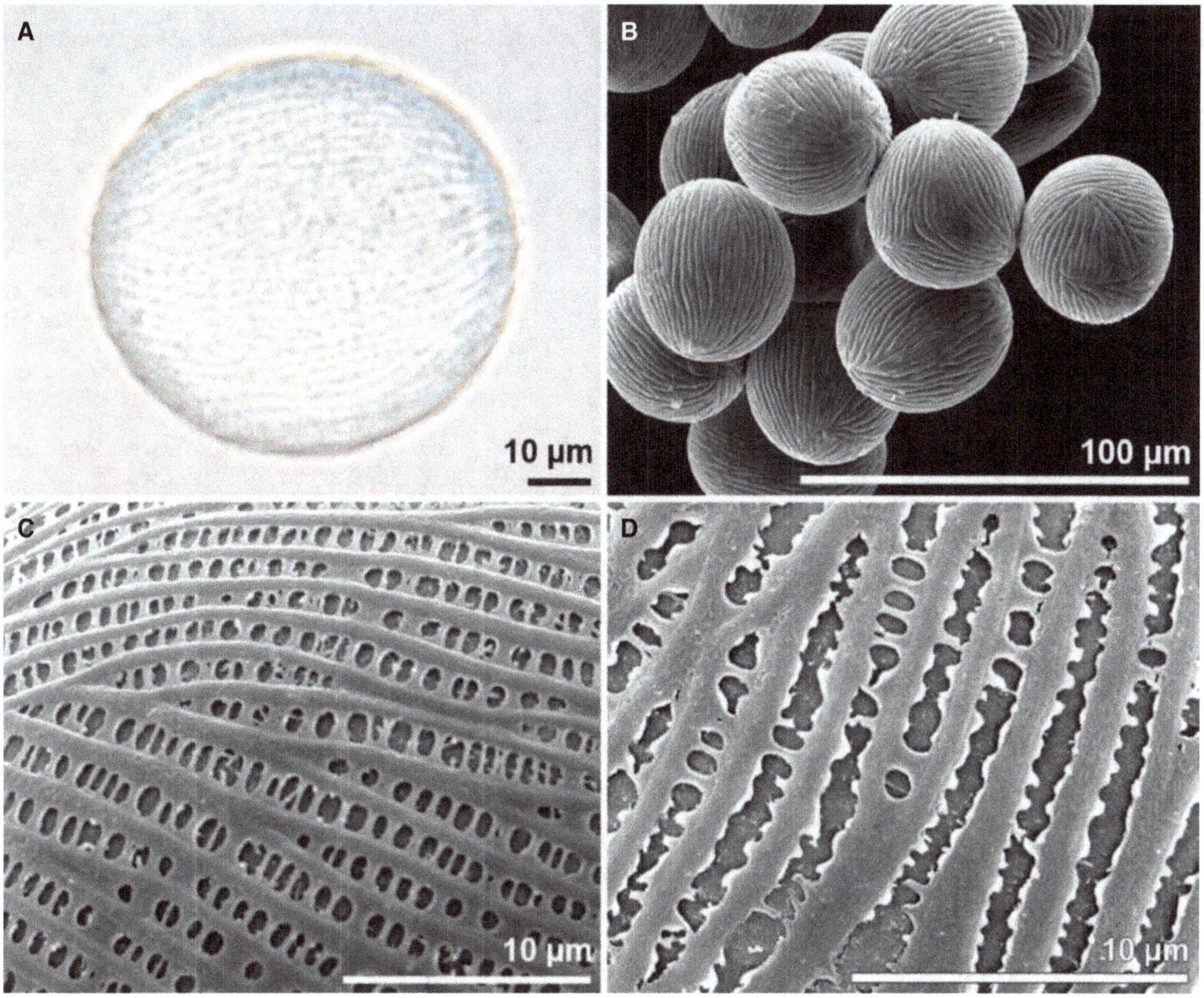

Fig. 21 Preparation effect on ornamentation. A-D. *Amorphophallus longituberosus*, Araceae, hydrated pollen in water with striate ornamentation, LM (**A**), dry pollen in SEM, striate (**B**), hydrated pollen in SEM, striate to reticulate (**C**), hydrated pollen in SEM, ornamentation striate with expanding thin surface layer (**D**)

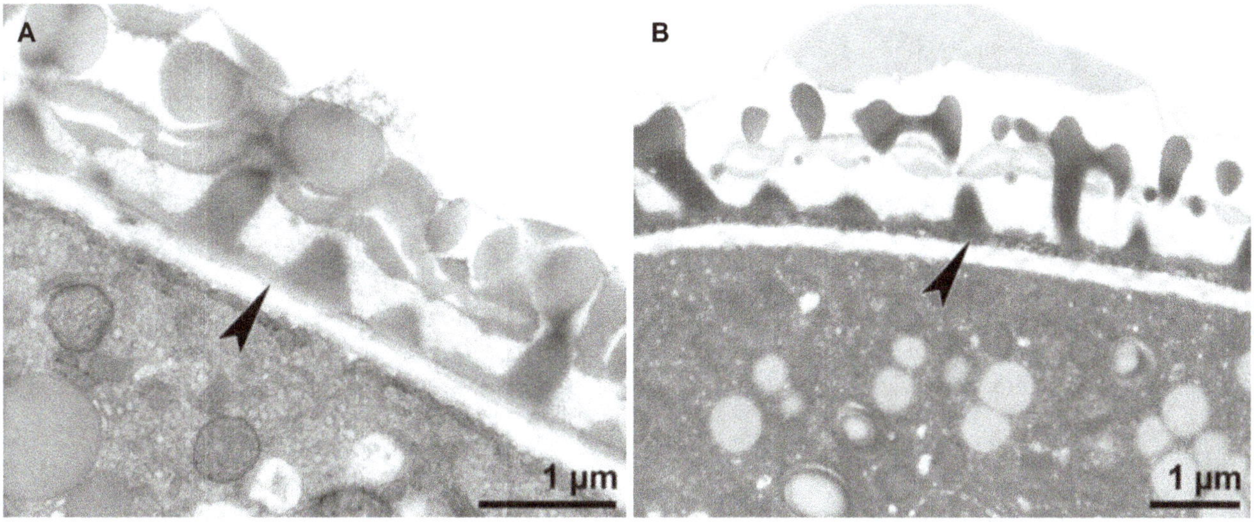

Fig. 22 Absence or presence of endexine. A-B. *Thymus odoratissimus*, Lamiaceae, U + Pb staining, endexine (arrowhead) not clearly visible (**A**), potassium permanganate staining, endexine (arrowhead) clearly visible (**B**)

References

Erdtman G (1952) Polelen Morphology and Plant Taxonomy. Angiosperms. Almqvist & Wiksell, Stockholm

Feuer SM (1977) Pollen morphology and evolution in the Santalales sen. str., a parasitic order of flowering plants. Thesis, University of Massachusetts

Feuer SM, Kuijt J (1985) Fine structure of mistletoe pollen VI. Small flowered neotropical Loranthaceae. Ann Missouri Bot Gard 72: 187–212

Gabarayeva NI, Rowley JR (1994) Exine development in *Nymphaea colorata* (Nymphaeaceae). Nordic J Bot 14: 671–691

Grayum MH (1992) Comparative external pollen ultrastructure of the Araceae and putatively related taxa. Monogr Syst Bot Missouri Bot Garden 43: 1–167

Grímsson F, Grimm, GW, Zetter R (2018) Evolution of pollen morphology in Loranthaceae. Grana 57: 16–116

Halbritter H (2016) Pontederia cordata. In: *PalDat* - A palynological database. https://www.paldat.org/pub/Pontederia_cordata/300430; [accessed 2018-08-07]

Halbritter H, Hesse M (1993) Sulcus morphology in some monocot families. Grana 32: 87–99

Harley MM (1999) Tetrad variation: its influence on pollen form and systematics in the Palmae. In: Kurmann MH, Hemsley AR (eds) The Evolution of Plant Architecture. Royal Botanic Gardens, Kew, p. 289–304

Harley MM (2004) Triaperturate pollen in the monocotyledons: configuration and conjecture. Plant Syst Evol 247: 75–122

Hesse M, Halbritter H, Weber M (2009) *Beschorneria yuccoides* and *Asimina triloba* (L.) Dun: examples for proximal polar germinating pollen in angiosperms. Grana 48: 1151–159

Hesse M, Zetter R (2005) Ultrastructure and diversity of recent and fossil zona-aperturate pollen grains. Plant Syst Evol 255: 145–176

Huynh KL (1976) Arrangement of some monosulcate, disulcate, trisulcate, dicolpate and tricolpate pollen types in the tetrads, and some aspects of evolution in the angiosperms. In: Ferguson IK, Muller M (eds) The evolutionary significance of the exine. Academic Press, London, p. 101–124

Iversen J, Troels-Smith J (1950) Pollenmorfologiske definitioner og typer. Pollenmorphologische Definitionen und Typen. Danm Geol Unders, ser 4, 3: 1–54

Kremp GOW (1968) Morphologic Encyclopedia of Palynology. 2nd edition, Arizona Press, Tucson

Punt W, Hoen PP, Blackmore S, Nilsson S, Le Thomas A (2007) Glossary of pollen and spore terminology. Rev Palaeobot Palynol 143: 1–81

Richard LC (1817) De Orchideis Europaeis annotationes, praesertim ad genera dilucidanda spectantes. Belin, Paris

Schols P, Furness CA, Merckx V, Wilkin P, Smets E (2005) Comparative pollen development in Dioscoreales. Int J Plant Sci 166: 909–924

Teppner H (2007) Notes on terminology for Mimosaceae polyads, especially in Calliandra. Phyton 46(2): 231–236

Traverse A (2007) Paleopalynology. 2nd ed, Springer, Dordrecht

Ulrich S, Hesse M, Bröderbauer D, Bogner J, Weber M, Halbritter H (2013) *Calla palustris* (Araceae): New insights with special regard to its controversial systematic position and to closely related genera. Taxon 62: 701–712

Ulrich S, Hesse M, Weber M, Halbritter H (2017) *Amorphophallus*: New insights into pollen morphology and the chemical nature of the pollen wall. Grana 56: 1–36

Wagenitz G (2003) Wörterbuch der Botanik. – 2nd edition, Spektrum, Heidelberg

Walker JW (1971) Pollen morphology, phytogeography, and phylogeny of the Annonaceae. Contributions from the Gray Herbarium 202: 1–131

Weber M, Halbritter H, Hesse M (1999) The basic pollen wall types in Araceae. Int J Plant Sci 160: 415–423

Weber M, Ulrich S (2010) The endexine: a frequently overlooked pollen wall layer and a simple method for detection. Grana 49: 83–90

Wettstein R (1907) Handbuch der systematischen Botanik 2(2/1): 161–394

Zetter R, Ferguson DK (2001) Trapaceae pollen in the Cenozoic. Acta Palaeobot 41: 321–339

How to Describe and Illustrate Pollen Grains

H. Halbritter et al., *Illustrated Pollen Terminology*, https://doi.org/10.1007/978-3-319-71365-6_5

For the description of a pollen grain, a number of features are used including size, polarity and shape, aperture condition, ornamentation, and pollen wall structure. Additional and often more specialized features depend on the group of plants under study, Gymnosperms (Cycadales, Ginkgoales, Pinales, Gnetales) vs. Angiosperms (magnoliids, monocots, commelinids, eudicots). These features can only be obtained by the application of a combined analysis with LM, SEM, and TEM (Fig. 1). In order to compare and categorize pollen, a common language and understanding of technical terms is necessary.

The description and illustration of a pollen grain depends on how the material is going to be presented and if one is describing a single fossil pollen grain, pollen of a particular extant species, pollen representing several species, a whole genus, several related genera, a complete family, or even a number of families. For future work it is important to provide both LM and SEM micrographs (even TEM), including incorporated scale bars, showing each taxon and close-ups of what are considered diagnostic features of pollen. When documenting the sculpture of pollen grains in SEM it has to be made sure that the magnification is high enough to distinguish the shape and outline of sculpture elements larger than 0.1 μm in diameter. LM- and SEM-diagnosis may be different from each other, due to the methods and techniques used. The methods used to prepare pollen grains for LM, SEM, and TEM must be mentioned along with the pollen descriptions, preferably in a material and method section.

Pollen from a Single Extant Taxon: Online Publication in *PalDat*

Pollen grains from single extant species have rarely been accepted by scientific journals. There is now a new online venue *PalDat*, for publishing pollen from a single species. *PalDat* is the world's most comprehensive pollen database (www.paldat.org) and contains tools for pollen identification as well as global, free online submission and publication with review and editorial process (Weber and Ulrich 2017). *PalDat* already provides a large amount of pollen data on a variety of plant families. Each taxon entry (online publication) ideally includes a detailed description and micrographs (LM, SEM, and TEM) of the pollen, as well as images of the plant/inflorescence/flower and information on relevant literature (Fig. 2). *PalDat* is freely accessible and following a free registration it is open for contributions from all those willing to publish their pollen descriptions and micrographs online. Registered authors may also contribute as co-authors to existing publications by submitting new images and/or new data to pollen diagnosis (with review and editorial process). All changes are recorded in the database history as links to previous versions of the publication. Each contribution is citable and accessible for all users. Registered users can download publications in pdf form. The terminology used in *PalDat* follows this book.

Groups of Extant Pollen

Many of the classical papers on pollen morphology and ultrastructure, covering a large number of extant taxa, provide only a general description of pollen types with pollen of different species lumped together. Furthermore, micrographs are showing selected taxa and usually not the same taxon photographed in both LM and SEM. This makes the data unreliable and not very useful for among others paleopalynologists that want to compare their fossil pollen grains very precisely to particular extant taxa. The decision on particular potential modern analogues of the fossil pollen grain can have major effects on the paleovegetation reconstruction and paleoecological and paleoclimate interpretations of the fossil assemblage, as well as on the paleophytogeographic signal of the taxon. It is recommended, disregardless of the description, that all species be fully illustrated by LM and SEM (and TEM when possible) and their basic and diagnostic morphological features compiled in a table so they can be easily compared (Table 1). The example shown here are Winteraceae pollen tetrads. When portraying tetrads it is useful to show their basal-, lateral-, as well as apical view, in both LM (Fig. 3) and SEM (Fig. 4). Pollen grains should be portrayed in polar and equatorial view. Illustrating pollen from different taxa together on a plate/figure with the same magnification makes it easier to realize size differences. The SEM close-ups are then used to highlight the main sculpture features or the dissimilarities of the taxa. Ideally all close-ups showing sculpture elements should have the same magnification for an easy comparison (Fig. 5).

Diagnosis					
Taxon				Family	
Shape, Size and Aperture					
Pollen unit		**Aperture**		**Polarity and shape**	
Pollen unit		Aperture number		Polarity	
Dispersal unit and peculiarities		Aperture type		Shape	
		Aperture Condition		Outline in polar view	
Size (pollen unit)		Aperture peculiarities		Shape (dry pollen)	
Pollen class				Outline in polar view (dry pollen)	
				Infoldings (dry pollen)	
Ornamentation and structure					
LM		**TEM**			
Ornamentation LM		Tectum			
Nexine		Infratectum			
Sexine		Foot layer			
SEM		Endexine			
Ornamentation SEM		Intine			
Peculiarities					
Wall peculiarities					
Miscellaneous					
Pollen coatings		Cell number			
Reserves in cytoplasm		Ubisch bodies			
Annotations					
Author(s) of diagnosis					

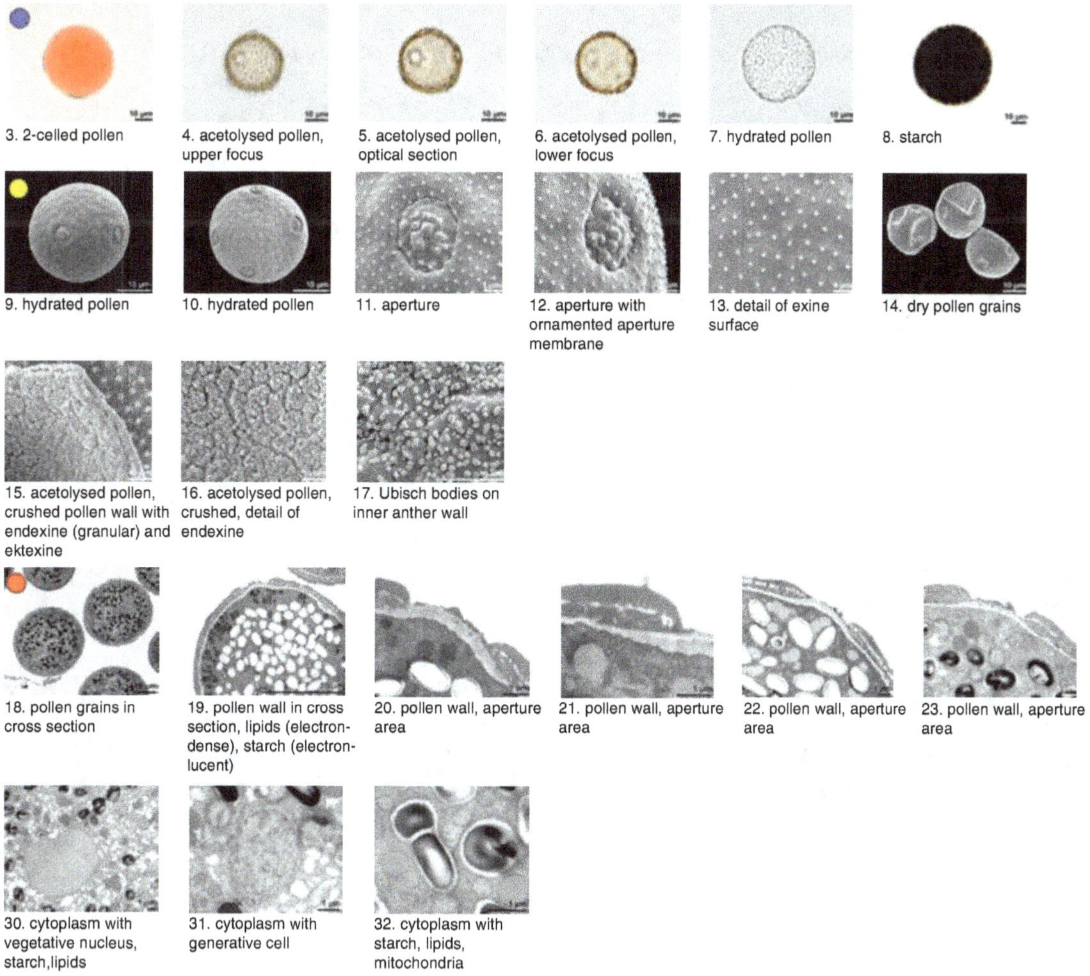

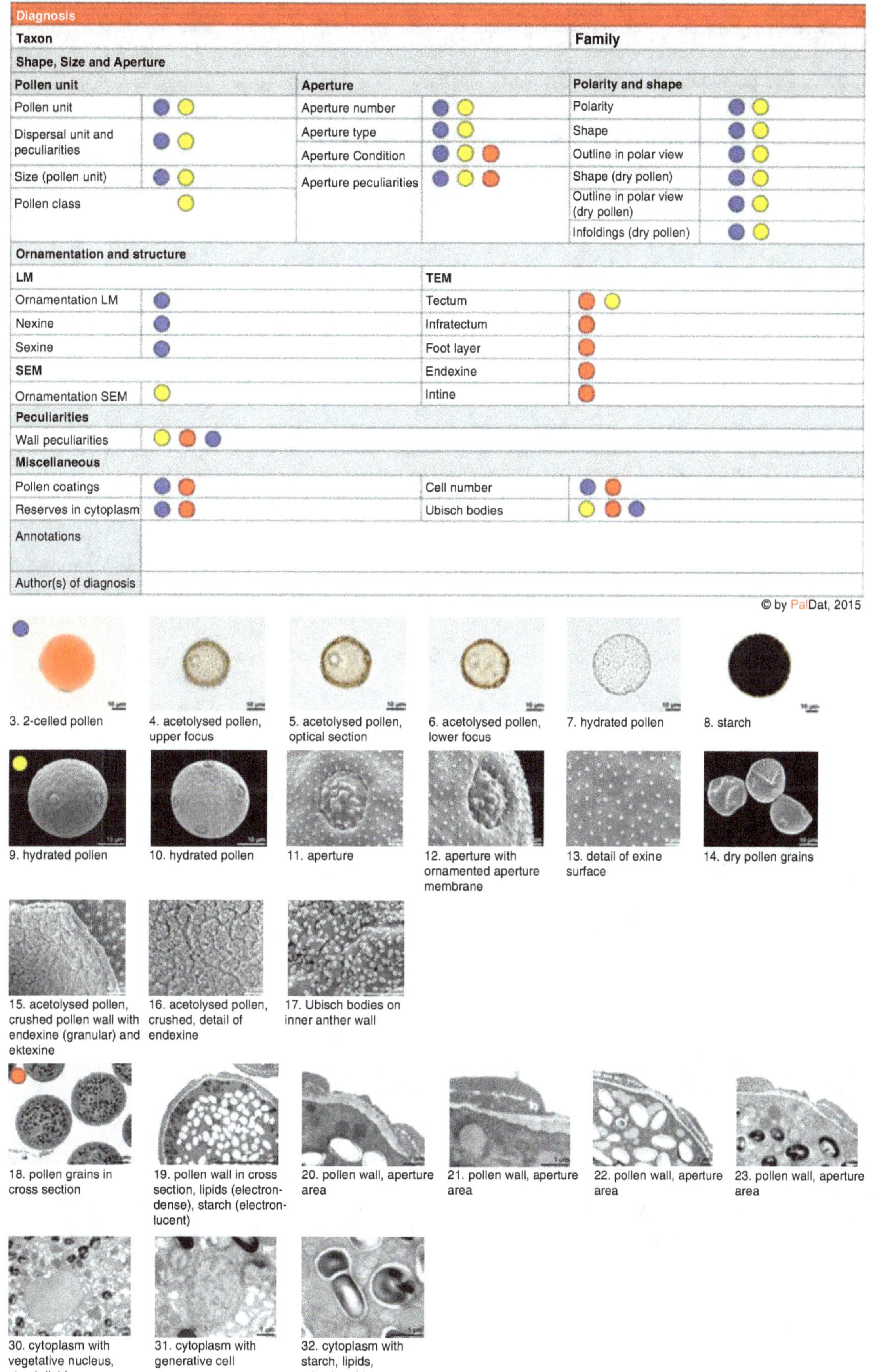

Fig. 1 Diagnosis worksheet. *PalDat* worksheet with all pollen features obtained by a combined analysis using LM, SEM, and TEM. Blue dots indicate LM-, yellow dots SEM-, and red dots TEM-based analyses. *PalDat* pictures showing *Plantago maritima*

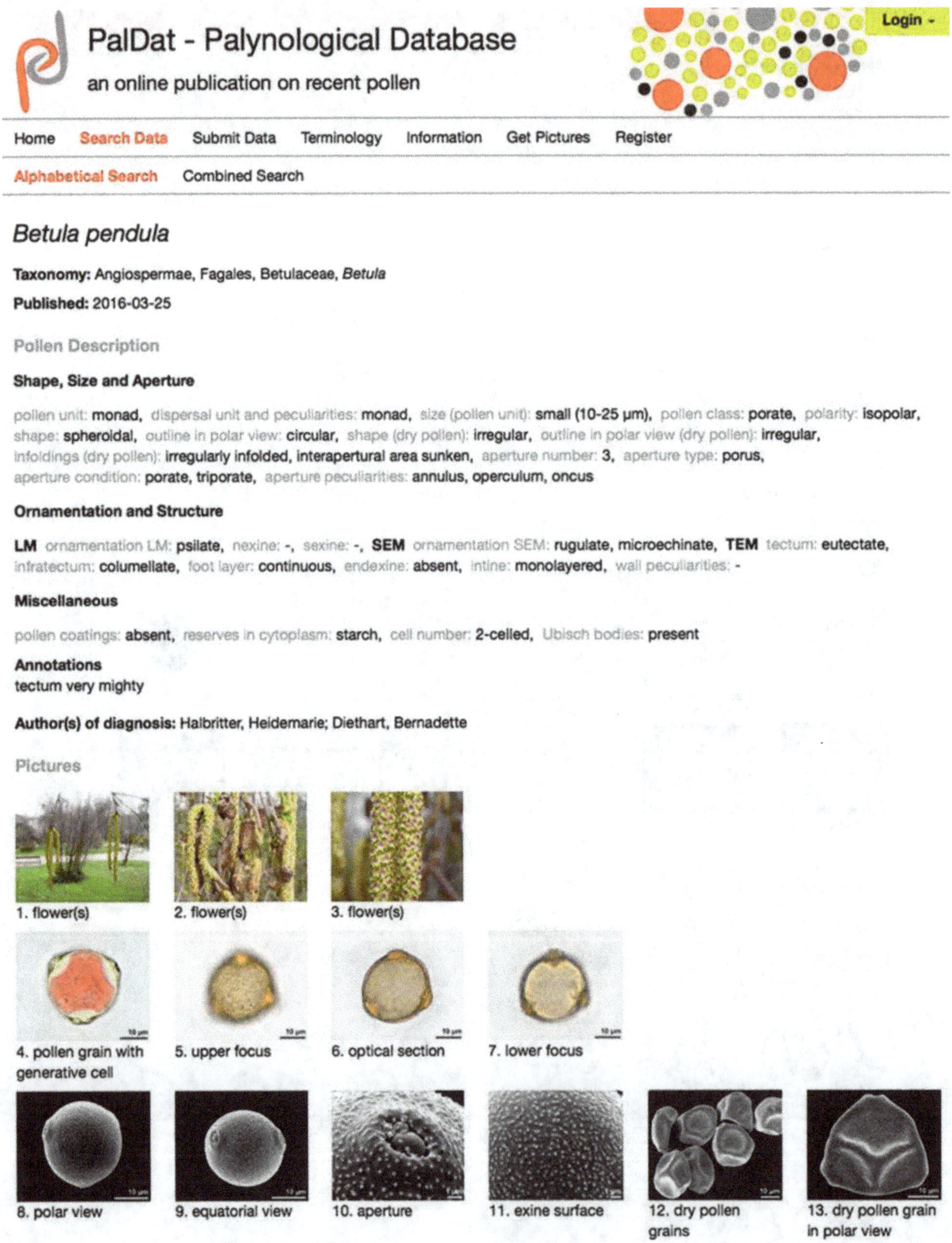

PalDat - Palynological Database

an online publication on recent pollen

Login -

Home | Search Data | Submit Data | Terminology | Information | Get Pictures | Register

Alphabetical Search | Combined Search

Betula pendula

Taxonomy: Angiospermae, Fagales, Betulaceae, *Betula*

Published: 2016-03-25

Pollen Description

Shape, Size and Aperture

pollen unit: **monad,** dispersal unit and peculiarities: **monad,** size (pollen unit): **small (10-25 µm),** pollen class: **porate,** polarity: **isopolar,** shape: **spheroidal,** outline in polar view: **circular,** shape (dry pollen): **irregular,** outline in polar view (dry pollen): **irregular,** infoldings (dry pollen): **irregularly infolded, interapertural area sunken,** aperture number: **3,** aperture type: **porus,** aperture condition: **porate, triporate,** aperture peculiarities: **annulus, operculum, oncus**

Ornamentation and Structure

LM ornamentation LM: **psilate,** nexine: **-,** sexine: **-,** **SEM** ornamentation SEM: **rugulate, microechinate,** **TEM** tectum: **eutectate,** infratectum: **columellate,** foot layer: **continuous,** endexine: **absent,** intine: **monolayered,** wall peculiarities: **-**

Miscellaneous

pollen coatings: **absent,** reserves in cytoplasm: **starch,** cell number: **2-celled,** Ubisch bodies: **present**

Annotations
tectum very mighty

Author(s) of diagnosis: Halbritter, Heidemarie; Diethart, Bernadette

Pictures

1. flower(s)
2. flower(s)
3. flower(s)
4. pollen grain with generative cell
5. upper focus
6. optical section
7. lower focus
8. polar view
9. equatorial view
10. aperture
11. exine surface
12. dry pollen grains
13. dry pollen grain in polar view

Fig. 2 Online publication in *PalDat*. Screenshot showing part of the online publication of *Betula pendula* (Halbritter and Diethart 2016)

Table 1 Winteraceae pollen tetrads

	Takhtajania perrieri	*Exospermum stipitatum*	*Tasmannia insipida*
Tetrad diameter (LM; µm)	58–65	32–38	28–33
Apertures surrounded by an annulus-like rim (width; µm)	Yes, 2.5–6	No	No
Width of aperture region (µm, longest axis)	12–17	5–6	7–11
Exine thickness (LM; µm)*	Max. 5.5	Max. 3	Max. 3.2
Nexine thickness (LM; µm)*	Max. 0.9	Max. 0.7	Max. 0.9
Sexine thickness (LM; µm)*	Max. 4.2	Max. 1.7	Max. 2.3
Sculpture (SEM)	Reticulate	Perforate to nanoreticulate	Reticulate
Muri	Broad and rounded	(Broad and rounded)	Narrow and crested
Diameter of (largest) lumina (µm; longest axis)	7–11	≤1	5–6
Number of lumina/perforations (one grain in lateral view)	c. 15/20	c. 120	c. 15–20
Height ratio columellae vs. muri	~1–1.5:1	?	~1:1
Columellae per µm	2 per 5 µm	2–3	1–2
Free-standing columellae	Frequent, mostly ≤1 µm; gemmae, bacula, and clavae	Absent	Rare, mostly ≤0.5 µm; verrucae, gemmae, and clavae
Ulcus membrane (SEM)	Granulate to microverrucate	Granulate to nanoverrucate	Granulate, nano- to microclavate

Main features of three different Winteraceae pollen tetrads

Annotation: Measurements like exine, nexine, and sexine thickness provided in Table 1 (asterisks) are commonly used in (paleo) palynological literature. Scientists should be aware that such measurements (e.g., 0.7 or 0.9 µm) vary highly, up to 30%, depending on the methods and tools used. Therefore, the measurements should not be overrated or used for taxonomic discrimination.

Fossil Pollen

From the birth of paleopalynology this branch of science has been plagued by the lack of taxonomic foundation when interpreting paleoenvironments. It is very unfortunate that numerous new "scientific" publications dealing with the subjects of paleoecology, paleovegetation, paleoclimate and various aspects of paleophytogeography still present only a list of taxa observed in LM. Some publications include LM micrographs of the most "common" taxa, but only in exceptional cases the LM micrographs are accompanied by SEM micrographs. The absence of illustrations makes it impossible for any reader to verify, or later revise, the taxonomic background and to conclude if the modern living relative or potential modern analogue of the fossil taxon is justified. Every proper scientific journal should make it a mandatory request that all pollen types are represented by at least one LM micrograph. Furthermore, all taxa that suggest some sort of different, abnormal or exceptional paleo-parameters, in an otherwise "homogeneous" assemblage, or taxa that are used to set any sort of boundaries (temperature, precipitation, biozone, time, etc.), should be illustrated using both LM and SEM (in some cases even TEM). These contrasting taxa might include a dry element in an otherwise humid assemblage, a tropical element in an otherwise temperate assemblage, or an African element in an otherwise North American-Eurasian assemblage. Even though the journal would not allow these illustrations in the printed version most of them now offer the possibility to archive online supplementary files where the pollen can be illustrated.

For those who want to produce a taxonomically valid study based on fossil material are advised to use the single-grain method when investigating fossil pollen and make sure not to sieve the sample

Fig. 3 LM micrographs of Winteraceae pollen tetrads. Tetrads shown in basal-(left), lateral-(middle), and apical (right) view at high focus (upper three rows) and in optical cross section (lower three rows). *Takhtajania perrieri* (first and fourth row), *Exospermum stipitatum* (second and fifth row), *Tasmannia insipida* (third and sixth row)

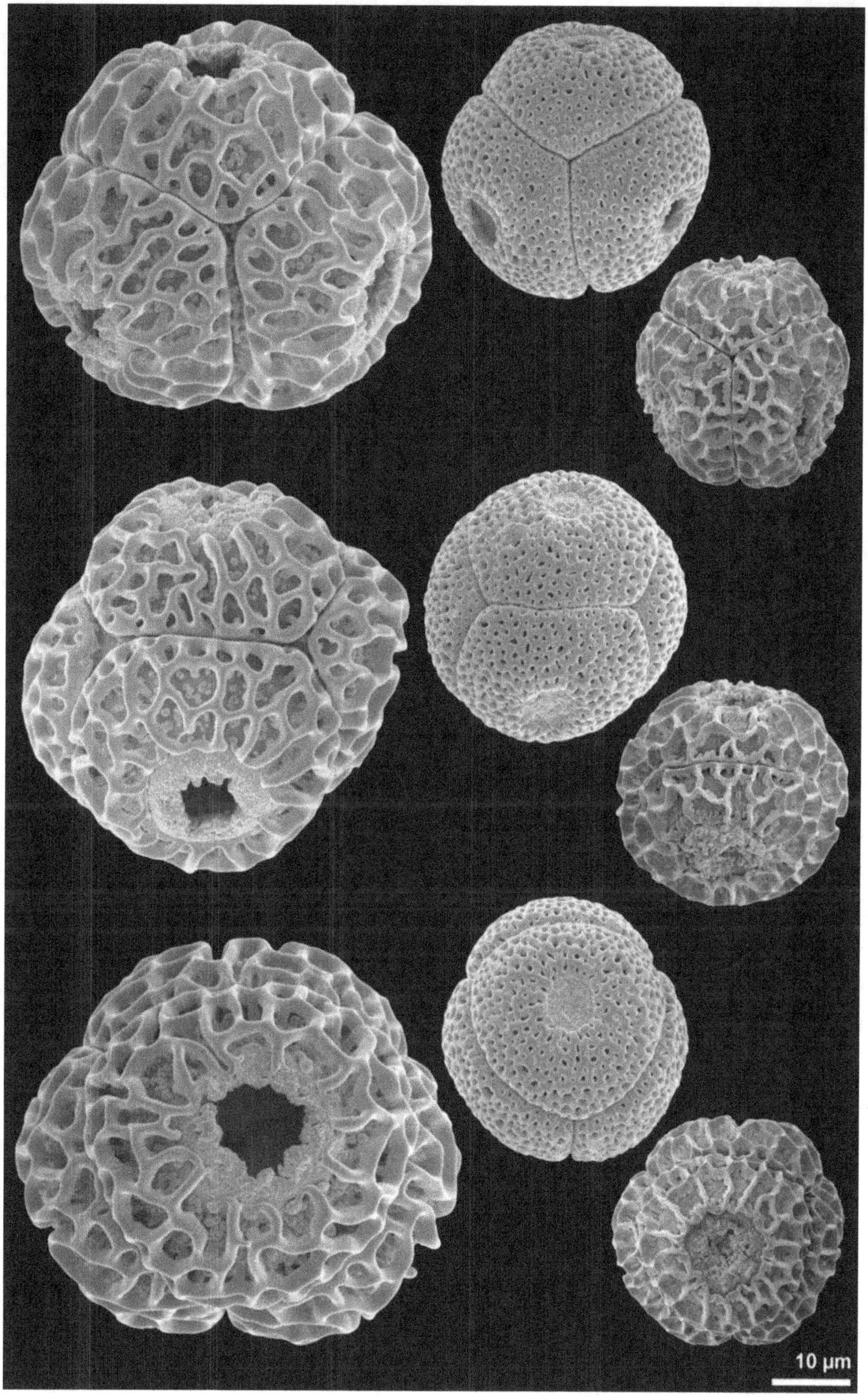

Fig. 4 SEM micrographs of Winteraceae pollen tetrads. Tetrads shown in basal view (upper row), lateral view (middle row) and apical view (lower row). *Takhtajania perrieri* (left), *Exospermum stipitatum* (middle), *Tasmannia insipida* (right)

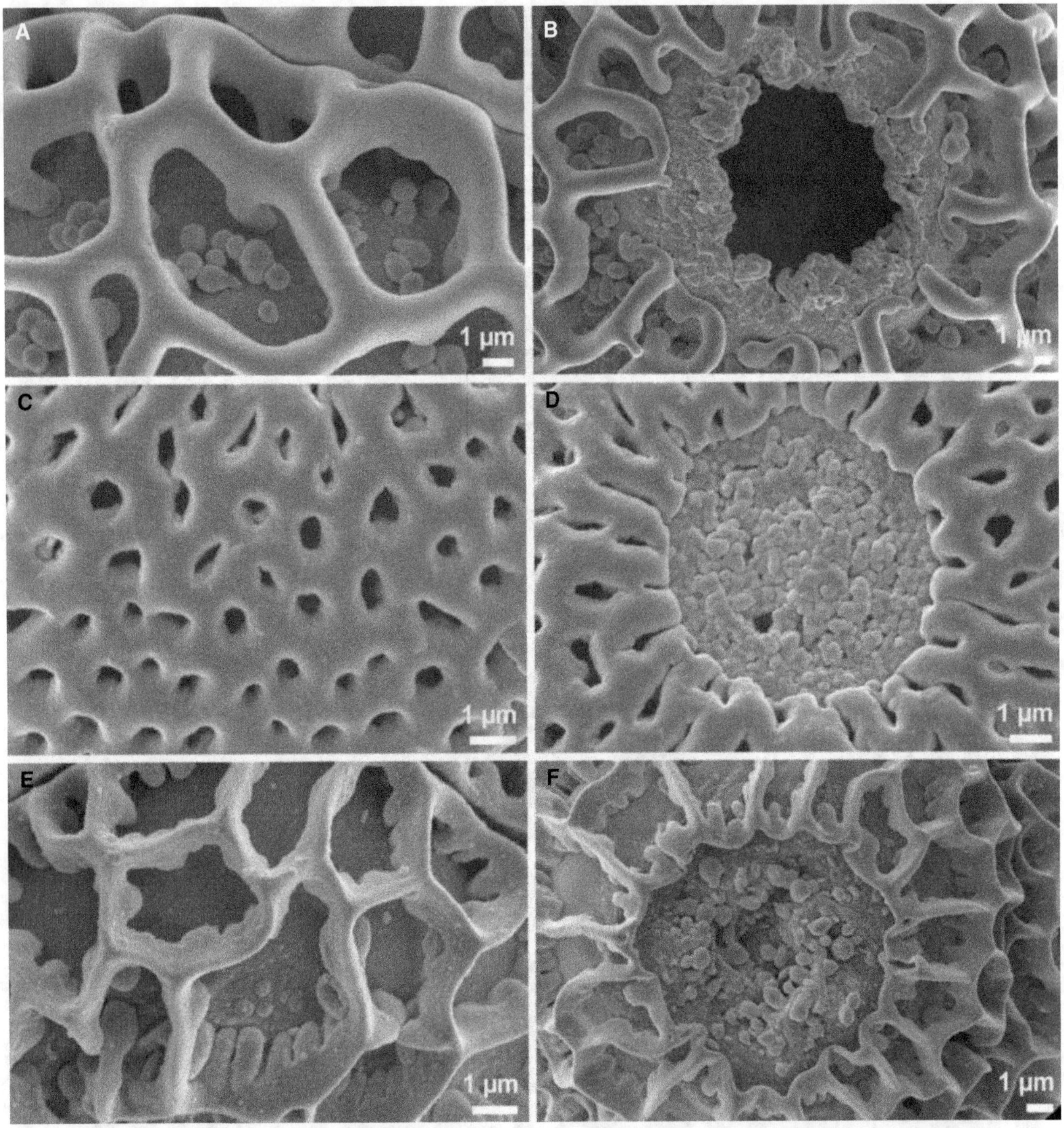

Fig. 5 Details of Winteraceae pollen tetrads. SEM close-ups of *Takhtajania perrieri* (**A-B**), *Exospermum stipitatum* (**C-D**) and *Tasmannia insipida* (**E-F**), showing sculpture on distal face of pollen (**A, C, E**) and the aperture region and ulcus membrane (**B, D, F**)

during preparation (see "Methods in Palynology"). This allows the researcher to study all elements occurring within a sample using both LM and SEM and to investigate even very small and/or rare pollen grains. The small and/or rare pollen (Fig. 6) would otherwise be overlooked during the old-fashion routine LM observation, where the researcher usually counts 300–600 grains. When illustrating fossil pollen it is important to show the grain in both LM and SEM. Close-ups taken with the SEM should have magnification high enough so all sculpture elements larger than 0.1 µm become distinguishable. Sculpture and suprasculpture elements smaller than 1 µm are not observed or hard to distinguish using LM only, but will be revealed using high magnification SEM (Fig. 7). Many pollen grains that look similar or the same in LM can be distinguished using SEM. In some cases it is beneficial to turn the pollen grain once it has been photographed in SEM, re-sputter and photograph again. This applies especially to heteropolar pollen grains (Fig. 8) as well as pollen dispersed in permanent tetrads. When single pollen

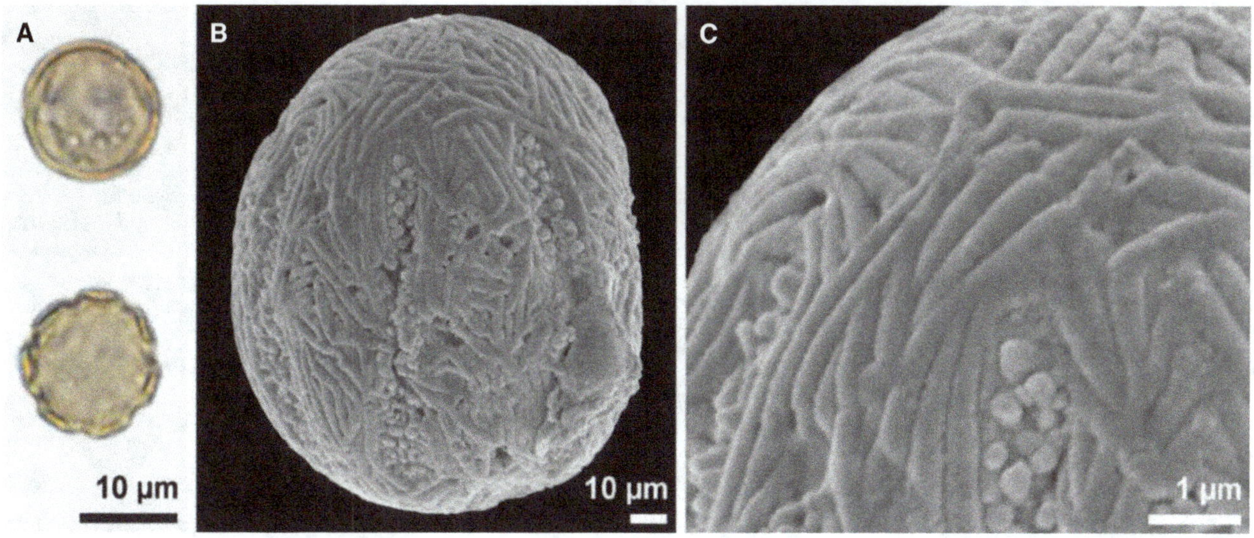

Fig. 6 Small and rare pollen, Paleocene, Western Greenland. A. small fossil grains (≤10 µm in diameter) usually absent in samples after sieving. LM micrographs (left) in equatorial (upper) and polar view (lower). **B.** pollen in equatorial view, SEM. **C.** striate sculpture not seen under low magnification LM

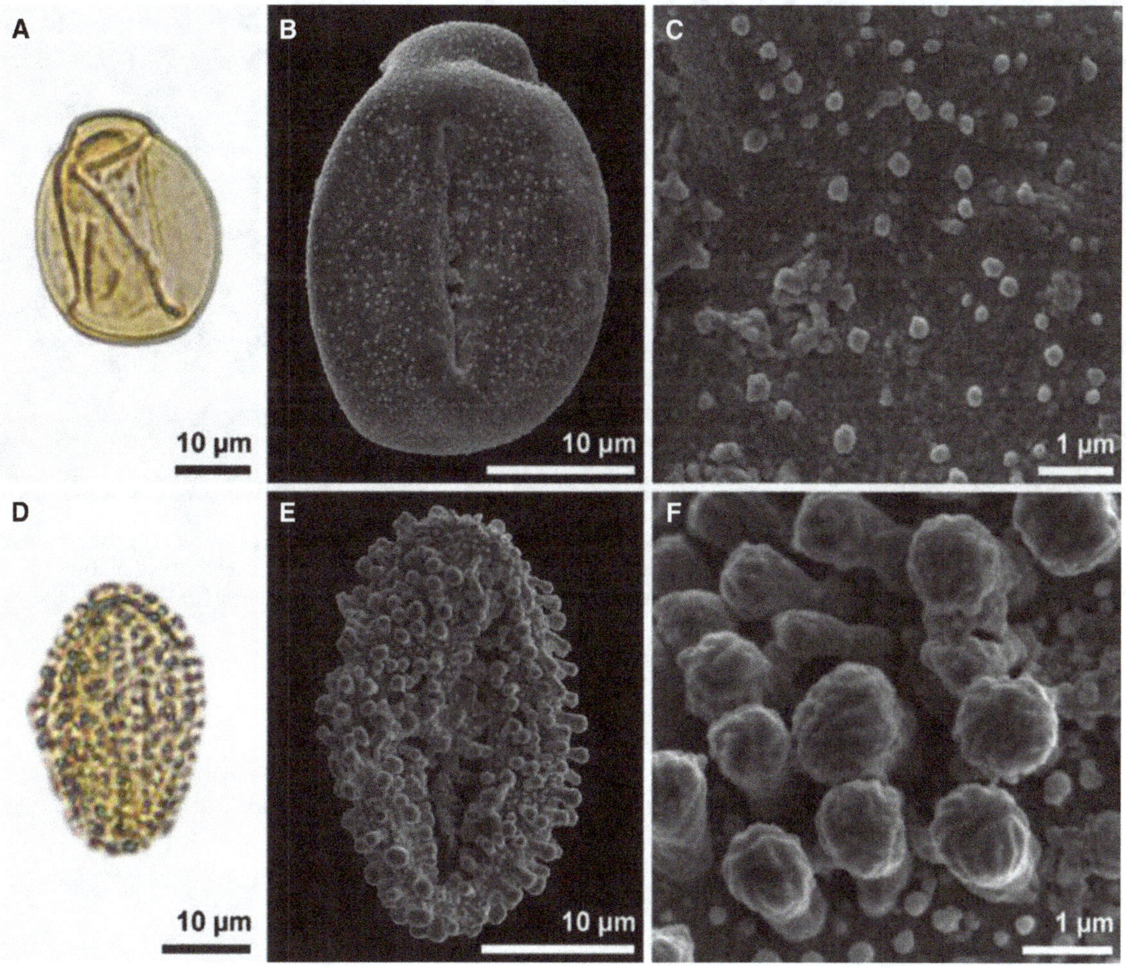

Fig. 7 Ornamentation LM vs. SEM, fossil, Middle Eocene, Western Greenland. A-C. *Eucommia* sp. **A.** Pollen psilate in LM. **B.** Pollen in SEM, equatorial view, note sculpture. **C.** Ornamentation nanoechinate (≤0.5 µm) and granulate. **D-F.** *Ilex sp.*, **E.** LM and SEM overviews show the typical clavate sculpture known for this genus. **F.** Microrugulate suprasculpture present on the distal part of the clavae, only observed using high magnification SEM

Fig. 8 Fossil heteropolar pollen grain, Paleocene, Western Greenland. A. LM micrographs showing proximal (left) and distal (middle) poles of pollen grain and equatorial view (right). **B.** SEM overviews showing both poles of the pollen grain and the different aperture arrangements. **C-D.** SEM close-ups of proximal (**C**) and distal poles (**D**) show that the muri are much broader on the proximal pole

grains or tetrads are studied using SEM, changes in sculpture over the pollen surface are often observed, for example polar vs. equatorial region, mesocolpium vs. aperture region vs. aperture membrane (Fig. 9). Some pollen or tetrads also have Ubisch bodies or viscin threads (Hesse et al. 2000). These differences in the sculpture of fossil pollen need to be documented and it is therefore often necessary to show more than a single close-up taken with the SEM.

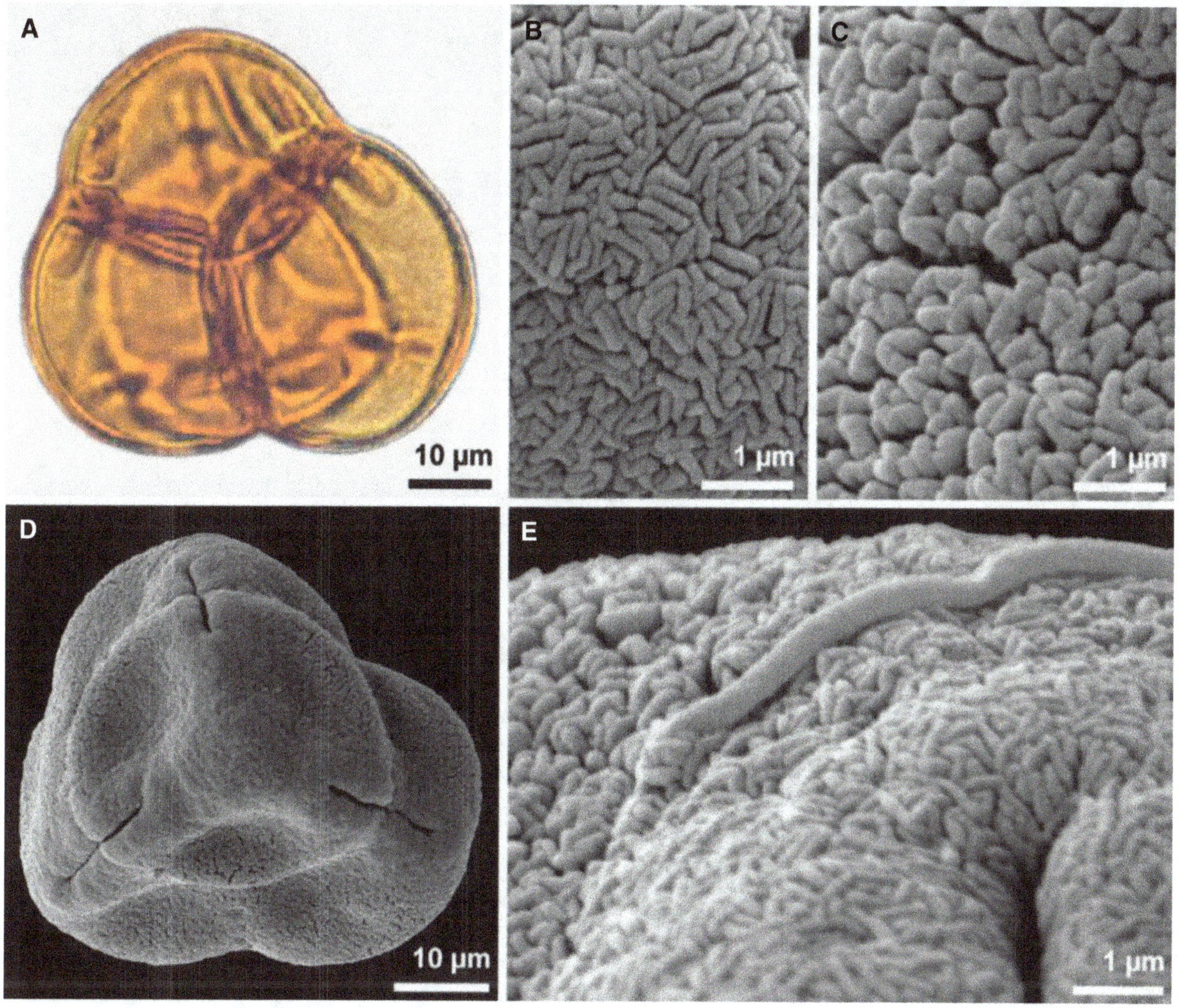

Fig. 9 Fossil tetrad, *Rhododendron* sp., Miocene, North-east China. A, D. Tetrad, overviews in LM vs. SEM. **B-C.** close-ups at same, magnification show difference in sculpture at polar region of pollen grain (**B**) vs. interapertural area (**C**). **E.** exine surface with viscin thread, SEM

References

Halbritter H, Diethart B (2016) *Betula pendula*. In: PalDat – a palynological database. Published on the Internet https://www.paldat.org/pub/Betula_pendula/300732 [accessed 2017-04-28]

Hesse M, Vogel S, Halbritter H (2000) Thread-forming structures in angiosperm anthers: their diverse role in pollination ecology. Plant Syst Evol 222: 281–292

PalDat – a palynological database (2000 onwards, www.paldat.org)

Weber M, Ulrich S (2017) PalDat 3.0 – second revision of the database, including a free online publication tool. Grana 56: 257–262

Methods in Palynology

H. Halbritter et al., *Illustrated Pollen Terminology*, https://doi.org/10.1007/978-3-319-71365-6_6

Preparation of Recent and Fossil Material for LM, SEM, and TEM

Multiple methods and techniques should be used when investigating pollen grains in order to provide comprehensive and accurate information about pollen morphology and ultrastructure (see also "Misinterpretations in Palynology"). The preparation methods used depend on the material to be studied, if the pollen grains are to be obtained from recent flower material (herbarium sheets, newly collected) or from various sedimentary rocks, sediments or soils (fossil to subfossil pollen). Recent and fossil pollen grains are easily studied using both LM and SEM, but recent pollen grains are also more often studied using TEM.

For an accurate description of any taxonomic value, it is important to study pollen grains in both LM and SEM. The LM will provide, among others, information on the endoaperture that cannot be obtained using SEM. Likewise the SEM will provide detailed information on the sculpture of the pollen grain that is not visible under the low magnification provided by the LM. For example, terms with "micro-" (like microreticulate) or "nano-" (like nanoechinate) can only be observed using SEM (Fig. 1).

Annotation: The methods described in this section are the standard palynological techniques applied by the authors of this book and may differ in other working groups/labs around the world. All LM, SEM, and TEM micrographs in this book are produced following these standard protocols. Recipes for preparations are included at the end of this section.

Light Microscopy

Pollen Hydration Status at Dispersal

To clarify the dehydration status of pollen grains at anthesis, pollen must be collected from newly opened anthers (Fig. 2). Fresh pollen grains are transferred immediately into a drop of pure glycerine and should be observed as soon as possible, as pollen grains expand in glycerine (within days or

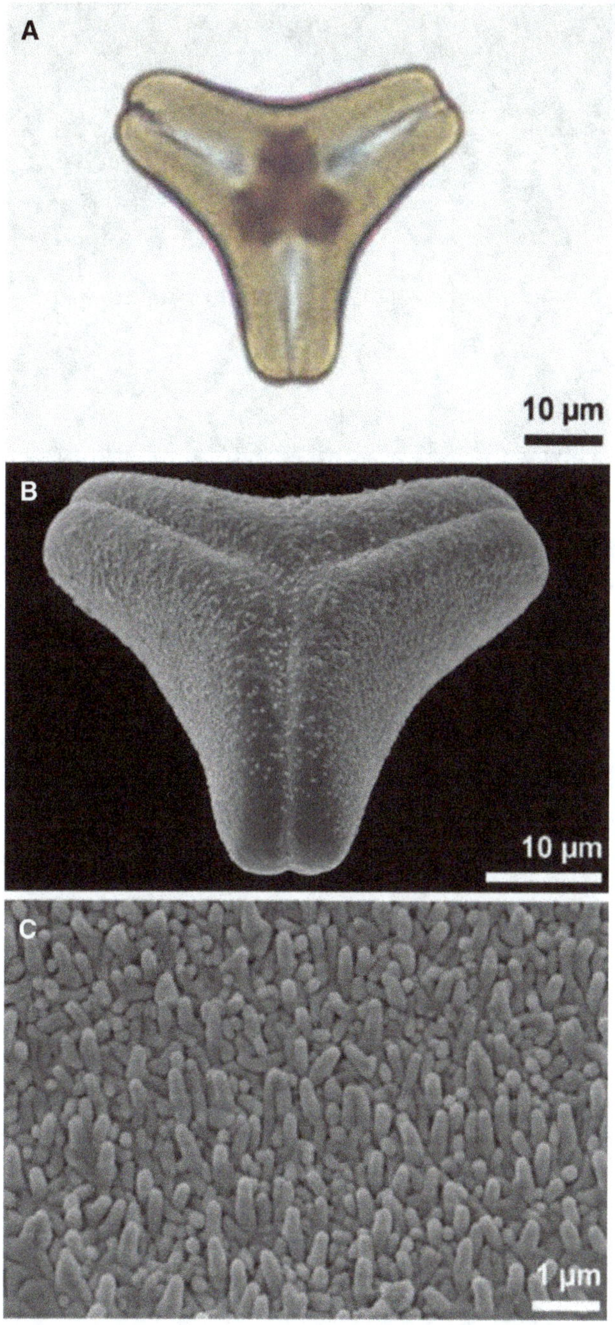

Fig. 1 **LM vs. SEM. A-C.** *Aetanthus coriaceus*, Loranthaceae. **A.** Pollen grain looks psilate or scabrate in LM. **B.** Sculpture elements become visible under SEM. **C.** The sculpture elements are nano- to microbaculate and only identifiable using high magnification

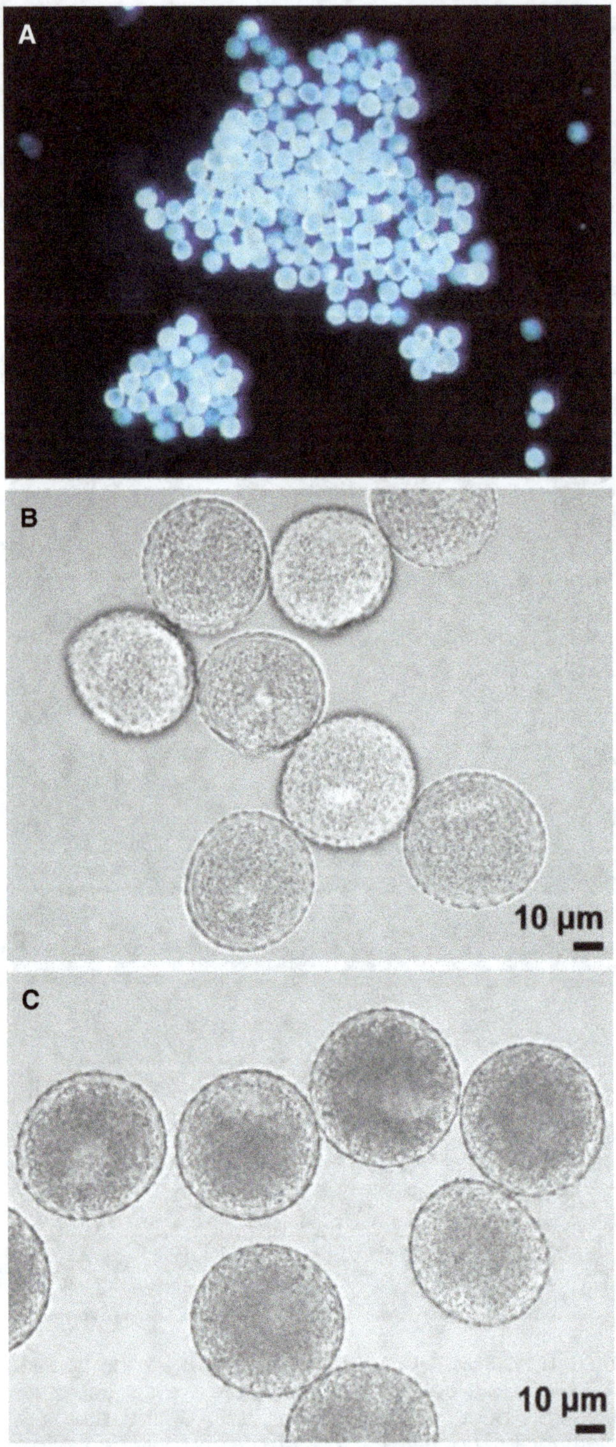

Fig. 2 Pollen hydration status at dispersal. A-C. *Alocasia* sp., Araceae. **A.** Pollen grains fully hydrated at anthesis, binocular microscope. **B.** Pollen in glycerine, LM. **C.** Pollen hydrated in water, LM

weeks). The water content of pollen grains at the time of dispersal varies and pollen can be fully hydrated, partially hydrated, or partially dehydrated (Heslop-Harrison 1979; Nepi et al. 2001; see also **harmomegathic effect** in "Pollen Morphology and Ultrastructure").

Pollen Hydrated in Water

Fresh or dry pollen grains are hydrated in a drop of water on a glass slide and observed in LM. This should be the first step before preparing pollen for SEM to get an impression about the quality of the collected material, to make sure that the material is not degenerated or contaminated by fungi (Fig. 2). Observations on pollen hydrated in water with the LM can reveal interesting aspects. One example is *Montrichardia* (Araceae), where a drop of water triggers a massive expansion of the thick intine resulting in an explosive opening of the pollen wall (Weber and Halbritter 2007).

Clarify the Pollen Polarity and Aperture Type

To clarify the pollen polarity and the aperture type, anthers with pollen tetrads must be collected before anthesis (usually found in flower buds). Pollen tetrads can be released from the anthers in a drop of water or in glycerine. Quite often different developmental stages can be found in one anther: microspores in early and late tetrad stages (with or without callose wall), but also young microspores (before first pollen mitosis) released from the tetrad as well as mature pollen grains (Fig. 3; see also Fig. 1 in "Pollen Development"). For the investigation it might be useful to stain the material, e.g. with toluidine blue or basic fuchsin (Siegel 1967).

Acetocarmine Staining: Detection of the Cellular Condition

For the detection of the cellular condition of pollen grains, fresh pollen are put into a drop of acetocarmine and warmed on a heating plate (up to

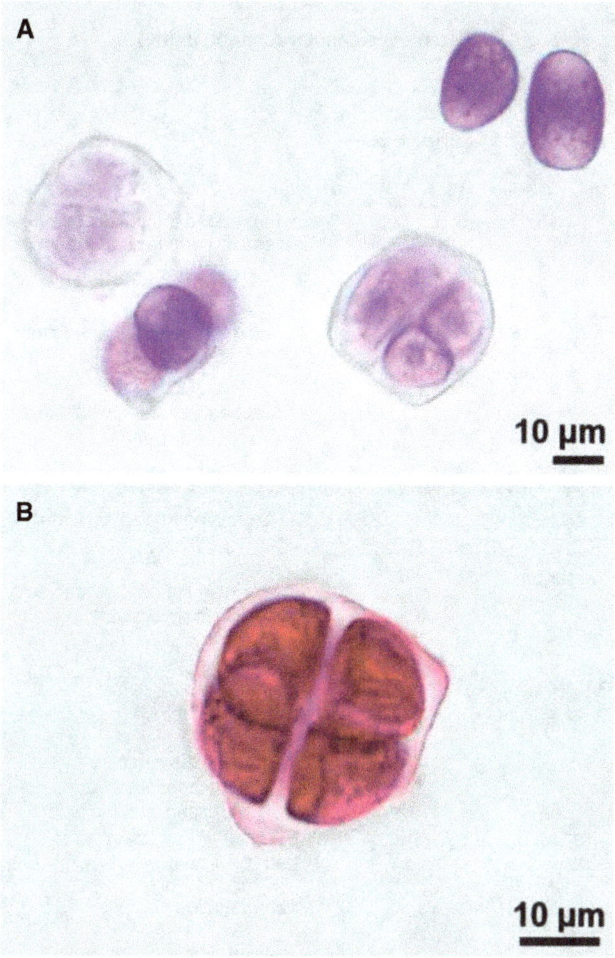

Fig. 3 Clarify the pollen polarity. A-B. *Calla palustris*, Araceae, tetrads in different stages as well as free microspores stained with toluidine blue (**A**) and basic fuchsin (**B**)

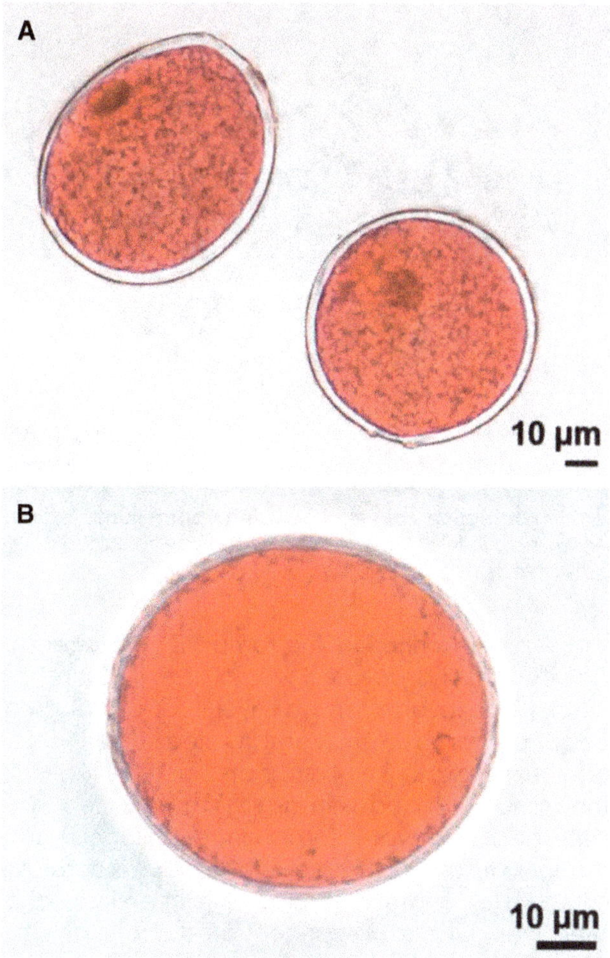

Fig. 4 Clarification of the cellular condition using aceto-carmine. A. Binucleate pollen of *Anchomanes welwitschii*, Araceae, generative nucleus stains intensive red. **B.** Trinucleate pollen of *Amorphophallus krausei*, Araceae, sperm nuclei stain intensive red

70 °C), for a few seconds to several minutes (species dependent), and observed under the LM (Gerlach 1984). The generative nucleus in binucleate pollen grains and the sperm nuclei in trinucleate pollen stain intensively red with aceto-carmine (Fig. 4). The generative nucleus usually stains less intensive.

Potassium Iodine: Detection of Starch

For the detection of starch as reserves in the cytoplasm, pollen grains are stained with aqueous potassium iodine (Gerlach 1984). Fresh or dry pollen grains are transferred into a drop of staining solution on a glass slide. Starch present in pollen grains will stain dark brown to black (Fig. 5).

Acetolysis: Visualizing Pollen Ornamentation and Aperture Number in Recent and Fossil Pollen

Acetolysis (Erdtman 1960) is a standard palynological preparation technique and an indispensable method for illustrating pollen grains with the LM. Untreated or stained pollen grains will hide much of the important information for the description of a pollen grain. The acetolysis treatment should remove the cellular content and the intine, but can also destroy the aperture membrane. Moreover, it cleans pollen surfaces and colors pollen grains brown, which makes it easier to observe all details of the pollen wall.

The normal preparation procedure is a combination of two steps, chlorination and acetolyzation

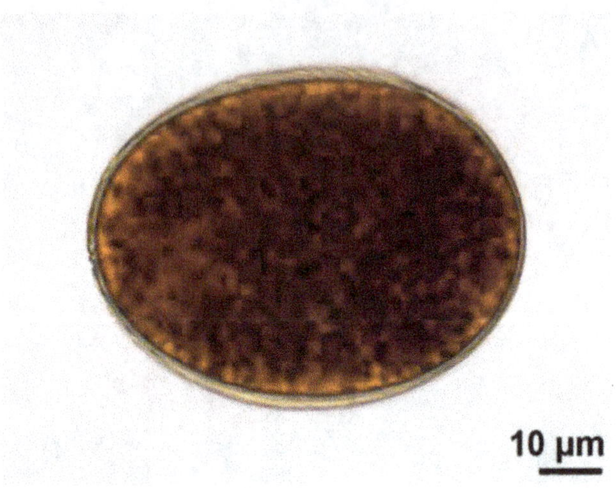

Fig. 5 Detection of starch using potassium iodine. *Amorphophallus interruptus*, Araceae, starch (in amyloplasts) stained with potassium iodine

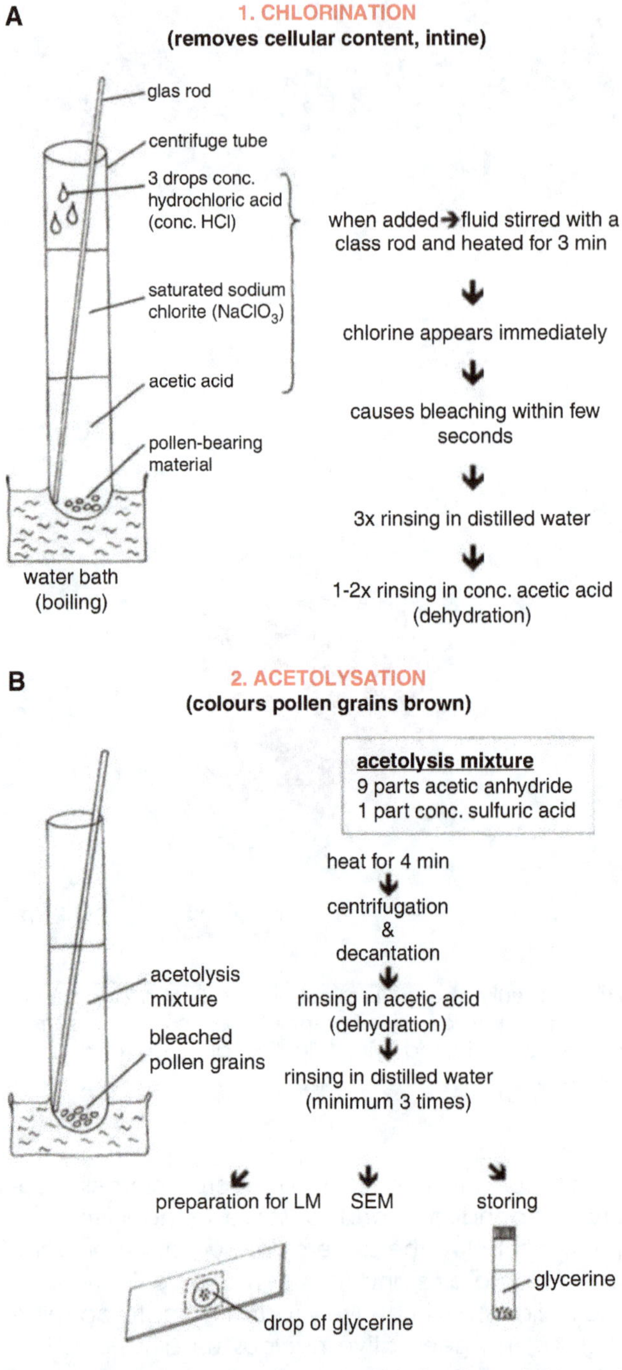

Fig. 6 Acetolysis treatment. Chlorination (**A**) and acetolyzation (**B**), the two steps of acetolysis

(Fig. 6). For **chlorination**, the sample is transferred to a test tube and covered with a layer (1.5 cm) of glacial acetic acid and a layer (ca. 3 cm) of a freshly prepared solution of saturated sodium chlorate. After adding 3 or 4 drops of concentrated HCl, the mixture is stirred with a glass rod, heated in a bath of boiling water for 3 min, centrifuged, and the liquid fraction decanted. The residue is carefully rinsed to eliminate any remaining chemicals and then finally washed in concentrated acetic acid or acetic anhydride to remove the water. For the **acetolyzation**, the sample is put into a mixture of 9 parts acetic anhydride and 1 part concentrated sulfuric acid and heated to 100 °C (at least 80 °C) for approximately 4 min (up to 10 min). The samples are ideally acetolyzed in an ultrasonic bath to avoid boiling retardation and to reduce water condensation. After the mixture has been centrifuged and the liquid fraction decanted, the residue is washed in acetic acid and 3 times with water. After washing, test tubes are turned upside down and the content dried. Glycerine is then added to the sample. For fossil pollen material both steps (chlorination and acetolyzation) are usually applied.

When **preparing recent material** (Fig. 7) it is routine to apply only the second step (acetolyzation). Traditionally, the term "acetolysis" is also used even when pollen grains have been acetolyzed only and not previously chlorinated. For acetolysis of recent pollen fresh or air dried pollen/anthers are transferred into test tubes and can be acetolyzed directly. For the analysis of soil, dust, honey, or any other samples, the material has to be washed in a beaker with about 200 ml distilled water (and detergent, e.g., Tween) and can be sieved to remove bigger parts (leaves, branches) from the sample. In order to prevent pollen loss, it is important to use sieves with big mashes (E-D-quick sieve "260 µm"). The material is then concentrated in test tubes by centrifuging at 3000 rpm and the water decanted. The residue is washed in concentrated acetic acid to remove the remaining water and

Fig. 7 Acetolyzation treatment of recent material. A. Washing the sample in a beaker. **B.** Washing with a detergent "Tween". **C.** Sieving the sample. **D.** Decanting water from the test tube after centrifuging; the organic fraction remains at the bottom. **E.** Fresh acetolysis mixture is added to the sample in the test tube. **F.** Samples are heated in an ultrasonic bath. **G.** During acetolyzation the solution turns brown. **H.** Residue washed in acetic acid followed by water. **I.** Drying of the acetolyzed sample. **J.** Acetolyzed material in glycerine stored in cryo tubes. **K.** acetolyzed pollen from honey in LM

subsequently acetolyzed (see description "acetolyzation" above). For light microscopy one part of the acetolyzed material is transferred into glycerine. For scanning electron microscopy, acetolyzed pollen is transferred into a drop of anhydrous ethanol on a SEM stub and sputter coated with gold (see also below "Preparation of fossil material").

Annotation: After rehydration or washing of the material (pollen/anthers) use acetic acid before and after the use of the acetolysis mixture, as it reacts intensively with water. Fresh acetolysis mixture is light yellow colored and highly reactive. Over time the mixture obtains a dark brown color and becomes less reactive.

Heavy Liquid Separation

Samples (recent and fossil) that still contain a very high mineral content after acetolysis should be treated with heavy liquid (e.g., zinc bromide solution; e.g., Eyring 1996, Traverse 2007; Fig. 8). Add ca. 2 cm of zinc bromide solution into the centrifuge tube and mix with the organic residue. Distilled

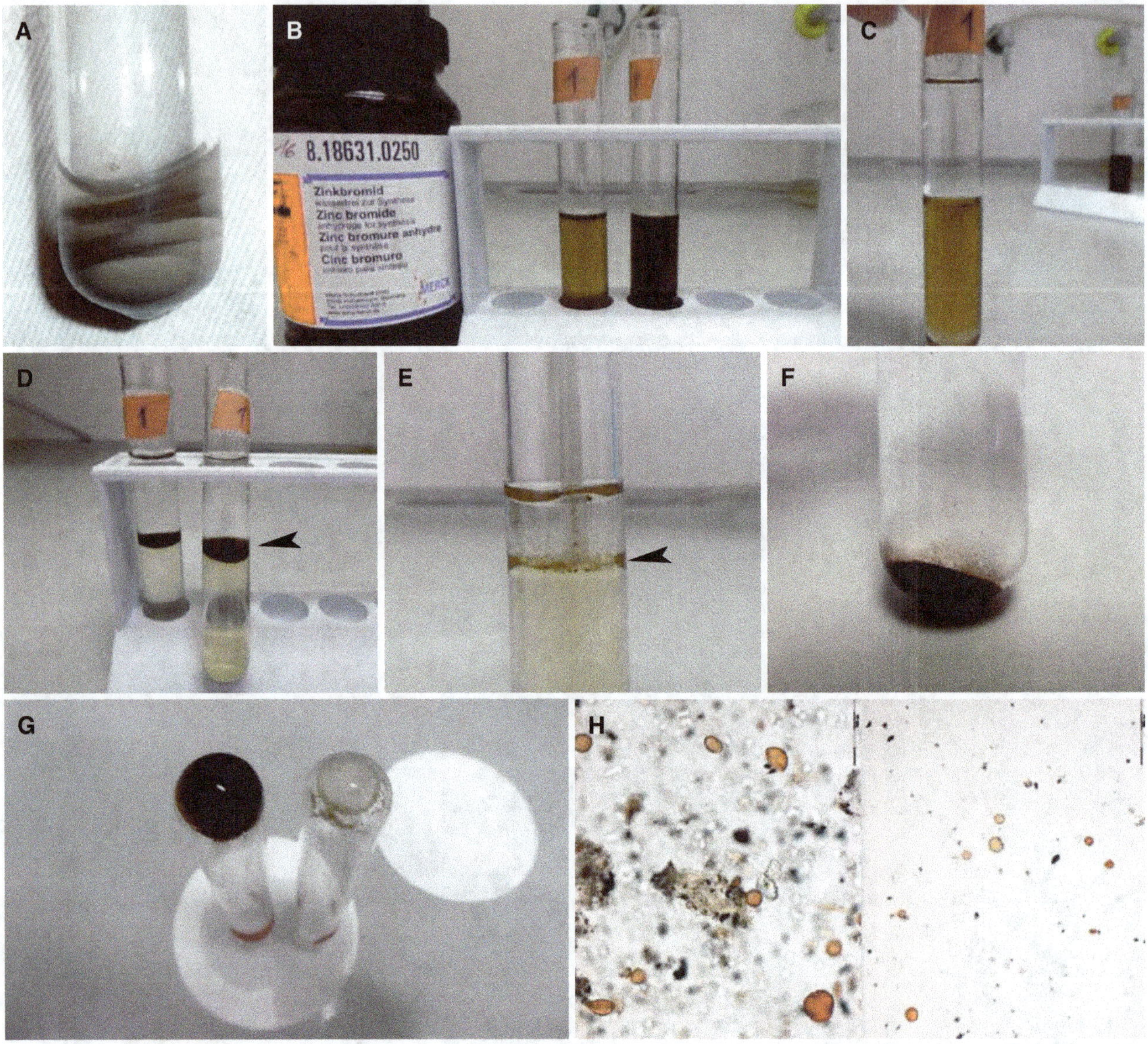

Fig. 8 Heavy liquid separation. A. Sample with high mineral content (light grey layers) after acetolysis. **B.** Mixing the sample with heavy liquid. **C.** Distilled water added without intermixing the liquids. **D.** Organic fraction (arrowhead) floating on the heavy liquid. **E.** Organic fraction (arrowhead) pipetted to a new test tube. **F.** Washing the organic fraction with water. **G.** Drying the acetolyzed sample (left) and the mineral fraction (right). **H.** Sample untreated (left) and treated with heavy liquid separation (right)

water is then carefully poured into the test tube (ca. 2 cm) and make sure that the two liquids do not intermix. After centrifuging for about 5–8 min at 3000 rpm the organic material is floating on the heavy liquid and below the distilled water. The organic material can then be transferred with a pipette into a new test tube for further washing. The inorganic parts remain at the bottom of the solution.

Acetolysis the Fast Way

A fast and easy way to prepare recent pollen grains for LM and SEM is to have a small glass bottle with a readymade acetolysis fluid (nine to one mix of 99% acetic anhydride and 95–97% sulfuric acid) at hand. Place a drop(s) of the acetolysis fluid on a glass slide. Remove anthers from the flowers and place them into the fluid on the glass slide (Fig. 9). To soften up the material let it lay in the liquid for some time and break the anther/flower material by squeezing and pressing it with the tip of a teasing needle. The slides are then heated over a candle flame for a short time to soften up the anthers, release the pollen grains from the anthers, dissolve extra organic material on pollen grain surfaces, "rehydrate" pollen grains and release their cell contents, and finally, to stain the pollen grains for LM photography. Make sure not to hold the slide over the flame for too long since it will make the pollen grains too dark. Best is to

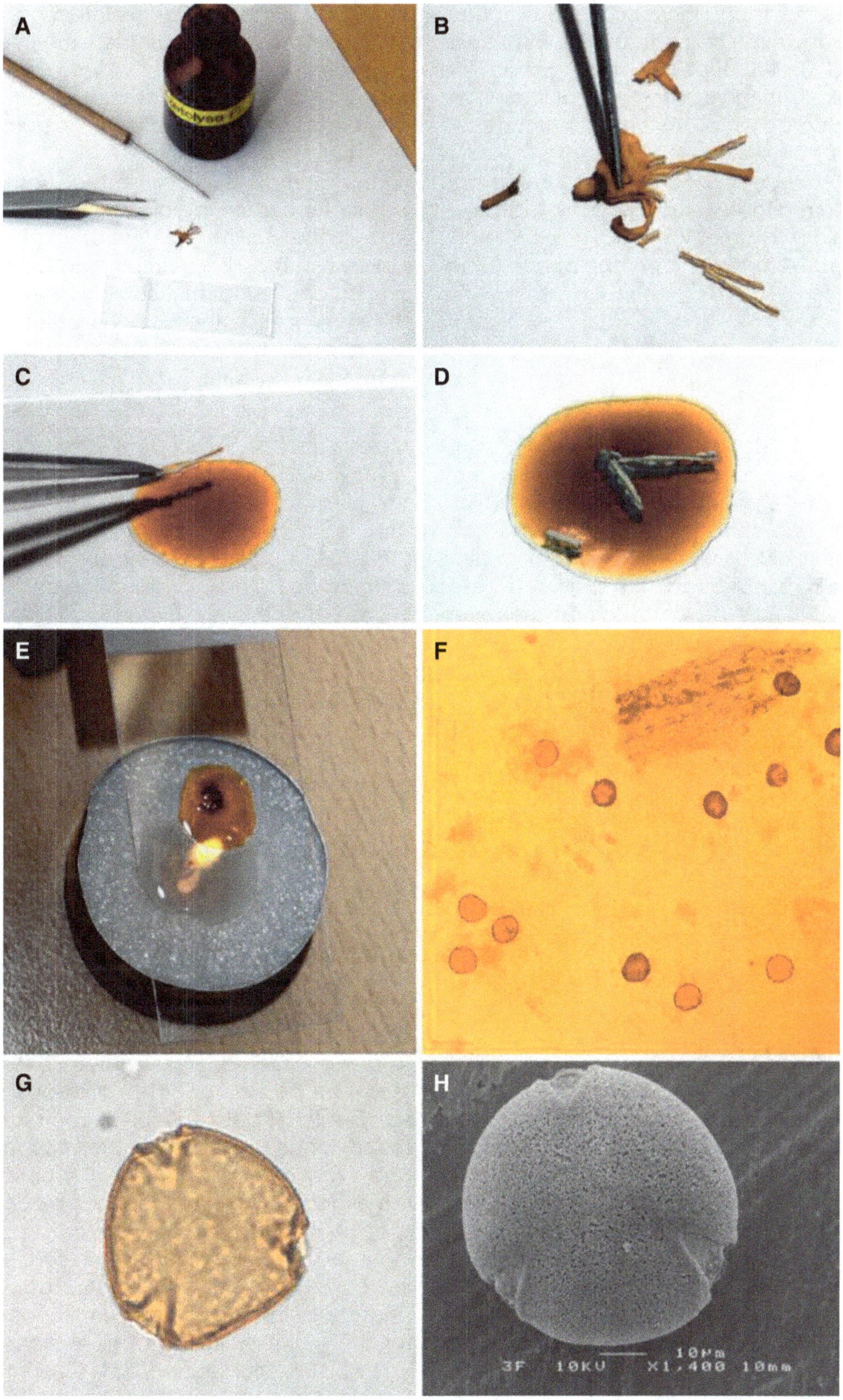

Fig. 9 Acetolysis the fast way. A. Flower and tools needed for preparation. **B.** Brake or cut off anthers. **C.** Transfer anthers into acetolysis fluid on glass slide. **D.** To soften up the material it can lay in the fluid for some time. **E.** Carefully heat the slides over a candle light. **F.** Readymade pollen grains in the acetolysis fluid. **G.** Transfer pollen grains to fresh drops of glycerine on new glass slides and photograph in LM. **H.** Same grain photographed in SEM using the "single-grain method"

heat the slides shortly and then use the teasing needle to break down the anther material. This should be repeated until the pollen have gained the required color. Using a micromanipulator (see below) selected pollen grains are then transferred into fresh drops of glycerine on new glass slides and photographed under LM. Some pollen grains can also be transferred to SEM stubs using the technique of the "single-grain method" described below, sputter coated with gold and photographed under the SEM.

Scanning Electron Microscopy: Preparation of Recent Pollen

SEM techniques cannot substitute LM, but they can provide a great deal more information, especially about ornamentation. Samples prepared for SEM should ideally reflect the fully hydrated condition of a living pollen grain. In addition, all types of pollen coatings must be removed from the pollen surface, not to obscure details of the pollen wall.

For scanning electron microscopy dehydration and drying techniques are of great importance. The principle of critical point drying (CPD) is to avoid any damaging to the pollen due to surface tension forces occurring during transition from the liquid to the vapor phase. Due to the slow penetration time of DMP, large samples (e.g., large anthers, whole parts of flowers) should be dehydrated in a series of alcohol (70–85–96%, each about 20 min) and acetone or dehydrated in 70% ethanol (3 days) and formaldehyde dimethyl acetal (FDA, 1 day or overnight).

The DMP Direct Method: Dimethoxypropane

With the DMP direct method (Halbritter 1998) important details of hydrated pollen grains, which may be lost by conventional methods (alcohol), are well preserved without shrinkage, distortion, or dissolution (Fig. 10). The best results are obtained using acidified dimethoxypropane (DMP) for dehydration. Anthers should be collected at anthesis. Take whole or parts of anthers, or loose pollen grains and put them into a pouch made of filter paper. For analyzing pollen in hydrated condition, moisture the filter pouch with a droplet of water and wait for a few seconds before transferring them into acidified 2,2-dimethoxypropane. After 20–30 min (or up to 24 h) in DMP samples are transferred into pure acetone for a few minutes and critical-point dried in CO_2 using acetone as the intermediate fluid. The CPD-pollen samples are then mounted on stubs using double-sided adhesive tape, sputter coated with gold and observed with an SEM. CPD samples can be stored, e.g., in a sealed plastic box to protect them from humidity.

This method can be used for fresh material as well as for herbarium samples (after rehydration in water). The chemical dehydration of unfixed plant material with DMP is a simple and fast method and can be applied to small samples only.

Unless stated otherwise, the pollen grains shown in this book are prepared using the DMP direct method by Halbritter (1998).

Transmission Electron Microscopy: Pollen Wall Stratification and Ultrastructure

For TEM studies of recent and fossil pollen, more than one protocol for fixation and staining may be needed.

Fixation and Embedding

Fixation of samples for TEM studies (Hayat 2000) is a time-consuming process that starts with fixation on the first day (Fig. 11), followed by dehydration and infiltration on the second and third day and ends with embedding on the fourth day (Fig. 12). For prefixation, the samples (closed anthers or pollen suspension) are placed in phosphate buffered glutaraldehyde (3%). In case of large specimens (flower/anther), the relevant parts of the sample are prepared/cut within the fixation solution under a binocular microscope (placed at the fume hood to prevent toxic substances from inhalation). Samples must be free of gaseous/air-bubbles. Transfer samples into Eppendorf tubes and make holes into the lid. Place the tubes into the vacuum desiccator and evacuate from air for 10–30 min. For pre-fixation the evacuated samples are then placed for 6 h in a specimen rotator (at room temperature). After rinsing in buffer and distilled water, samples are postfixed in 2% osmium tetroxide plus 0.8% phosphate-buffered potassium ferrocyanide (2:1) for 8–12 h at 6 °C (for osmium storage see also Fig. 28). On the second day osmium tetroxide is removed and samples are washed in distilled water (3 times for 5 min each) followed by dehydration in 2,2-dimethoxypropane (3 times, for 10 min each) and finally by pure acetone (2 times for 15 min each). The infiltration process starts by adding a few drops of the embedding media (1:2) to the samples

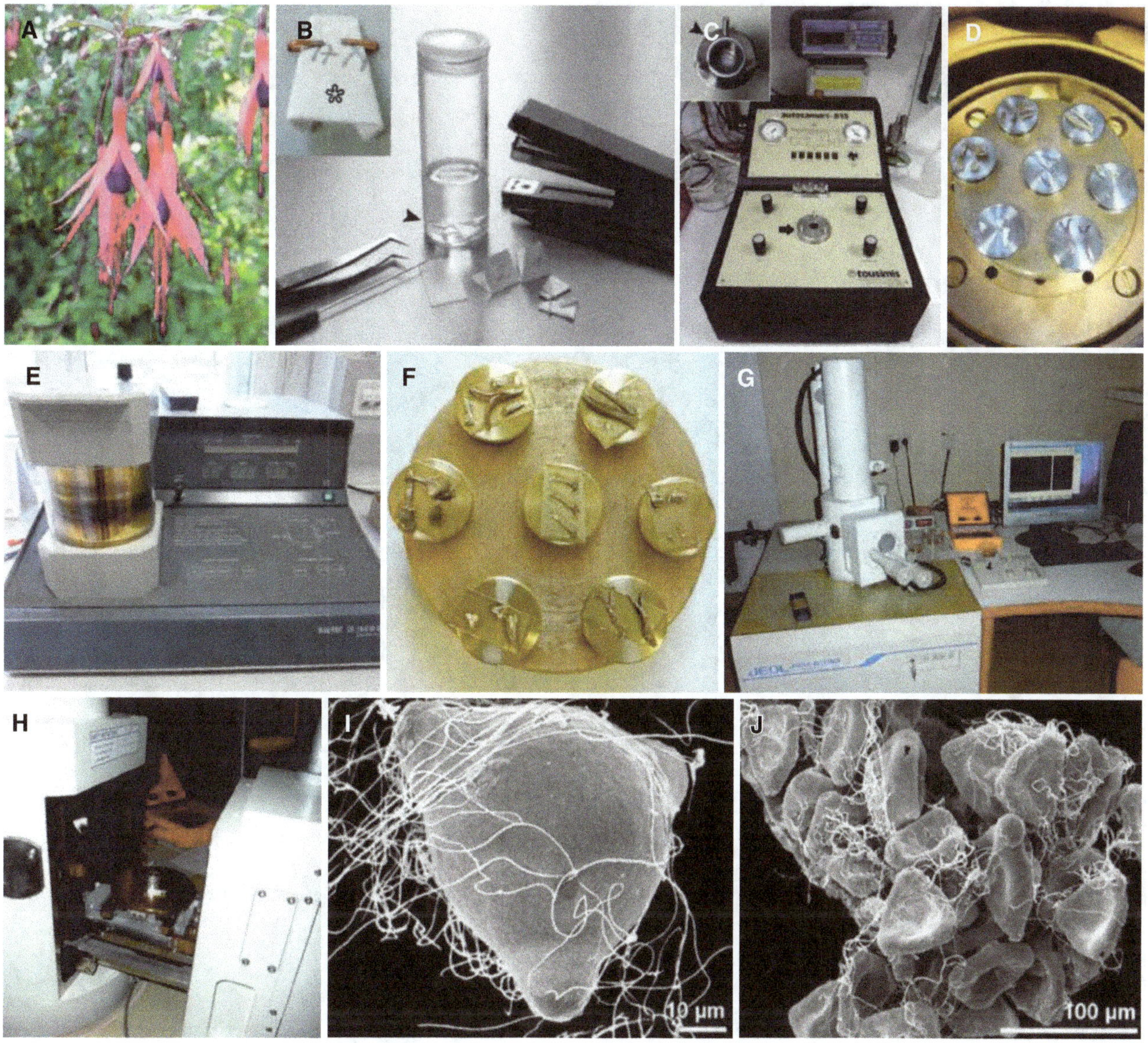

Fig. 10 The DMP direct method. A. Pollen collected at anthesis, *Fuchsia magellanica*, Onagraceae. **B.** Filter pouches for pollen preparation; moisture filter pouches (pollen samples) with a droplet of water (asterisk) before dehydration in DMP (arrowhead). **C.** Critical point dryer (CPD) with closed chamber and upper view on open chamber (arrowhead). **D.** CPD-pollen samples mounted on stubs using double-sided adhesive tape. **E.** Sputter coater. **F.** Samples sputter-coated with gold. **G.** SEM. **H.** Open chamber. **I.** Pollen in hydrated condition, SEM. **J.** Pollen in dry condition, SEM

and swirl the mixture. Repeat the procedure in 6–7 h, then let samples infiltrate overnight. This process has to be repeated on the third day. On the fourth day, acetone has to be removed before embedding the material: extract half of the acetone-resin-mixture with a pipette and wait for 2–3 h until the remaining acetone evaporates. After the fixation process the material should be stained intensive black (due to osmium), if not start from the beginning with new material.

The fixed material can now be transferred into embedding forms filled with fresh **embedding media** (Agar low-viscosity resin, see section "Recipes for TEM"). Polymerization takes place in an oven for about 12 h at 70 °C. After polymerization the specimen blocks can be stored in small plastic bags and are ready for ultrathin sectioning.

Annotation: For fixation of pollen, the material must be centrifuged after each step and the fixation mixture/water/DMP must be extracted with a pipette.

Ultramicrotomy

A lot of equipment and preliminary steps are involved in the ultramicrotomy process: **preparation of formvar film-coated grids, section-manipulators**

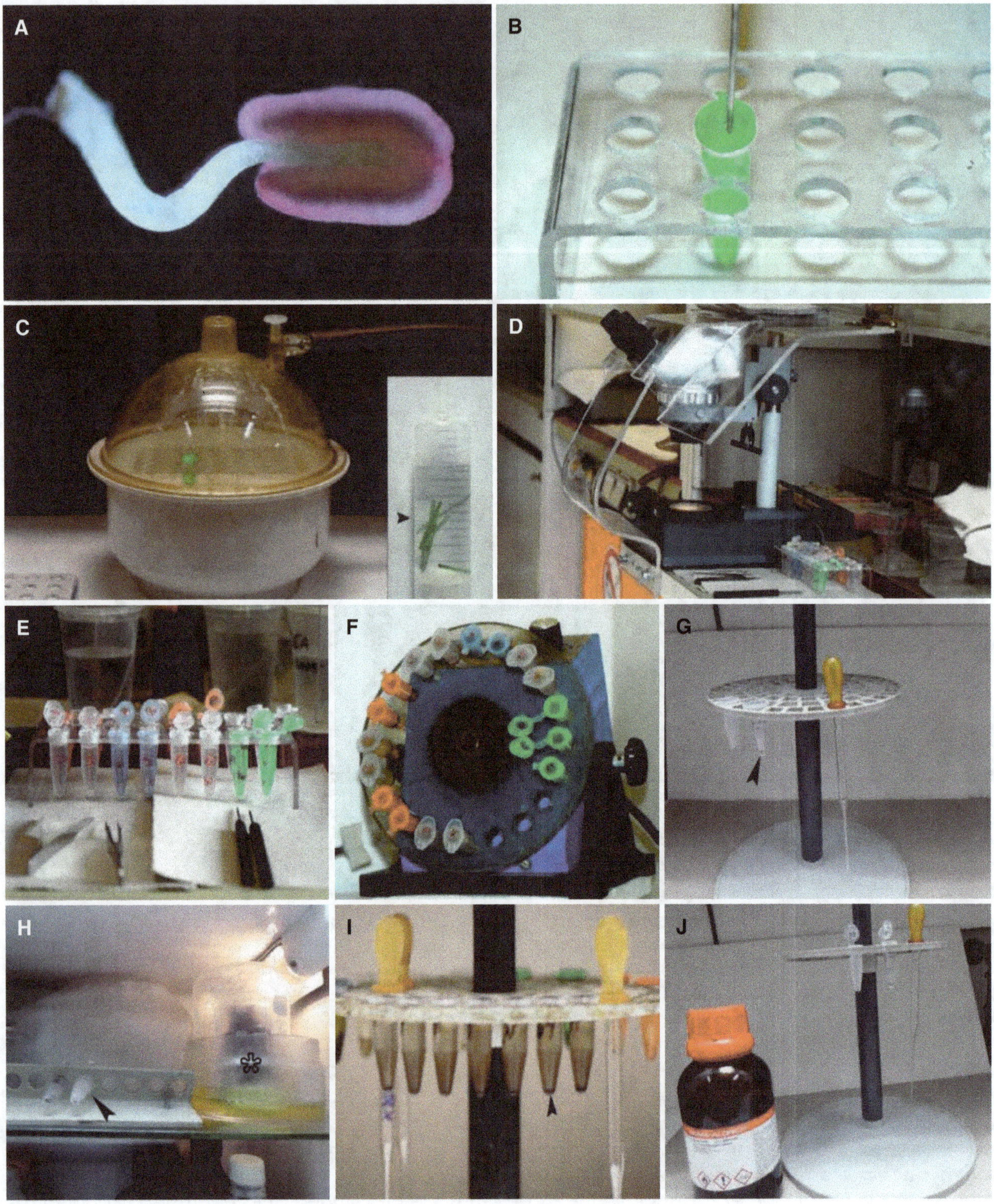

Fig. 11 Fixation and embedding day 1–2. A. Closed anther for pre-fixation. **B.** Material in Eppendorf tube with fixation solution, make holes in lid before evaporation. **C.** Evacuation in vacuum desiccator (left) or manually in a syringe (right). **D.** Preparation/cutting of samples within the fixation solution under a binocular microscope (placed at the fume hood). **E.** Transfer of selected parts of the sample into small Eppendorf tubes with fixation solution (3% GA). **F.** Samples in specimen rotator. **G.** Post-fixation; arrowhead indicates sample with osmium solution. **H.** Post-fixation of samples (arrowhead) for 8–12 h at 6 °C (fridge in a fume hood) in Eppendorf tubes; Note: osmium solution stored in fridge (asterisk). **I.** Samples after 8–12 h: material blackened due to osmium (arrowhead). **J.** After removal of osmium, samples are dehydrated, followed by pure acetone

Fig. 12 Fixation and embedding day 2–4. A. Infiltration starts by repeatedly adding few drops of embedding media. **B.** Embedding solution (Agar low-viscosity resin) mixed using a magnetic stirrer. **C.** Final embedding into adequate embedding forms under binocular microscope. **D.** Polymerization at 70 °C in a thermostat oven (arrow). **E.** Examples of various embedding forms. **F.** Polymerized samples. **G.** Specimen blocks stored in small plastic bags

and preparation of loops, specimen block trimming, semi-thin sectioning, making of glass knives, diamond knives, and **ultra-thin sectioning**. Another indispensable equipment for ultramicrotomy are tweezers with an ultra fine pointed, curved, and angled precision tip.

Formvar Film-Coated Grids

Coated grids are made with a formvar solution (see "Recipes for TEM"; Fig. 13). New and cleaned glass slides are dipped with a special self-made "filming machine" into the formvar solution (minimize evaporation of the chloroform). The extraction speed of the slide influences the thickness of the formvar film: a thin film is produced by a slow, steady movement. After 1–2 min remove the glass slide steadily from the solution and dry for 2–3 min. The film can then be transferred onto a clean water surface (use distilled water in a clean staining cuvette). To loosen the film, cut the film with a scalpel along the edges of the slide and blow moist air (with a straw from your mouth) onto the film. In the same instance, dip the slide into the water at an angle of 45° to remove the film from the glass slide. When the film is floating on the water surface, don't pull out the slide, but let it slowly set into the cuvette. The quality of the film is indicated by the color: a thin film is grey to silver, whereas gold is too thick. Grids cleaned with chloroform are placed using fine pointed tweezers onto the film. To know which side of the grid is coated, always put one side (either

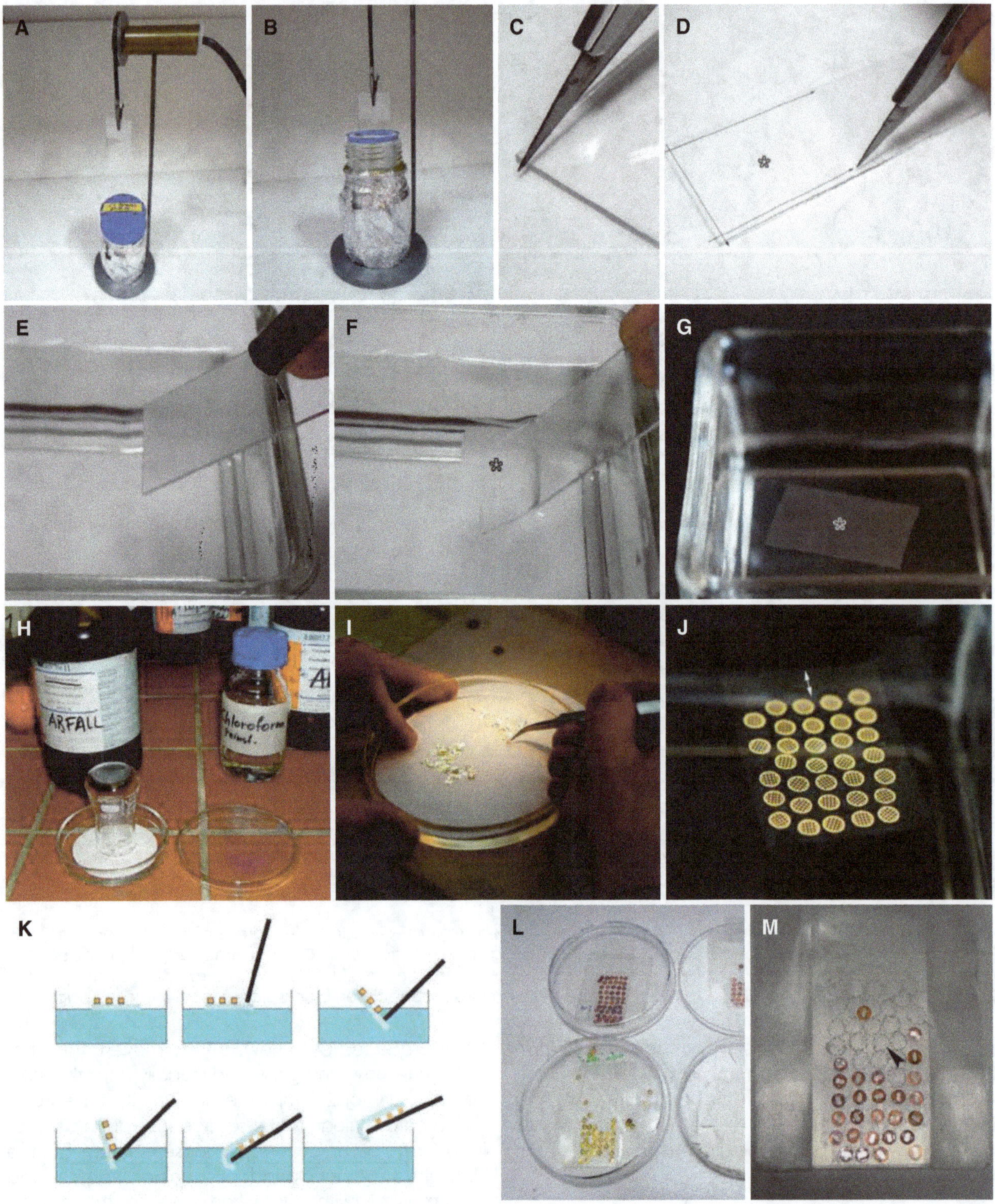

Fig. 13 Making formvar film-coated grids. **A.** Filming machine with holder for glass slide; filming solution should be protected from light. **B.** Glass slide dipped into formvar solution (under fume hood). **C-D.** Film cut along edges (arrowhead); arrows indicate cutting line on film (asterisk). **E.** Moisture film before dipping the slide under water; arrowhead indicates straw. **F.** Dipping the slide into the water at an angle of 45°; film partly floating on water (asterisk). **G.** Thin film floating on water surface (silver colored). **H.** Clean grids with chloroform. **I.** Shiny or dull side of the grid is visible under binocular. **J.** Grids on thin floating formvar film, arrow indicates space left for film extraction. **K.** Extraction of coated grids from the water surface using a parafilm-coated glass slide. **L.** Coated grids dried and stored in petri dish. **M.** Parafilm-coated slide with formvar coated grids, perforations (arrowhead) outline removed grids

shiny or dull side) of the grid down on the film. Make sure to leave enough space between grids and along one short margin to extract the film from the water surface. Use a parafilm-coated glass slide to extract the filmed grids: place the slide on free space of the film and dip with quick and steady motion at about 45° angle into the water and then pull out the slide again (Fig. 13K). Place the slide on a filter paper in a petri dish and let it dry. Formvar film-coated grids should be stored protected from light and dust-free (e.g., in the petri dish). To isolate the grids, use a needle to make perforations around the grids and remove them carefully with a forceps. Before ultra-thin sectioning, check the formvar film-coated grids for defects (e.g., holes, dust) under binocular microscope and place them with the filmed side up on a filter paper (see also Fig. 20 "Section pick up").

Section Manipulators (Eyelash or Other Adequate Type of Hair)

To separate and move semi-thin and ultra-thin sections floating on the water surface an **eyelash manipulator** is used. Usually a human eyelash (untreated) is fixed with glue or wax on a short glass pipette or wooden stick. The eyelashes should be cleaned with alcohol each time used and stored dust-free (e.g., covered with the back end of a bigger plastic pipette) (Fig. 14).

Loops

Loops are used to transfer ultra-thin sections onto formvar-coated grids (see Fig. 20 "Section pick up"). A loop should take up a droplet of water accurately and should fit exactly onto the grid. Therefore, two types of loops are produced (1) **circular loops**, that fit onto mesh-grids and (2) **oval** loops, used for slot-grids (Fig. 15).

Loops are made with wires from conventional electric cables (wires should not be too thick or thin). For making a circular loop a small piece of wire can be twisted around a circular object with appropriate diameter (e.g., screw driver). To produce an oval loop make a smaller circle and press it from two sides with a plier into an oval shape (fitting the grid slot). More ideally wrap the wire around a self-made model form fitting the grid size/slot. The wire of the loop is finally flattened with a hammer and the twisted (non-flattened) appendices fixed with glue or wax, e.g., on a short glass pipette. The loop should be cleaned before use with alcohol and stored free of dust.

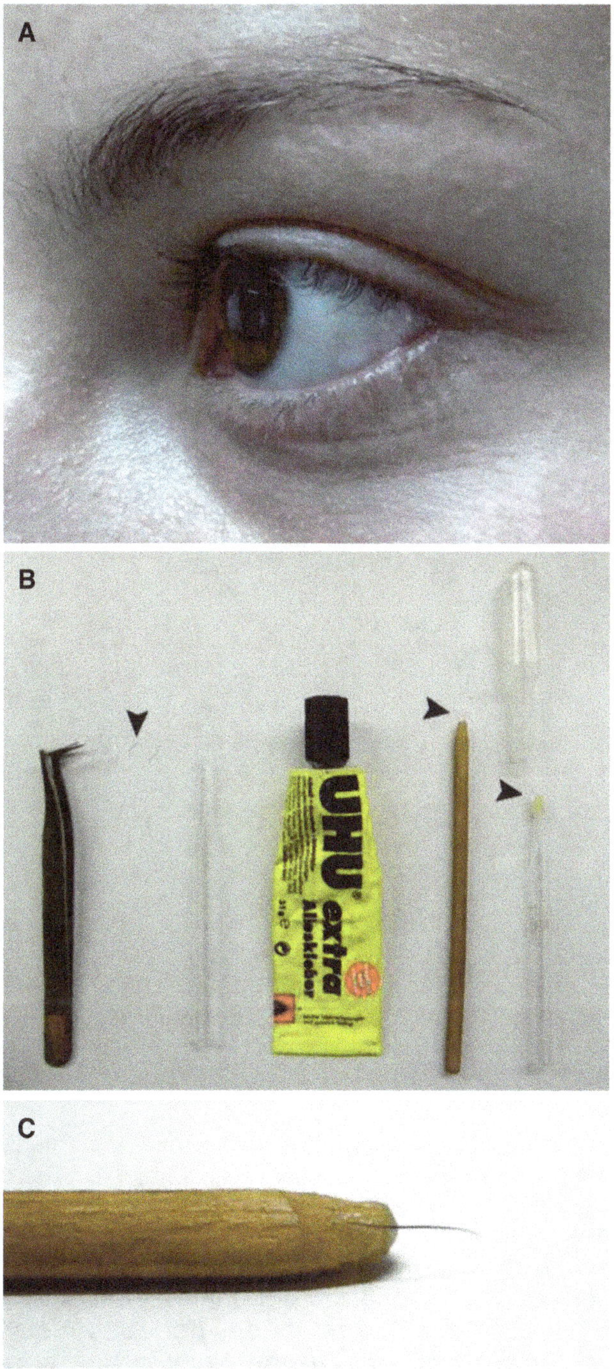

Fig. 14 Making a section manipulator. A. Human eyelash. **B.** Technical equipment for making a section manipulator; arrowheads indicate eyelashes. **C.** eyelash fixed with glue on wooden stick

Specimen Block Trimming

Criteria for block trimming are: (1) a small sample size, (2) the location of the sample should be in the center of the block-face (trapezoid) and surrounded by resin, (3) the straightness of the block-face edges (parallel edges).

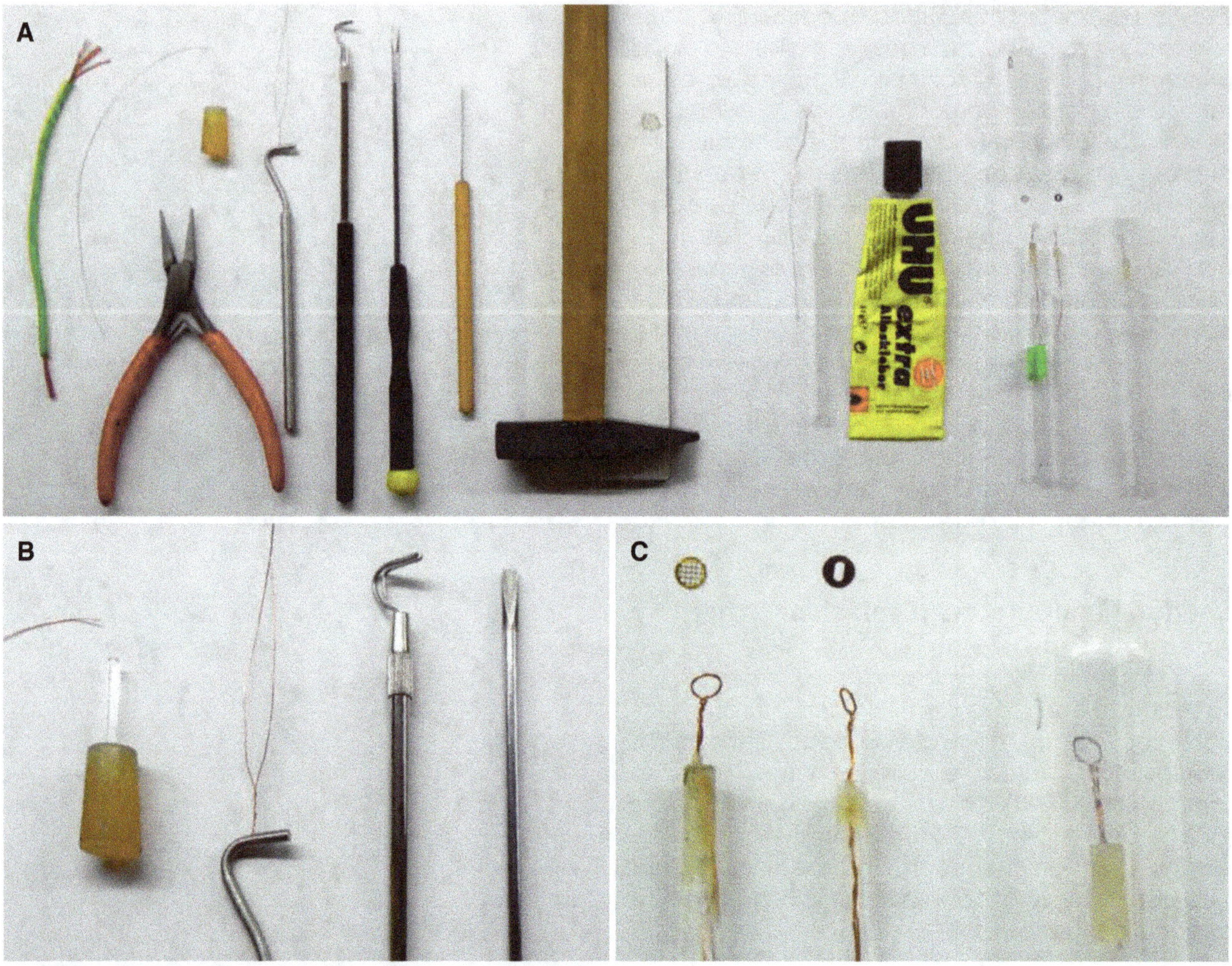

Fig. 15 Making loops. A. Technical equipment for making loops. **B.** Wire twisted around a circular object (model form) for mesh grids. **C.** Loops fixed with wax on glass pipettes (left), for storage loop covered with the back end of a disposable plastic pipette (right)

A specimen block must be trimmed (cut) to get small sections with a block-face of 4 mm by 4 mm in size (Fig. 16 O). A small block-face ensures good sectioning performance. Trimming is conducted with razor blades (for each block use a new razor blade). The block is fixed in a specimen holder and trimmed under a binocular microscope. The specimen block is trimmed into a pyramid with a trapezoid-shaped block-face. The tip of the pyramid should be cut away until you reach the appropriate level within the sample. A glass knife is used for initial cuts. If the specimen is rather big, the block-face can be larger for semi-thin sectioning (max. 4 mm^2) to ensure that the area of interest is preserved. Such a large block must be trimmed further to reach the final required block-face for ultra-thin sectioning.

Glass Knives

Glass knives are generally used for semi-thin sectioning and are replaced by diamond knives for ultra-thin sectioning. Glass knives are produced with a "knife-maker" (Fig. 17). Specially produced glass strips (e.g., 6.4 × 25 mm) are first cut into squares. The squares are then cut diagonally into two triangles, each with a knife edge (Fig. 17 E). The breaking line (stress line) indicates the quality of the knives. The left side of the glass knife is sharper and can also be used for ultra-thin sectioning, whereas the right side is used for semi-thin sectioning only. "**Glass knife boats**" (disposable plastic forms) are attached and sealed with hot melted dental wax (hot plate and ethanol burner) to the glass knife (see also Fig. 18). Glass knives should be stored dust-free and safe in a "glass knife box."

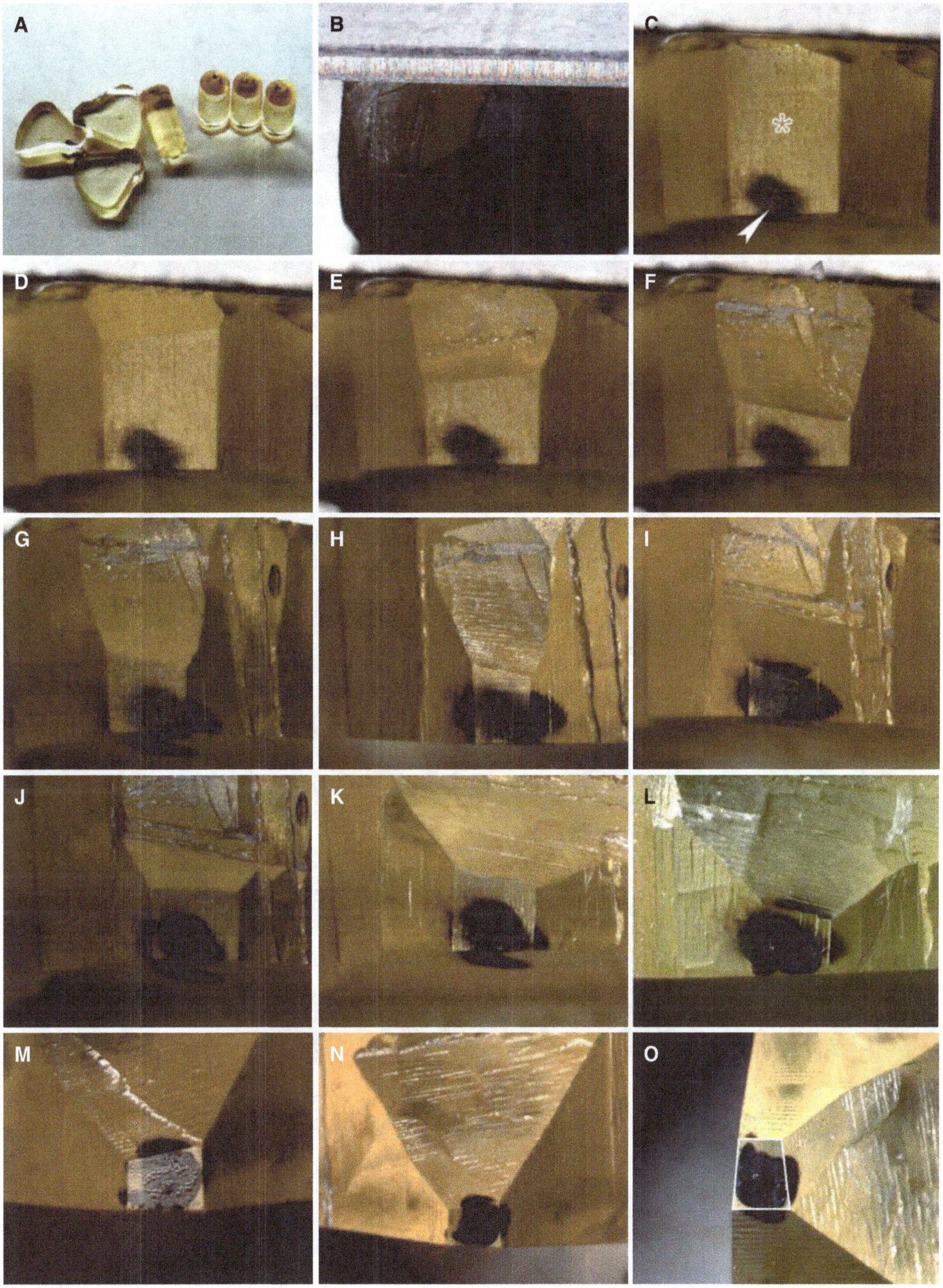

Fig. 16 Specimen block trimming. A. Specimen blocks of various shapes. **B**. Trimming is conducted with razor blades. **C**. Untrimmed specimen block with view on block-face (asterisk), arrowhead indicates position of specimen inside block. **D-N**. Blocks are trimmed into a pyramid with a +/– trapezoid shaped block-face and parallel edges. **O**. Final block-face with trapezoid form (white trapeze)

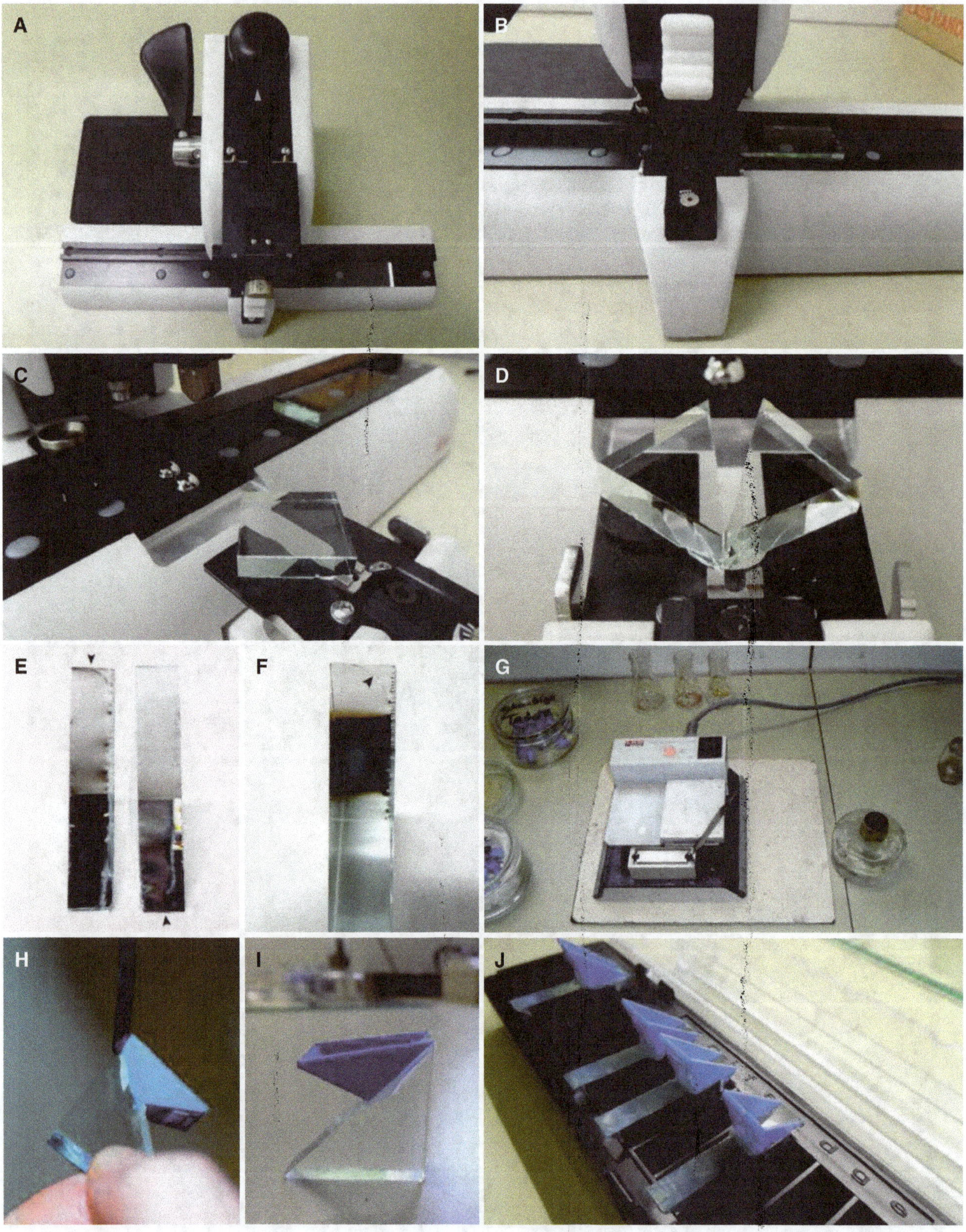

Fig. 17 Making glass knives. **A**. Knife maker. **B**. Glass stripes cut into squares. **C**. Squares are cut into two triangles. **D**. Two triangles (glass knives). **E**. Each triangle has a knife edge (arrowhead). **F**. Detail of triangle with knife edge, arrowhead indicating breaking (stress) line. **G**. Hot plate and ethanol burner for melting dental wax. **H**. Glass knife boats attached and sealed with hot wax using a spatula. **I**. Readymade glass knife. **J**. Knives stored in glass knife box

Semi-Thin Sectioning

Before selecting an area of the specimen block for ultra-thin sectioning, semi-thin sections are cut with an ultramicrotome, using a glass knife (Fig. 18). The settings for semi-thin sectioning are: section thickness between 0.5 and 2 µm (interference color purple to blue) and cutting speed 2 mm per second. Semi-thick sections are transferred with a loop into a drop of water on a glass slide. For a fast drying process put the slide on a hot plate (approx. 70 °C). While the water evaporates the sections will stretch. The dried sections are stained with toluidine blue on the glass slide, which can be sped up by placing the slide for max. 5 s on the hot plate. Carefully wash the slide with water and dry the glass slide in a filter paper block. The stained semi-thin sections are controlled with the LM to determine the quality of the fixation and to ensure that the appropriate area of the specimen is in the correct position for ultra-thin sectioning.

Ultra-Thin Sectioning

Ultra-thin sections between 60 and 90 nm (interference color silver to pale gold) are cut using an ultramicrotome (Fig. 19). **Diamond knives** are more suitable for cutting plant material, as e.g., crystals in cells destroy the cutting edge of glass knives, generating scratches within the sections or even splitting the sections.

The knife is placed in the knife holder and the knife boat filled with distilled water. The knife should be clean, free of dust and moistened with water. The specimen block has to be placed in the specimen arm in the upper position. Then the block has to be positioned parallel to the knife-edge by rotational or lateral adjustments of block as well as the knife. By moving the block up and down in front of the knife a slit of reflected light helps to adjust the block to the knife. A narrow slit of light indicates that the block is close to the knife and a constant thickness of the slit, along the whole block-face, indicates that the block face and the knife-edge are parallel. This is the ideal position for sectioning. The settings for ultra-thin sectioning are: section thickness between 60 and 90 nm and cutting speed 1 mm per second. The section settings can be adjusted while cutting until pale gold to silver sections are produced. Sections are floating on the water surface and can be manipulated with an eye-lash. Before the ultra-thin sections can be transferred to grids, sections must be stretched to remove compressions due to cutting. For stretching a solvent (e.g., xylol, chloroform, acetone vapor) or a hot pen can be used. For the vapor method use a thin, wedge-shaped piece of filter paper moistened with a drop of solvent, hold it closely above the sections while moving it back and forth.

Section Pick-Up

The stretched sections are picked up from the water surface with a loop (Fig. 20). Depending on the size of the sections between 3 and 10 sections can be picked up at once. Center the loop above the selected sections, dip it on to the water surface, lift the sections up within a droplet of water and transfer onto a grid under a binocular microscope. Center the loop above the grid and lower it onto the grid surface. Lift up the loop and the attached grid. The water is removed slowly with a filter paper touching the first twist by the loop (Fig. 20 D). Transfer the grid with a forceps into a grid-box (sections should face the same side). Make a section protocol. Store the grid box away from light.

Staining Methods

The application of different TEM staining techniques for one and the same sample is very important and highly recommended to avoid misinterpretations of the pollen wall structure. Therefore, sections of pollen grains are routinely stained using the several different staining methods (Figs. 21 and 22). Most staining solutions are harmful or even toxic and therefore applied under fume hood.

Annotation: In electron microscopy there is no grey-scale terminology from white to black. Use "electron dense" for black or darkly colored structures and "electron translucent" for white to light grey colored.

Uranyl Acetate-Lead Citrate Staining: U + Pb

Uranyl acetate-lead citrate staining is a conventional staining method (Hayat 2000; Figs. 21 and 22). Ultra-thin sections are usually collected on copper grids. Sections are stained in uranyl acetate solution (Leica Ultrastain-1) for 45 min followed by lead citrate staining (Leica Ultrastain-2) for 1–5 min at room temperature. Use of sodium hydroxide pellets for lead citrate staining prevents crystalline precipitation by absorbing moisture and carbon dioxide from the air. Sections are thoroughly washed in distilled water after each staining step (3 times for 5 min in a row of water drops).

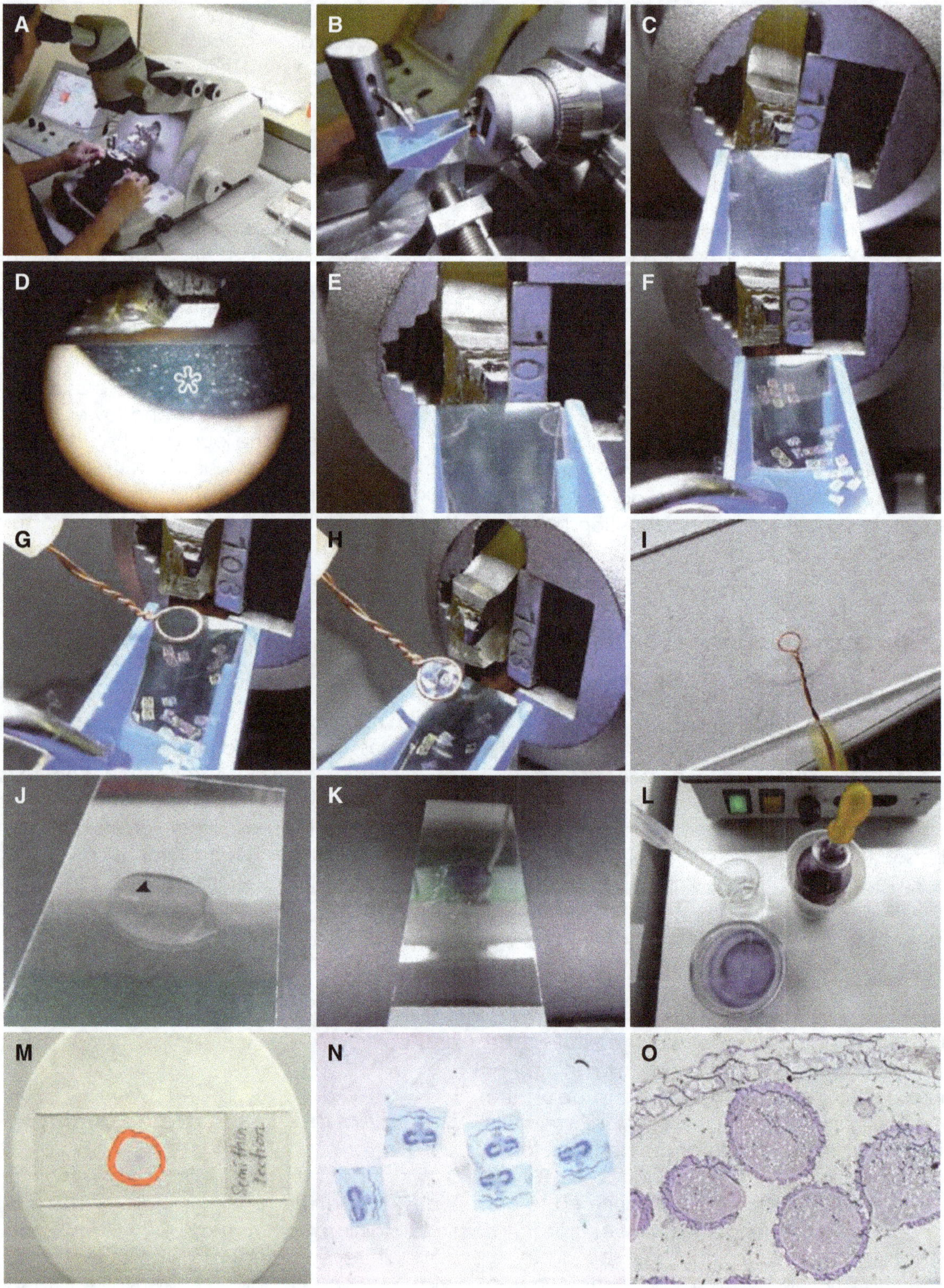

Fig. 18 Semi-thin sectioning. **A.** Ultramicrotome. **B.** Glass knife positioned in the knife holder and knife boat filled with distilled water; specimen block fixed within the specimen holder. **C.** Block adjustment parallel to knife-edge by use of reflecting light. **D.** Knife should be clean, free of dust and moistened with water, asterisk indicates slightly lowered water level at knife edge for sectioning, but still moistened. **E.** Block must be close enough to knife (until slit of light almost disappears) to start sectioning. **F.** Semi-thin sections between 0.5 and 2 µm (interference color purple to blue) floating on water. **G-H.** Section pick-up with a loop (see "Section pick-up"). **I.** Transfer of sections in a drop of water on a glass slide. **J.** Slide on a hot plate (arrowhead indicates semi-thin sections). **K.** Staining sections with toluidine blue on hot plate. **L.** Rinsing the stained sections with water. **M.** Stained semi-thin sections ready for LM. **N.** Toluidine blue sections seen under LM. **O.** Final quality check before ultra-thin sectioning

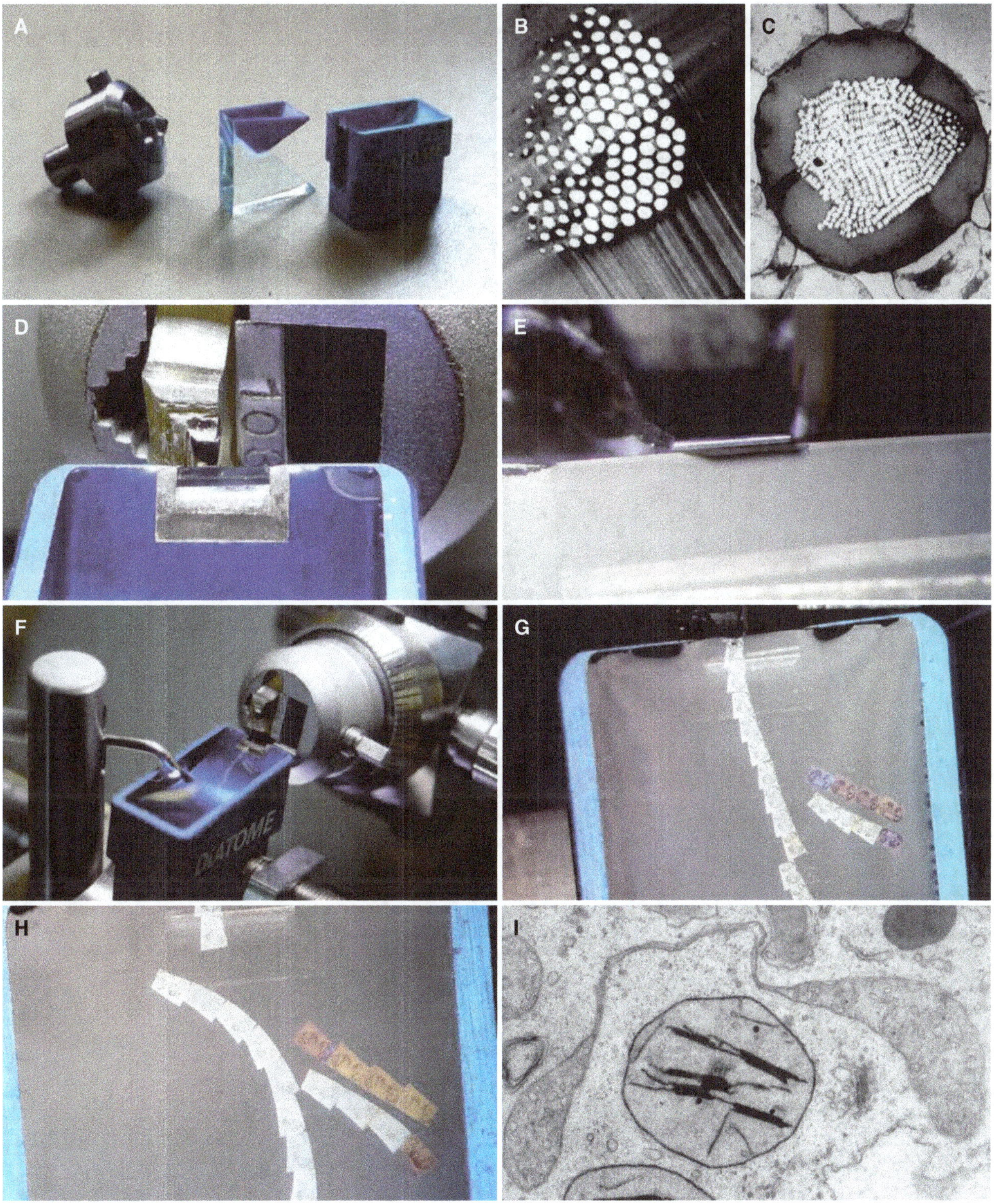

Fig. 19 Ultra-thin sectioning. A. Specimen block holder with trimmed block, glass and diamond knife. **B.** Crystals in plant cells cut with glass knife, note scratches. **C.** Crystals in plant cells cut with diamond knife. **D-E.** Block adjusted parallel to knife-edge by use of reflecting light. **F-G.** Sections between 60 and 90°nm (interference color silver to pale gold). **H.** Stretched sections, note the change in size and thickness (for color change compare to picture **G**). **I.** Ultrastructure of a plant cell showing high quality fixation of several organelles in TEM

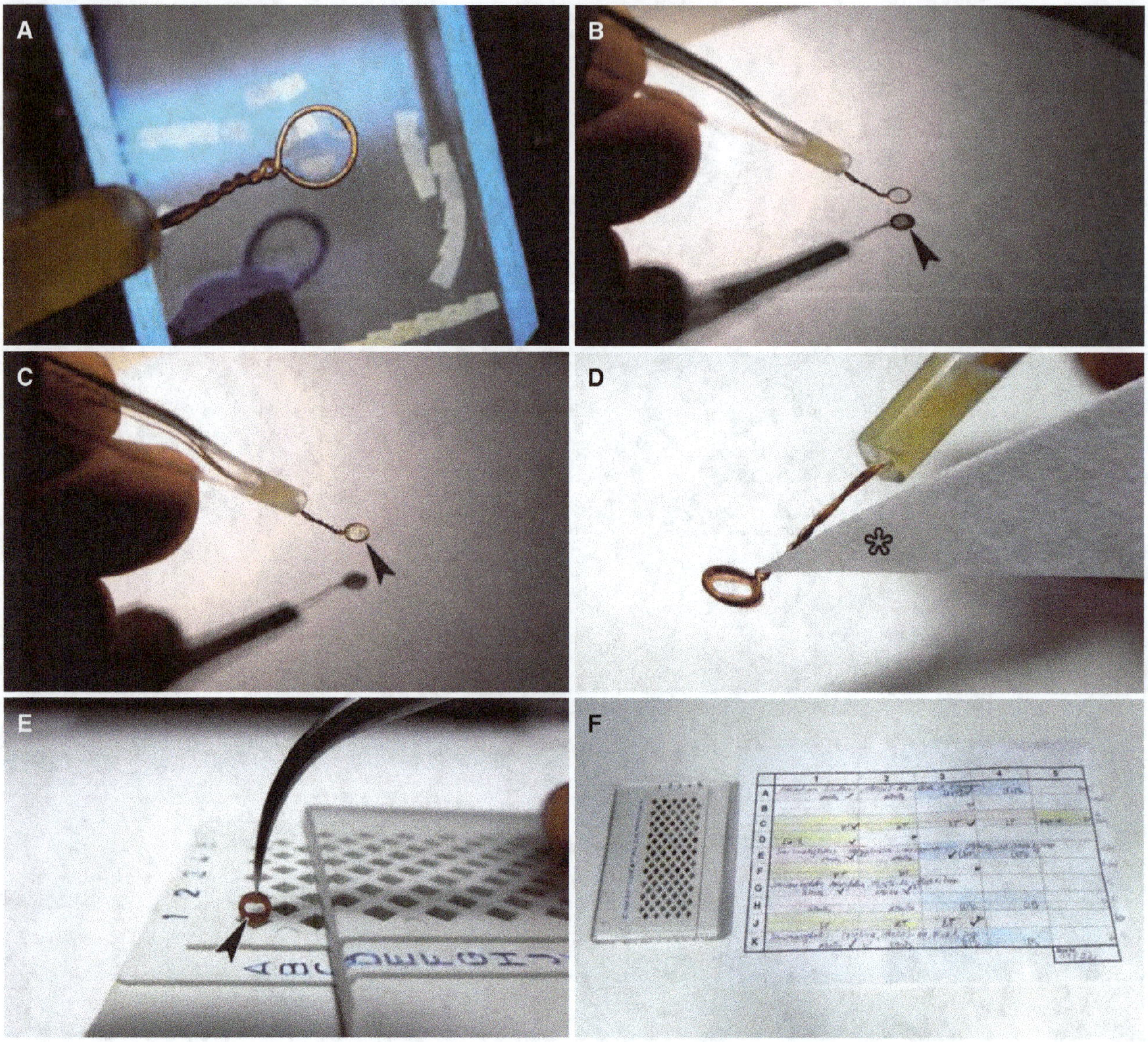

Fig. 20 Section pick-up. A. Loop for section pick-up. **B.** Loop centered above grid (arrowhead). **C.** Grid attached to loop (arrowhead). **D.** Water removed from grid, asterisk indicates wet filter paper. **E.** Dry grid placed into grid-box, sections on the left side (arrowhead). **F.** Grid-box and section protocol with color code used for different staining methods

The Lipid Test for the Detection of Unsaturated Lipids: TCH + SP

The endexine can be differentiated from the ektexine and the intine by thiocarbohydrazide-silver proteinate (TCH+SP) staining in osmium-fixed material. The endexine stains electron dense after the lipid test, indicating lipidic compounds (Fig. 22 B).

Ultra-thin sections on gold grids are treated with 0.2% TCH for 8–15 h and 1% SP for 30 min and thoroughly washed in water (3 times for 5 min in a row of water drops) (Rowley and Dahl 1977; Weber 1992).

Thiéry-Test: PA + TCH + SP

The Thiéry-test is used for the detection of neutral polysaccharides in osmium-free material (Thiéry 1967). Ultra-thin sections from osmium-free material are placed on gold grids and treated with 1% periodic acid (PA) for 45 min, 0.2% thiocarbohydrazide (TCH) for 8–15 h, and 1% silver proteinate (SP) for 30 min (Thiéry 1967). The polysaccharide intine and starch grains in amyloplasts stain electron dense (Fig. 22 C). For control samples leave out the thiocarbohydrazide step. If osmium fixed material is

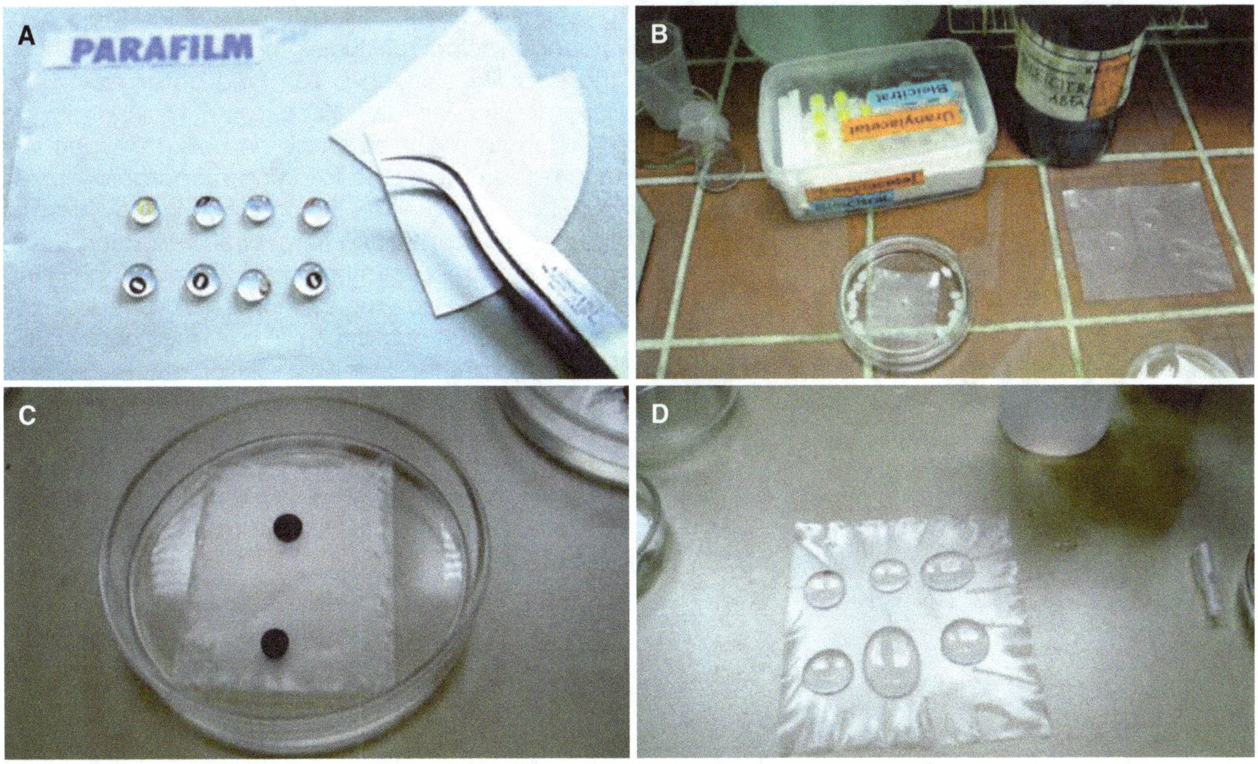

Fig. 21 Staining methods for ultra-thin sections. A. Ultra-thin sections on copper or gold grids stained in a small drop of uranyl acetate on parafilm. **B.** Small drops of lead citrate on parafilm and sodium hydroxide pellets in a closed petri dish. **C.** Small drops of potassium permanganate on parafilm. **D.** Row of large water drops for washing placed on parafilm

used for the Thiéry-test, the staining time for 1% periodic acid has to be prolonged up to 60 min (instead of 30 min), to remove the osmium tetroxide from the material.

Modified Thiéry-Test: PA + TCH + SP (short)

The modified (short) Thiéry-test (Weber and Frosch 1995) is especially effective after fixation of specimens with osmium and potassium ferrocyanide and is a good method for general enhancement of contrast in the cytoplasm and the pollen wall (Fig. 22 D). Ultra-thin sections are collected on gold grids and stained with 1% periodic acid (PA) for 10 min, 0.2% thiocarbohydrazide (TCH) for 15 min, and 1% silver proteinate (SP) for 10 min (at room temperature). After all steps the sections are thoroughly washed in distilled water (3 times for 5 min in a row of water drops), and following the TCH first washed in 3% acetic acid.

Potassium Permanganate: $KMnO_4$

Potassium permanganate staining is a simple method for the detection of the endexine. Using uranyl acetate and lead citrate, ektexine and endexine may differ in their electron opaqueness in that the endexine is higher in electron density than the ektexine, or vice versa. When the endexine is thin and less compact or discontinuous, the differentiation of the two layers may be insufficient. Typical for the endexine is its increasing thickness close to the aperture. Potassium permanganate stains the endexine electron dense, producing a distinct contrast (Weber and Ulrich 2010; Fig. 22 E). Ultra-thin sections from osmified material on copper grids are treated with 1% aqueous potassium permanganate solution for 7 min and thoroughly washed in water (3 times for 5 min in a row of water drops).

Preparation of Fossil Pollen

There are numerous methods currently used to extract organic material, including fossil and subfossil pollen, from all different types of sediments (rocks) and soils. These methods have been summarized in detail by, e.g., Erdtman (1943), Brown (1960), Fægri and Iversen (1989), Moore et al. (1991), Wood et al. (1996), and Traverse (2007). Most of these preparation methods involve sieving of some sort and the final production of palynomorphs enclosed in

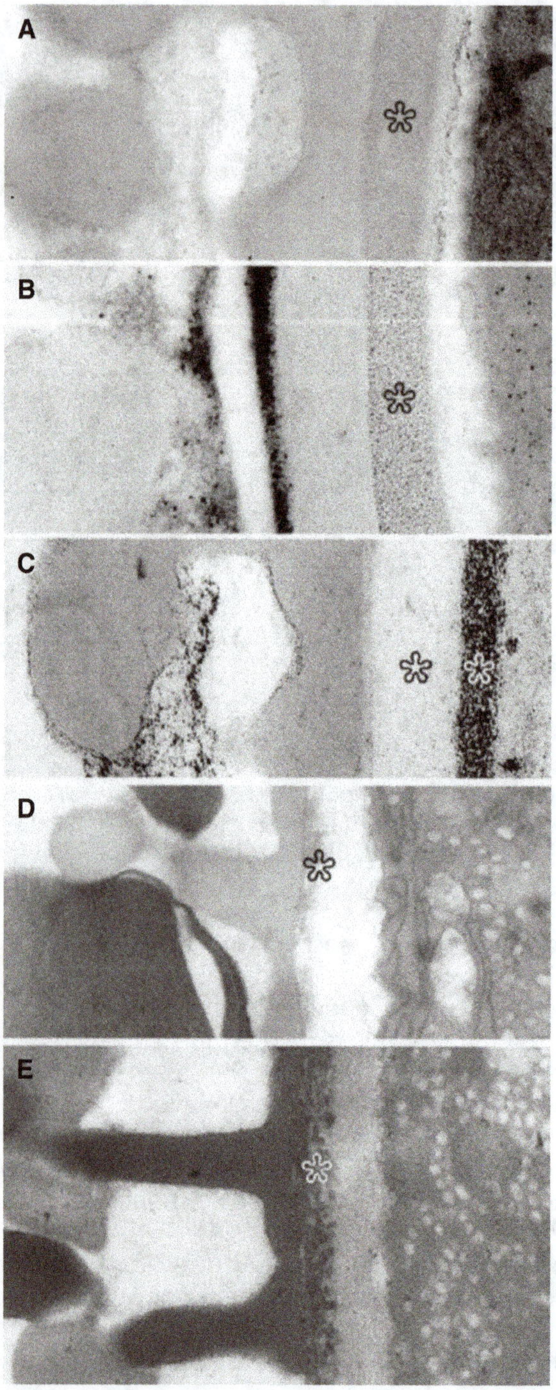

Fig. 22 Stained pollen walls and behavior of endexine (cross-section, TEM). A-C. *Apium nodiflorum*, Apiaceae. **A.** Uranyl acetate + lead citrate (U + Pb), compact-continuous endexine (asterisk). **B.** Lipid test (TCH+SP). compact-continuous endexine (asterisk) stains electron dense. **C.** Thiéry-Test (PA + TCH + SP), compact-continuous endexine (asterisk) stains electron translucent, intine electron dense (white asterisk). **D-E.** *Mentha aquatica*, Lamiaceae. **D.** Modified Thiéry-Test (PA + TCH + SP), thin compact-continuous endexine (asterisk) only slightly visible. **E.** Potassium permanganate ($KMnO_4$), thin compact-continuous endexine (asterisk) electron dense

glycerine gelatine on sealed glass-slides. Majority of paleopalynological studies then focus on counting the quantity of each pollen type observed on the slides (often between 300 and 600 grains), with an unfortunate minor emphasis on pollen morphology and ultrastructure. The following preparation procedure has been used by the paleopalynology team at the University of Vienna for over 30 years and is suitable for most sedimentary rocks with minor variations. During preparation the solution is not sieved at any stage, so not to lose any small or exceptionally large palynomorphs, and the final solution is stored in glycerine suspension in small sample tubes so the palynomorphs can be studied using the so-called "single grain method." This method has been evolved to able researchers to obtain pollen characters from single fossil grains using both LM and SEM and sometimes TEM.

Preparation Method: From Rock to Palynomorphs

Sedimentary rock samples (20–50 g) are washed and dried and hand ground in a mortar with a pestle (Fig. 23). Using a glass beaker the resulting powder is boiled in ≥200 ml of concentrated hydrochloride acid (HCl) for 5–10 min; this should remove all carbonates. Let the solution stand and when the residue has settled, decant most of the HCl liquid. Transfer the remainder of the solution into a copper pan or pot and add ≥150 ml of hydrofluoric acid (HF) and boil for approx. 10 min while stirring with a copper stick or spoon (or let stand in cold HF for 3–5 days, stir regularly, use acid-resistant plastic containers and tools); this should remove all the silicates. The solution is then poured slowly into a 4 L plastic beaker filled with water. After settling, the liquid is decanted and the remainder solution poured into glass beakers along with ≥200 ml of HCl and boiled again for 5–10 min; this prevents the formation of fluorite crystals. After cooling and settling decant most of the HCl and pour the remainder of the solution into two separate test tubes (glass centrifuge type). Wash the solutions 4 times with water and centrifuge and decant the liquid following each wash. Fill one large glass tube with cold water and add 1–2 teaspoons of sodium chlorate (pure crystalline powder; $NaClO_3$). Shake this large tube and when there are crystals that cannot be dissolved in the water the solution is ready. Pour ca 1 ml of acetic acid glacial (100%, CH_3COOH) and 3–4 ml of the sodium chlorate solution into the two original test tubes, then add five drops of HCl. Place the tubes in boiling water for at least 5 min and

Fig. 23 From rock to palynomorphs. A. Different types of sedimentary rocks: reddish, yellowish, and white-greyish samples usually contain few and/or badly preserved palynomorphs (back row), brown, dark-grey to blackish samples often contain well preserved pollen (front row). **B**. Sedimentary rock sample (ca 30 g) hand grounded in a mortar with a pestle. **C**. Sample boiling in ≥200 ml of HCl. **D**. Sample boiled in a copper pan with ≥150 ml of HF. **E**. The HF solution is poured slowly into a large plastic beaker filled with water. **F**. Organic material settled on the bottom of the beaker. **G**. Acetolysis, test tube in boiling water, note the stirring glass stick. **H**. Acetolyzed sample before decanting of water following the final wash

have a stirring glass stick in it at all times. The color of the sample solutions should change from dark blackish to brown or reddish. Centrifuge the test tubes and decant the liquid. Wash the residue 3–4 times with water and one last time with acetic acid glacial. Prepare a new solution in a clean and dry measuring glass-tube with 9 parts acetic anhydride (99%, $(CH_3CO)_2O$) and one part sulfuric acid (95–97%, H_2SO_4). Make sure to produce at least 10 ml of this solution for each original (fossil) test tube you process. Pour ca 10 ml of the new solution into each test tube. Direct tube away from your face and make sure no water comes into contact with the solution. Place the tubes again into the boiling water bath for at least 5 min. Then centrifuge and decant the liquid (again avoid contact with water). First wash the remaining residue once with acetic acid glacial, centrifuge and descant liquid, and then wash them 3–4 times with water. The remaining organic material in the test tube is finally mixed with glycerine and transferred, using pipettes, into small closable plastic test tubes. Test tubes are labelled accordingly.

The Single-Grain Method

A combined method for the investigation of fossil pollen grains was initiated by Daghlian (1982), suggesting that the same individual fossil grain should/could be observed in LM, SEM, and even TEM. This idea of how to properly investigate fossil pollen grains in a taxonomically valid way was taken further by Zetter (1989) who evolved a relatively easy method to investigate the same single fossil grain using the so-called "single-grain method," also described in Ferguson et al. (2007). To apply this method the following equipment and tools are necessary: samples prepared in the way described above, narrow glass-pipettes (see below, Fig. 24), teasing needle with an attached human nasal hair (see below, Fig. 25), an erect image compound microscope with a photographing unit, 10 and/or 20× objective lens with a minimum 10 mm working distance, glass slides, ethanol absolute, SEM stubs, sputter coater, and a functional scanning electron microscope.

Making Glass-Pipettes

It is important to have enough cheap and dispensable glass-pipettes to transfer parts of the sample from the storage tubes onto the glass slides for primary LM investigations. These pipettes are also used to make very small drops of ethanol on the surface of the SEM stubs when transferring pollen from glycerine drops using the micromanipulator (see below

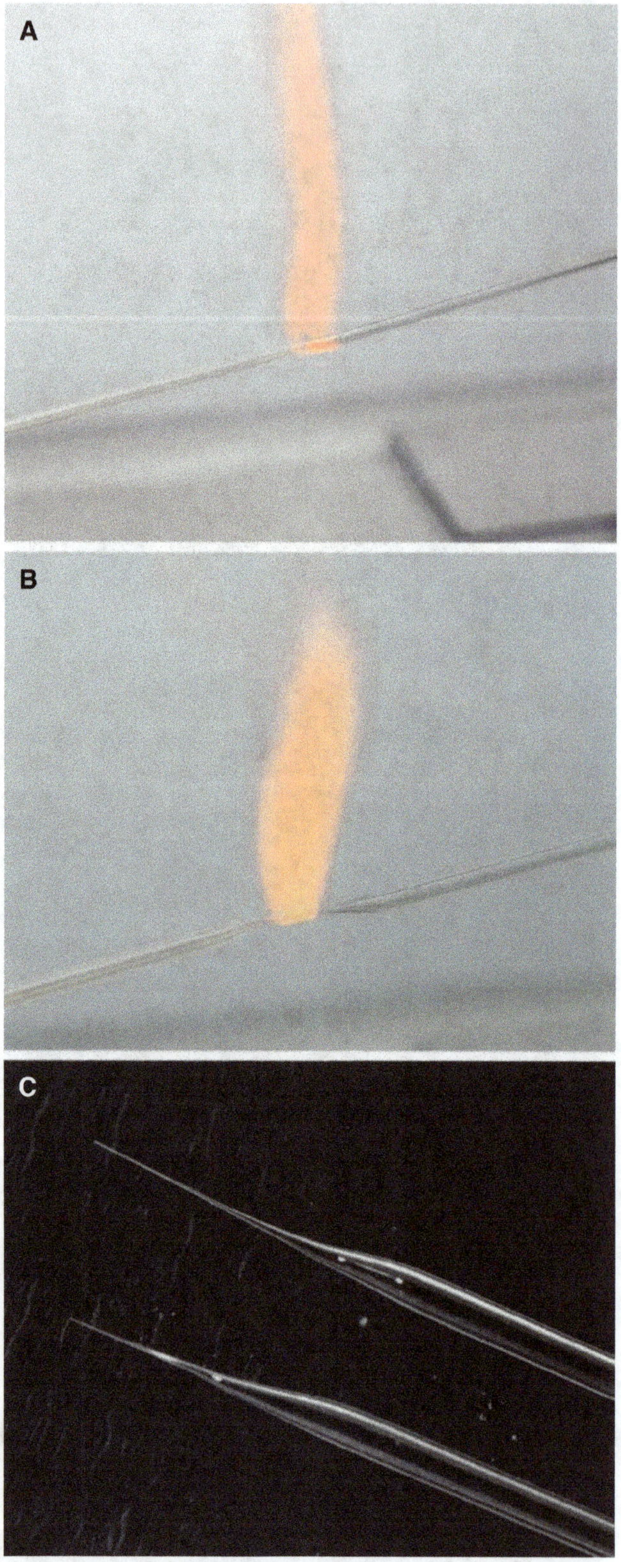

Fig. 24 Making glass-pipettes. A. Glass pipe held in burning gas flame starting to melt. **B.** Melting glass pulled very slowly and gently apart. **C.** Two freshly made pipettes ready for use

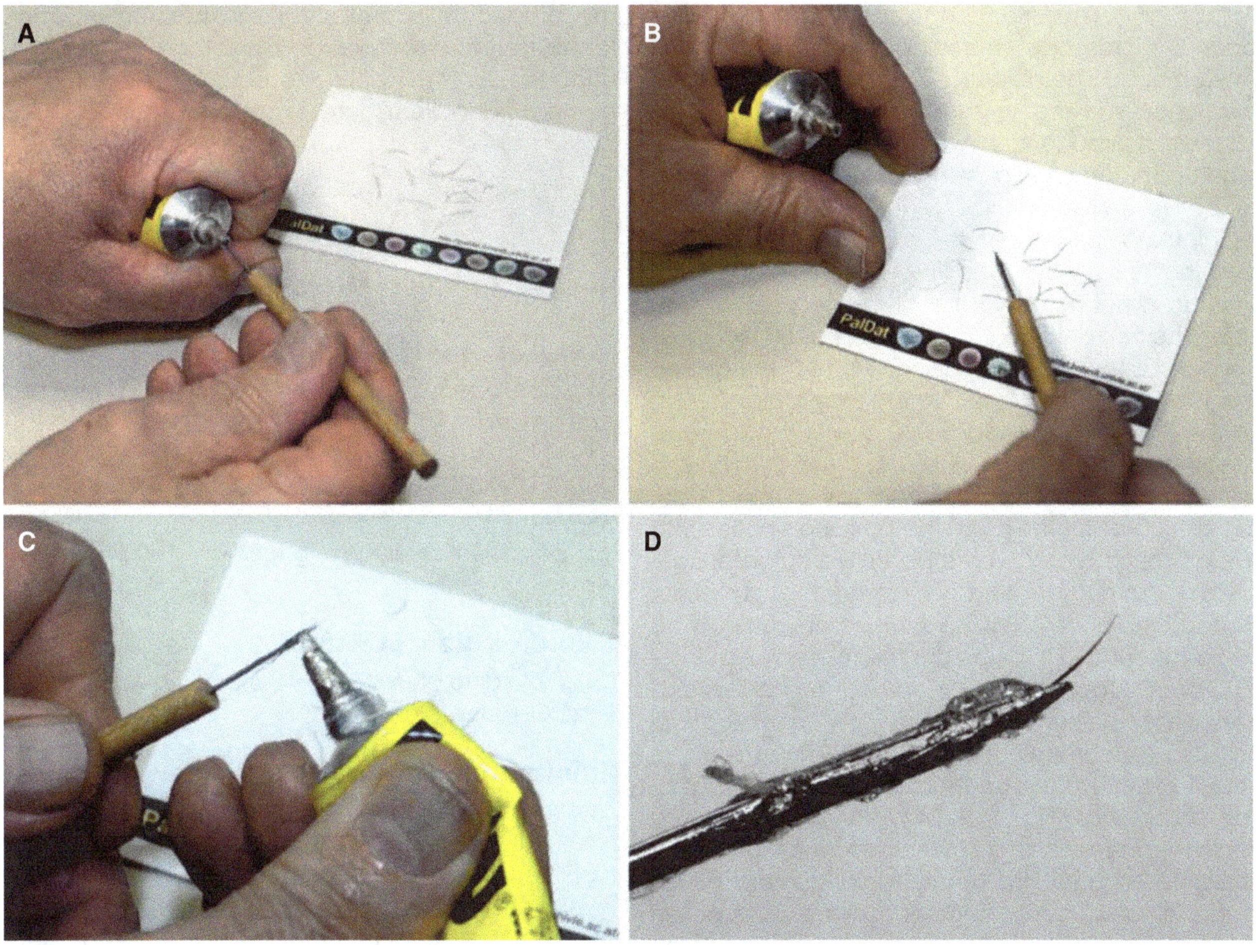

Fig. 25 Producing a micromanipulator. A. Needle pushed ca 1 cm into glue tube and turn in circles. **B.** Needle pressed onto a nasal hair. **C.** Extra glue added around the proximal part of the hair. **D.** Readymade micromanipulator

"Producing a Micromanipulator"). One possibility is to make your own pipettes (Fig. 24) by cutting down 4 mm wide glass pipes (cylinders) into ca 30 cm long units. The middle part of these is then held in/over a burning gas flame. While the glass starts to heat and melt you pull it apart from each end. The pipes will quickly give away in the middle as the glass melts. When pulled apart the glass will form two very long and narrow cones until they finally detach and one holds a perfect pipette in each hand.

Producing a Micromanipulator

The easiest way to make a really good and functional micromanipulator, that can be used to push around and pick out single pollen grains, is to attach a human nasal hair to a teasing needle (Fig. 25). Collect fresh nasal hairs from your professor or senior scientist (avoid the grey and white hairs) and lay them on a sheet of paper. Take a teasing needle and push it ca 1 cm into a glue tube while squeezing gently and turning the needle in circles. Pull out the needle and press onto one of the hairs already laying on the sheet of paper. Make sure that the distal end of the hair is facing the same way as the distal end of the needle and that it extends a few mm longer than the needle. When the hair is attached to the glue, add a little extra glue to cover the proximal part of the hair. Place the needle across the small opening of the glue tube, then press the tube gently for additional glue and at the same time turn the needle in circles while moving it back and forward.

Applying the Single-Grain Method

Use one of the self-made glass-pipettes to stir the sample and blow air through it to mix up the particles real good. Then suck up a tiny portion of the sample using the pipette and transfer onto a glass slide. When the tip of the pipette touches the glass slide drag the pipette along the middle section of the slide (left to right) to produce a long and relatively narrow glycerine strip. Using an erect image compound microscope (meaning when something is moved under the objective lens from left to right it is also seen moving in that same direction when observed through the eyepiece) place the glass slide under the special working distance 10× or 20× objective lens and move the distal end of the micromanipulator in-between the glass slide and the lens and then gently press the tip of the micromanipulator (the nasal hair) into the glycerine (Fig. 26 A-B). Using the micromanipulator grains of particular interest are brushed or pushed to the edge of the glycerine, then out of the glycerine until they are attached to the nasal hair and can be picked up and transferred to another glass slide (Fig. 26 C-H). Have a fresh drop of glycerine ready on a new glass slide. Dip the tip of the hair with the attached pollen into the glycerine drop and the pollen will automatically detach from the hair and rest in the glycerine. Because no cover slip is used this pollen can now be turned around with help from the micromanipulator and photographed in polar and equatorial views as well under different foci (high-, low focus, optical section), documenting important features such as sculpture, apertures, and thickenings or thinnings of the pollen wall (Fig. 27 A-D). After this, the pollen grain is transferred to a SEM stub to which a drop of absolute ethanol has been added to remove all traces of the glycerine from the surface of the pollen grains (Fig. 27 E-G). For this, the best way is to position the light microscope close beside a binocular stereoscope. Place a single SEM stub under the stereoscope and have a small container with fresh ethanol at your side as well as one of the glass-pipettes mentioned above. First pick out a pollen grain with the micromanipulator from the glycerine drop and slowly move over to the stereoscope. Dip the tip of the pipette into the ethanol container and it will automatically suck up a small portion of the ethanol. Press the tip of the pipette on the surface of the SEM stub to leave a tiny drop of ethanol. Then gently press the tip of the nasal hair with the attached pollen into the drop of ethanol and the pollen will be detached from the hair, float a bit in the drop and finally rest on the stub surface when the ethanol evaporates. Try to make the ethanol drops small and close to the center of the SEM stub. Up to 10 different types of grains can be placed on a single stub and additional ethanol drops can be added to clean the glycerine thoroughly off the pollen grains. The stub is then sputter coated with gold and the pollen photographed using a SEM (overviews and close-ups). Pollen of particular interests can be turned. Add a drop of ethanol to the sputtered sample and flip the grain over using the micromanipulator before the ethanol evaporates (under the stereoscope). Re-sputter the sample and photograph it again using the SEM. This applies especially to any kind of heteropolar pollen/spores or tetrads of some sort.

Recipes

Recipes for Light Microscopy (LM)

Acetocarmine (Staining)

30 g acetocarmine + 2 L 45% acetic acid, 4 h boiled and filtered.

Potassium Iodine (Lugol's Iodine, Detection of Starch)

2 g potassium iodine + 1 g iodine + 100 ml distilled water

Toluidine Blue (Staining)

0.1 g Toluidine blue + 100 ml 2.5% sodium carbonate ($NaCO_3$); durable at +4 °C

Chlorination Mixture

Acetic acid (CH_3COOH) + saturated sodium chlorate ($NaClO_3$)* + 3–5 drops hydrochloric acid (conc. HCl)

*Saturated sodium chlorate solution: about 10 g of $NaClO_3$ in 10 ml distilled water (25 °C); the solution is saturated when crystals are still present.

Annotation: solubility of sodium chlorate is depending on the temperature of water.

Acetolysis Mixture

Acetolysis mixture: 9 parts acetic anhydride (99%) are mixed with 1 part concentrated sulfuric acid (96%).

Zinc Bromide Solution (Heavy Liquid Separation for Samples with a High Mineral Content)

250 g zinc bromide (Merck 8.18631.0250) + 25 ml 10% HCl*, mix until all zinc bromide is solved (takes some time!), then add 100 ml distilled water.

*10% HCl-Lösung: 27 ml H_2O + 10 ml HCl (37%) = 37 ml 10% HCl

Recipes for Scanning Electron Microscopy (SEM)

Dimethoxypropane (Dehydration)

30 ml 2,2-dimethoxypropane (DMP) + 1 drop 0.2 n hydrochloric acid (HCl)

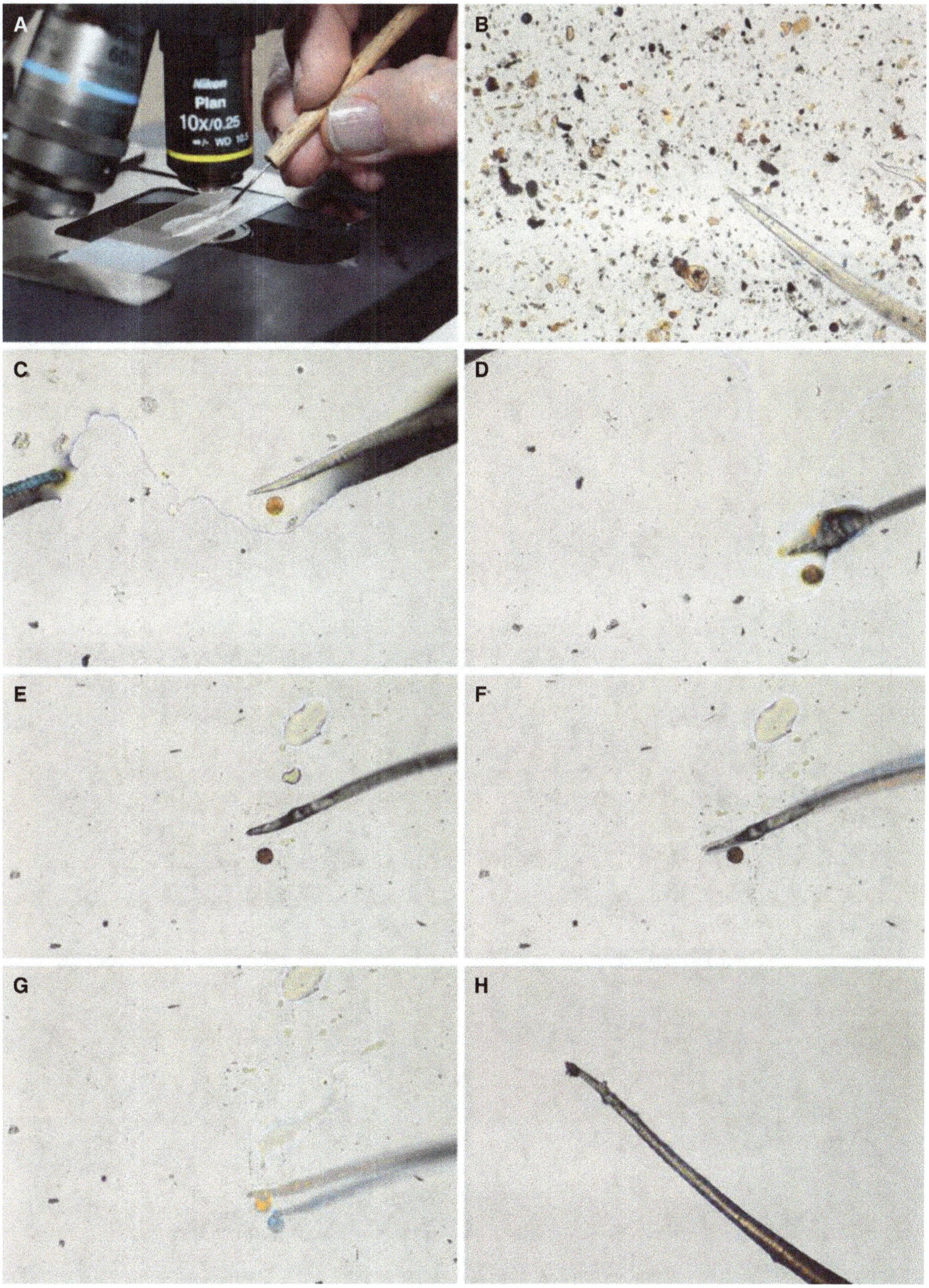

Fig. 26 Applying the single-grain method — Part 1. A. Sample on a glass slide under LM, working distance from sample to objective approx. 1 cm. **B.** Organic-rich sample and the tip of a nasal hair seen through the LM. **C.** Fossil pollen grain being brushed/pushed towards the margin of the glycerine. **D.** Fossil pollen grain pushed further away from the glycerine. **E.** Grain out of glycerine and ready to be picked up by the nasal hair. **F.** Pollen pushed a bit further. **G.** In a pushing or brushing motion the pollen is picked up from the glass slide. **H.** Single fossil pollen grain attached to tip of nasal hair

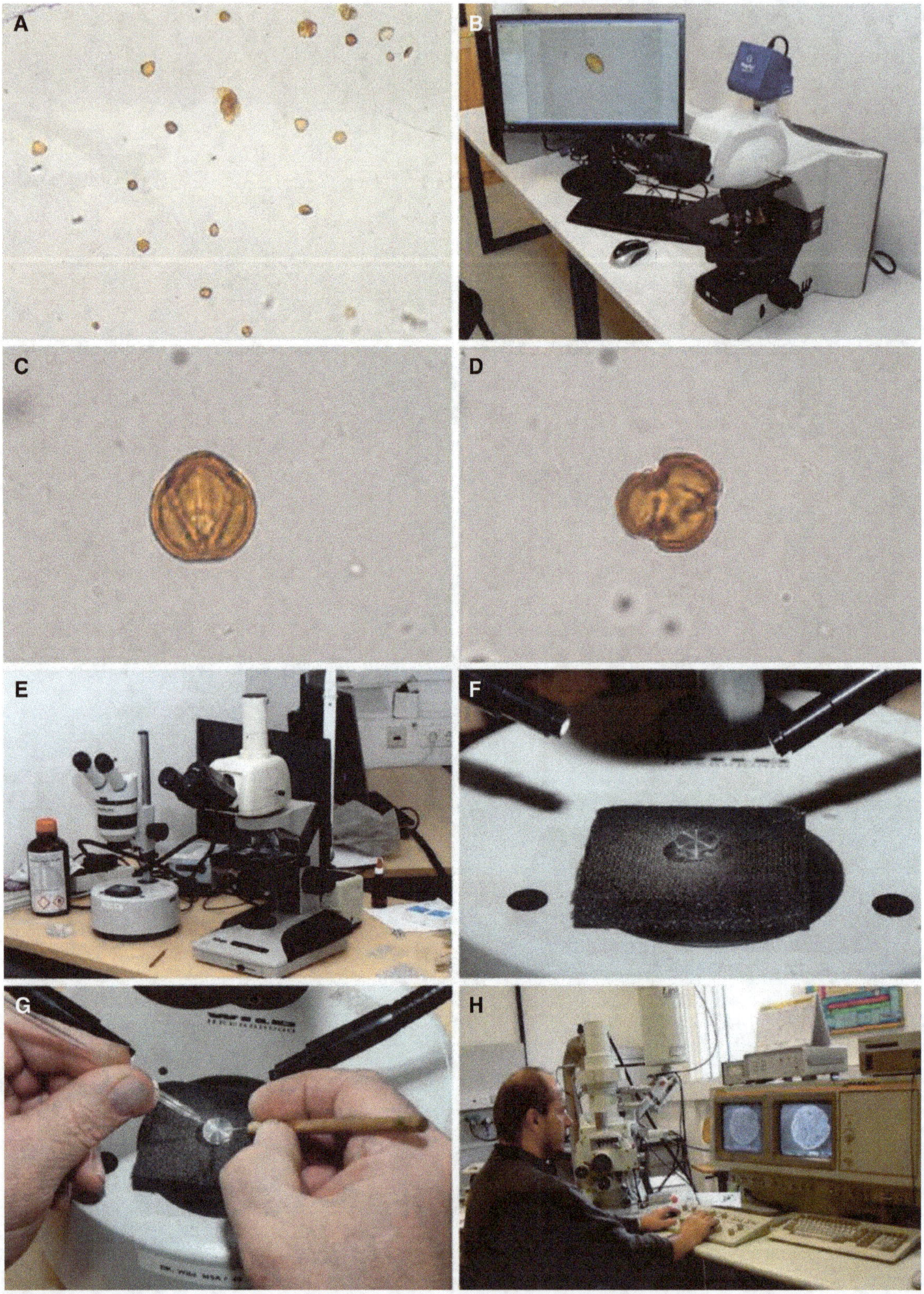

Fig. 27 Applying the single-grain method — Part 2. A. Selected well-preserved pollen grains in a fresh drop of glycerine. **B**. Light microscope equipped with a photographic unit to document pollen grains and their diagnostic features. **C**. Pollen turned and photographed in equatorial view. **D**. Same pollen grain turned and photographed in polar view. **E**. Arrangement of the light microscope and stereomicroscope along with a bottle of ethanol and other tools used when transporting pollen grains from glass slides over to SEM stubs. **F**. Cleaned SEM stub under a stereomicroscope waiting for fossil pollen grains. **G**. How to hold the pipette with the ethanol (left) and the micromanipulator (right) when transferring fossil pollen grains onto SEM stubs. **H**. Photographing fossil pollen using SEM

Recipes for Transmission Electron Microscopy (TEM)

3% Glutaraldehyde (Fixation)

100 ml glutaraldehyde: 12 ml glutaraldehyde (GA, 25%) + 88 ml phosphate buffer (pH 7.2).

1 % Osmium Tetroxide (Fixation)

0.1 g osmium tetroxide (OsO_4) + 10 ml distilled water.

Osmium can be acquired in crystalline form within glass ampullae. The osmium crystals usually adhere inside the ampulla and can be loosened by dipping the ampulla in liquid nitrogen (in a styrofoam box). The ampulla can then be opened and the osmium crystals transferred into distilled water in a vial. Close the vial and seal it with parafilm. For faster dissolution, place the vial in an ultrasonic bath. Mix the osmium solution and pipette it into a vapor-tight bottle. Store it at 6 °C.

*Annotation: osmium is volatile and toxic, use in fume hood only; for storage, use oil with high percentage of unsaturated fatty acids (e.g. corn oil) to bind volatiles of osmium tetroxide (*Fig. 28*).*

Phosphate Buffer pH 7.2 (Fixation)

1 phosphate buffer saline tablet (phosphate buffer saline tablets, $Na_2HPO_4{\cdot}2H_2O$, sodium hydrogen phosphate) + 200 ml distilled water (dispense tablet in ultrasonic bath).

0.8% Potassium Ferrocyanide (Accelerator for Osmium)

0.1 g potassium hexacyanoferrate (II) ($K_4Fe(CN)_6{\cdot}3H_2O$) + 12.5 ml distilled water (dispense in ultrasonic bath).

Annotation: the fresh solution is uncolored and becomes yellow after a few days.

Agar Low Viscosity Resin Kit (Embedding)

LV-resin (Agar Scientific): 48 g LV Resin + 8 g hardener VH1 + 44 g hardener VH2 + 2.5 ml accelerator.

Annotation: Mix the embedding solution in a disposable plastic beaker by using a magnetic stirrer. The first two components must be mixed first before adding the remaining ingredients, then mix well again. The mixture can be used immediately for infiltration and then for embedding. Embedding solution can be stored in a freezer.

Potassium Iodine (Staining)

3 g potassium iodide + 7 g iodine + 100 ml ethanol (92%).

Fig. 28 Osmium storage. A. Osmium solution stored at 6 °C (fridge placed in a fume hood). **B-D.** Osmium solution in a sealed bottle and stored in a plastic container, plastic container placed in glass vessel containing oil. **C.** Arrowheads showing osmium contamination from volatiles. **D.** Second glass vessel placed over the osmium containers, osmium vapor is bound to the oil and cannot escape into the atmosphere

1% Potassium Permanganate (Fixation and Staining)

1% potassium permanganate: 1 g potassium permanganate in 100 ml distilled water

1% Periodic Acid (Staining)

1 g periodic acid (PA, Firma Fluka) + 100 ml distilled water

0.2% Thiocarbohydrazide (Staining)

0.2 g thiocarbohydrazide (TCH, by *Serva*) + 100 ml 20% acetic acid (20 ml 100 % CH_3COOH + 80 ml distilled water)

1% Silver Proteinate (Staining)

0.25 g silver proteinate (SP, by *Merck*) + 25 ml distilled water

Uranyl Acetate (Staining)

Prefabricated solution: "Ultrostain 1" by Leica

Lead Citrate (Staining)

Prefabricated solution: "Ultrostain 2" by Leica; used with potassium hydroxide pellets

Formvar Filming Solution (Film-Coated Grids)

2 g formvar (15/45 E) + 100 ml chloroform (pure); mix with a magnetic stirrer

References

Brown CA (1960) Palynological techniques. Lousiana State University, Baton Rouge, La

Daghlian CP (1982) A simple method for combined light, scanning and transmission electron microscope observation of single pollen grains from dispersed pollen samples. Pollen Spores 24: 537–545

Erdtman G (1943) An introduction to pollen analysis. Chronica Botanica, Waltham, Mass

Erdtman G (1960) The acetolysis method. Svensk Bot Tidskr 54: 561–564

Eyring MB (1996) Soil pollen analysis from a forensic point of view. Microscope 44: 81–97

Fægri K, Iversen J (1989) Textbook of Pollen analysis. 4th edition, John Wiley & Sons, Chichester

Ferguson DF, Zetter R, Paudayal KN (2007) The need for the SEM in paleopalynology. C R Palevol 6: 423–430

Gerlach D (1984) Botanische Mikrotechnik. 3rd edition, Thieme, Stuttgart

Halbritter H (1998) Preparing living pollen material for scanning electron microscopy using 2,2-dimethoxypropane (DMP) and critical-point drying. Biotech Histochem 73: 137–143

Hayat MA (2000) Principles and techniques of electron microscopy: Biological applications. Cambridge University Press, Cambridge

Heslop-Harrison J(1979) An interpretation of the hydrodynamics of pollen. Am J Bot 66: 737–743

Moore PD, Webb JA, Collinson ME (1991) Pollen analysis. 2nd edition. Blackwell Scientific Publication, Oxford

Nepi M, Franchi GG, Pacini E (2001) Pollen hydration status at dispersal: cytophysiological features and strategies. Protoplasma 216: 171–180

Rowley JR, Dahl AO (1977) Pollen development in *Artemisia vulgaris* with special reference to Glycocalyx material. Pollen Spores 19: 169–284

Siegel I (1967) Toluidine blue O and naphthol yellow S; a highly polychromatic general stain. Stain Technol 42: 29–30

Thiéry J-P (1967) Mise en évidence des polysaccharides sur coupes fines en microscopie électronique. J Microscopie 6: 987–1018

Traverse A (2007) Paleopalynology. 2nd ed, Springer, Dordrecht

Weber M (1992) Nature and distribution of the exine-held material in mature pollen grains of *Apium nodiflorum* L. (Apiaceae). Grana 31: 17–24

Weber M, Frosch A (1995) The development of the transmitting tract in the pistil of *Haquetia epipactis* (Apiaceae). Int J Plant Sci 156: 615–621

Weber M, Halbritter H (2007) Exploding pollen in *Montrichardia arborescens* (Araceae). Plant Syst Evol 263: 51–57

Weber M, Ulrich S (2010) The endexine: a frequently overlooked pollen wall layer and a simple method for detection. Grana 49: 83–90

Wood GD, Gabriel AM, Lawson JC (1996) Chapter 3. Palynological techniques – processing and microscopy. In: Jansonius J, McGregor DC (eds) Palynology: principles and applications. American Association of Stratigraphy Palynologists Foundation, Vol. 1. Publishers Press, Salt Lake City, Utah, USA, p. 29-50

Zetter R (1989) Methodik und Bedeutung einer routinemäßig kombinierten lichtmikroskopischen und rasterelektronenmikroskopischen Untersuchung fossiler Mikrofloren. Cour Forsch-Inst Senckenberg 109: 41–50

Illustrated Pollen Terms

This part is divided into 6 topic related chapters: "Pollen- and Dispersal Units," "Shape and Polarity," "Aperture," "Ornamentation," "Pollen Wall," and "Pollen Class." Terms are either morphologically or alphabetically grouped depending on practical use. When a term is illustrated by numerous images (one or more plates), the definition of the term occurs along with the first image. Features are often highlighted (colored) for easy recognition. Each image is accompanied by the name of the plant species illustrated, the current family name, and a short relevant description. The majority of the micrographs are SEM pictures, but include also LM and TEM. The SEM micrographs usually represent the turgescent (hydrated) state of recent pollen, but they can also be in dry condition or fossilized. The LM micrographs usually show acetolyzed pollen, but pollen grains can also be hydrated in water, glycerine, or stained with biological stains. Exceptions from the standard method (LM, SEM) are specified in the picture legend. For TEM micrographs the staining method is provided when necessary.

Contents

Pollen- and Dispersal Units

H. Halbritter et al., *Illustrated Pollen Terminology*, https://doi.org/10.1007/978-3-319-71365-6_7

monad

unit consisting of a single pollen grain

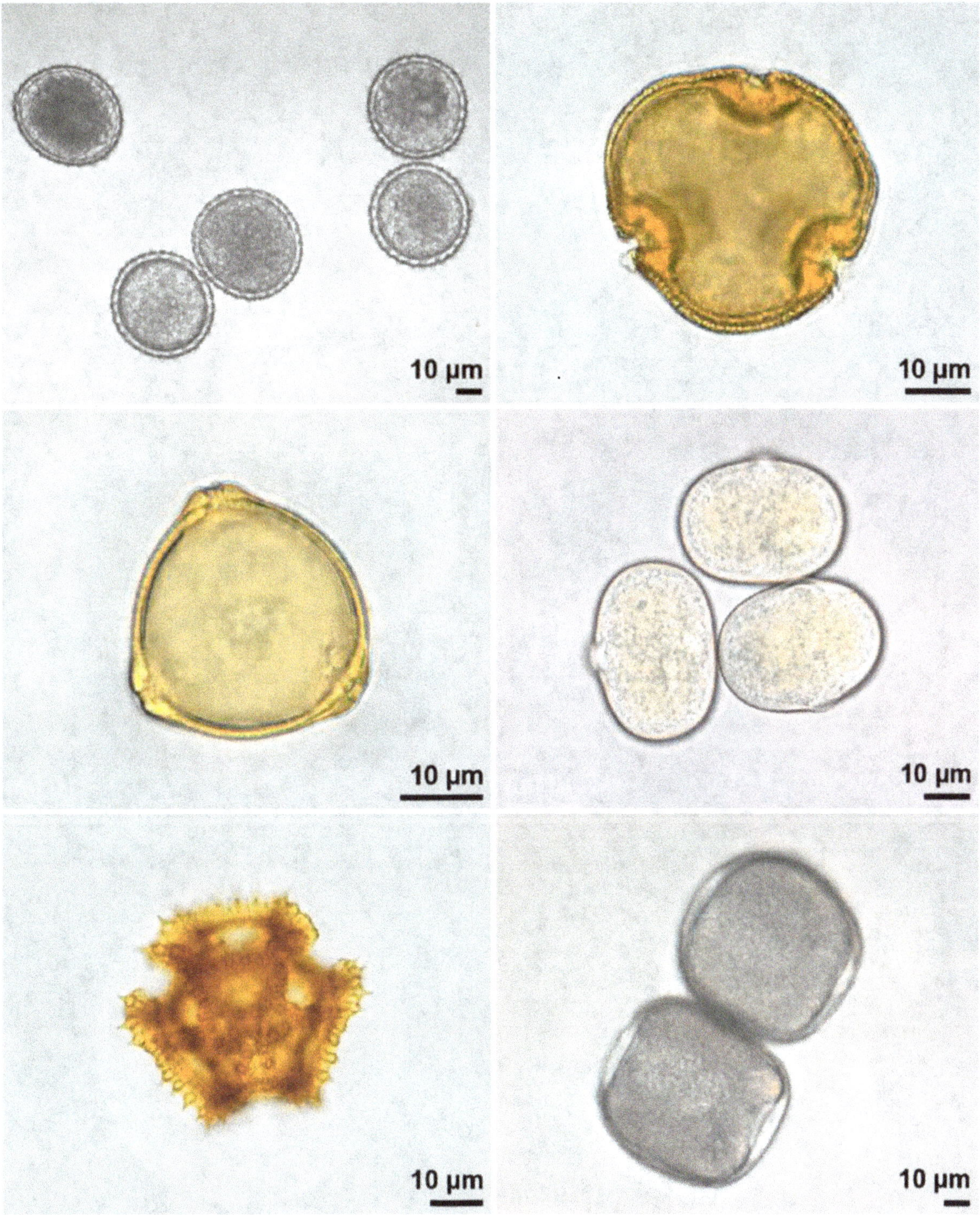

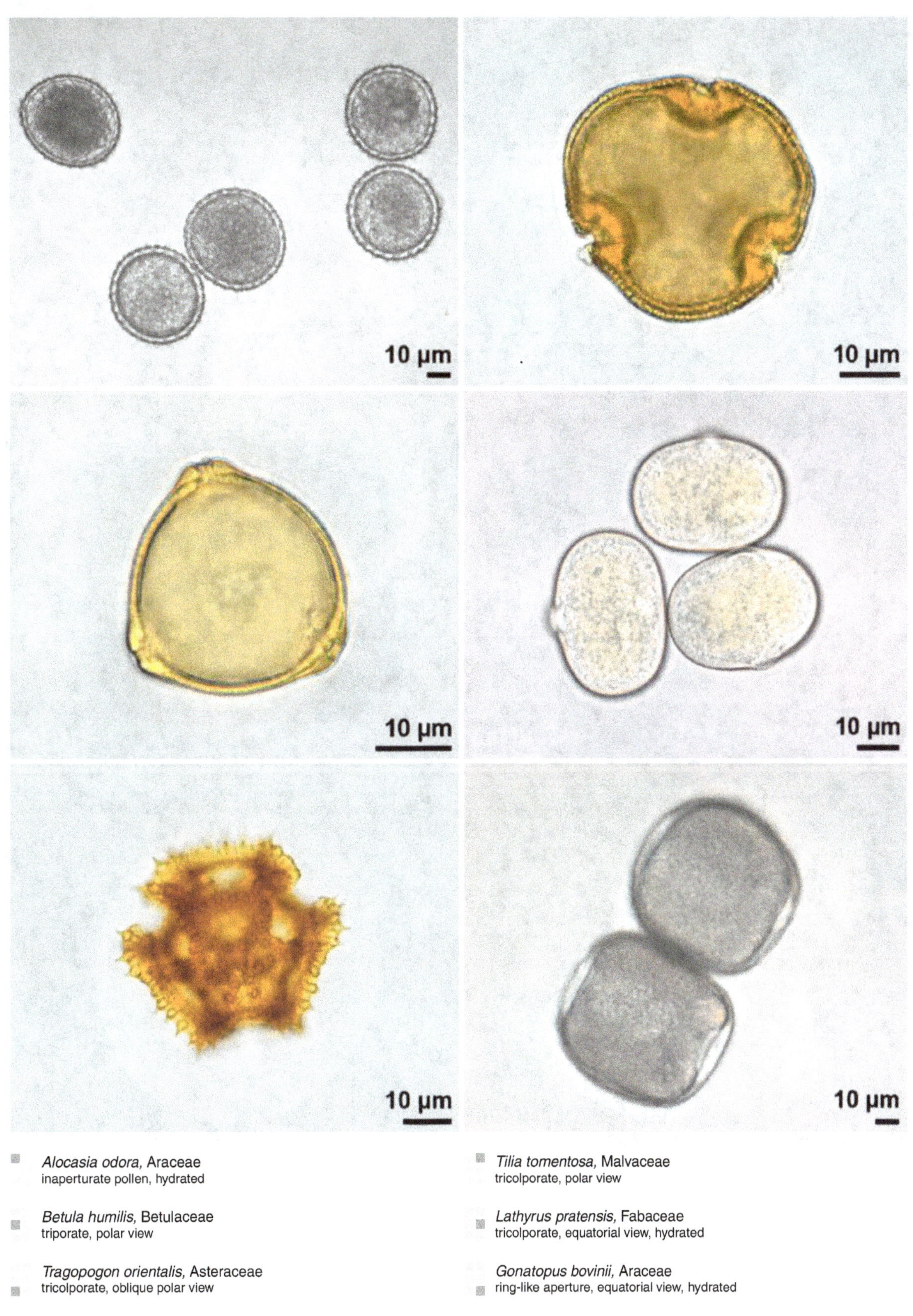

- *Alocasia odora,* Araceae
  inaperturate pollen, hydrated
- *Betula humilis,* Betulaceae
  triporate, polar view
- *Tragopogon orientalis,* Asteraceae
  tricolporate, oblique polar view
- *Tilia tomentosa,* Malvaceae
  tricolporate, polar view
- *Lathyrus pratensis,* Fabaceae
  tricolporate, equatorial view, hydrated
- *Gonatopus bovinii,* Araceae
  ring-like aperture, equatorial view, hydrated

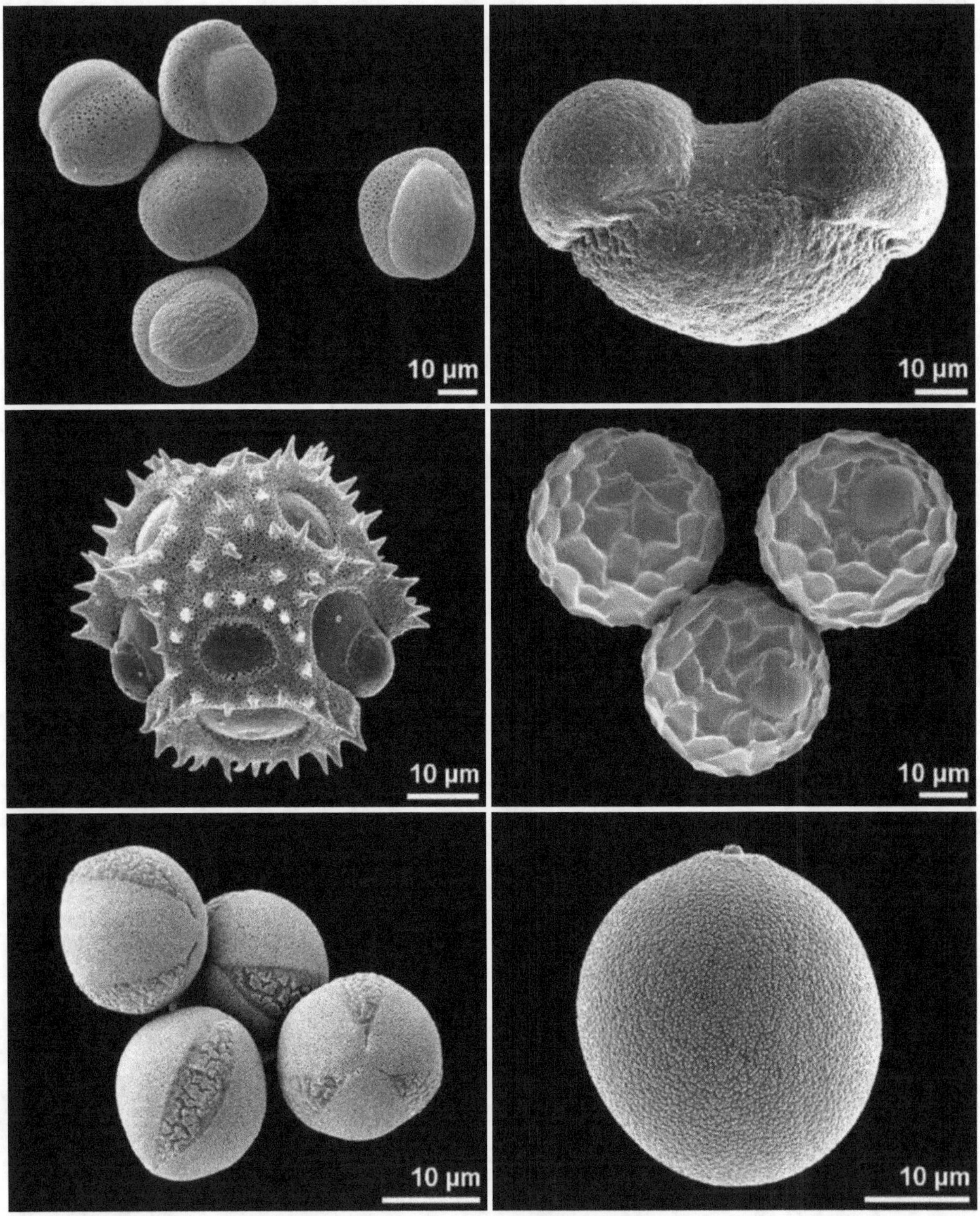

Haworthia herbacea, Xanthorrhoeaceae
sulcate

Tragopogon orientalis, Asteraceae
tricolporate, oblique polar view

Lamium galeobdolon, Lamiaceae
tricolpate

Pinus strobus, Pinaceae
bisaccate, equatorial view

Bituminaria bituminosa, Fabaceae
tricolporate, dry pollen

Dactylis glomerata, Poaceae
ulcerate, equatorial view

dyad

unit of two pollen grains

Polypleurum stylosum, Podostemaceae

Polypleurum stylosum, Podostemaceae
pollen collapsed

Zeylanidium olivaceum, Podostemaceae
equatorial view

Zeylanidium subulatum, Podostemaceae

Thelethylax minutiflora, Podostemaceae
equatorial view

Scheuchzeria palustris, Scheuchzeriaceae

pseudomonad

unit of a permanent tetrad with three rudimentary pollen grains

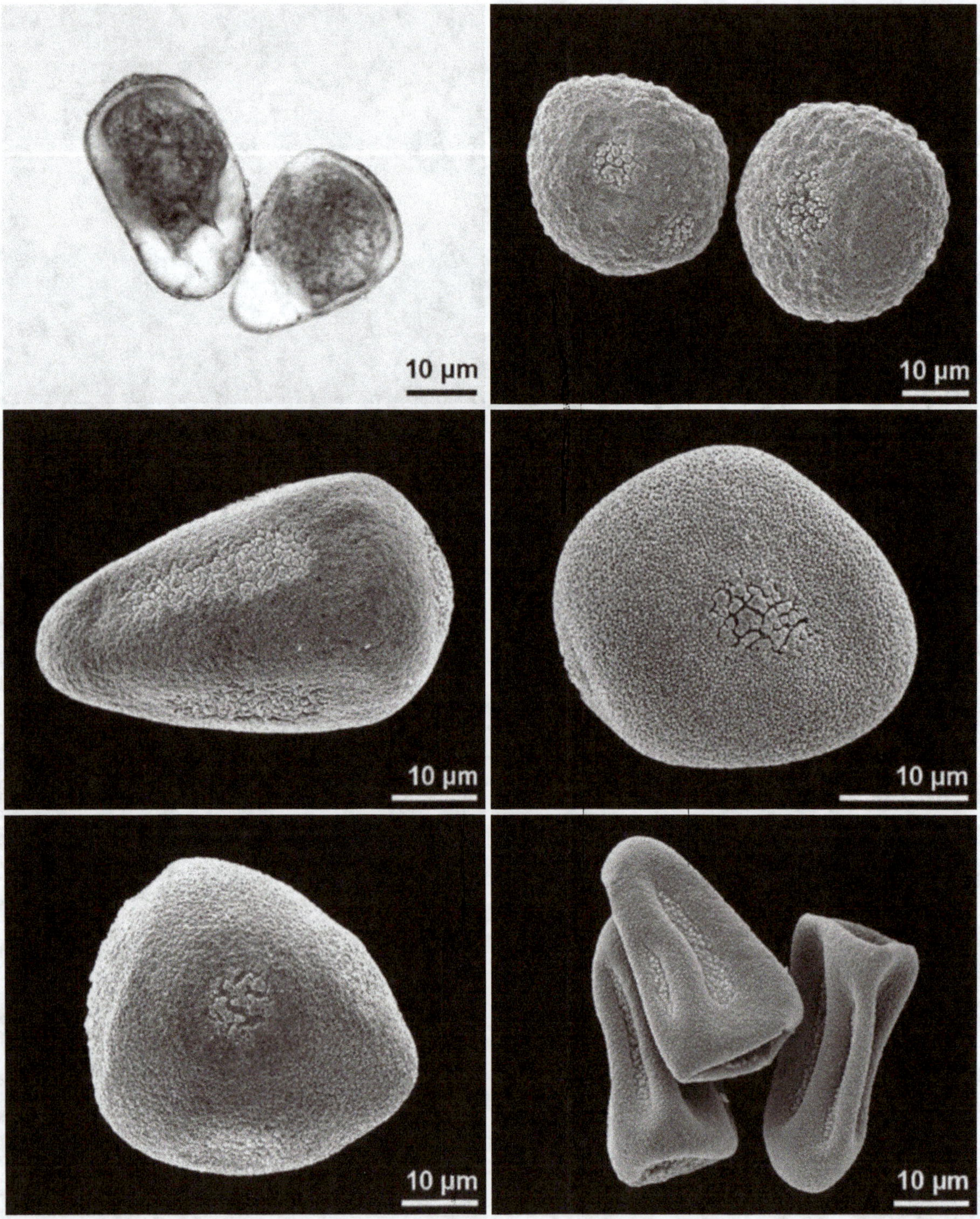

Carex sp., Cyperaceae
hydrated pseudomonads, note remnants of aborted pollen at proximal pole (white zone)

Carex atrata, Cyperaceae

Schoenoplectus lacustris, Cyperaceae
equatorial view

Scirpus sylvaticus, Cyperaceae
equatorial view

Carex distans, Cyperaceae
equatorial view

Cyperus conglomeratus, Cyperaceae
dry pollen

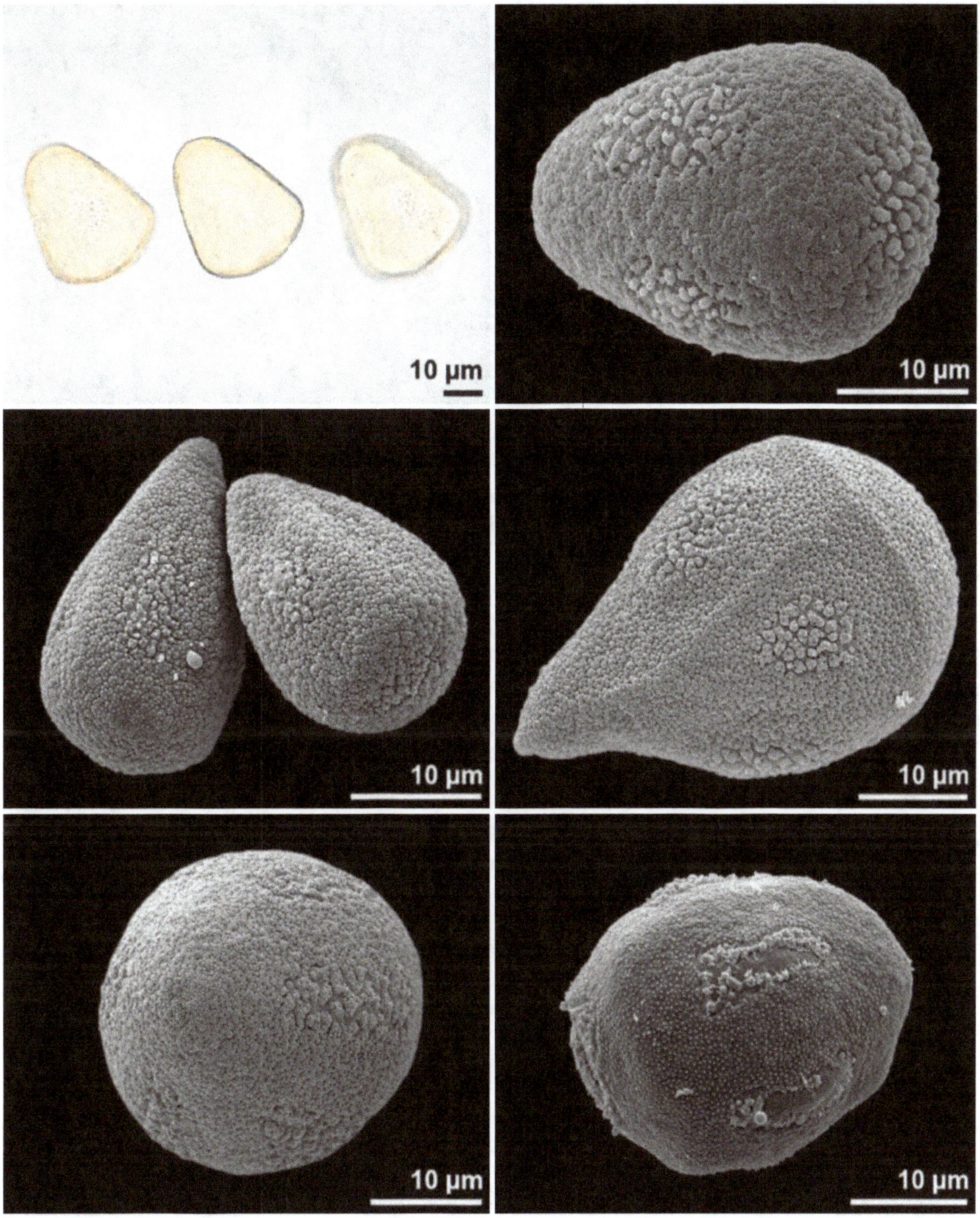

Carex digitata, Cyperaceae
pollen at different focal depths

Cyperus diffusus, Cyperaceae
equatorial view

Eleocharis palustris, Cyperaceae
proximal polar view

Trichophorum cespitosum, Cyperaceae
equatorial view

Scirpoides holoschoenus, Cyperaceae
equatorial view

Cyperus papyrus, Cyperaceae
equatorial view

tetrad

unit of four pollen grains

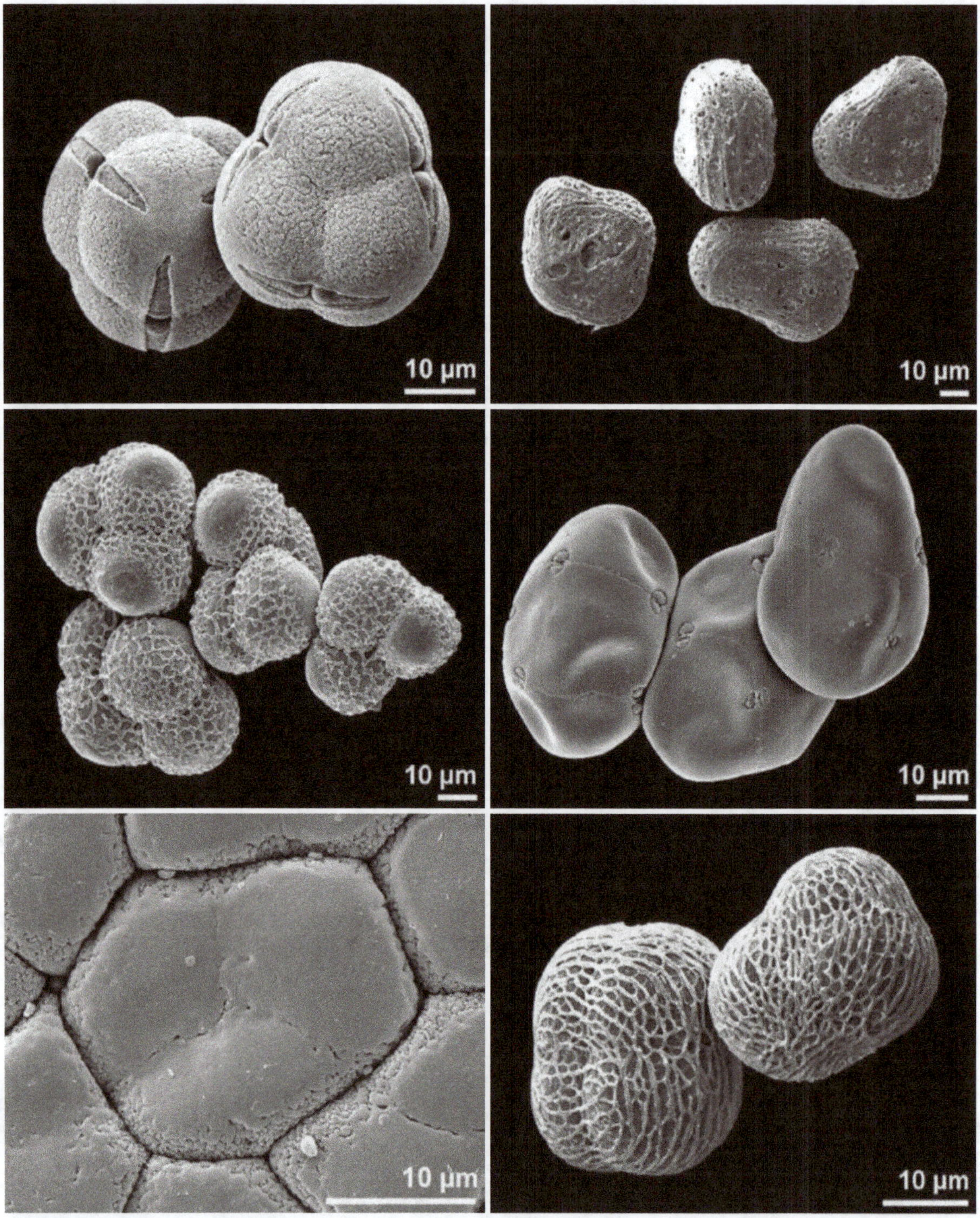

Erica herbacea, Ericaceae
tetrahedral tetrads

Epipactis helleborine, Orchidaceae
decussate tetrads

Epidendrum centropetalum, Orchidaceae
tetrad, part of pollinium

Chlorospatha kolbii, Araceae
differnt planar tetrads

Cyprinia gracilis, Apocynaceae
planar tetrads, dry pollen

Xanthosoma ceronii, Araceae
tetrad planar (left) and decussate (right)

tetrad tetrahedral

unit of four pollen grains in which the centers of the grains define a tetrahedron

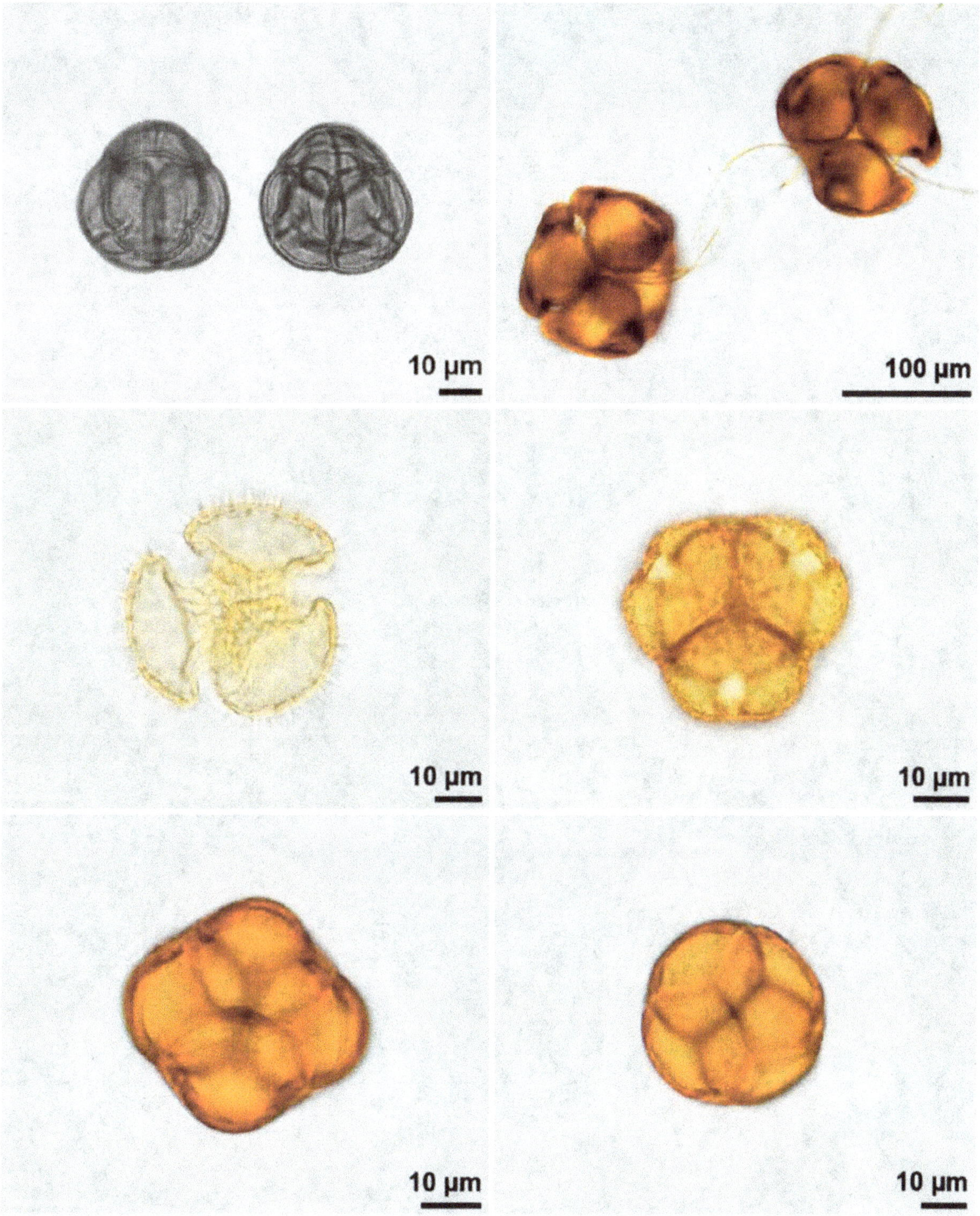

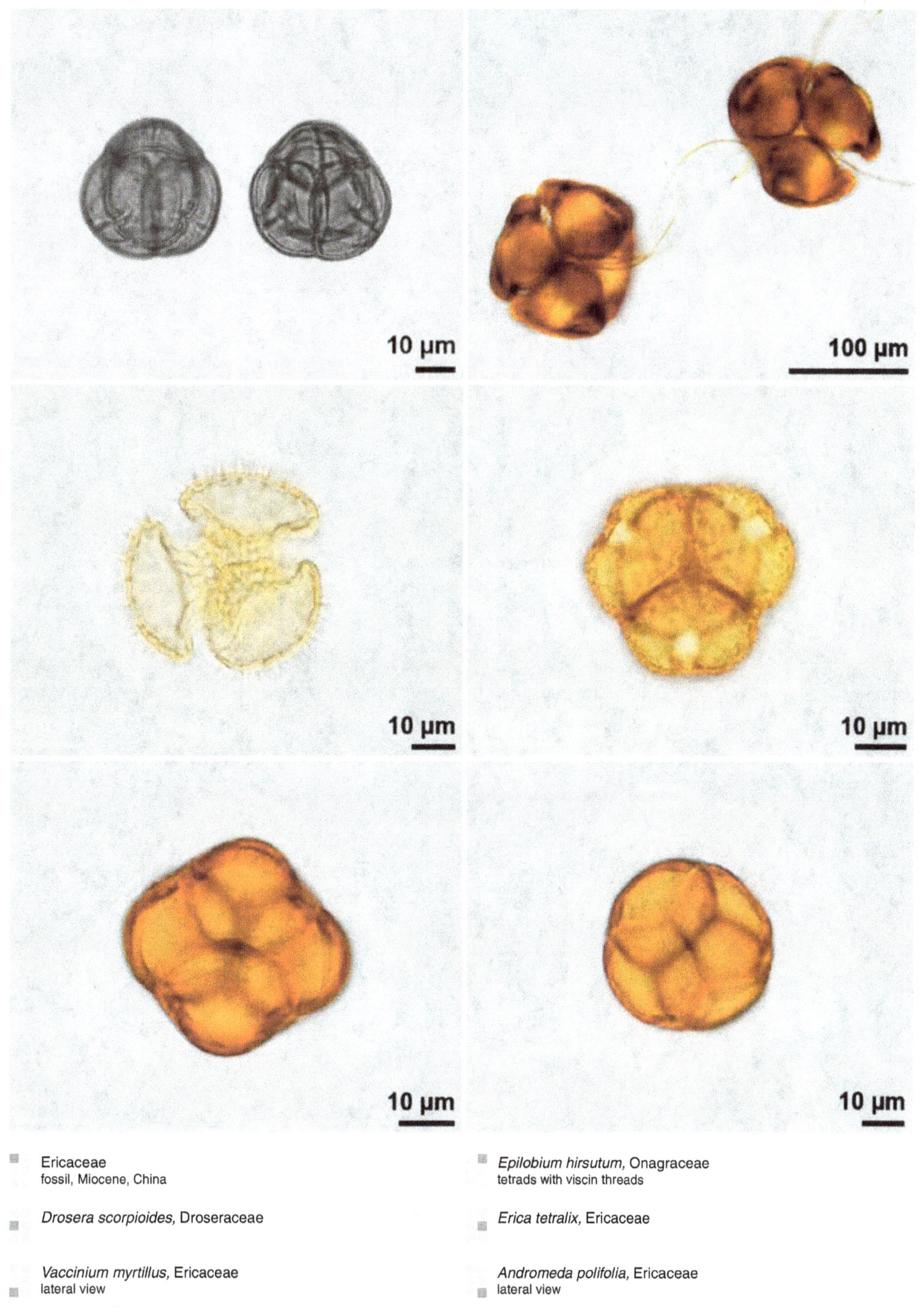

Ericaceae
fossil, Miocene, China

Epilobium hirsutum, Onagraceae
tetrads with viscin threads

Drosera scorpioides, Droseraceae

Erica tetralix, Ericaceae

Vaccinium myrtillus, Ericaceae
lateral view

Andromeda polifolia, Ericaceae
lateral view

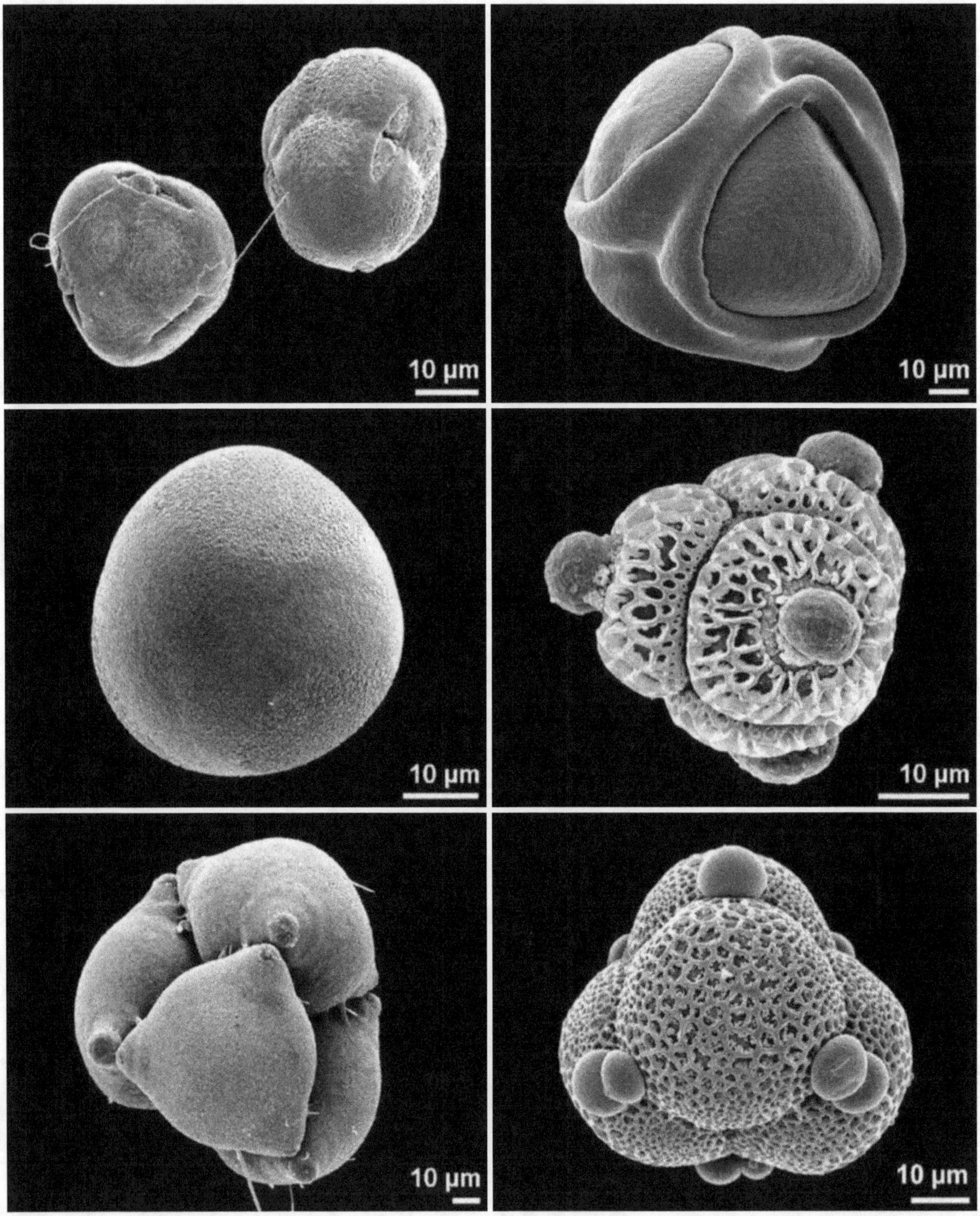

Rhododendron hirsutum, Ericaceae
tetrads with viscin threads

Victoria regia, Nymphaeaceae
dry pollen

Luzula campestris, Juncaceae
ulcerate pollen

Drimys granatensis, Winteraceae
ulcerate pollen

Epilobium montanum, Onagraceae
viscin threads

Oxyanthus subpunctatus, Rubiaceae
apical view

Epilobium parviflorum, Onagraceae
tetrads with viscin threads, dry pollen

Chelonanthus alatus, Gentianaceae
basal view

Ludwigia octovalvis, Onagraceae
pollen with viscin threads

Arbutus unedo, Ericaceae

Mimosa pudica, Mimosaceae

Agapetes macrantha, Ericaceae
tetrads with aborted monad ("pseudotriads")

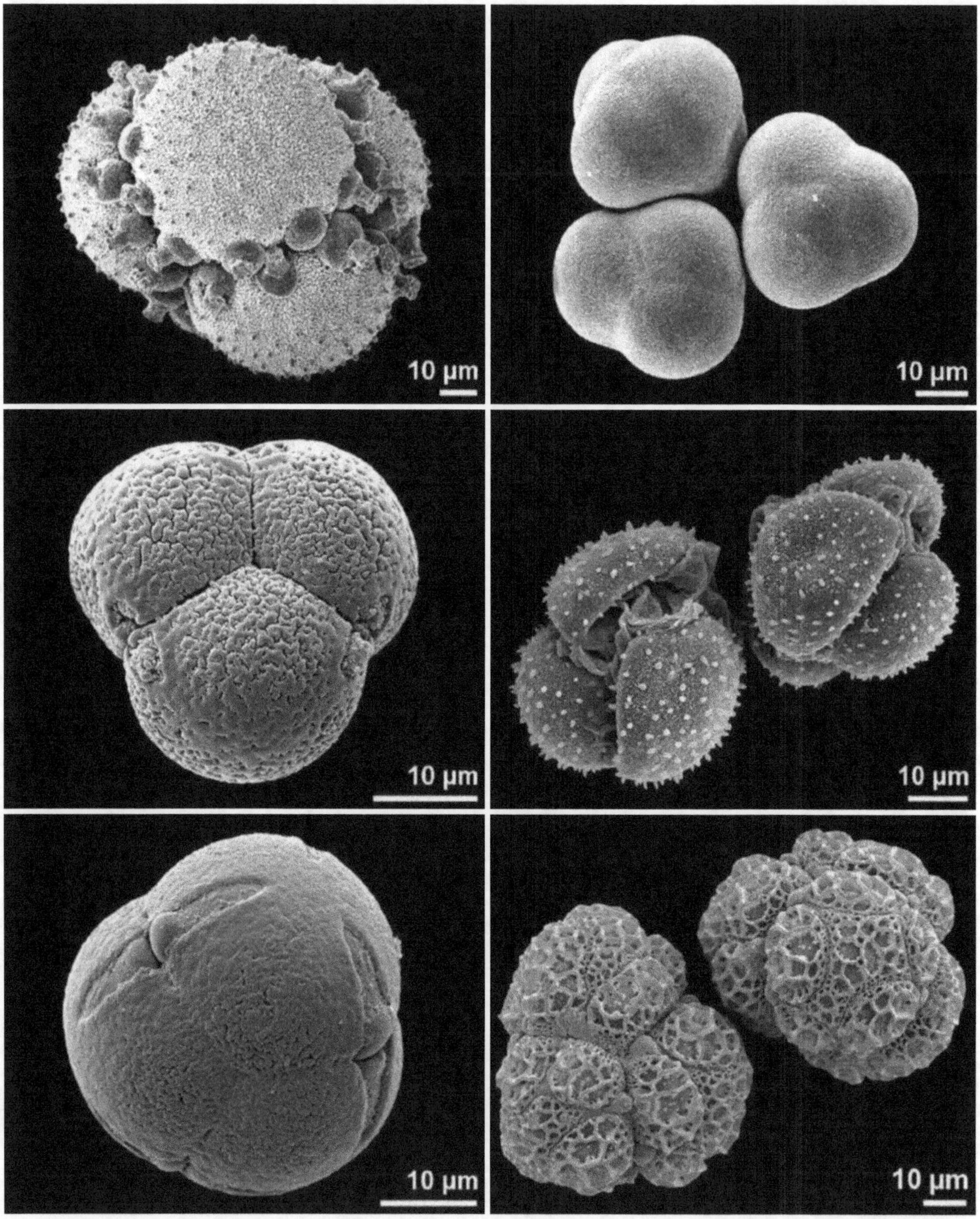

Dionaea muscipula, Droseraceae

Moneses uniflora, Ericaceae
basal view

Vaccinium vitis-idaea, Ericaceae

Juncus effusus, Juncaceae
ulcerate pollen

Drosera binata, Droseraceae
dry pollen

Calolisianthus pendulus, Gentianaceae

tetrad decussate

unit of four pollen grains arranged in two pairs in two different plains

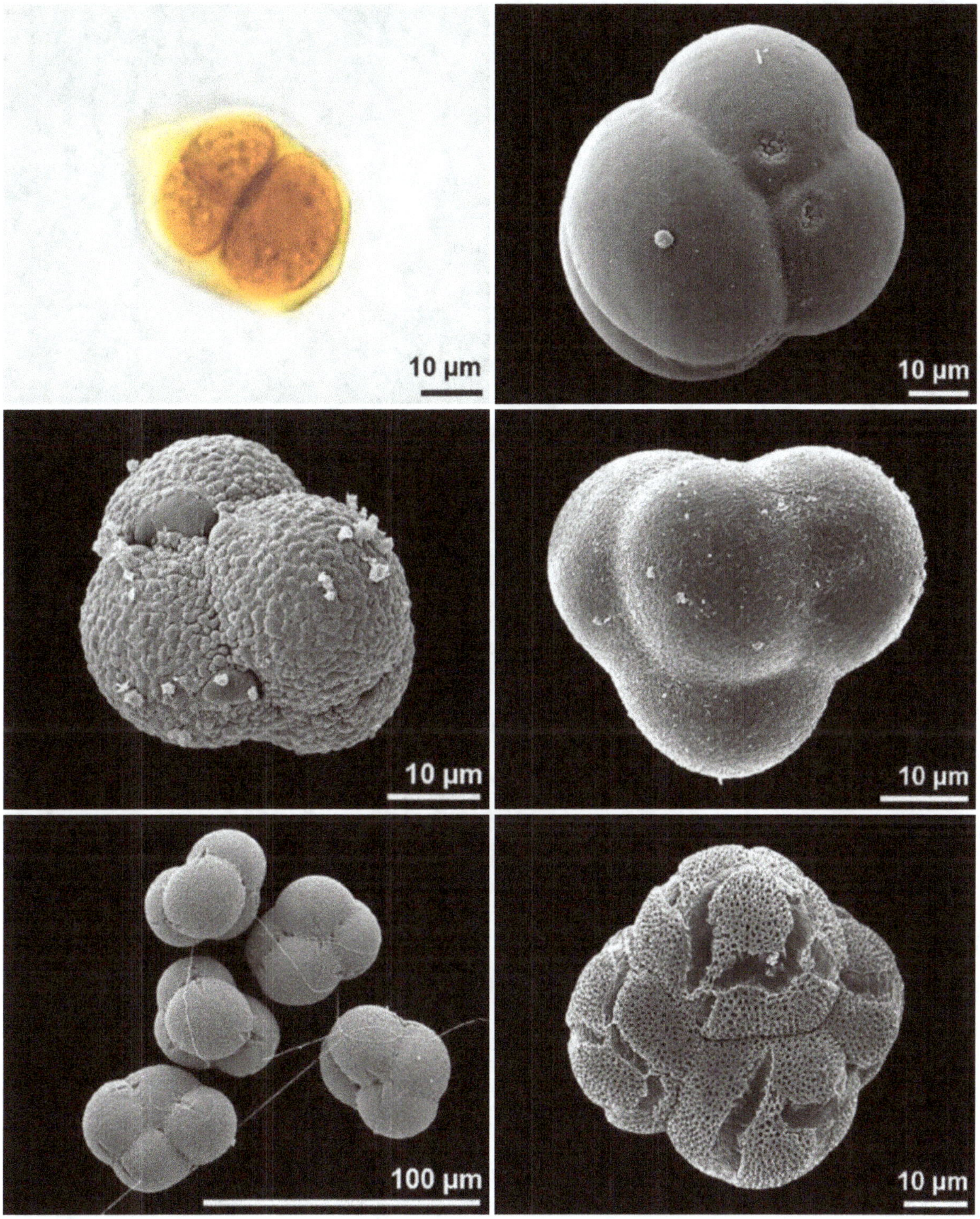

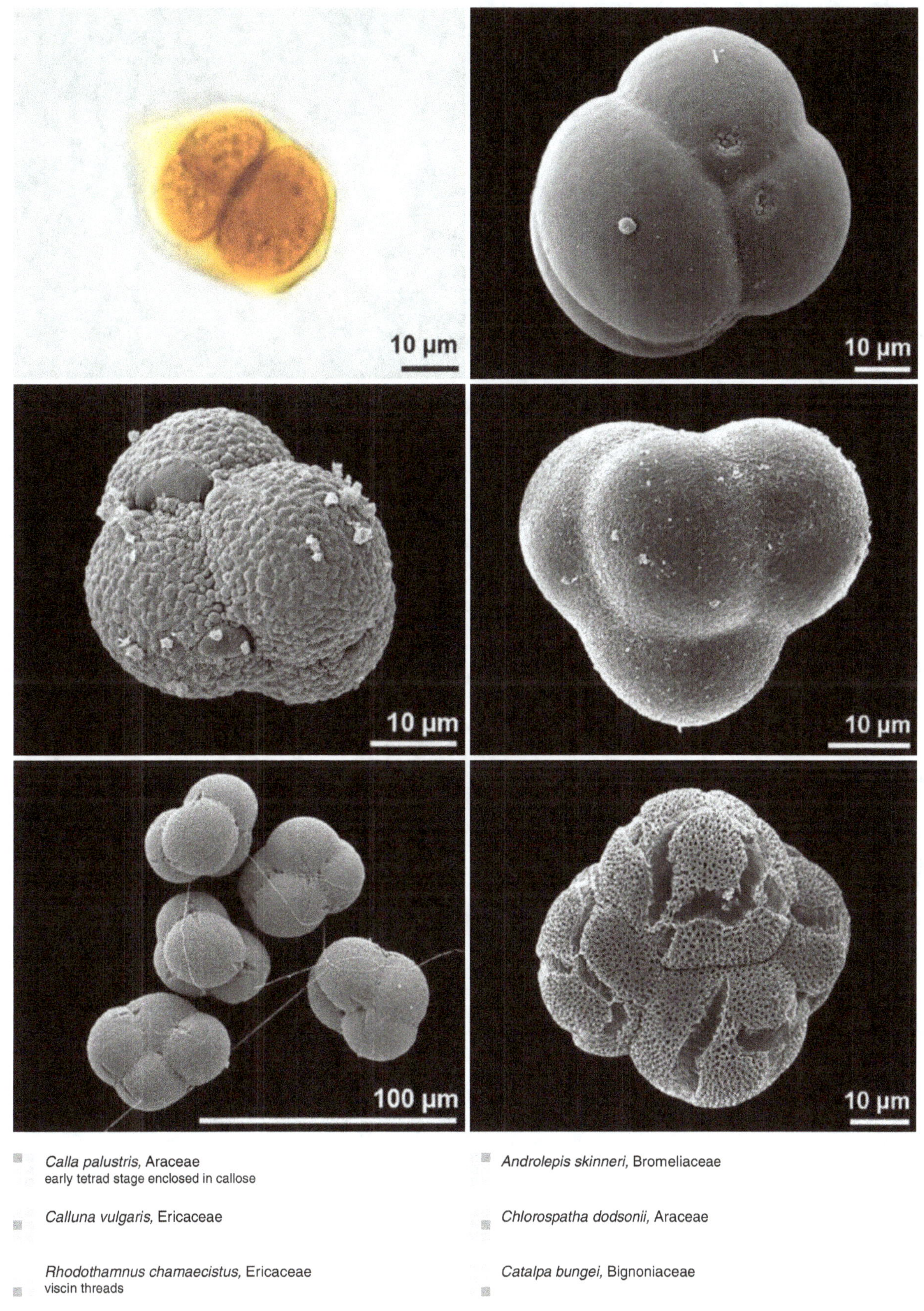

Calla palustris, Araceae
early tetrad stage enclosed in callose

Androlepis skinneri, Bromeliaceae

Calluna vulgaris, Ericaceae

Chlorospatha dodsonii, Araceae

Rhodothamnus chamaecistus, Ericaceae
viscin threads

Catalpa bungei, Bignoniaceae

tetrad planar

unit of four pollen grains arranged in one plane: tetragonal, T-shaped, linear

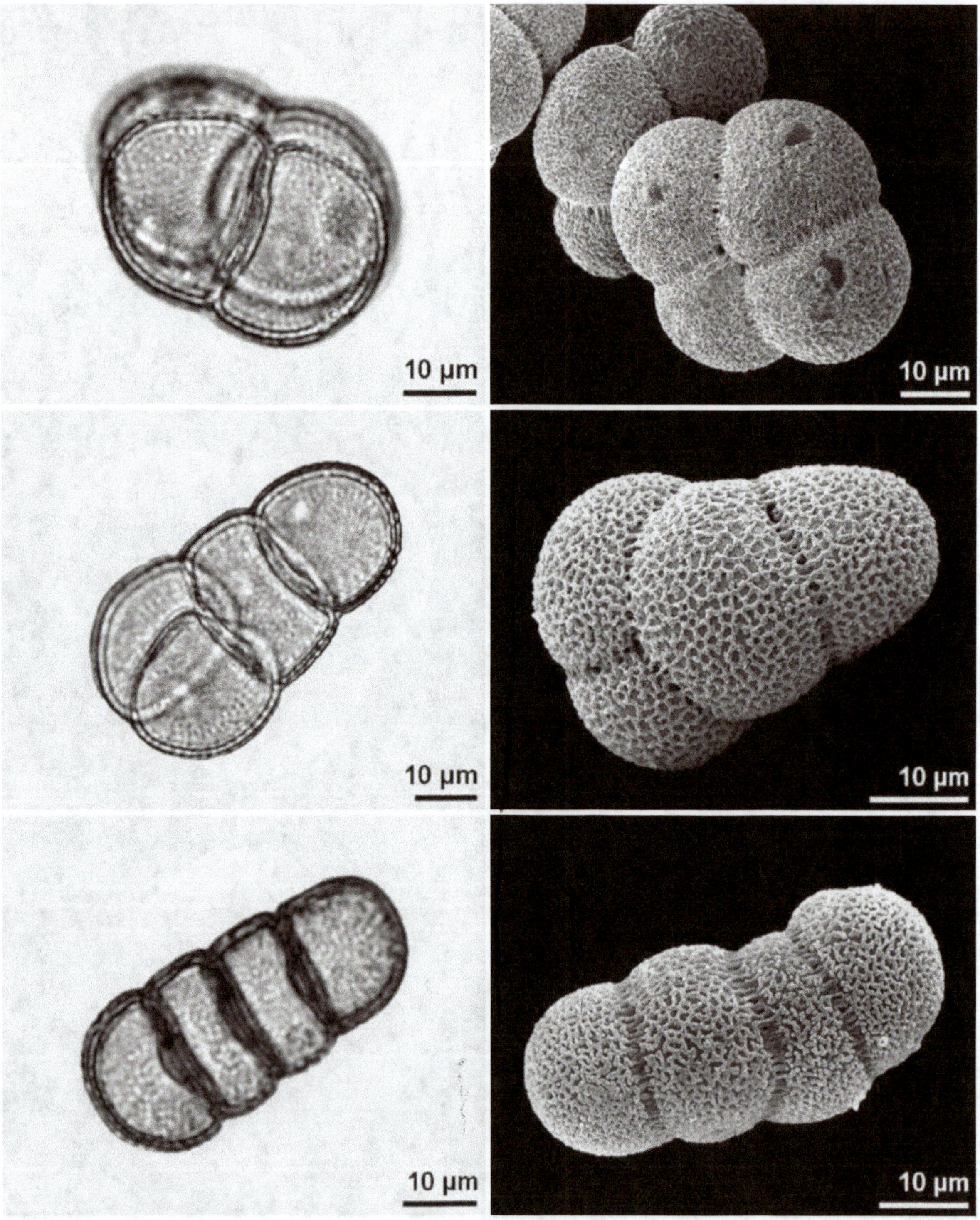

Typha latifolia, Typhaceae
tetrad tetragonal

Typha latifolia, Typhaceae
tetrad T-shaped

Typha latifolia, Typhaceae
tetrad linear

Typha latifolia, Typhaceae
tetrad tetragonal

Typha latifolia, Typhaceae
tetrad T-shaped

Typha latifolia, Typhaceae
tetrad linear

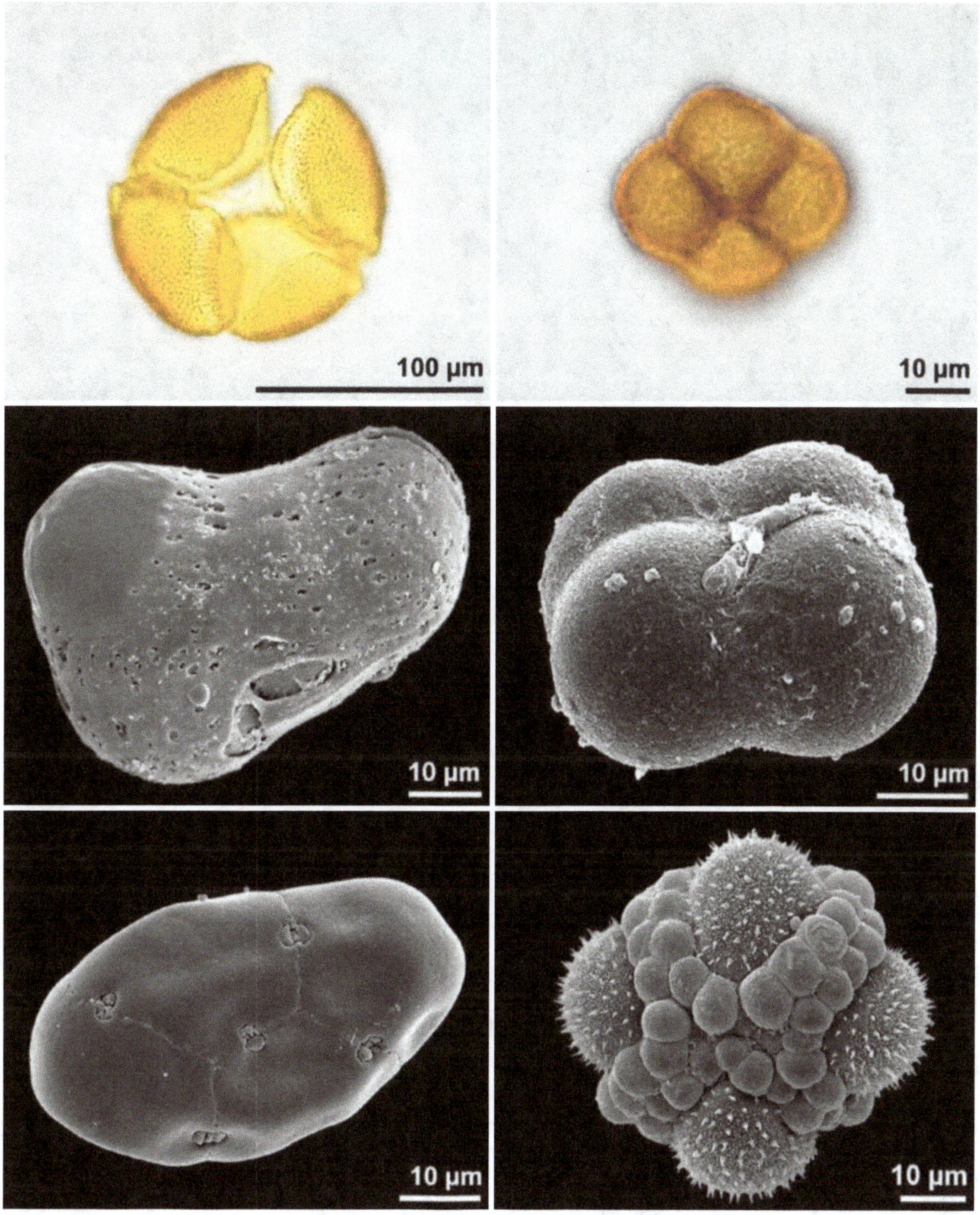

Asimina triloba, Annonaceae
tetrad tetragonal

Calluna vulgaris, Ericaceae
tetrad tetragonal

Chlorospatha kolbii, Araceae
tetrad T-shaped

Chlorospatha dodsonii, Araceae
tetrad tetragonal

Cyprinia gracilis, Apocynaceae
tetrad planar, dry pollen

Drosera peltata, Droseraceae
tetrad tetragonal

polyad

unit of more than four pollen grains (multiple of 4)

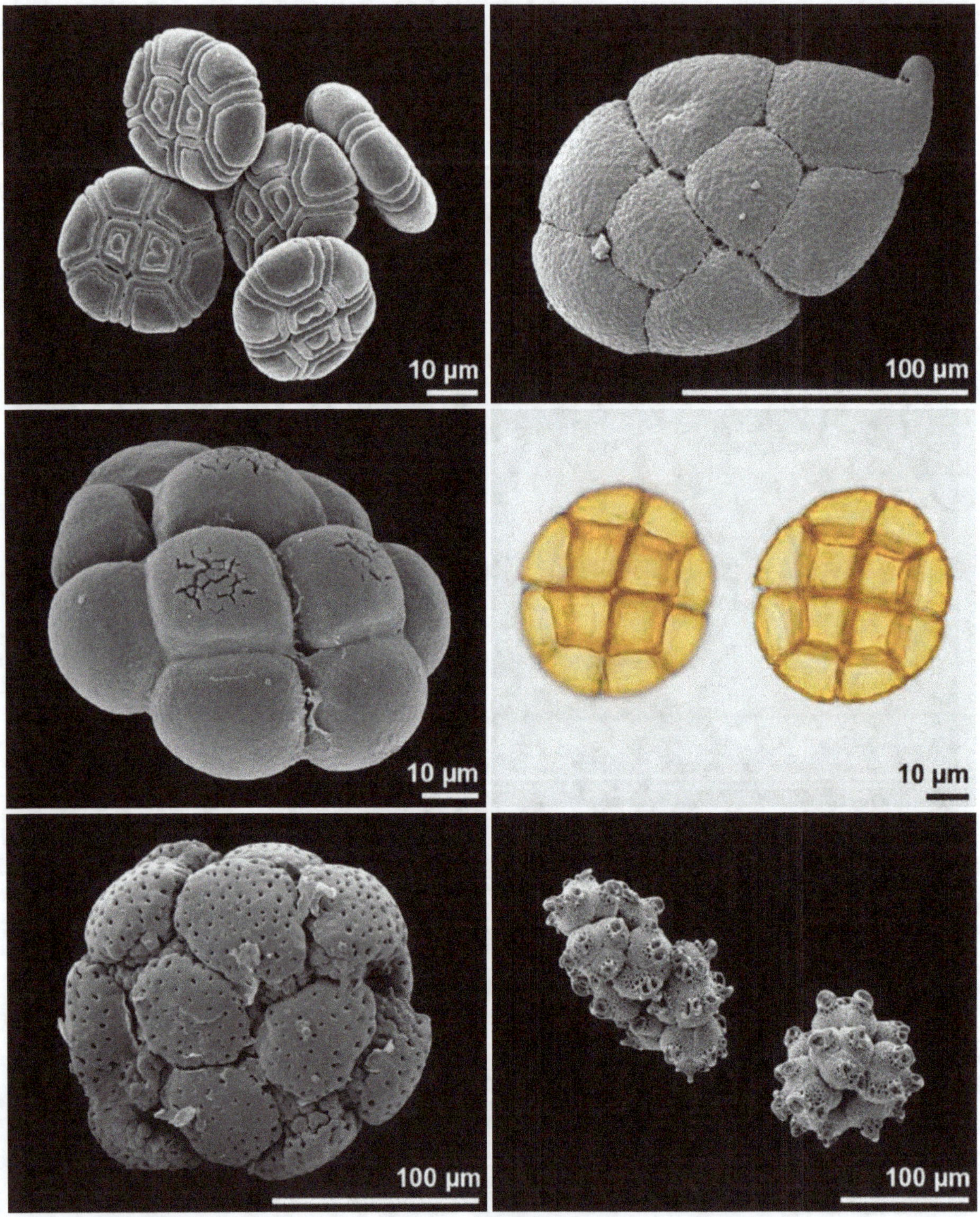

Acacia myrtifolia, Fabaceae
polyad with 8 monads, dry pollen

Calliandra tergemina, Fabaceae
polyad with 8 monads

Pithecellobium dulce, Fabaceae
polyad with 16 monads

Acacia sp., Fabaceae
polyad with 16 monads

Cymbopetalum aequale, Annonaceae
polyad with 16 monads

Chelonanthus purpurascens, Gentianaceae
polyad with 32 monads (8 united tetrads)

Acacia sp., Fabaceae
polyad of 8 monads, irregularly arranged, dry pollen

Acacia karroo, Fabaceae
polyad with 16 monads

Albizia julibrissin, Fabaceae
polyad with 16 monads

Acacia dealbata, Fabaceae
polyad with 16 monads

Acacia karroo, Fabaceae
polyad with 16 monads, dry pollen

Albizia saman, Fabaceae
polyad with 32 monads

massula

unit of more than four pollen grains but less than the locular content of a theca
Comment: In angiosperms only used for Orchidaceae with sectile pollinia

Traunsteinera globosa, Orchidaceae

Epipogium aphyllum, Orchidaceae

Herminium monorchis, Orchidaceae
massulae forming pollinium

Gennaria diphylla, Orchidaceae
massulae forming pollinium

Orchis italica, Orchidaceae

Orchis purpurea, Orchidaceae

pollinium

unit of a more or less interconnected loculiform pollen mass
Comment: loculi may be subdivided by septae, thus resulting in more than two pollinia

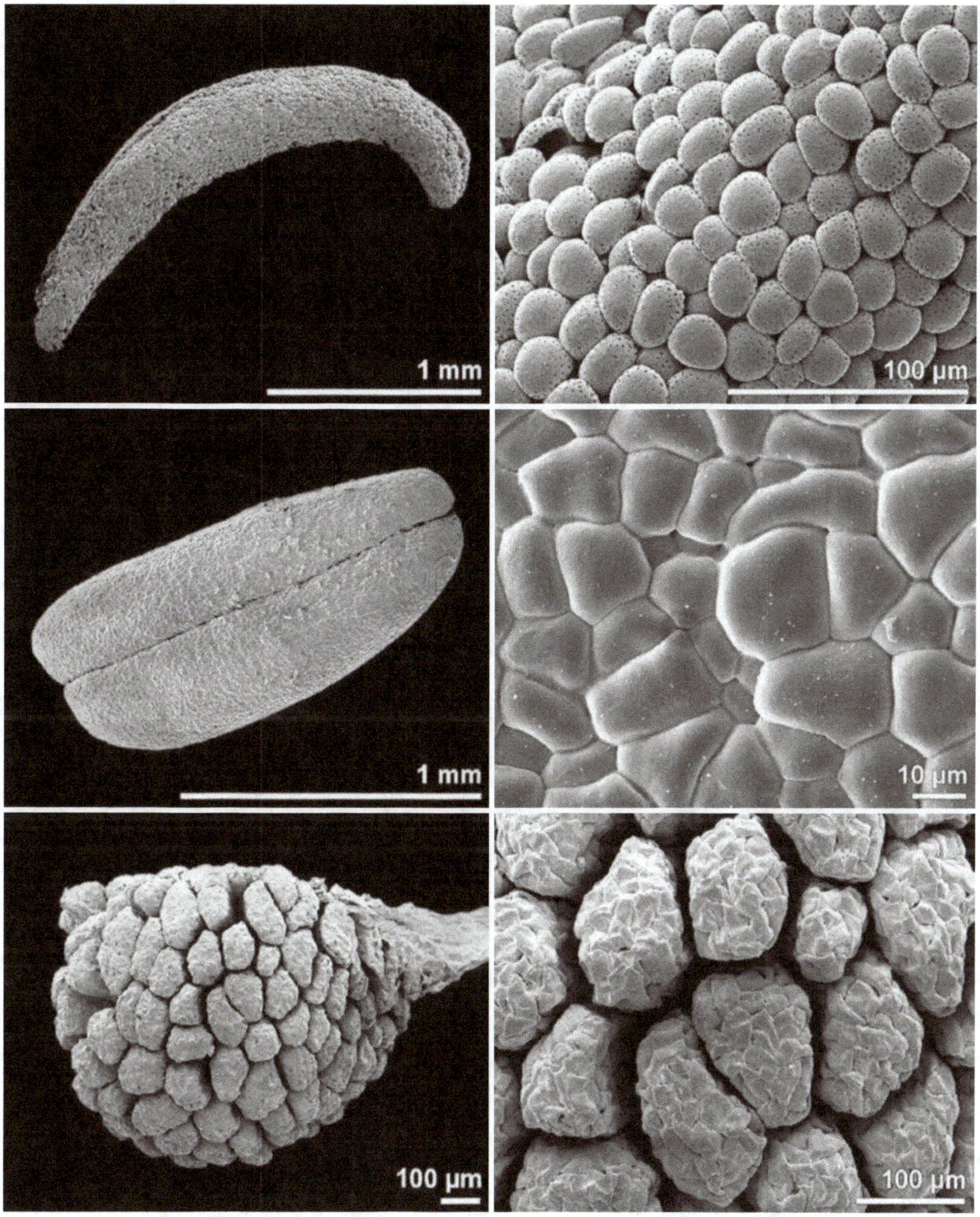

Cephalanthera longifolia, Orchidaceae

Cephalanthera longifolia, Orchidaceae
detail (monads)

Dendrobium farmeri, Orchidaceae

Dendrobium farmeri, Orchidaceae
detail (tetrads)

Steveniella satyrioides, Orchidaceae
sectile pollinium

Steveniella satyrioides, Orchidaceae
sectile pollinium, detail (massulae)

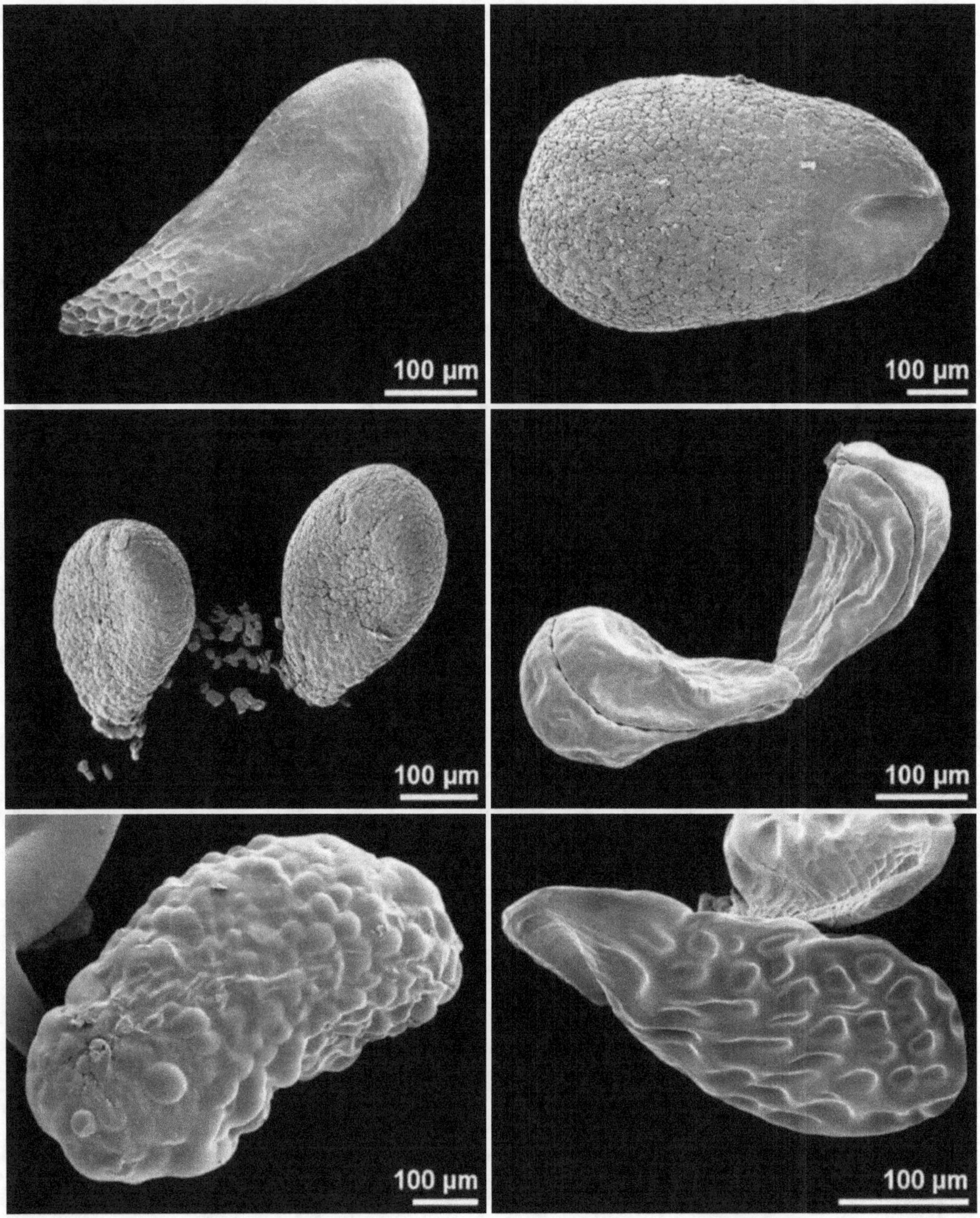

Hammarbya paludosa, Orchidaceae

Plectrophora cultrifolia, Orchidaceae

Restrepia muscifera, Orchidaceae
2 pollinia

Malaxis monophyllos, Orchidaceae
4 pollinia

Stephanotis floribunda, Asclepiadaceae

Hoodia flava, Asclepiadaceae

pollinarium

dispersal unit of pollinium (or pollinia) plus secretions and/or tissues that aid in the removal of the structure from the flower

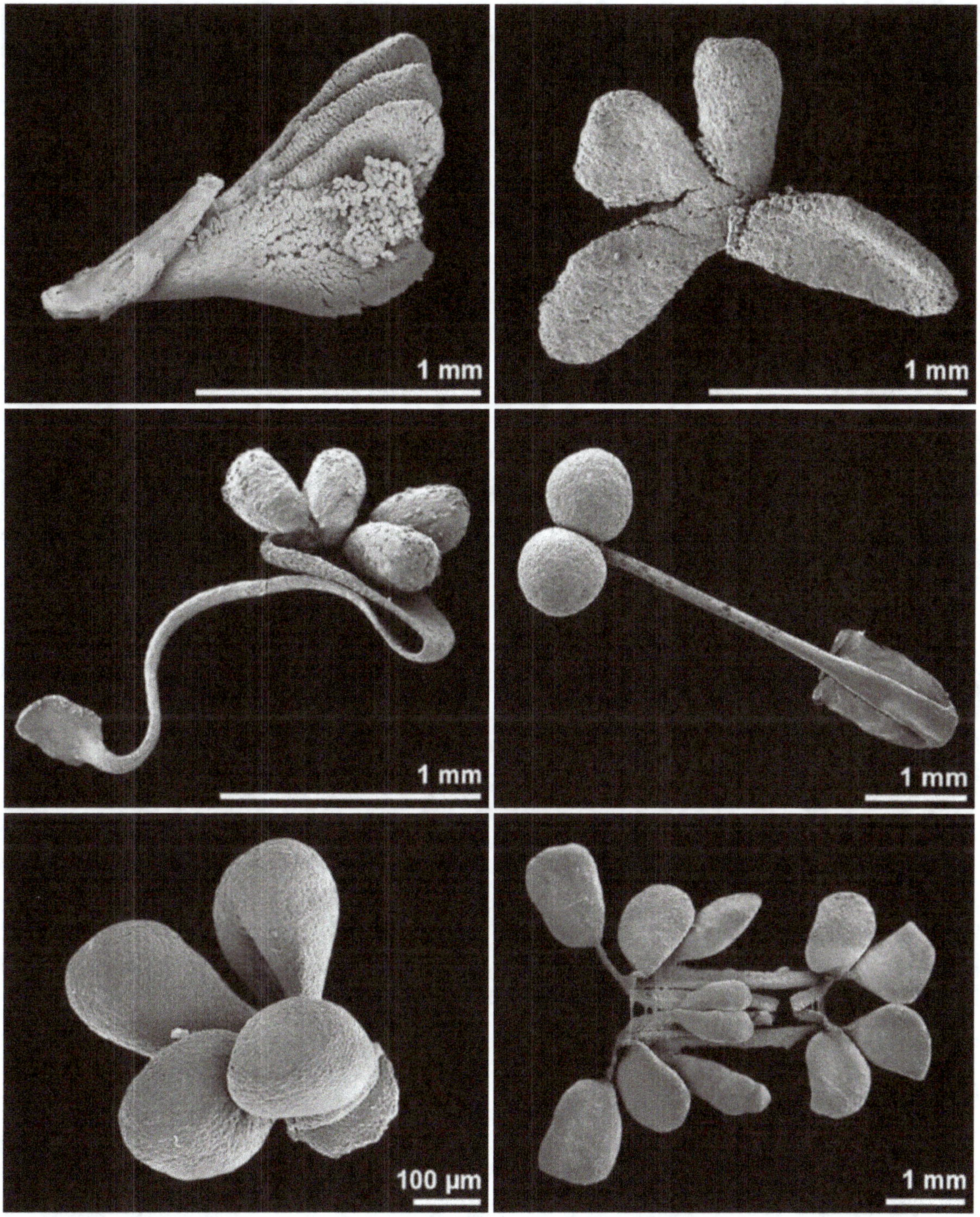

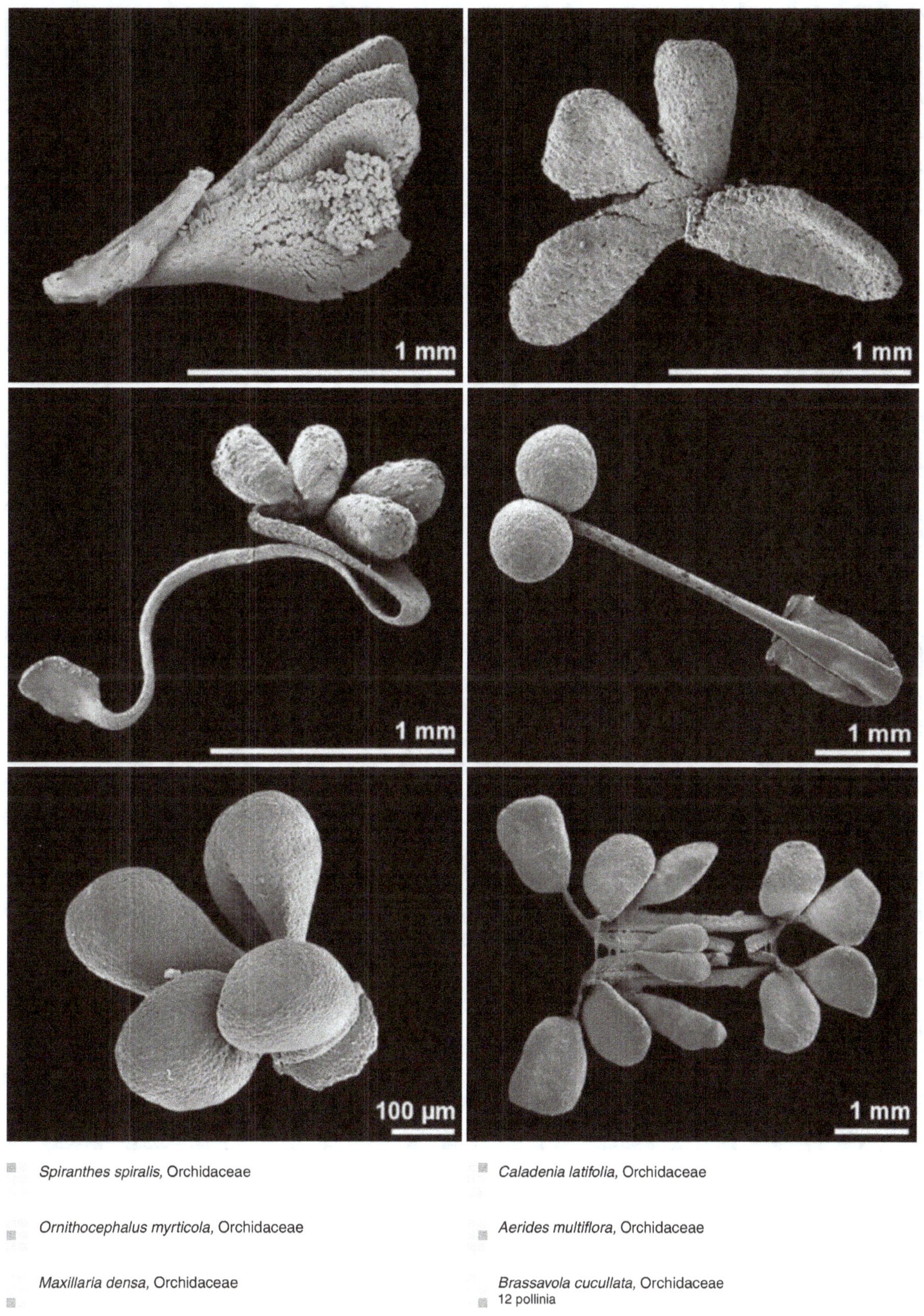

Spiranthes spiralis, Orchidaceae

Caladenia latifolia, Orchidaceae

Ornithocephalus myrticola, Orchidaceae

Aerides multiflora, Orchidaceae

Maxillaria densa, Orchidaceae

Brassavola cucullata, Orchidaceae
12 pollinia

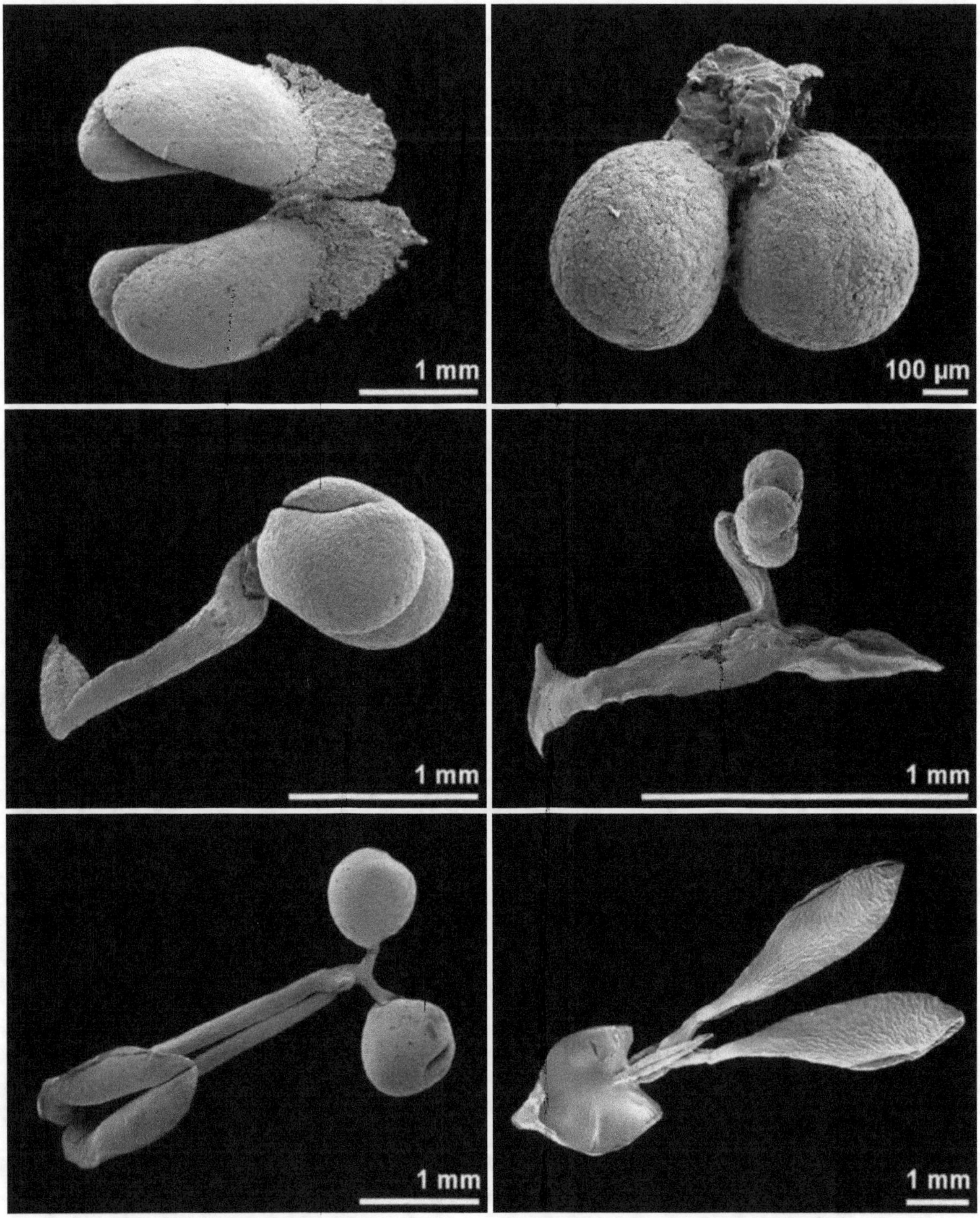

Coelogyne fimbriata, Orchidaceae

Oncidium maizaefolium, Orchidaceae

Haraella odorata, Orchidaceae

Pleurothallis loranthophylla, Orchidaceae

Schoenorchis fragrans, Orchidaceae

Stanhopea oculata, Orchidaceae

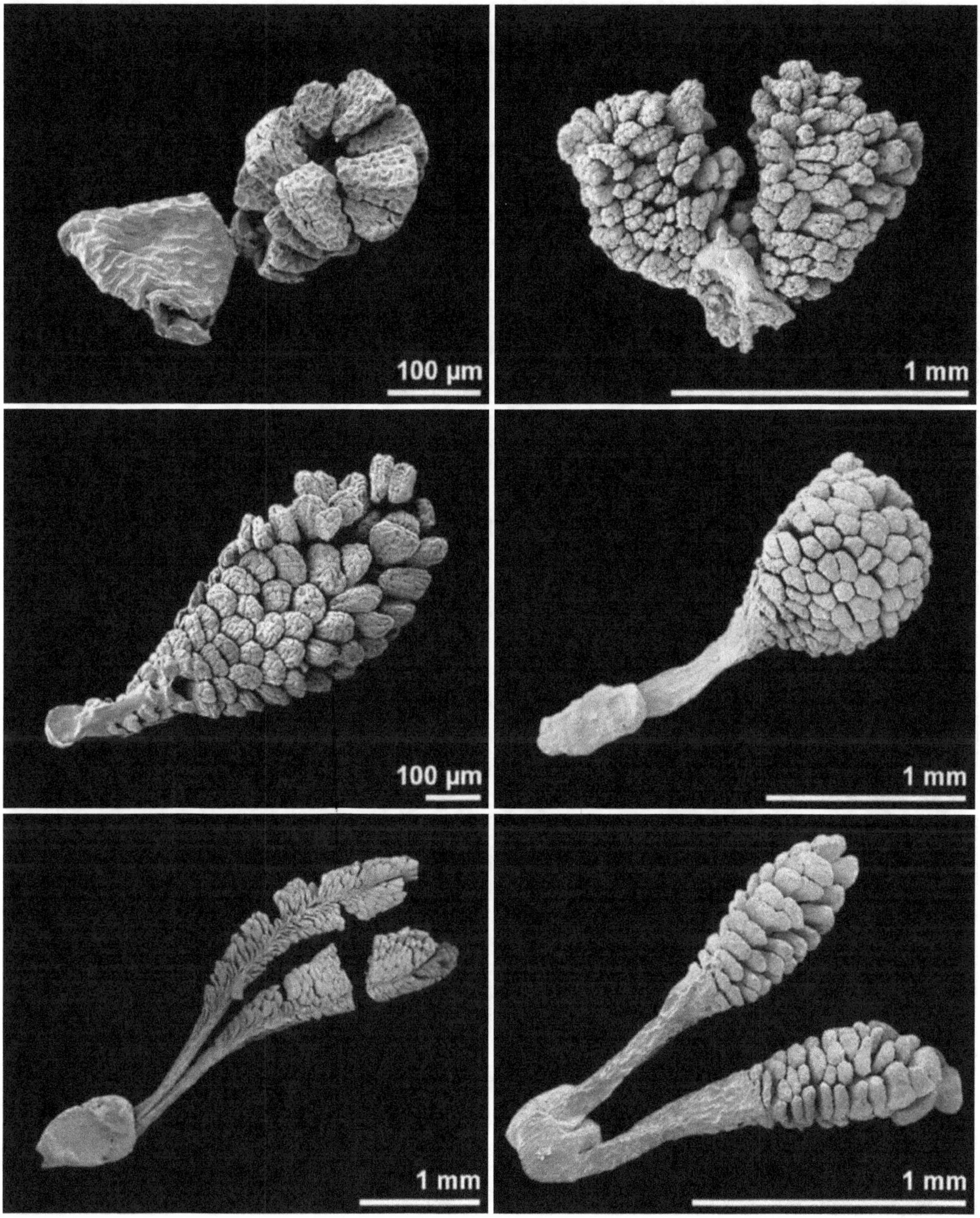

Herminium monorchis, Orchidaceae

Goodyera repens, Orchidaceae

Gennaria diphylla, Orchidaceae

Steveniella satyrioides, Orchidaceae

Ludisia discolor, Orchidaceae

Anacamptis pyramidalis, Orchidaceae

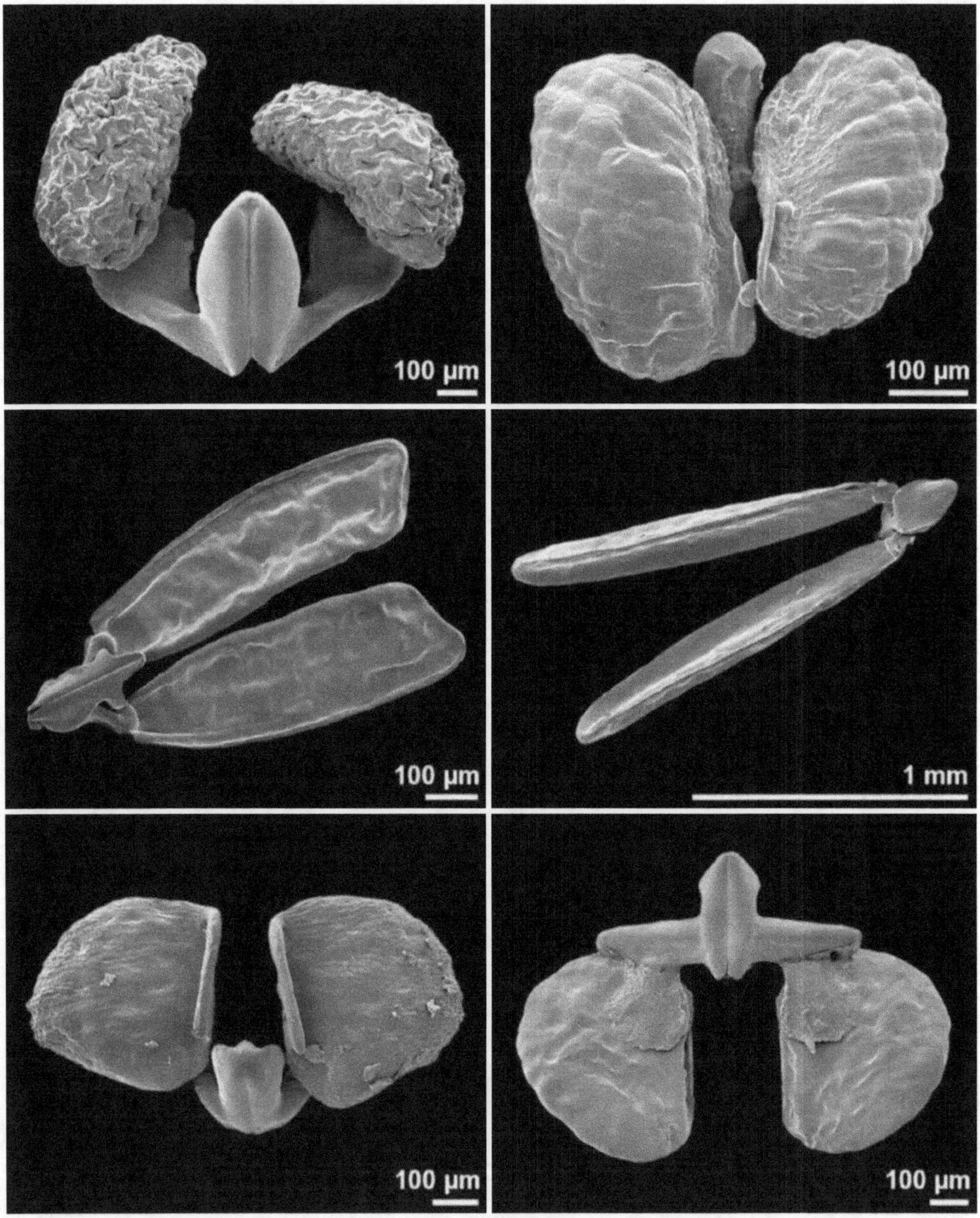

Stephanotis floribunda, Asclepiadaceae

Ceropegia sandersonii, Asclepiadaceae

Hoya carnosa, Asclepiadaceae

Hoya multiflora, Asclepiadaceae

Frerea indica, Asclepiadaceae

Orbeanthus hardyi, Asclepiadaceae

Shape and Polarity

H. Halbritter et al., *Illustrated Pollen Terminology*, https://doi.org/10.1007/978-3-319-71365-6_8

outline circular

outline describes the contour of pollen grains in polar and/or equatorial view

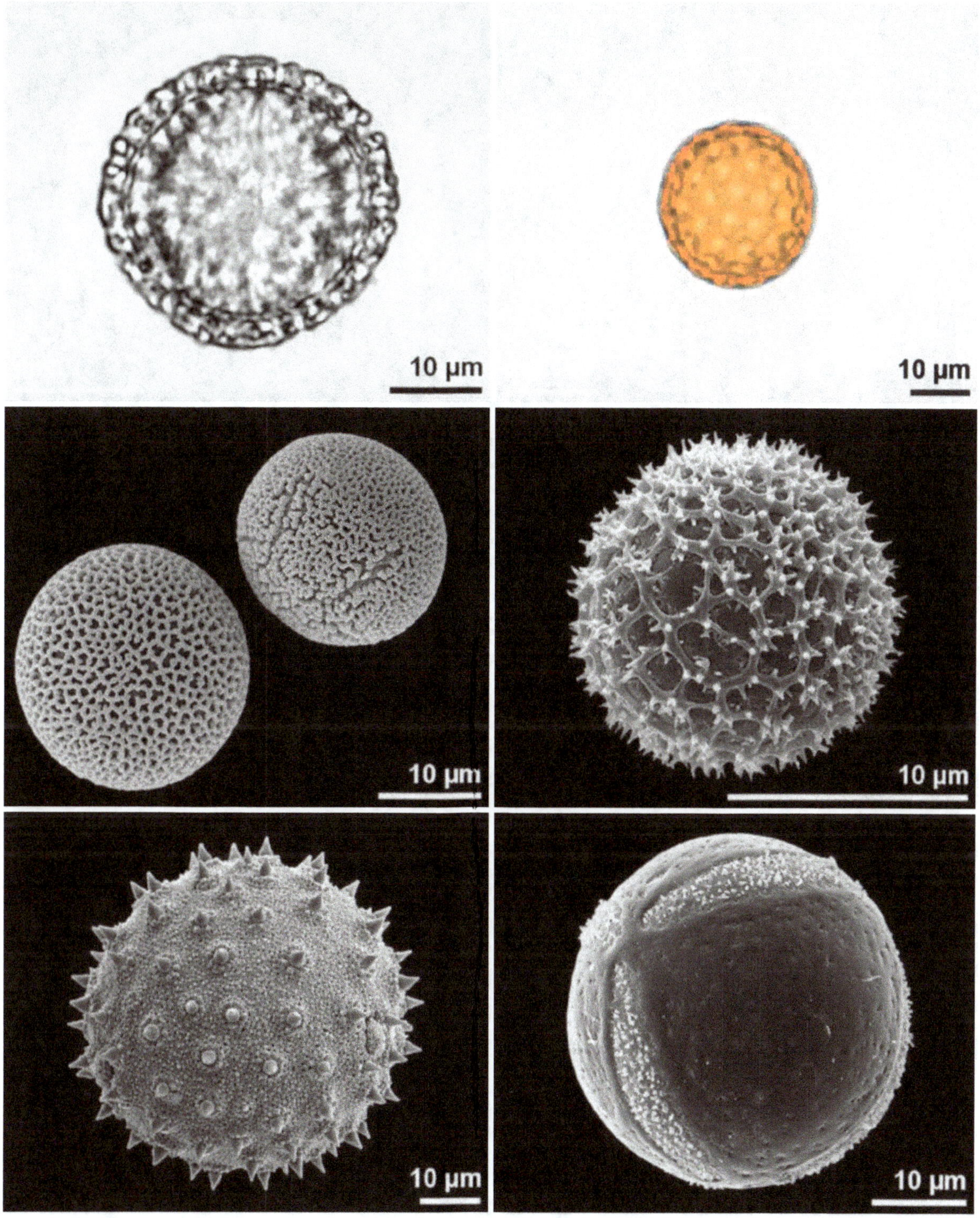

Ligustrum sp., Oleaceae
fossil, middle Miocene, Austria, equatorial view

Chenopodiaceae

Hedyosmum scaberrimum, Chloranthaceae
polychotomosulcate

Anthurium ovatifolium, Araceae
porate

Abutilon theophrasti, Malvaceae
triporate, equatorial view

Corydalis ophiocarpa, Papaveraceae
hexacolpate

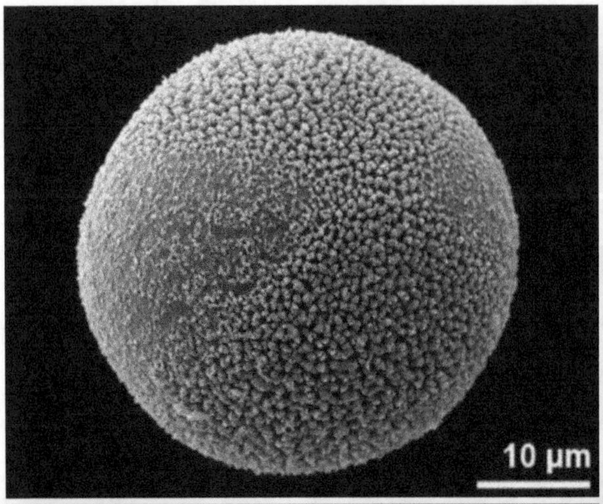

Mayna odorata, Achariaceae
dicolpate, polar view

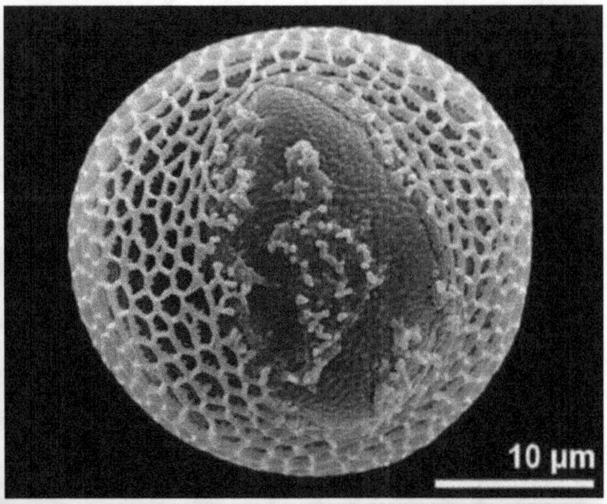

Saruma henryi, Aristolochiaceae
sulcate, distal polar view

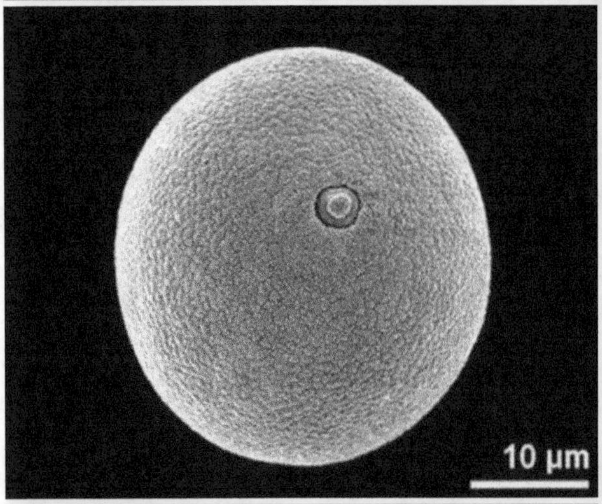

Phleum pratense, Poaceae
ulcerate, distal polar view

Fraxinus ornus, Oleaceae
tricolporate, polar view

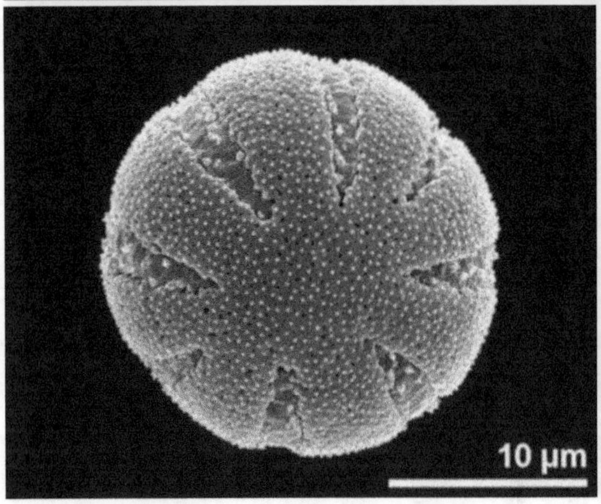

Galium lucidum, Rubiaceae
stephanocolpate, polar view

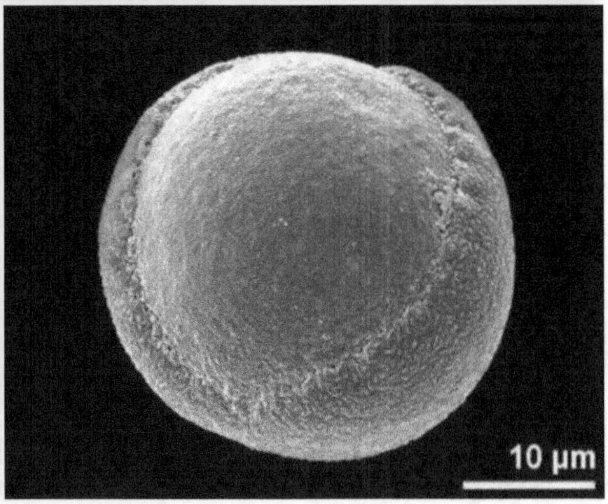

Ginkgo biloba, Ginkgoaceae
sulcate, oblique distal polar view

outline elliptic

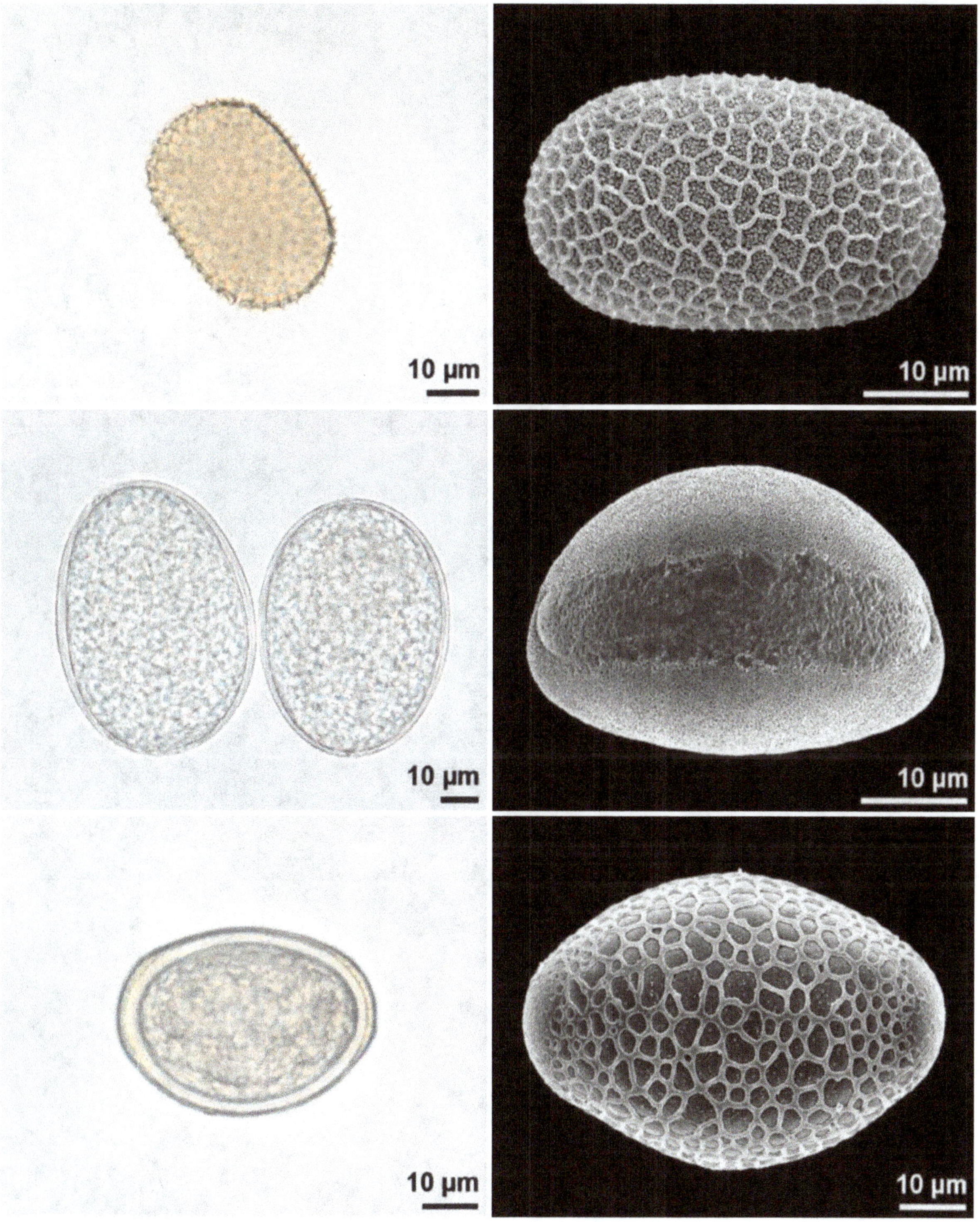

Impatiens parviflora, Balsaminaceae
tetracolpate, polar view

Impatiens parviflora, Balsaminaceae
tetracolpate, polar view

Amorphophallus interruptus, Araceae
inaperturate, hydrated pollen

Allium oleraceum, Amaryllidaceae
sulcate, distal polar view

Ambrosina bassi, Araceae
inaperturate, hydrated pollen

Aechmea dealbata, Bromeliaceae
diporate

Salvia coccinea, Lamiaceae
hexacolpate, polar view

Billbergia porteana, Bromeliaceae
sulcate, distal polar view

Galeopsis tetrahit, Lamiaceae
tricolpate, dry pollen

Commelina erecta, Commelinaceae
sulcate, proximal polar view

Zamia loddigesii, Zamiaceae
sulcate, dry pollen

Physostegia virginiana, Lamiaceae
tricolpate, dry pollen

outline lobate

outline in polar view of a pollen grain with bulged interapertural areas (mainly in dry pollen grains)

Acer pseudoplatanus, Sapindaceae
tricolpate, dry pollen

Artemisia pontica, Asteraceae
tricolporate, polar view

Sanguisorba officinalis, Rosaceae
hexacolporate, dry pollen

Orthilia secunda, Ericaceae
tricolporate, dry pollen

Gunnera tinctoria, Gunneraceae
tricolpate, polar view

Gunnera tinctoria, Gunneraceae
dry pollen, equatorial (left) and polar view (right)

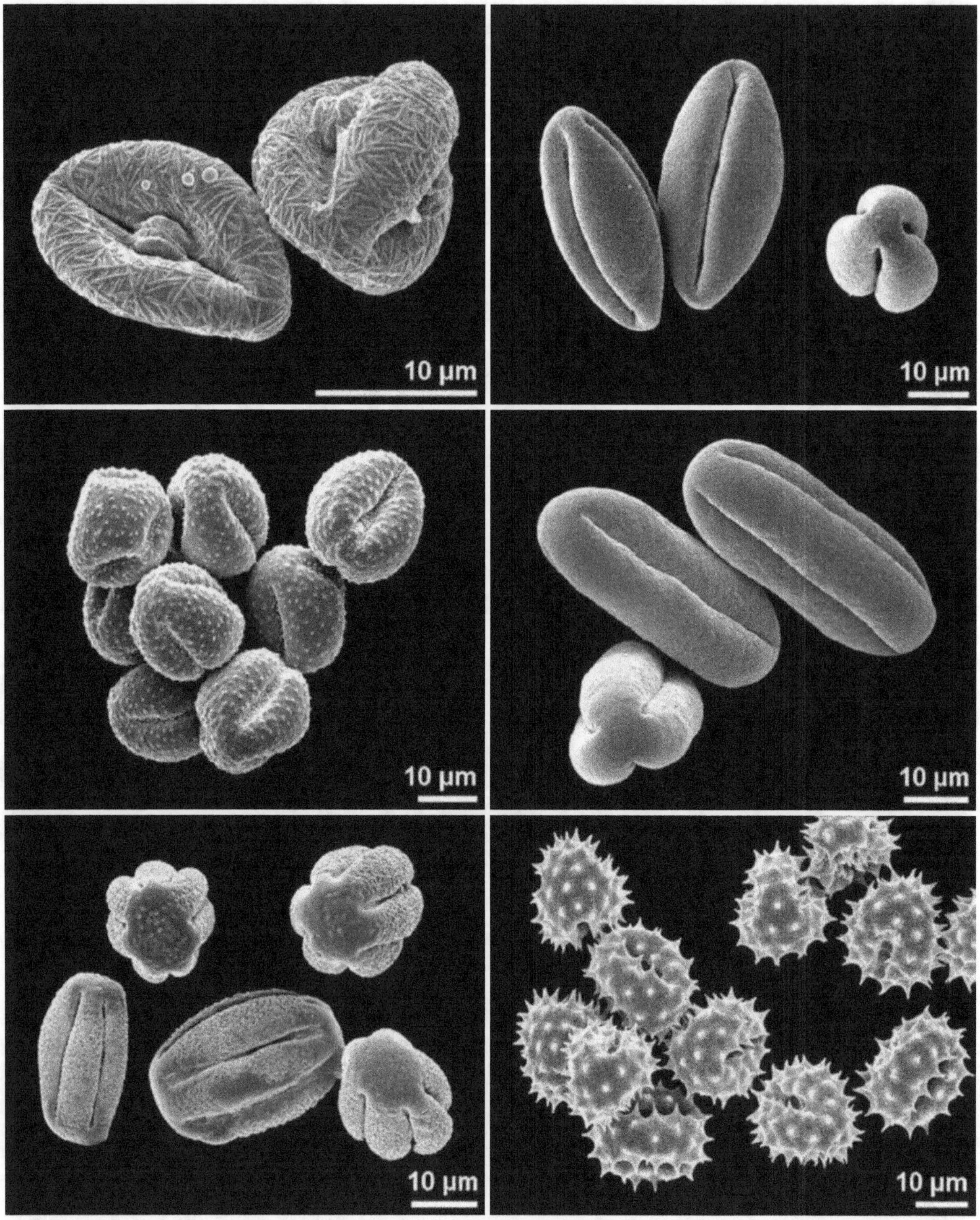

Sedum rupestre, Crassulaceae
tricolporate, dry pollen

Clematis heracleifolia, Ranunculaceae
tricolpate, dry pollen

Pinguicula ehlersiae, Lentibulariaceae
stephanocolporate, dry pollen

Viola alba, Violaceae
tricolporate, dry pollen

Sanicula europaea, Apiaceae
tricolporate, dry pollen

Bellis perennis, Asteraceae
tricolporate, dry pollen

Artemisia sp., Asteraceae
tricolporate, polar view

Hypecoum imberbe, Papaveraceae
dicolpate, polar view

Pelargonium punctatum, Geraniaceae
tricolporate, dry pollen

Nicotiana tabacum, Solanaceae
tetracolporate, polar view

Barringtonia asiatica, Lecythidaceae
tricolpate, dry pollen

Viola riviniana, Violaceae
tetracolporate, dry pollen

outline triangular

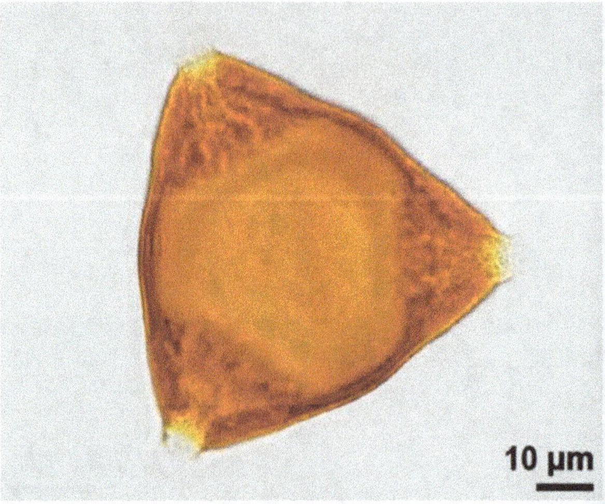

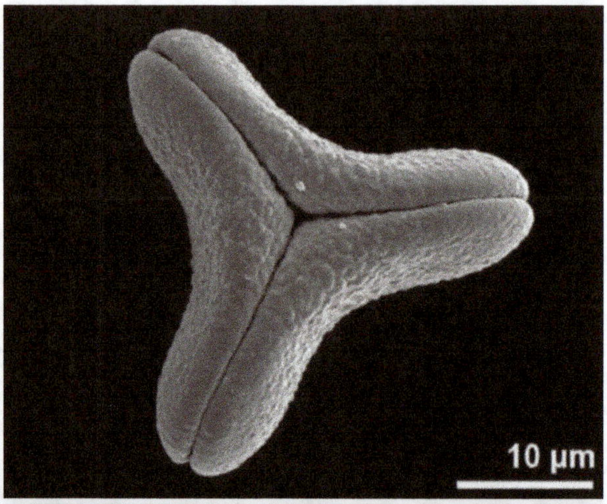

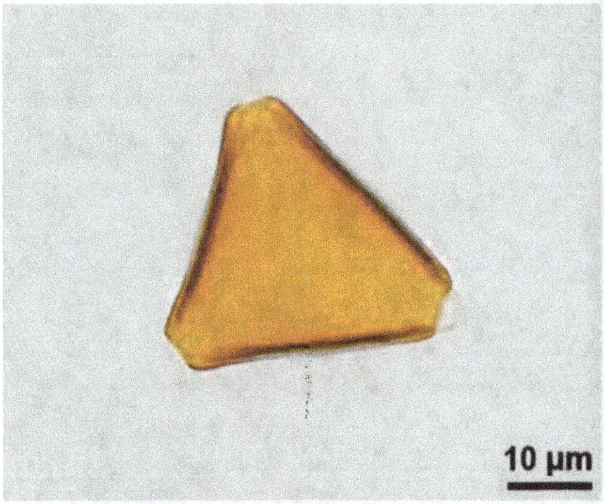

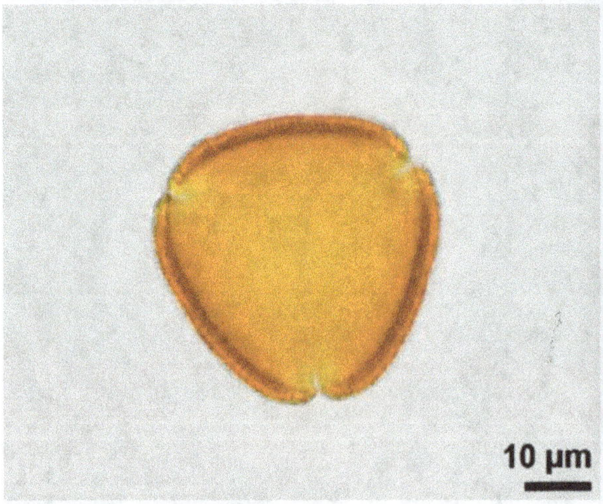

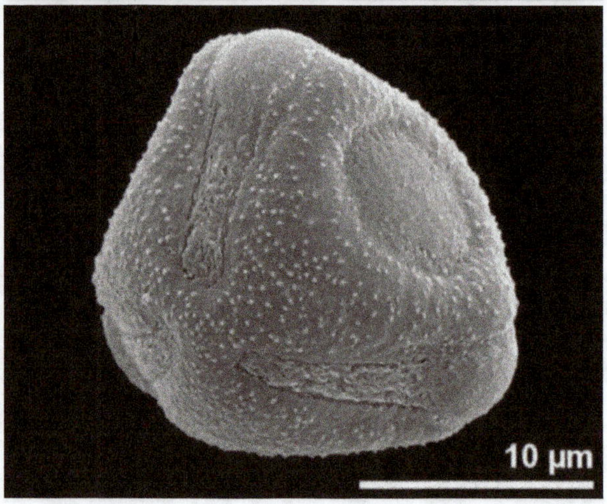

Lopezia racemosa, Onagraceae
triporate, polar view

Macadamia ternifolia, Proteaceae

Kolkwitzia amabilis, Caprifoliaceae

Loranthus europaeus, Loranthaceae
syncolpate, polar view, dry pollen

Tropaeolum emarginatum, Tropaeolaceae
tricolpate, polar view

Acicarpha tribuloides, Calyceraceae
tricolporate, oblique polar view

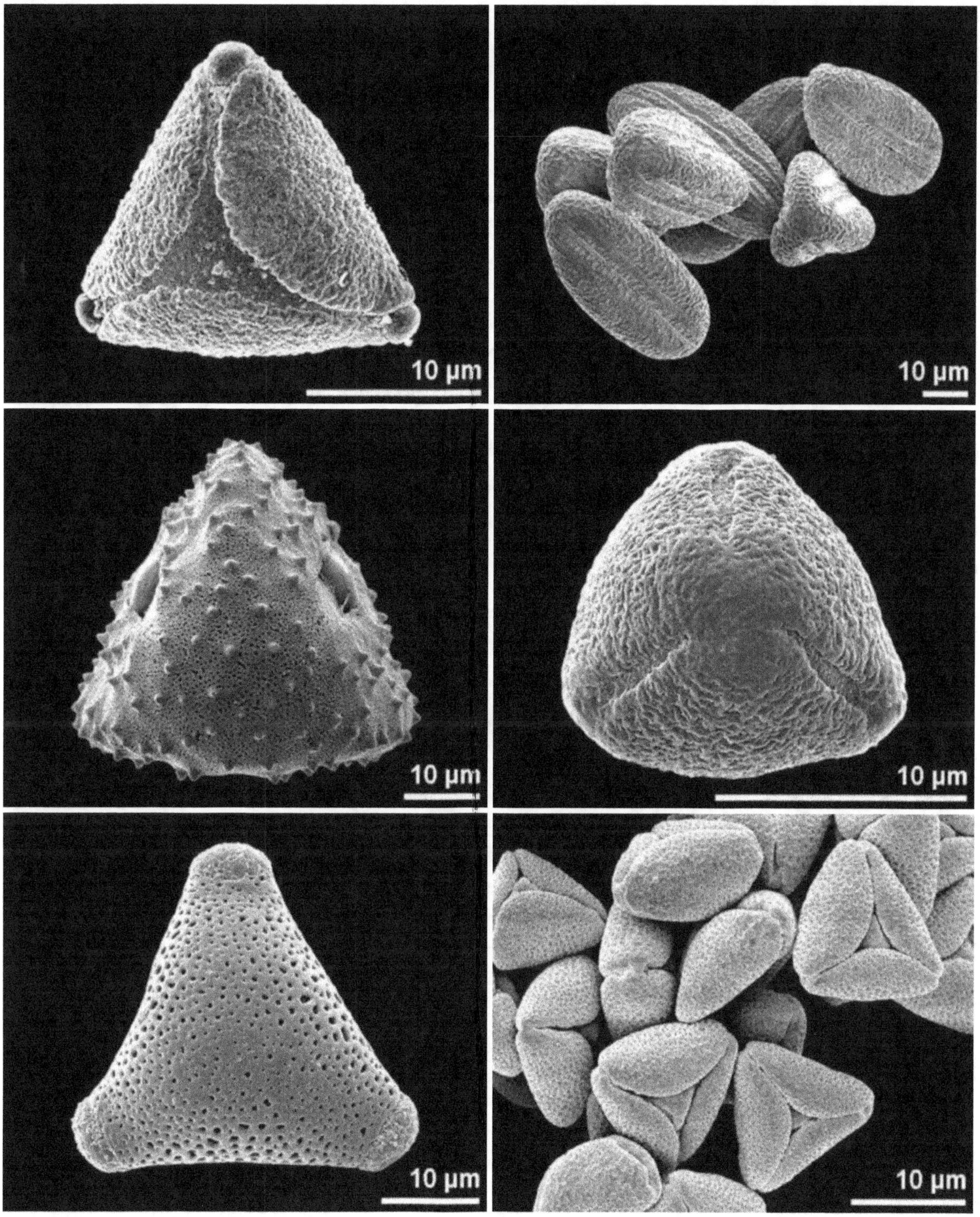

Callistemon coccineus, Myrtaceae
tricolporate, synaperturate, polar view

Echinops ritro, Asteraceae
tricolporate, polar view

Paullinia tomentosa, Sapindaceae
triporate, polar view

Hypoestes phyllostachya, Acanthaceae
tricolpate, dry pollen

Bupleurum rotundifolium, Apiaceae
tricolporate, polar view

Primula denticulata, Primulaceae
tricolporate, synaperturate, dry pollen

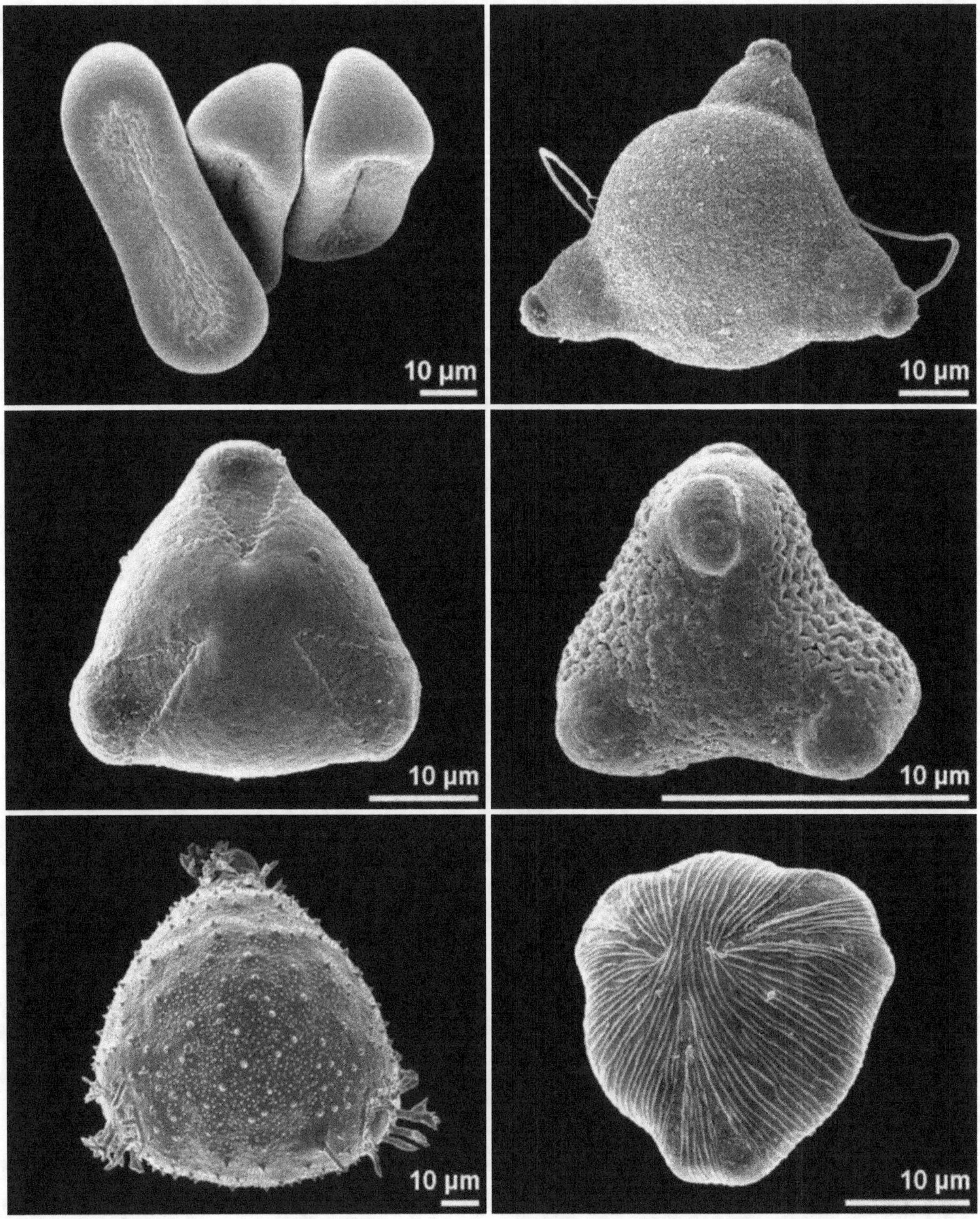

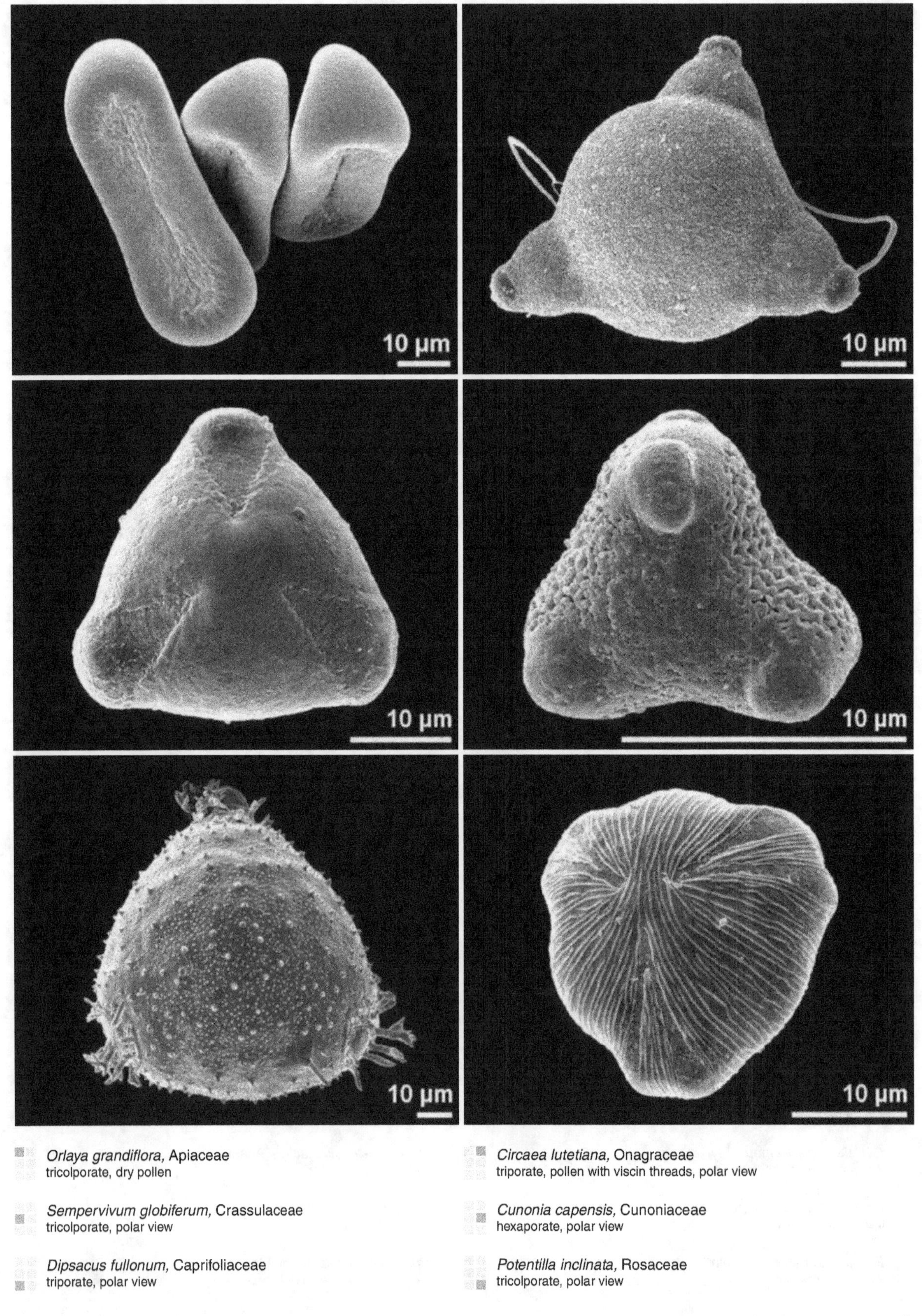

Orlaya grandiflora, Apiaceae
tricolporate, dry pollen

Sempervivum globiferum, Crassulaceae
tricolporate, polar view

Dipsacus fullonum, Caprifoliaceae
triporate, polar view

Circaea lutetiana, Onagraceae
triporate, pollen with viscin threads, polar view

Cunonia capensis, Cunoniaceae
hexaporate, polar view

Potentilla inclinata, Rosaceae
tricolporate, polar view

outline quadrangular

Anchusa officinalis, Boraginaceae
tetracolporate, dry pollen

Viola tricolor, Violaceae
tetracolporate, polar view

Herniaria glabra, Caryophyllaceae
hexaporate

Nonea pulla, Boraginaceae
tetracolporate, polar view

Eremurus robustus, Xanthorrhoeaceae
sulcate, distal polar view

Sideritis romana, Lamiaceae
tetracolpate, dry pollen

outline polygonal

Viola arvensis, Violaceae
pentacolporate, polar view

Opuntia basilaris, Cactaceae
pantocolpate, dry pollen

Sarracenia alata, Sarraceniaceae
stephanocolporate, polar view

Viola arvensis, Violaceae
pentacolporate, polar view

Talinum paniculatum, Talinaceae
pantocolpate, dry pollen

Stellaria holostea, Caryophyllaceae
pantoporate, dry pollen

P/E-ratio, oblate

P/E-ratio refers to the length of the polar axis between the two poles compared to the equatorial diameter
oblate: pollen grain with a polar axis shorter than the equatorial diameter

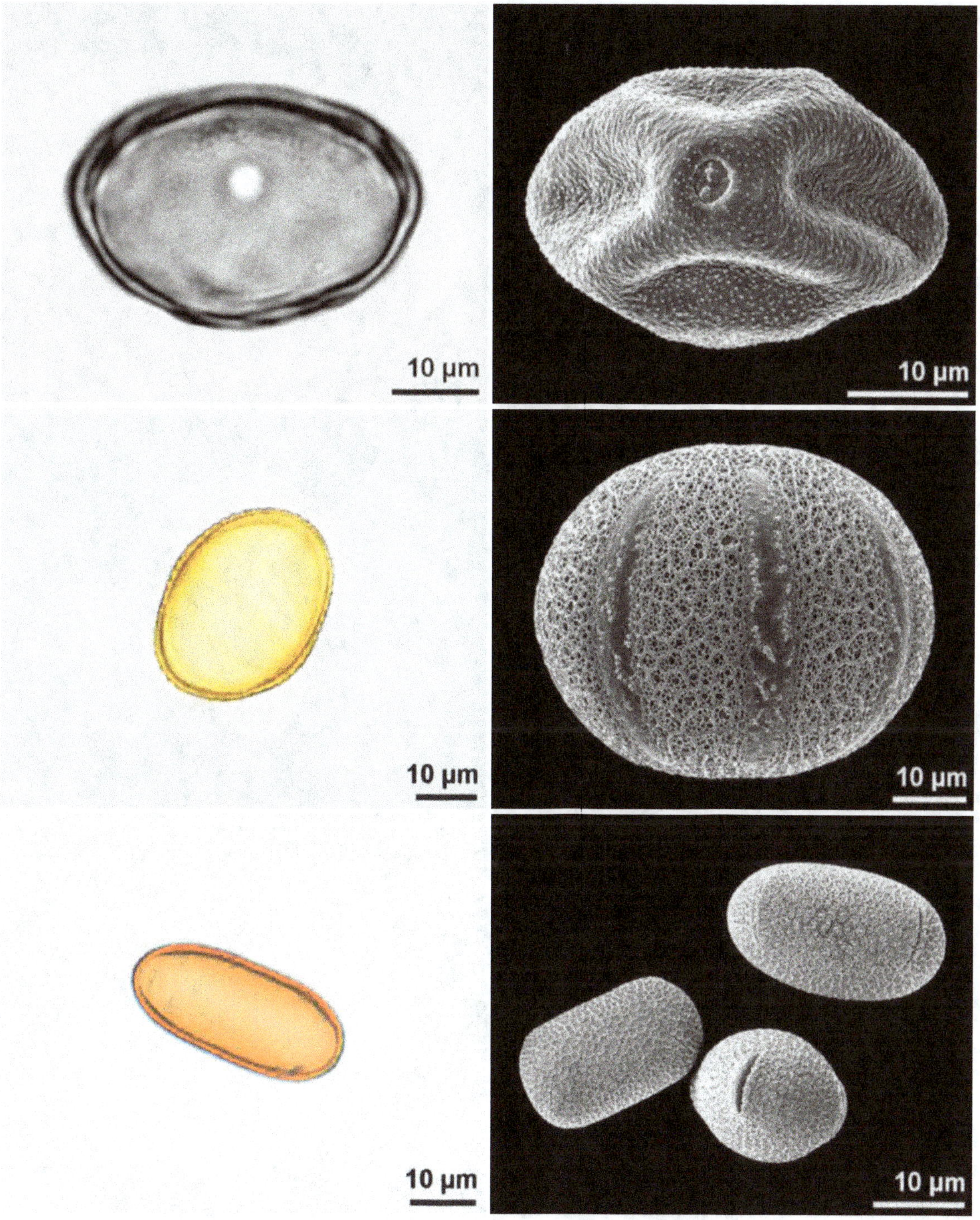

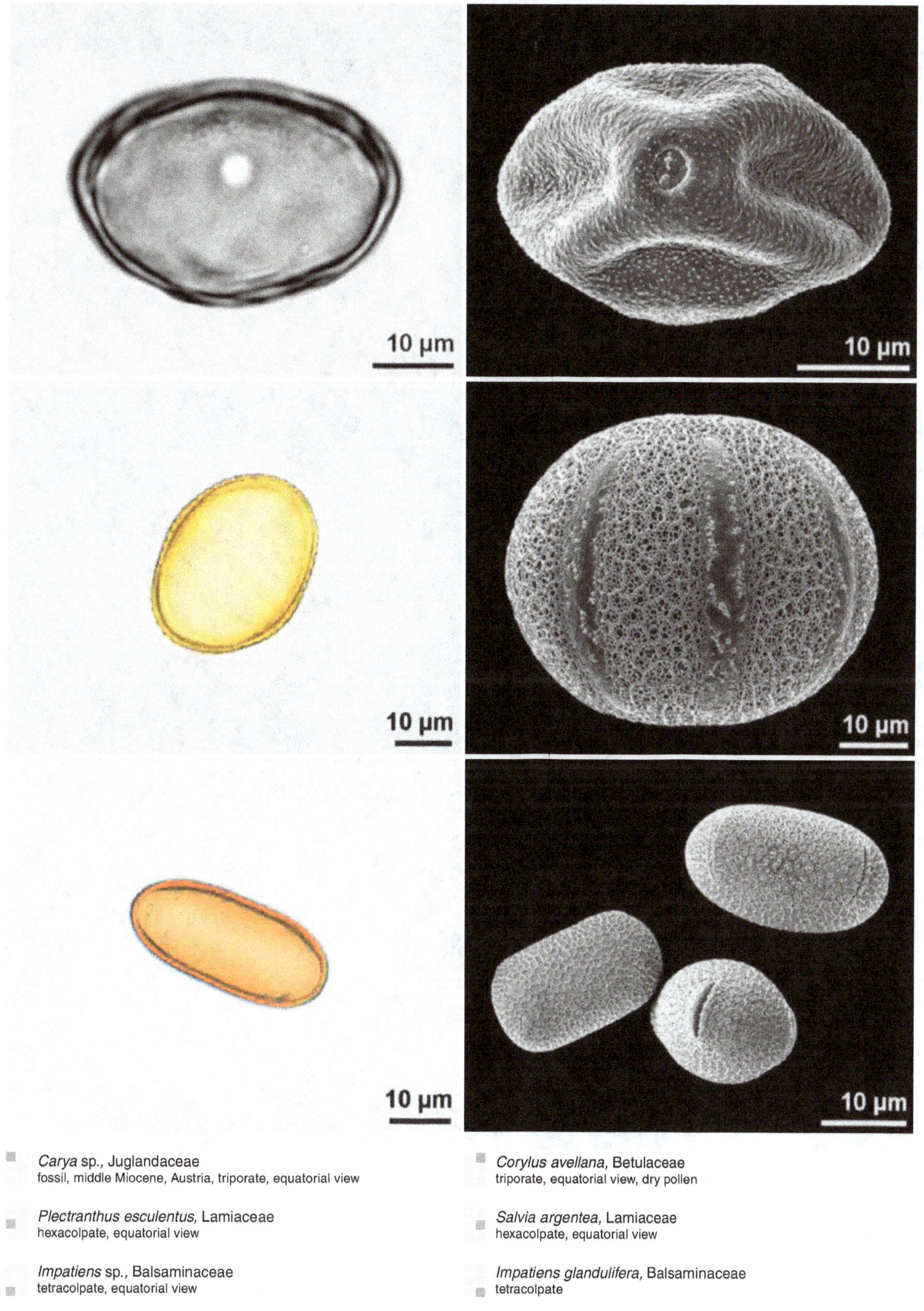

Carya sp., Juglandaceae
fossil, middle Miocene, Austria, triporate, equatorial view

Corylus avellana, Betulaceae
triporate, equatorial view, dry pollen

Plectranthus esculentus, Lamiaceae
hexacolpate, equatorial view

Salvia argentea, Lamiaceae
hexacolpate, equatorial view

Impatiens sp., Balsaminaceae
tetracolpate, equatorial view

Impatiens glandulifera, Balsaminaceae
tetracolpate

Sarracenia alata, Sarraceniaceae
stephanocolporate

Knautia drymeia, Caprifoliaceae
triporate, dry pollen

Hakea kippistiana, Proteaceae
triporate

Hymenocallis tubiflora, Amaryllidaceae
sulcate, proximal polar view

Cuphea procumbens, Lythraceae
tricolporate, dry pollen

Roridula gorgonias, Roridulaceae
tricolpate, equatorial view

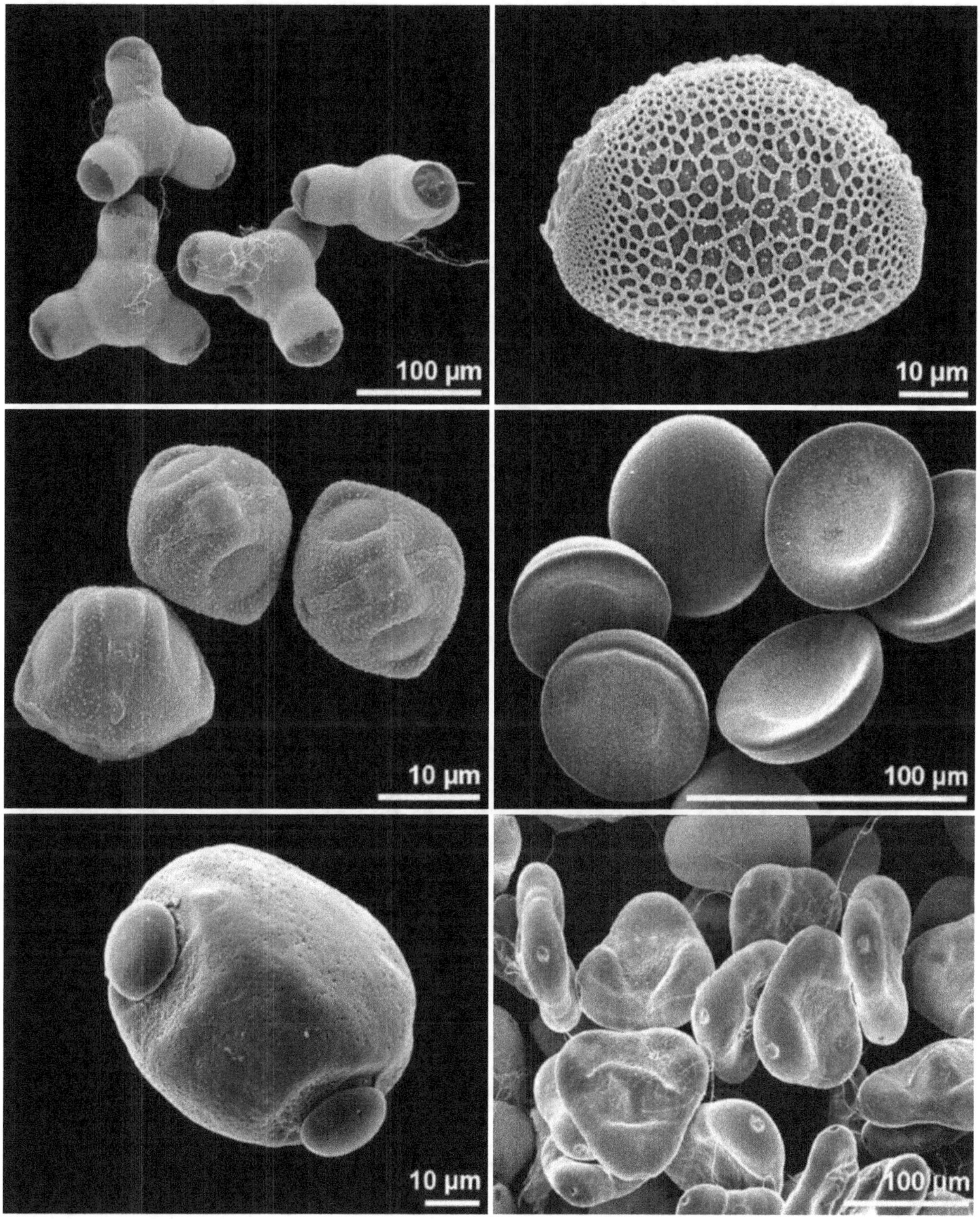

Clarkia unguiculata, Onagraceae
triporate, pollen with viscin threads

Acicarpha tribuloides, Calyceraceae
tricolporate

Amsonia ciliata, Apocynaceae
tricolporate, oblique equatorial view

Vriesea neoglutinosa, Bromeliaceae
sulcate, equatorial view

Heliconia sp., Heliconiaceae
ulcerate, dry pollen

Clarkia purpurea, Onagraceae
triporate, dry pollen

P/E-ratio, isodiametric

isodiametric: pollen grain with a polar axis equal to the equatorial diameter

Parnassia palustris, Celastraceae
spheroidal, tricolporate, equatorial view

Roemeria hybrida, Papaveraceae
spheroidal, pantoporate

Iris pumila, Iridaceae
spheroidal, sulcate, distal polar view

Campanula fenestrellata, Campanulaceae
spheroidal, stephanoporate, oblique polar view

Silene nutans, Caryophyllaceae
polygonal, pantoporate, dry pollen

Sarcocapnos enneaphylla, Papaveraceae
hexacolpate

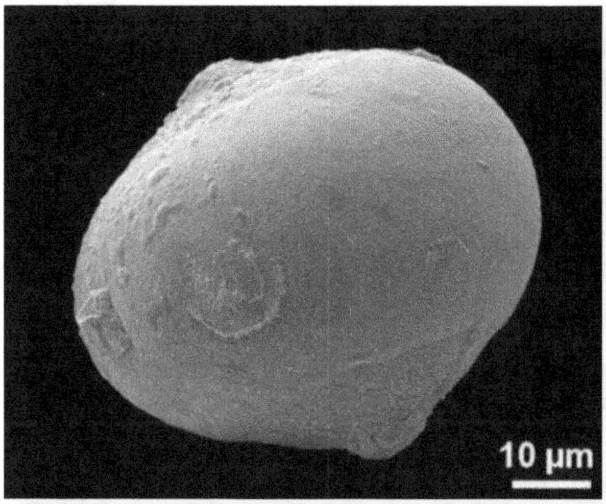

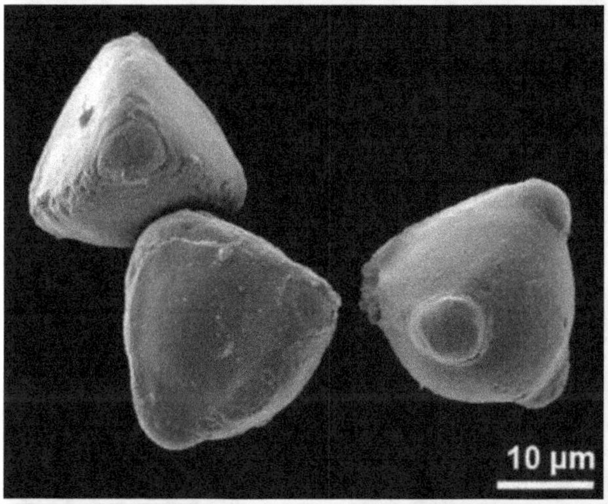

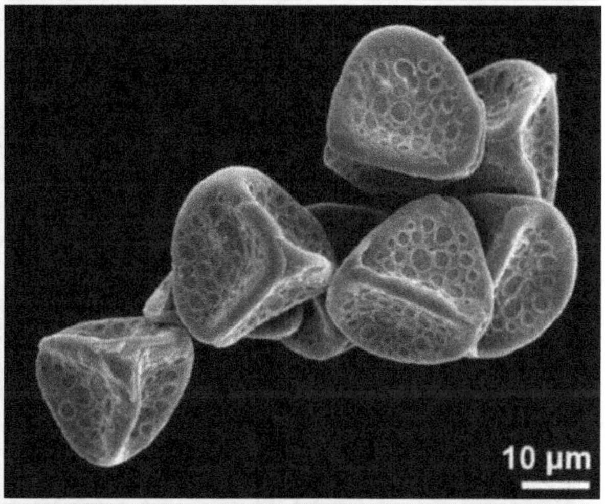

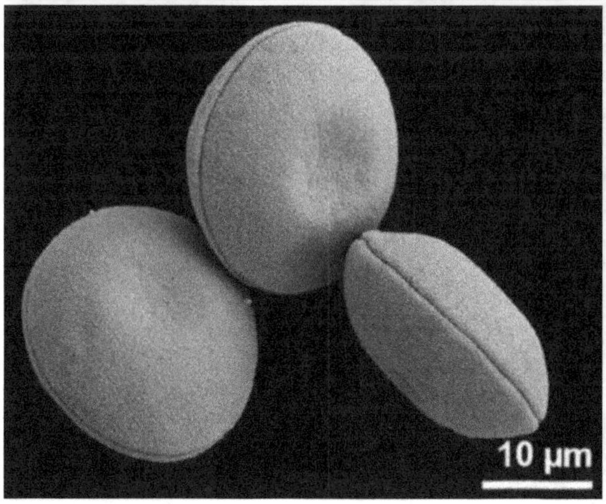

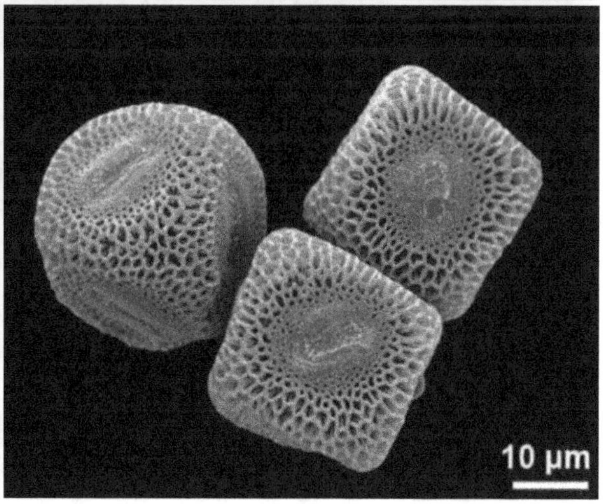

- *Whitfieldia lateritia*, Acanthaceae
  diporate
- *Schoepfia schreberi*, Schoepfiaceae
  tetraaperturate
- *Pedicularis gyroflexa*, Orobanchaceae
  ring-like aperture, dry pollen
- *Whitfieldia lateritia*, Acanthaceae
  dry pollen
- *Thesium arvense*, Santalaceae
  triradiate colpi, dry pollen
- *Basella alba*, Basellaceae
  hexacolpate

P/E-ratio, prolate

prolate: pollen grain with a polar axis longer than the equatorial diameter

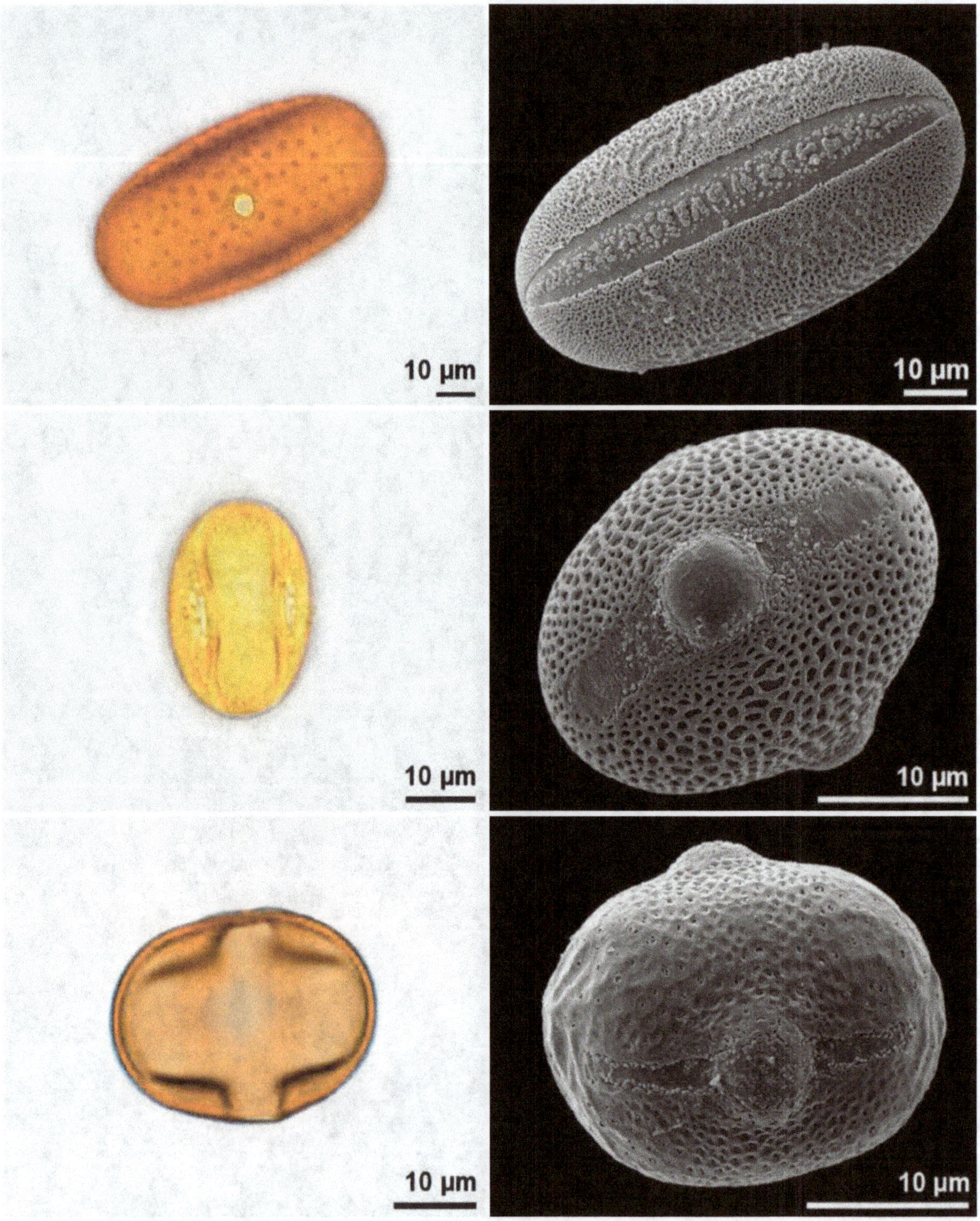

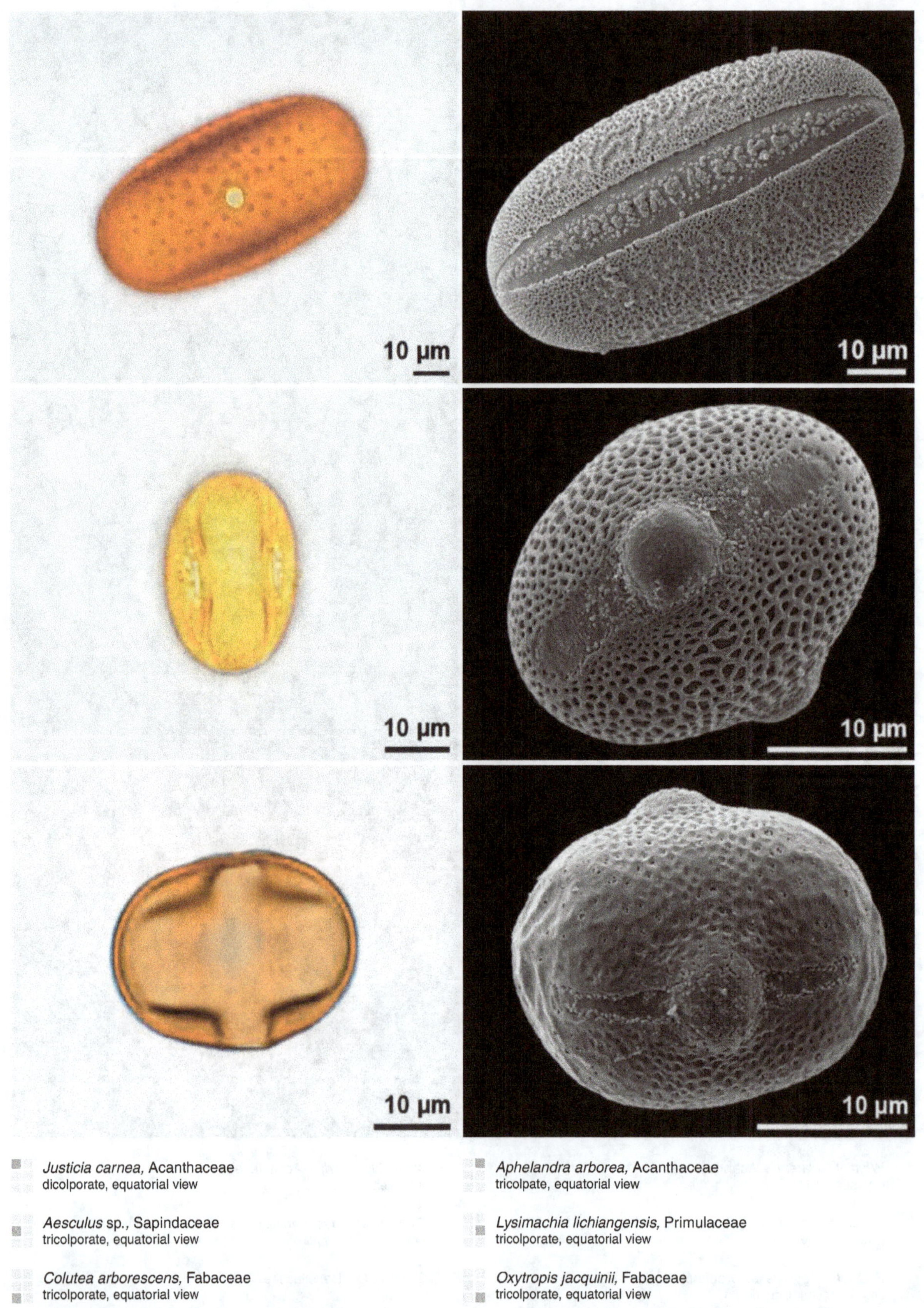

Justicia carnea, Acanthaceae
dicolporate, equatorial view

Aesculus sp., Sapindaceae
tricolporate, equatorial view

Colutea arborescens, Fabaceae
tricolporate, equatorial view

Aphelandra arborea, Acanthaceae
tricolpate, equatorial view

Lysimachia lichiangensis, Primulaceae
tricolporate, equatorial view

Oxytropis jacquinii, Fabaceae
tricolporate, equatorial view

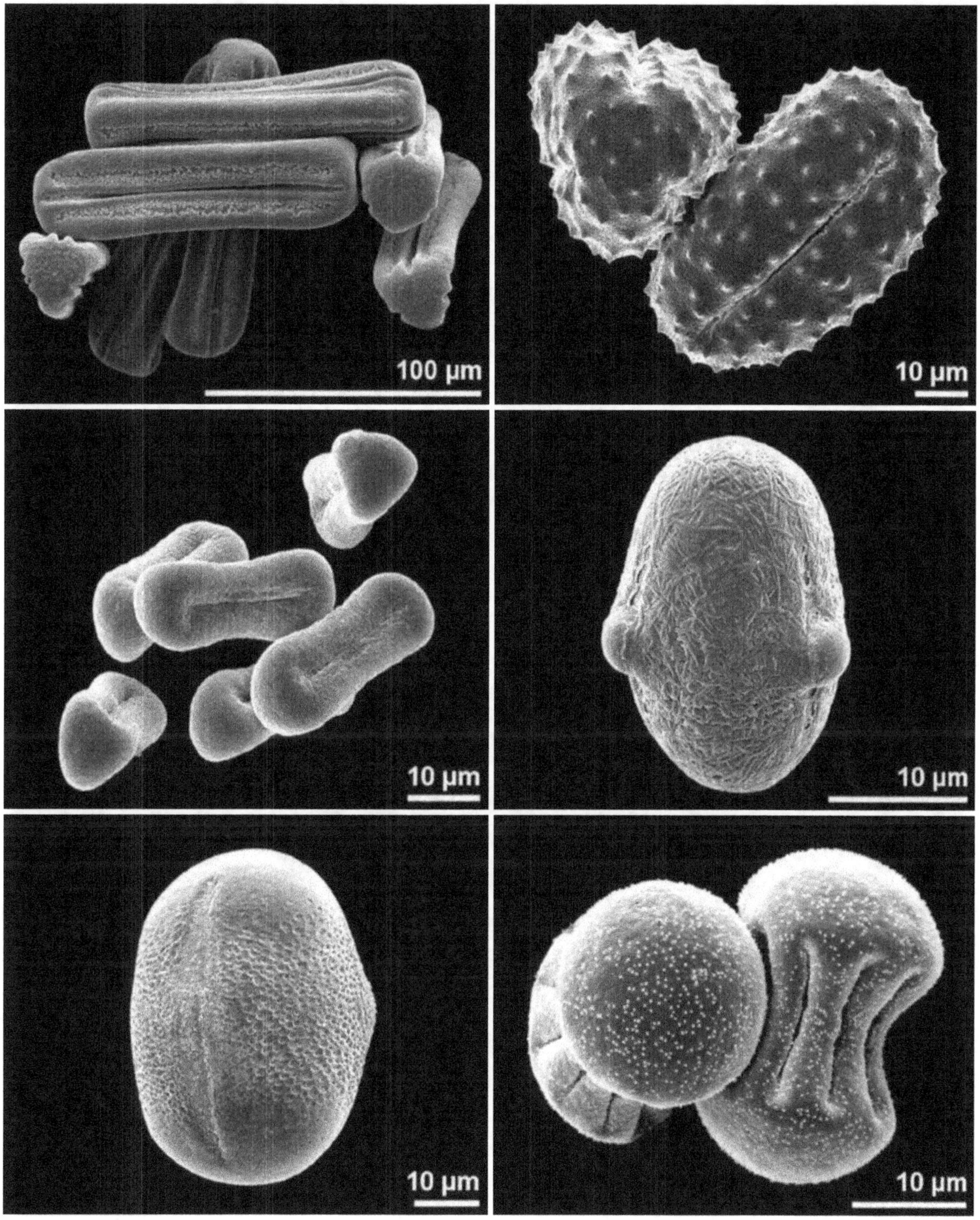

Crossandra flava, Acanthaceae
tricolpate, dry pollen

Torilis arvensis, Apiaceae
tricolporate, dry pollen

Astragalus onobrychis, Fabaceae
tricolporate, equatorial view

Jurinea mollis, Asteraceae
tricolporate, dry pollen

Peucedanum cervaria, Apiaceae
tricolporate, equatorial view

Symphytum officinale, Boraginaceae
stephanocolporate, dry pollen

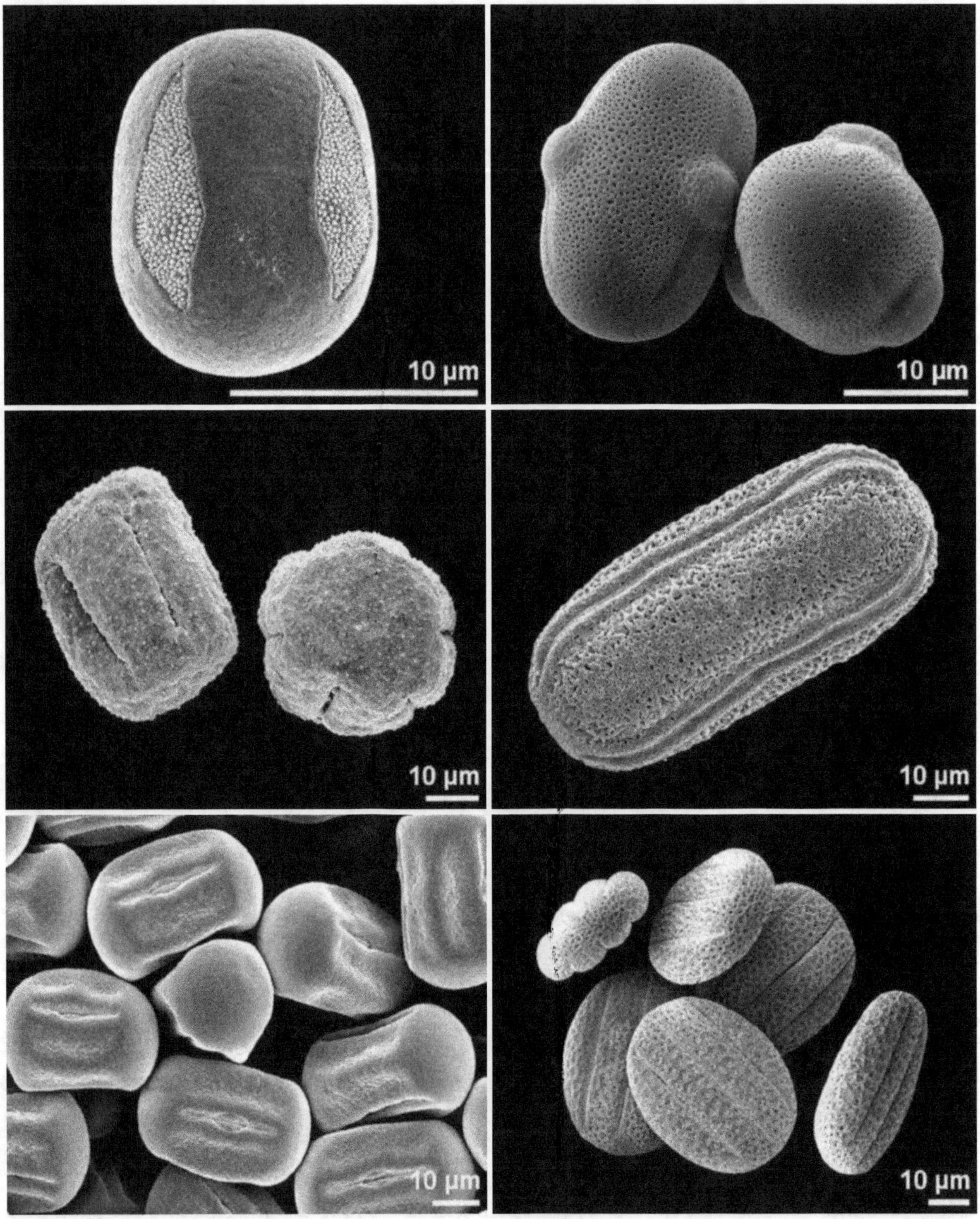

Buglossoides purpurocaerulea, Boraginaceae
tetracolporate, equatorial view

Platycodon grandiflorus, Campanulaceae
stephanocolpate, dry pollen

Lathyrus tuberosus, Fabaceae
tricolporate, dry pollen

Vitaliana primuliflora, Primulaceae
tricolporate

Stenandrium guineense, Acanthaceae
tricolpate, equatorial view

Salvia sclarea, Lamiaceae
hexacolpate, dry pollen

isopolar

pollen grain with identical proximal and distal faces

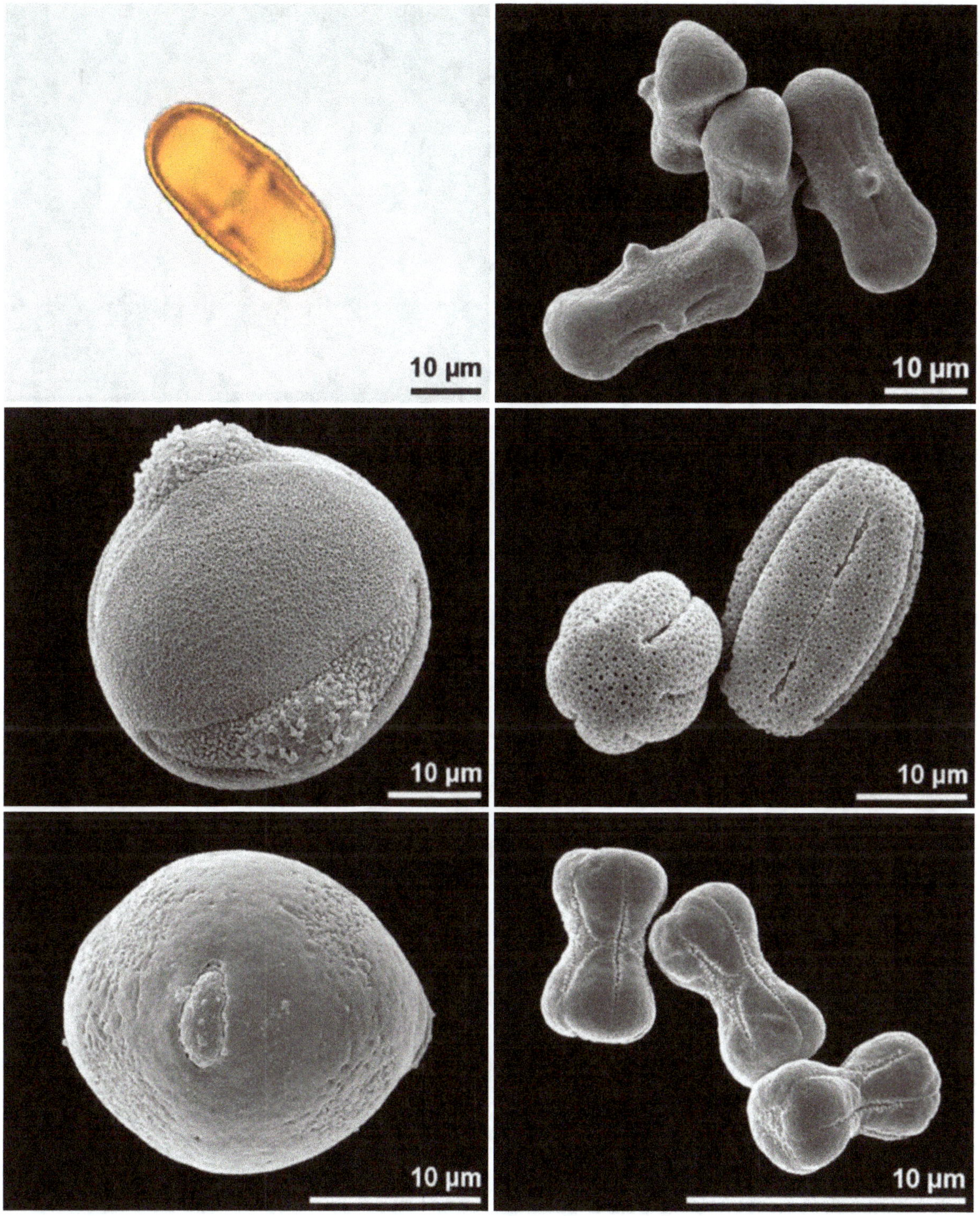

Apiaceae
equatorial view

Columnea magnifica, Gesneriaceae
equatorial view

Monotropa hypopitys, Ericaceae
equatorial view

Bifora radians, Apiaceae

Asperula tinctoria, Rubiaceae
dry pollen

Myosotis scorpioides, Boraginaceae
dry pollen

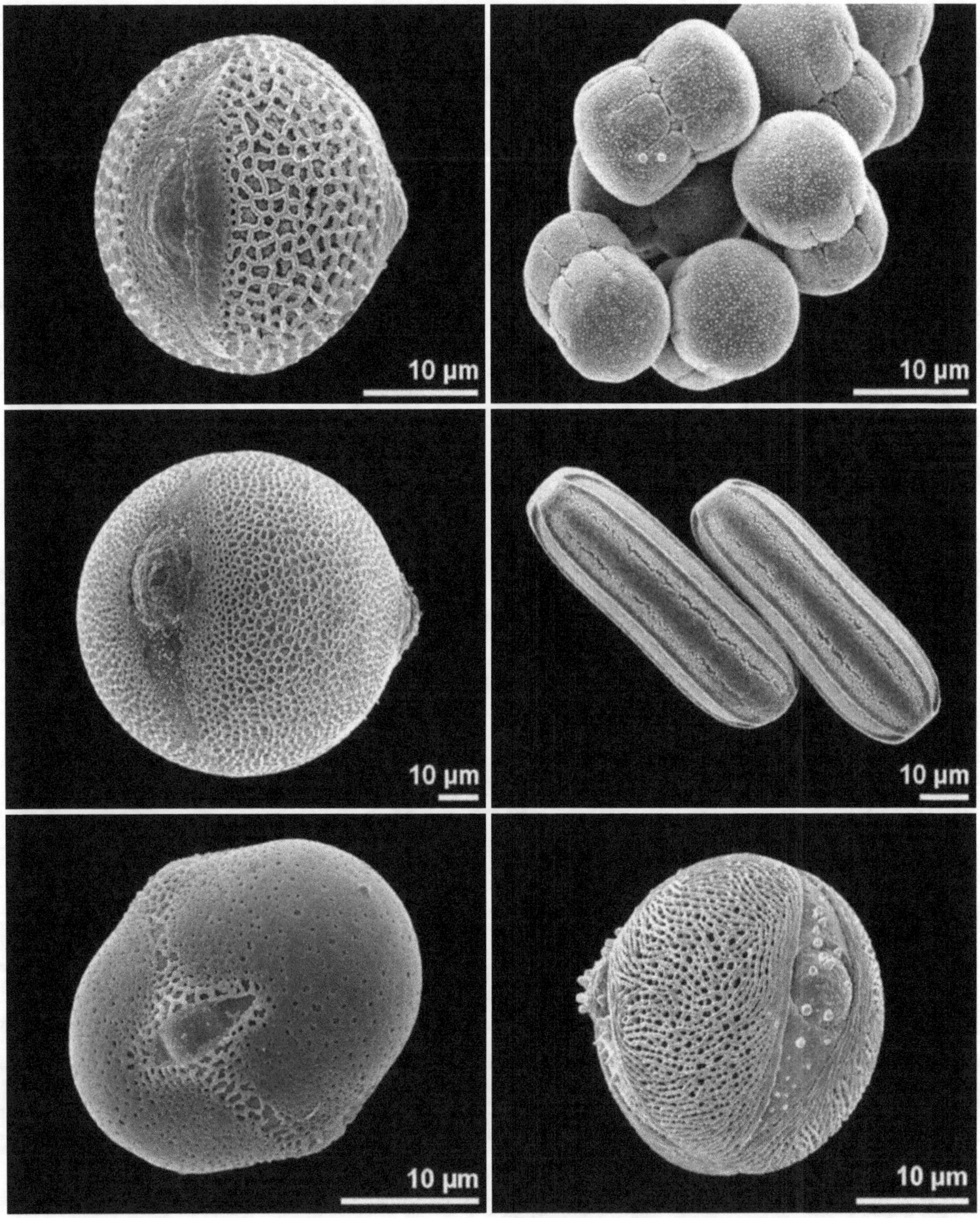

Viburnum tinus, Adoxaceae
equatorial view

Luffa cylindrica, Cucurbitaceae
equatorial view

Pulmonaria angustifolia, Boraginaceae
equatorial view

Cerinthe minor, Boraginaceae
dry pollen

Crossandra flava, Acanthaceae
equatorial view

Aesculus flava, Sapindaceae
equatorial view

heteropolar

pollen grain with different proximal and distal faces

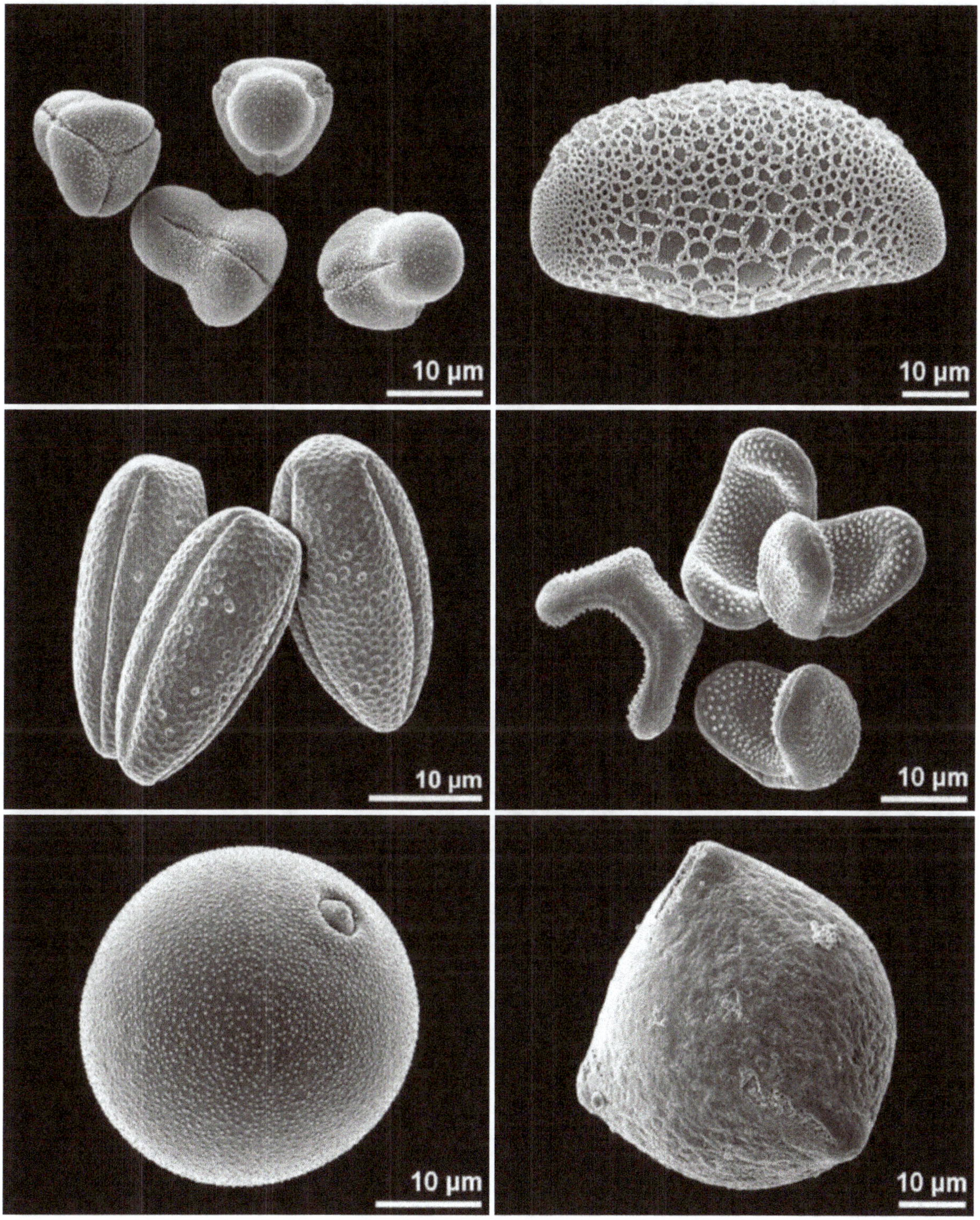

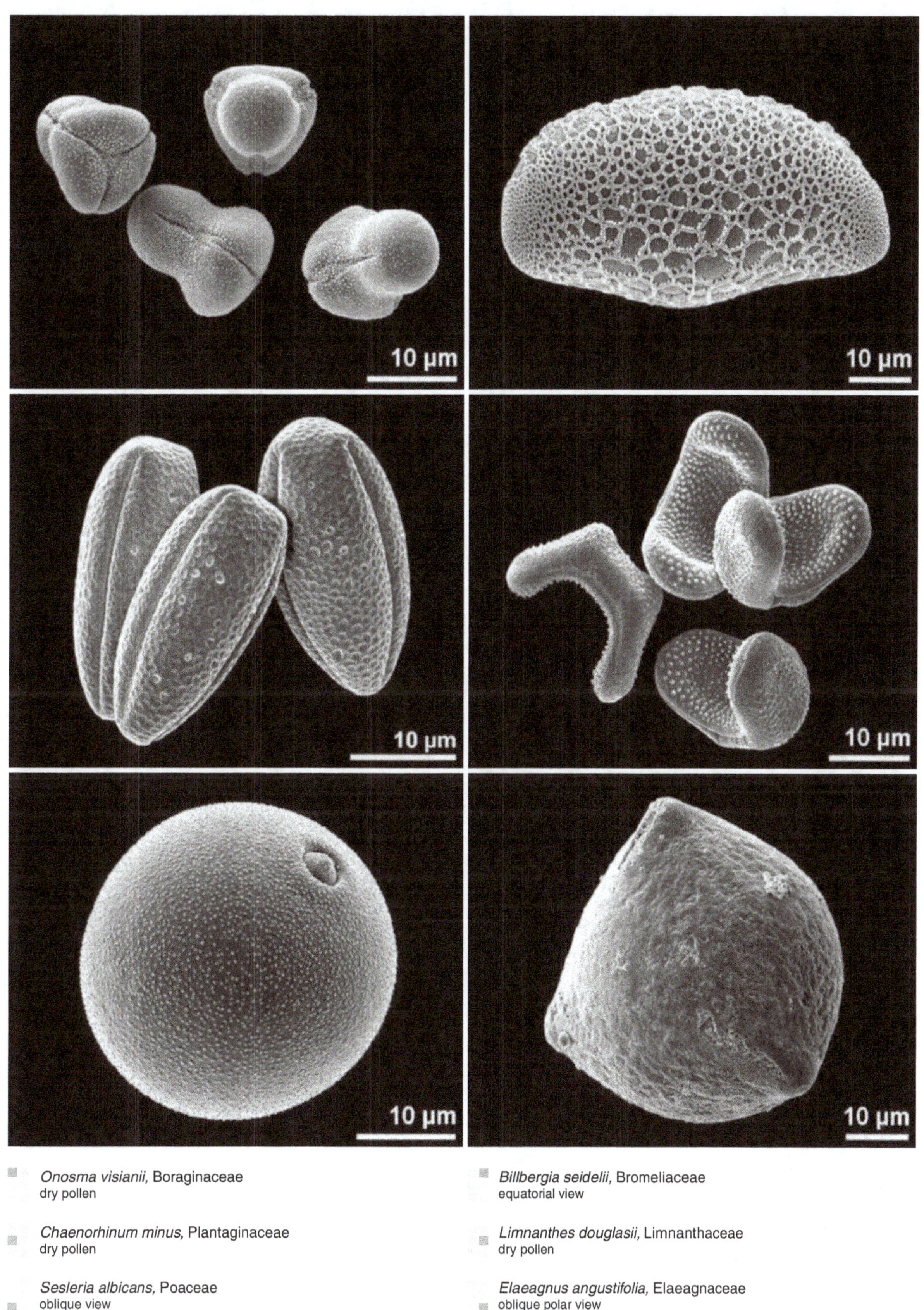

Onosma visianii, Boraginaceae
dry pollen

Billbergia seidelii, Bromeliaceae
equatorial view

Chaenorhinum minus, Plantaginaceae
dry pollen

Limnanthes douglasii, Limnanthaceae
dry pollen

Sesleria albicans, Poaceae
oblique view

Elaeagnus angustifolia, Elaeagnaceae
oblique polar view

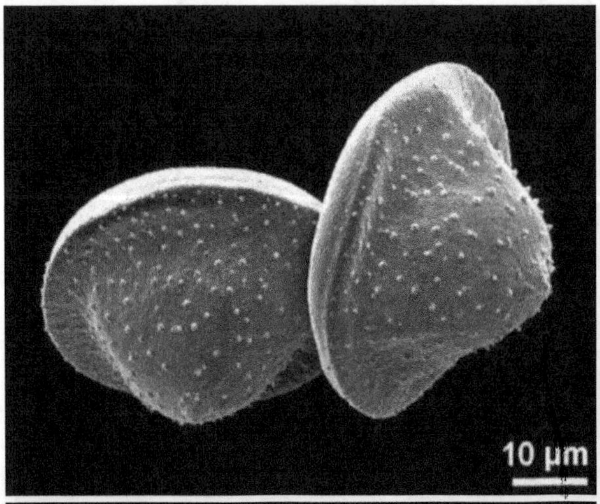

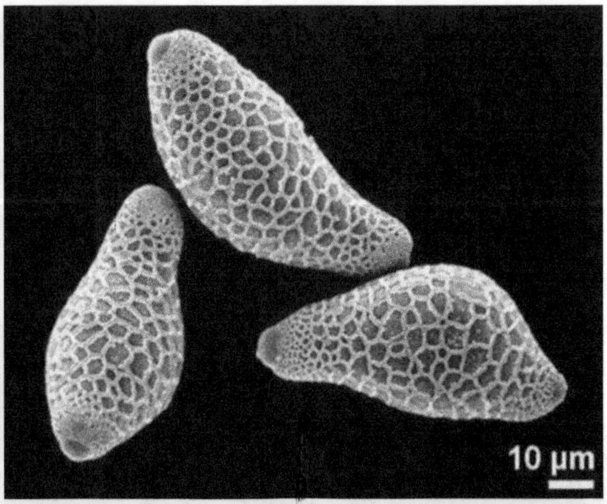

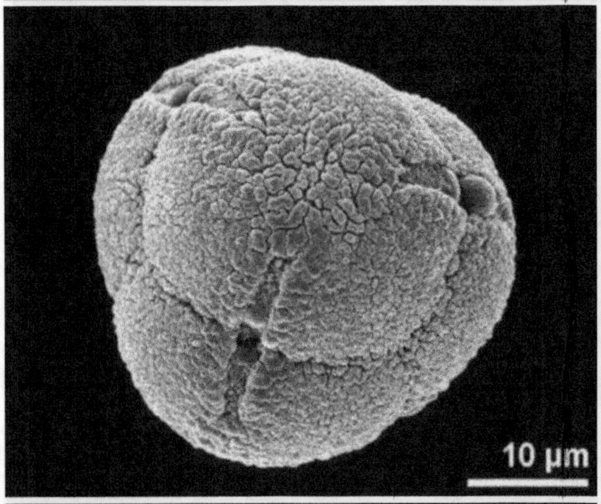

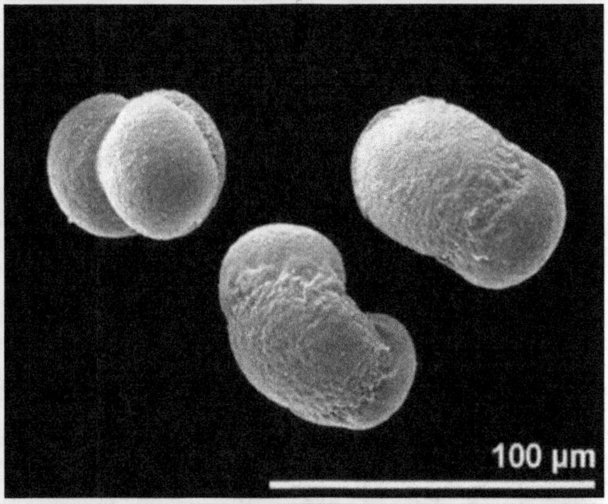

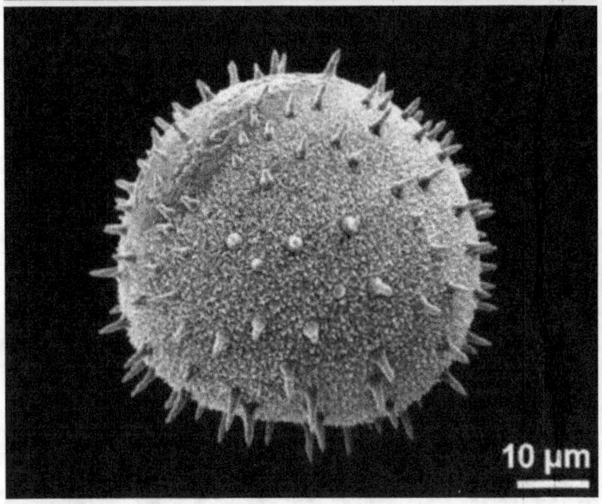

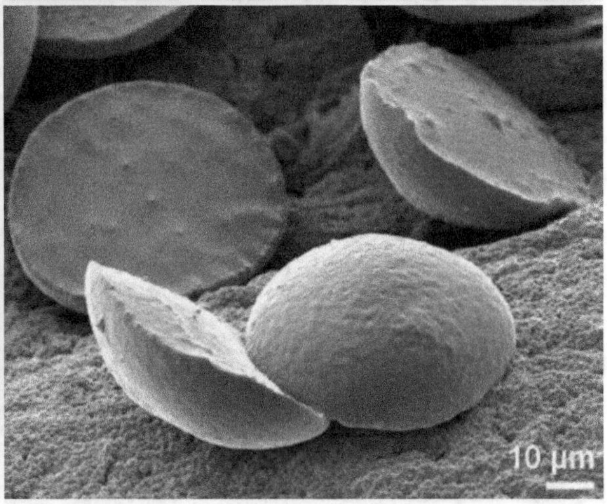

Heliconia sp., Heliconiaceae
dry pollen

Erica arborea, Ericaceae
tetrad

Nuphar lutea, Nymphaeaceae
equatorial view

Quesnelia augusto-coburgii, Bromeliaceae

Pinus strobus, Pinaceae
saccate

Sansevieria parva, Asparagaceae
dry pollen

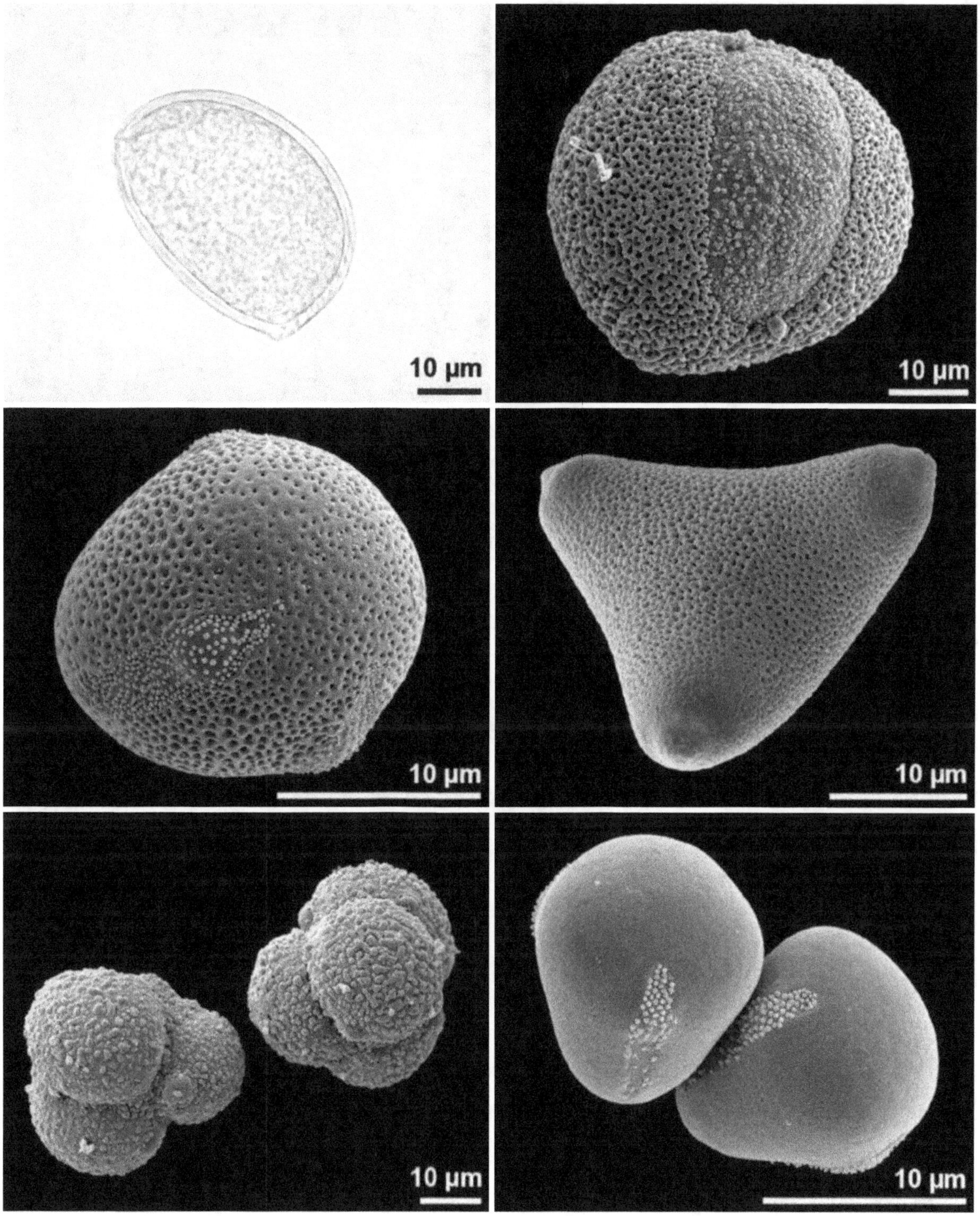

Amorphophallus yunnanensis, Araceae
hydrated, equatorial view

Austrobaileya scandens, Austrobaileyaceae
distal polar view

Echium italicum, Boraginaceae
oblique view

Adenanthos sericeus, Proteaceae
polar view

Calluna vulgaris, Ericaceae
tetrads

Alkanna corcyrensis, Boraginaceae

shape

3-dimensional form of a pollen grain in relation to the P/E-ratio

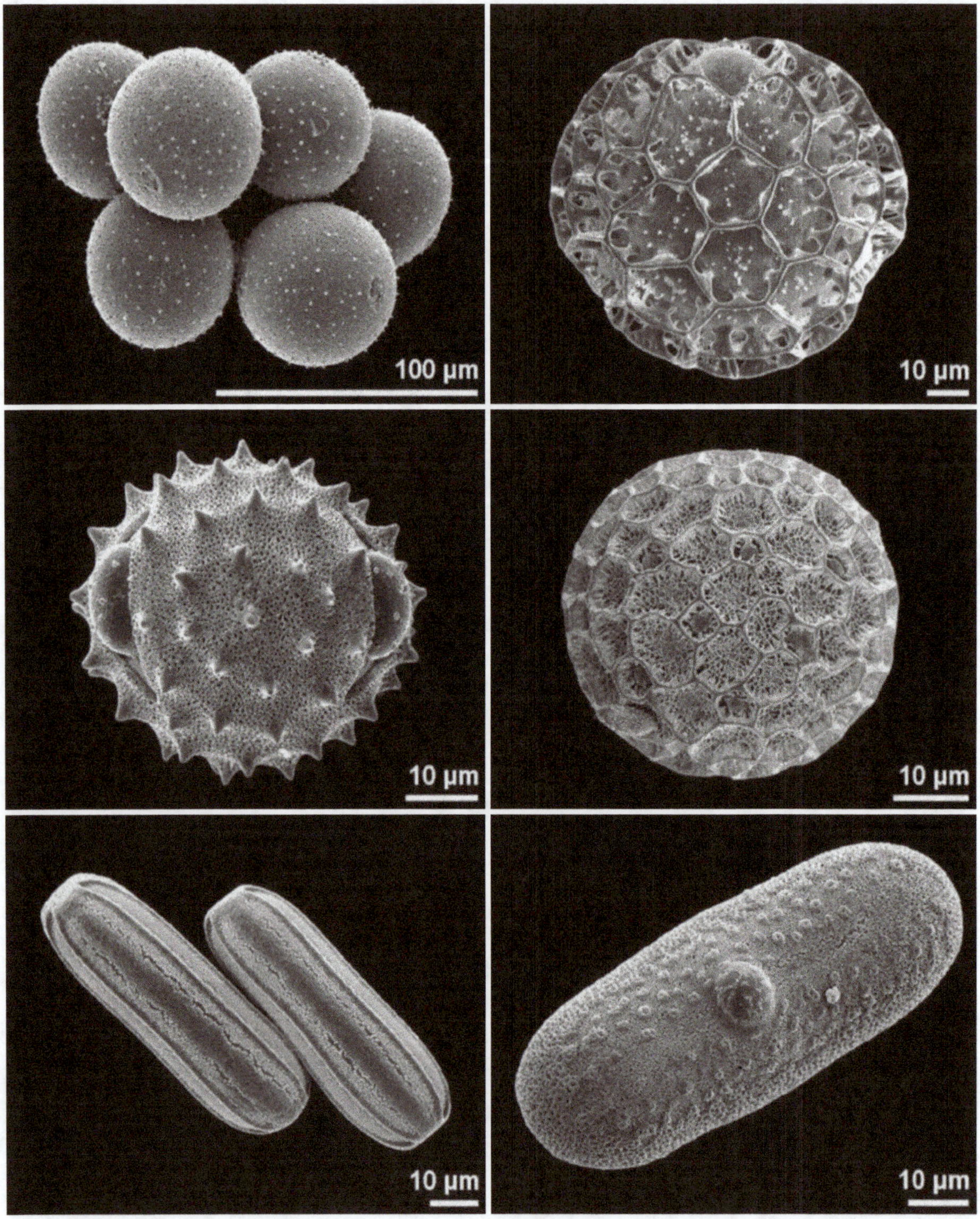

Adansonia gregorii, Malvaceae
spheroidal

Ruellia macrantha, Acanthaceae
spheroidal, polar view

Cirsium oleraceum, Asteraceae
spheroidal, equatorial view

Phlox paniculata, Polemoniaceae
spheroidal

Crossandra flava, Acanthaceae
cylindric, equatorial view

Justicia carnea, Acanthaceae
cylindric, equatorial view

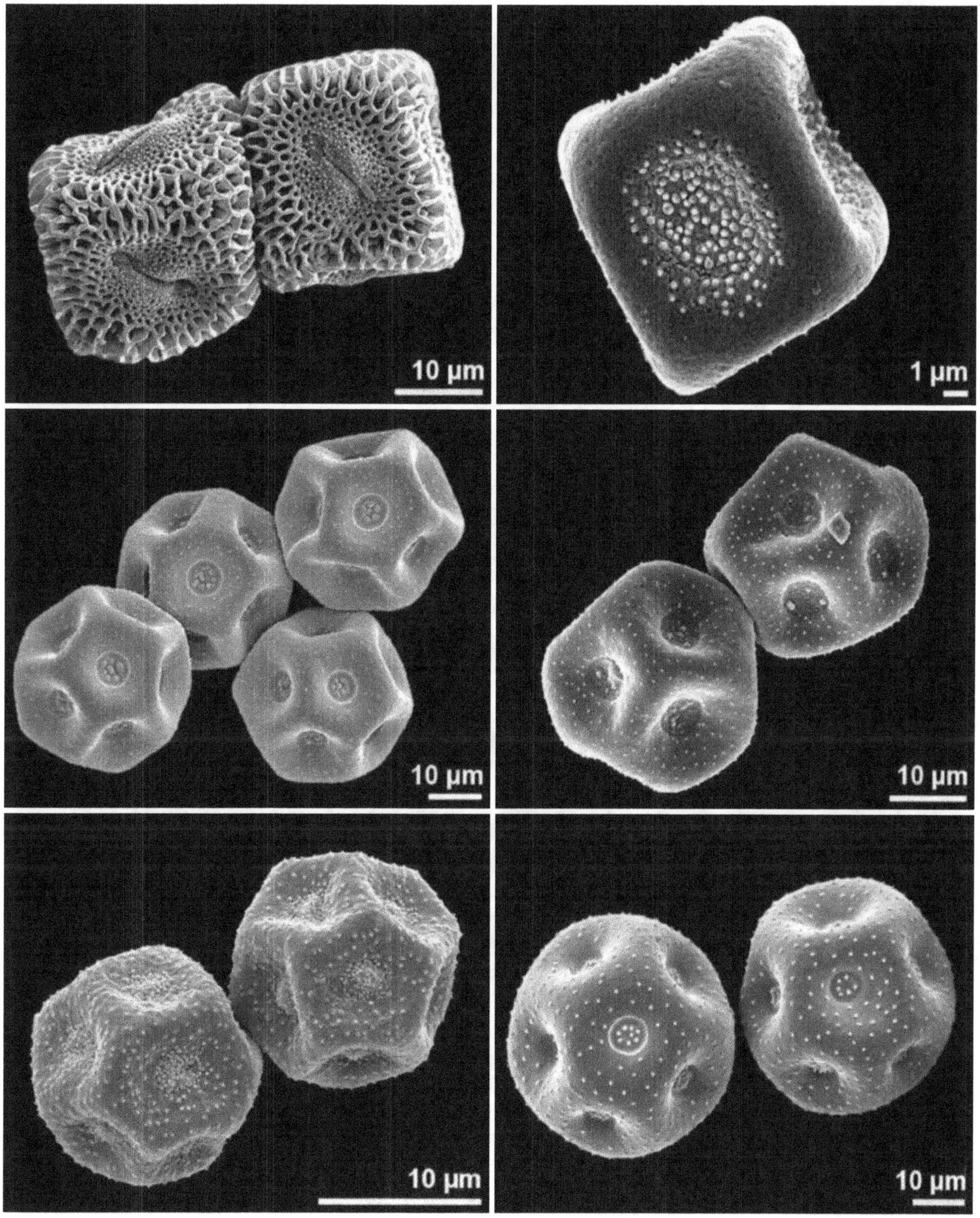

Basella alba, Basellaceae
cubical, dry pollen

Cerastium dubium, Caryophyllaceae
polygonal, dry pollen

Paronychia polygonifolia, Caryophyllaceae
polygonal, dry pollen

Herniaria glabra, Caryophyllaceae
cubical, dry pollen

Eremogone procera, Caryophyllaceae
polygonal, dry pollen

Paronychia polygonifolia, Caryophyllaceae
polygonal, dry pollen

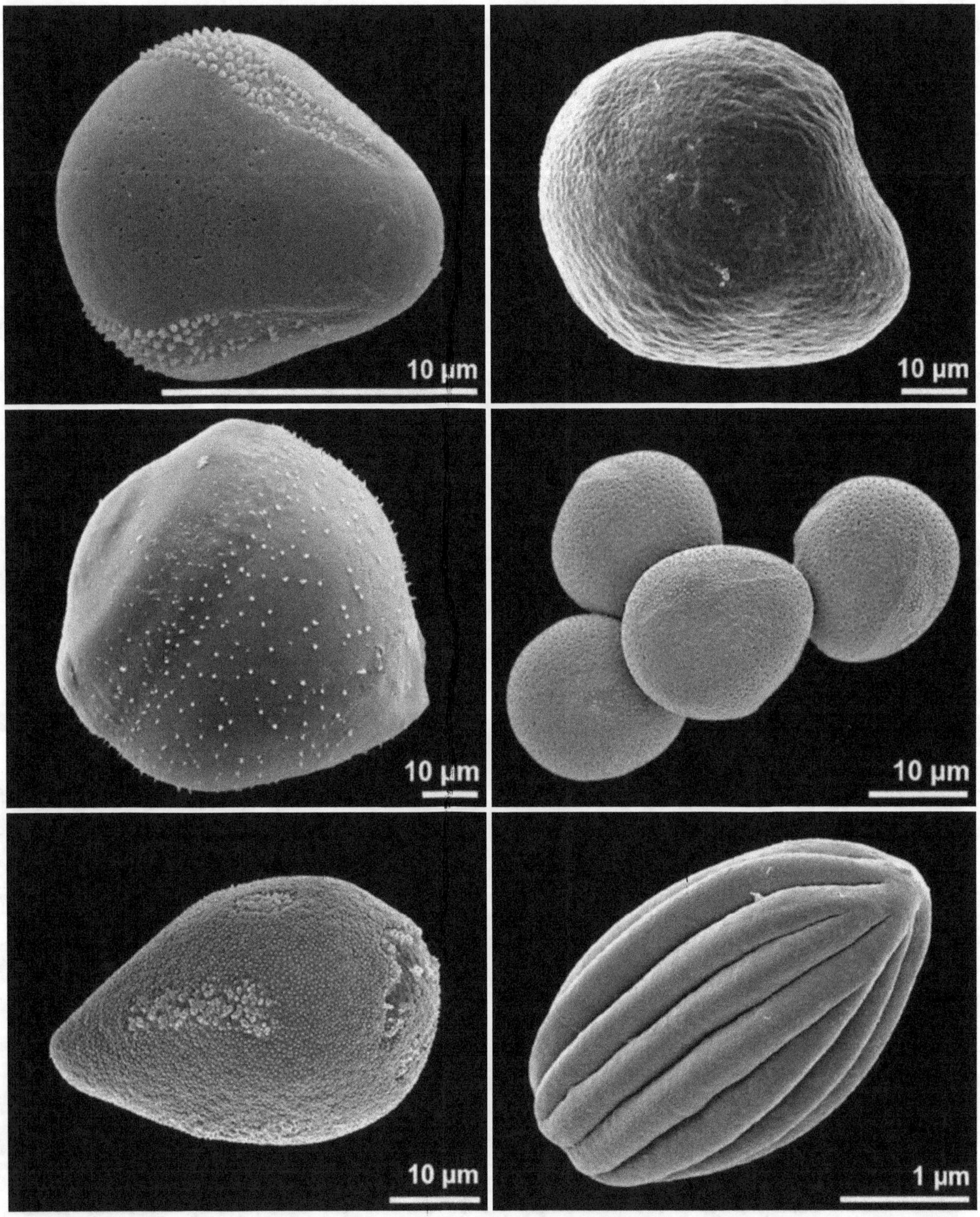

Alkanna corcyrensis, Boraginaceae
pear-shaped, equatorial view

Heliconia rostrata, Heliconiaceae
pear-shaped, equatorial view

Cyperus longus, Cyperaceae
pear-shaped, equatorial view, pseudomonad

Montrichardia arborescens, Araceae
pear-shaped, equatorial view

Echium plantagineum, Boraginaceae
pear-shaped

Ephedra foeminea, Ephedraceae
spindle-shaped, dry pollen

Brugmansia suaveolens, Solanaceae
barrel-shaped, dry pollen

Anthyllis vulneraria, Fabaceae
barrel-shaped, dry pollen

Corydalis cheilanthifolia, Papaveraceae
triangular pyramid, dry pollen

Schoepfia schreberi, Schoepfiaceae
triangular pyramid

Aechmea drakeana, Bromeliaceae
wedge-shaped

Cardiospermum halicacabum, Sapindaceae
convex triangular

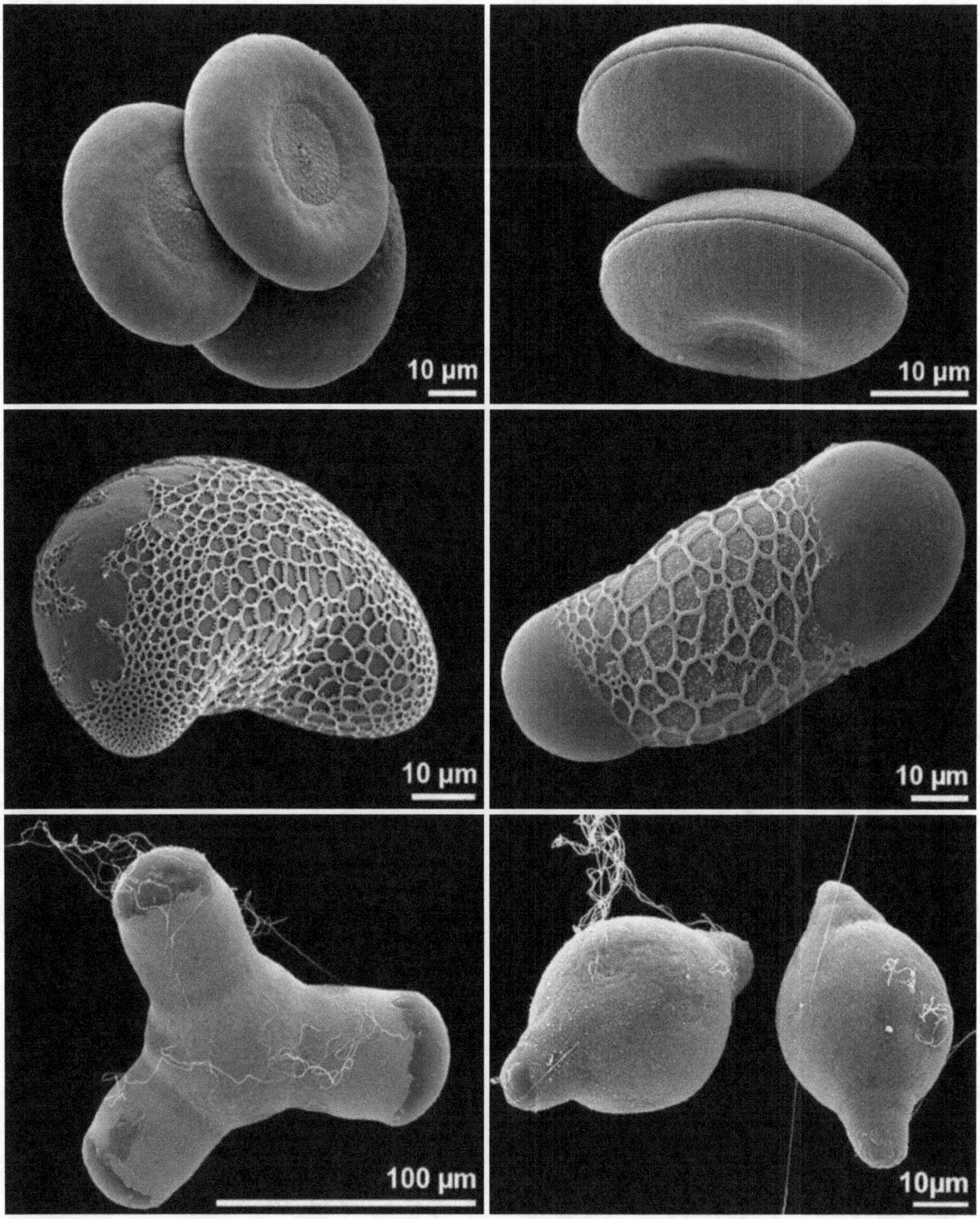

Whitfieldia lateritia, Acanthaceae
disc-shaped, dry pollen

Pedicularis portenschlagii, Orobanchaceae
disc-shaped, dry pollen

Billbergia pyramidalis, Bromeliaceae
bean-shaped

Quesnelia imbricata, Bromeliaceae
cylindric

Clarkia unguiculata, Onagraceae
triangular star

Fuchsia paniculata, Onagraceae
lemon-shaped

Acicarpha tribuloides, Calyceraceae
triangular dipyramid, dry pollen

Nicandra physalodes, Solanaceae
dry pollen

Myosotis alpestris, Boraginaceae
dumbbell-shaped, dry pollen

Loranthus europaeus, Loranthaceae
dry pollen

Sansevieria suffruticosa, Asparagaceae
cup-shaped, dry pollen

Juncus jacquinii, Juncaceae
tetrad, monads cup-shaped, dry pollen

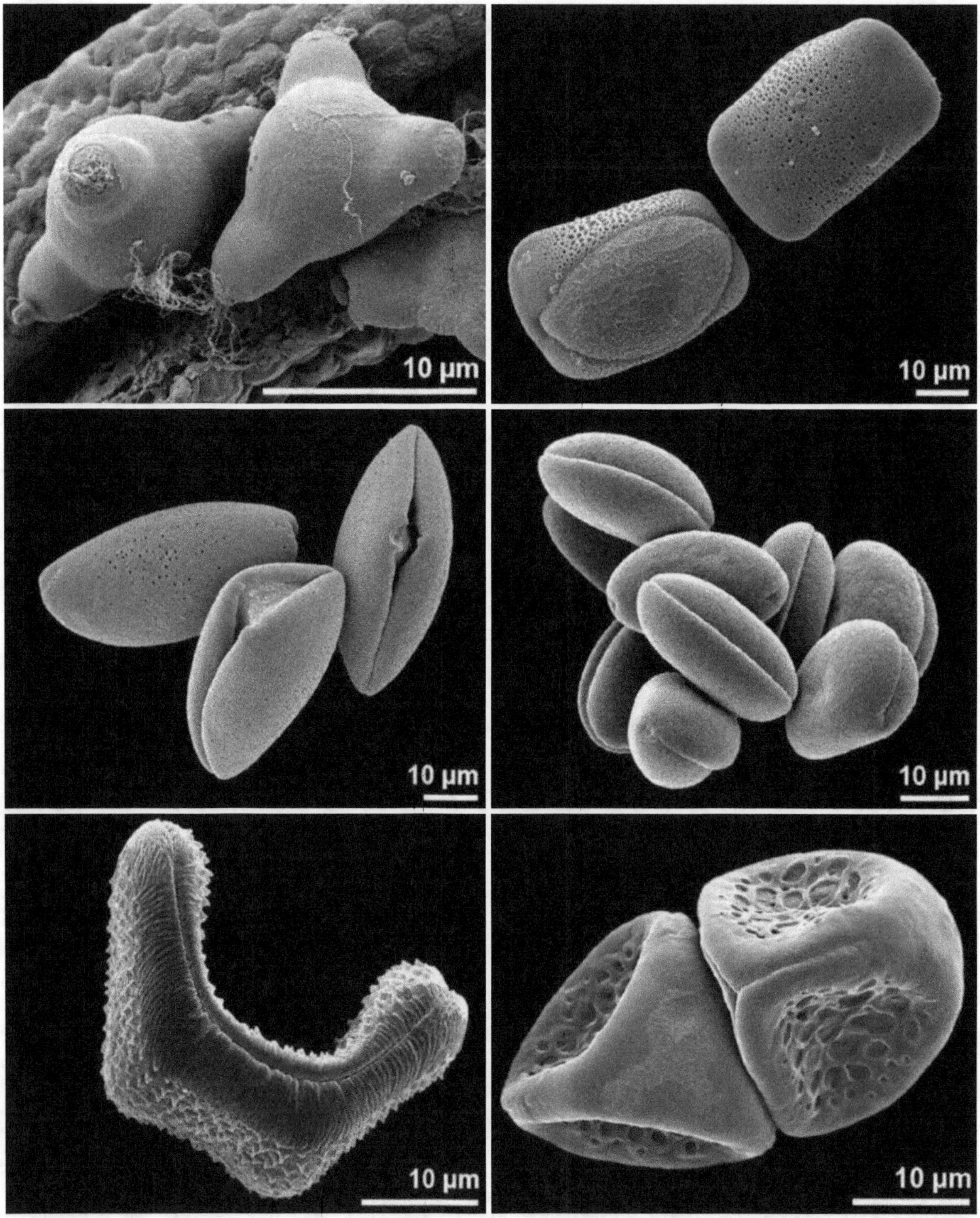

Gaura lindheimeri, Onagraceae

Eremurus robustus, Xanthorrhoeaceae
rounded cuboid

Hyacinthoides italica, Asparagaceae
boat-shaped, dry pollen

Galanthus nivalis, Amaryllidaceae
boat-shaped, dry pollen

Limnanthes douglasii, Limnanthaceae
U-shaped, dry pollen

Thesium dollineri, Santalaceae
convex triangular pyramid, dry pollen

saccus/saccate, corpus

saccus: exinous expansion forming an air sac
corpus: body of a saccate pollen grain

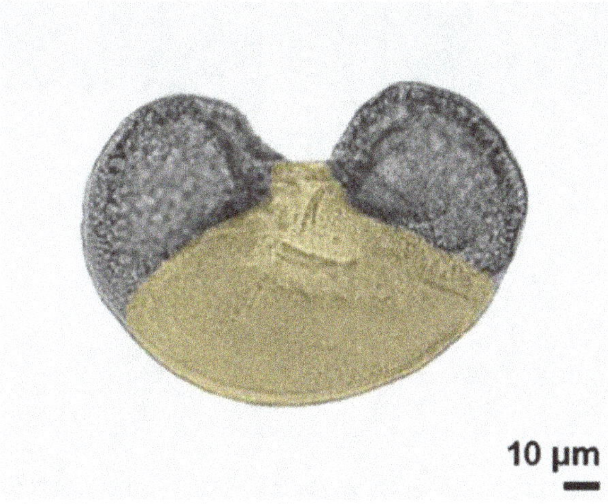

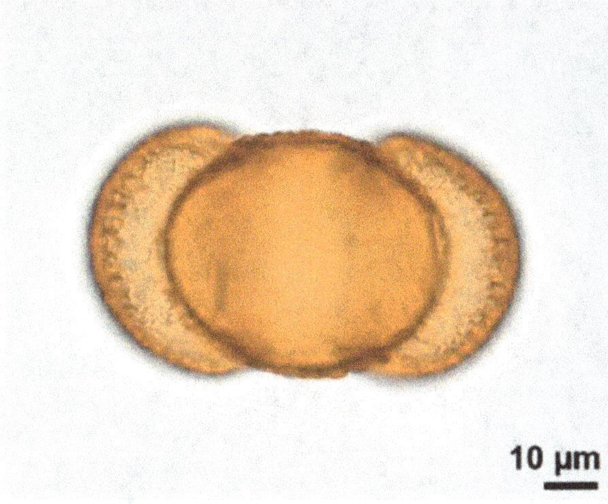

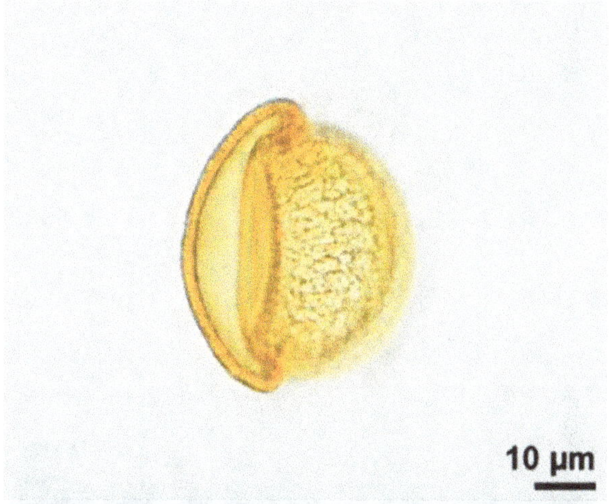

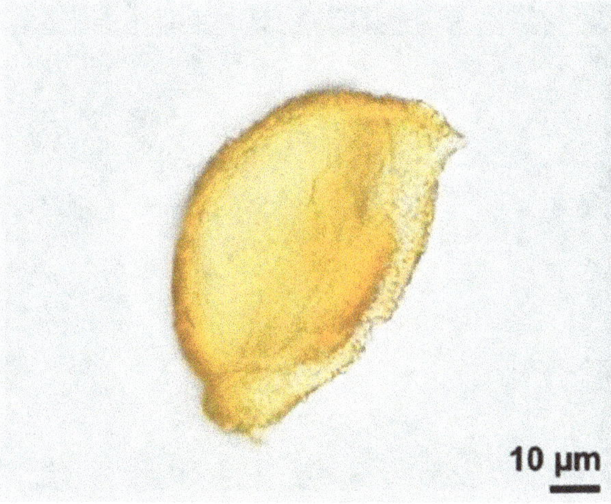

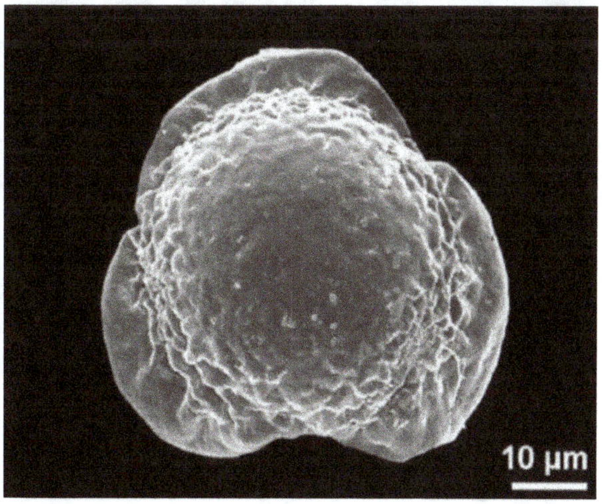

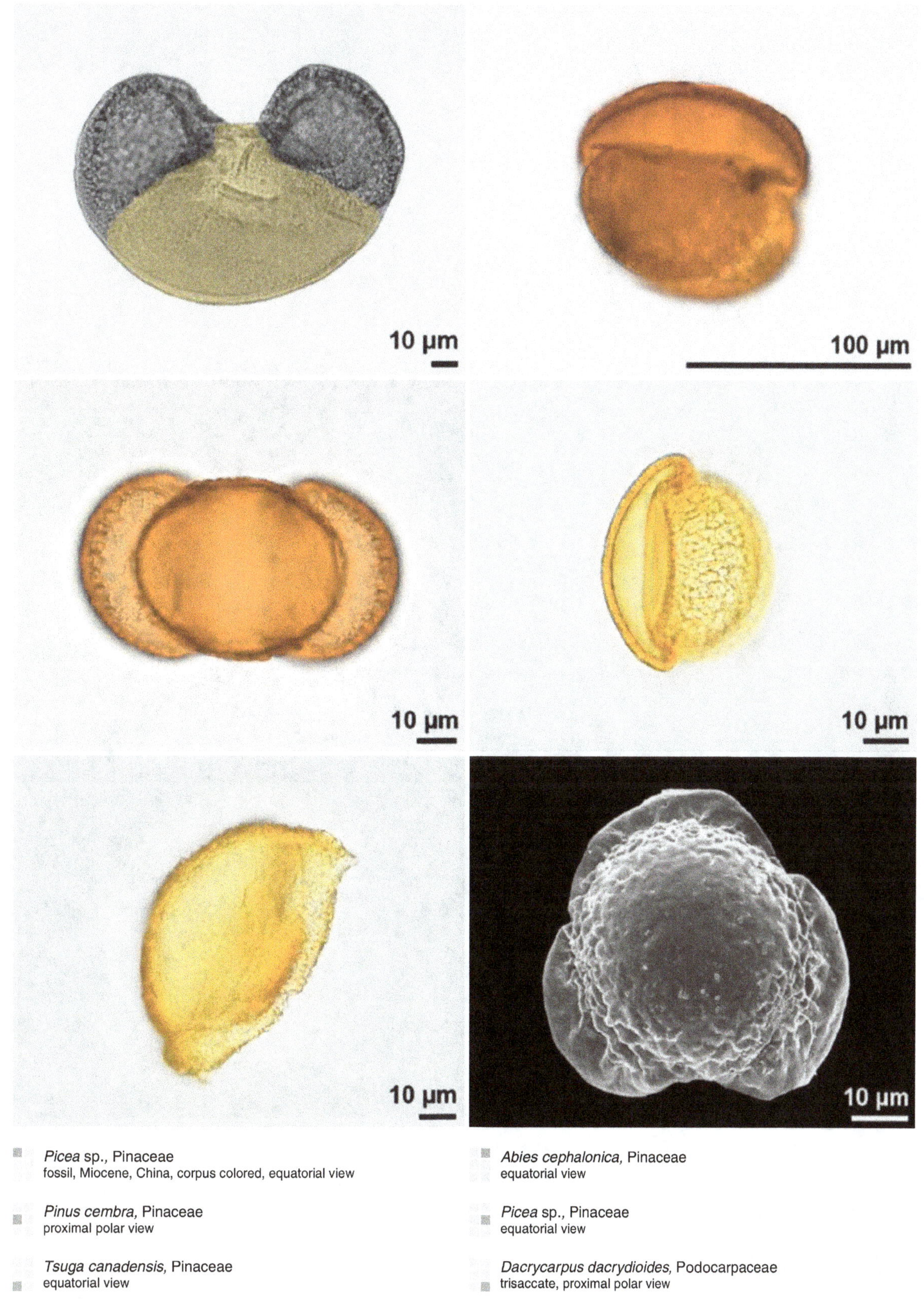

Picea sp., Pinaceae
fossil, Miocene, China, corpus colored, equatorial view

Pinus cembra, Pinaceae
proximal polar view

Tsuga canadensis, Pinaceae
equatorial view

Abies cephalonica, Pinaceae
equatorial view

Picea sp., Pinaceae
equatorial view

Dacrycarpus dacrydioides, Podocarpaceae
trisaccate, proximal polar view

saccus, monosaccate

monosaccate: pollen grain with a single saccus

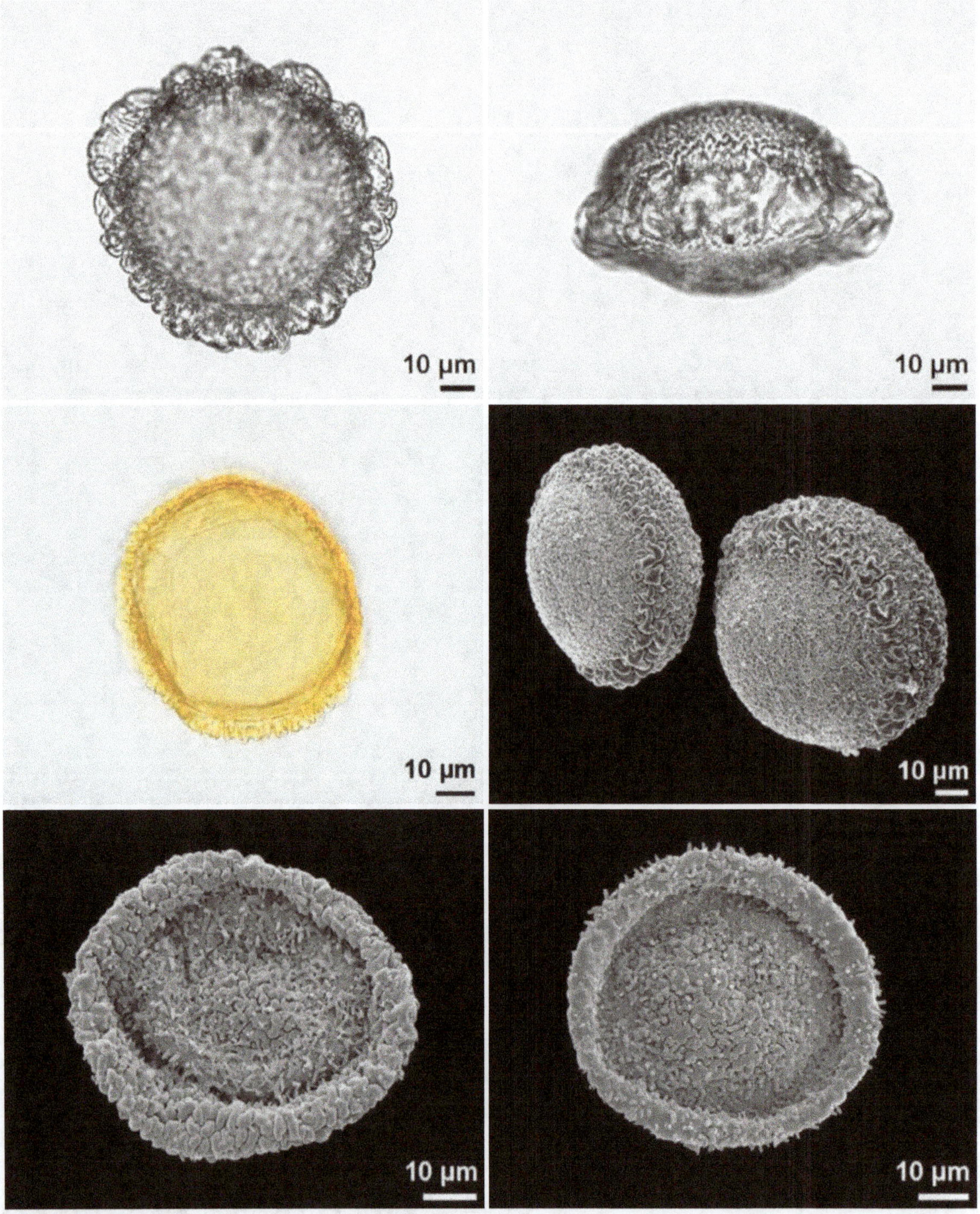

Tsuga sp., Pinaceae
fossil, middle Miocene, Austria, polar view

Tsuga sp., Pinaceae
fossil, middle Miocene, Austria, equatorial view

Tsuga canadensis, Pinaceae
polar view

Tsuga canadensis, Pinaceae
equatorial view

Tsuga sp., Pinaceae
fossil, Miocene, China, polar view

Tsuga sp., Pinaceae
fossil, Miocene, China, polar view

saccus, bisaccate

bisaccate: pollen grain with two sacci

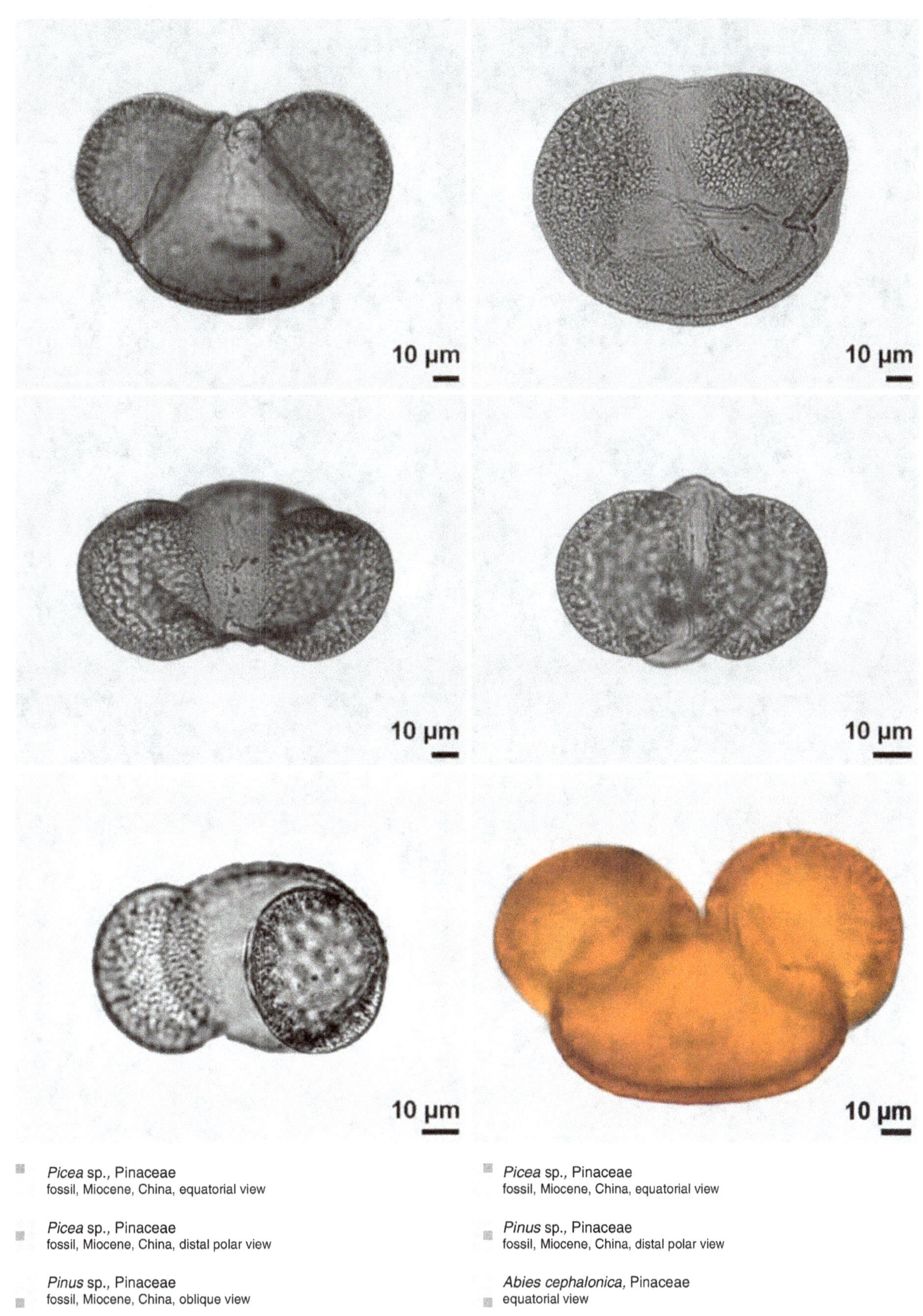

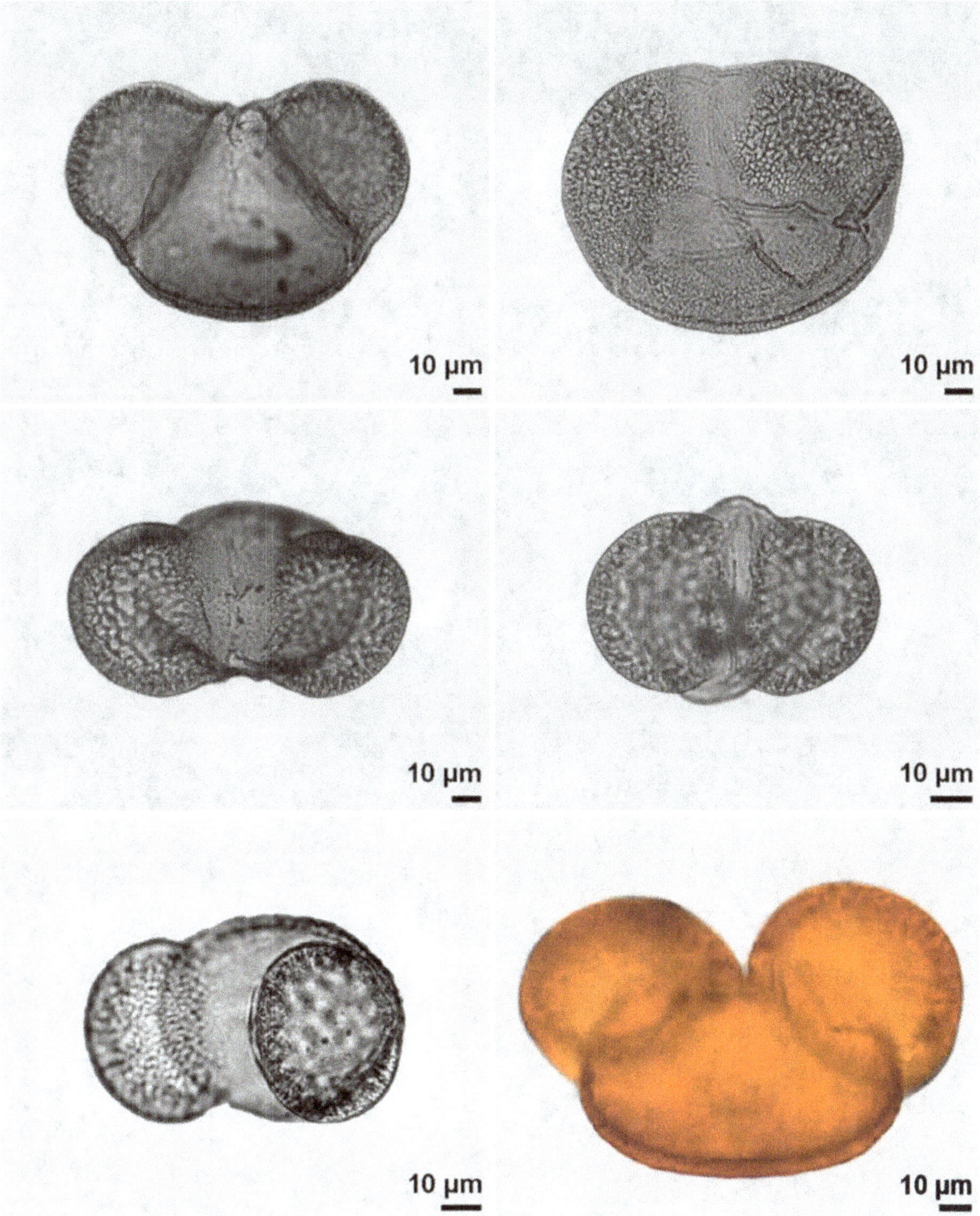

Picea sp., Pinaceae
fossil, Miocene, China, equatorial view

Picea sp., Pinaceae
fossil, Miocene, China, distal polar view

Pinus sp., Pinaceae
fossil, Miocene, China, oblique view

Picea sp., Pinaceae
fossil, Miocene, China, equatorial view

Pinus sp., Pinaceae
fossil, Miocene, China, distal polar view

Abies cephalonica, Pinaceae
equatorial view

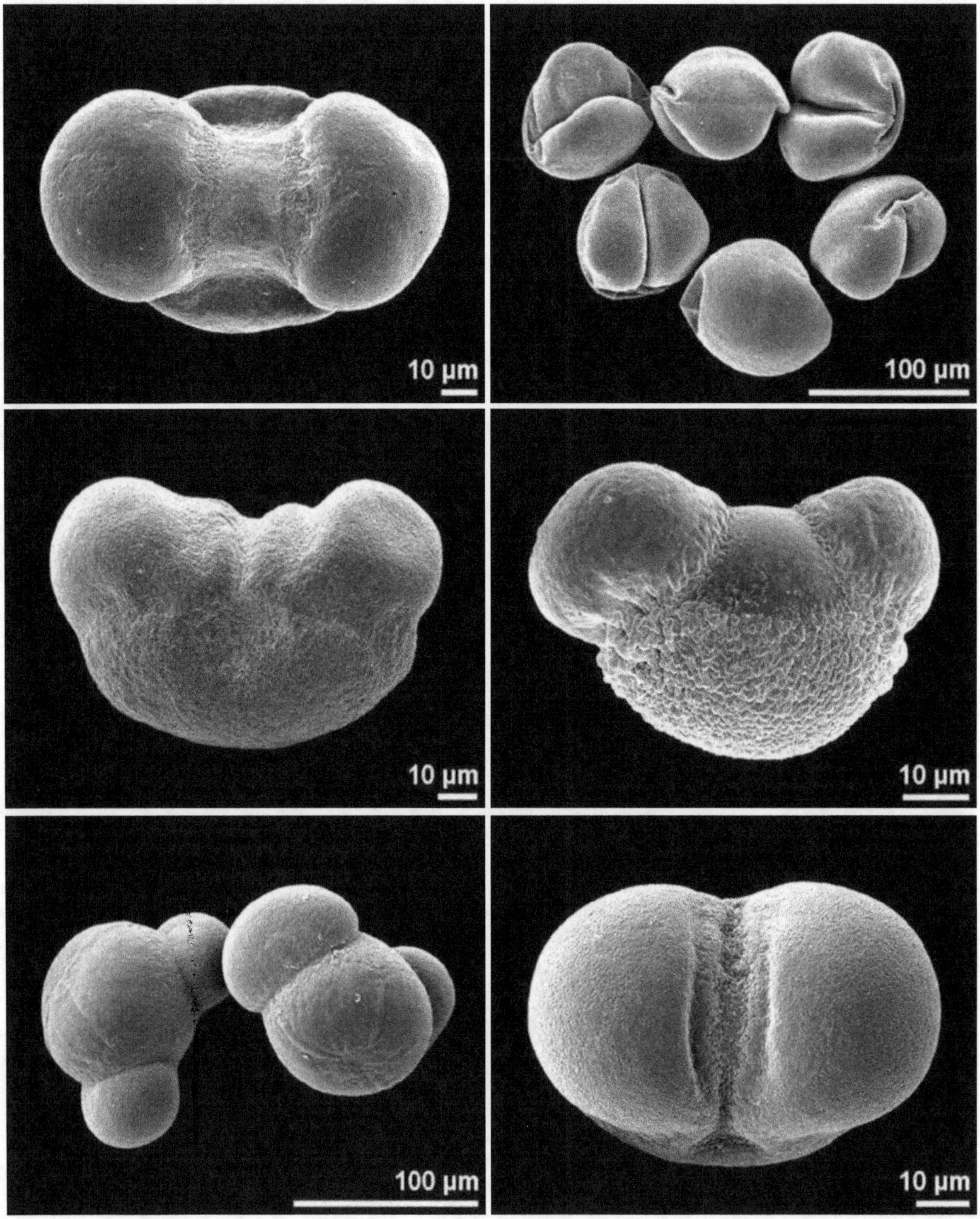

Abies cephalonica, Pinaceae
distal polar view

Picea abies, Pinaceae
equatorial view

Abies nordmanniana, Pinaceae

Picea abies, Pinaceae
dry pollen

Pinus mugo, Pinaceae
equatorial view

Picea pungens, Pinaceae
distal polar view

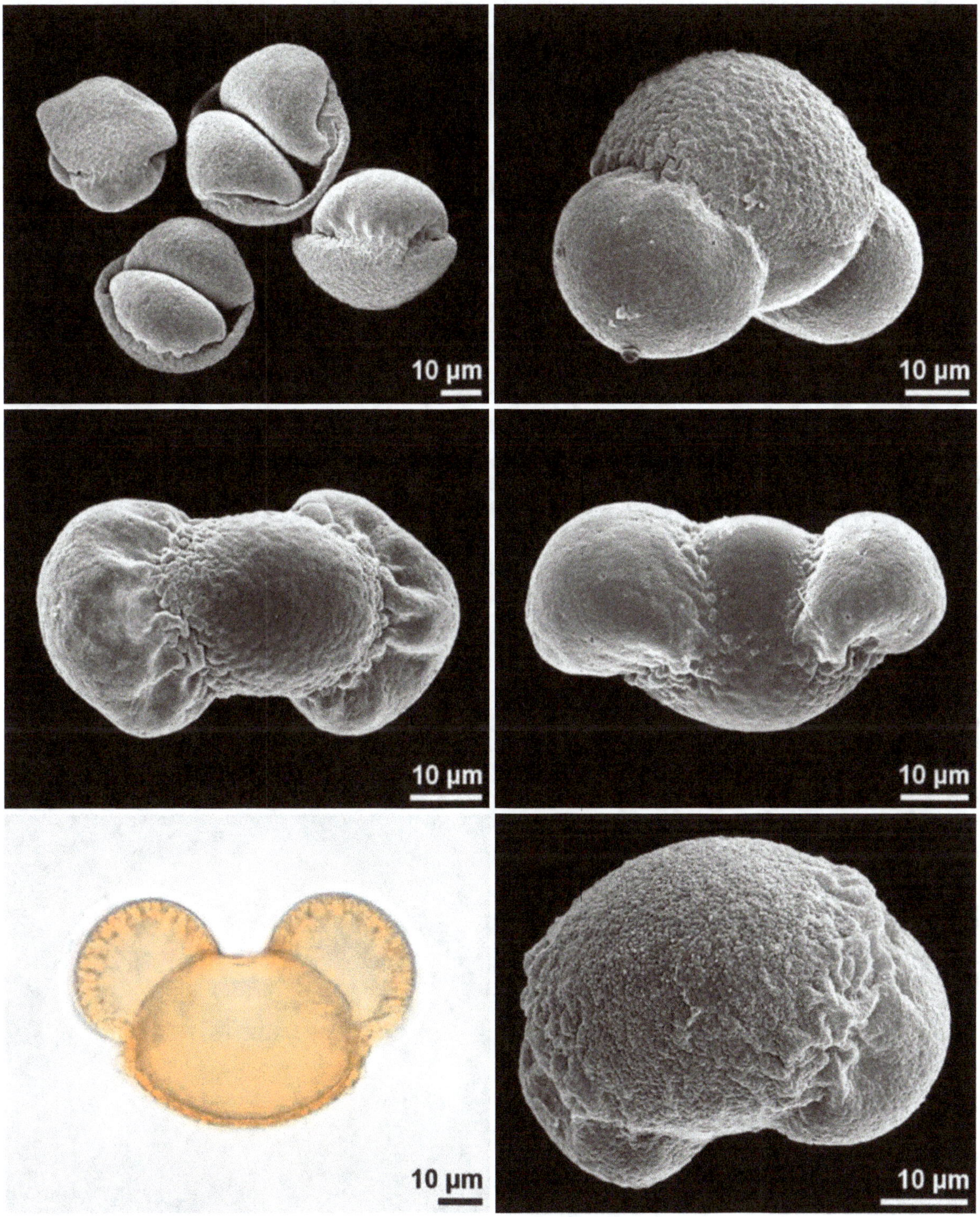

Pinus heldreichii, Pinaceae
dry pollen

Podocarpus sp., Podocarpaceae
proximal polar view

Pinus mugo, Pinaceae
equatorial view

Pinus nigra, Pinaceae
equatorial view

Podocarpus sp., Podocarpaceae
equatorial view

Pinus contorta, Pinaceae
oblique proximal polar view

saccus, trisaccate

trisaccate: pollen grain with three sacci

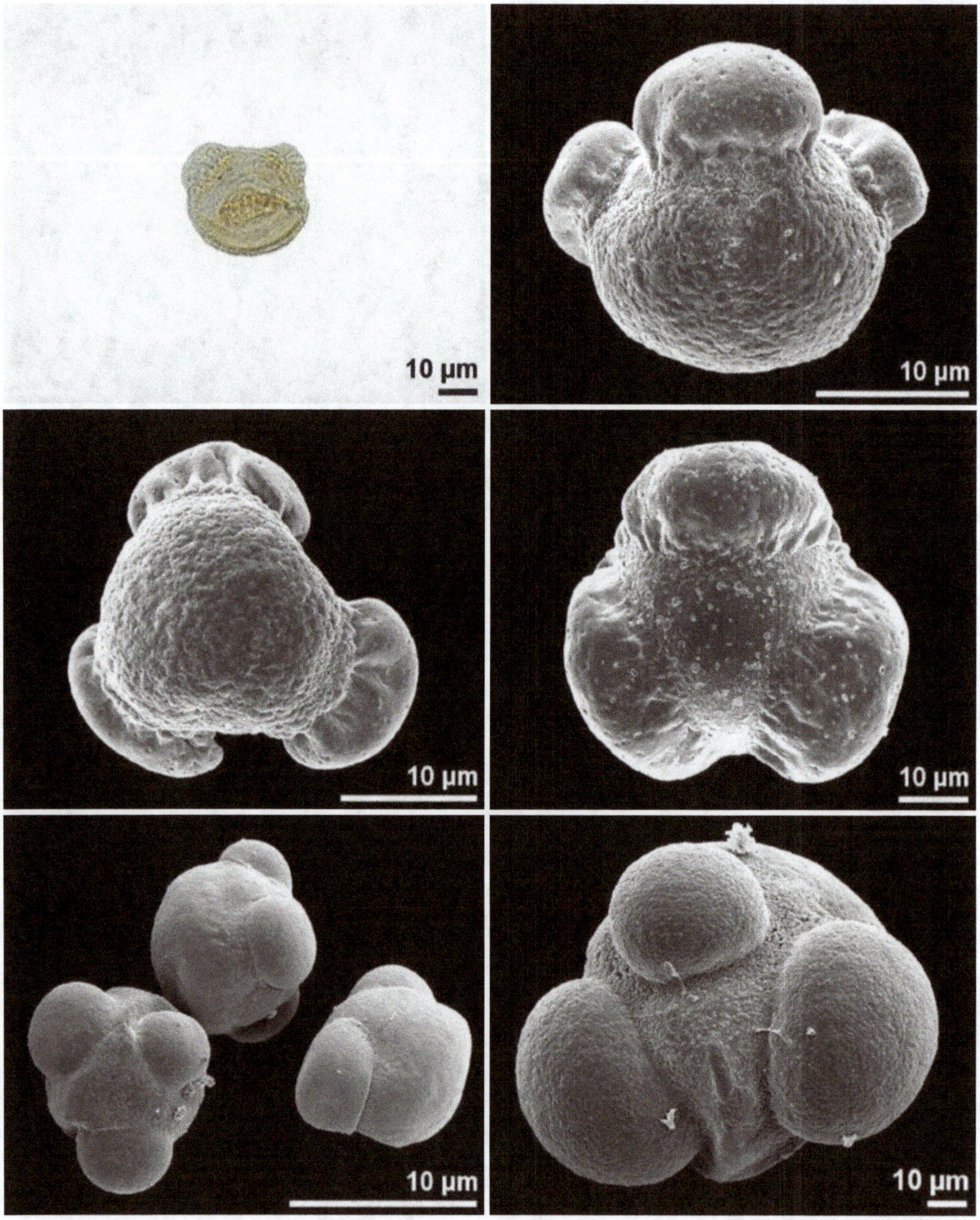

Podocarpaceae
fossil, early Miocene, South-Africa, oblique view

Pherosphaera hookeriana, Podocarpaceae
equatorial view

Pherosphaera hookeriana, Podocarpaceae
proximal polar view

Dacrycarpus dacrydioides, Podocarpaceae
distal polar view

Abies concolor, Pinaceae

Abies concolor, Pinaceae
distal polar view

infoldings, apertures sunken

infoldings (dry pollen): consequence of harmomegathy in dry condition

Fritillaria pontica, Liliaceae
sulcate

Galium odoratum, Rubiaceae
colpate

Roemeria hybrida, Papaveraceae
pantoporate

Veratrum album, Melanthiaceae
sulcate

Sparmannia africana, Tiliaceae
colporate

Bifora radians, Apiaceae
tricolporate

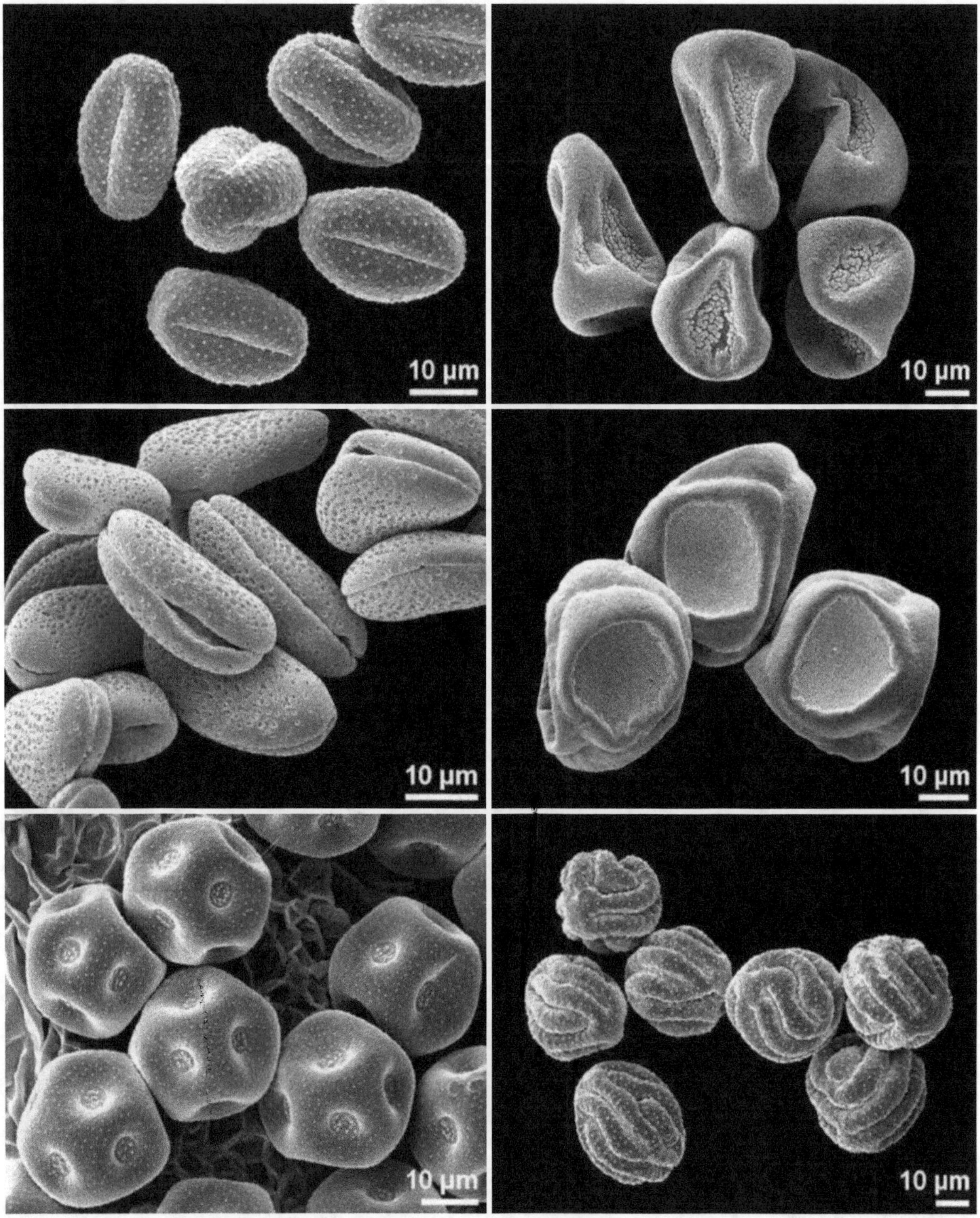

Artemisia pontica, Asteraceae
colporate

Lachenalia aloides, Asparagaceae
sulcate

Moehringia muscosa, Caryophyllaceae
pantoporate

Carex alba, Cyperaceae
pseudomonads, poroidate

Luzula sylvatica, Juncaceae
tetrads, ulcerate

Anemone hortensis, Ranunculaceae
spiraperturate

infoldings, boat-shaped

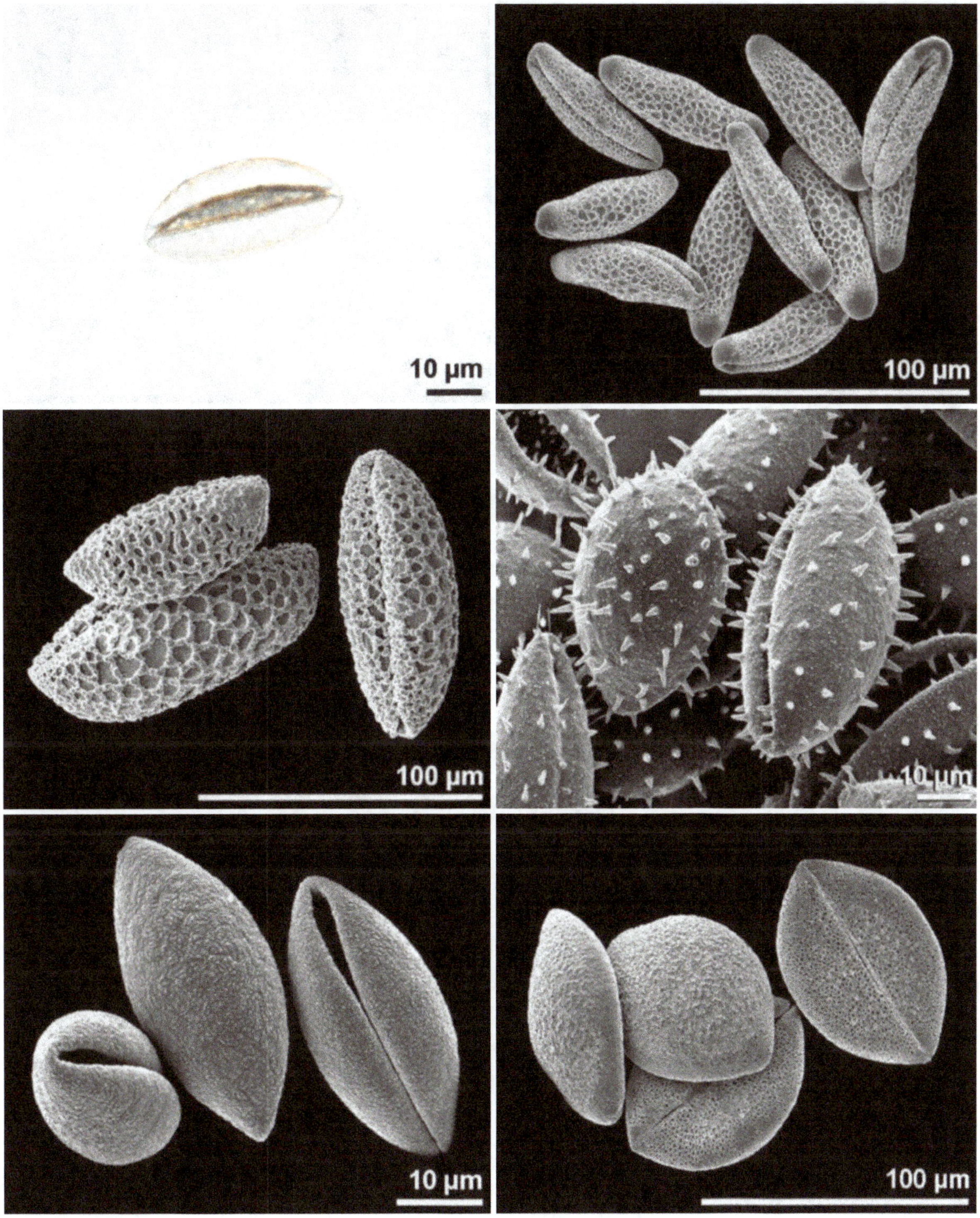

Dracontium asperum, Araceae
sulcate

Lilium candidum, Liliaceae
sulcate

Ginkgo biloba, Ginkgoaceae
sulcate

Billbergia seidelii, Bromeliaceae
sulcate

Nuphar lutea, Nymphaeaceae
sulcate

Asphodeline lutea, Xanthorrhoeaceae
sulcate

Lysichiton americanus, Araceae
sulcate

Gagea lutea, Liliaceae
sulcate

Dioon edule, Zamiaceae
sulcate

Piper nigrum, Piperaceae
sulcate

Sparganium erectum, Typhaceae
ulcerate

Symplocarpus foetidus, Araceae
sulcate

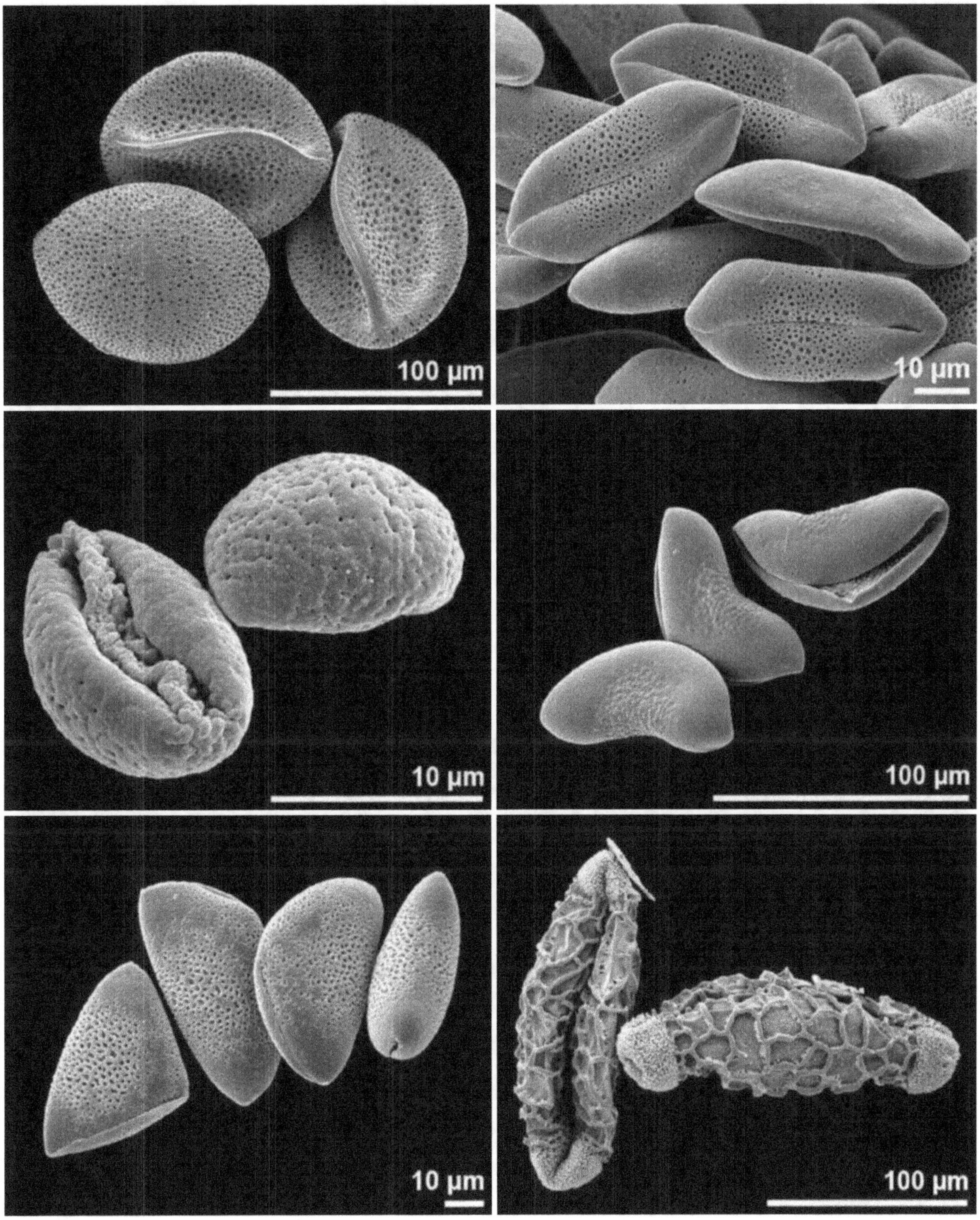

Asphodelus fistulosus, Xanthorrhoeaceae
sulcate

Eremurus thiodanthus, Xanthorrhoeaceae
sulcate

Piper auritum, Piperaceae
sulcate

Wachendorfia thyrsiflora, Haemodoraceae
sulcate

Haemanthus coccineus, Amaryllidaceae
sulcate

Hymenocallis tubiflora, Amaryllidaceae
sulcate

infoldings, cup-shaped

Bougainvillea sp., Nyctaginaceae
colpate

Tilia euchlora, Malvaceae
colporate

Luzula campestris, Juncaceae
tetrads, ulcerate

Heliconia sp., Heliconiaceae
ulcerate

Elaeagnus angustifolia, Elaeagnaceae
colporate

Tsuga canadensis, Pinaceae
leptoma

Adenanthos sericeus, Proteaceae
porate

Petrea volubilis, Verbenaceae
brevicolpate

Cunninghamia lanceolata, Cupressaceae
leptoma

Heliconia stricta, Heliconiaceae
ulcerate

Leucadendron brunioides, Proteaceae
porate

Hibiscus schizopetalus, Malvaceae
porate

infoldings, interapertural area sunken

Alnus glutinosa, Betulaceae
porate

Bupleurum rotundifolium, Apiaceae
colporate

Leucadendron discolor, Proteaceae
porate

Erica arborea, Ericaceae
tetrads, colporate

Melampyrum arvense, Orobanchaceae
colpate

Melastoma sanguineum, Melastomataceae
colporate, heteroaperturate

Verbena officinalis, Verbenaceae
colporate

Ardisia crenata, Primulaceae
syncolporate

Grevillea banksii, Proteaceae
porate

Tsusiophyllum tanakae, Ericaceae
tetrads, colporate

Thesium arvense, Santalaceae
colpate, triradiate colpus

Tropaeolum moritzianum, Tropaeolaceae
colpate

infoldings, irregular

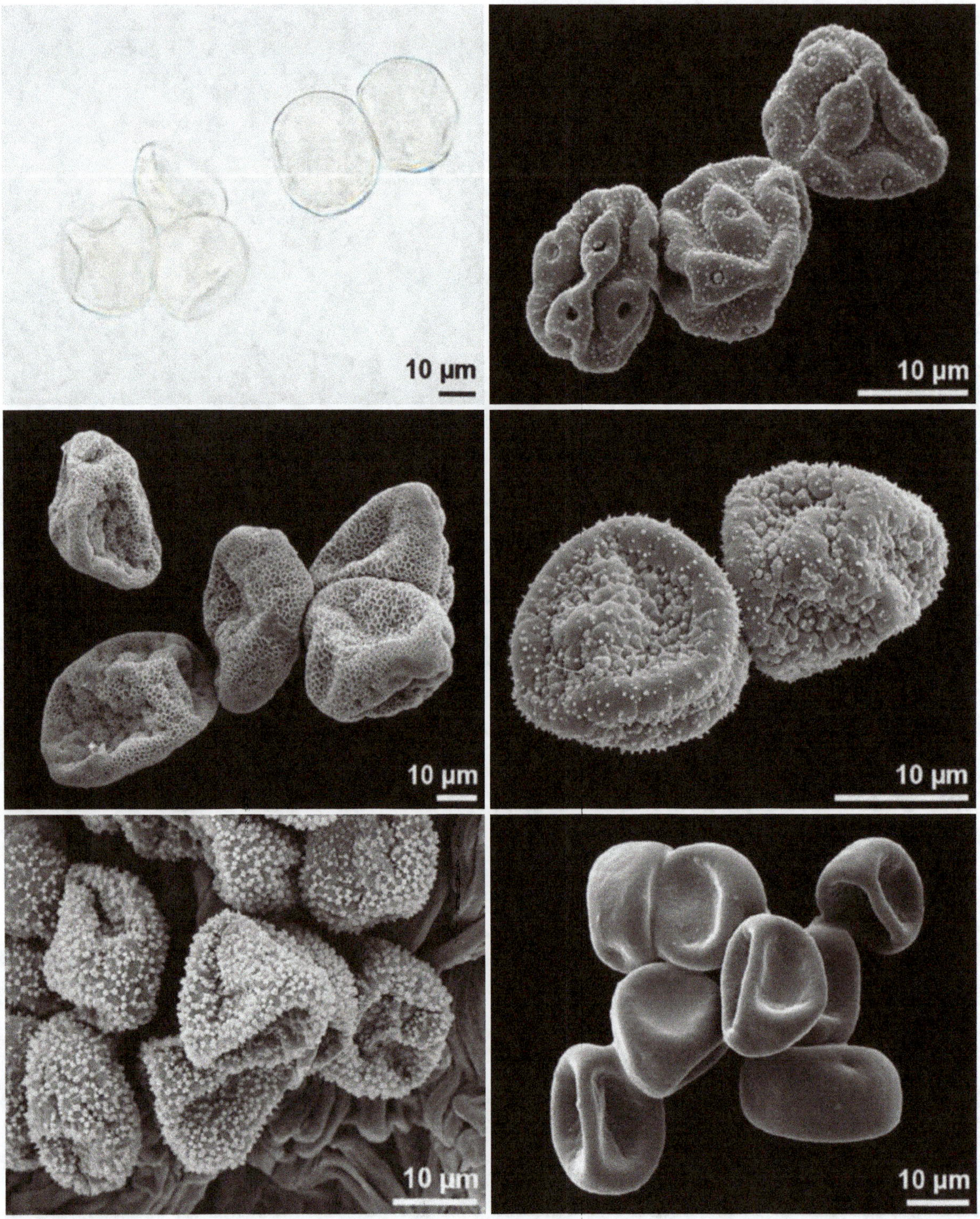

Poaceae
ulcerate

Guzmania elvallensis, Bromeliaceae
sulcate

Callitriche stagnalis, Plantaginaceae
inaperturate

Dorstenia contrajerva, Moraceae
pantoporate

Smilax spinosa, Smilacaceae
aperture condition unclear

Vanilla planifolia, Orchidaceae
porate

Urtica dioica, Urticaceae
porate

Sesleria albicans, Poaceae
ulcerate

Coriaria nepalensis, Coriariaceae
porate

Populus alba, Salicaceae
inaperturate

Anthurium radicans, Araceae
porate

Orobanche hederae, Orobanchaceae
inaperturate

Aperture

H. Halbritter et al., *Illustrated Pollen Terminology*, https://doi.org/10.1007/978-3-319-71365-6_9

angulaperturate

pollen grain with an angular outline where the apertures are located at the angles

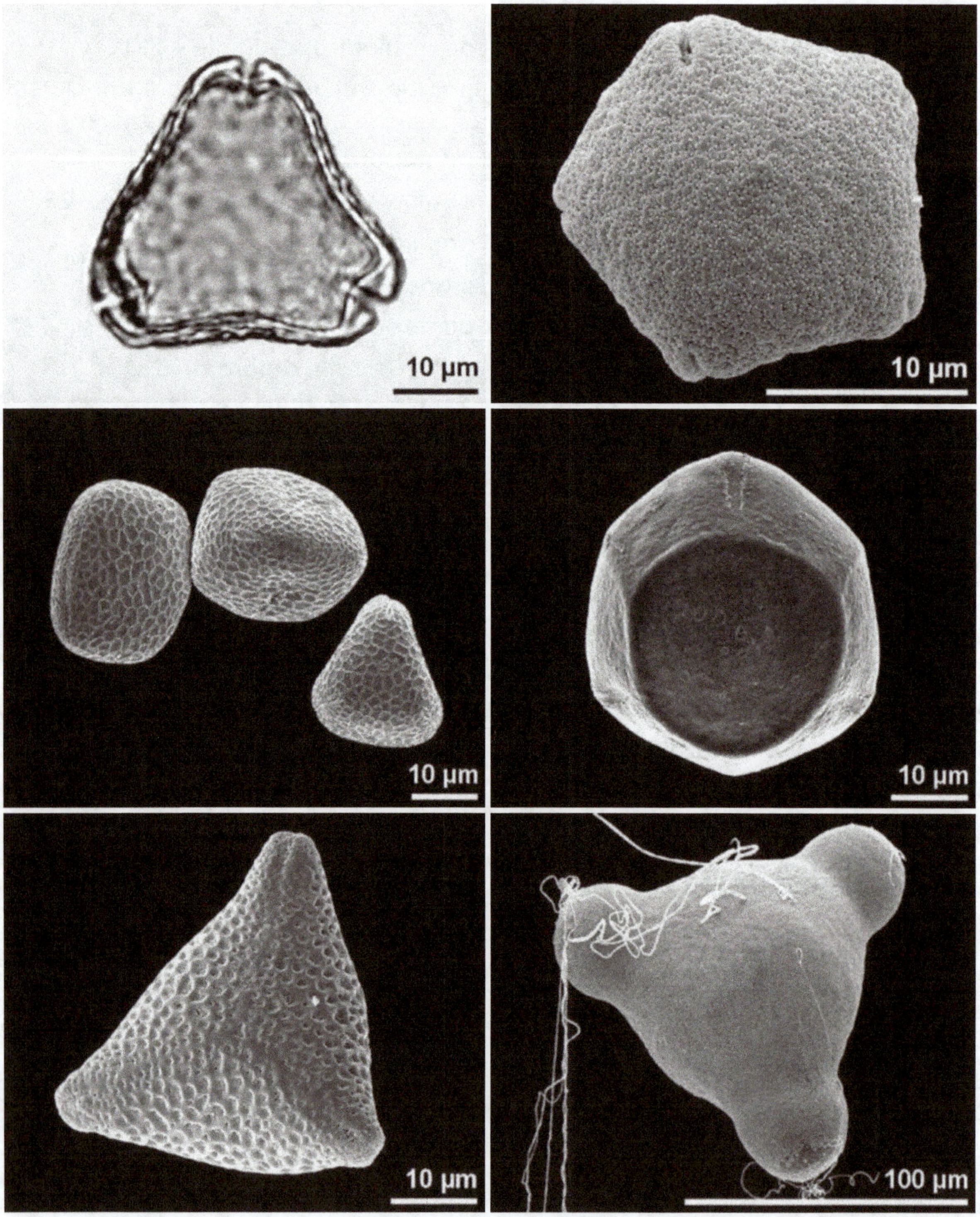

Symplocos sp., Symplocaceae
fossil, middle Miocene, Austria, polar view

Tropaeolum majus, Tropaeolaceae
dry pollen

Cardiospermum halicacabum, Sapindaceae
polar view

Acalypha macrostachya, Euphorbiaceae
polar view

Anthyllis vulneraria, Fabaceae
polar view

Oenothera biennis, Onagraceae
polar view, viscin threads

annulus/annulate

ring like wall thickening surrounding a porus or ulcus

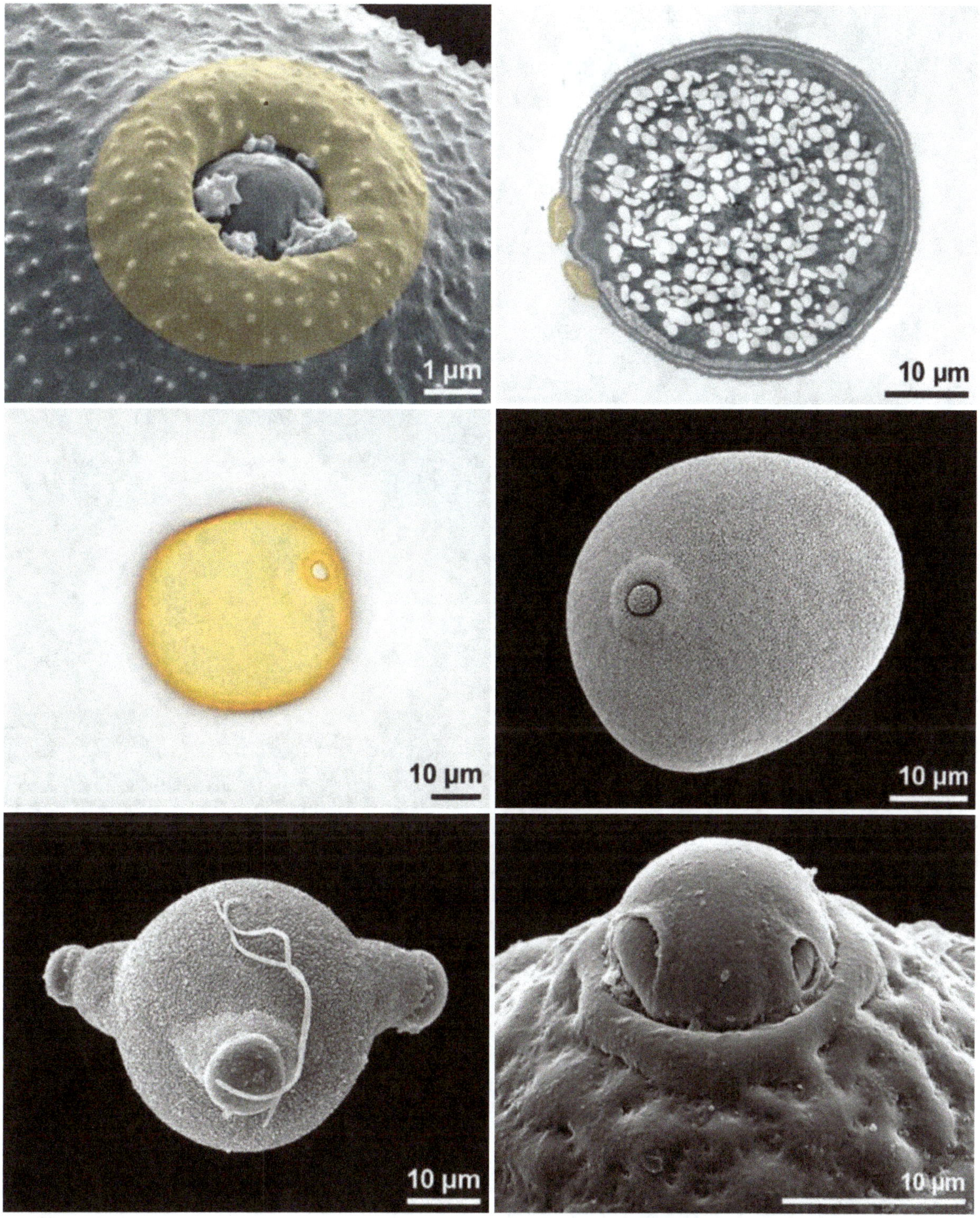

Betula humilis, Betulaceae
annulus colored

Triticum aestivum, Poaceae
cross section, annulus colored

Poaceae
oblique view

Secale cereale, Poaceae
oblique distal polar view

Circaea lutetiana, Onagraceae
equatorial view, viscin threads

Fumaria officinalis, Papaveraceae

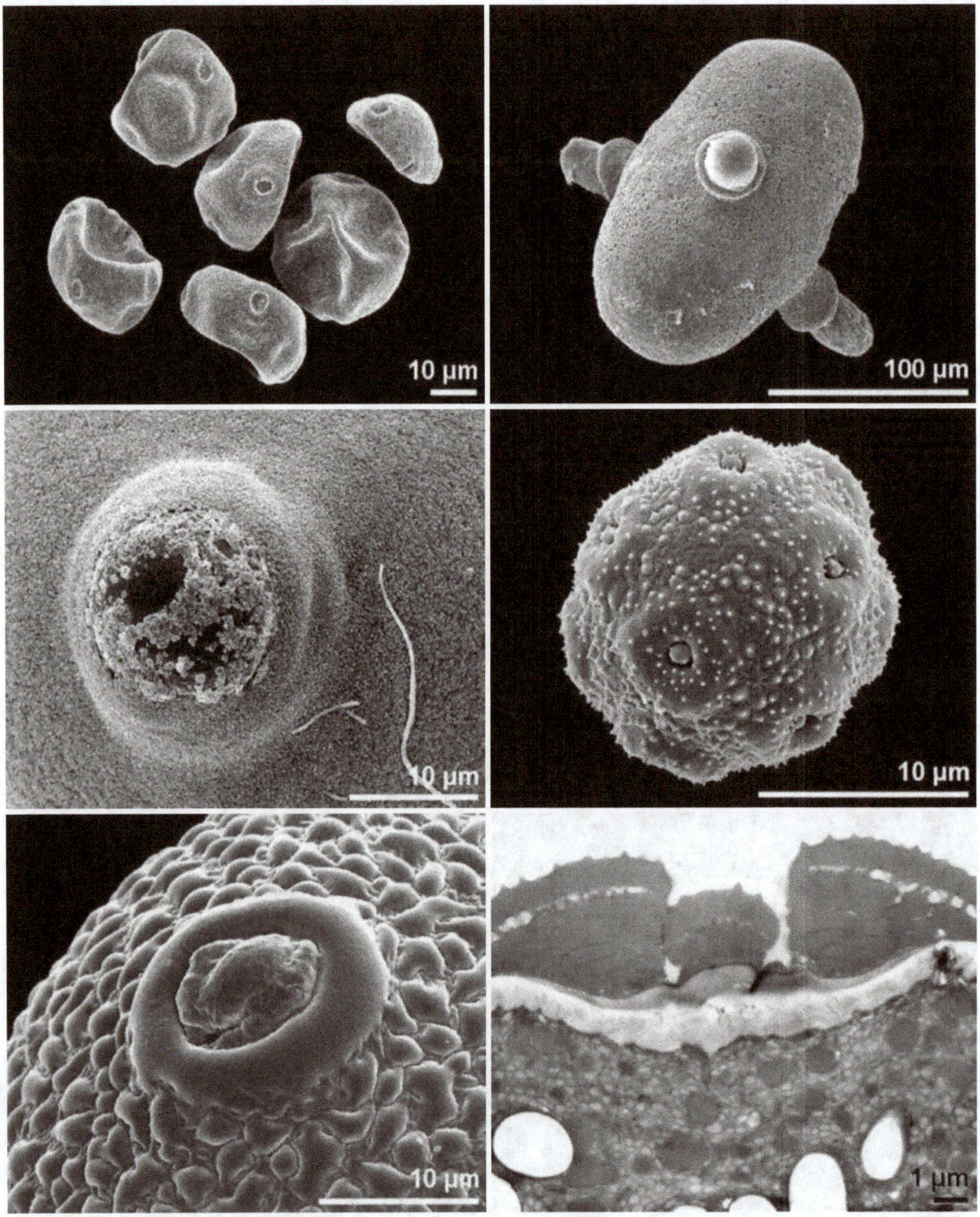

Myriophyllum spicatum, Haloragaceae
dry pollen

Morina longifolia, Caprifoliaceae
equatorial view

Epilobium angustifolium, Onagraceae

Dorstenia contrajerva, Moraceae

Trichosanthes cucumerina, Cucurbitaceae

Zea mays, Poaceae
cross section of aperture

aperture/aperturate

region of the pollen wall that differs morphologically and/or anatomically significantly from the rest of the pollen wall

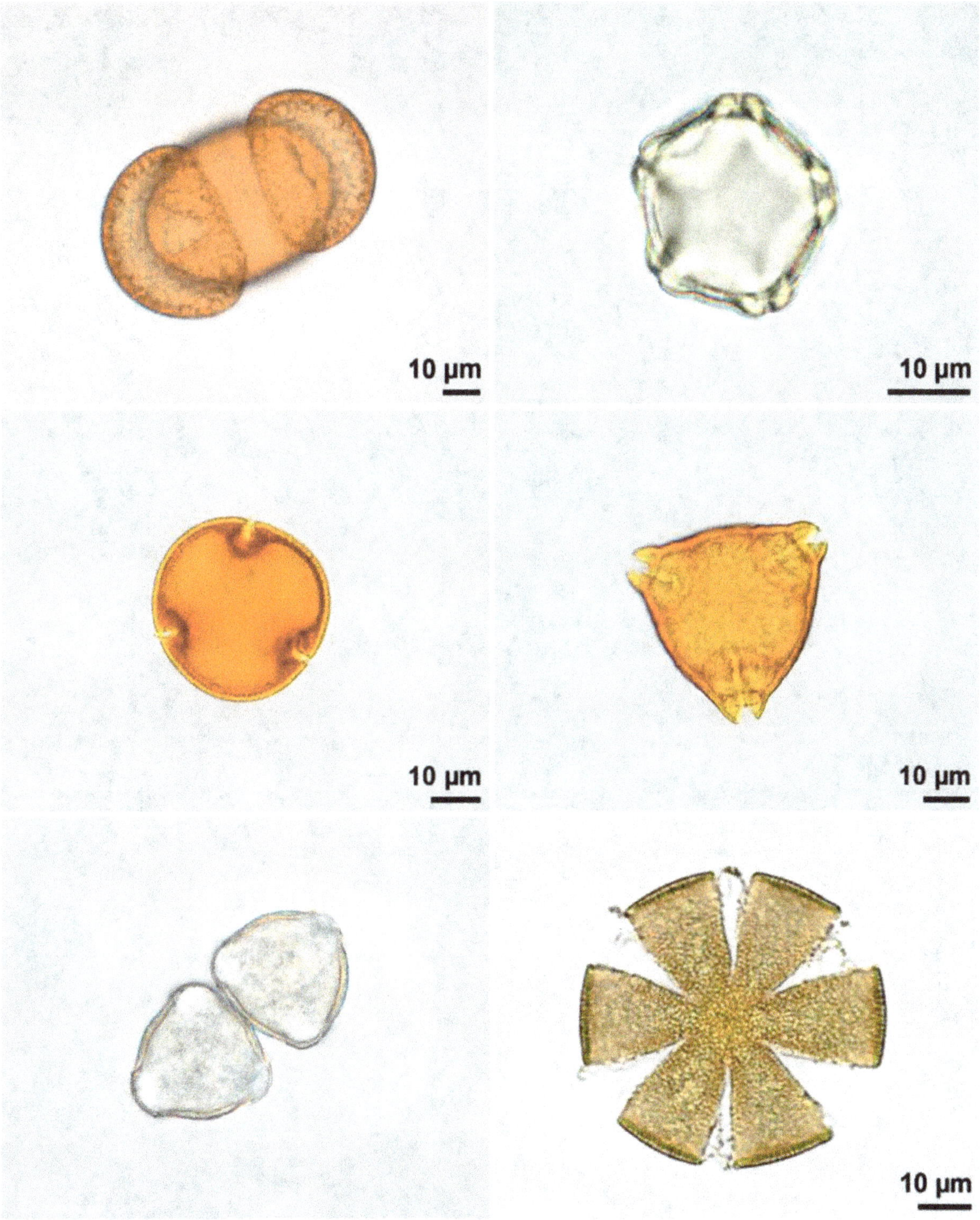

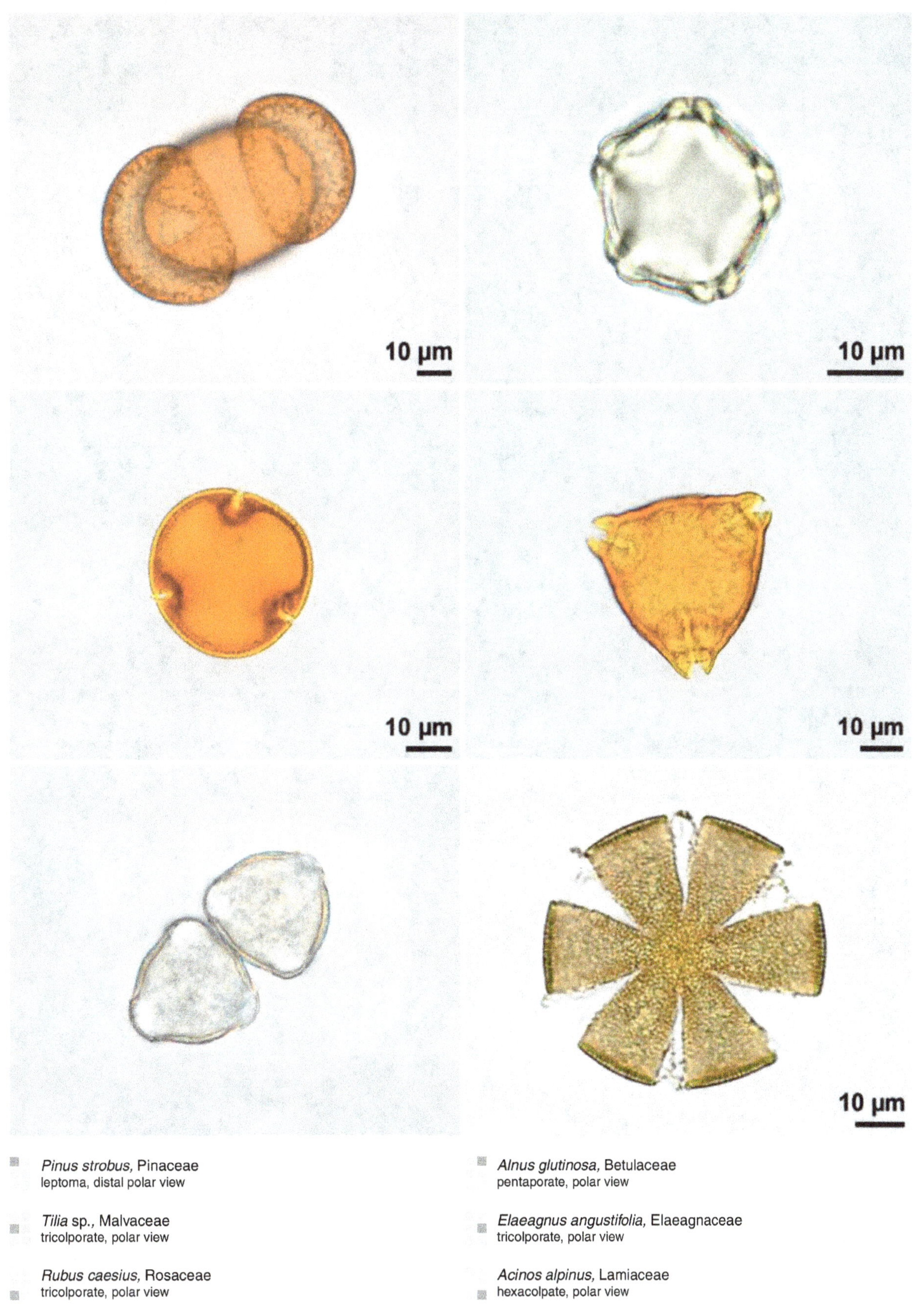

Pinus strobus, Pinaceae
leptoma, distal polar view

Tilia sp., Malvaceae
tricolporate, polar view

Rubus caesius, Rosaceae
tricolporate, polar view

Alnus glutinosa, Betulaceae
pentaporate, polar view

Elaeagnus angustifolia, Elaeagnaceae
tricolporate, polar view

Acinos alpinus, Lamiaceae
hexacolpate, polar view

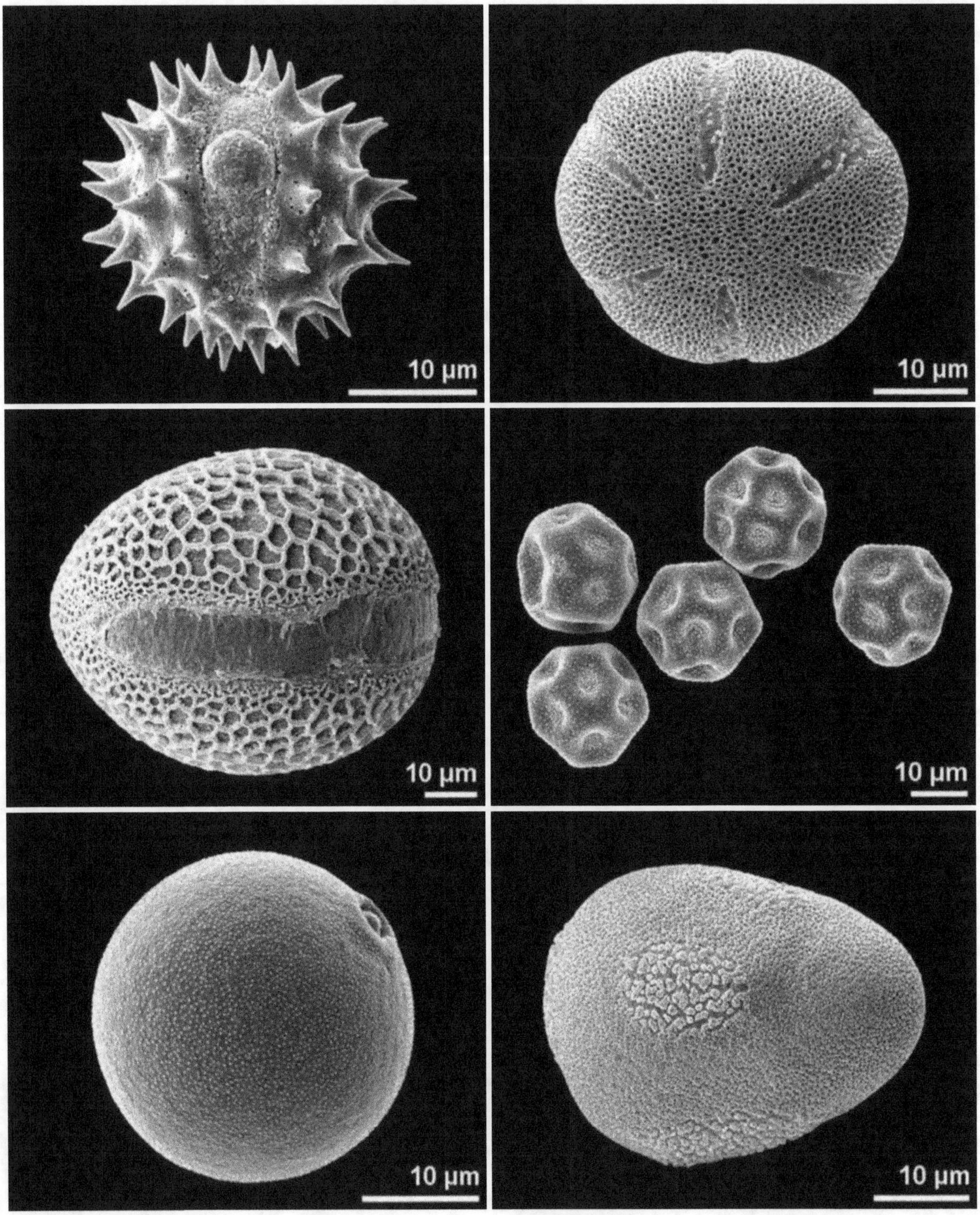

Eupatorium cannabinum, Asteraceae
tricolporate, equatorial view

Clinopodium vulgare, Lamiaceae
hexacolpate, polar view

Eucharis grandiflora, Amaryllidaceae
sulcate, distal polar view

Alisma plantago-aquatica, Alismataceae
pantoporate, dry pollen

Sesleria sadleriana, Poaceae
ulcerate, equatorial view

Carex alba, Cyperaceae
poroidate, pseudomonad

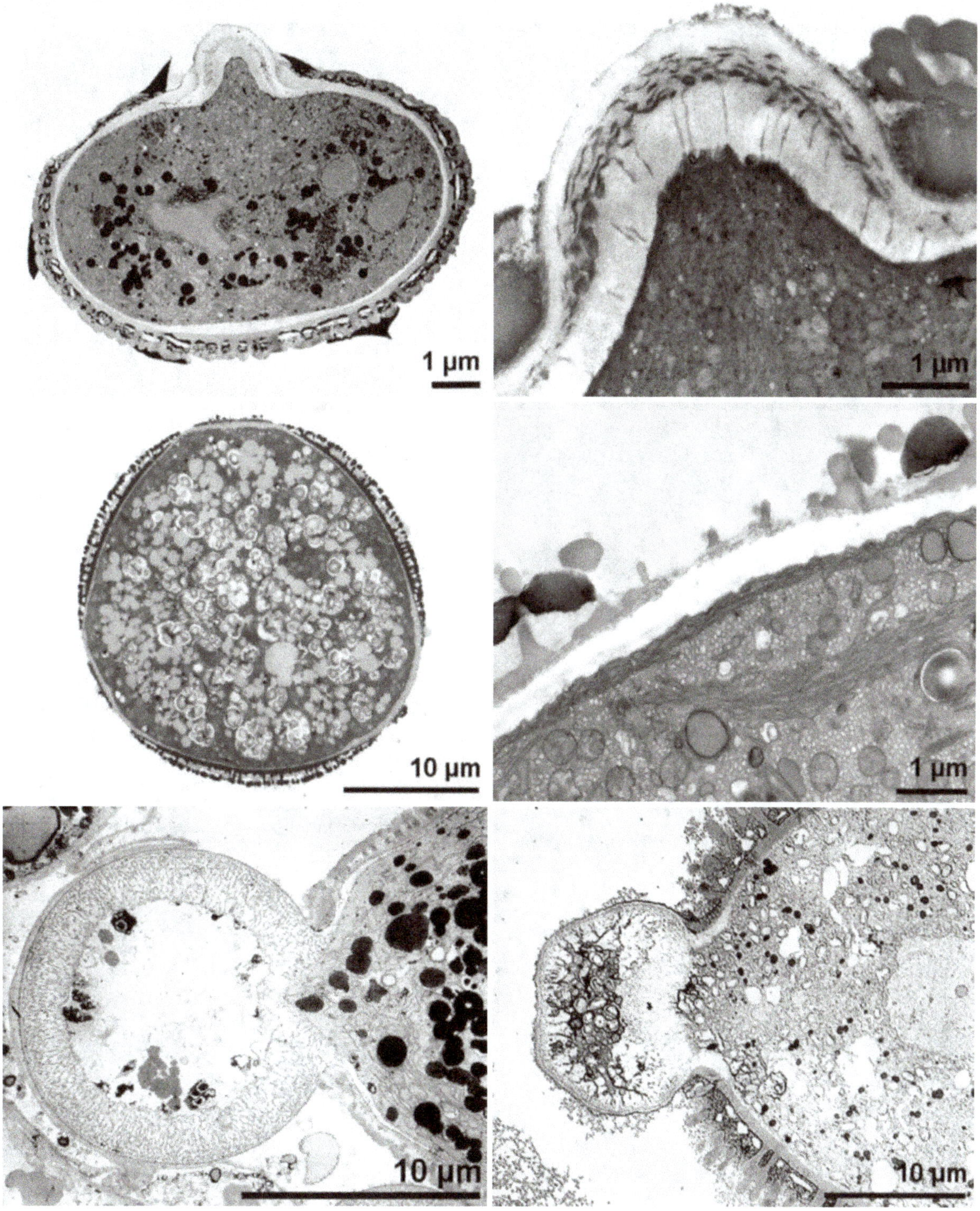

Apium nodiflorum, Apiaceae
cross section

Apium nodiflorum, Apiaceae
cross section of aperture

Veronica spicata, Plantaginaceae
cross section

Mentha aquatica, Lamiaceae
cross section of aperture

Ophiorrhiza sp., Rubiaceae
cross section of aperture, apertural intine protrusion

Geranium robertianum, Geraniaceae
cross section of aperture, apertural intine protrusion

aperture membrane

exine layer covering an aperture
aperture membrane psilate

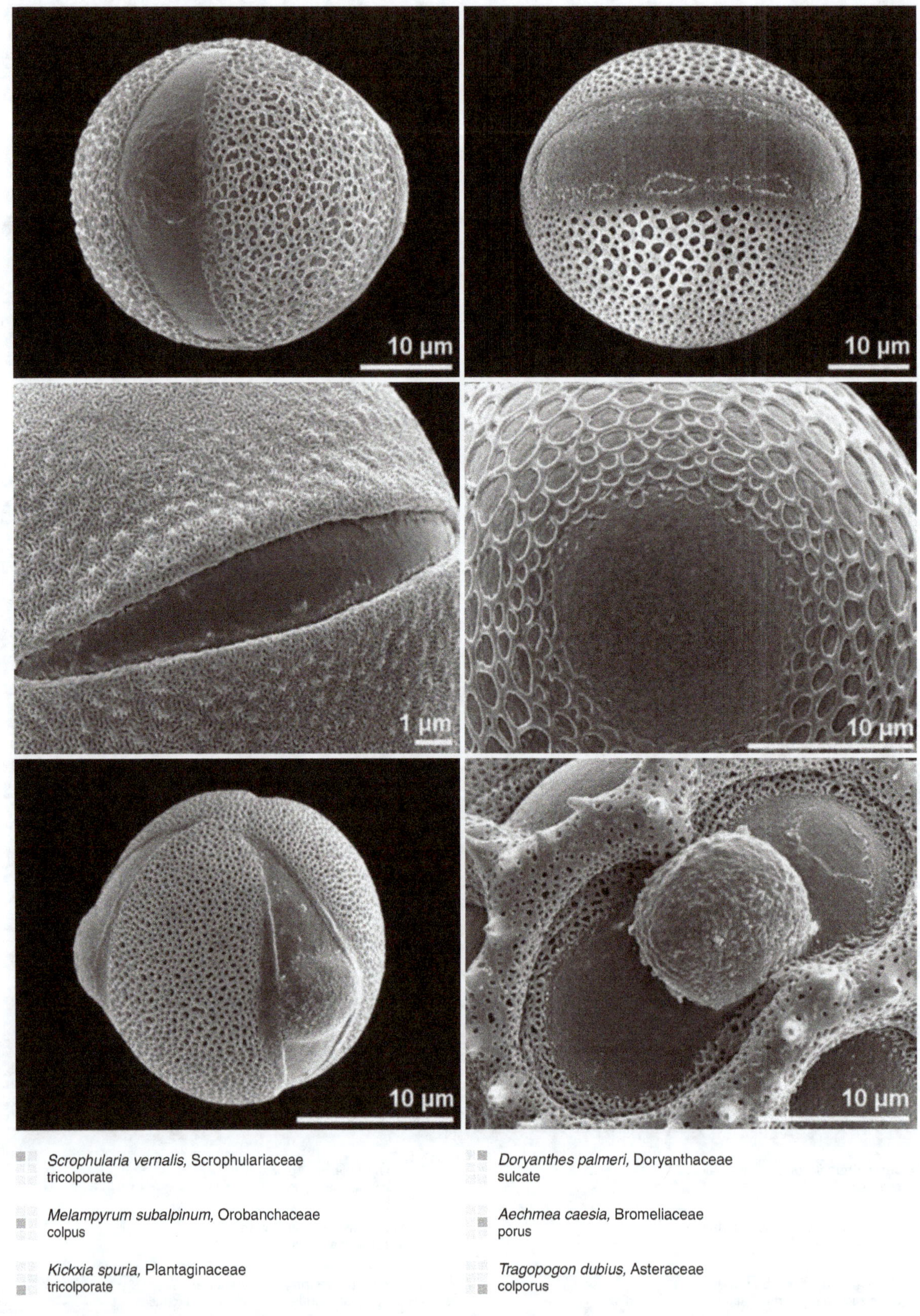

Scrophularia vernalis, Scrophulariaceae
tricolporate

Doryanthes palmeri, Doryanthaceae
sulcate

Melampyrum subalpinum, Orobanchaceae
colpus

Aechmea caesia, Bromeliaceae
porus

Kickxia spuria, Plantaginaceae
tricolporate

Tragopogon dubius, Asteraceae
colporus

aperture membrane ornamented

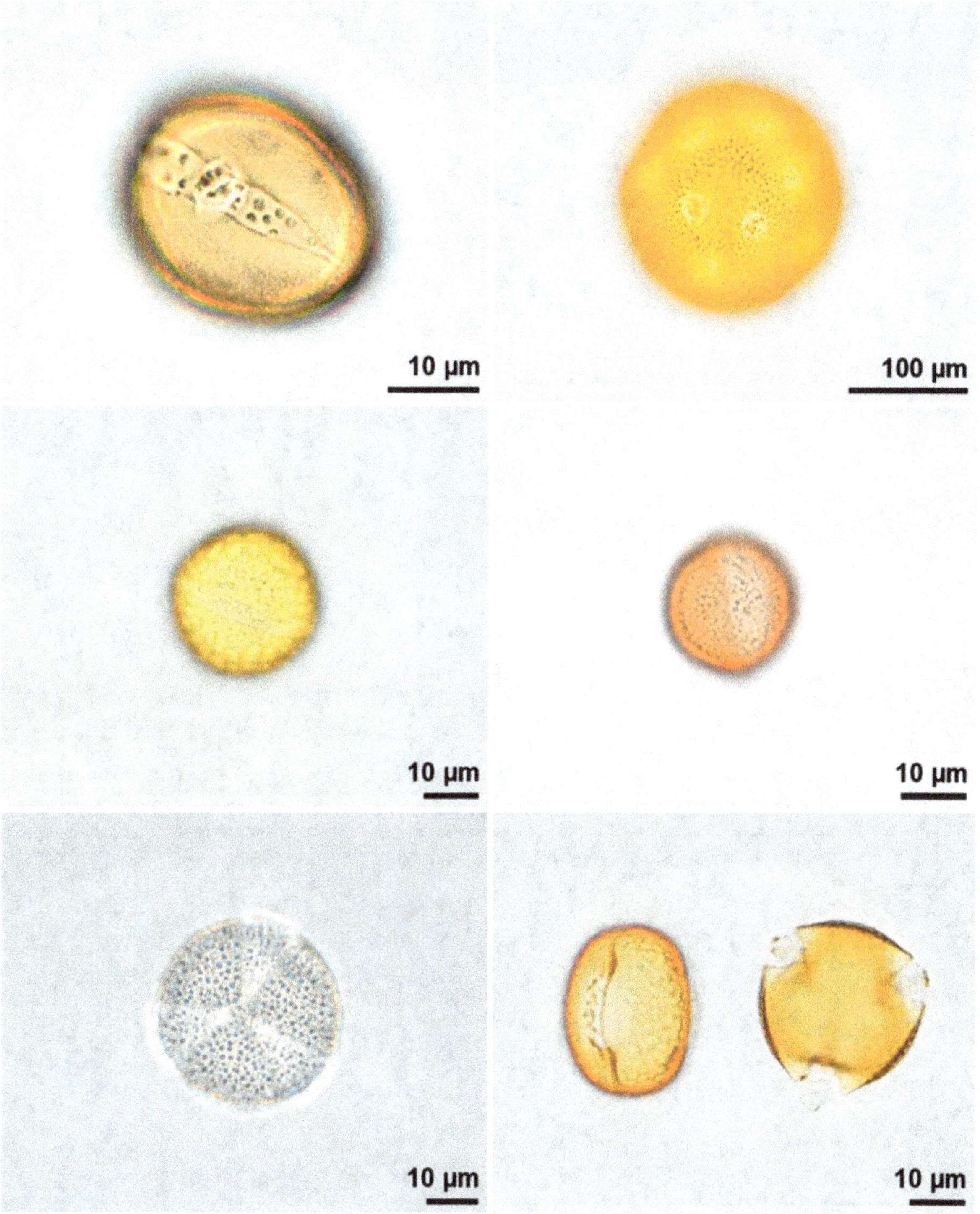

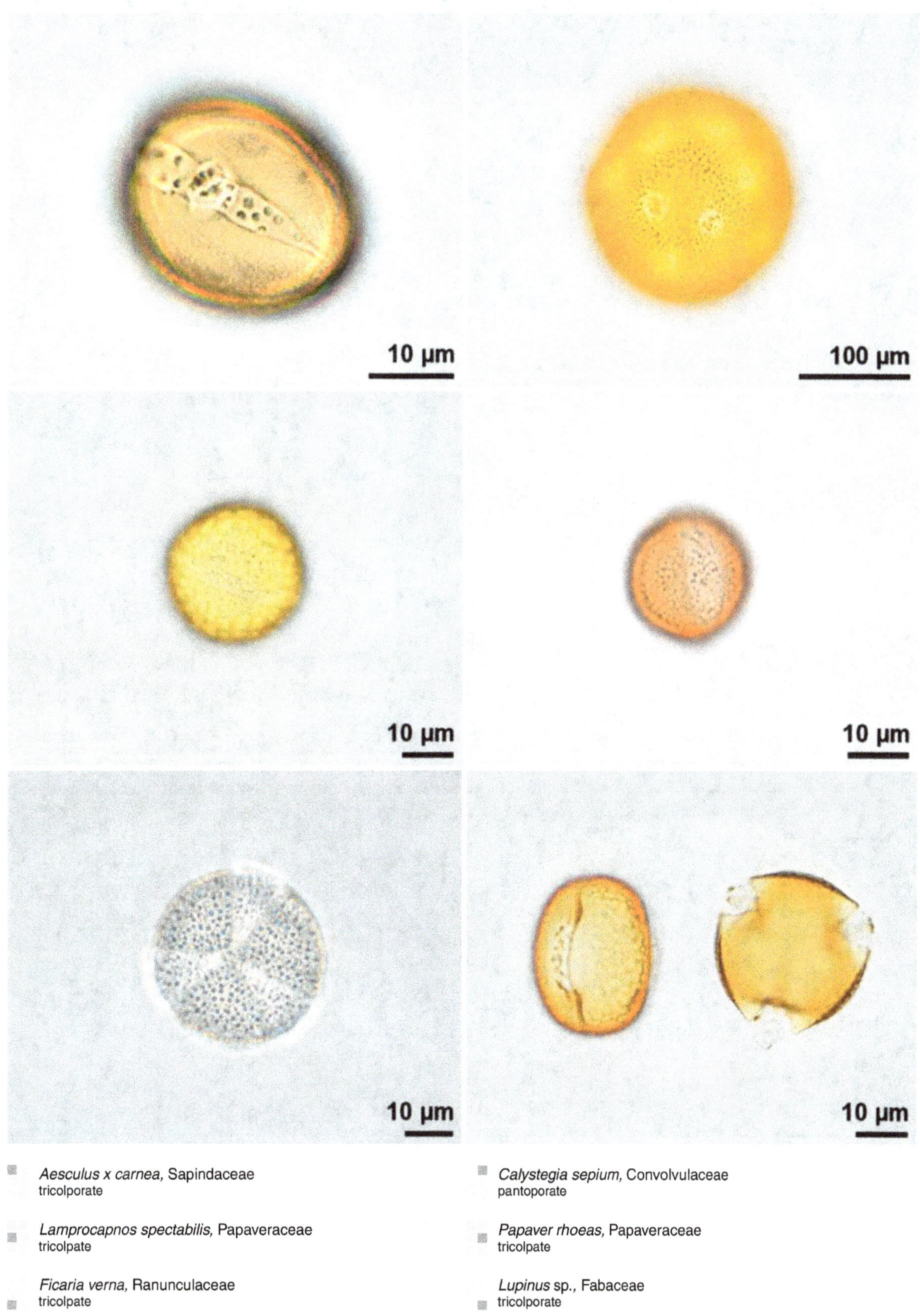

Aesculus x carnea, Sapindaceae
tricolporate

Lamprocapnos spectabilis, Papaveraceae
tricolpate

Ficaria verna, Ranunculaceae
tricolpate

Calystegia sepium, Convolvulaceae
pantoporate

Papaver rhoeas, Papaveraceae
tricolpate

Lupinus sp., Fabaceae
tricolporate

aperture membrane ornamented

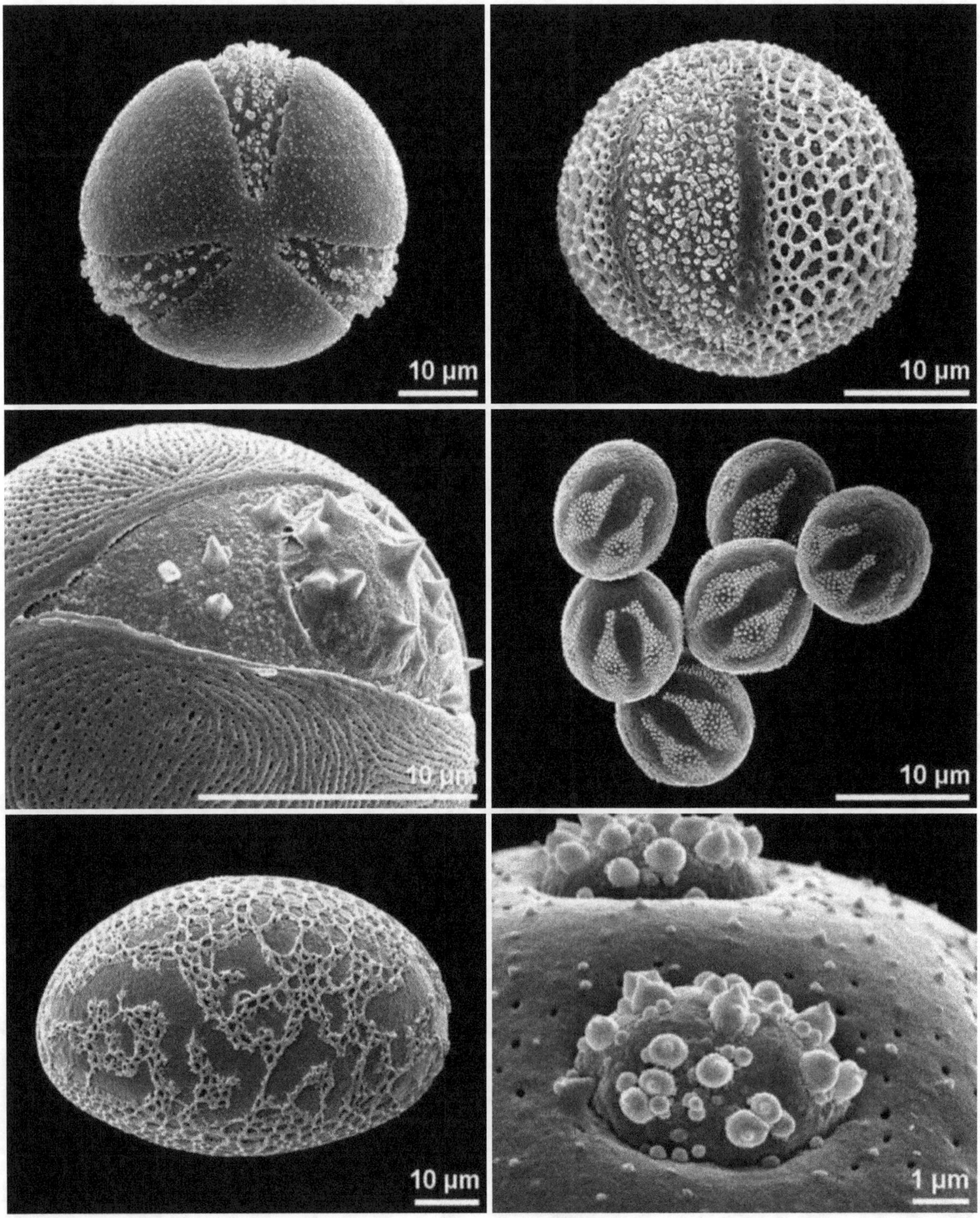

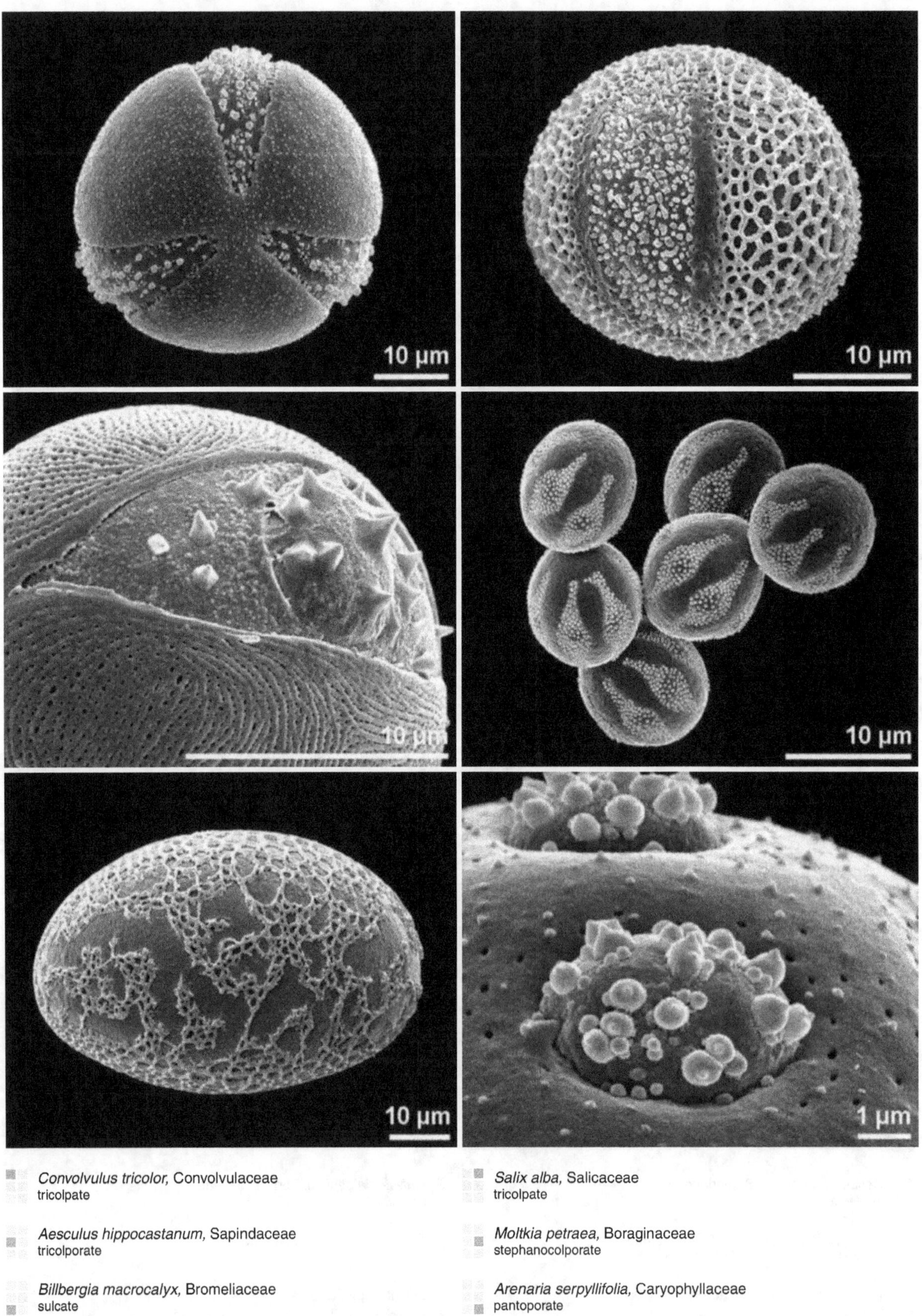

Convolvulus tricolor, Convolvulaceae
tricolpate

Salix alba, Salicaceae
tricolpate

Aesculus hippocastanum, Sapindaceae
tricolporate

Moltkia petraea, Boraginaceae
stephanocolporate

Billbergia macrocalyx, Bromeliaceae
sulcate

Arenaria serpyllifolia, Caryophyllaceae
pantoporate

aperture membrane ornamented

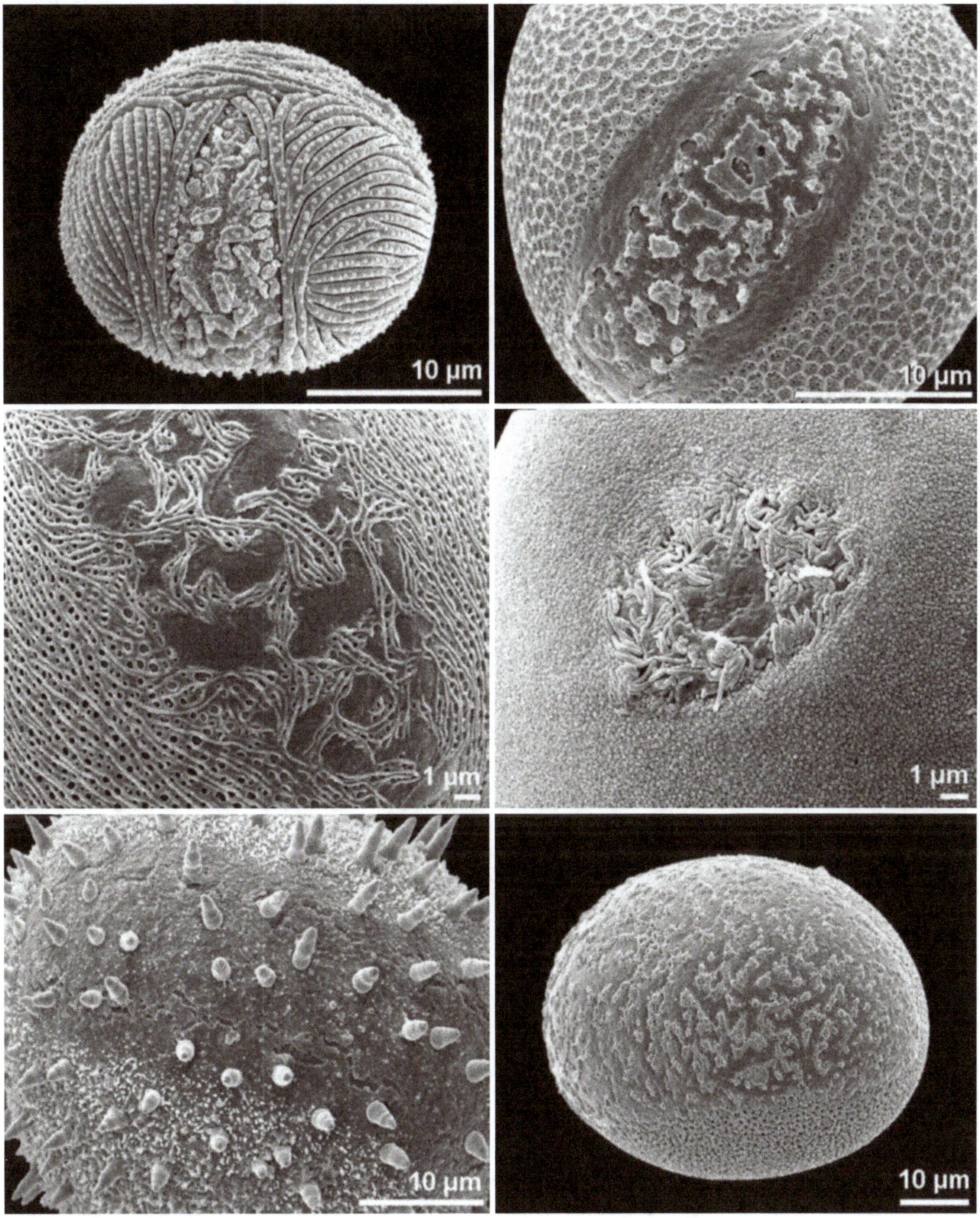

Saxifraga vandellii, Saxifragaceae
tricolpate

Galeopsis tetrahit, Lamiaceae
tricolpate

Veronica wyomingensis, Plantaginaceae
tricolpate

Clarkia pulchella, Onagraceae
triporate

Nuphar lutea, Nymphaeaceae
sulcate

Gagea villosa, Liliaceae
sulcate

aperture membrane ornamented

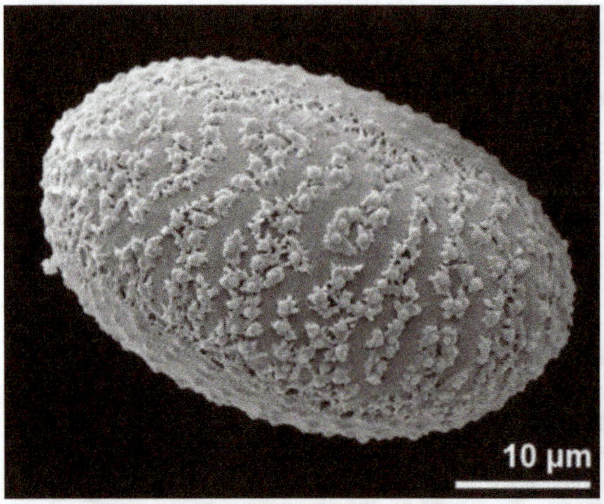

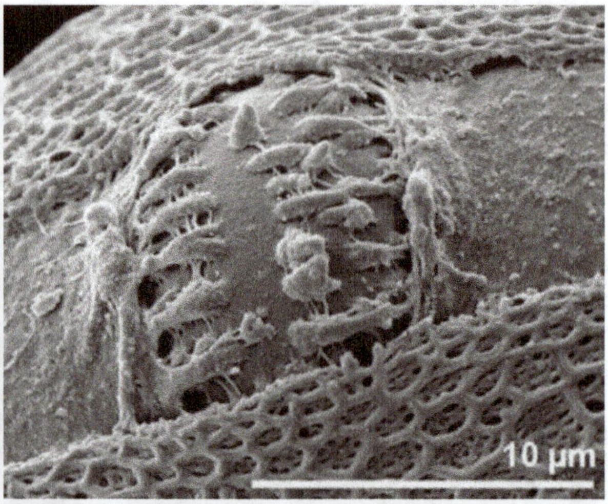

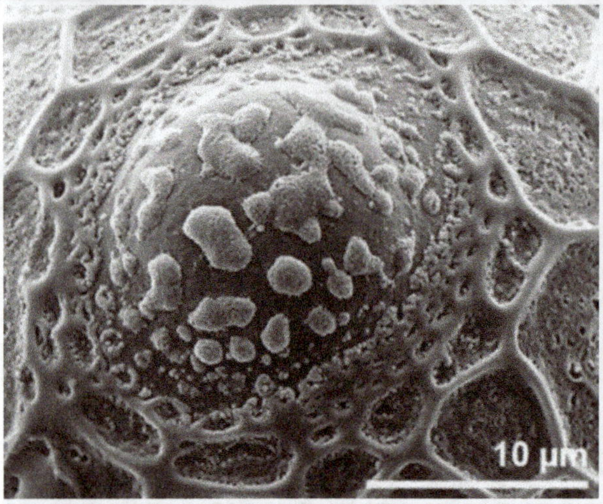

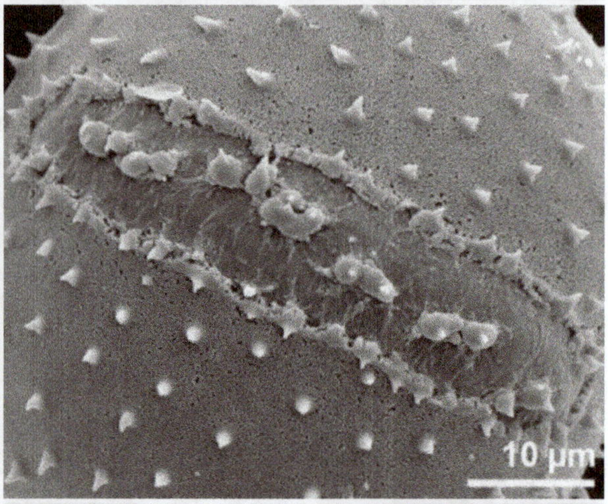

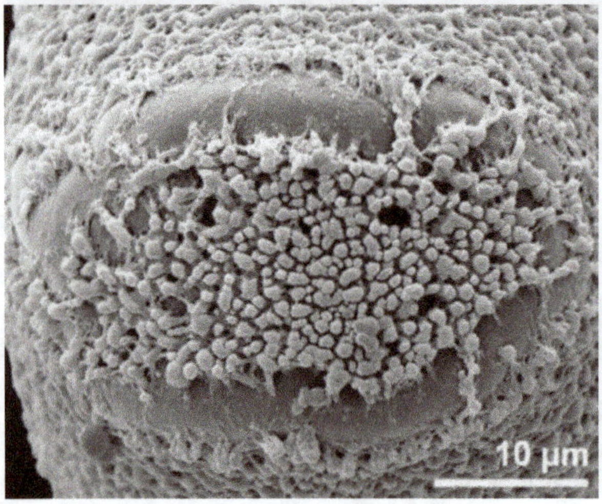

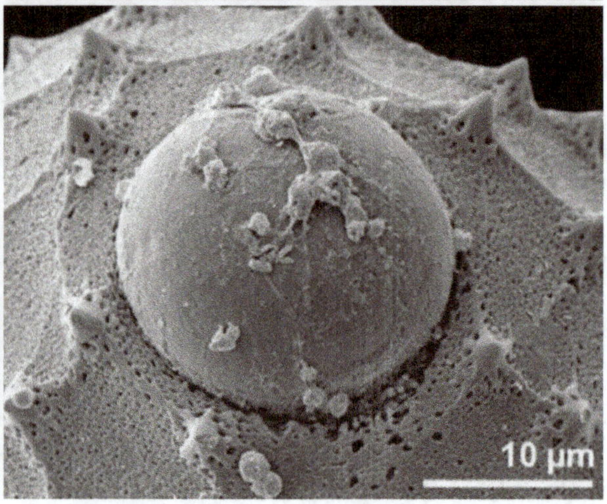

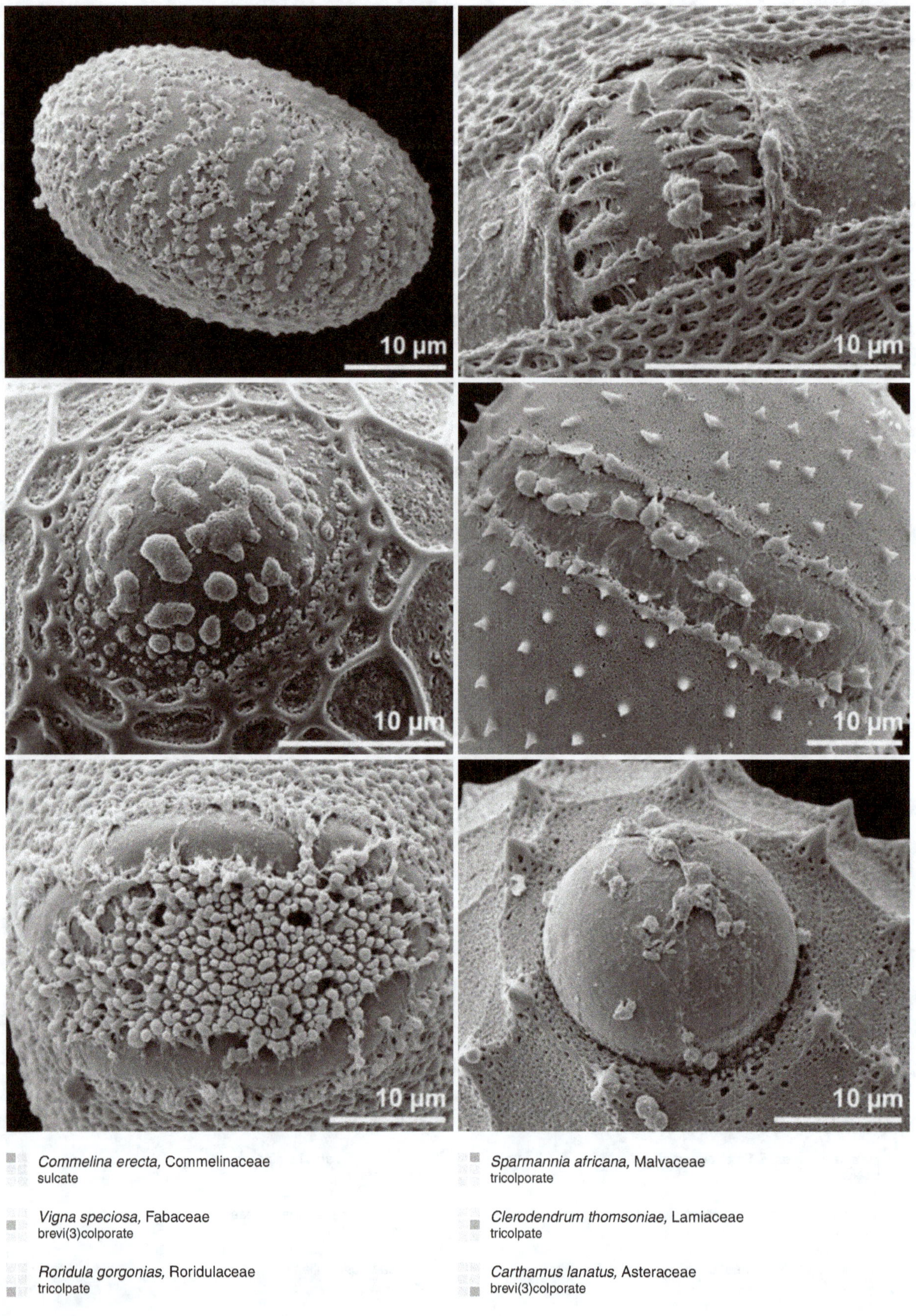

Commelina erecta, Commelinaceae
sulcate

Vigna speciosa, Fabaceae
brevi(3)colporate

Roridula gorgonias, Roridulaceae
tricolpate

Sparmannia africana, Malvaceae
tricolporate

Clerodendrum thomsoniae, Lamiaceae
tricolpate

Carthamus lanatus, Asteraceae
brevi(3)colporate

aperture membrane ornamented

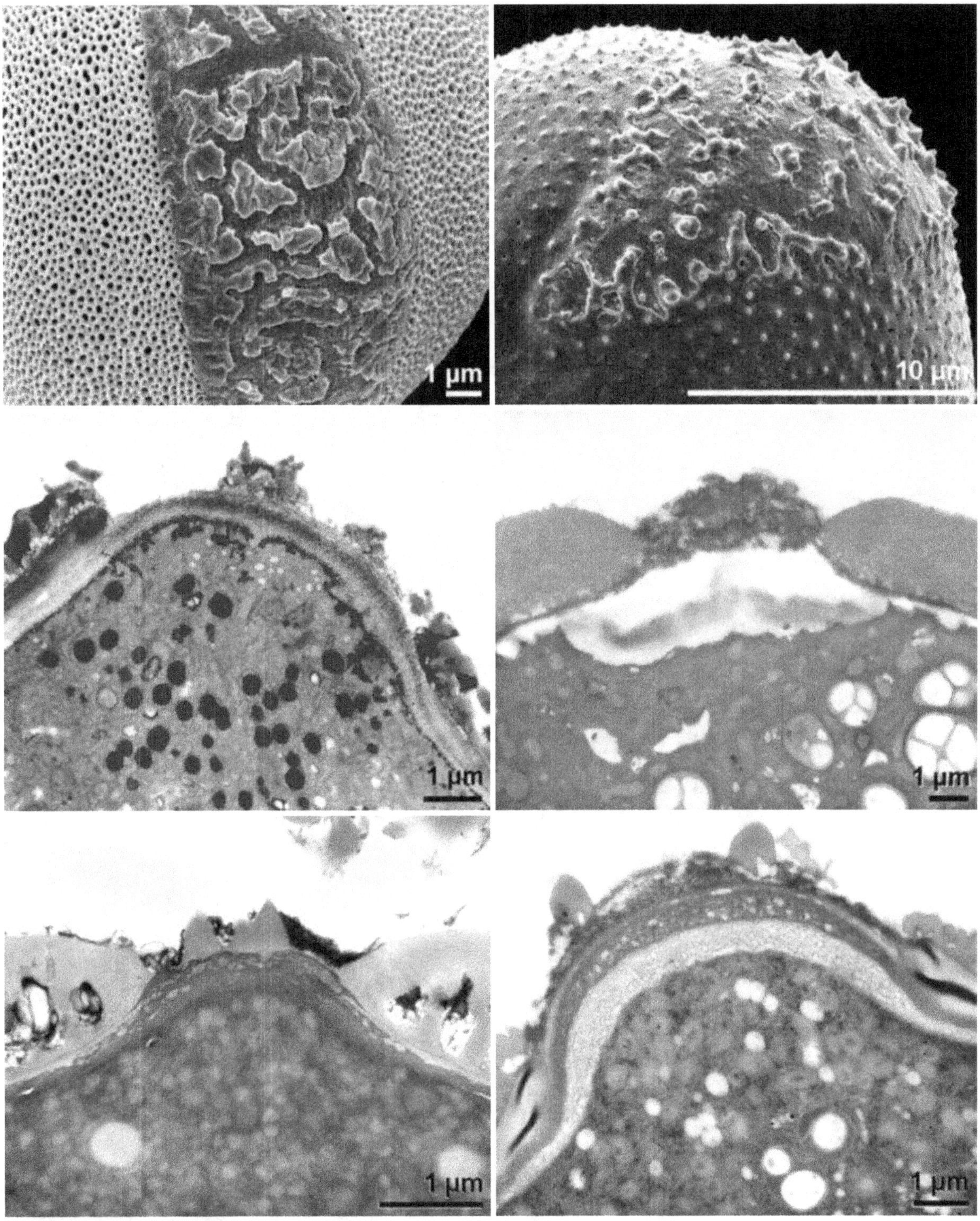

- *Lamiastrum galeobdolon*, Lamiaceae
  tricolpate
- *Aconitum lycoctonum*, Ranunculaceae
  tricolpate
- *Mercurialis perennis*, Euphorbiaceae
  cross section of aperture
- *Ulmus minor*, Ulmaceae
  cross section of aperture
- *Chenopodium hybridum*, Amaranthaceae
  cross section of aperture
- *Aesculus carnea*, Sapindaceae
  cross section of aperture

atrium, oncus

atrium: space between diverging exine layers within the aperture
oncus: lens-shaped body located beneath the aperture, not resistant to acetolysis

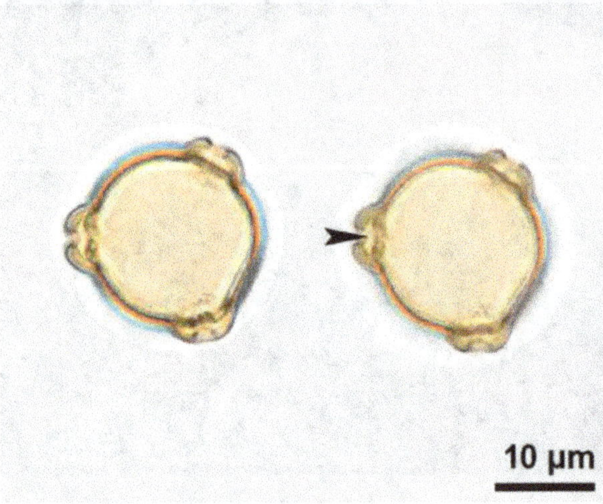

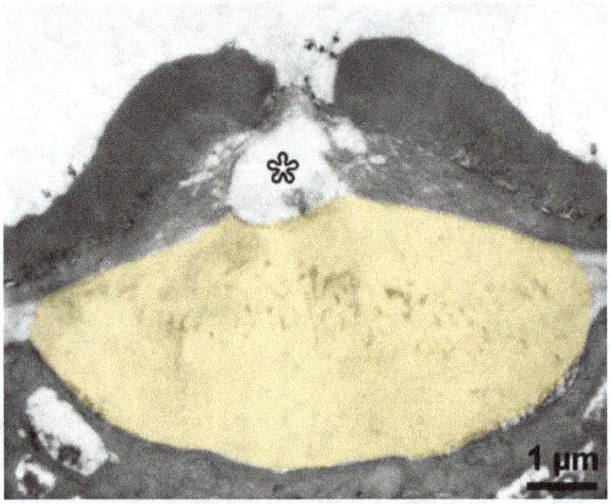

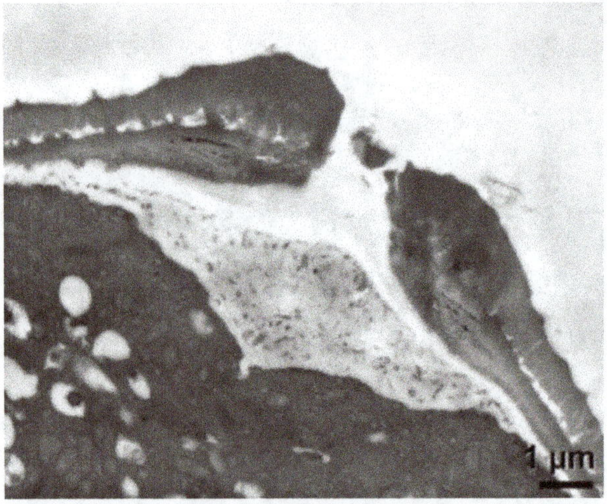

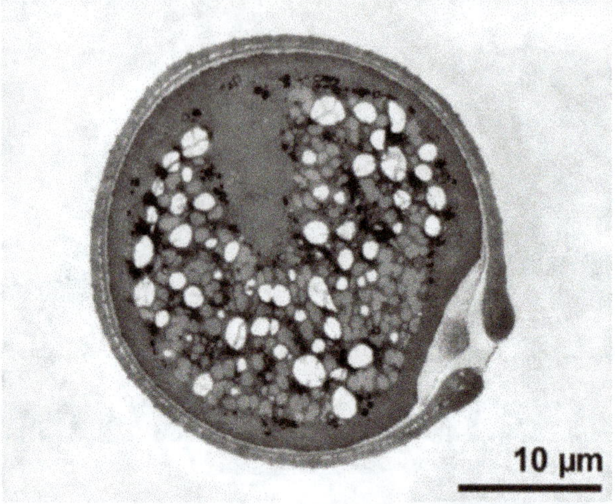

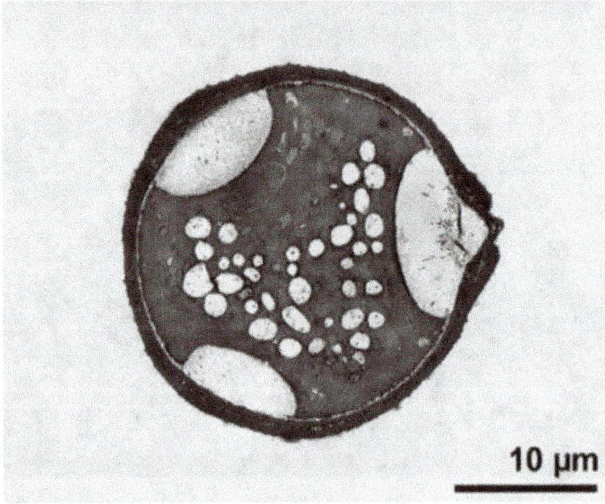

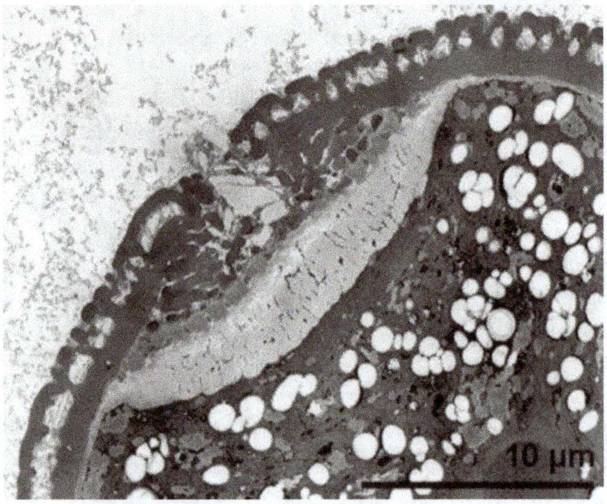

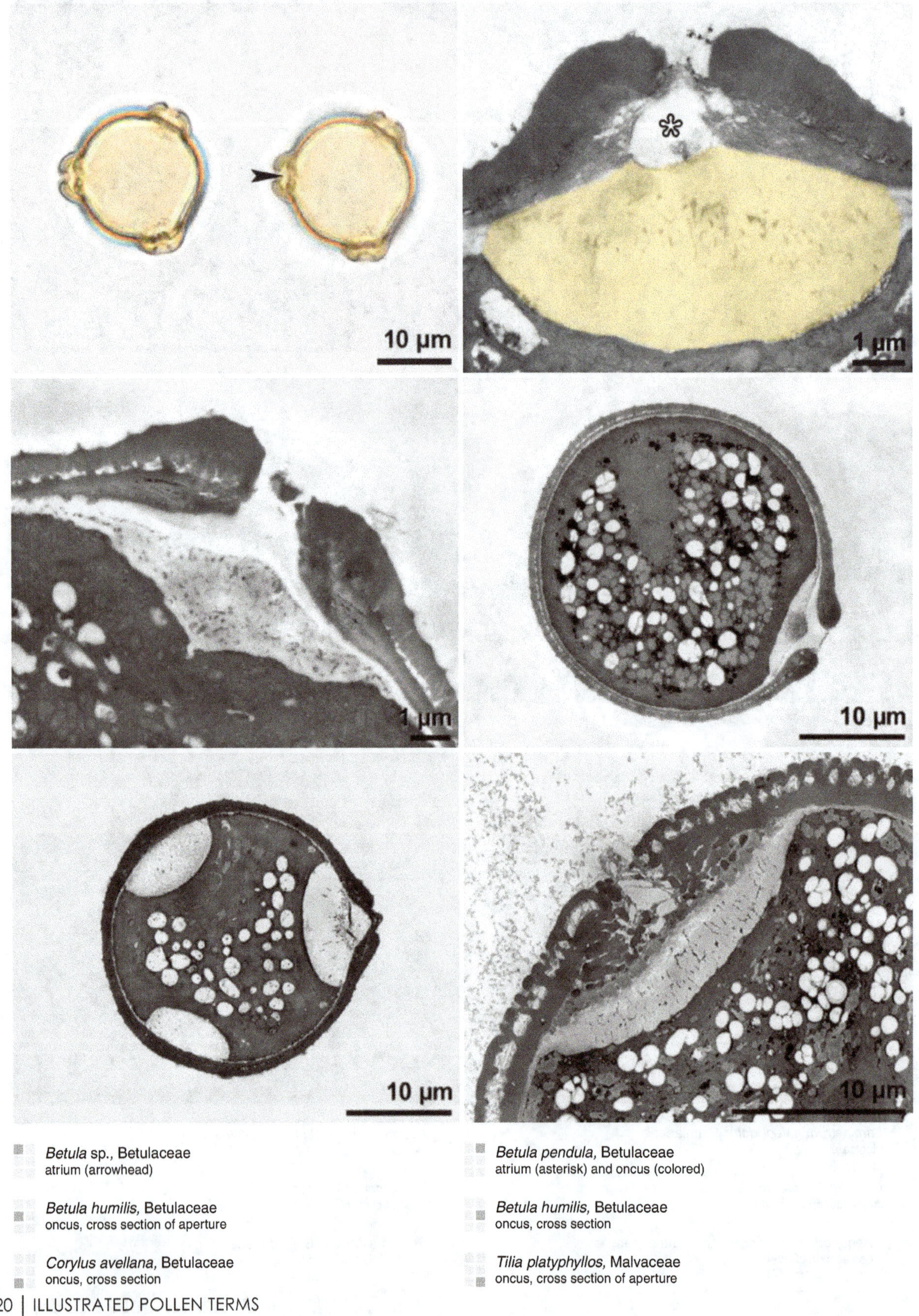

Betula sp., Betulaceae
atrium (arrowhead)

Betula humilis, Betulaceae
oncus, cross section of aperture

Corylus avellana, Betulaceae
oncus, cross section

Betula pendula, Betulaceae
atrium (asterisk) and oncus (colored)

Betula humilis, Betulaceae
oncus, cross section

Tilia platyphyllos, Malvaceae
oncus, cross section of aperture

brevicolpus/brevicolpate

short colpus situated equatorially

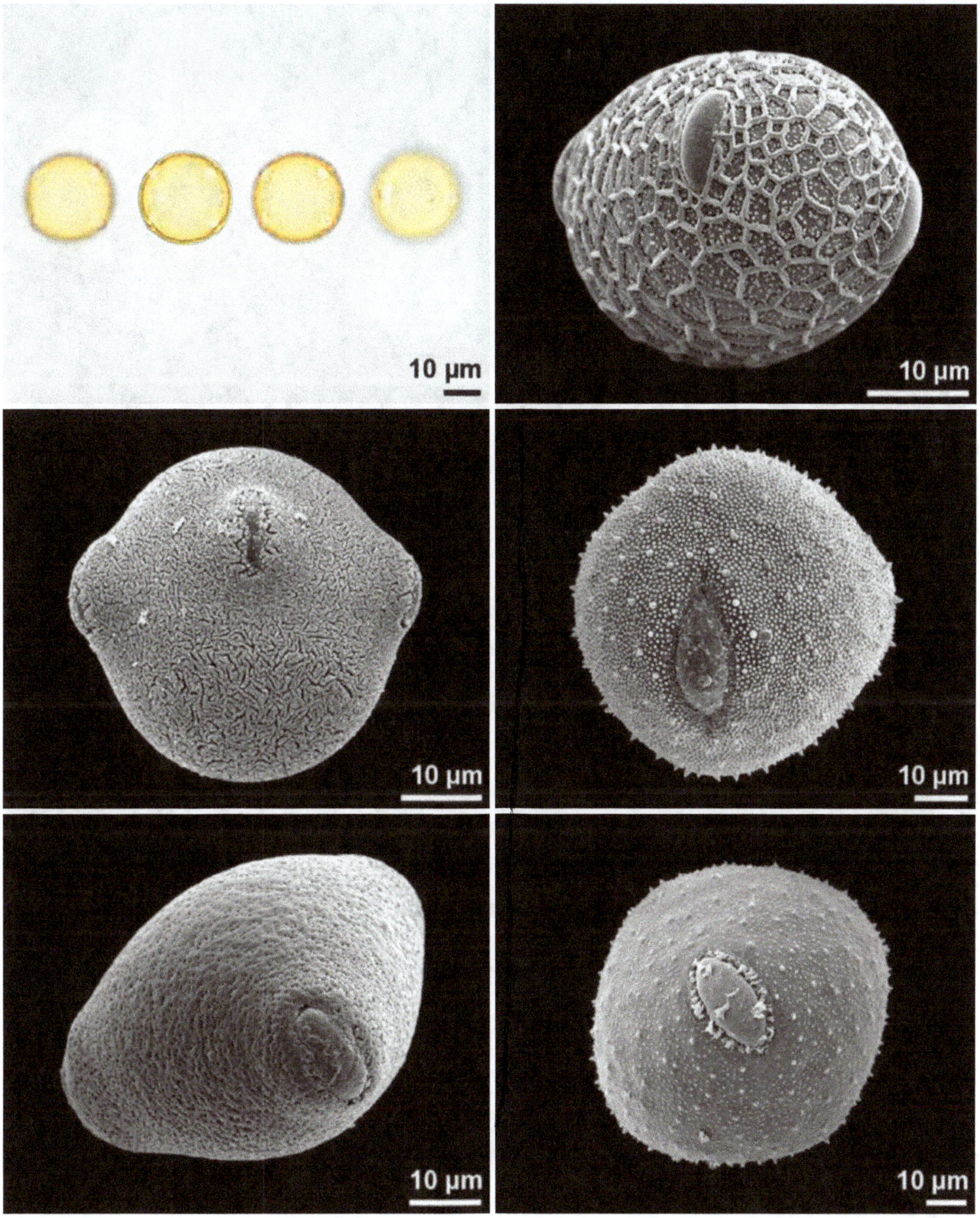

Pistacia sp., Anacardiaceae
tetracolpate

Impatiens columbaria, Balsaminaceae
tetracolpate, equatorial view

Mendoncia albida, Acanthaceae
pentacolpate, oblique equatorial view

Scabiosa ochroleuca, Caprifoliaceae
tricolpate, equatorial view

Petrea volubilis, Verbenaceae
tricolpate, oblique equatorial view

Succisa pratensis, Caprifoliaceae
tricolpate, equatorial view

brevicolporus/brevicolporate

short colpus in a compound aperture situated equatorially

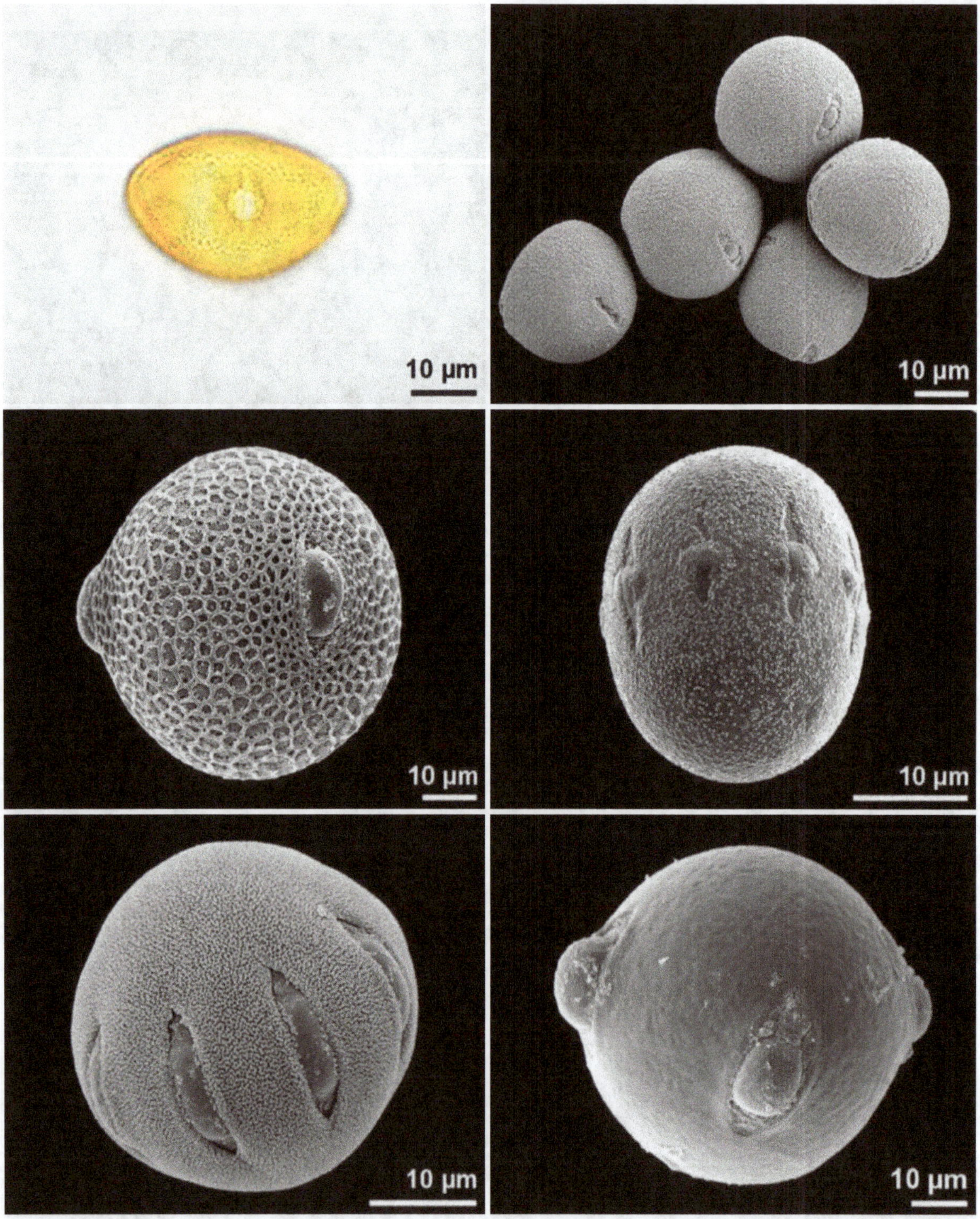

Tilia platyphyllos, Malvaceae
tricolporate, equatorial view

Coriaria myrtifolia, Coriariaceae
tricolporate

Dalechampia spathulata, Euphorbiaceae
tricolporate, equatorial view

Symphytum orientale, Boraginaceae
stephanocolporate, equatorial view

Borago pygmaea, Boraginaceae
stephanocolporate, equatorial view

Tabernaemontana simulans, Apocynaceae
tetracolporate, equatorial view

bridge

exine connection(s) between the margins of an aperture

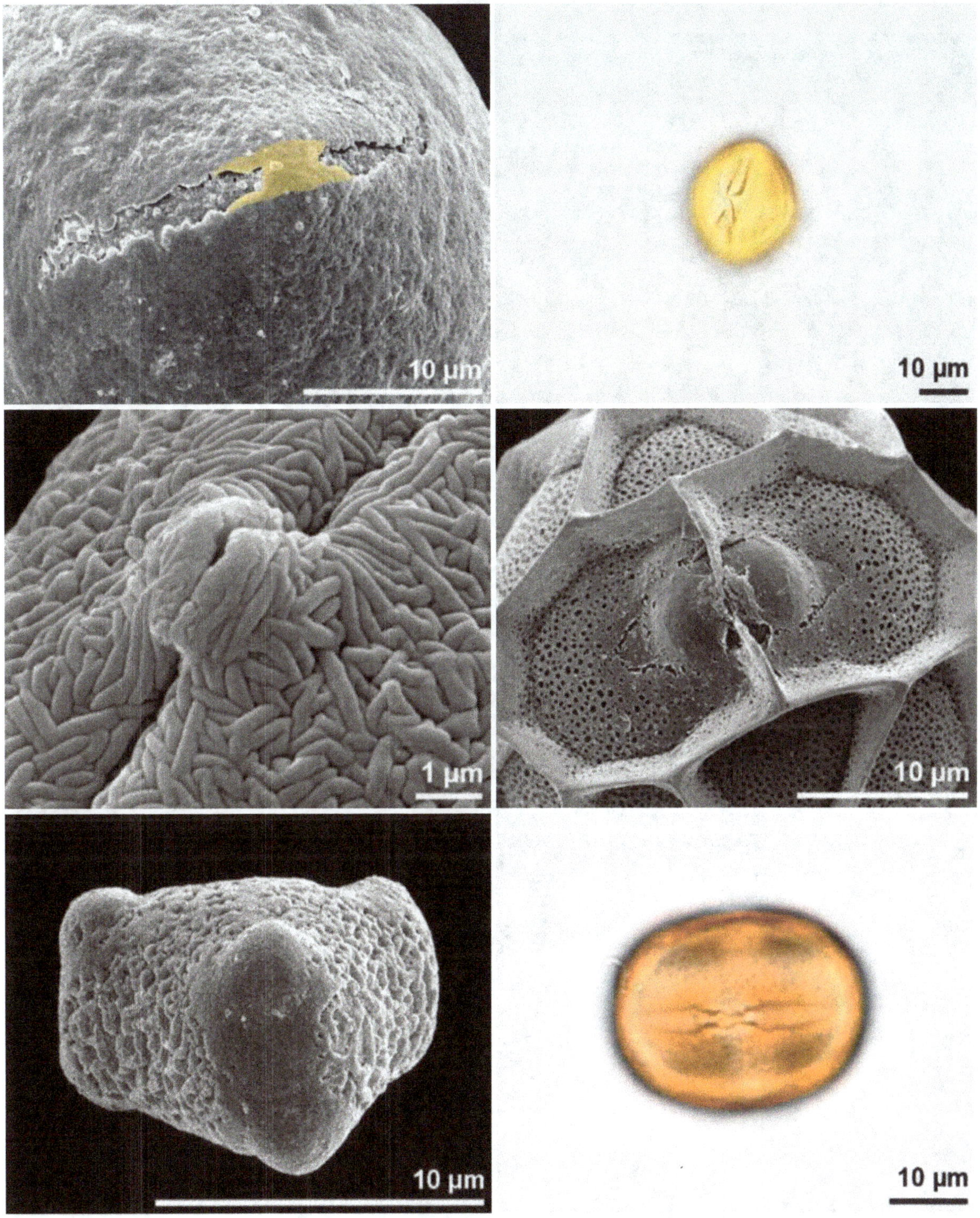

Elaeagnus angustifolia, Elaeagnaceae
tricolporate, bridge colored

Bifora radians, Apiaceae
tricolporate

Cunonia capensis, Cunoniaceae
hexaporate with bridge, equatorial view

Malus baccata, Rosaceae
tricolporate, equatorial view

Gazania rigens, Asteraceae
tricolporate

Colutea arborescens, Fabaceae
tricolporate, equatorial view

the term bridge is used in a more general context, e.g., for exine connections within tetrads

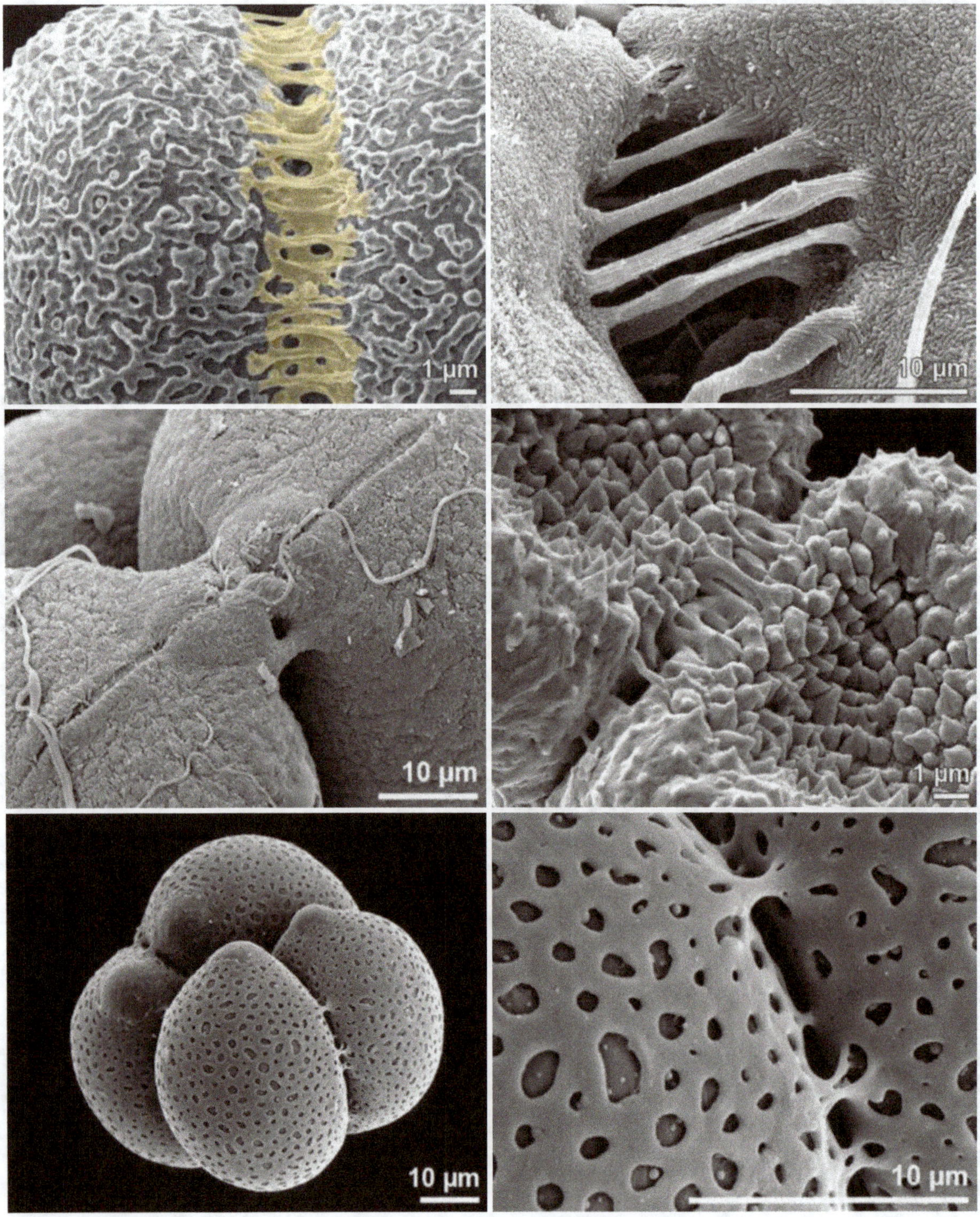

Typha latifolia, Typhaceae
tetrad, bridge (colored) between monads

Epilobium hirsutum, Onagraceae
tetrad, bridge between monads

Ludwigia octovalvis, Onagraceae
tetrad, monads also adnate in aperture region

Thelethylax minutiflora, Podostemaceae
dyad, bridge between monads

Aechmea subintegerrima, Bromeliaceae
tetrad, bridge between monads

Aechmea subintegerrima, Bromeliaceae

colpus/colpate, dicolpate

colpus: elongated aperture (length/width ratio >2) situated at the equator or globally distributed
dicolpate: pollen grain with two colpi

Hypecoum imberbe, Papaveraceae
polar view

Hypecoum procumbens, Papaveraceae
oblique polar view

Mayna odorata, Achariaceae
oblique polar view

Mayna odorata, Achariaceae
dry pollen

Pedicularis elongata, Orobanchaceae
oblique polar view

Pedicularis elongata, Orobanchaceae
dry pollen

colpus/colpate, tricolpate

tricolpate: pollen grain with three colpi

Lamium maculatum, Lamiaceae

Erysimum odoratum, Brassicaceae
polar view

Nelumbo nucifera, Nelumbonaceae
oblique polar view

Lonicera fragrantissima, Caprifoliaceae
equatorial view

Stachys palustris, Lamiaceae
dry pollen

Ceratostigma plumbaginoides, Plumbaginaceae
polar view

Fraxinus excelsior, Oleaceae
equatorial view

Odontites luteus, Orobanchaceae
dry pollen

Nandina domestica, Berberidaceae
polar view

Corylopsis platypetala, Hamamelidaceae
equatorial view

Trollius europaeus, Ranunculaceae
polar view

Veronica serpyllifolia, Plantaginaceae
equatorial view

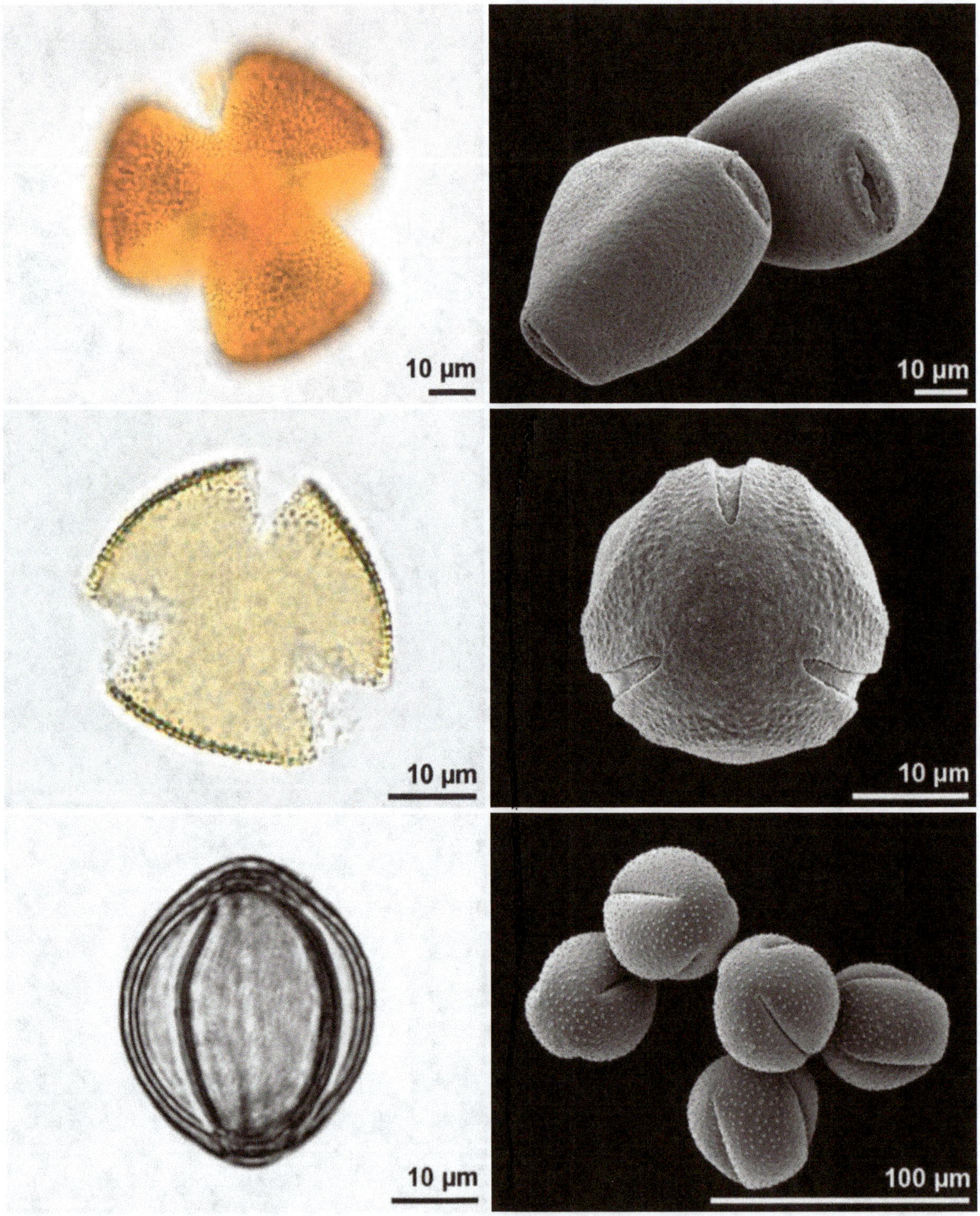

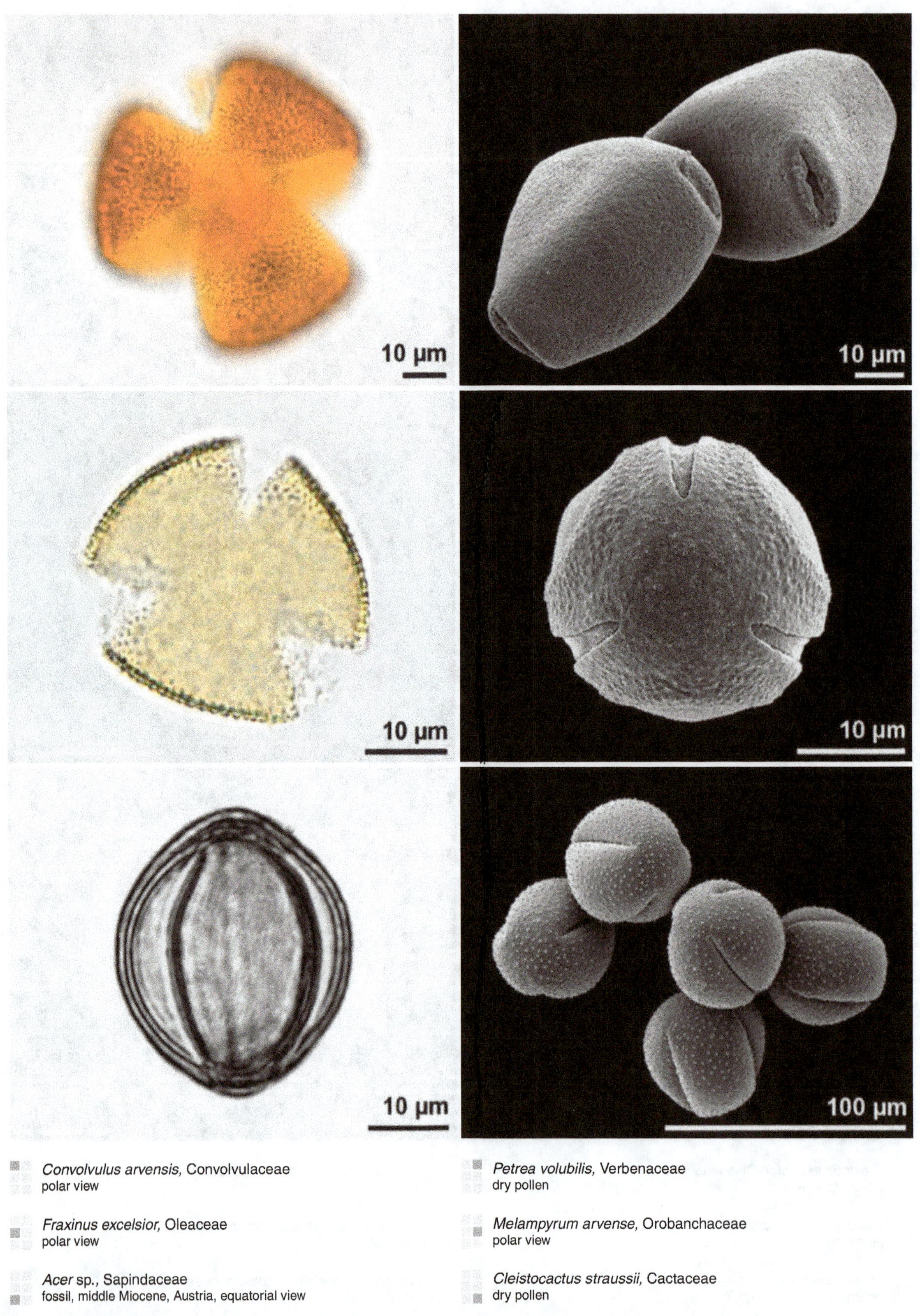

Convolvulus arvensis, Convolvulaceae
polar view

Fraxinus excelsior, Oleaceae
polar view

Acer sp., Sapindaceae
fossil, middle Miocene, Austria, equatorial view

Petrea volubilis, Verbenaceae
dry pollen

Melampyrum arvense, Orobanchaceae
polar view

Cleistocactus straussii, Cactaceae
dry pollen

colpus/colpate, tetracolpate, pentacolpate

tetra- and pentacolpate: pollen grain with four or five colpi

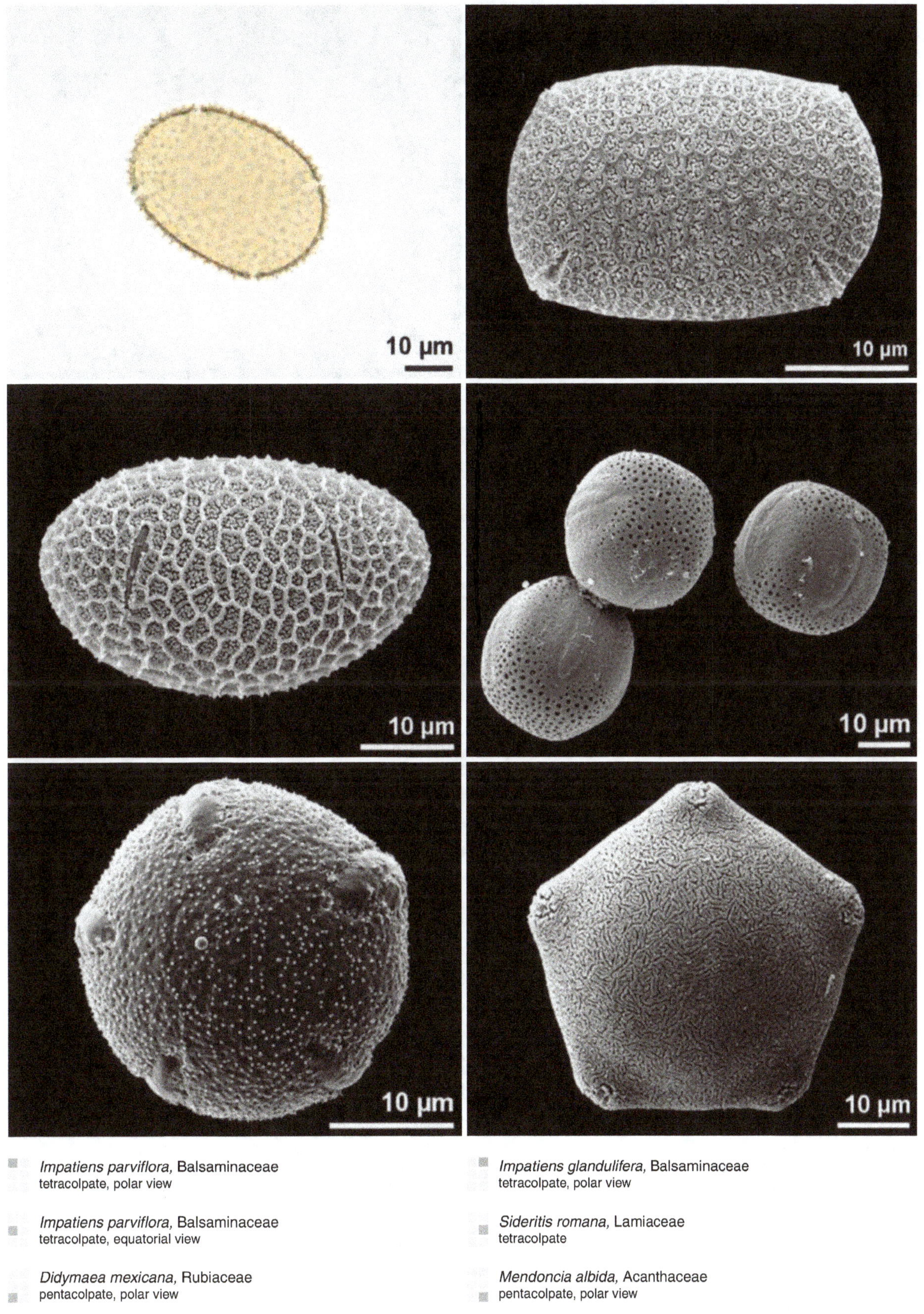

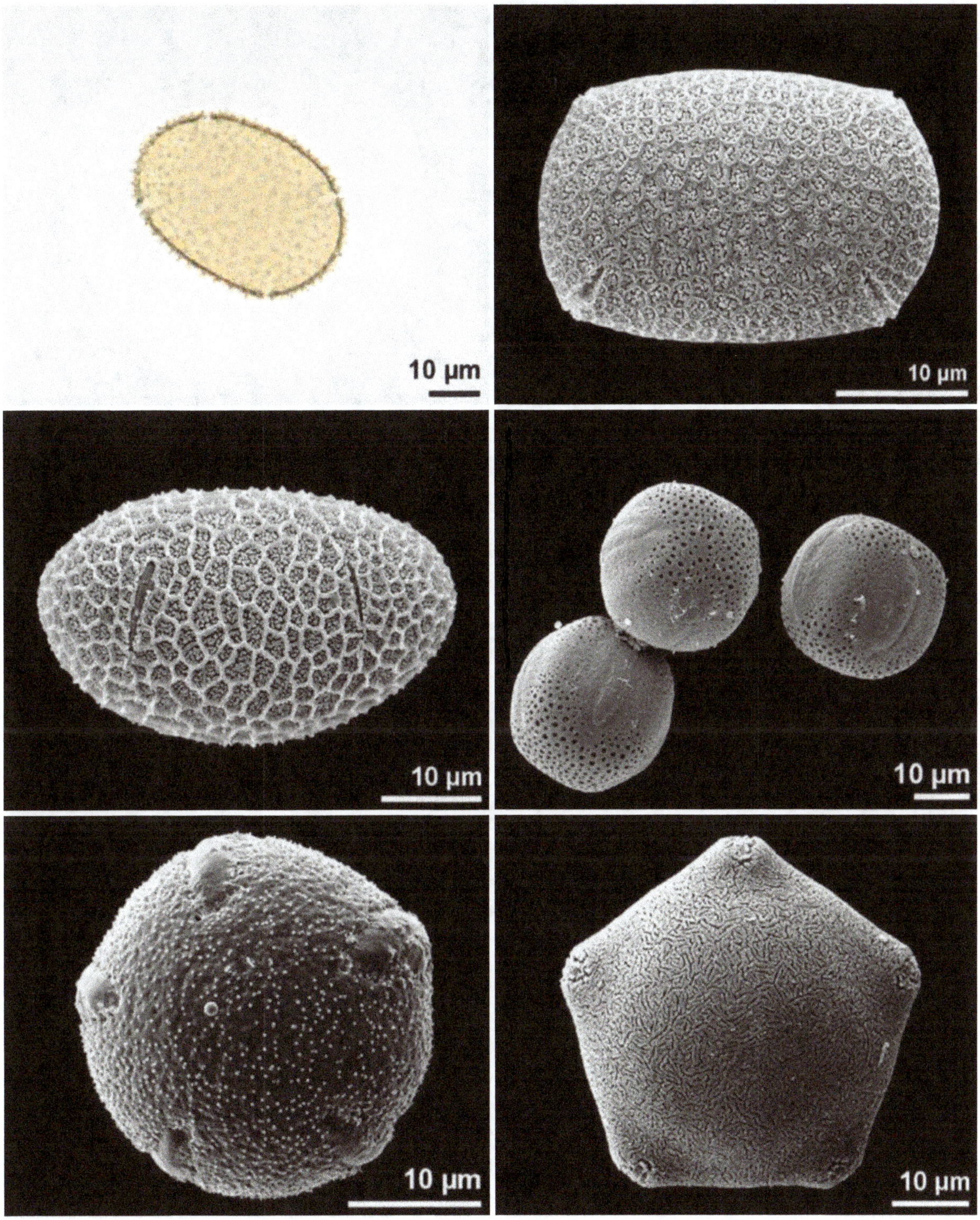

Impatiens parviflora, Balsaminaceae
tetracolpate, polar view

Impatiens glandulifera, Balsaminaceae
tetracolpate, polar view

Impatiens parviflora, Balsaminaceae
tetracolpate, equatorial view

Sideritis romana, Lamiaceae
tetracolpate

Didymaea mexicana, Rubiaceae
pentacolpate, polar view

Mendoncia albida, Acanthaceae
pentacolpate, polar view

colpus/colpate, hexacolpate

hexacolpate: pollen grain with six colpi

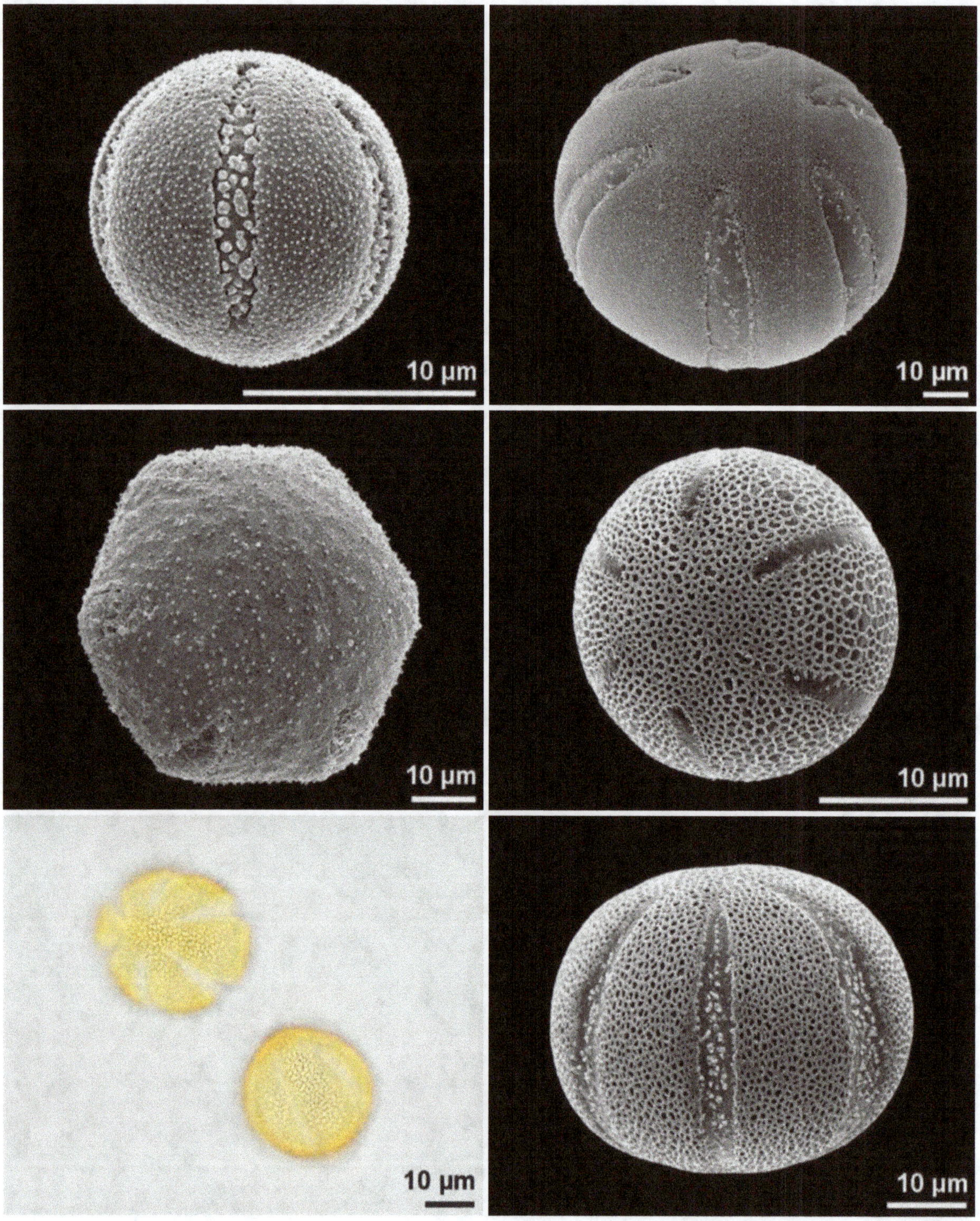

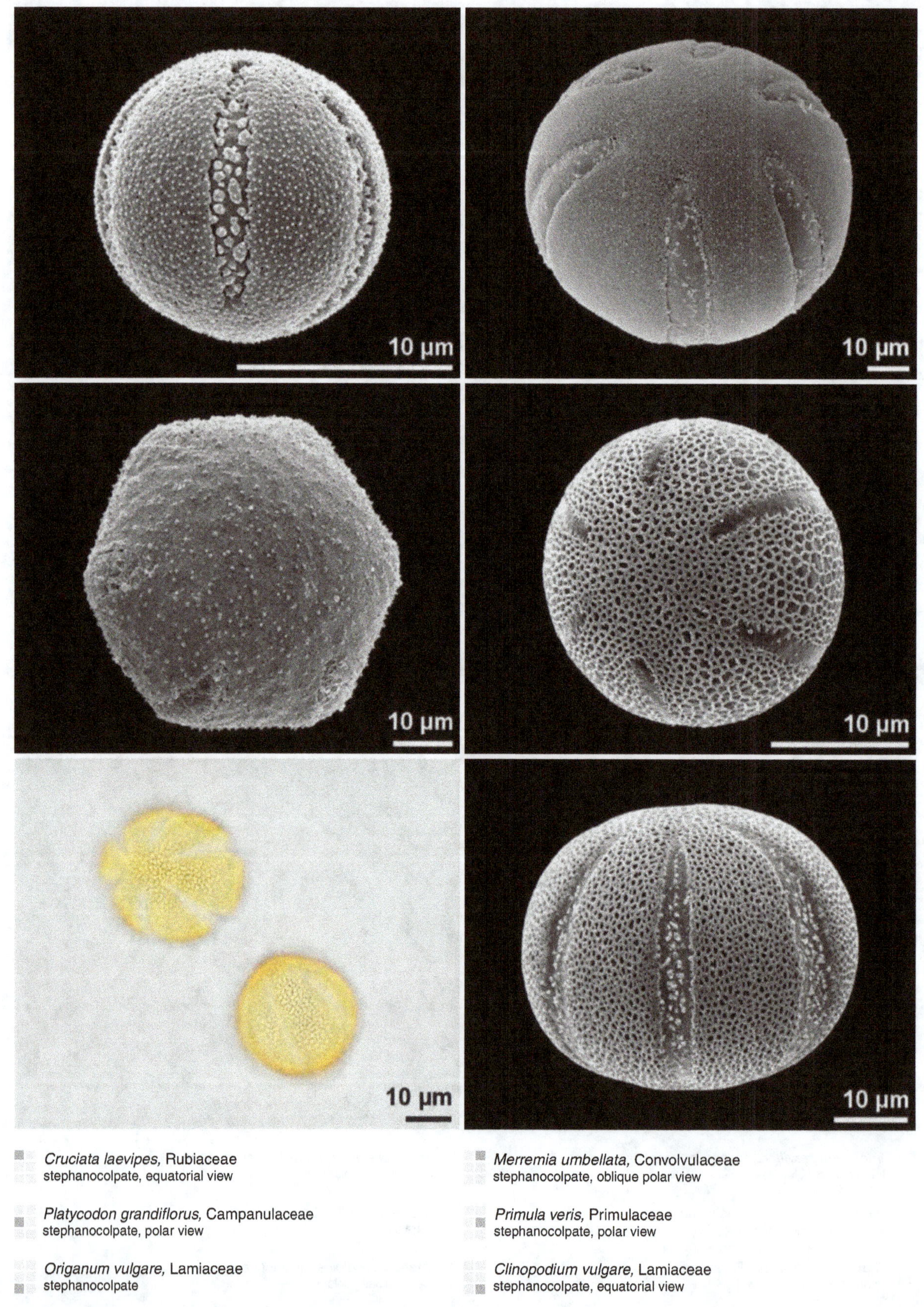

Cruciata laevipes, Rubiaceae
stephanocolpate, equatorial view

Merremia umbellata, Convolvulaceae
stephanocolpate, oblique polar view

Platycodon grandiflorus, Campanulaceae
stephanocolpate, polar view

Primula veris, Primulaceae
stephanocolpate, polar view

Origanum vulgare, Lamiaceae
stephanocolpate

Clinopodium vulgare, Lamiaceae
stephanocolpate, equatorial view

colpus/colpate, stephanocolpate

stephanocolpate: colpi situated at the equator (term usually used for six or more apertures)

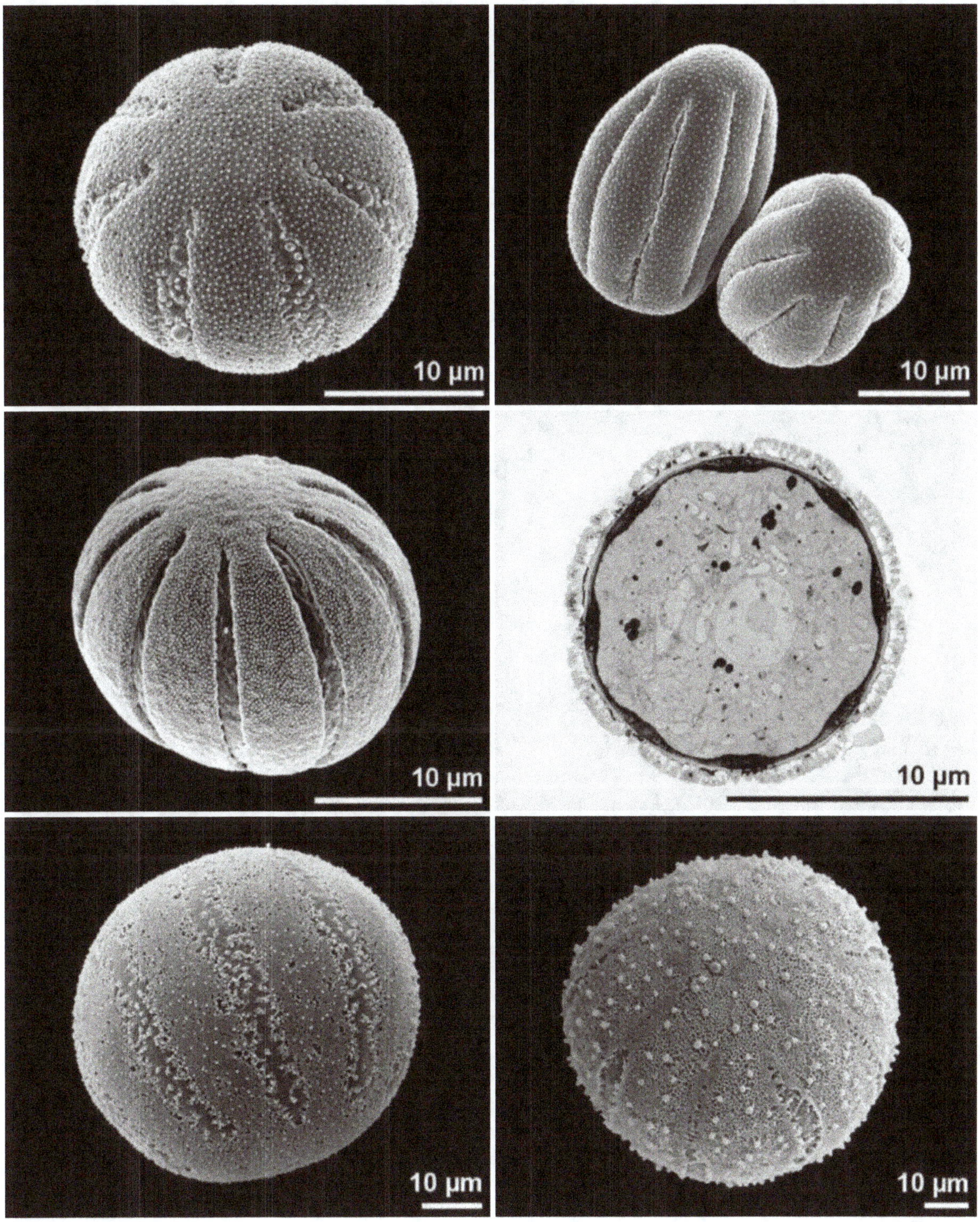

Galium glaucum, Rubiaceae
oblique polar view

Galium lucidum, Rubiaceae
dry pollen

Sherardia arvensis, Rubiaceae
oblique equatorial view

Galium odoratum, Rubiaceae
cross section

Codonopsis pilosula, Campanulaceae
equatorial view

Sechium edule, Cucurbitaceae
oblique polar view

colpus/colpate, pantocolpate

pantocolpate: pollen grain with colpi distributed more or less regularly over the surface

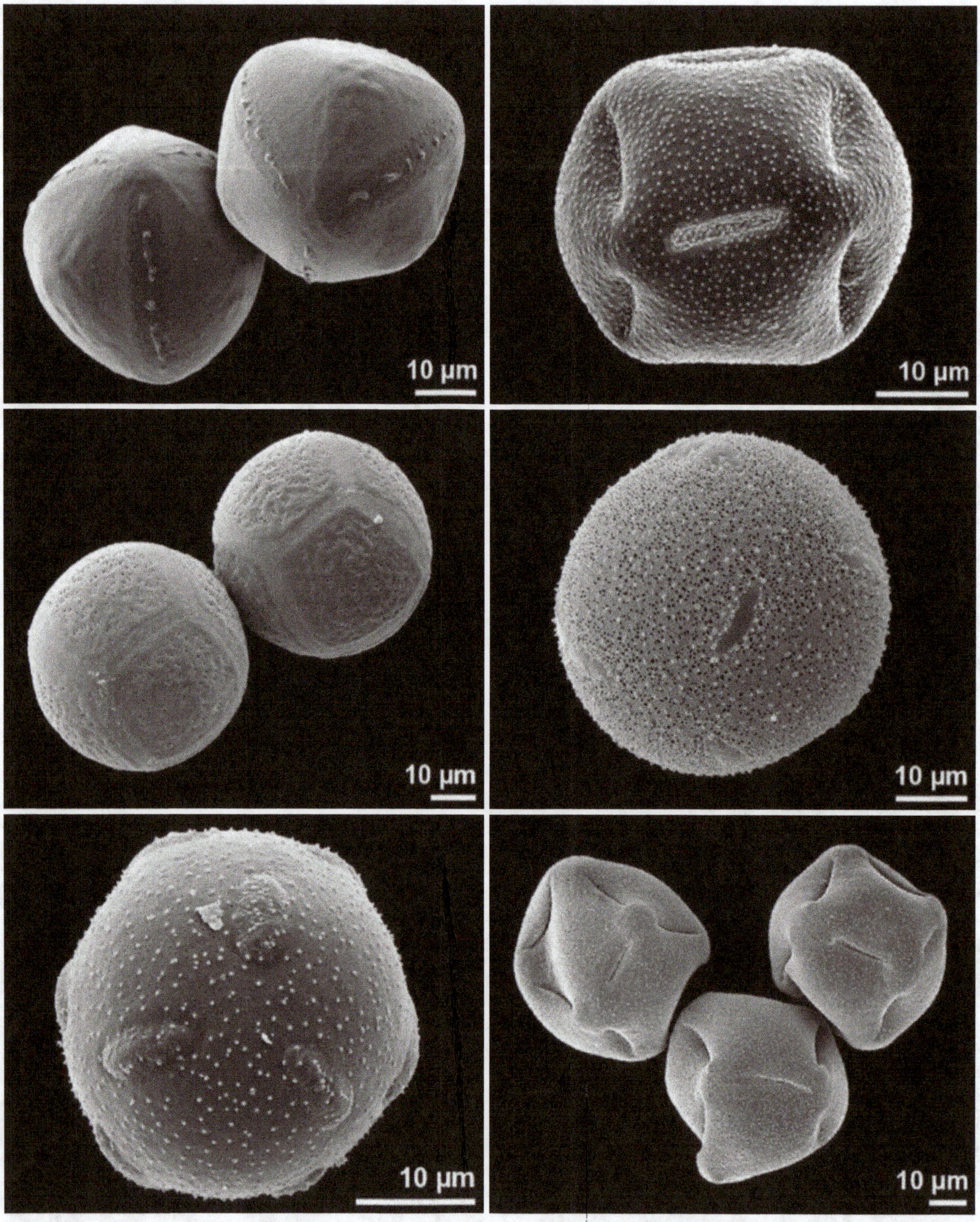

Sarcocapnos enneaphylla, Papaveraceae
panto(6)colpate

Pseudofumaria lutea, Papaveraceae
panto(syn)colpate

Mollugo verticillata, Molluginaceae

Talinum paniculatum, Talinaceae
dry pollen

Maripa nicaraguensis, Convolvulaceae

Turbinicarpus pseudomacrochele, Cactaceae
dry pollen

colporus/colporate, dicolporate

colporus: compound aperture composed of a colpus (ektoaperture) combined with an endoaperture of variable size and shape
dicolporate: pollen grain with two colpori

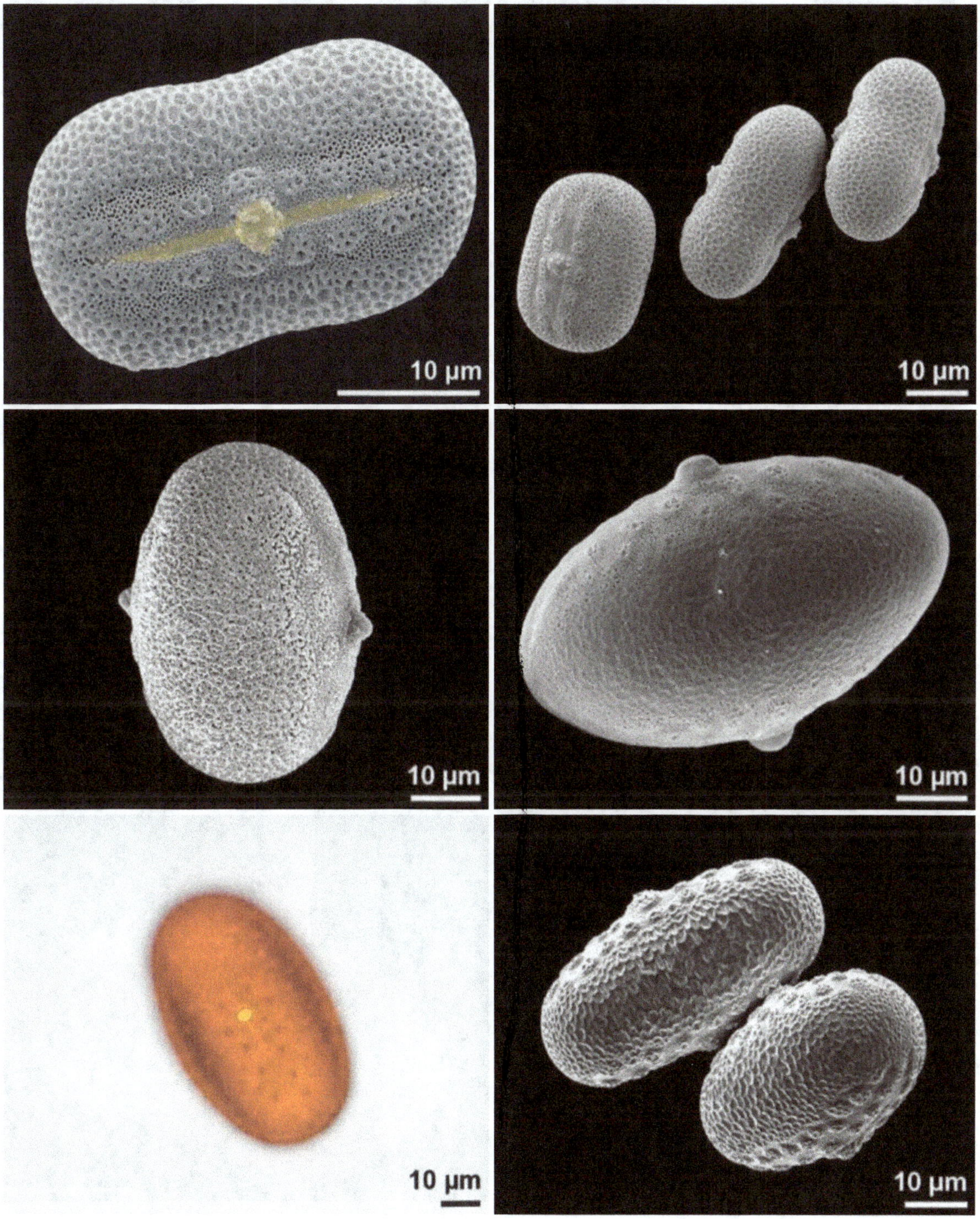

Justicia procumbens, Acanthaceae
colporus colored, equatorial view

Adhatoda schimperiana, Acanthaceae
equatorial view

Justicia carnea, Acanthaceae
equatorial view

Justicia procumbens, Acanthaceae
equatorial view

Justicia macrantha, Acanthaceae
equatorial view

Justicia xylosteoides, Acanthaceae

colporus/colporate, tricolporate

tricolporate: pollen grain with three colpori

Lathyrus vernus, Fabaceae
equatorial view

Kraussia floribunda, Rubiaceae
equatorial view

Hieracium hoppeanum, Asteraceae
equatorial view

Erica herbacea, Ericaceae
tetrad

Rumex acetosa, Polygonaceae
equatorial view

Aruncus dioicus, Rosaceae
germinating pollen

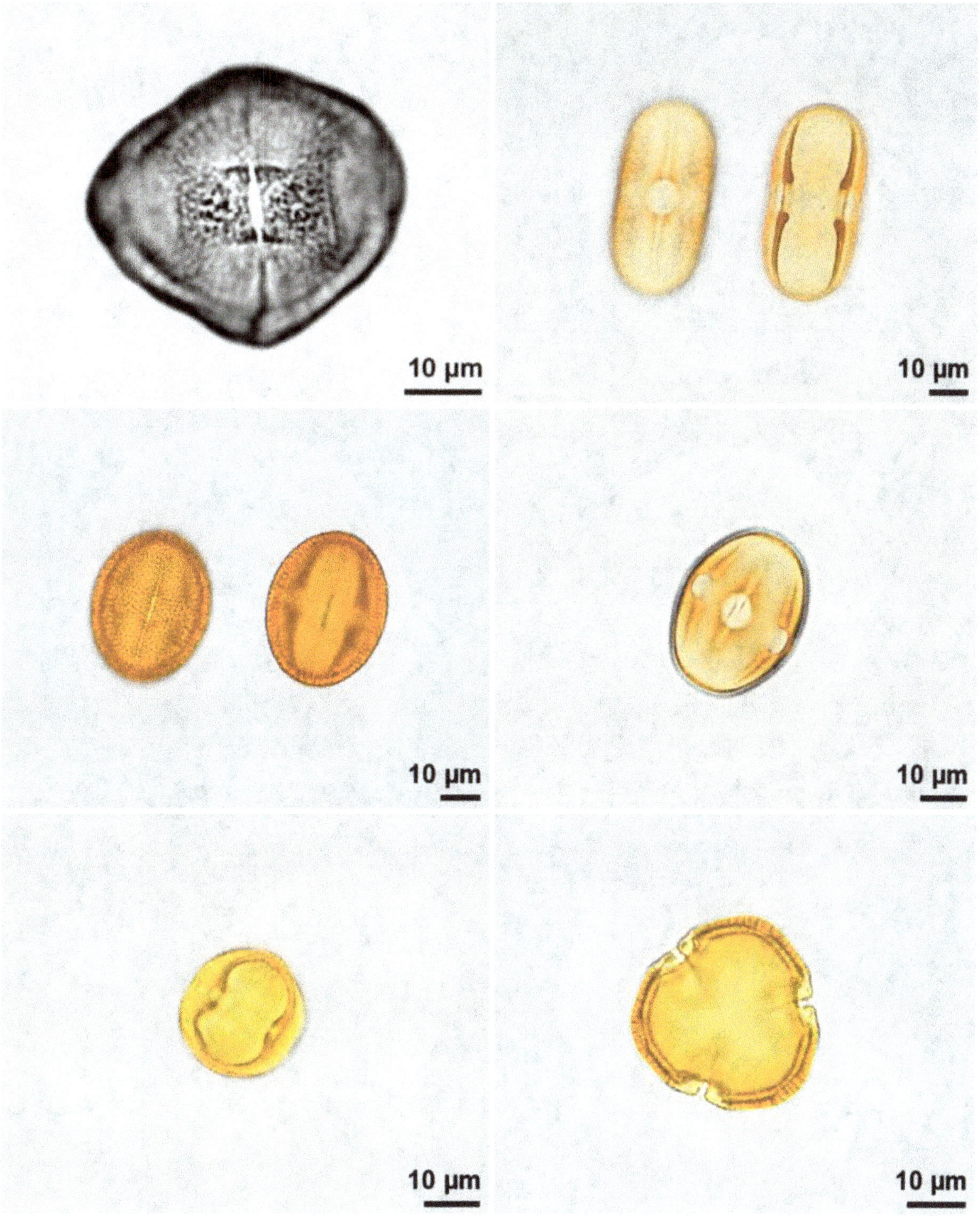

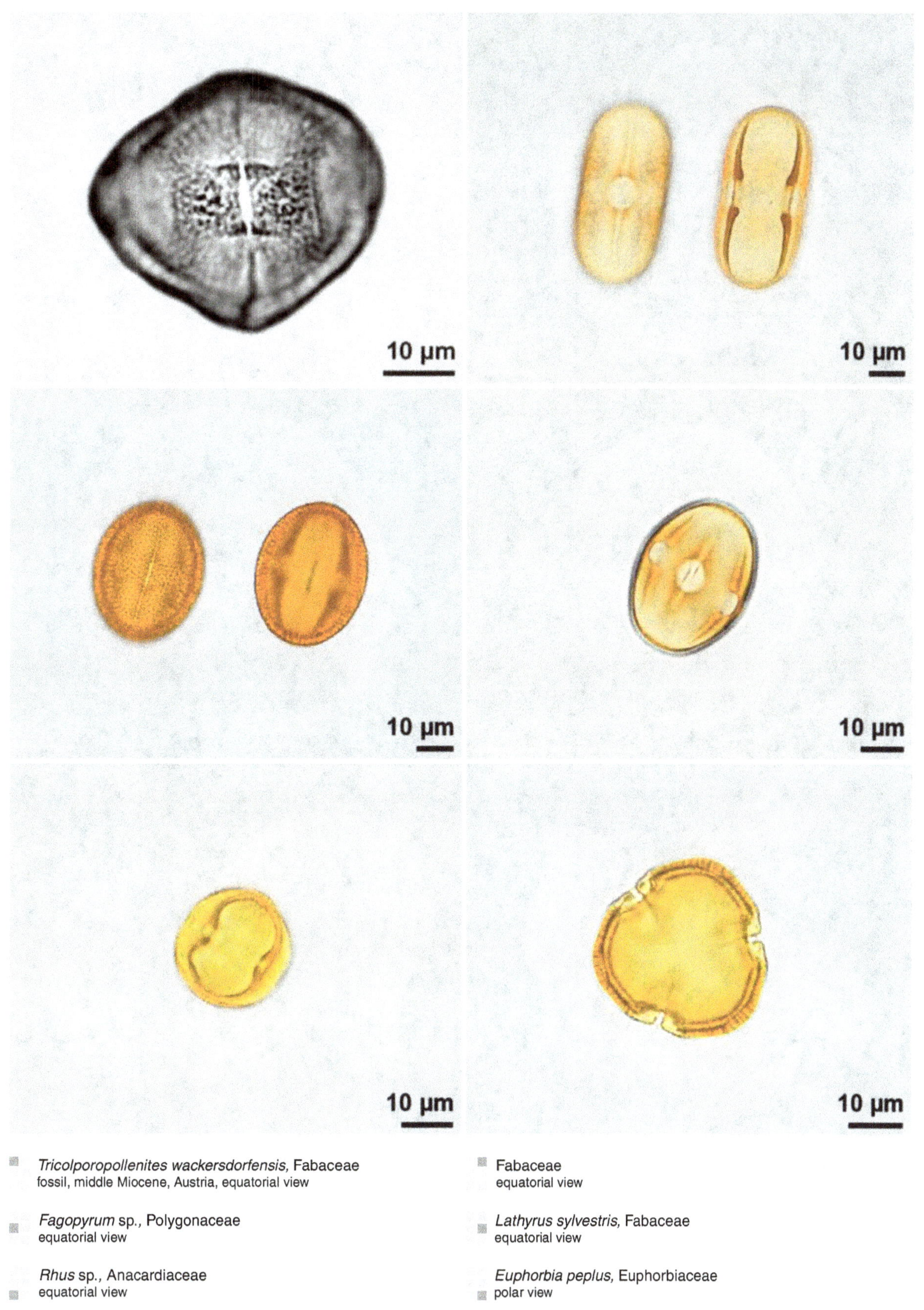

- *Tricolporopollenites wackersdorfensis,* Fabaceae
  fossil, middle Miocene, Austria, equatorial view
- *Fagopyrum* sp., Polygonaceae
  equatorial view
- *Rhus* sp., Anacardiaceae
  equatorial view
- Fabaceae
  equatorial view
- *Lathyrus sylvestris,* Fabaceae
  equatorial view
- *Euphorbia peplus,* Euphorbiaceae
  polar view

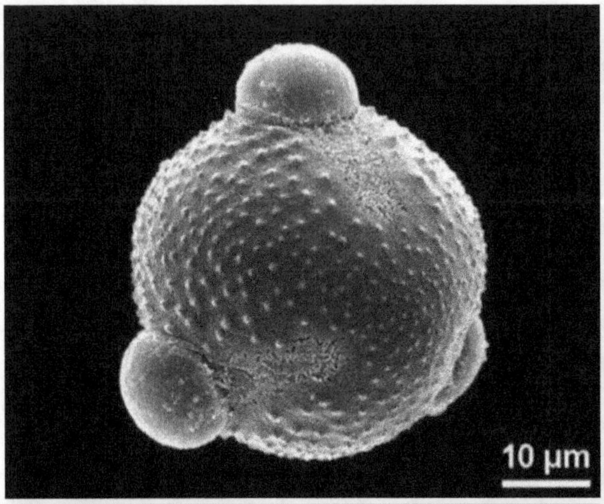

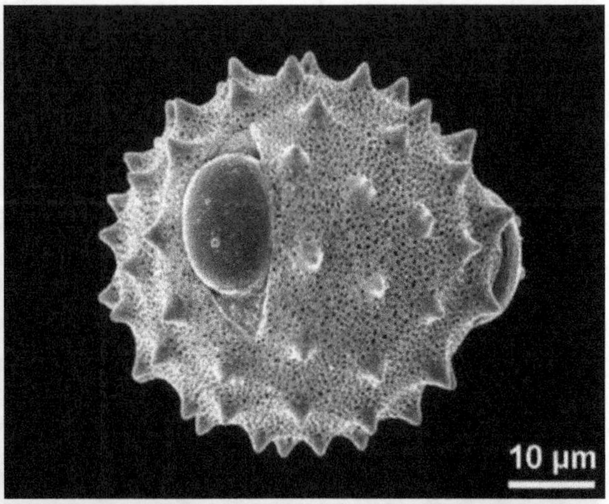

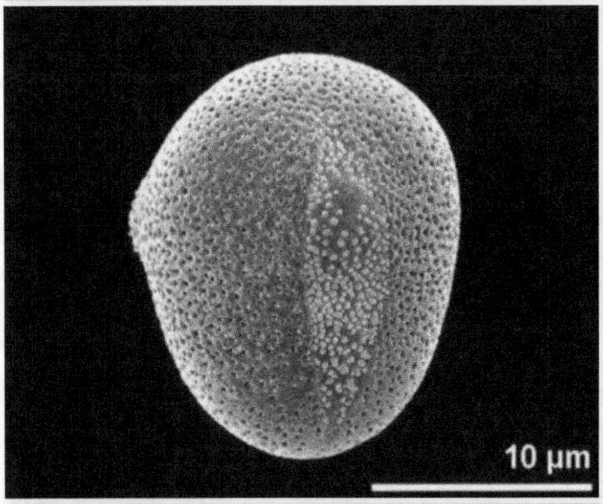

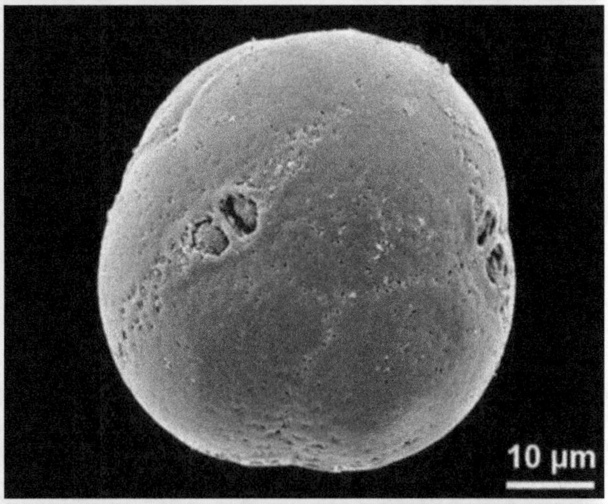

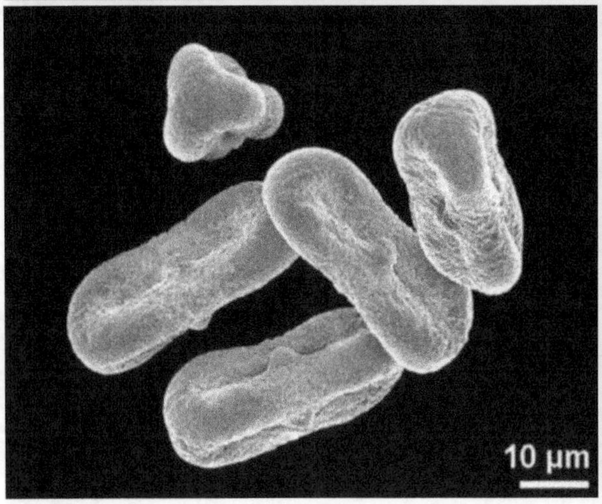

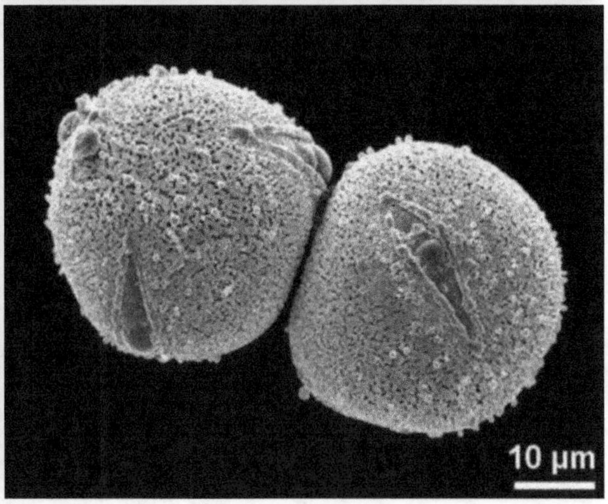

Centaurea scabiosa, Asteraceae
polar view

Cirsium oleraceum, Asteraceae
equatorial view

Echium vulgare, Boraginaceae
equatorial view

Gardenia thunbergia, Rubiaceae
tetrad

Myrrhis odorata, Apiaceae
dry pollen

Fatsia japonica, Araliaceae

colporus/colporate, tetracolporate

tetracolporate: pollen grain with 4 colpori

Citrus swinglei, Rutaceae

Nicotiana tabacum, Solanaceae
oblique polar view

Poncirus trifoliata, Rutaceae
oblique polar view

Pulmonaria mollis, Boraginaceae
equatorial view

Genlisea violacea, Lentibulariaceae

Tridax procumbens, Asteraceae

colporus/colporate, pentacolporate, stephanocolporate

pentacolporate: pollen grain with five colpori

stephanocolporate: colpori situated at the equator (term usually used for 6 or more apertures)

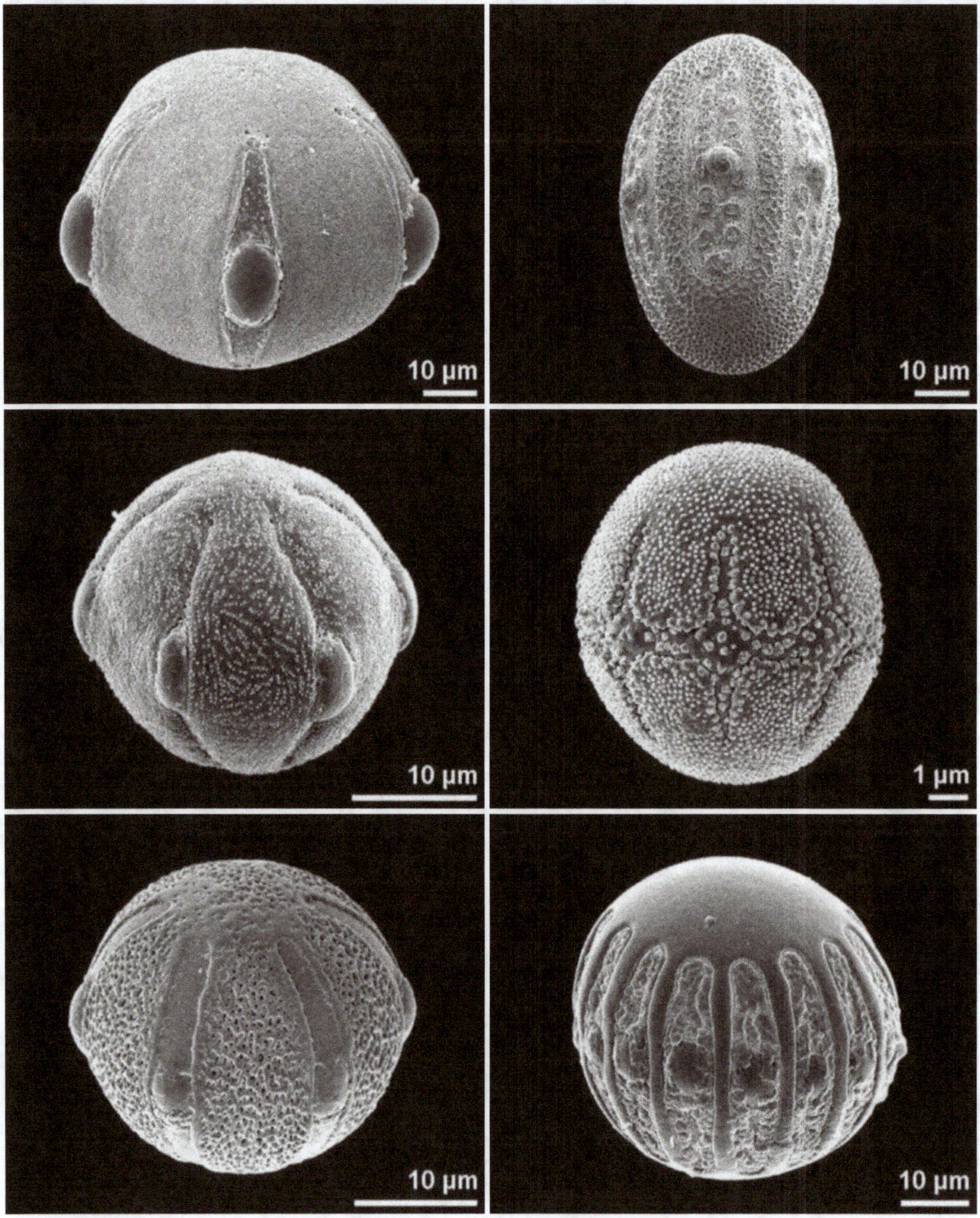

Viola arvensis, Violaceae
pentacolporate, equatorial view

Justicia menesii, Acanthaceae
hexacolporate, oblique equatorial view

Sanguisorba officinalis, Rosaceae
stephanocolporate, equatorial view

Cerinthe minor, Boraginaceae
stephanocolporate, equatorial view

Pinguicula ehlersiae, Lentibulariaceae
stephanocolporate, equatorial view

Polygala chamaebuxus, Polygalaceae
stephanocolporate, equatorial view

colporus/colporate, stephanocolporate

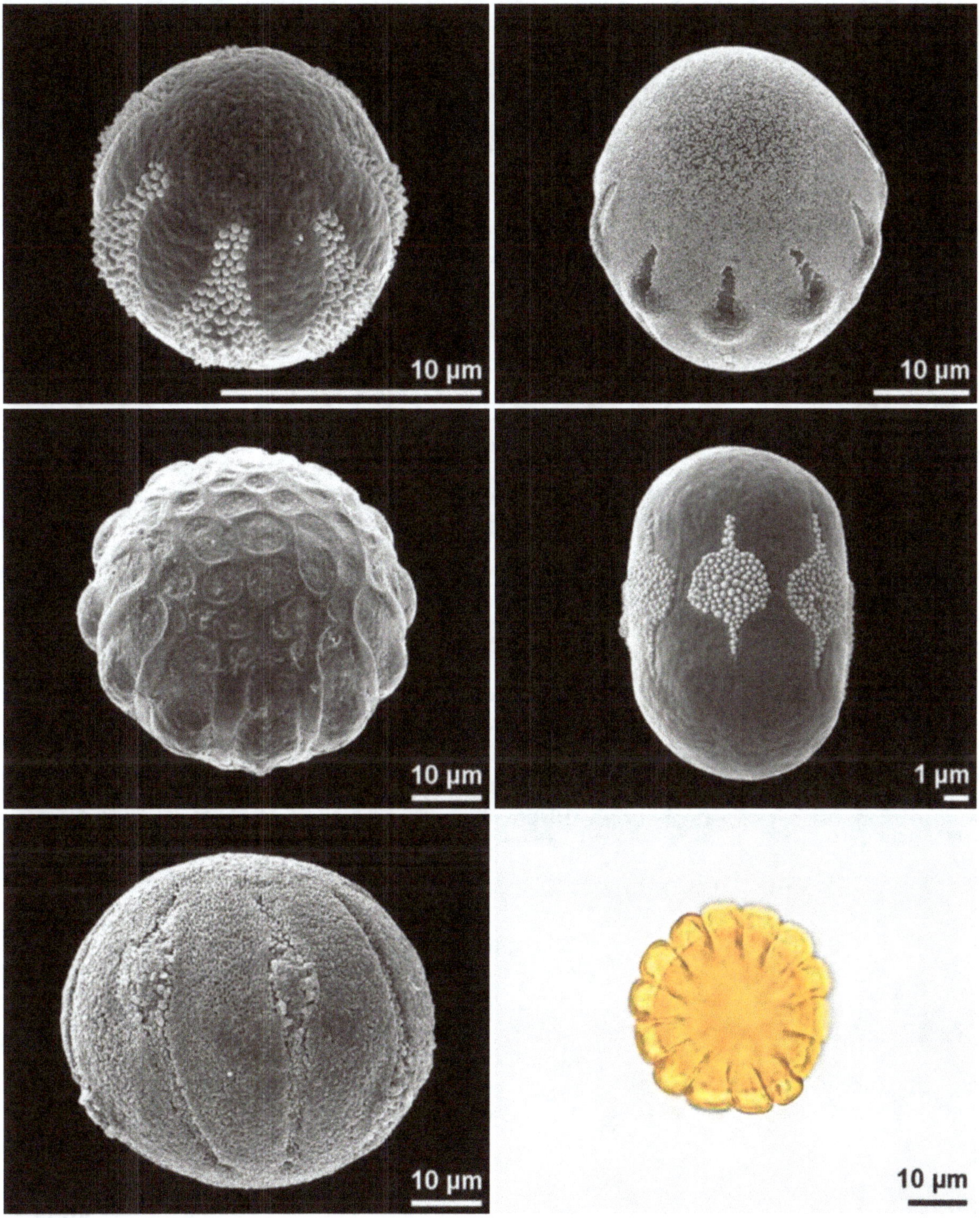

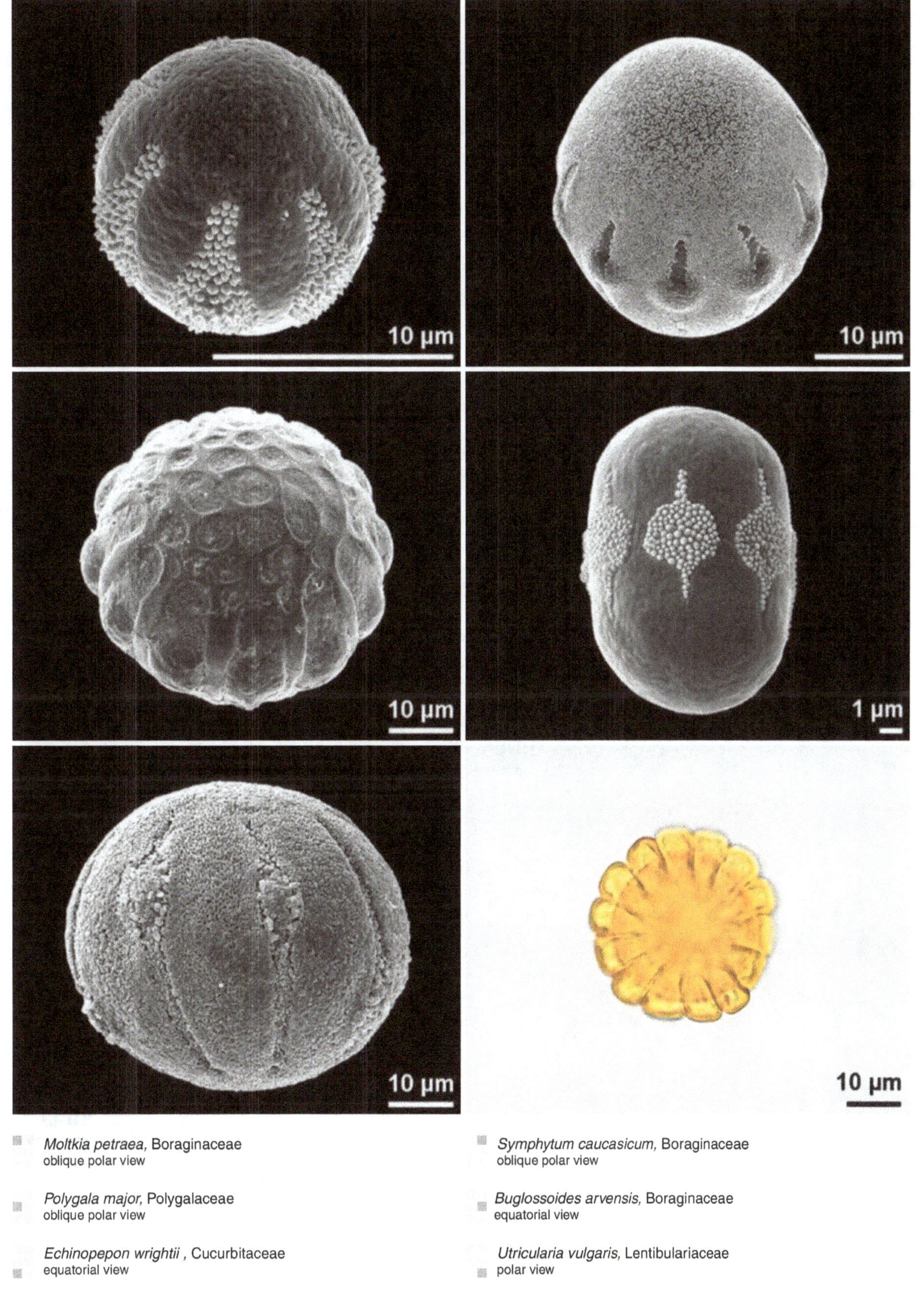

Moltkia petraea, Boraginaceae
oblique polar view

Symphytum caucasicum, Boraginaceae
oblique polar view

Polygala major, Polygalaceae
oblique polar view

Buglossoides arvensis, Boraginaceae
equatorial view

Echinopepon wrightii , Cucurbitaceae
equatorial view

Utricularia vulgaris, Lentibulariaceae
polar view

ektoaperture, endoaperture, lalongate, lolongate

ectoaperture: outer part of a compound aperture
endoaperture: inner part of a compound aperture

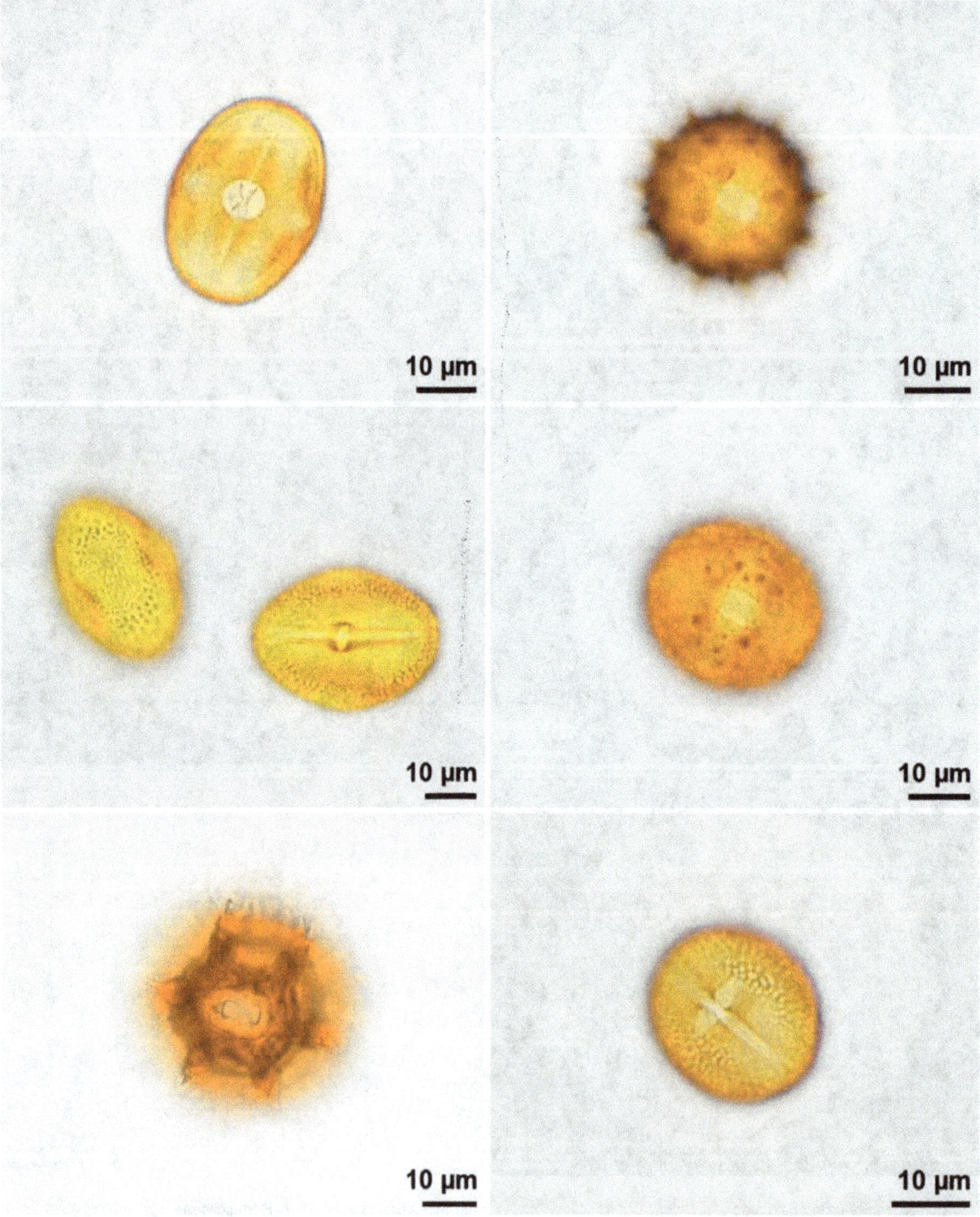

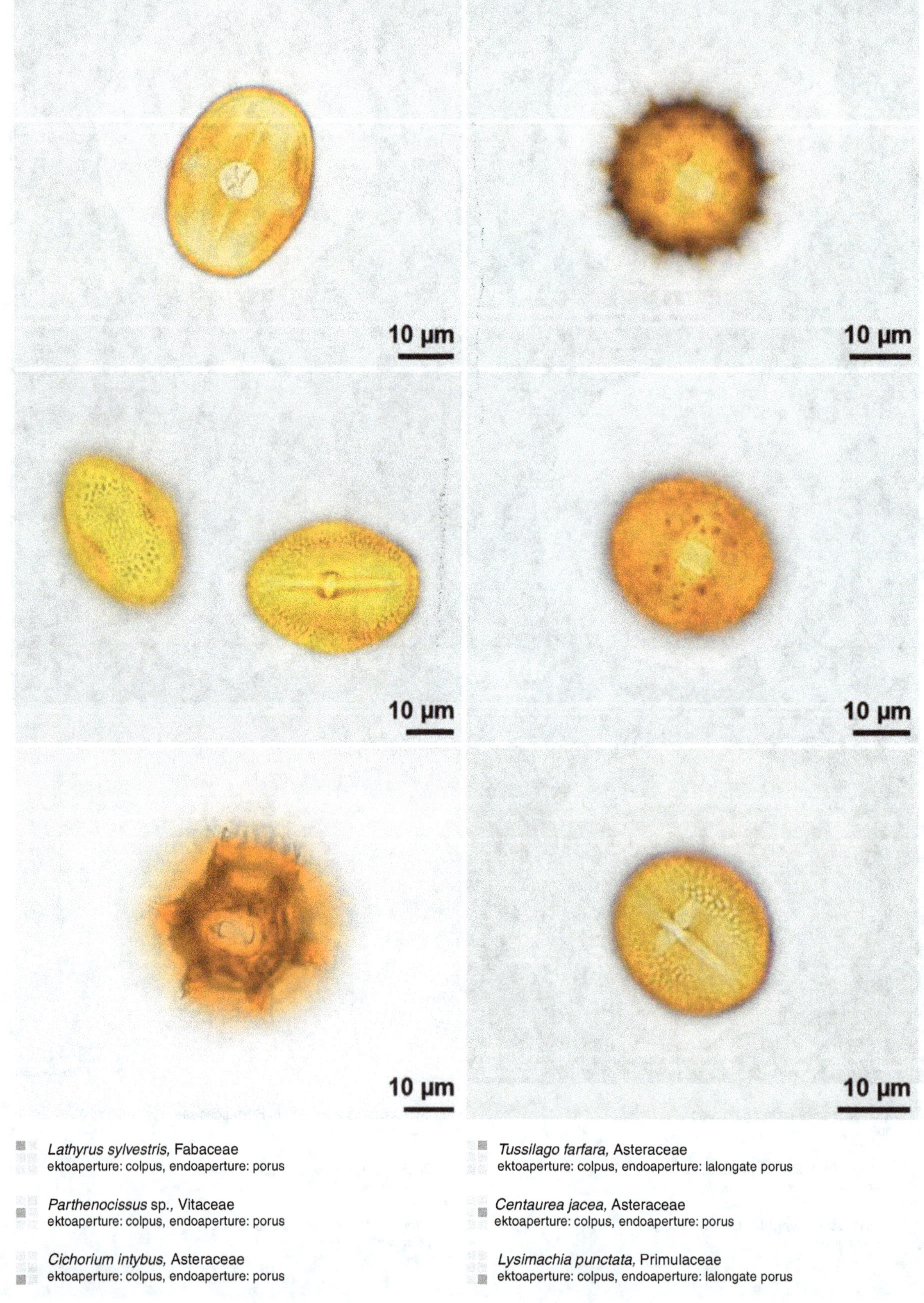

Lathyrus sylvestris, Fabaceae
ektoaperture: colpus, endoaperture: porus

Parthenocissus sp., Vitaceae
ektoaperture: colpus, endoaperture: porus

Cichorium intybus, Asteraceae
ektoaperture: colpus, endoaperture: porus

Tussilago farfara, Asteraceae
ektoaperture: colpus, endoaperture: lalongate porus

Centaurea jacea, Asteraceae
ektoaperture: colpus, endoaperture: porus

Lysimachia punctata, Primulaceae
ektoaperture: colpus, endoaperture: lalongate porus

lalongate: ectoaperture elongated equatorially
lolongate: ectoaperture alongated meridionally

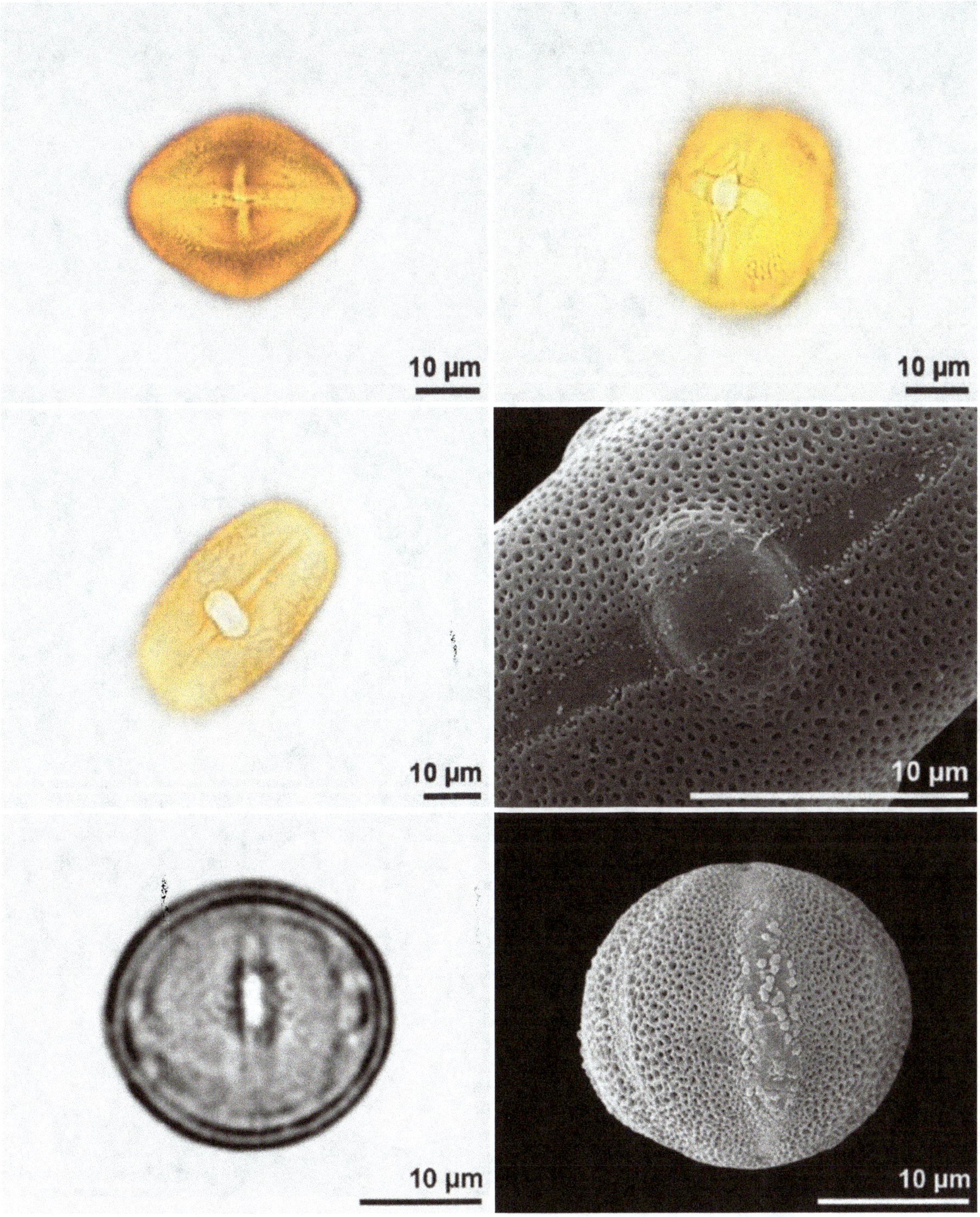

Dictamnus albus, Rutaceae
ektoaperture: colpus, endoaperture: lalongate porus

Vicia faba, Fabaceae
ektoaperture: colpus, endoaperture: lalongate porus

Rumex sp., Polygonaceae
fossil, Quaternary Austria, ektoaperture: colpus, endoaperture: lolongate porus

Scaevola aemula, Goodeniaceae
ektoaperture: colpus, endoaperture: lalongate porus

Vitaliana primuliflora, Primulaceae
ektoaperture: colpus, endoaperture: lalongate porus

Phacelia campanularia, Boraginaceae
heteroaperturate pollen, ektoaperture: colpus, endoaperture lolongate porus

heteroaperturate

pollen grain with different types of apertures; only one type presumed to function as germination site

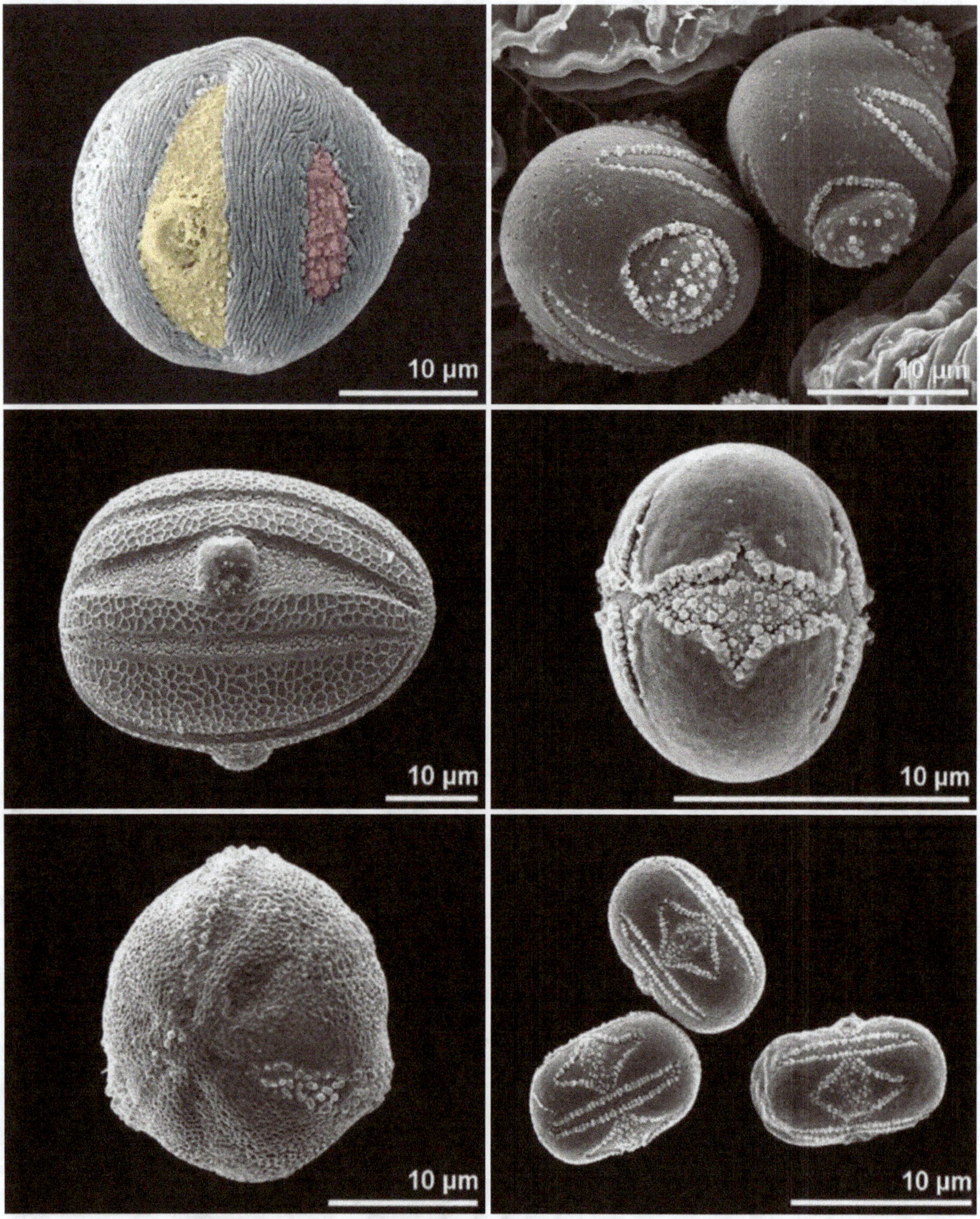

Lythrum hyssopifolia, Lythraceae
colporus (yellow) and pseudocolpus (purple), equatorial view

Tetramerium nervosum, Acanthaceae
equatorial view

Phacelia tanacetifolia, Boraginaceae
polar view

Pardoglossum sp., Boraginaceae

Cynoglossum officinale, Boraginaceae
equatorial view

Myosotis ramosissima, Boraginaceae

Pseuderanthemum alatum, Acanthaceae

Omphalodes linifolia, Boraginaceae

Meriania selvaflorensis, Melastomataceae

Combretum fruticosum, Combretaceae
equatorial view

Medinilla scortechinii, Melastomataceae

Phacelia campanularia, Boraginaceae

inaperturate

pollen grain without distinct apertures

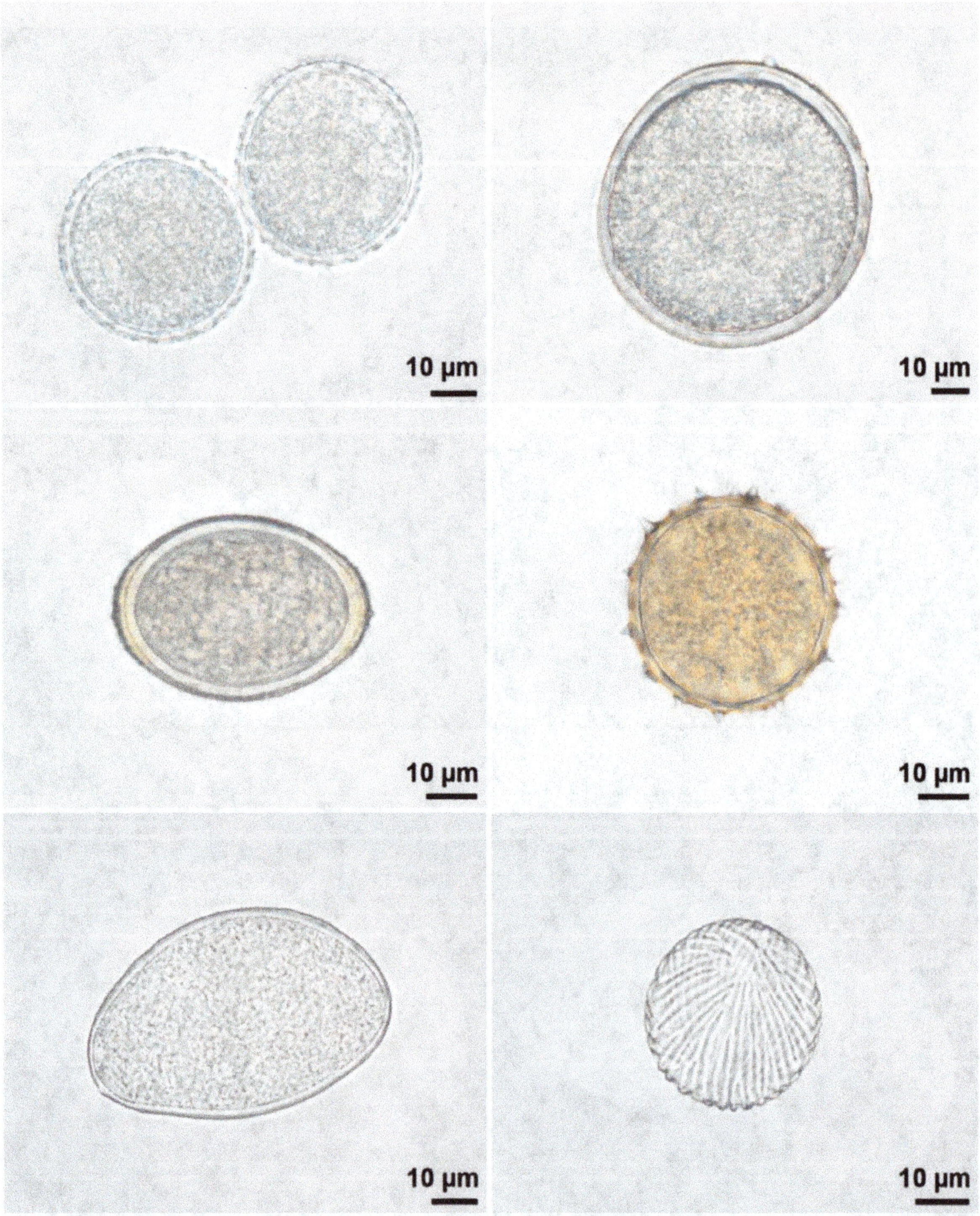

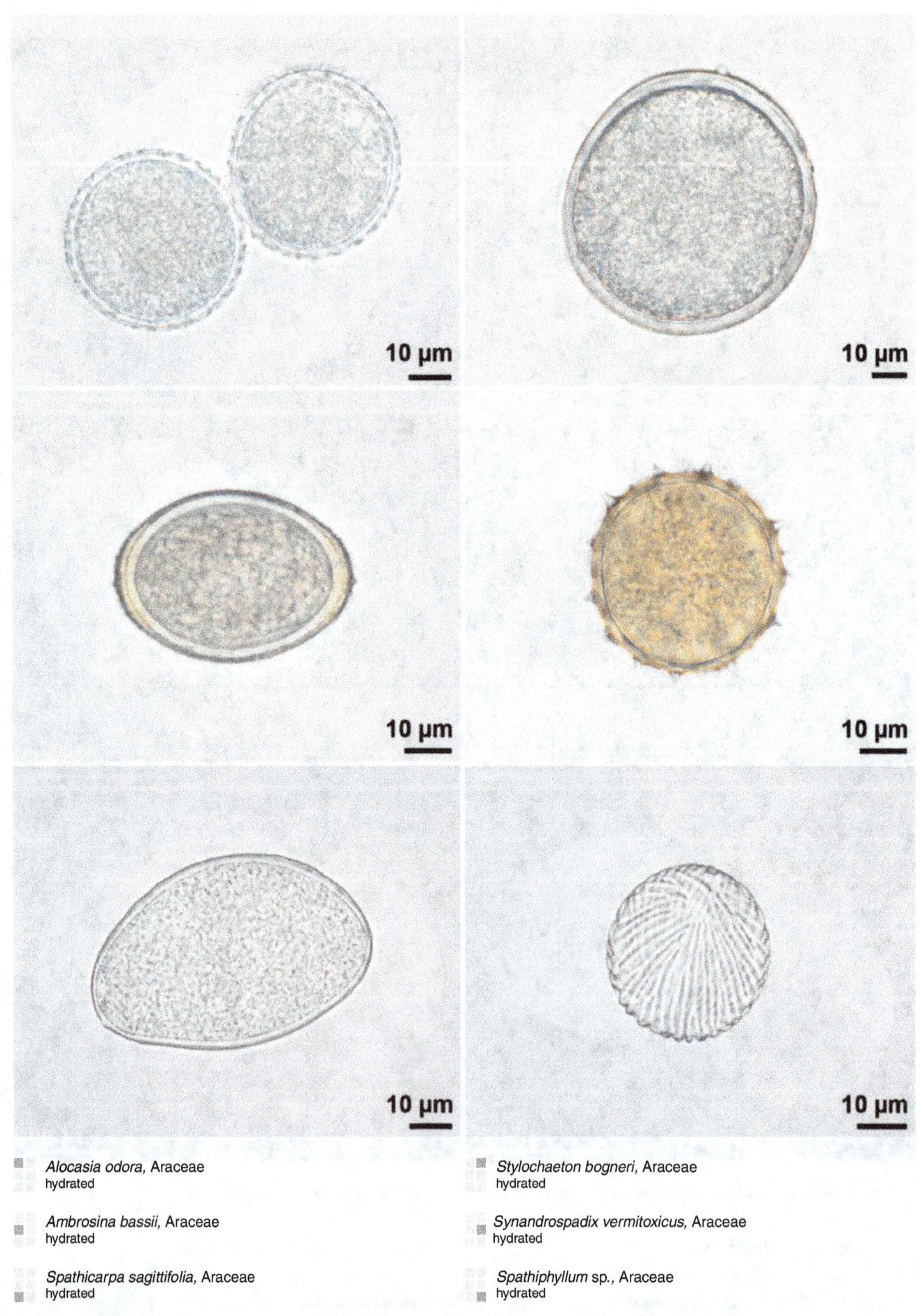

- *Alocasia odora*, Araceae
  hydrated
- *Ambrosina bassii*, Araceae
  hydrated
- *Spathicarpa sagittifolia*, Araceae
  hydrated
- *Stylochaeton bogneri*, Araceae
  hydrated
- *Synandrospadix vermitoxicus*, Araceae
  hydrated
- *Spathiphyllum* sp., Araceae
  hydrated

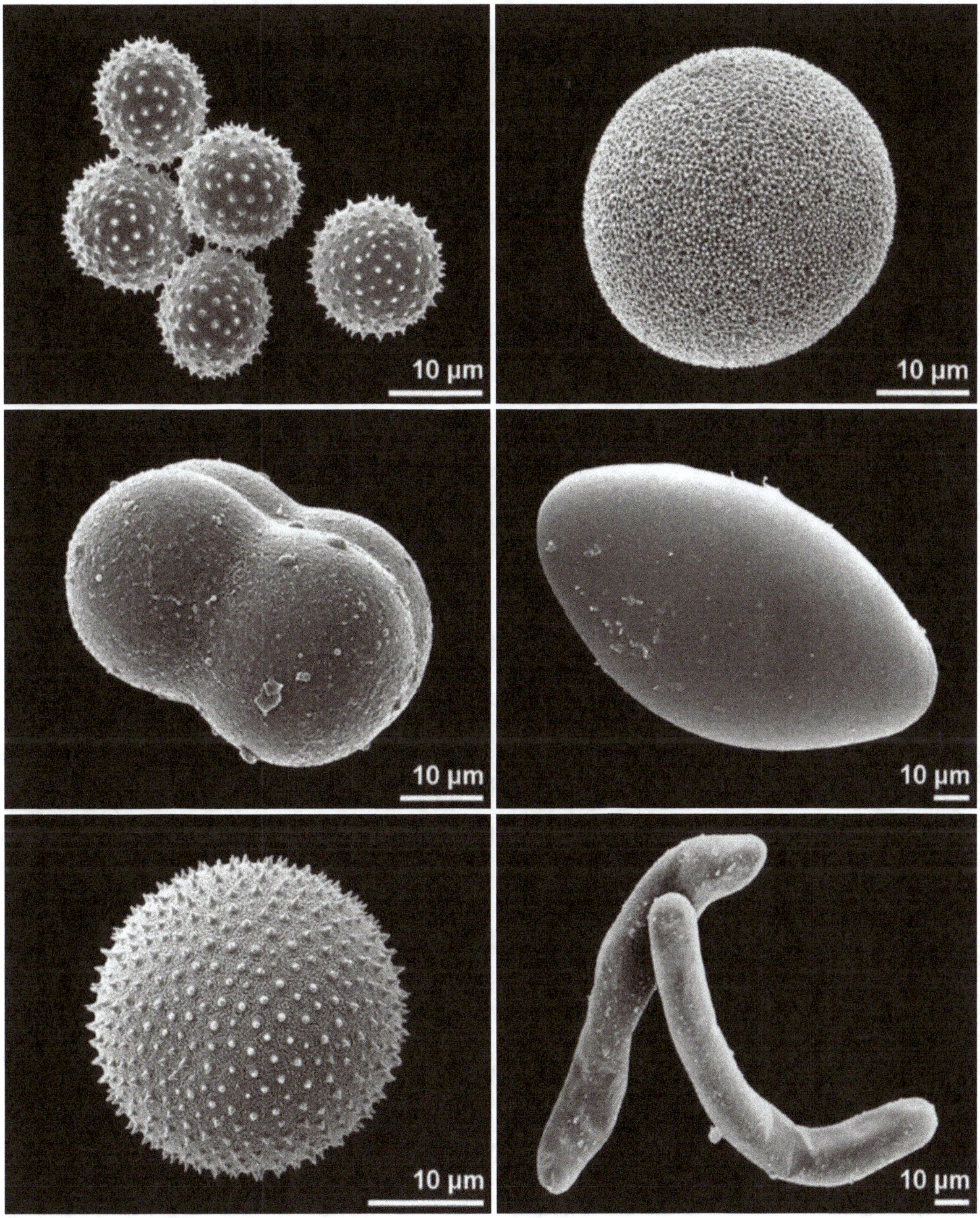

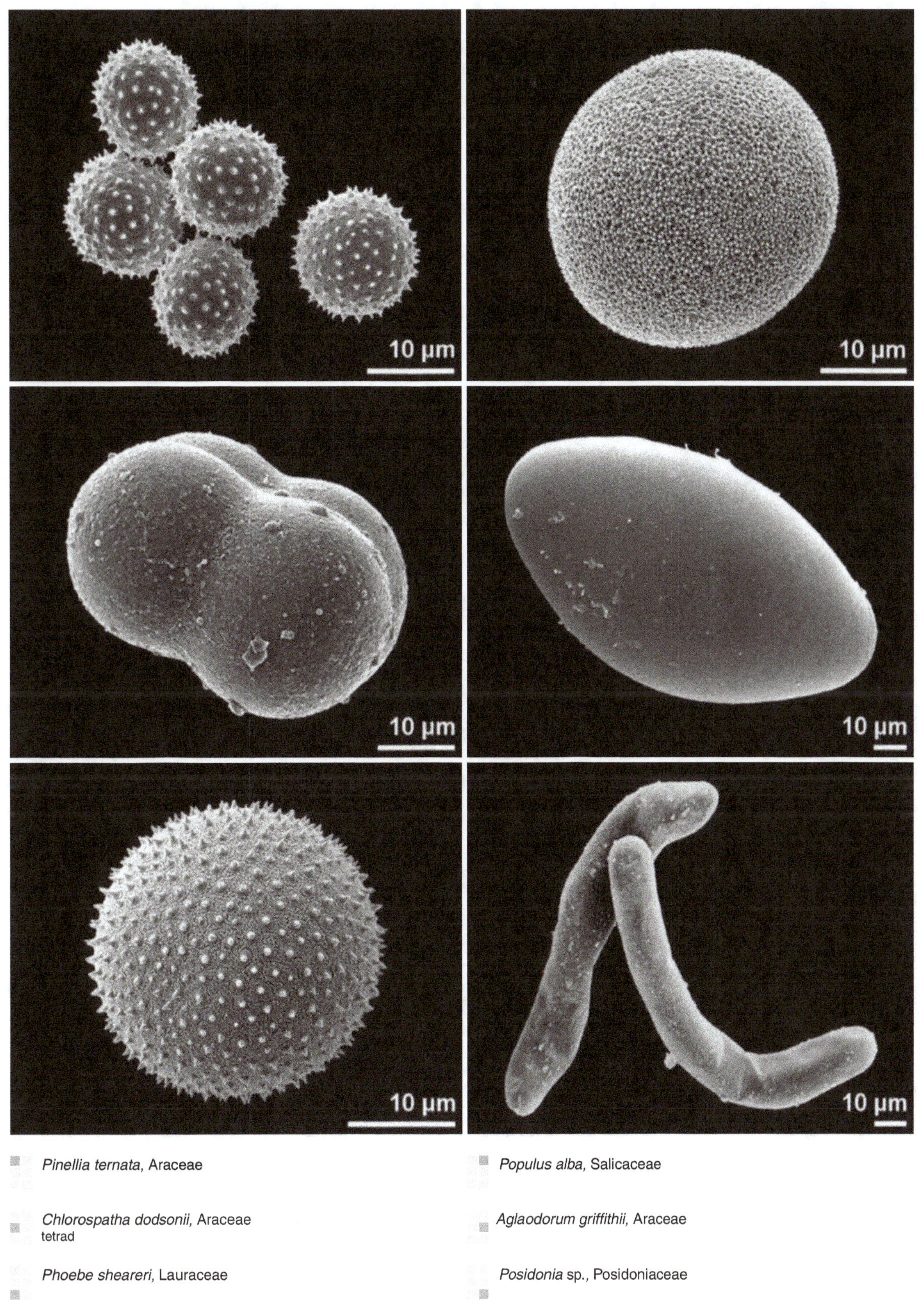

Pinellia ternata, Araceae

Populus alba, Salicaceae

Chlorospatha dodsonii, Araceae
tetrad

Aglaodorum griffithii, Araceae

Phoebe sheareri, Lauraceae

Posidonia sp., Posidoniaceae

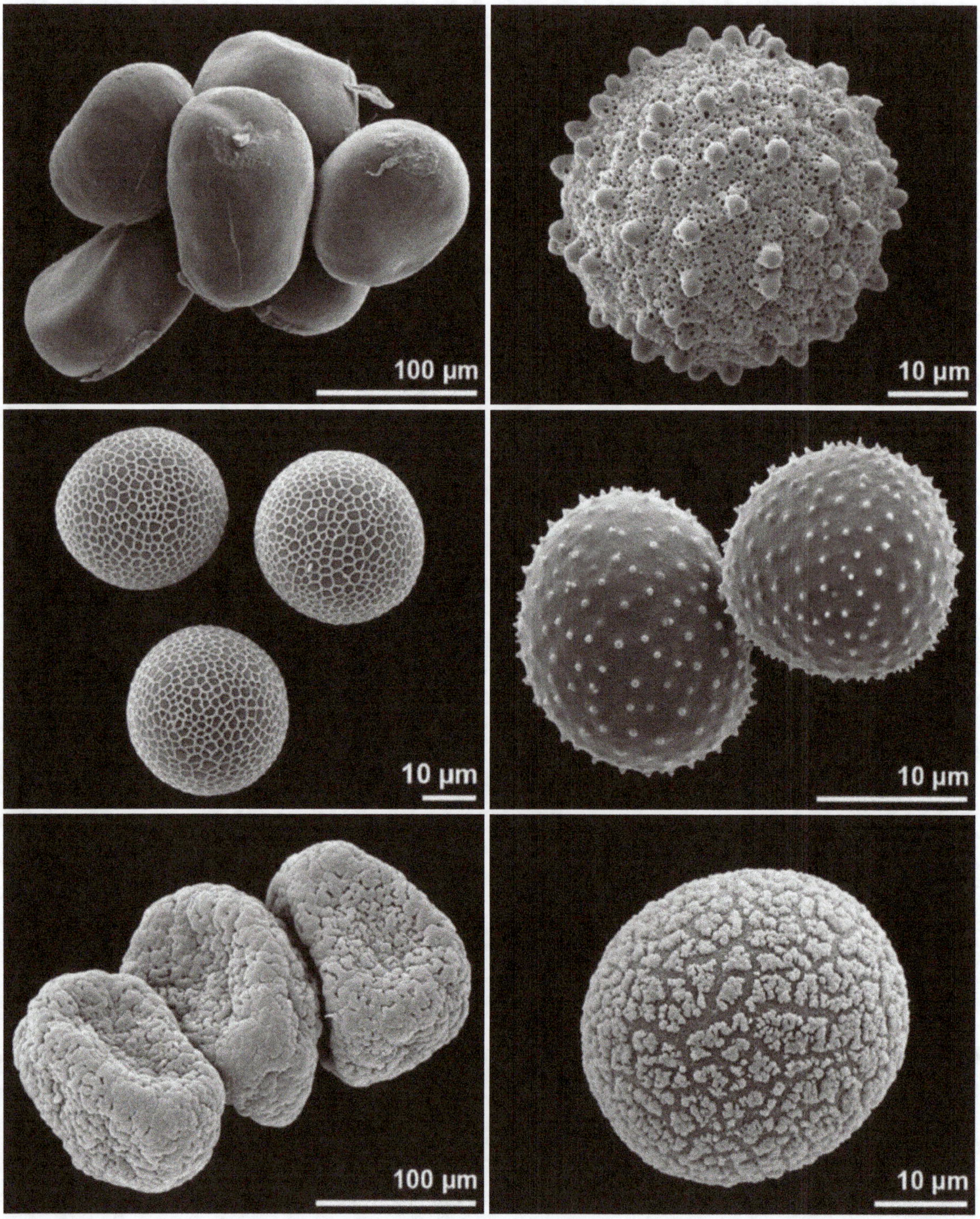

Orchidantha maxillarioides, Lowiaceae
dry pollen

Aristolochia arborea, Aristolochiaceae

Triglochin maritima, Juncaginaceae

Gnetum gnemon, Gnetaceae

Cytinus hypocistis, Cytinaceae
dry pollen

Trillium chloropetalum, Melanthiaceae

leptoma

thinning of the pollen wall on the distal face in conifers, presumed to function as germination area

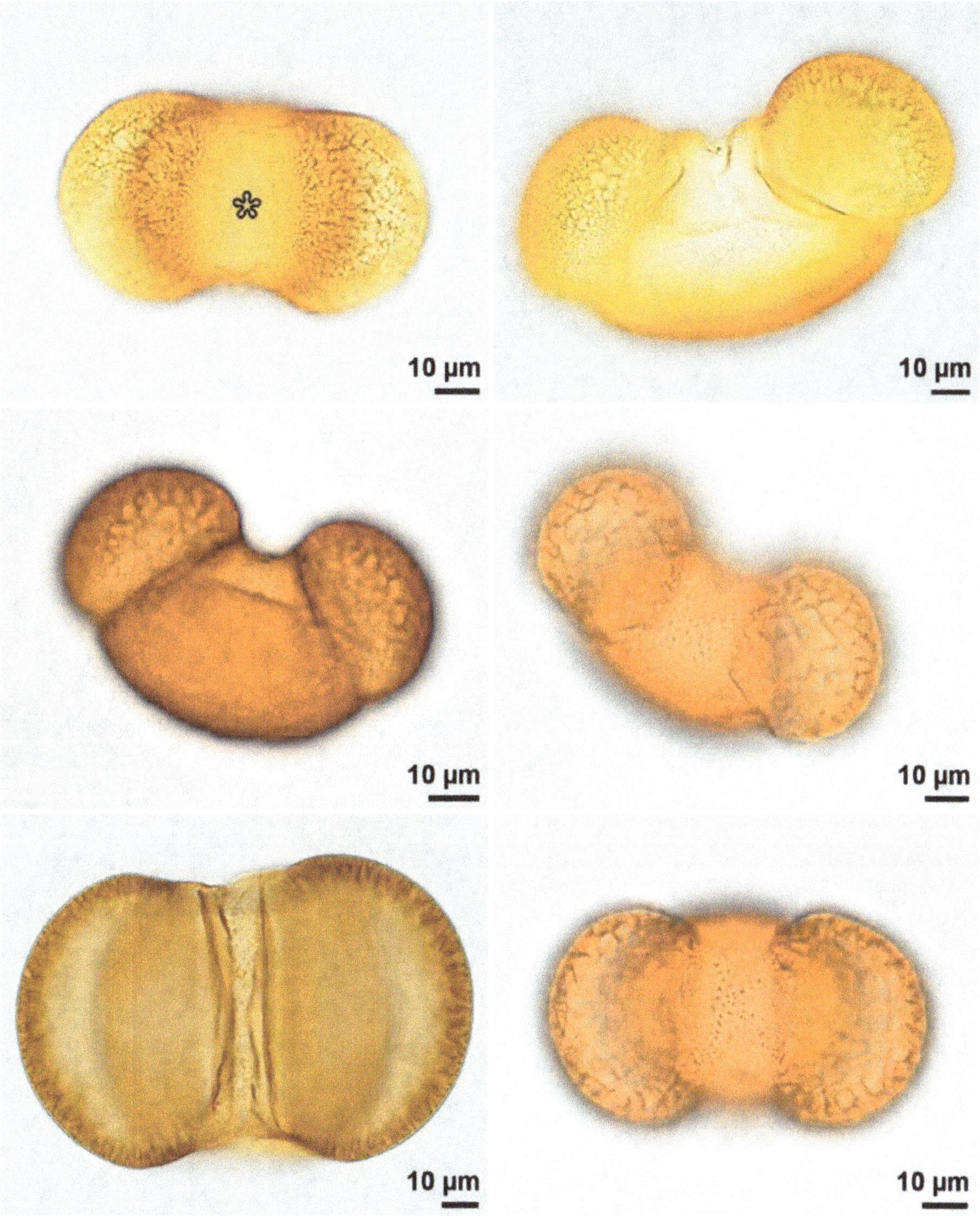

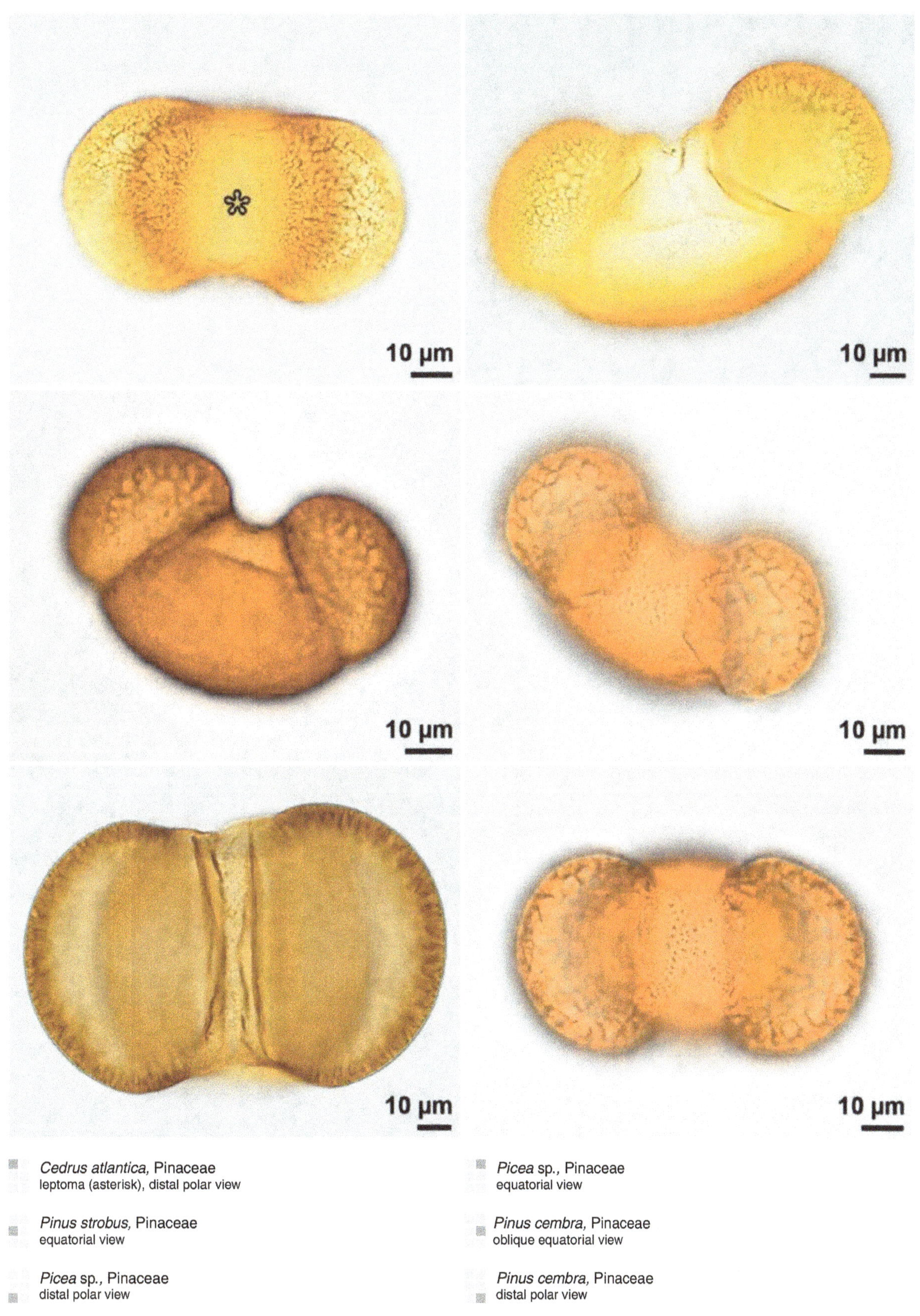

Cedrus atlantica, Pinaceae
leptoma (asterisk), distal polar view

Picea sp., Pinaceae
equatorial view

Pinus strobus, Pinaceae
equatorial view

Pinus cembra, Pinaceae
oblique equatorial view

Picea sp., Pinaceae
distal polar view

Pinus cembra, Pinaceae
distal polar view

margo

exine area with different ornamentation bordering a colpus/colporus/sulcus

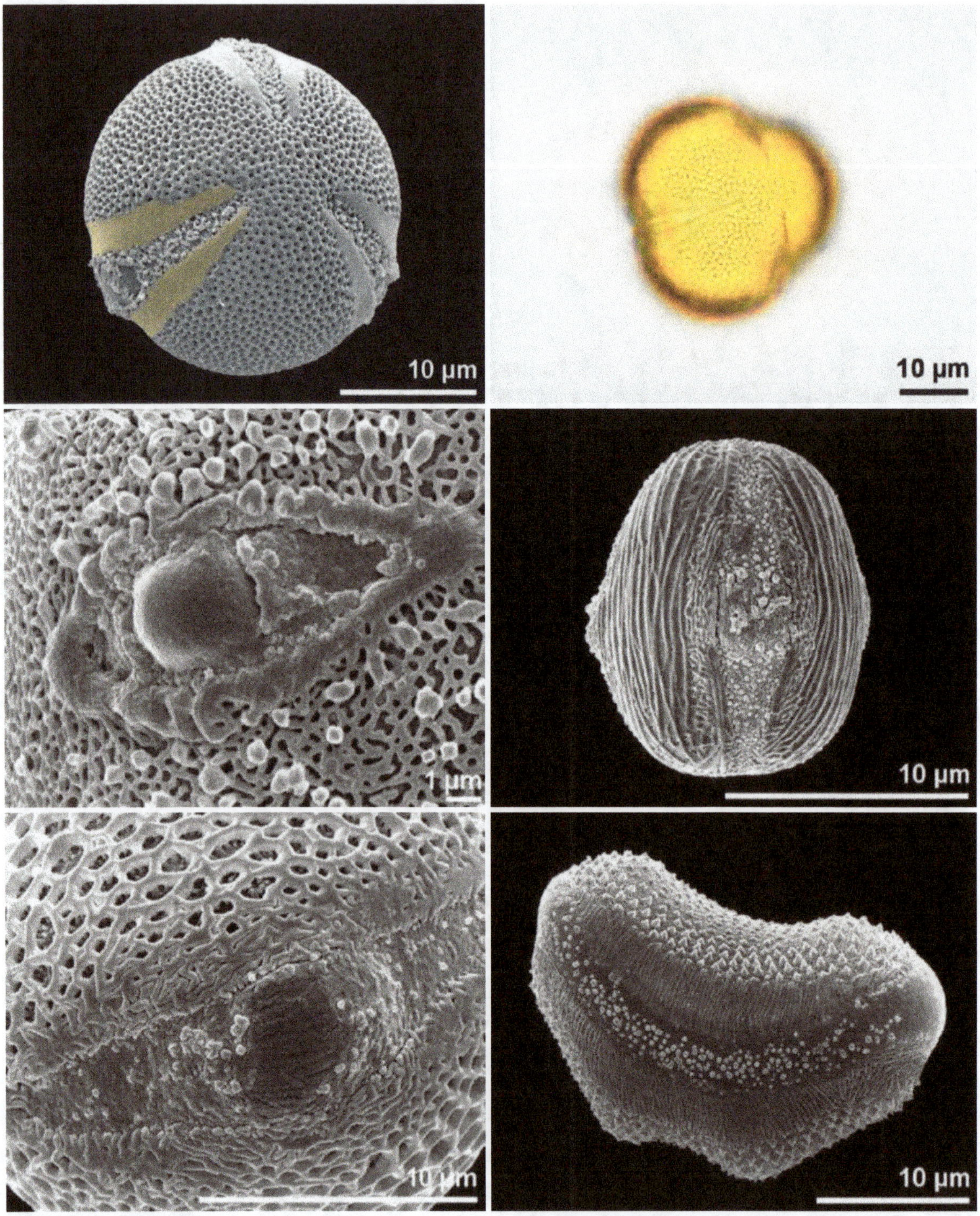

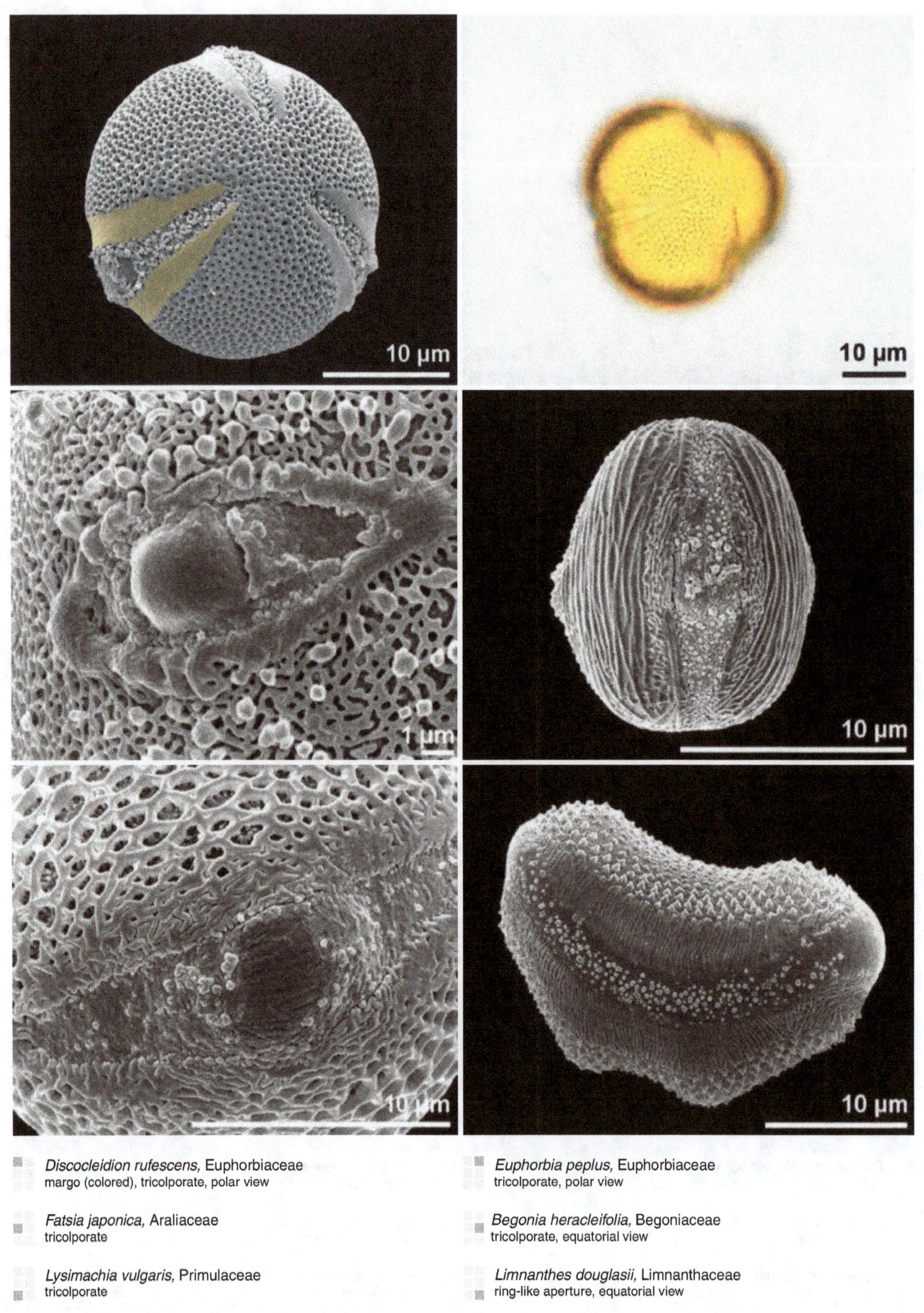

Discocleidion rufescens, Euphorbiaceae
margo (colored), tricolporate, polar view

Euphorbia peplus, Euphorbiaceae
tricolporate, polar view

Fatsia japonica, Araliaceae
tricolporate

Begonia heracleifolia, Begoniaceae
tricolporate, equatorial view

Lysimachia vulgaris, Primulaceae
tricolporate

Limnanthes douglasii, Limnanthaceae
ring-like aperture, equatorial view

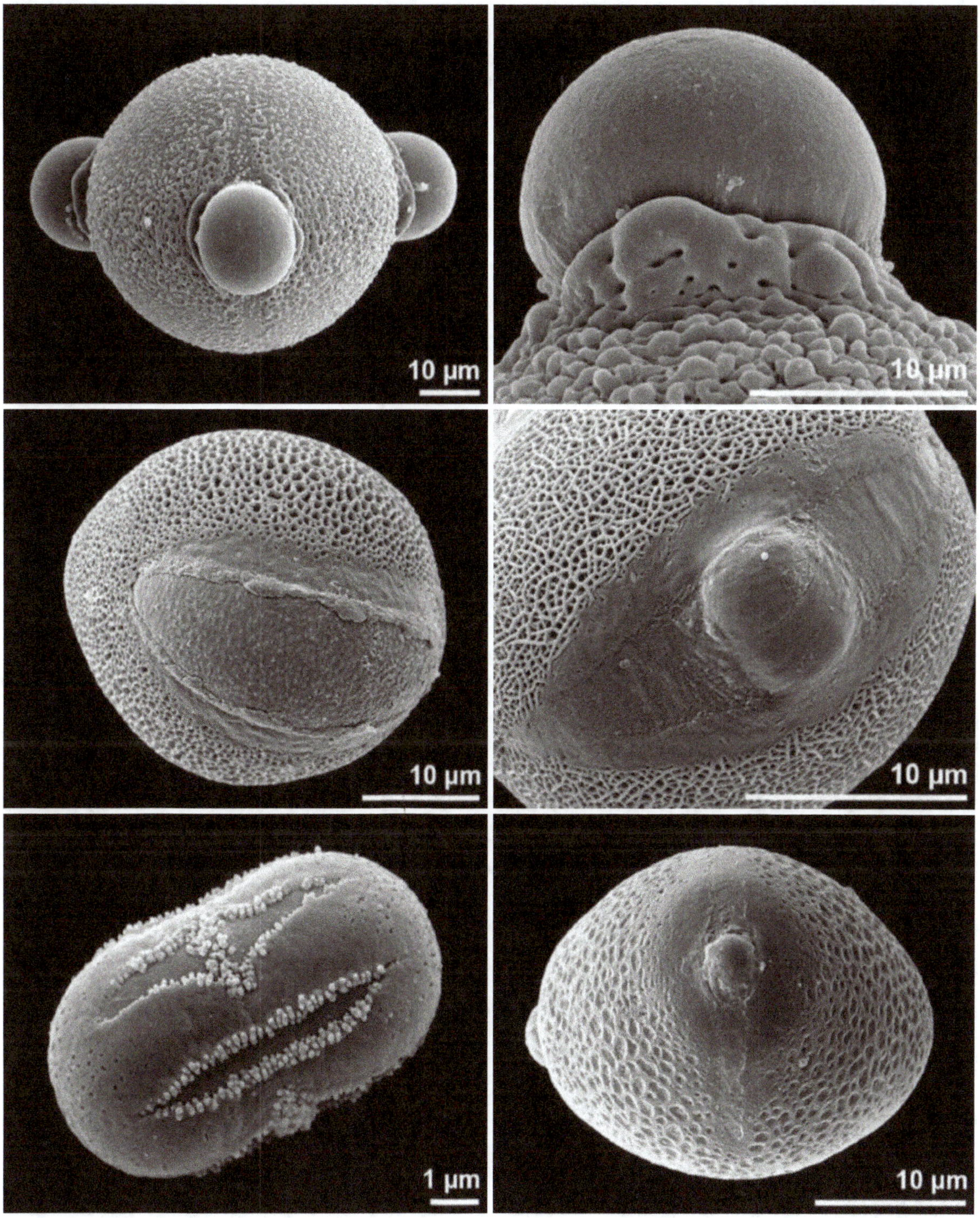

Merinthopodium neuranthum, Solanaceae
tricolporate, equatorial view

Butomus umbellatus, Butomaceae
sulcate, distal polar view

Omphalodes verna, Boraginaceae
heteroaperturate, equatorial view

Merinthopodium neuranthum, Solanaceae
close-up

Blumenbachia hieronymi, Loasaceae
tricolporate

Rhamnus cathartica, Rhamnaceae
tricolporate, equatorial view

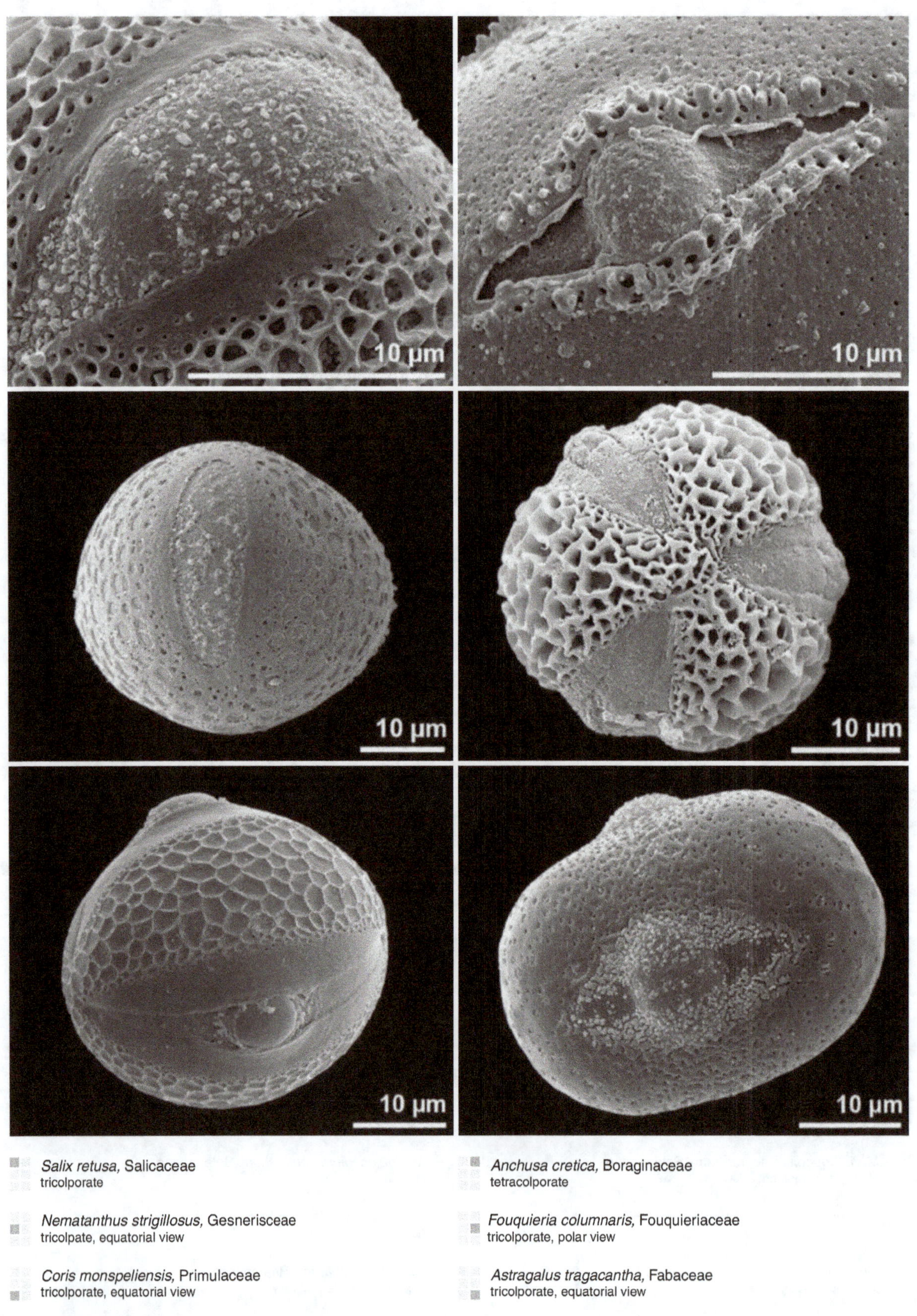

Salix retusa, Salicaceae
tricolporate

Anchusa cretica, Boraginaceae
tetracolporate

Nematanthus strigillosus, Gesnerisceae
tricolpate, equatorial view

Fouquieria columnaris, Fouquieriaceae
tricolporate, polar view

Coris monspeliensis, Primulaceae
tricolporate, equatorial view

Astragalus tragacantha, Fabaceae
tricolporate, equatorial view

operculum/operculate

distinctly delimited exine structure covering an aperture

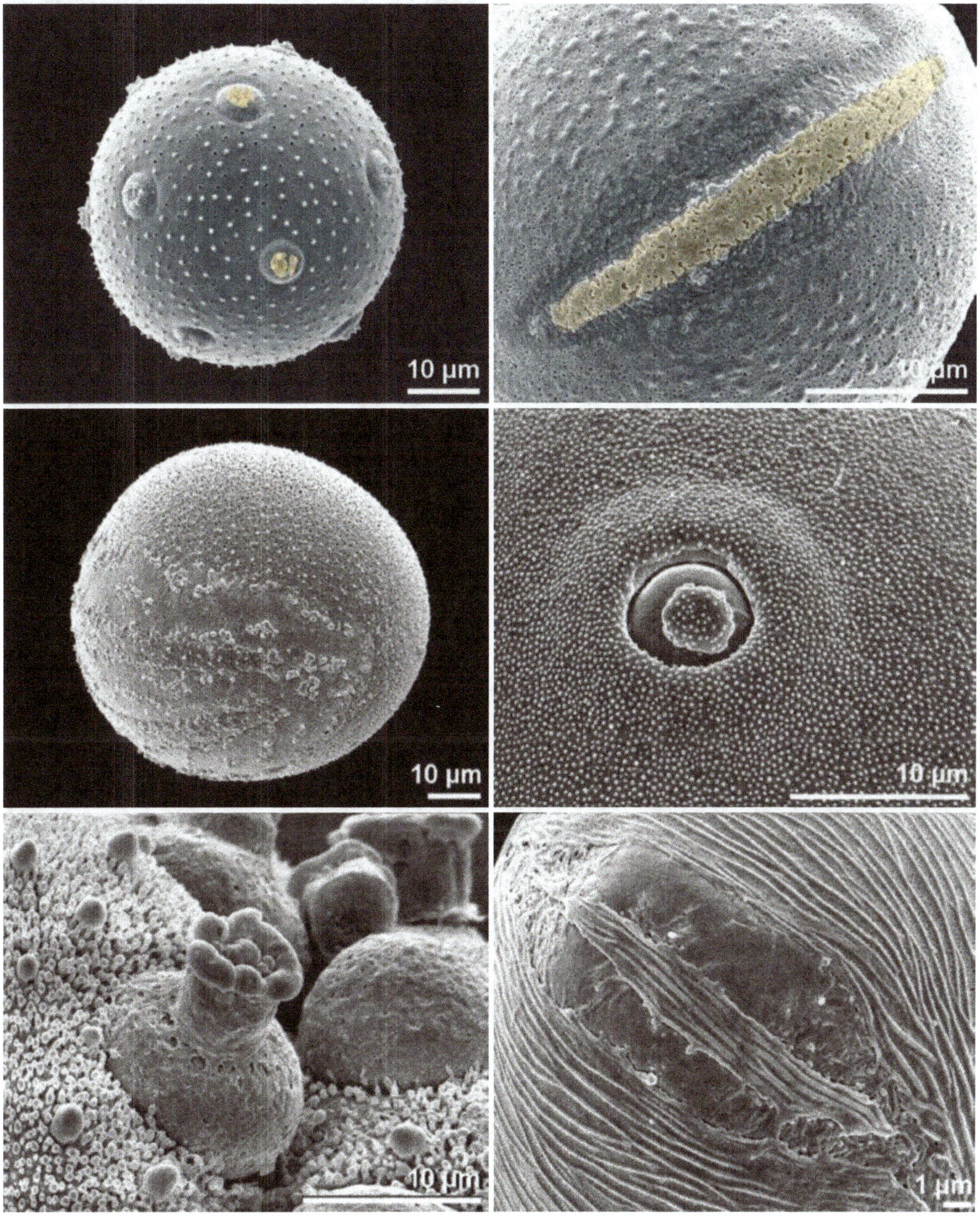

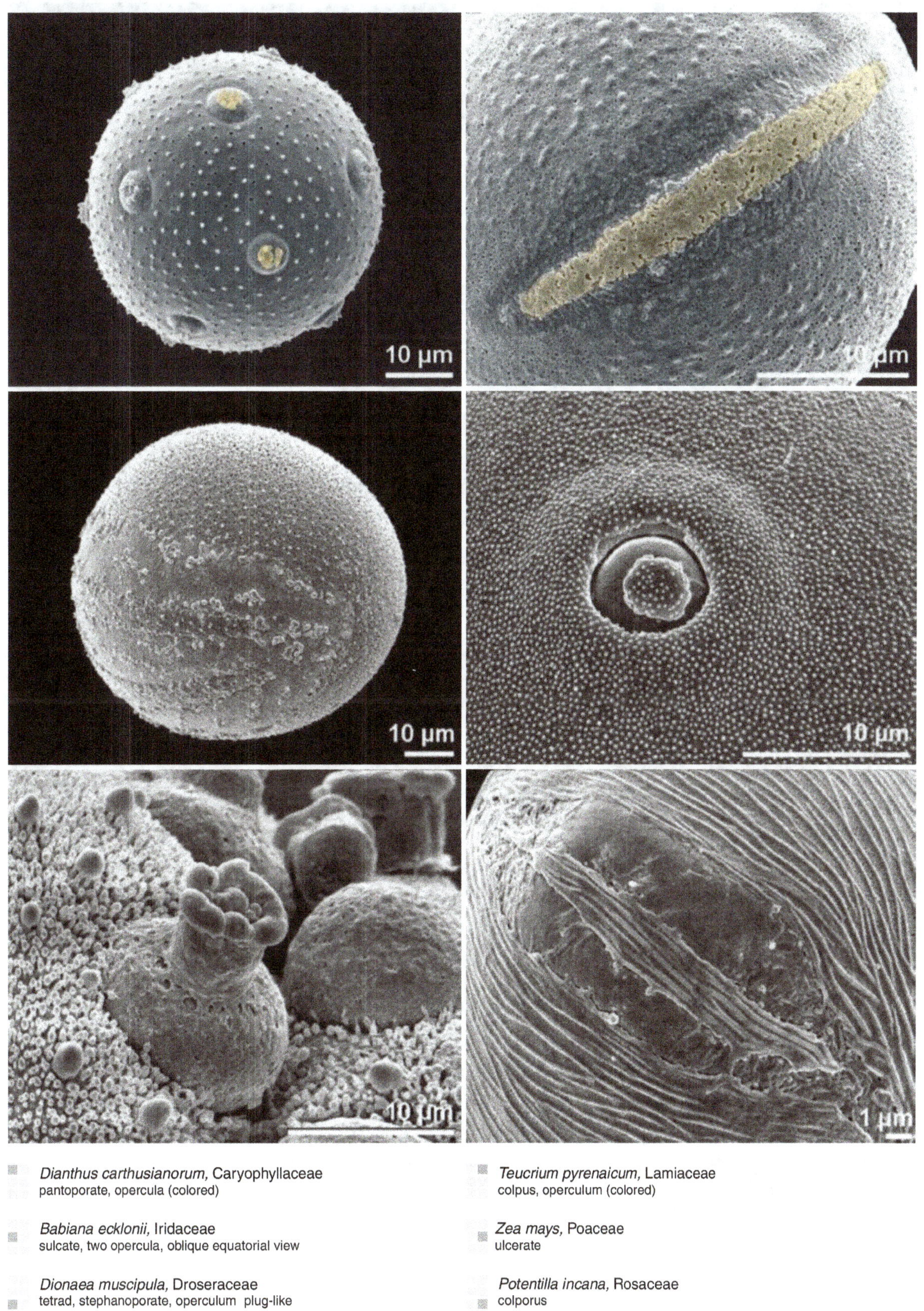

Dianthus carthusianorum, Caryophyllaceae
pantoporate, opercula (colored)

Babiana ecklonii, Iridaceae
sulcate, two opercula, oblique equatorial view

Dionaea muscipula, Droseraceae
tetrad, stephanoporate, operculum plug-like

Teucrium pyrenaicum, Lamiaceae
colpus, operculum (colored)

Zea mays, Poaceae
ulcerate

Potentilla incana, Rosaceae
colporus

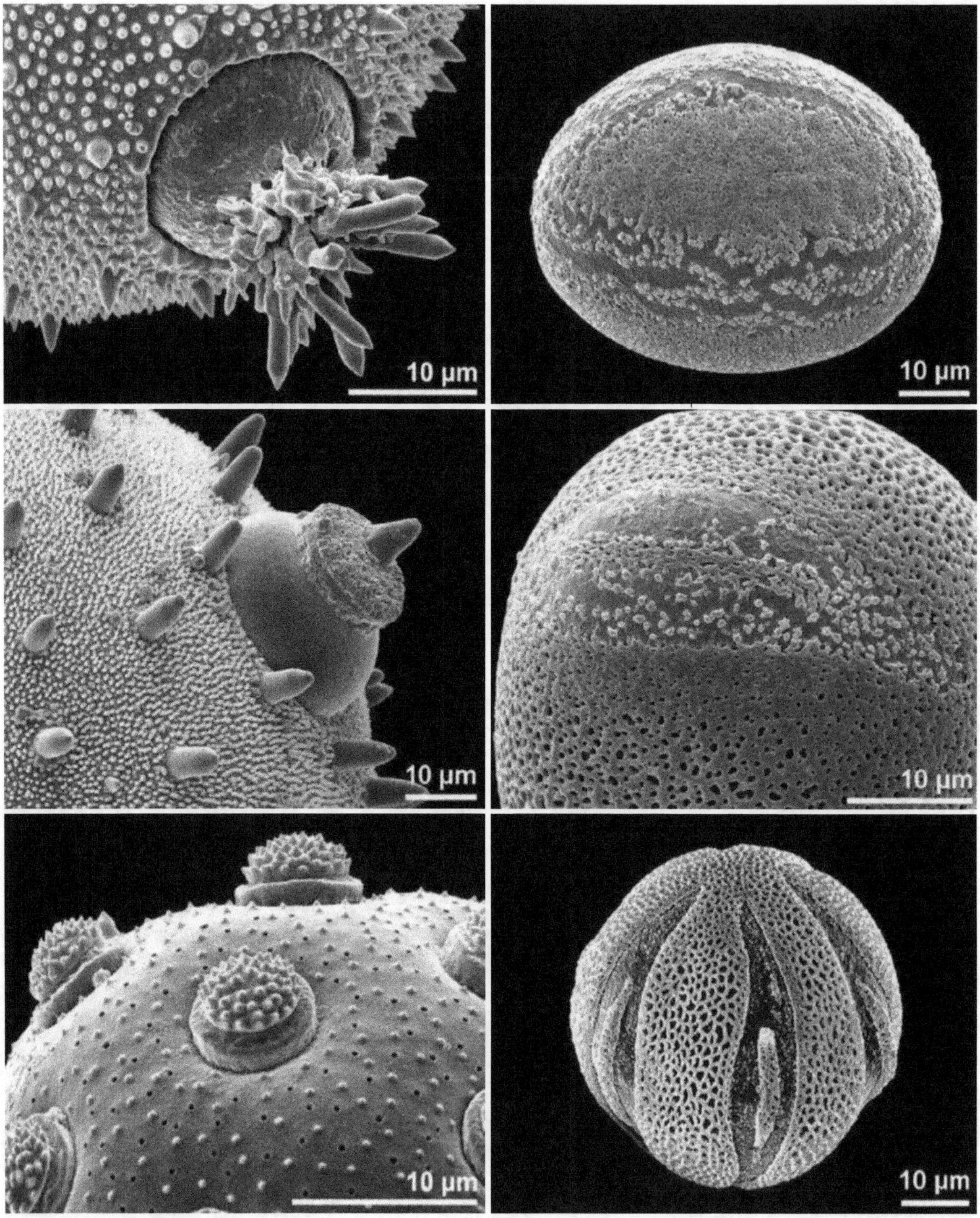

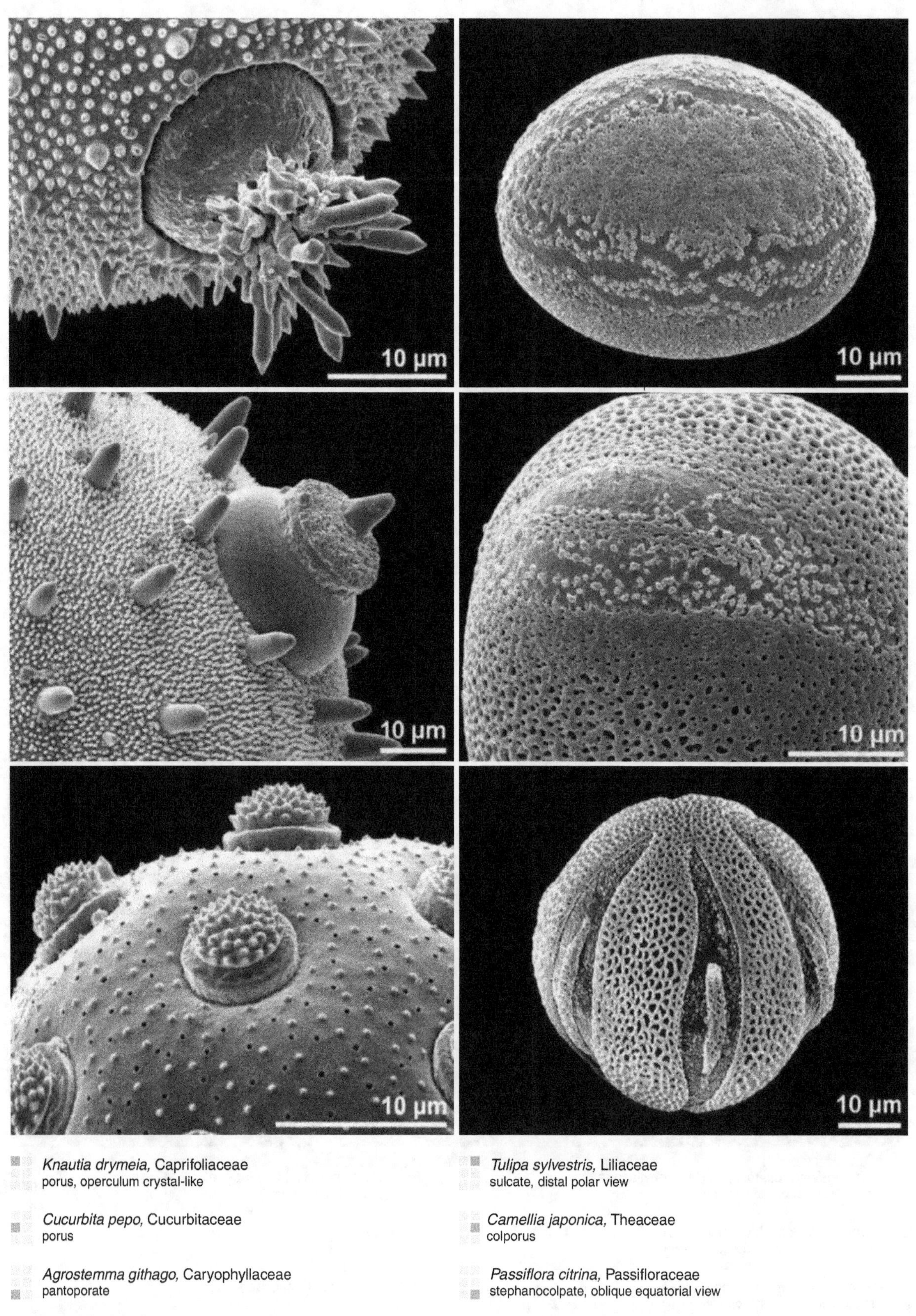

Knautia drymeia, Caprifoliaceae
porus, operculum crystal-like

Cucurbita pepo, Cucurbitaceae
porus

Agrostemma githago, Caryophyllaceae
pantoporate

Tulipa sylvestris, Liliaceae
sulcate, distal polar view

Camellia japonica, Theaceae
colporus

Passiflora citrina, Passifloraceae
stephanocolpate, oblique equatorial view

Rosa pendulina, Rosaceae
colporus

Passiflora suberosa, Passifloraceae
stephanocolpate, equatorial view

Agave asperrima, Asparagaceae
sulcate, distal polar view

Erythronium dens-canis, Liliaceae
sulcate, distal polar view

Erythronium dens-canis, Liliaceae
sulcate, dry pollen

Potentilla erecta, Rosaceae
tricolporate, equatorial view

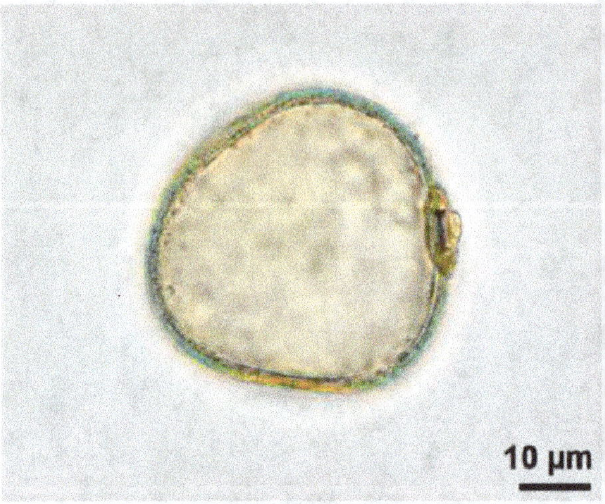

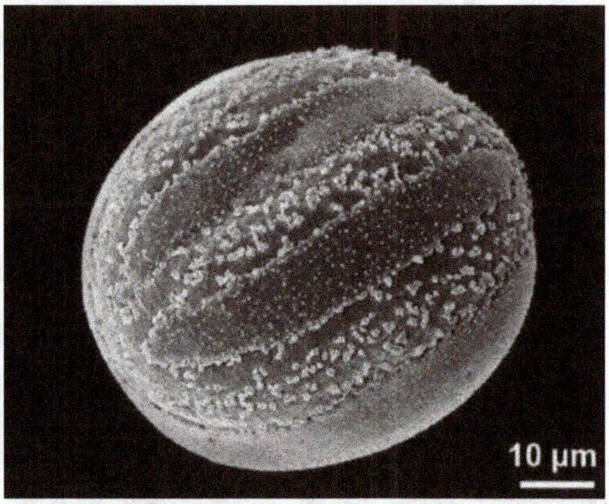

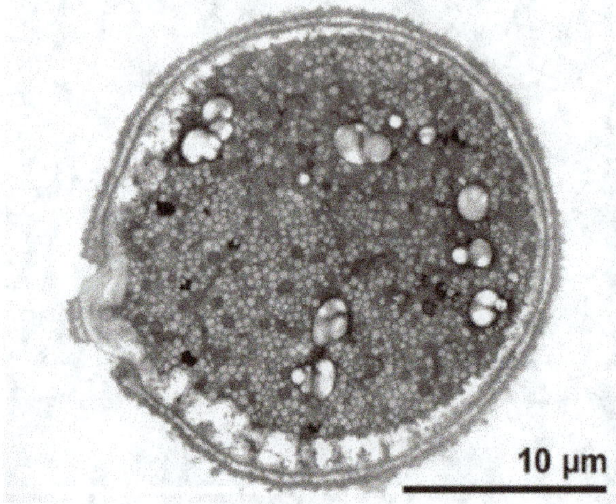

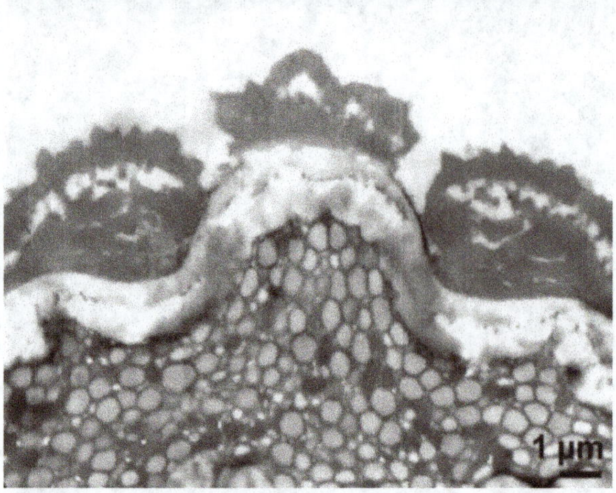

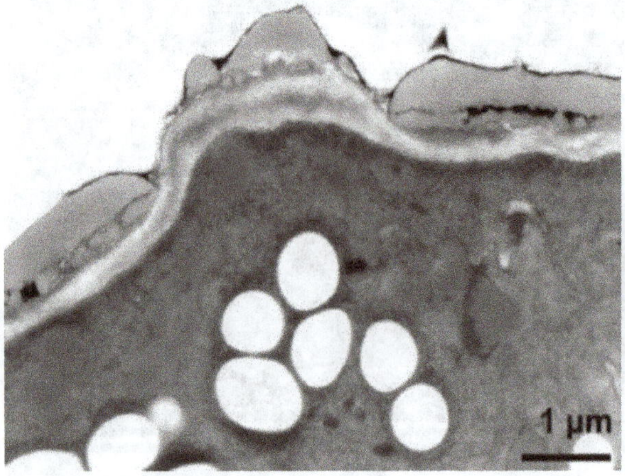

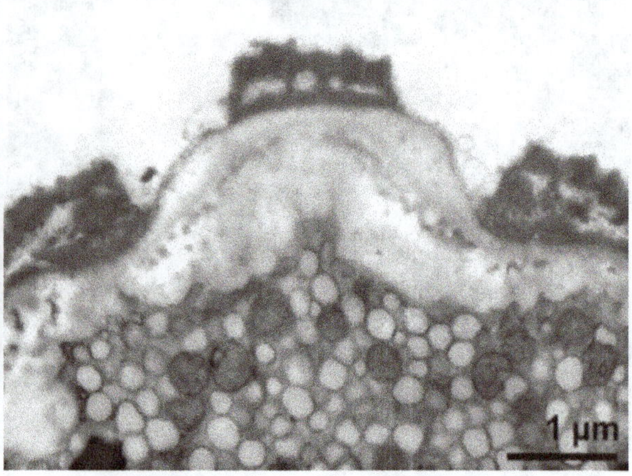

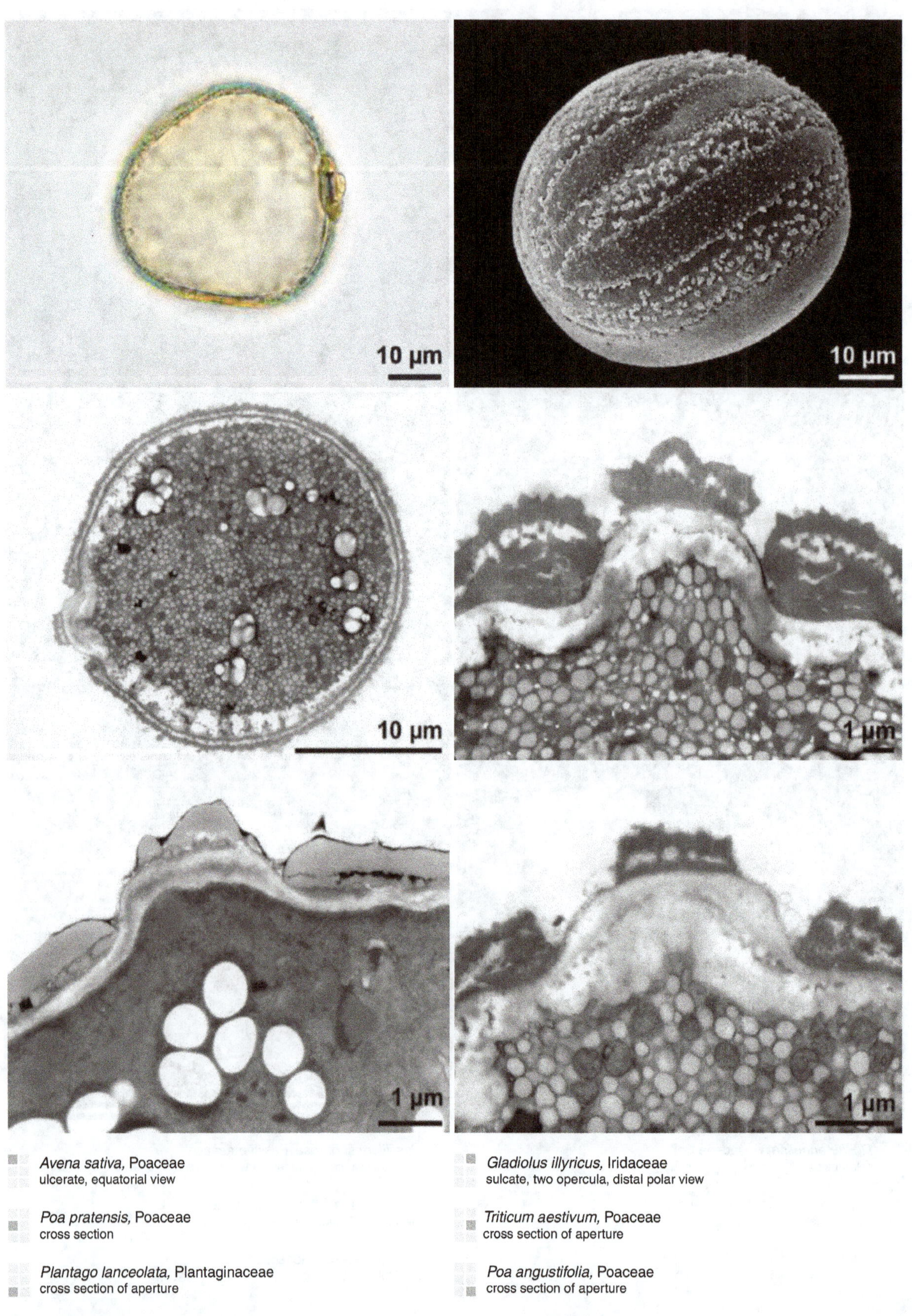

Avena sativa, Poaceae
ulcerate, equatorial view

Poa pratensis, Poaceae
cross section

Plantago lanceolata, Plantaginaceae
cross section of aperture

Gladiolus illyricus, Iridaceae
sulcate, two opercula, distal polar view

Triticum aestivum, Poaceae
cross section of aperture

Poa angustifolia, Poaceae
cross section of aperture

pantoaperturate, pantocolpate

pollen grain with apertures distributed more or less regularly over the surface

Anemone transsilvanica, Ranunculaceae

Opuntia basilaris, Cactaceae

Ranunculus lanuginosus, Ranunculaceae
dry pollen

Portulaca grandiflora, Portulacaceae

Sideritis syriaca, Lamiaceae
hexacolpate, dry pollen

Corydalis cava, Papaveraceae
hexacolpate

pantoaperturate, pantocolporate, pantoporate

pantocolporate, pantoporate

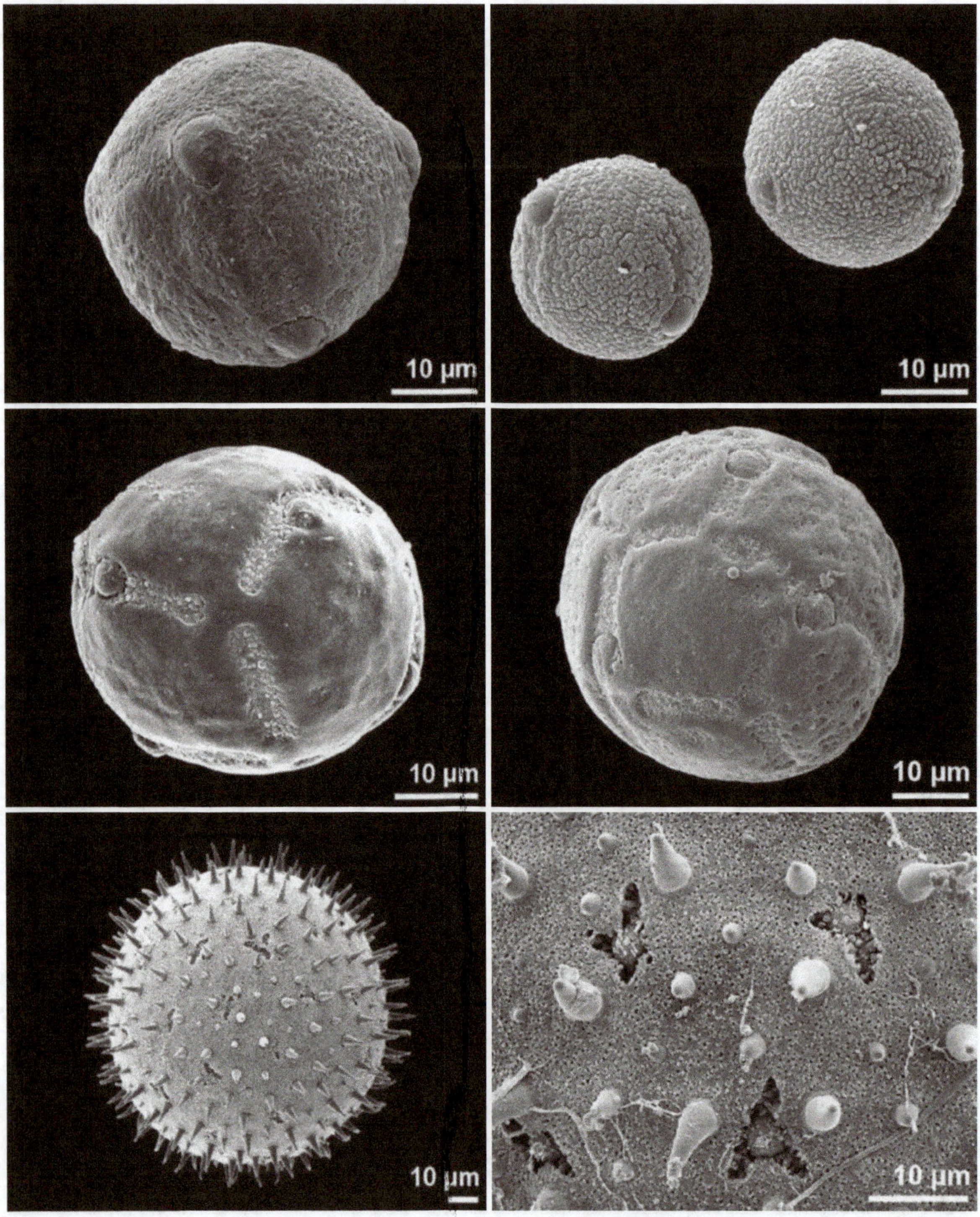

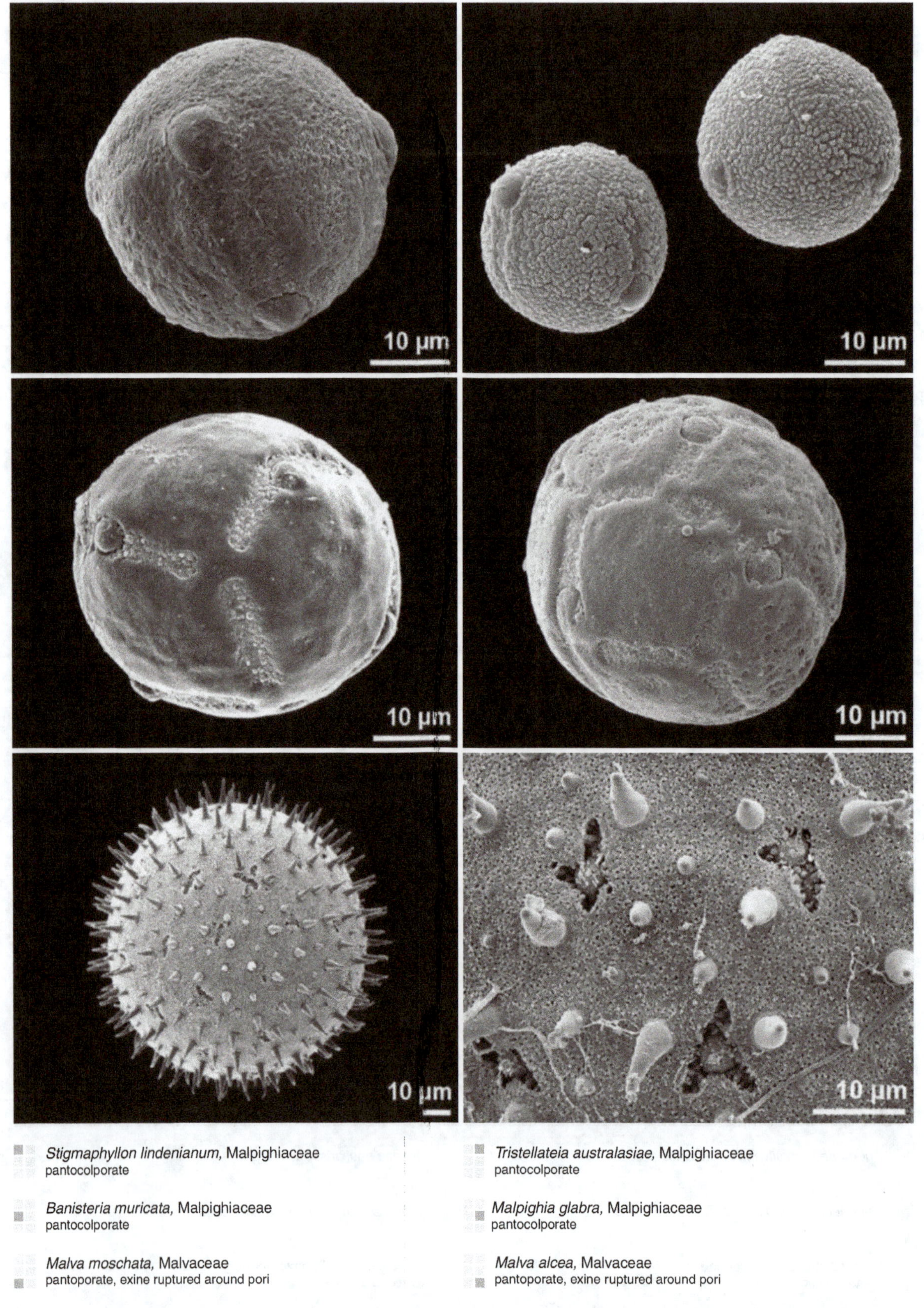

Stigmaphyllon lindenianum, Malpighiaceae
pantocolporate

Tristellateia australasiae, Malpighiaceae
pantocolporate

Banisteria muricata, Malpighiaceae
pantocolporate

Malpighia glabra, Malpighiaceae
pantocolporate

Malva moschata, Malvaceae
pantoporate, exine ruptured around pori

Malva alcea, Malvaceae
pantoporate, exine ruptured around pori

pantoaperturate, pantoporate

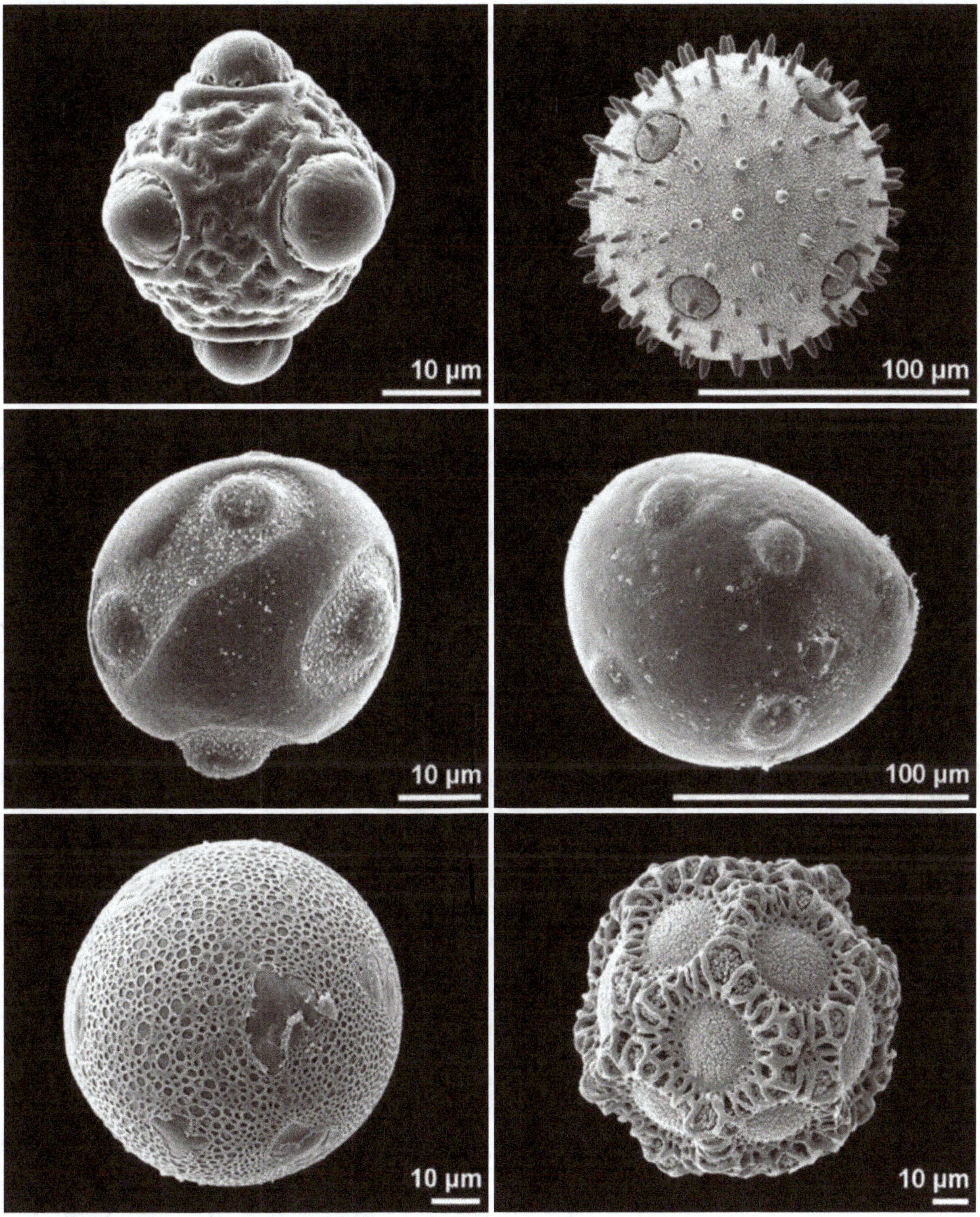

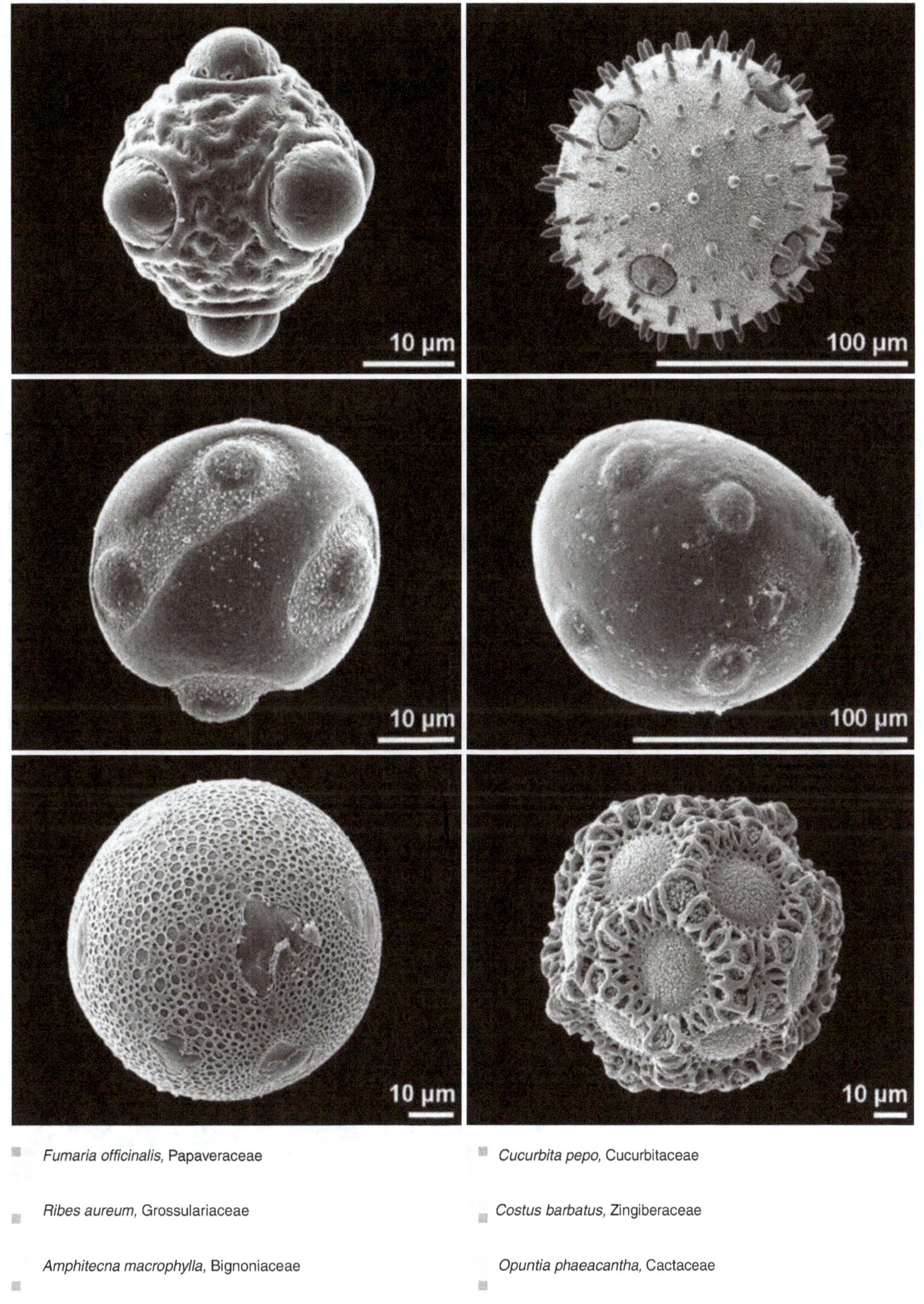

Fumaria officinalis, Papaveraceae

Cucurbita pepo, Cucurbitaceae

Ribes aureum, Grossulariaceae

Costus barbatus, Zingiberaceae

Amphitecna macrophylla, Bignoniaceae

Opuntia phaeacantha, Cactaceae

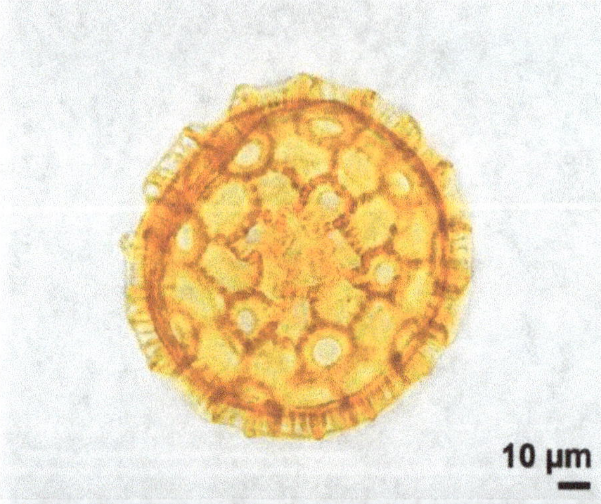

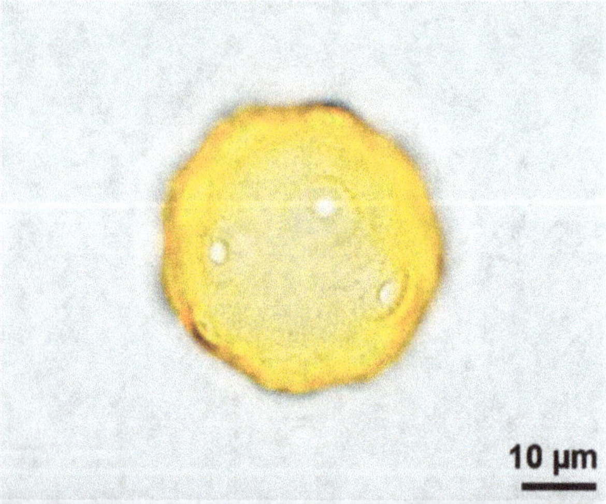

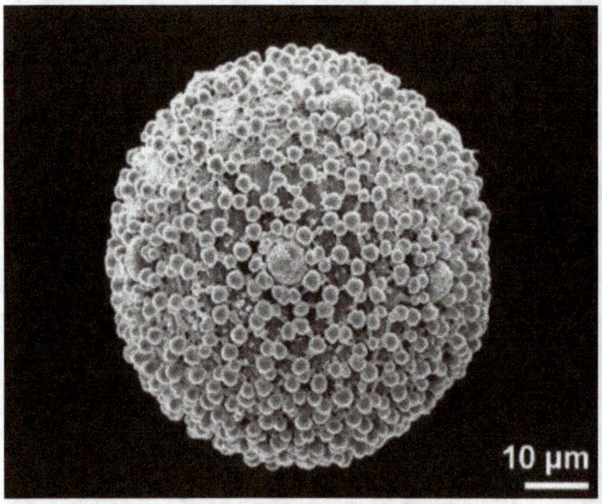

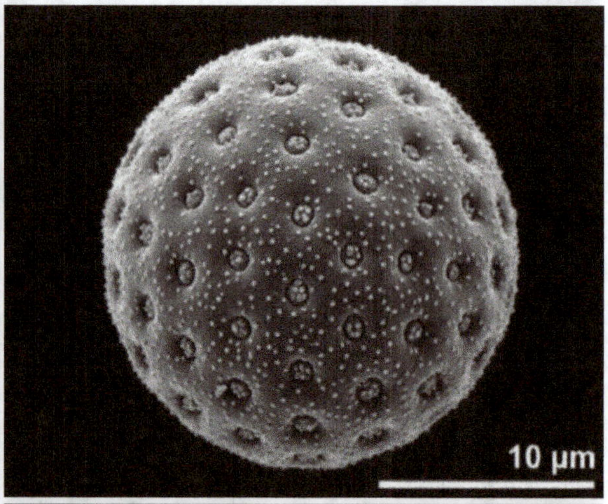

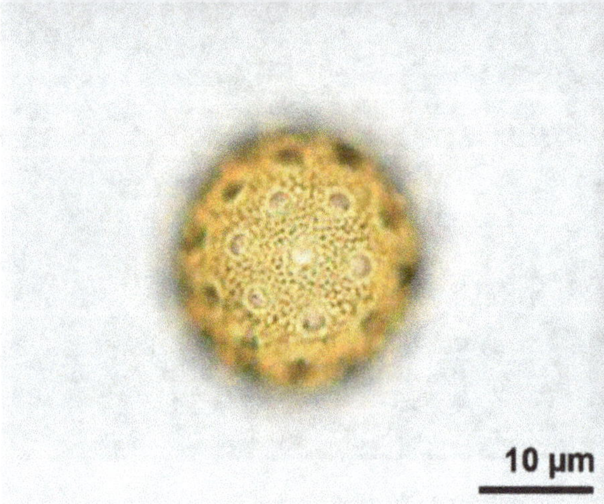

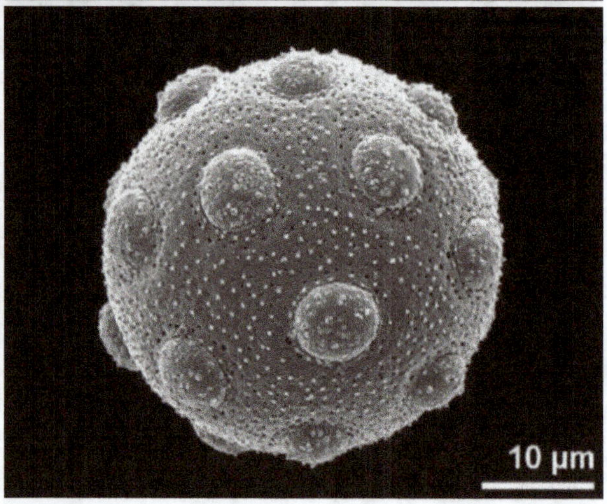

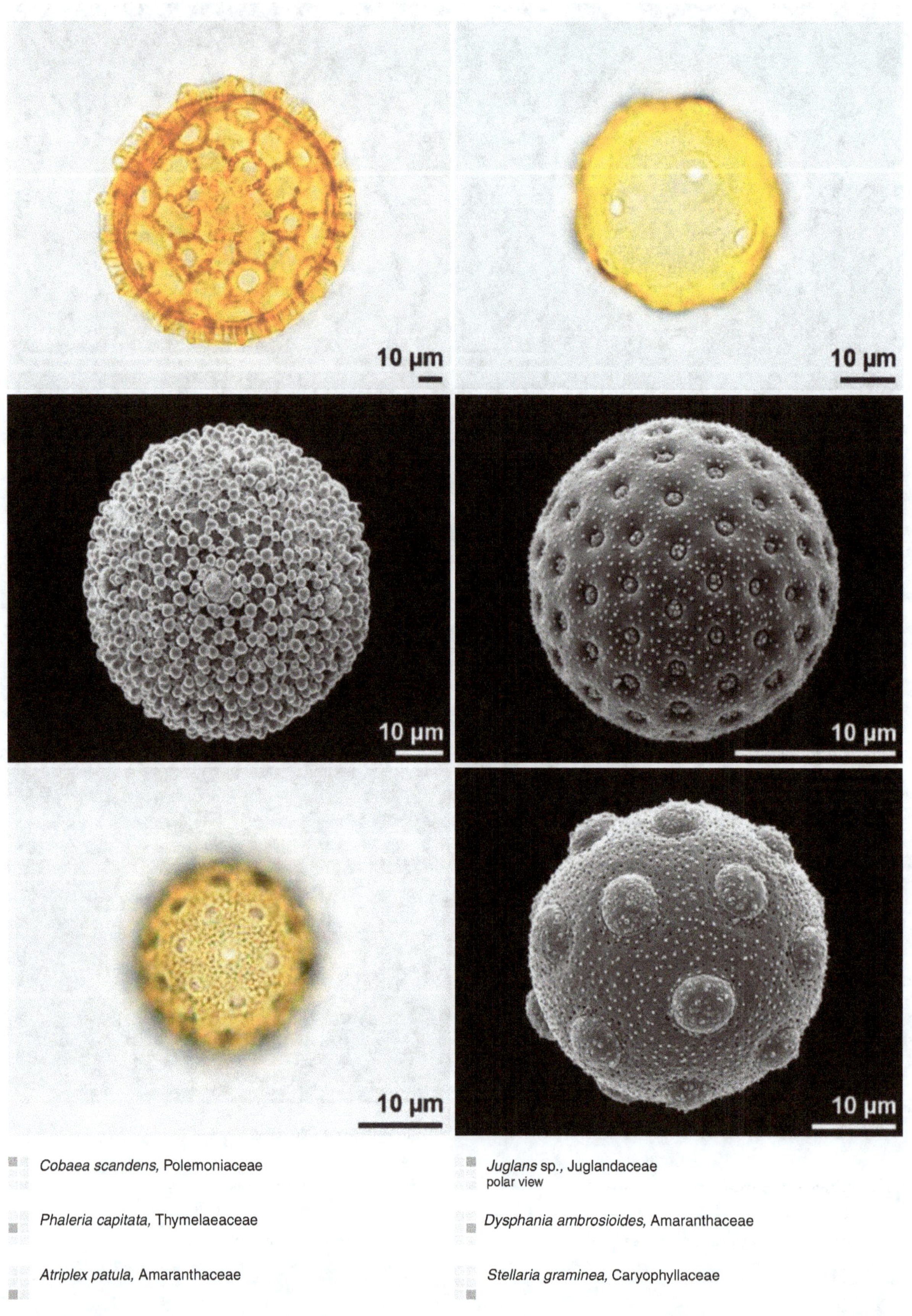

Cobaea scandens, Polemoniaceae

Juglans sp., Juglandaceae
polar view

Phaleria capitata, Thymelaeaceae

Dysphania ambrosioides, Amaranthaceae

Atriplex patula, Amaranthaceae

Stellaria graminea, Caryophyllaceae

papilla

small protuberance typical for Taxodioideae pollen located distally

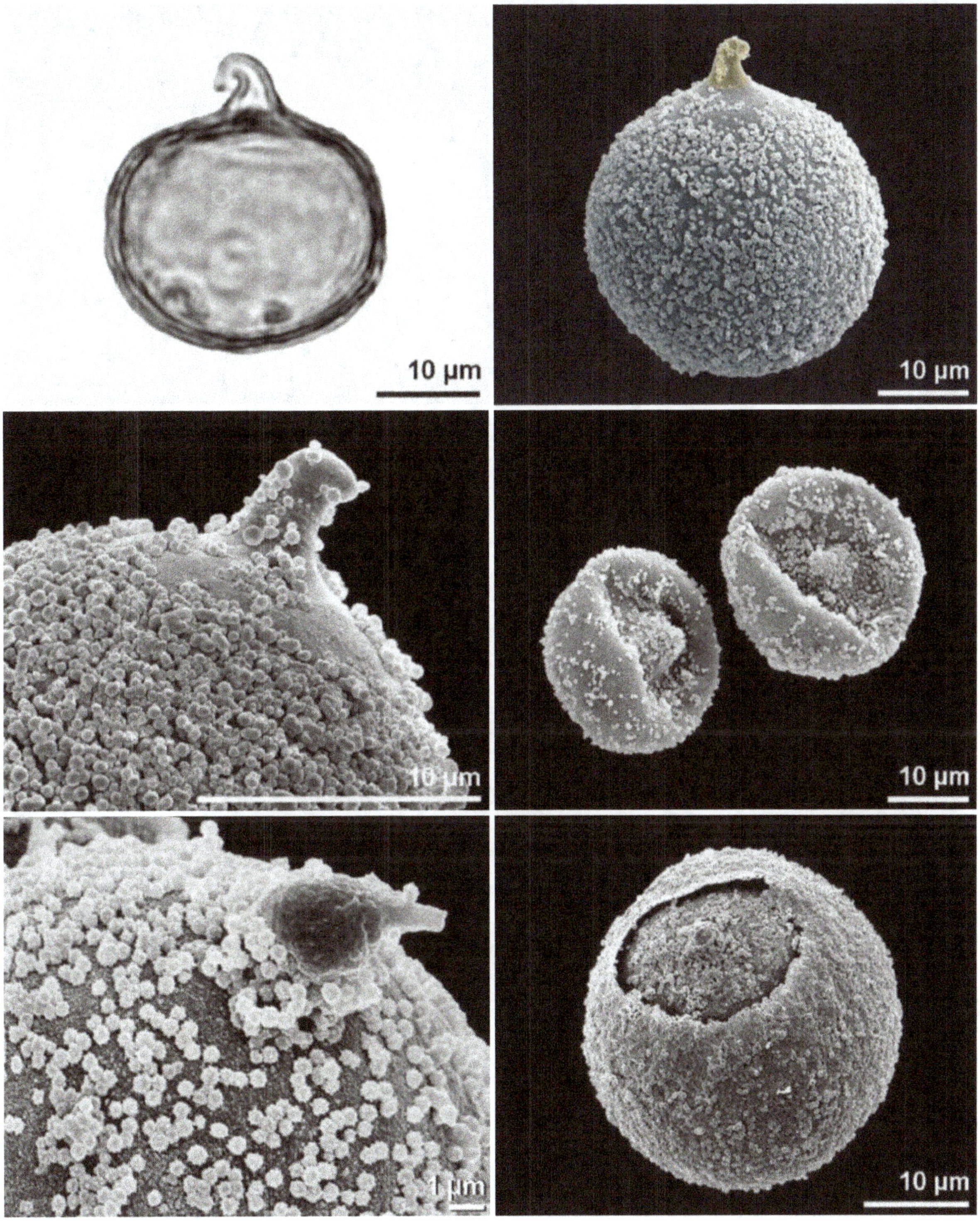

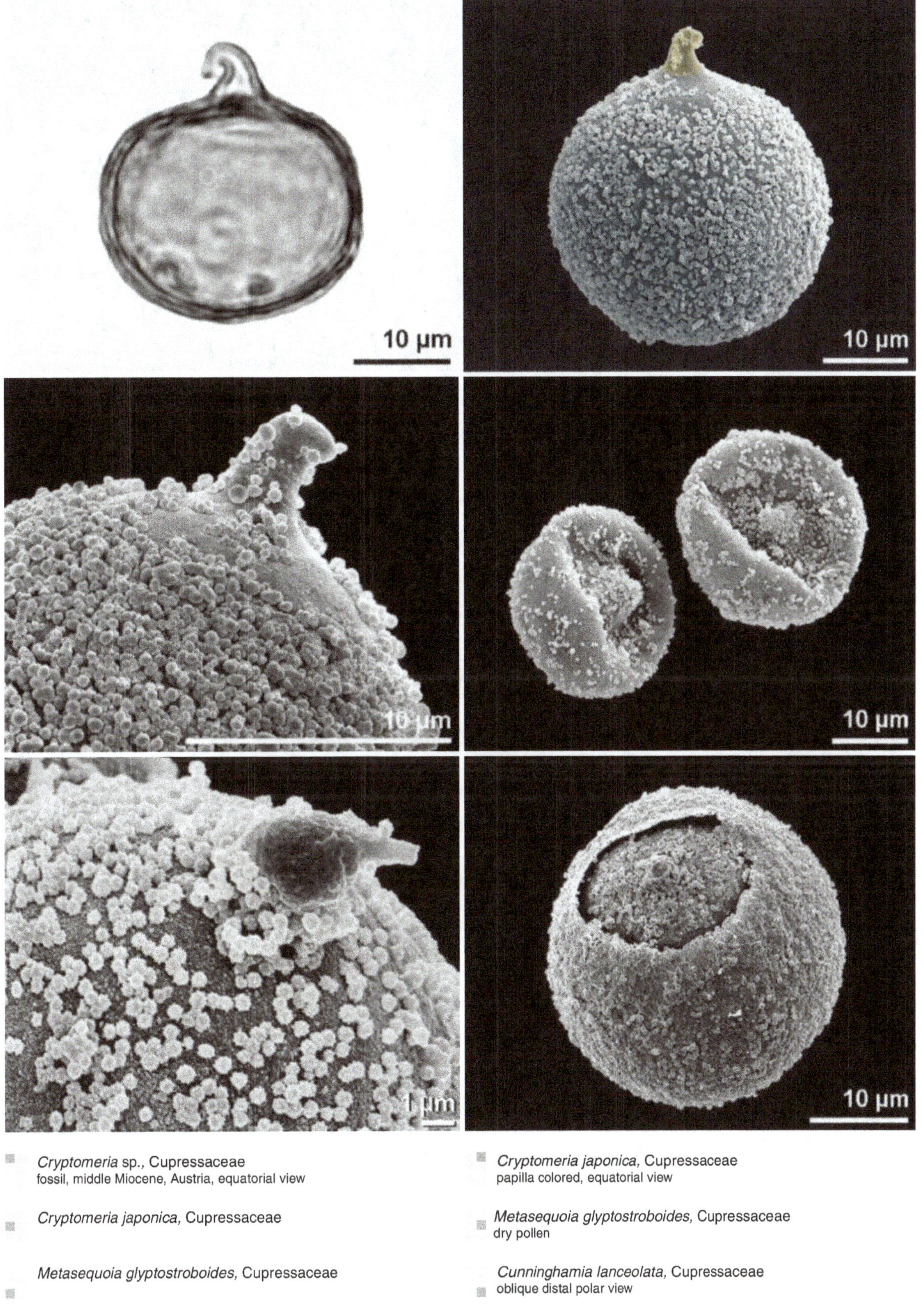

Cryptomeria sp., Cupressaceae
fossil, middle Miocene, Austria, equatorial view

Cryptomeria japonica, Cupressaceae

Metasequoia glyptostroboides, Cupressaceae

Cryptomeria japonica, Cupressaceae
papilla colored, equatorial view

Metasequoia glyptostroboides, Cupressaceae
dry pollen

Cunninghamia lanceolata, Cupressaceae
oblique distal polar view

planaperturate

pollen grain with an angular outline, where the apertures are situated between the angles

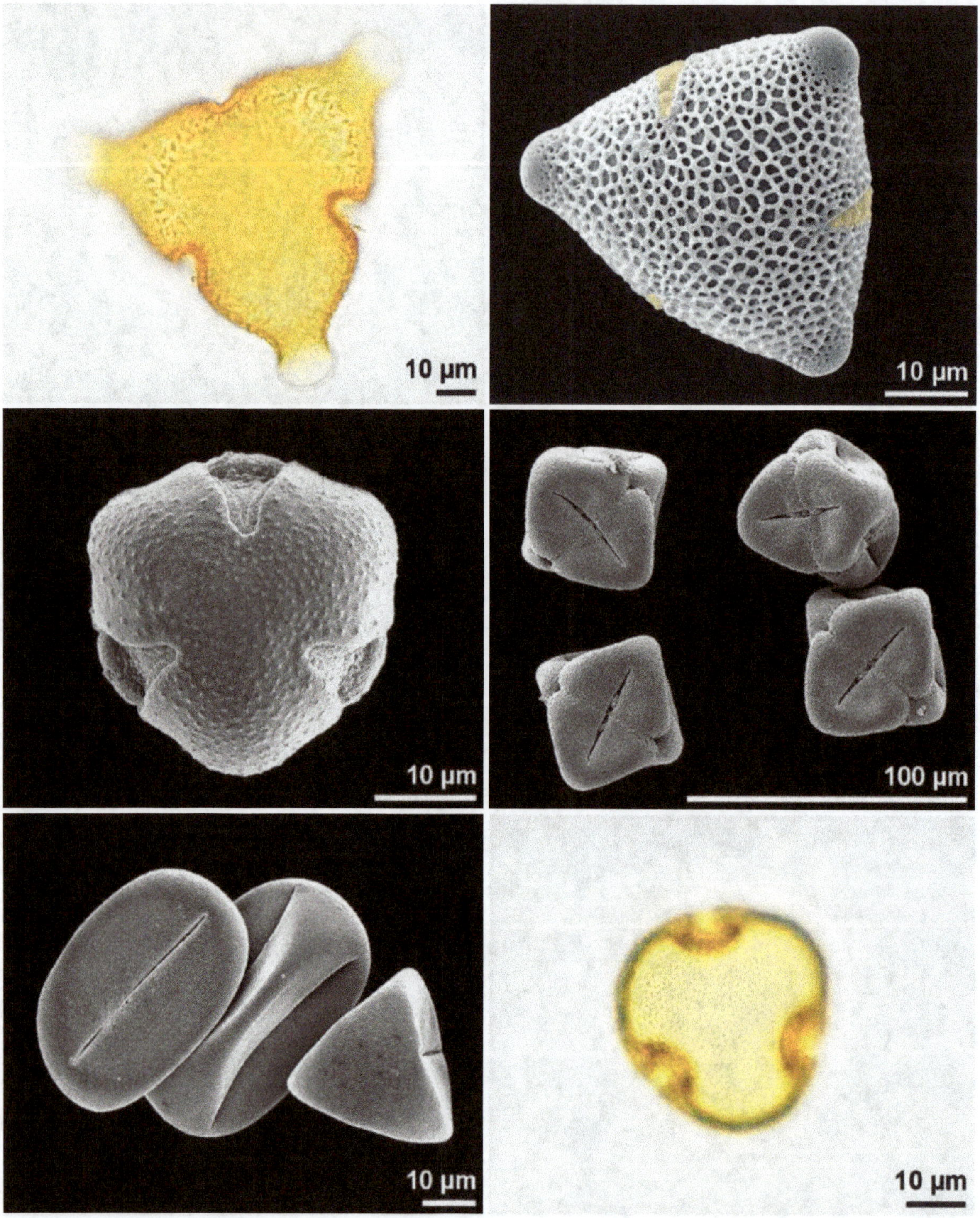

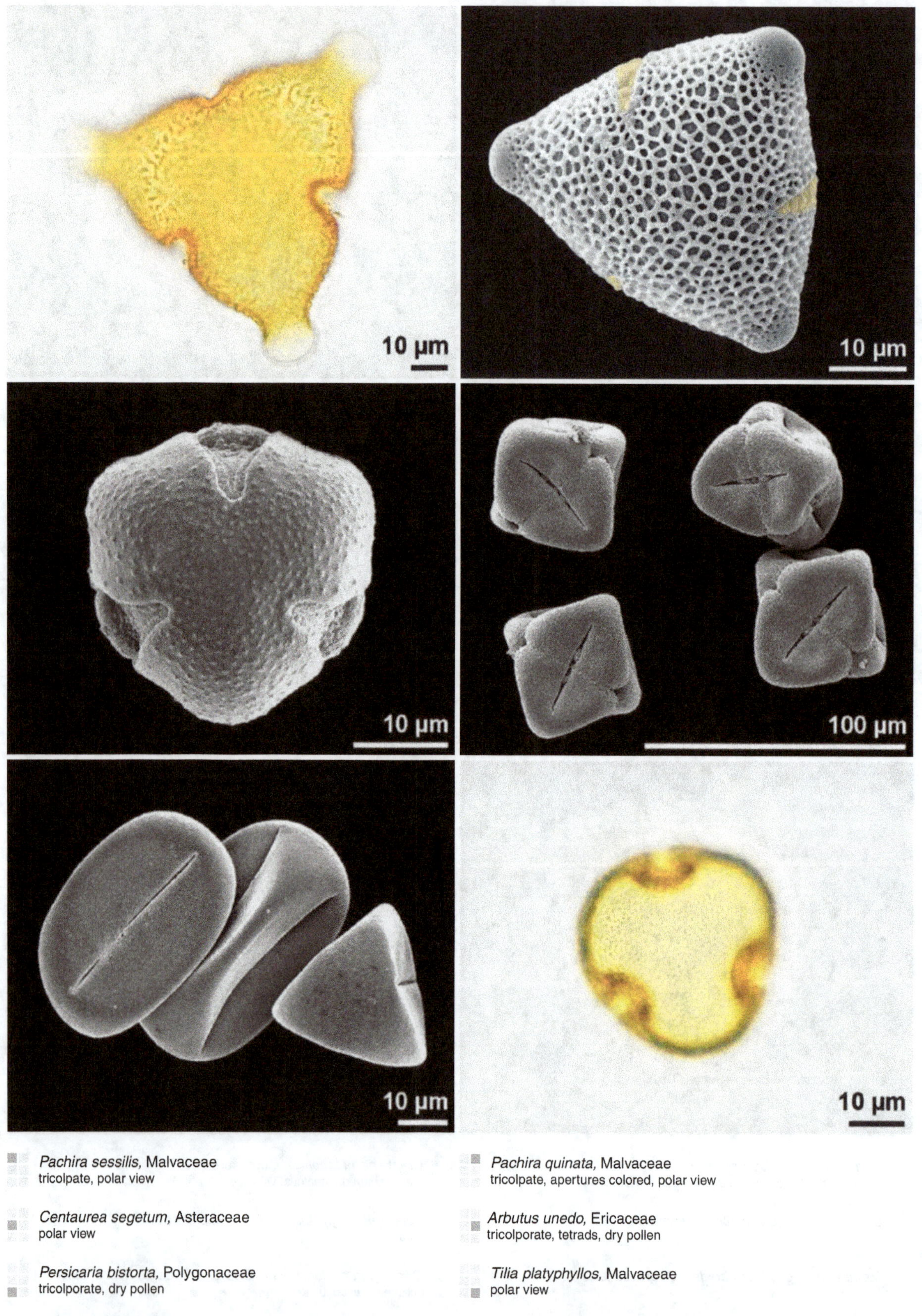

Pachira sessilis, Malvaceae
tricolpate, polar view

Pachira quinata, Malvaceae
tricolpate, apertures colored, polar view

Centaurea segetum, Asteraceae
polar view

Arbutus unedo, Ericaceae
tricolporate, tetrads, dry pollen

Persicaria bistorta, Polygonaceae
tricolporate, dry pollen

Tilia platyphyllos, Malvaceae
polar view

Euphorbia tithymaloides, Euphorbiaceae
tricolporate, polar view

Justicia brandegeeana, Acanthaceae
tricolporate, dry pollen

Schaueria flavicoma, Acanthaceae
tricolporate

Echinops exaltatus, Asteraceae
tricolporate

Pachira aquatica, Malvaceae
tricolpate

Pachira aquatica, Malvaceae
tricolpate, polar view

pontoperculum/pontoperculate

elongated operculum linked to the ends of the aperture

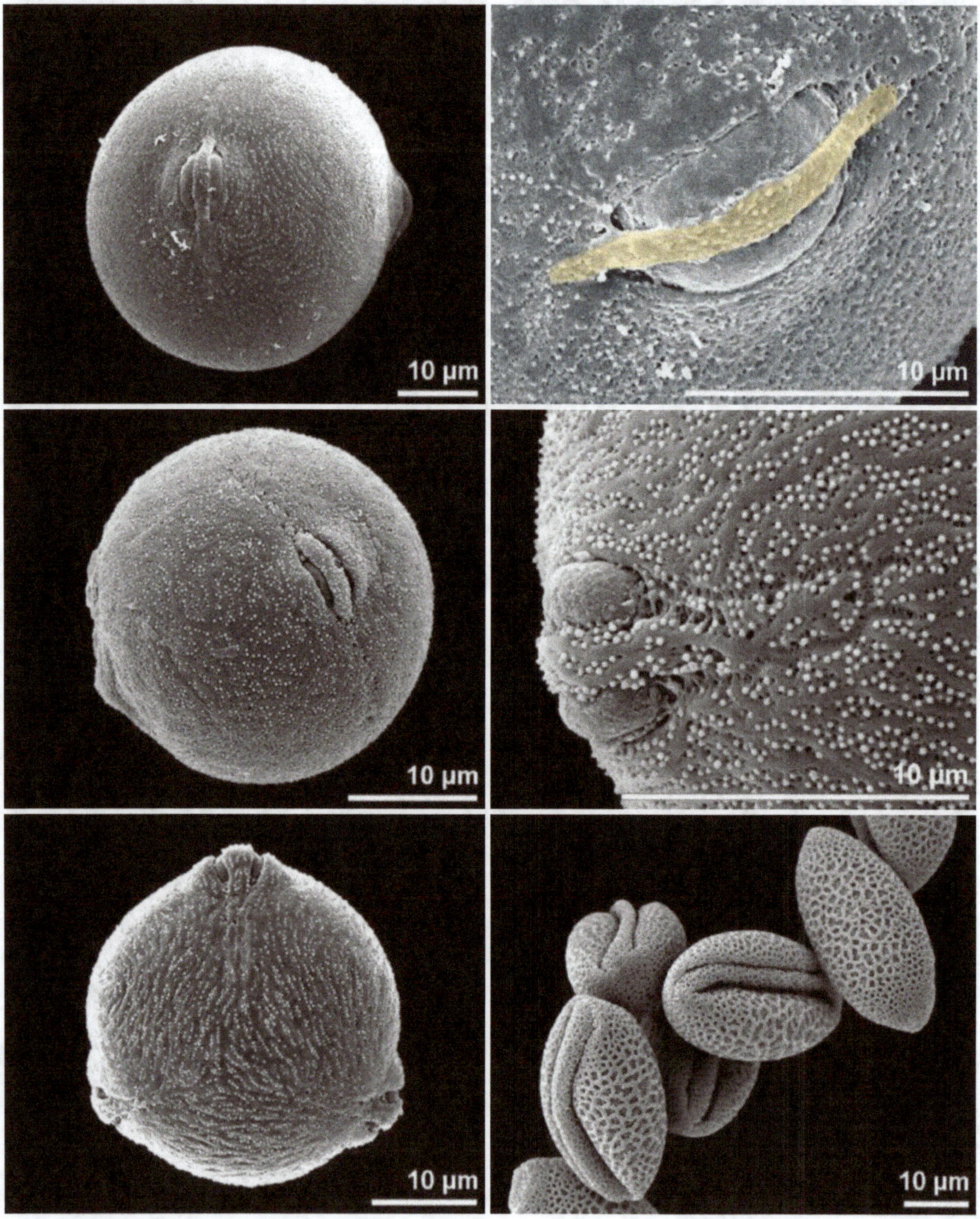

Sanguisorba cretica, Rosaceae
tricolporate, equatorial view

Sanguisorba cretica, Rosaceae
pontoperculum colored, close-up

Sarcopoterium spinosum, Rosaceae
tricolporate, equatorial view

Sarcopoterium spinosum, Rosaceae
close-up

Sanguisorba minor, Rosaceae
tricolporate, polar view

Veratrum nigrum, Melanthiaceae
sulcate, dry pollen

porus/porate, diporate

porus: more or less circular aperture; pori located at the equator or regularly spread over the pollen grain
diporate: pollen grains with two pori

Aechmea allenii, Bromeliaceae
equatorial view, pori colored

Sanchezia nobilis, Acanthaceae

Broussonetia papyrifera, Moraceae
equatorial view

Colchicum autumnale, Colchicaceae

Whitfieldia lateritia, Acanthaceae
dry (left) and turgescent pollen (right)

Quesnelia lateralis, Bromeliaceae
equatorial view

porus/porate, triporate

triporate: pollen grain with three pori

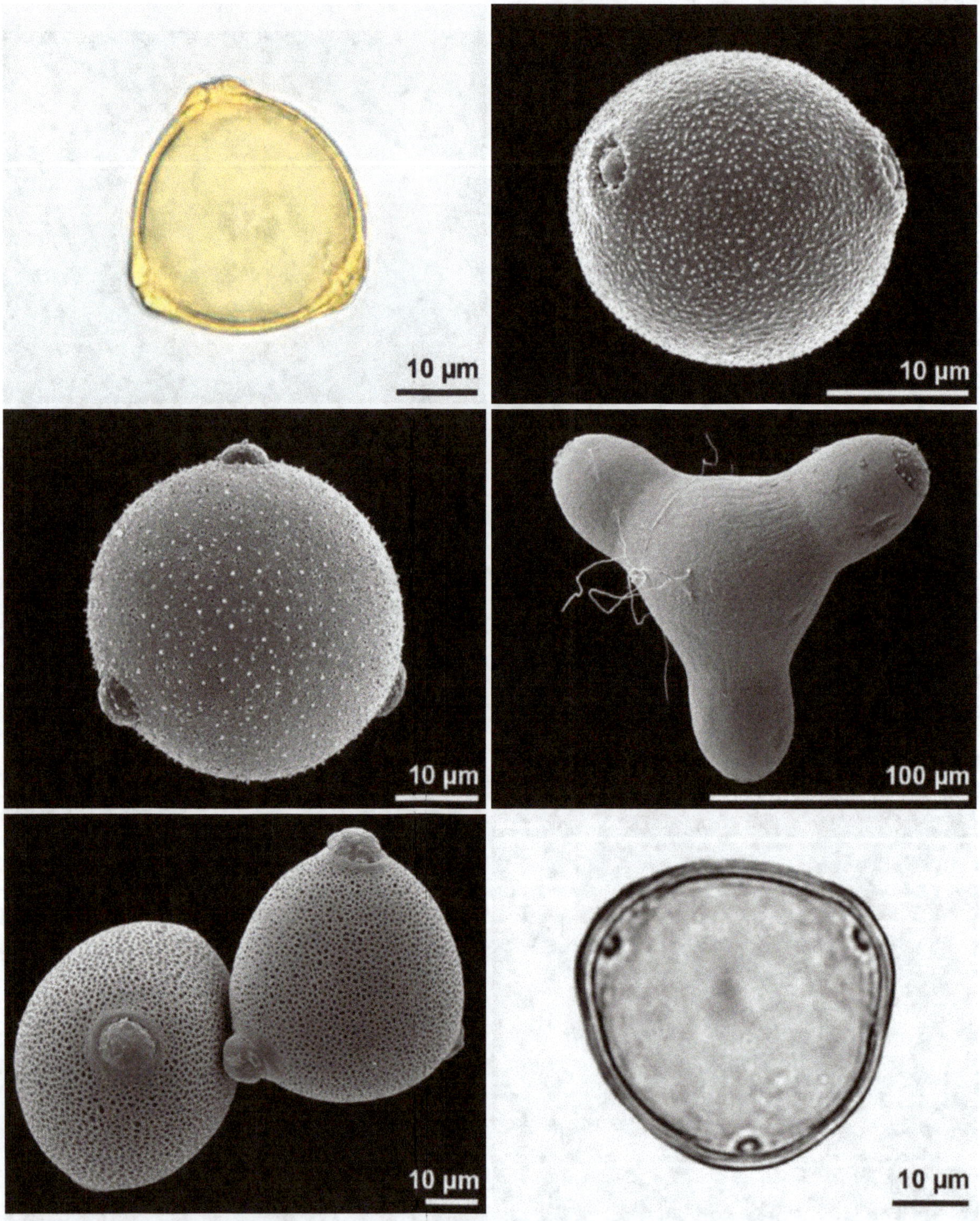

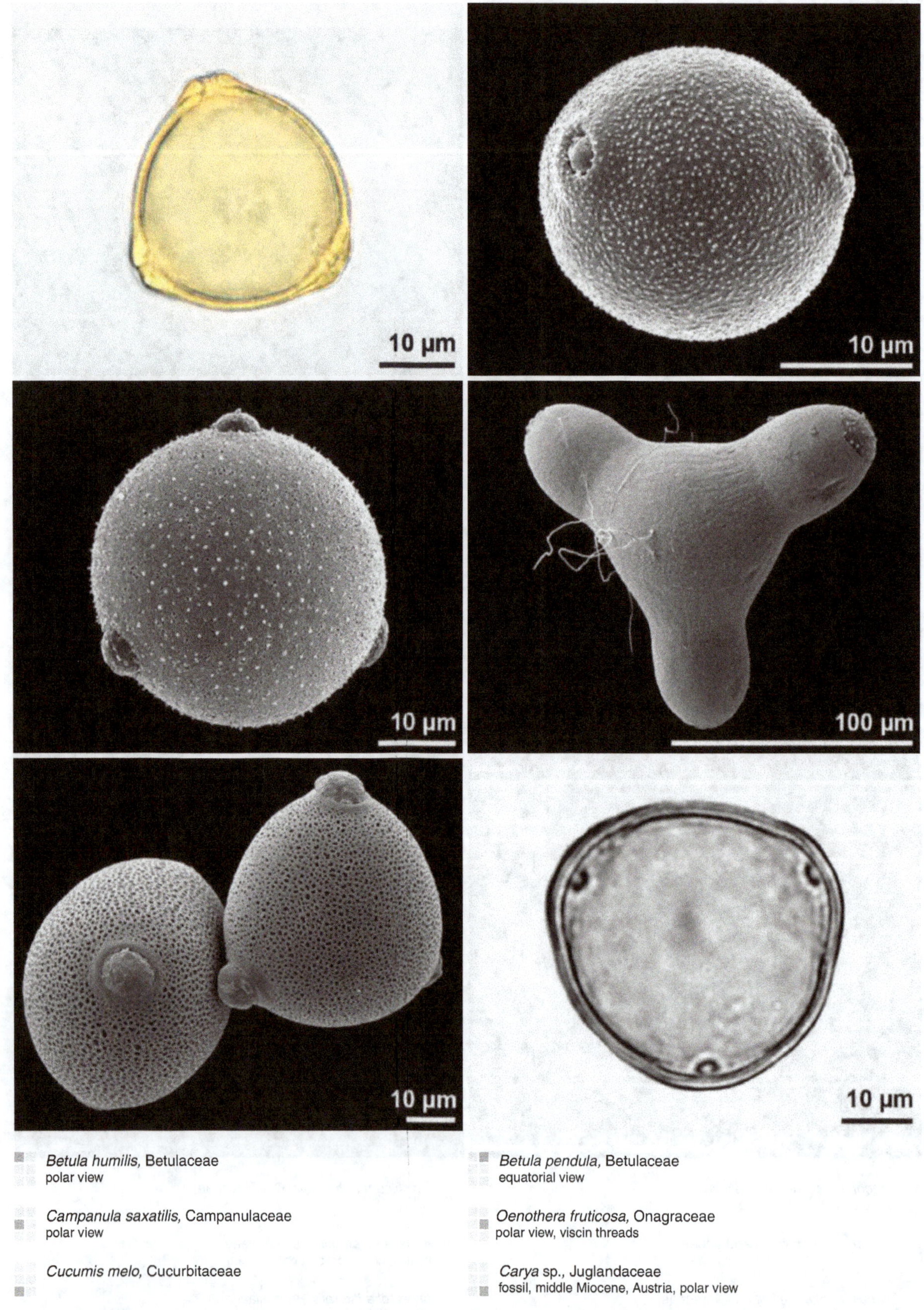

Betula humilis, Betulaceae
polar view

Campanula saxatilis, Campanulaceae
polar view

Cucumis melo, Cucurbitaceae

Betula pendula, Betulaceae
equatorial view

Oenothera fruticosa, Onagraceae
polar view, viscin threads

Carya sp., Juglandaceae
fossil, middle Miocene, Austria, polar view

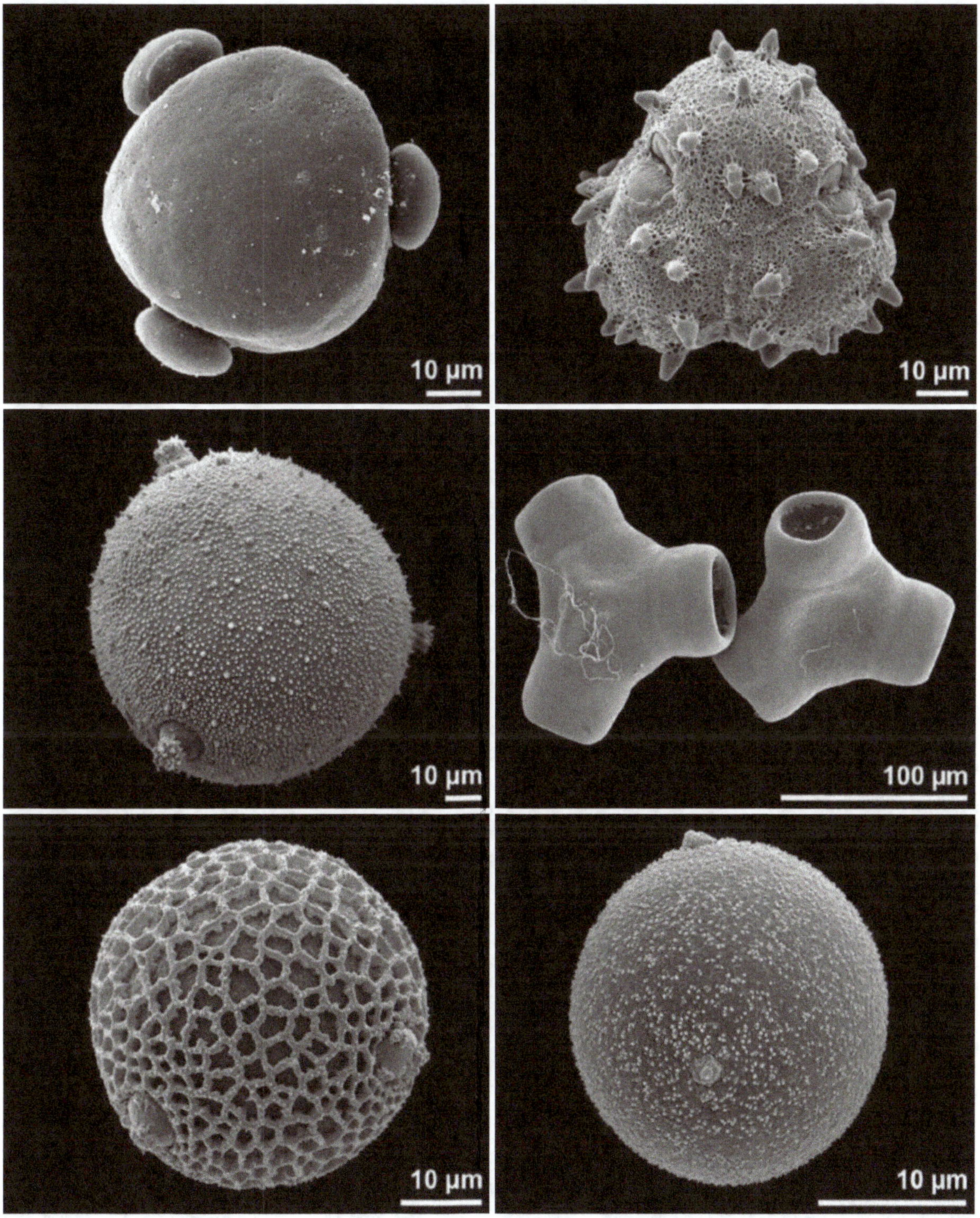

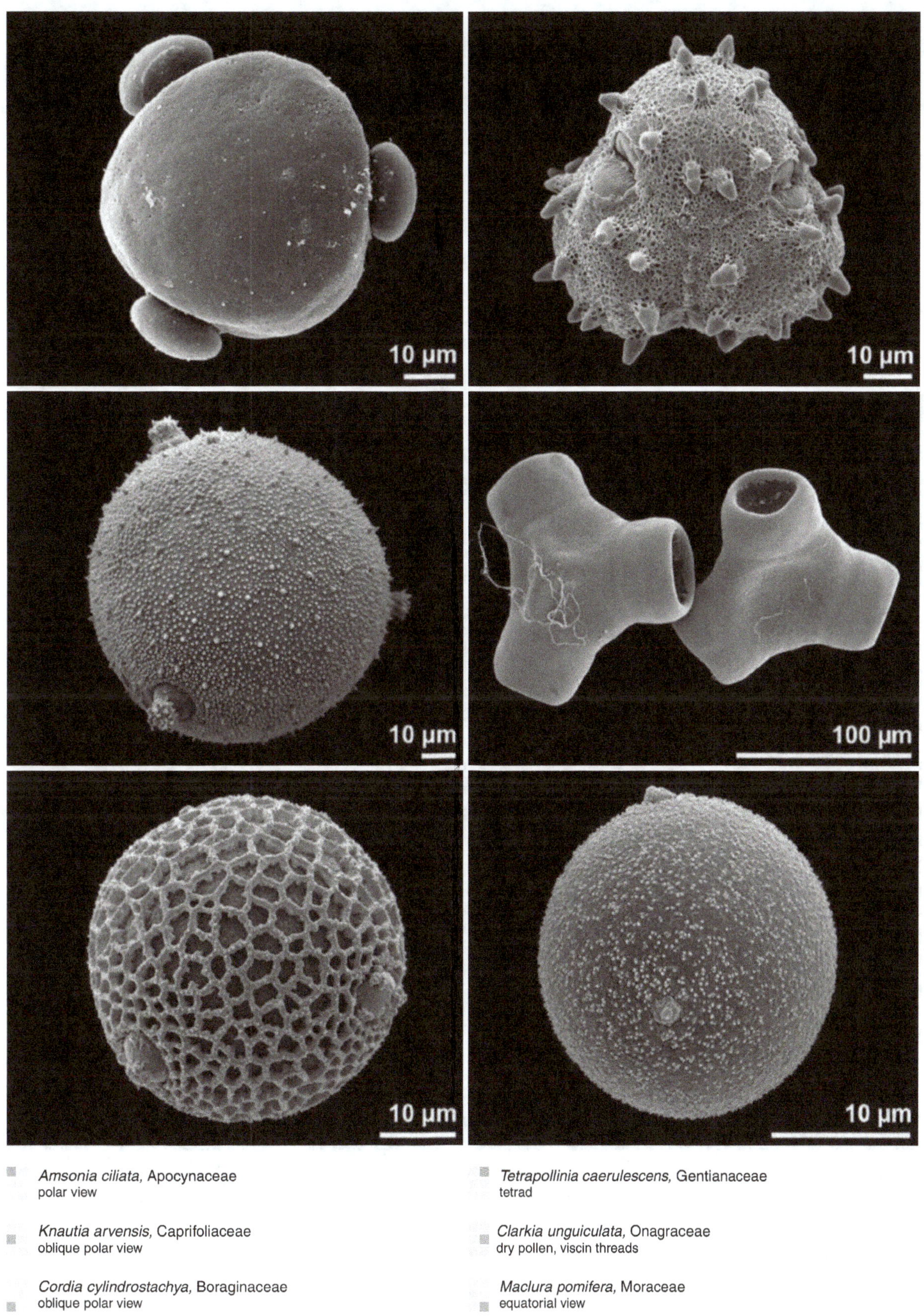

Amsonia ciliata, Apocynaceae
polar view

Knautia arvensis, Caprifoliaceae
oblique polar view

Cordia cylindrostachya, Boraginaceae
oblique polar view

Tetrapollinia caerulescens, Gentianaceae
tetrad

Clarkia unguiculata, Onagraceae
dry pollen, viscin threads

Maclura pomifera, Moraceae
equatorial view

porus/porate, tetraporate

tetraporate: pollen grain with four pori

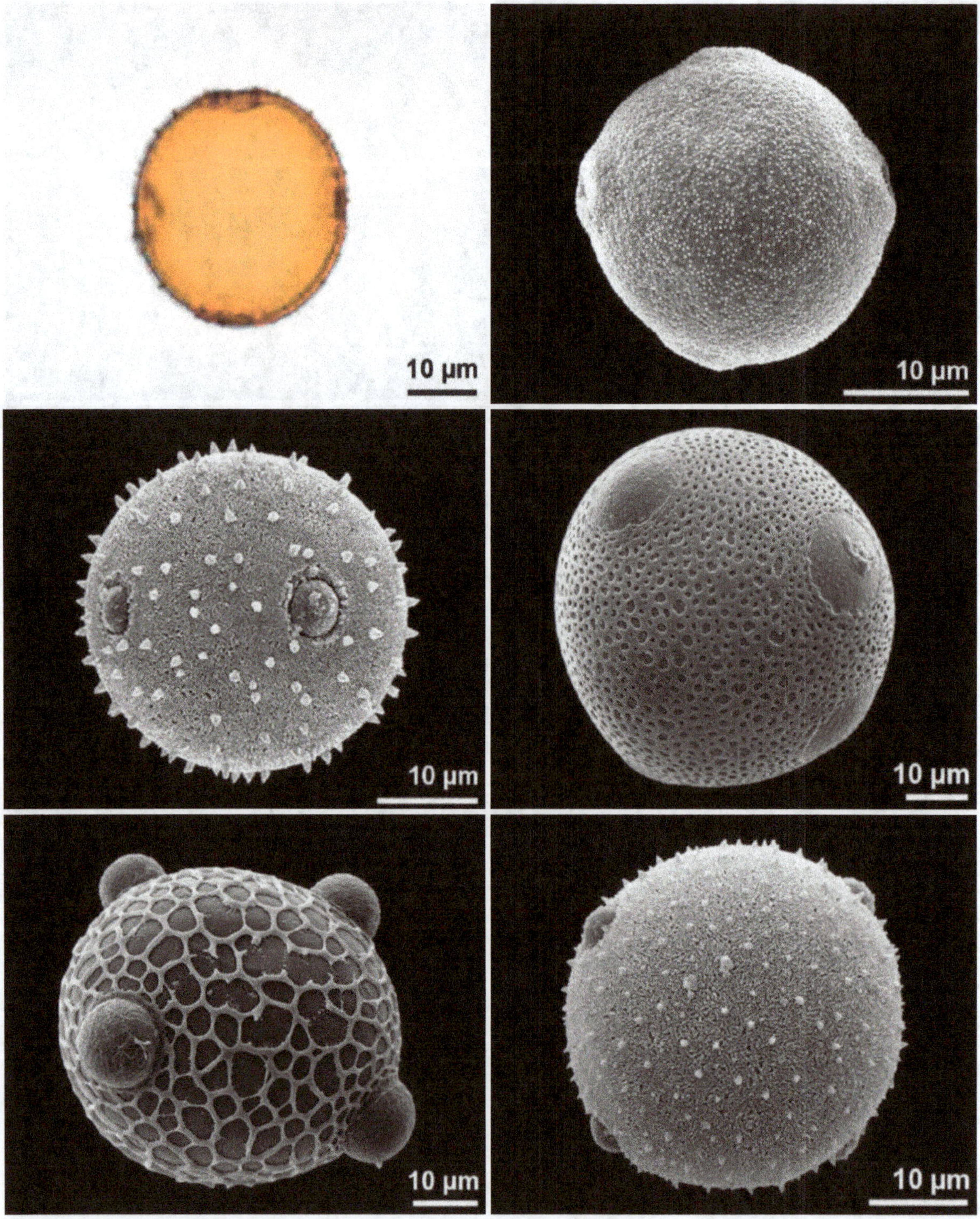

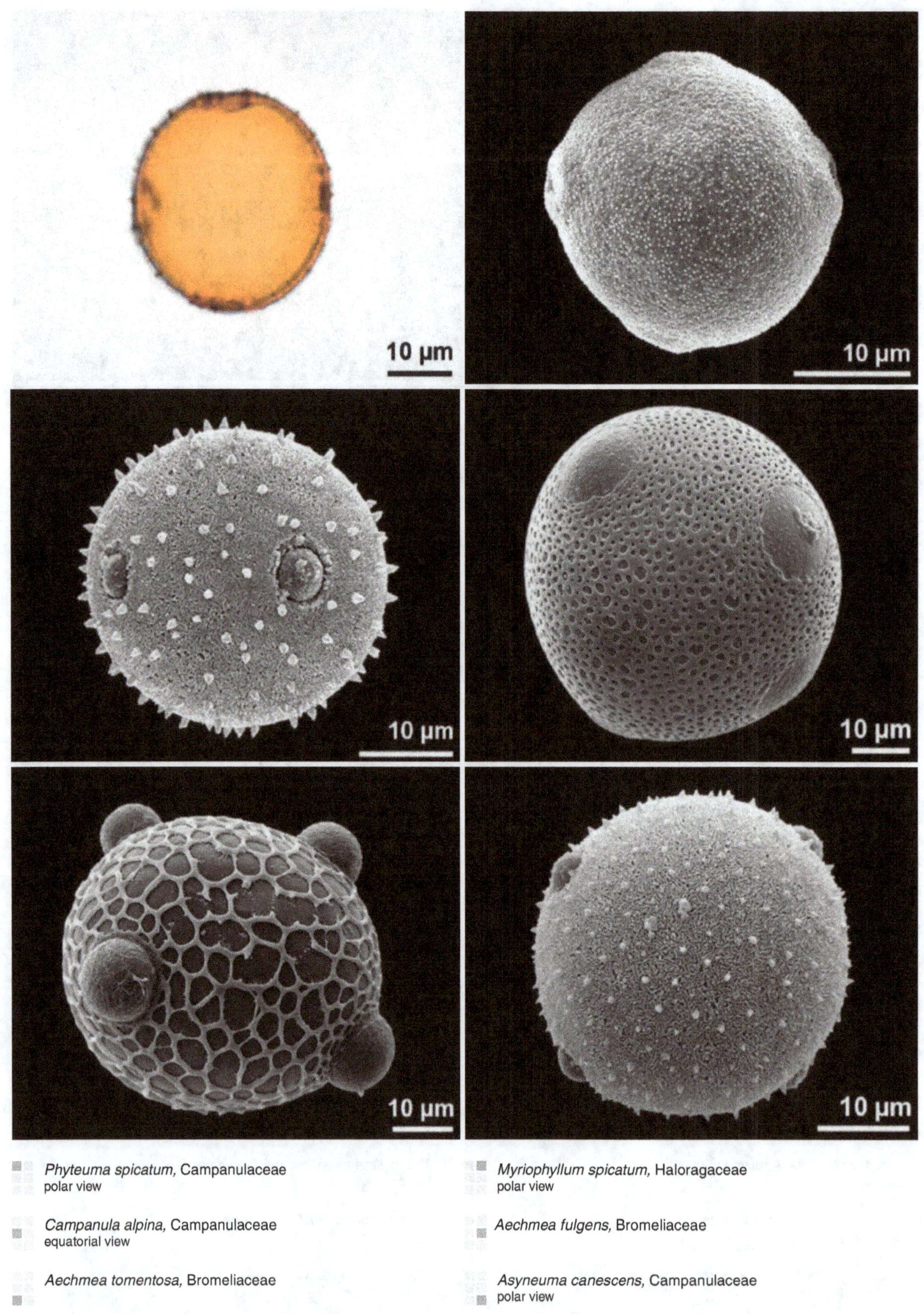

Phyteuma spicatum, Campanulaceae
polar view

Myriophyllum spicatum, Haloragaceae
polar view

Campanula alpina, Campanulaceae
equatorial view

Aechmea fulgens, Bromeliaceae

Aechmea tomentosa, Bromeliaceae

Asyneuma canescens, Campanulaceae
polar view

porus/porate, pentaporate

pentaporate: pollen grain with five pori

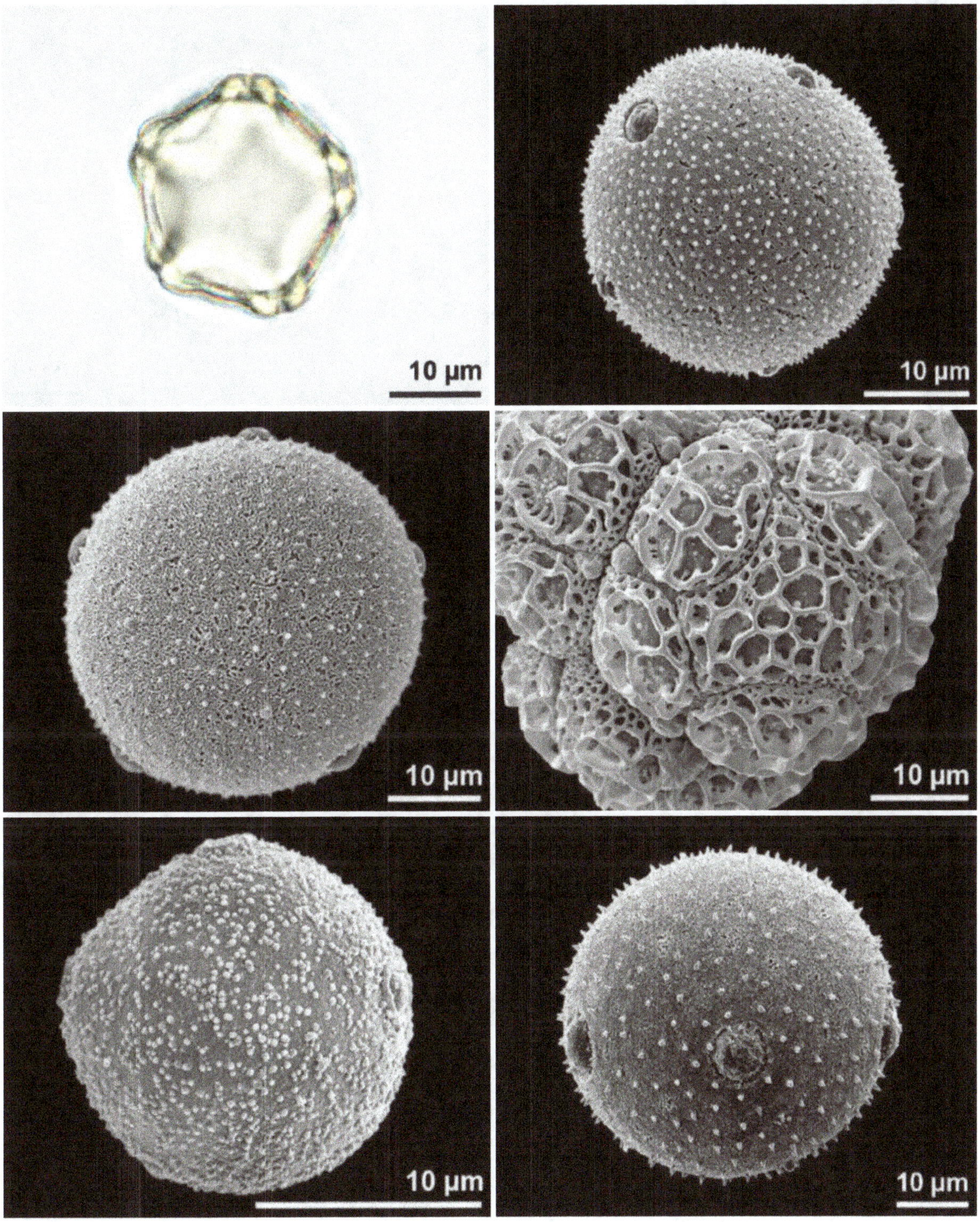

Alnus glutinosa, Betulaceae
polar view

Campanula garganica, Campanulaceae
polar view

Elatostema ambiguum, Urticaceae
polar view

Legousia speculum-veneris, Campanulaceae
polar view

Irlbachia pendula, Gentianaceae
tetrad

Campanula rapunculoides, Campanulaceae
equatorial view

porus/porate, stephanoporate

stephanoporate: pori situated at the equator (term usually used for six or more apertures)

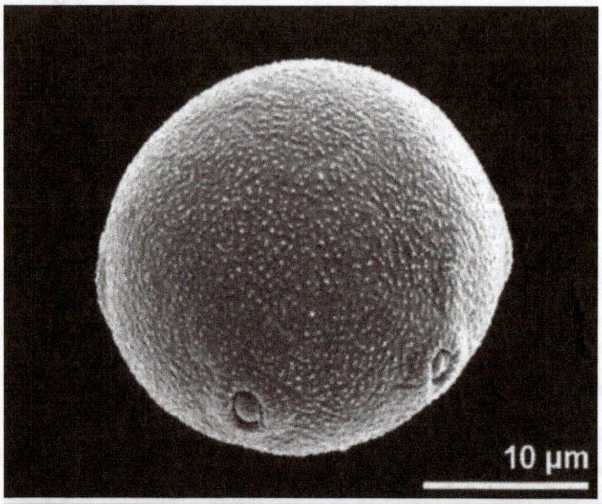

Alnus viridis, Betulaceae
oblique polar view

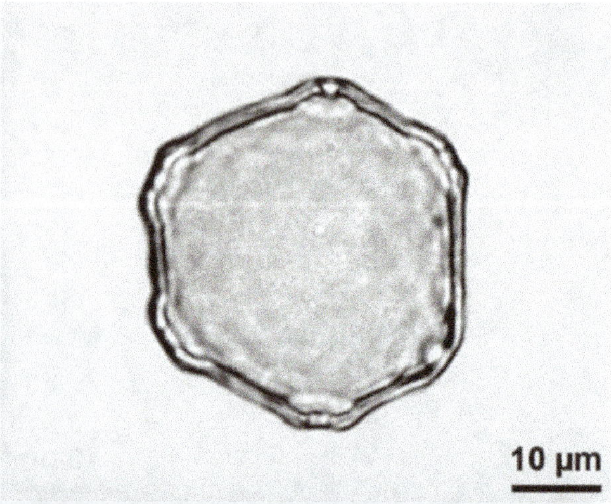

Pterocarya sp., Juglandaceae
fossil, middle Miocene, Austria, polar view

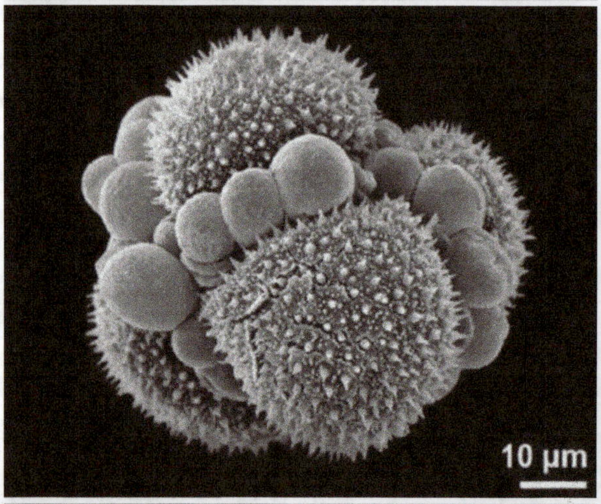

Drosera kansaiensis, Droseraceae
tetrad

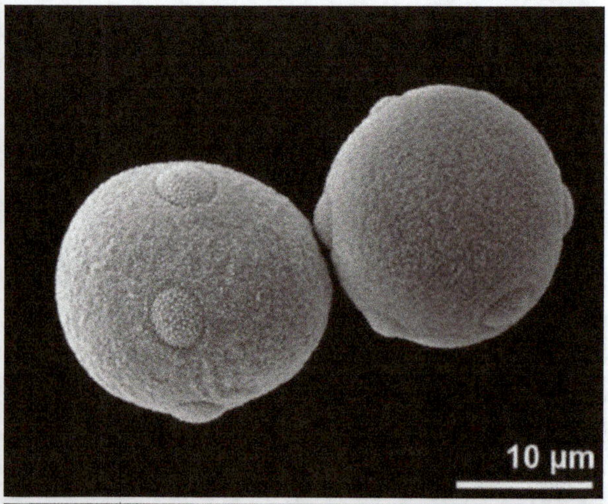

Paramoltkia doerfleri, Boraginaceae

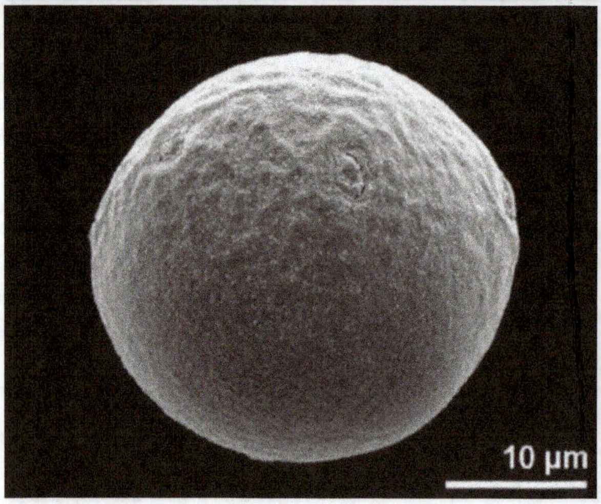

Ulmus minor, Ulmaceae
oblique equatorial view

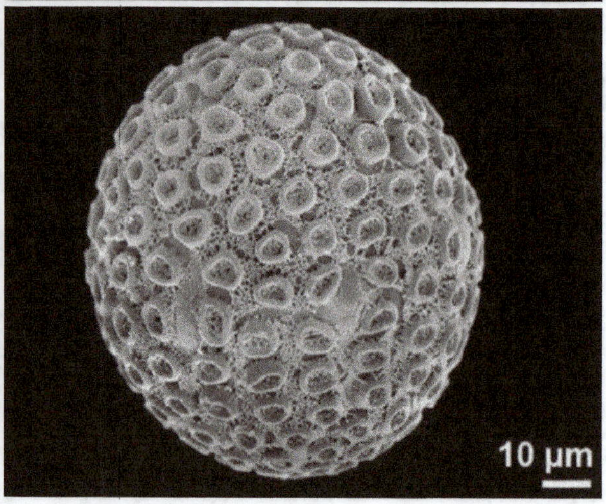

Megaskepasma erythrochlamys, Acanthaceae
equatorial view

porus/porate, pantoporate

pantoporate: pori distributed more or less regularly over the surface

Stellaria holostea, Caryophyllaceae

Cobaea scandens, Polemoniaceae

Ipomoea batatas, Convolvulaceae

Helianthium bolivianum, Alismataceae

Calystegia sepium, Convolvulaceae

Aechmea azurea, Bromeliaceae

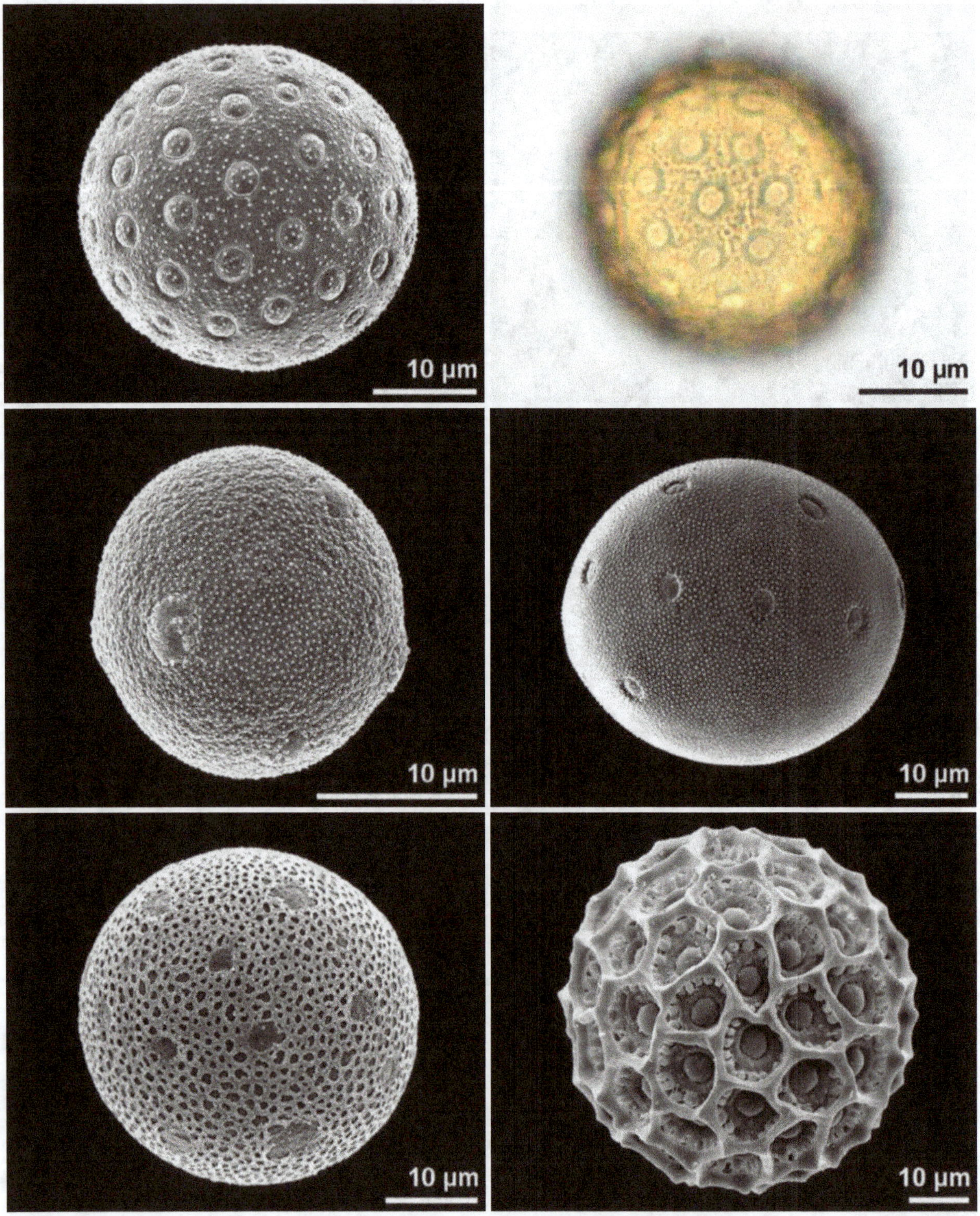

Bassia scoparia, Amaranthaceae

Bassia scoparia, Amaranthaceae

Plantago major, Plantaginaceae

Juglans regia, Juglandaceae
equatorial view, pores irregularly spread (exception)

Buxus sempervirens, Buxaceae

Kallstroemia maxima, Zygophyllaceae

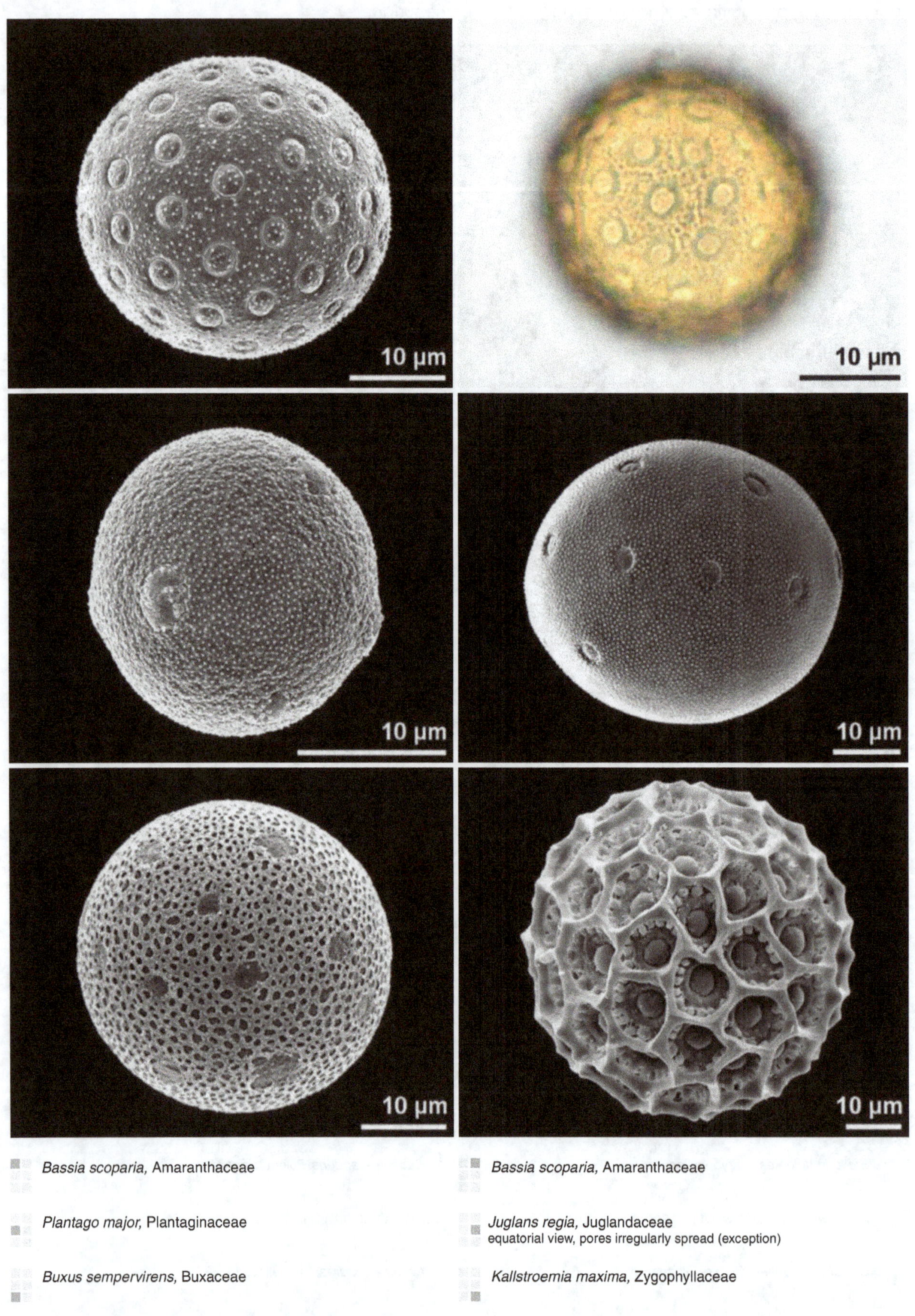

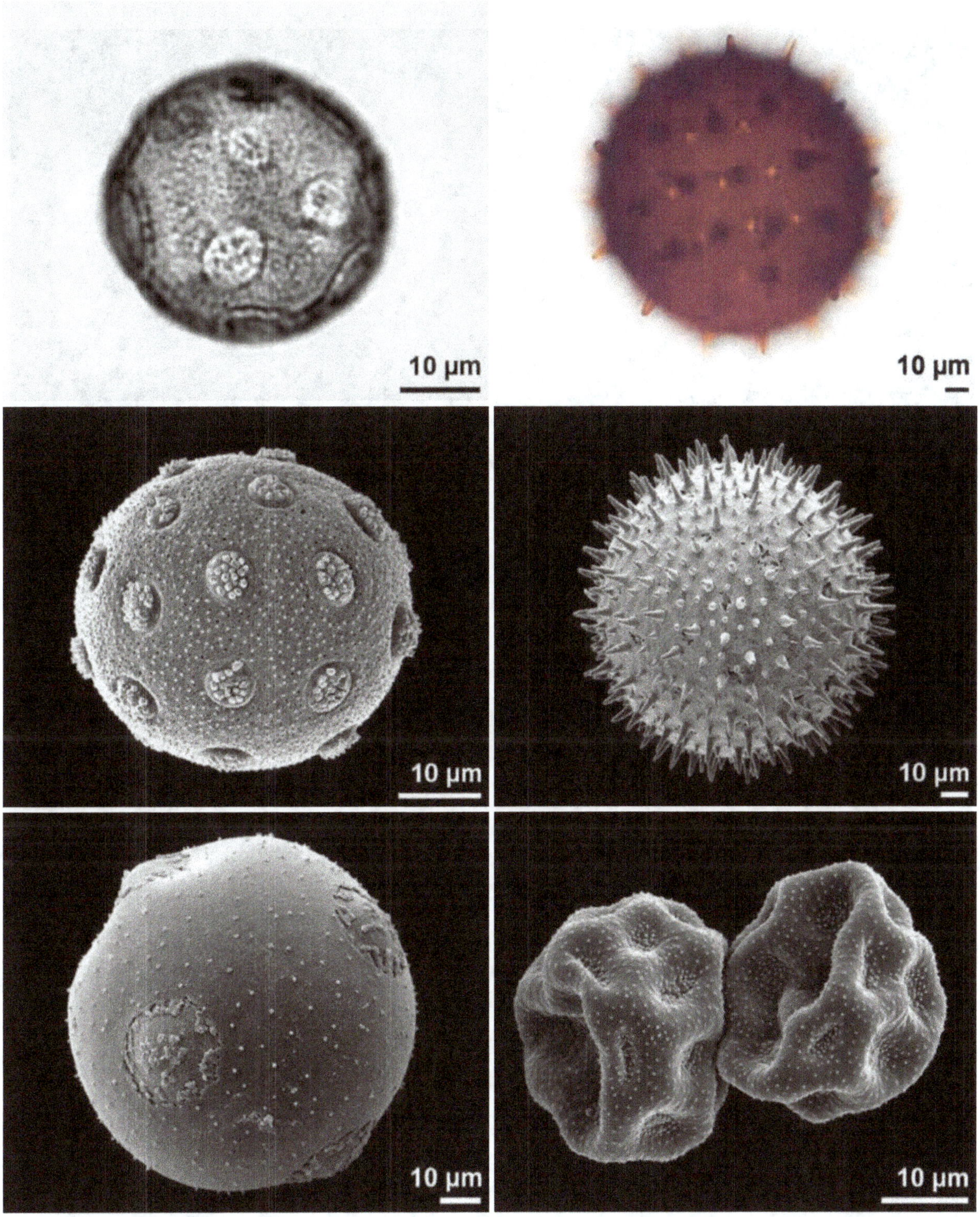

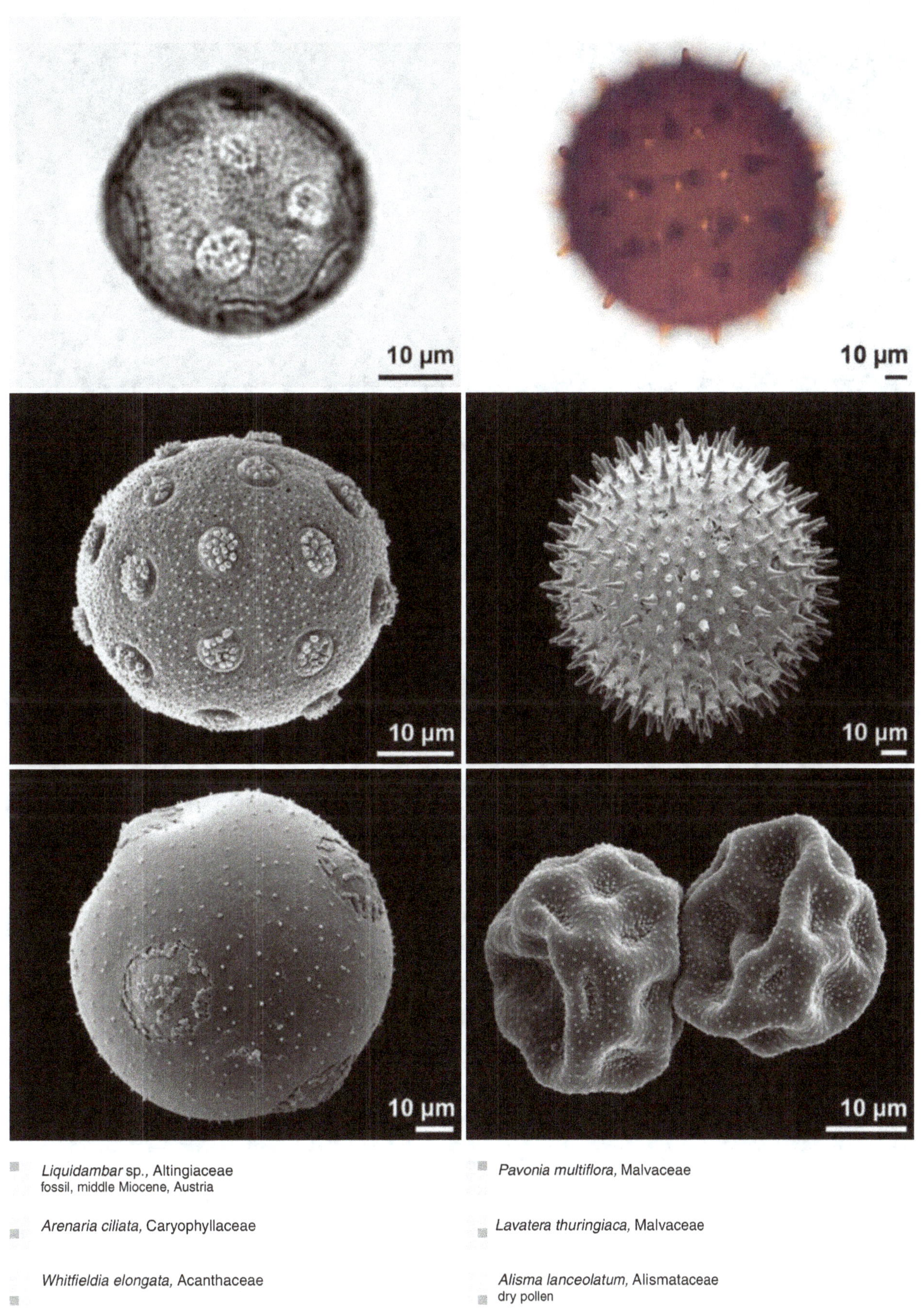

Liquidambar sp., Altingiaceae
fossil, middle Miocene, Austria

Pavonia multiflora, Malvaceae

Arenaria ciliata, Caryophyllaceae

Lavatera thuringiaca, Malvaceae

Whitfieldia elongata, Acanthaceae

Alisma lanceolatum, Alismataceae
dry pollen

poroid/poroidate

indistinct circular or elliptic aperture

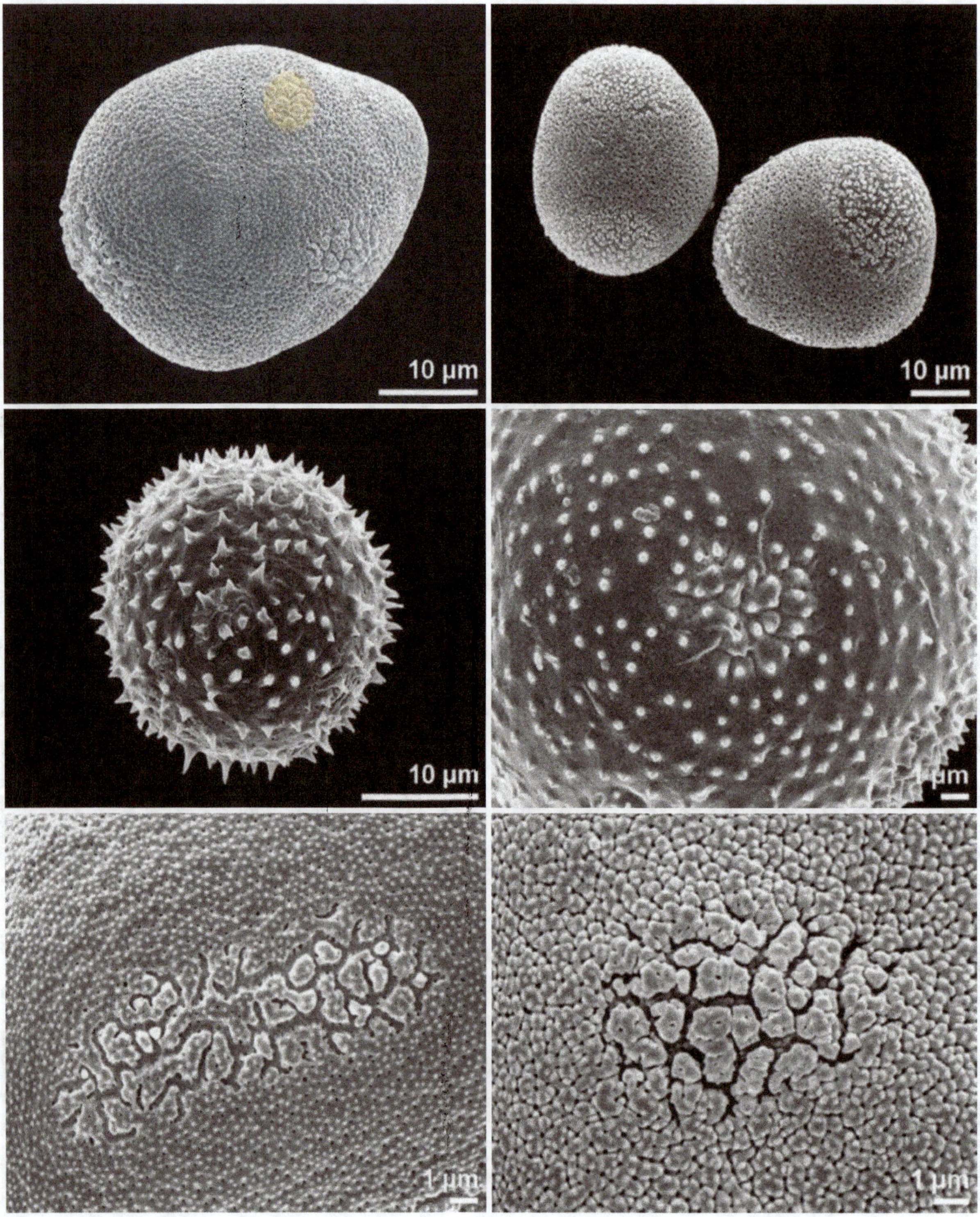

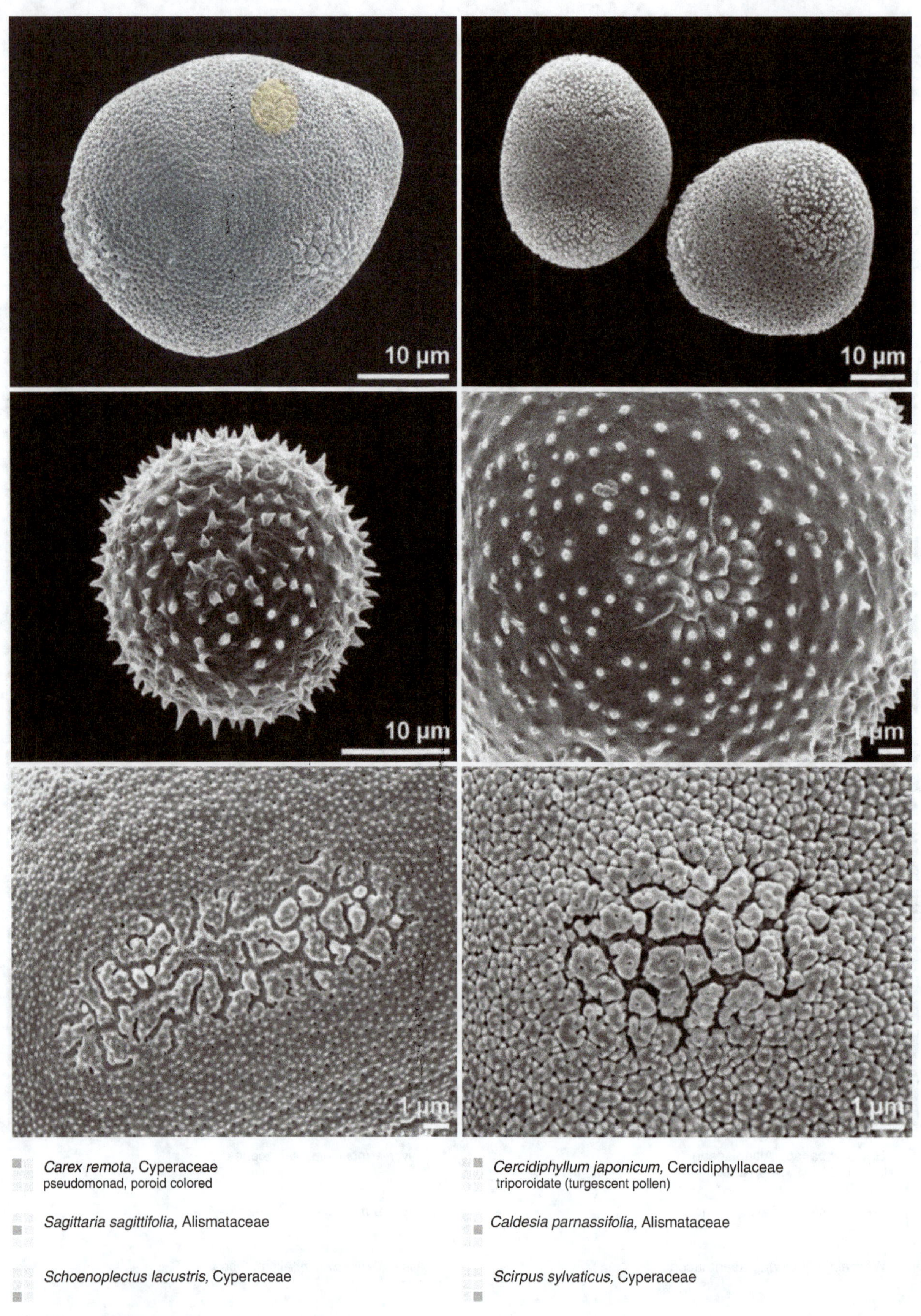

Carex remota, Cyperaceae
pseudomonad, poroid colored

Cercidiphyllum japonicum, Cercidiphyllaceae
triporoidate (turgescent pollen)

Sagittaria sagittifolia, Alismataceae

Caldesia parnassifolia, Alismataceae

Schoenoplectus lacustris, Cyperaceae

Scirpus sylvaticus, Cyperaceae

pseudocolpus

colpus in a heteroaperturate pollen grain, presumed not to function as germination site

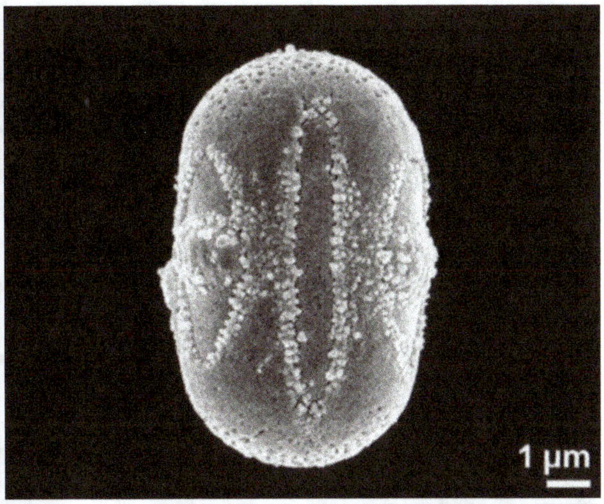

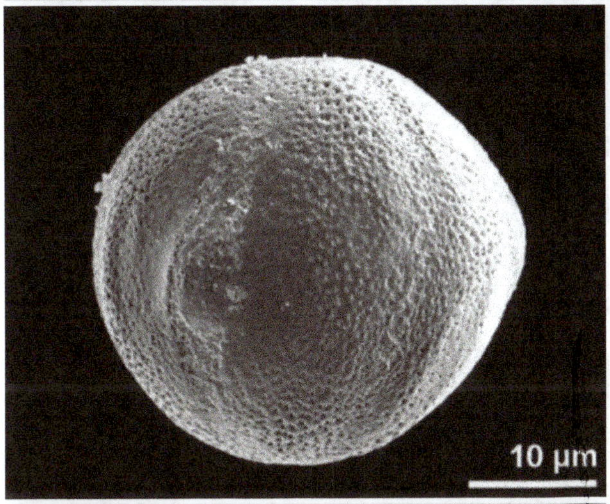

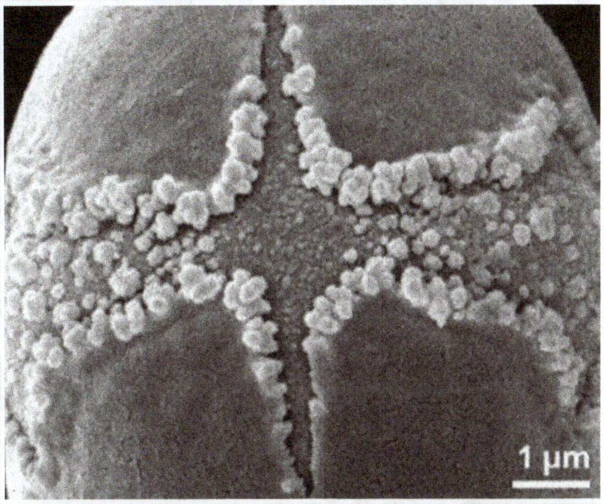

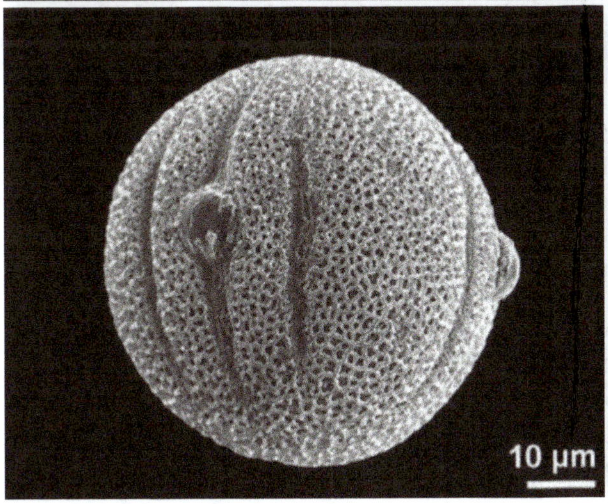

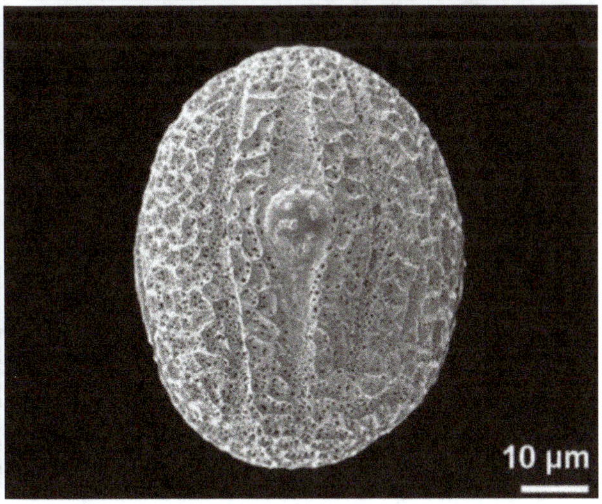

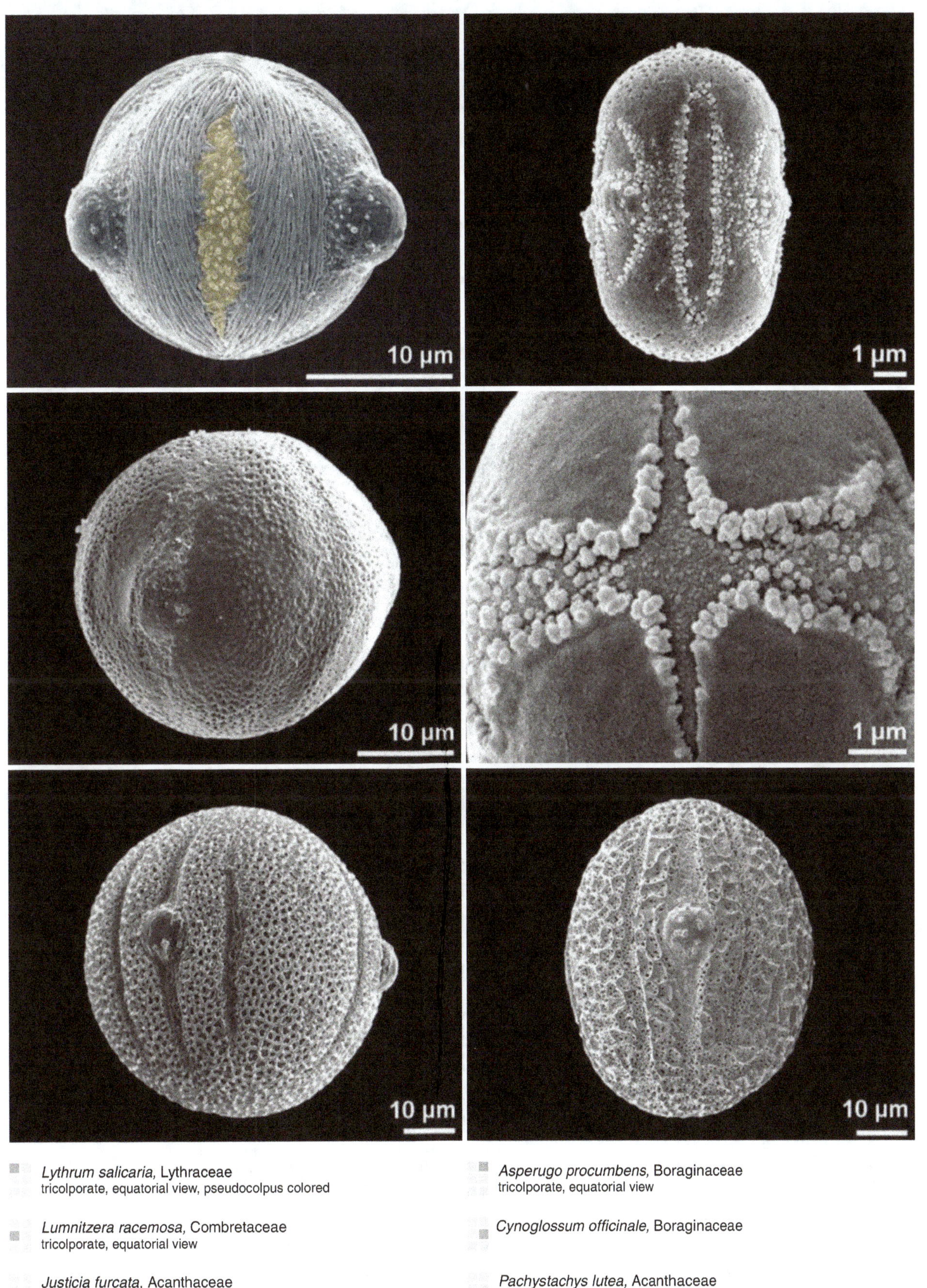

Lythrum salicaria, Lythraceae
tricolporate, equatorial view, pseudocolpus colored

Lumnitzera racemosa, Combretaceae
tricolporate, equatorial view

Justicia furcata, Acanthaceae
equatorial view, two pseudocolpi flanking colporus

Asperugo procumbens, Boraginaceae
tricolporate, equatorial view

Cynoglossum officinale, Boraginaceae

Pachystachys lutea, Acanthaceae
equatorial view, two pseudocolpi flanking colporus

ring-like aperture

circumferential aperture (situated more or less equatorially or, rarely, meridionally)

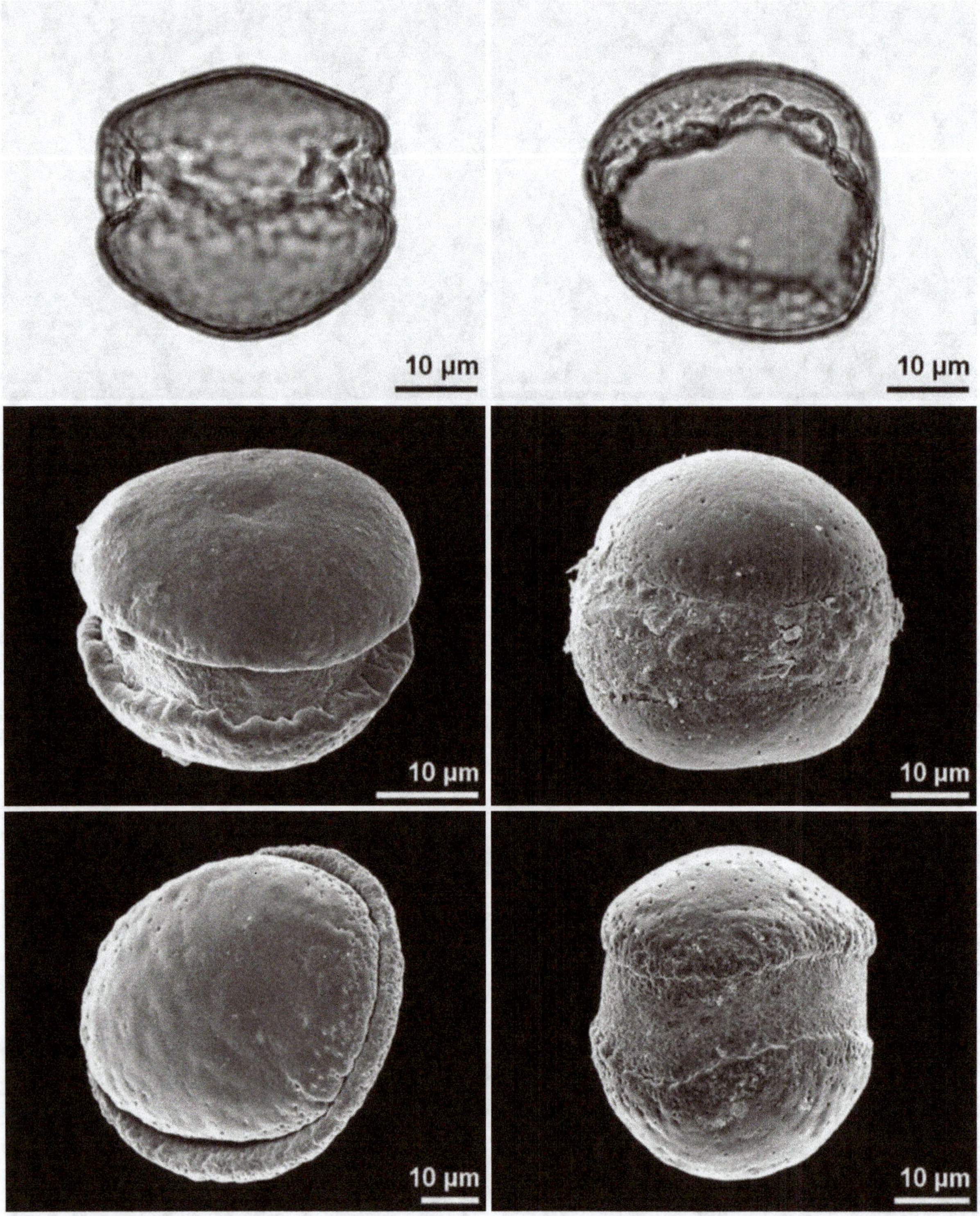

Zamioculcas zamiifolia, Araceae
equatorial view

Zamioculcas zamiifolia, Araceae
oblique polar view, artificially ruptured

Zamioculcas zamiifolia, Araceae
equatorial view

Monstera deliciosa, Araceae
equatorial view

Gonatopus angustus, Araceae
oblique polar view, dry pollen

Gonatopus angustus, Araceae
equatorial view

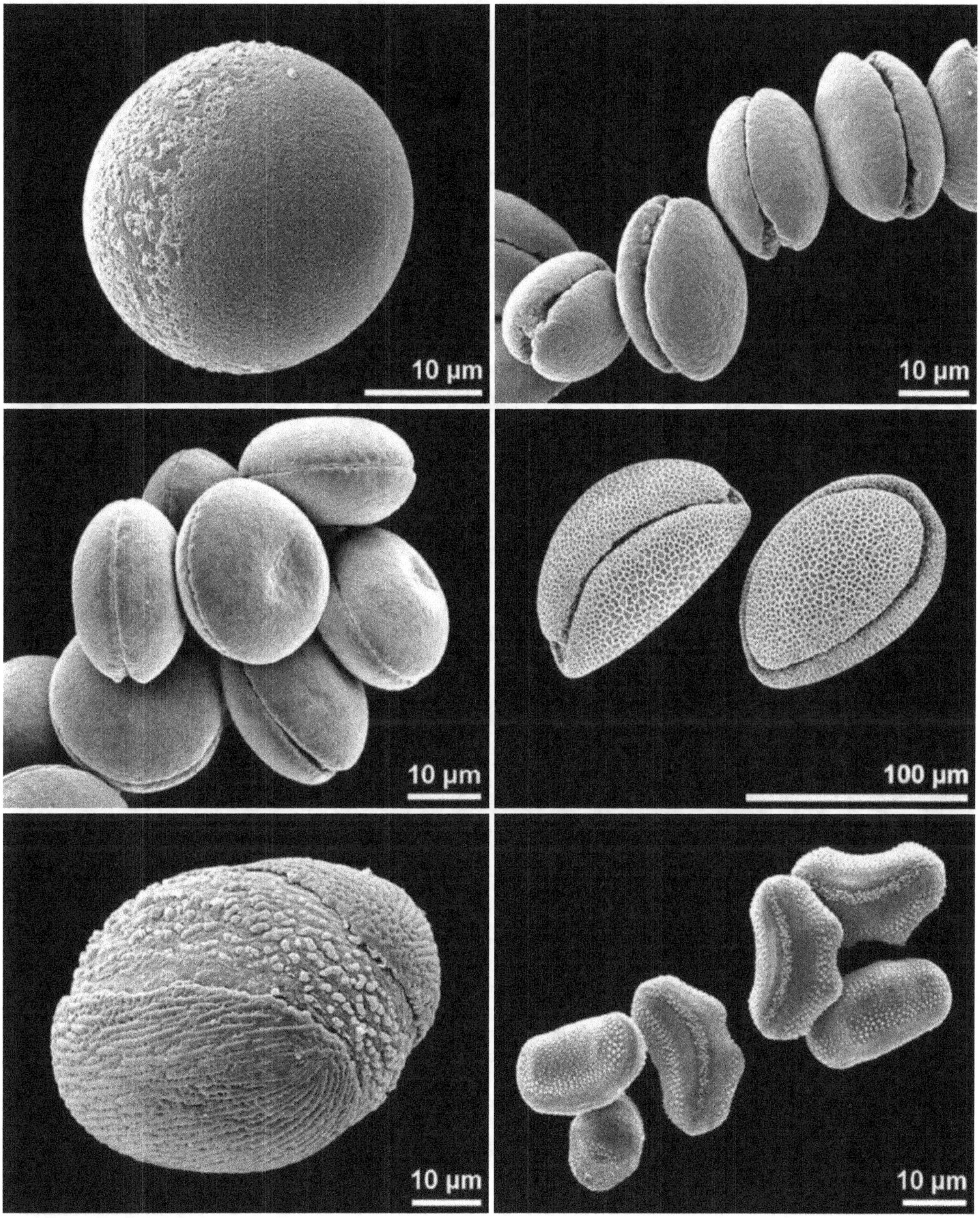

Pedicularis palustris, Orobanchaceae
aperture running meridionally

Pedicularis rostratocapitata, Orobanchaceae
aperture running meridionally, dry pollen

Cephalostemon riedelianus, Rapateaceae
aperture running equatorially, oblique equatorial view

Pedicularis palustris, Orobanchaceae
aperture running meridionally, dry pollen

Iris histrioides, Iridaceae
aperture running equatorially, dry pollen

Limnanthes douglasii, Limnanthaceae
aperture running equatorially

Gonatopus boivinii, Araceae

Victoria regia, Nymphaeaceae
tetrad

Pedicularis gyroflexa, Orobanchaceae
aperture running meridionally

Passiflora amethystina, Passifloraceae
debatable: 3 ring-like apertures or triporate with opercula

Acacia dealbata, Fabaceae
debatable: polyad, monads with ring-like structure/thinning

Acacia dealbata, Fabaceae
close-up of polyad

spiral aperture/spiraperturate

elongated, coiled aperture

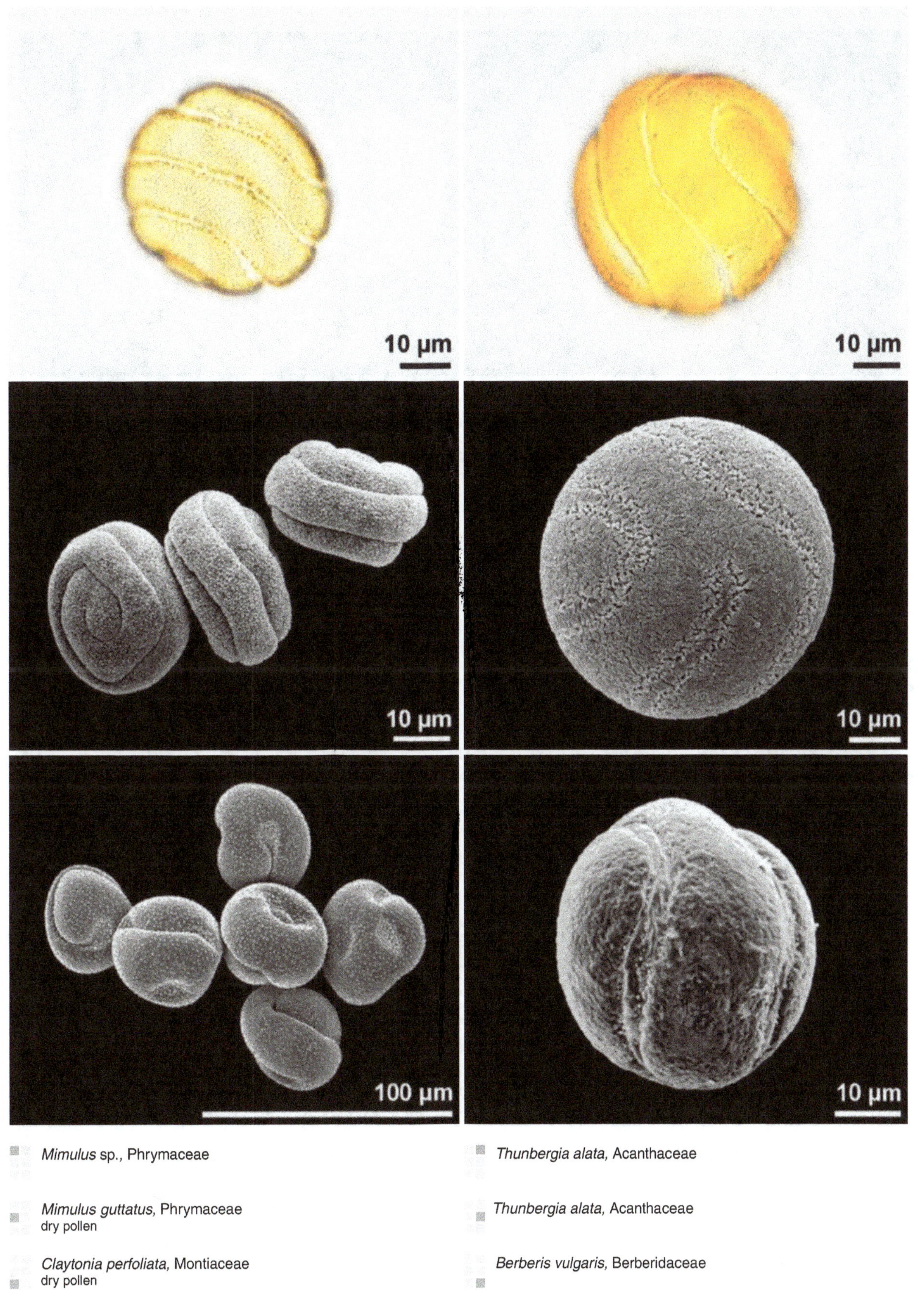

Mimulus sp., Phrymaceae

Thunbergia alata, Acanthaceae

Mimulus guttatus, Phrymaceae
dry pollen

Thunbergia alata, Acanthaceae

Claytonia perfoliata, Montiaceae
dry pollen

Berberis vulgaris, Berberidaceae

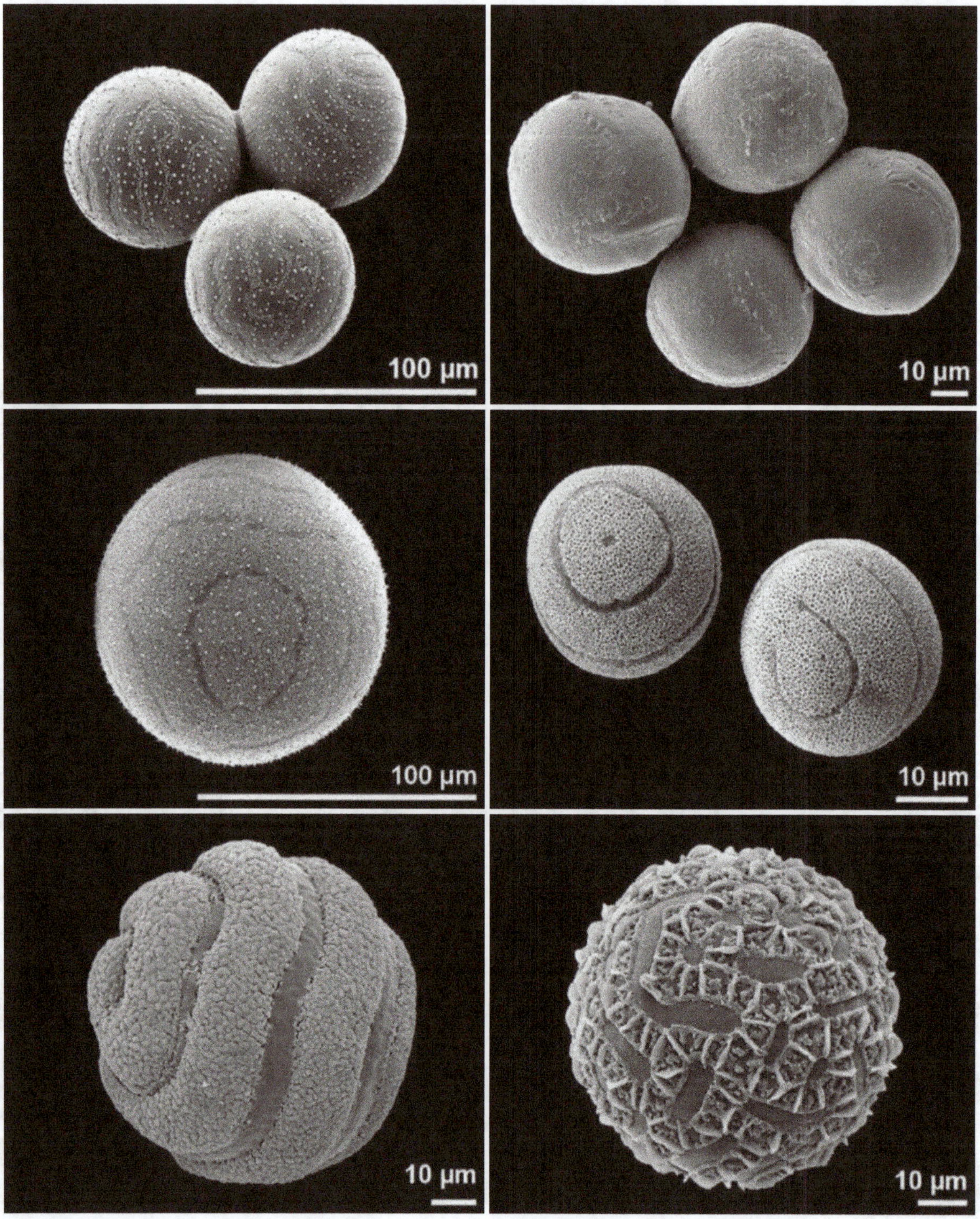

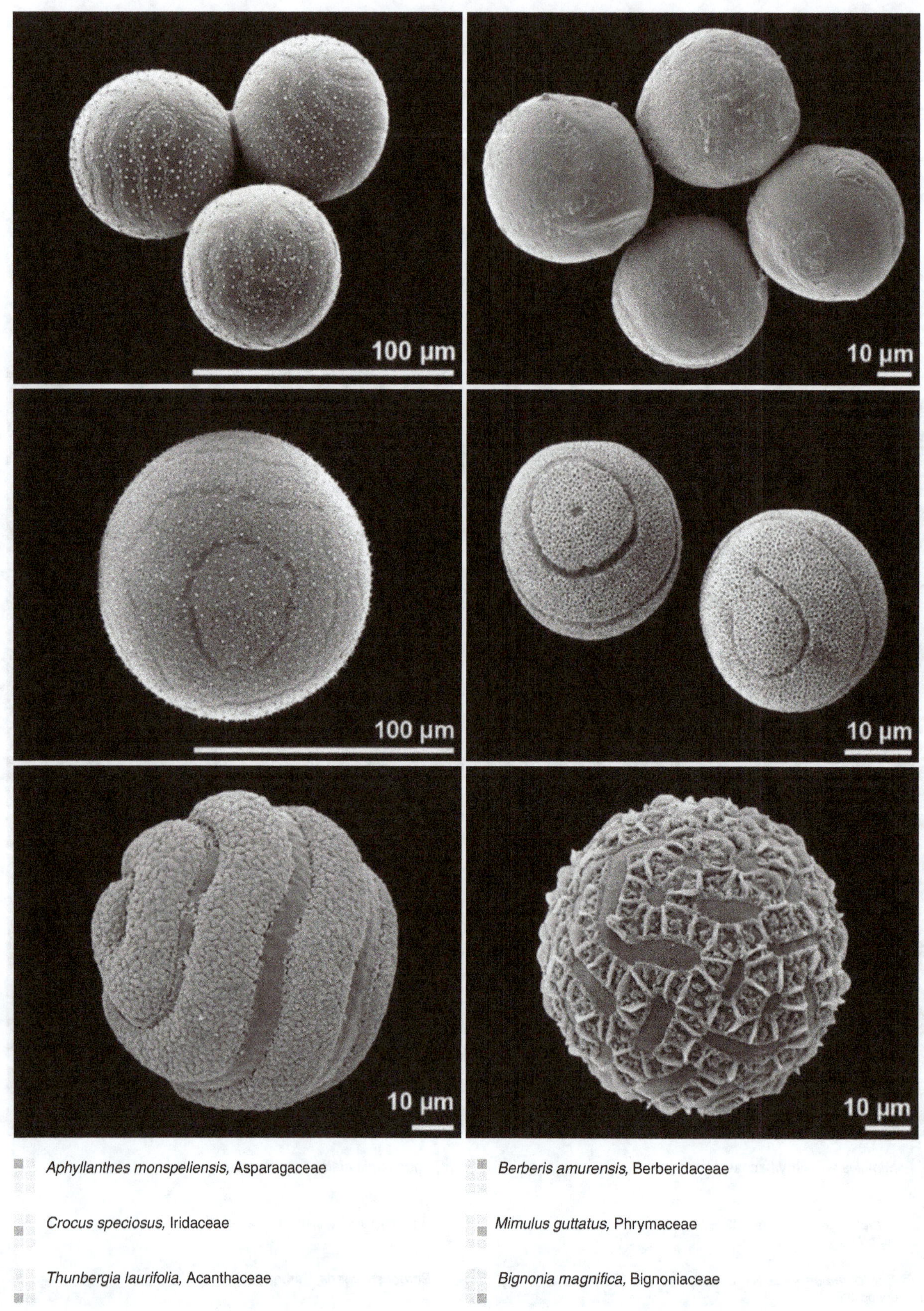

Aphyllanthes monspeliensis, Asparagaceae

Berberis amurensis, Berberidaceae

Crocus speciosus, Iridaceae

Mimulus guttatus, Phrymaceae

Thunbergia laurifolia, Acanthaceae

Bignonia magnifica, Bignoniaceae

stephanoaperturate

apertures situated at the equator (term usually used for more than six apertures): stephanocolpate, stephanocolporate, stephanoporate

Dracocephalum austriacum, Lamiaceae
hexacolpate, polar view

Plectranthus esculentus, Lamiaceae
hexacolpate, polar view

Polygala myrtifolia, Polygalaceae
colporate, equatorial view

Salvia glutinosa, Lamiaceae
hexacolpate, polar view

Sanguisorba officinalis, Rosaceae
colporate, equatorial view

Alnus glutinosa, Betulaceae
porate

stephanocolpate, stephanocolporate, stephanoporate

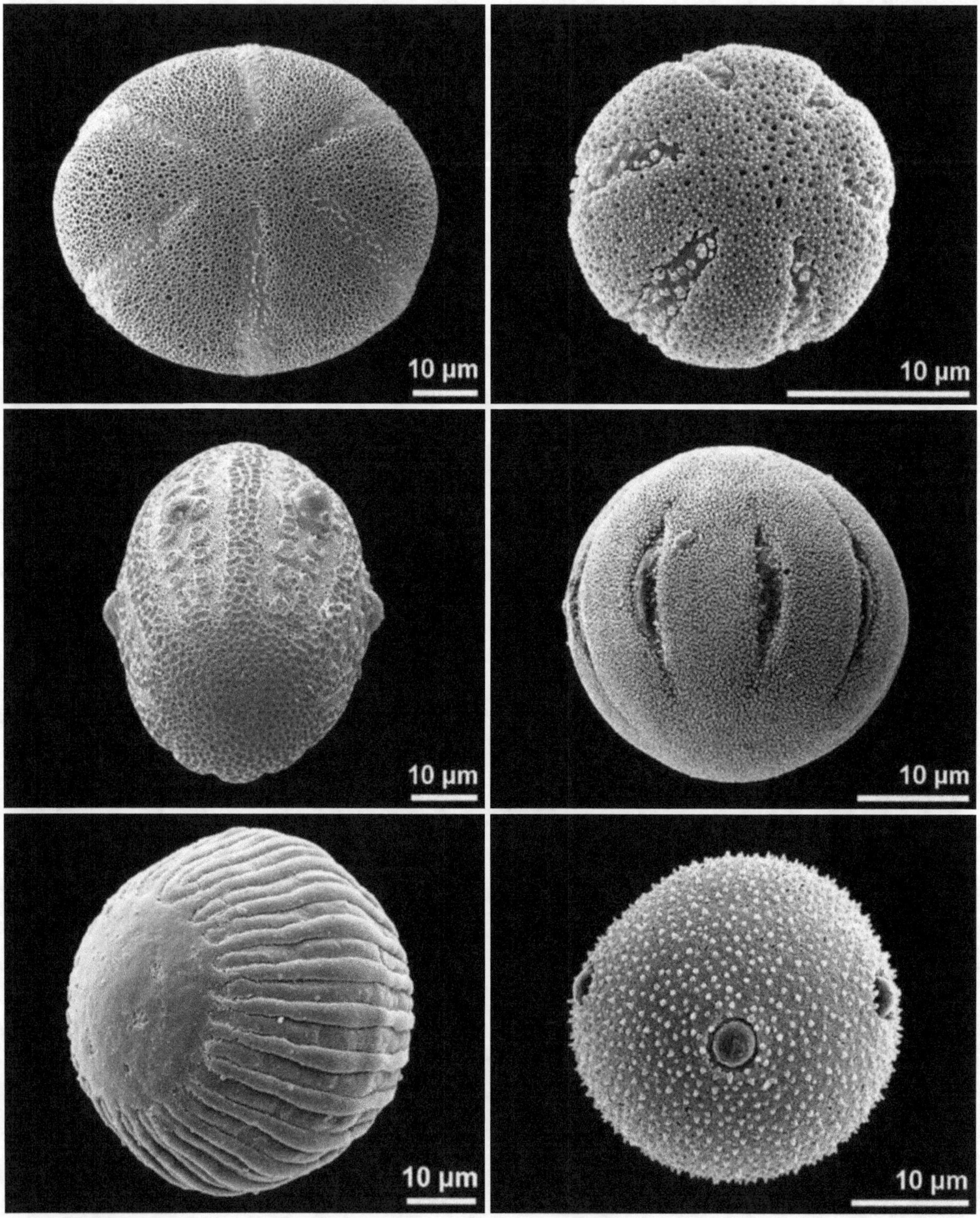

Dracocephalum austriacum, Lamiaceae
hexacolpate, polar view

Asperula tinctoria, Rubiaceae
hexacolpate, polar view

Justicia menesii, Acanthaceae
colporate, oblique polar view

Borago officinalis, Boraginaceae
colporate, equatorial view

Polygala myrtifolia, Polygalaceae
colporate, oblique equatorial view

Legousia speculum-veneris, Campanulaceae
porate, equatorial view

sulcus/sulcate

elongated aperture located distally

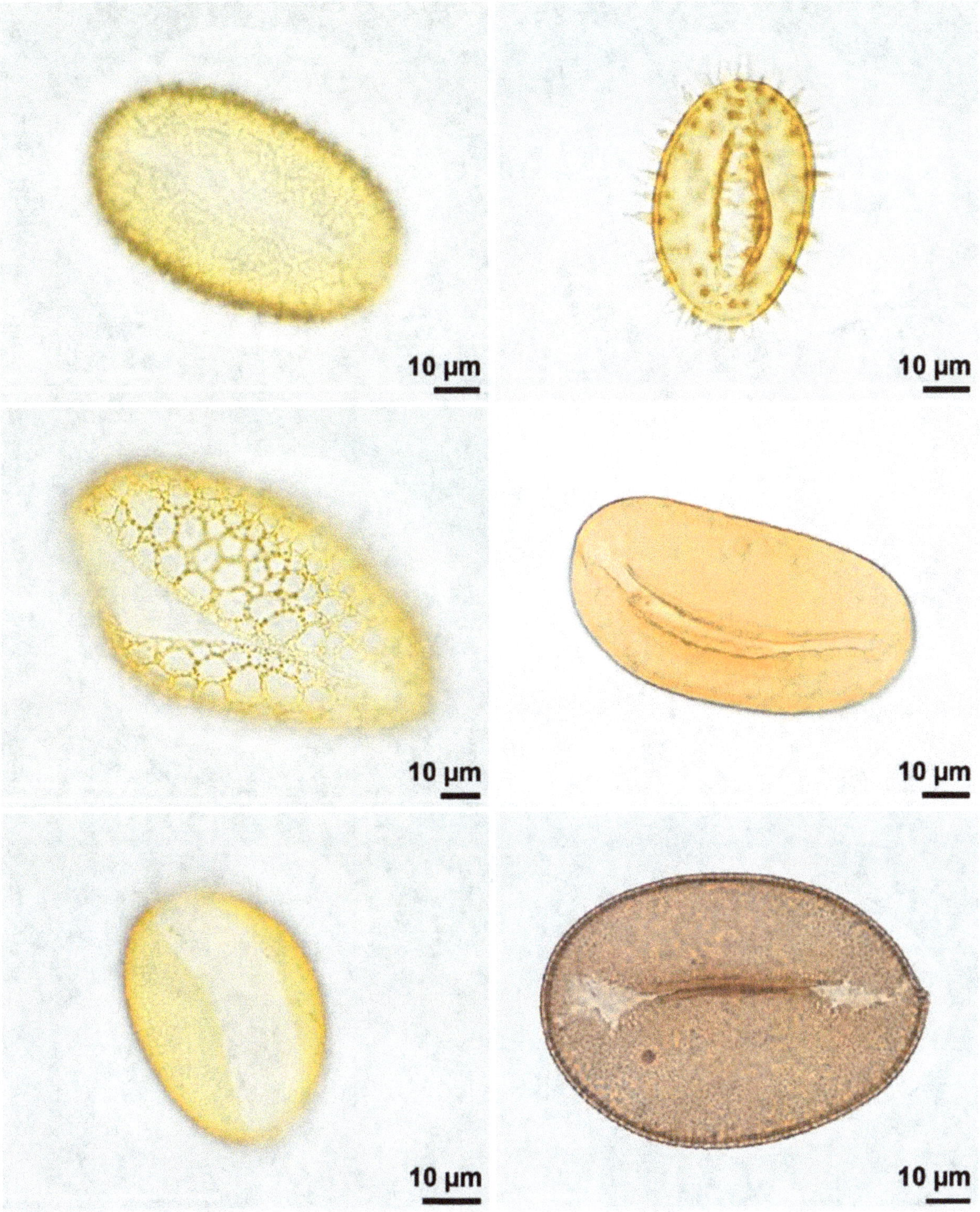

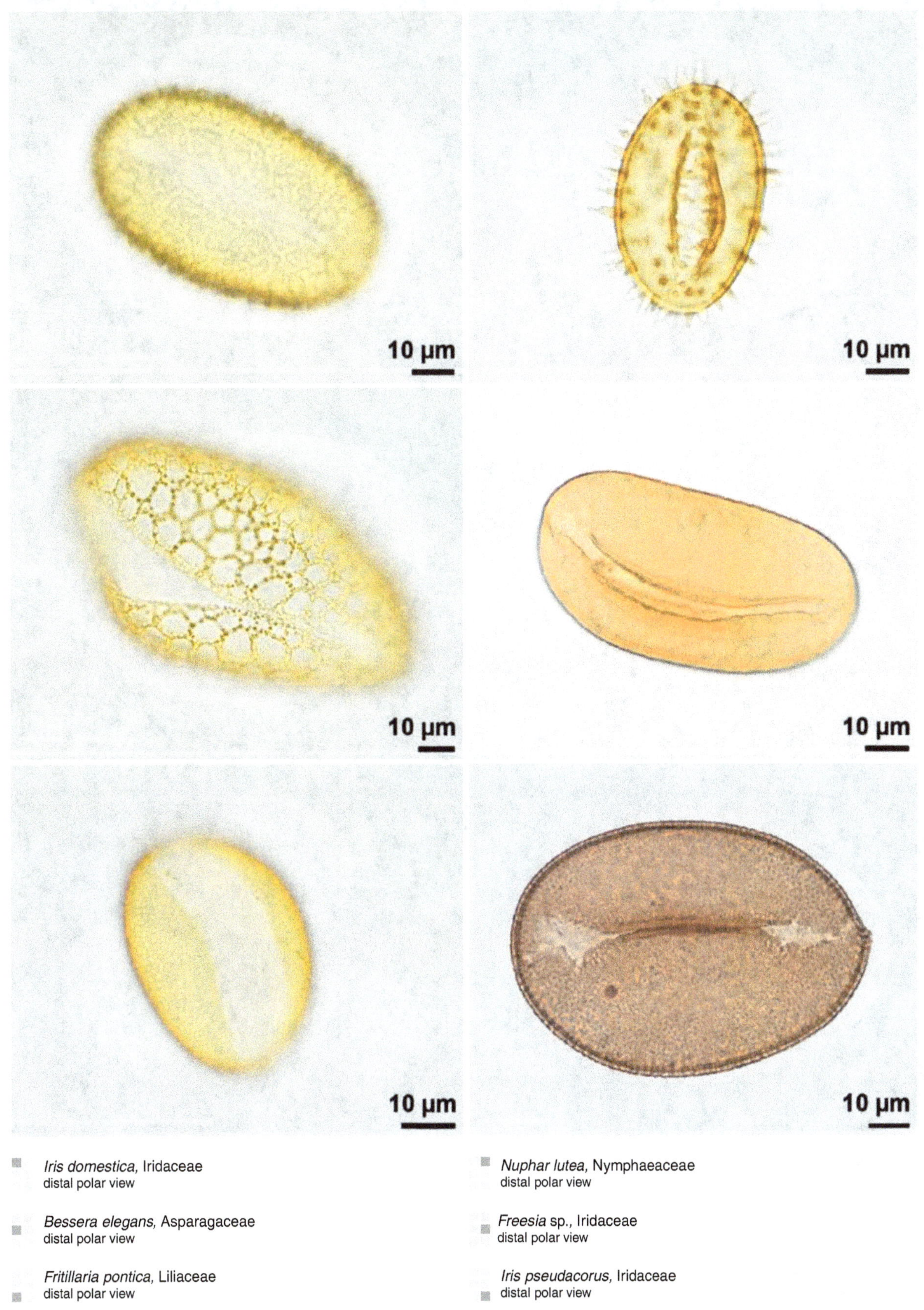

Iris domestica, Iridaceae
distal polar view

Bessera elegans, Asparagaceae
distal polar view

Fritillaria pontica, Liliaceae
distal polar view

Nuphar lutea, Nymphaeaceae
distal polar view

Freesia sp., Iridaceae
distal polar view

Iris pseudacorus, Iridaceae
distal polar view

Lilium martagon, Liliaceae
distal polar view

Doryanthes palmeri, Doryanthaceae
dry pollen

Cabomba palaeformis, Cabombaceae
oblique distal polar view

Galanthus nivalis, Amaryllidaceae
distal polar view

Allium ursinum, Amaryllidaceae
distal polar view

Asphodeline lutea, Xanthorrhoeaceae

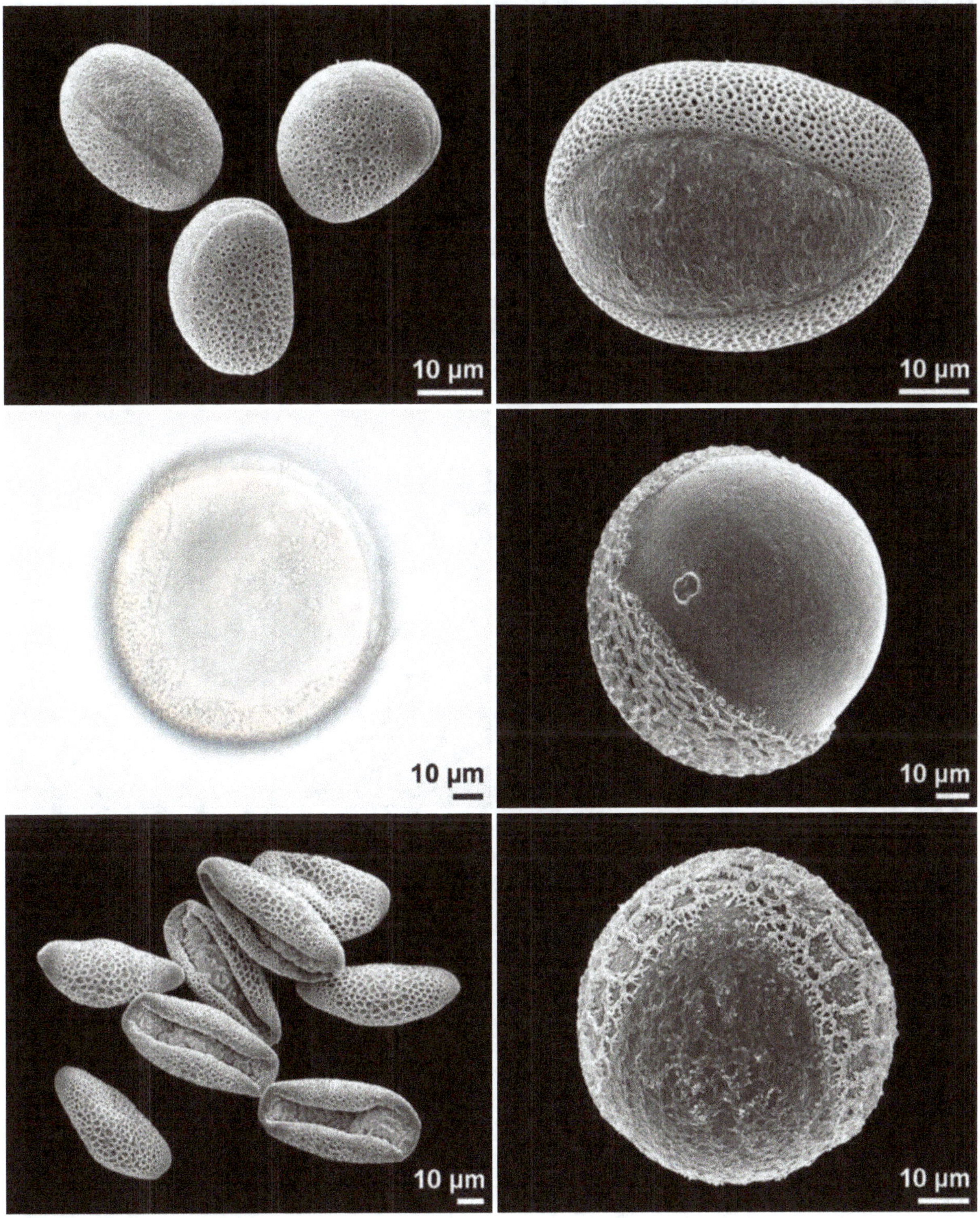

Lachenalia aloides, Asparagaceae

Catopsis floribunda, Bromeliaceae
distal polar view

Iris pseudacorus, Iridaceae
hydrated, equatorial view

Iris reichenbachii, Iridaceae
oblique distal polar view

Vriesea neoglutinosa, Bromeliaceae
dry pollen

Paradisea liliastrum, Asparagaceae
equatorial view

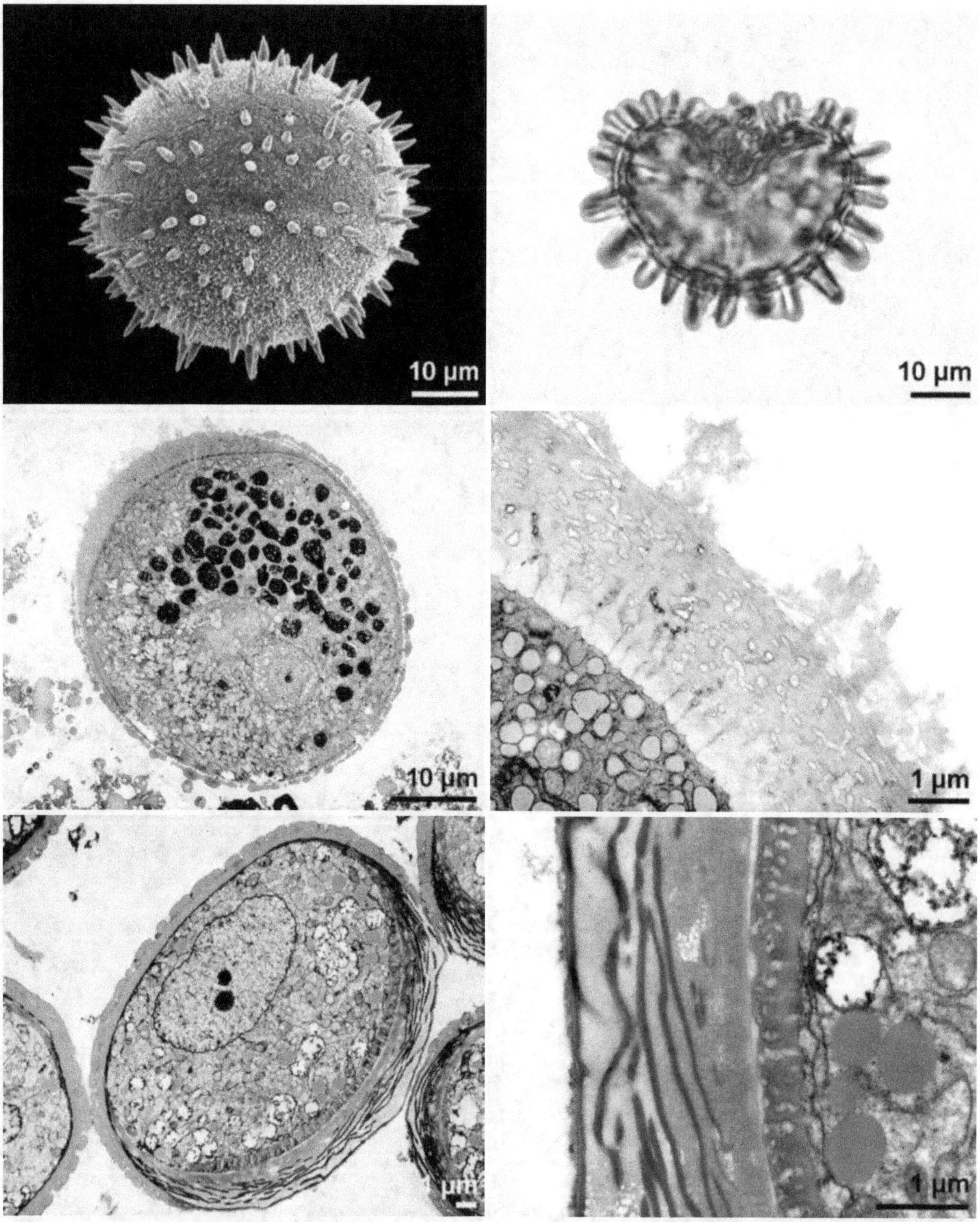

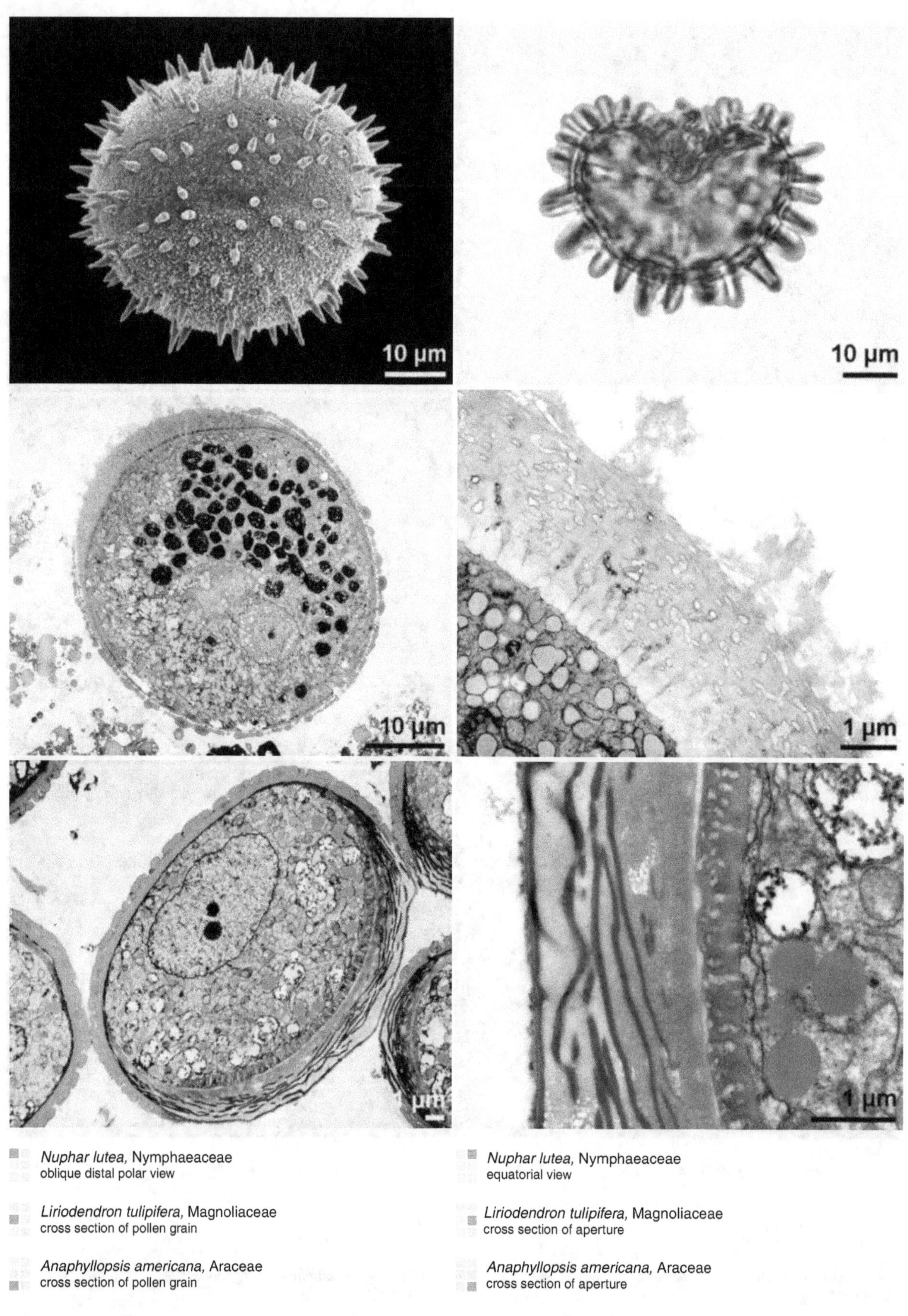

Nuphar lutea, Nymphaeaceae
oblique distal polar view

Nuphar lutea, Nymphaeaceae
equatorial view

Liriodendron tulipifera, Magnoliaceae
cross section of pollen grain

Liriodendron tulipifera, Magnoliaceae
cross section of aperture

Anaphyllopsis americana, Araceae
cross section of pollen grain

Anaphyllopsis americana, Araceae
cross section of aperture

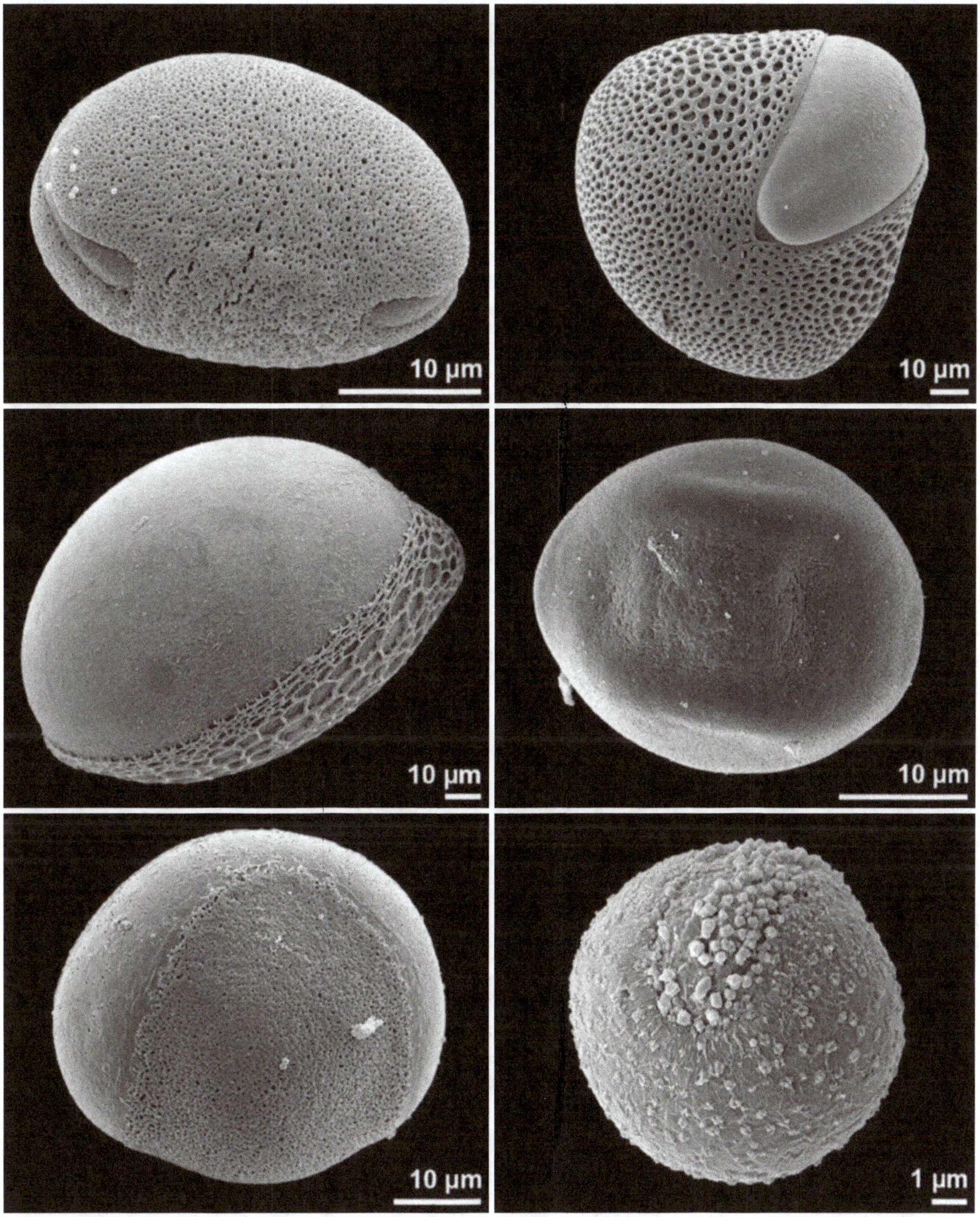

Allium sphaerocephalum, Amaryllidaceae
sulcus extended to proximal face, oblique proximal polar view

Asphodelus fistulosus, Xanthorrhoeaceae
equatorial view

Bessera elegans, Asparagaceae
oblique distal polar view

Ceratozamia kuesteriana, Zamiaceae
distal polar view

Cordyline fruticosa, Asparagaceae
extended sulcus, equatorial view

Saururus cernuus, Saururaceae
equatorial view

Guzmania elvallensis, Bromeliaceae
inconspicuous sulcus with broken reticulum, distal polar view

Chamaedorea microspadix, Arecaceae
dry pollen

Tradescantia zebrina, Commelinaceae
additional 2 tenuitates, proximal polar view, dry pollen

Wachendorfia thyrsiflora, Haemodoraceae
equatorial view

Tradescantia zebrina, Commelinaceae
additional 2 proximal situated tenuitates, dry pollen

Veratrum album, Melianthaceae
dry pollen

sulcus/sulcate, disulcate

disulcate: pollen grain with two sulci

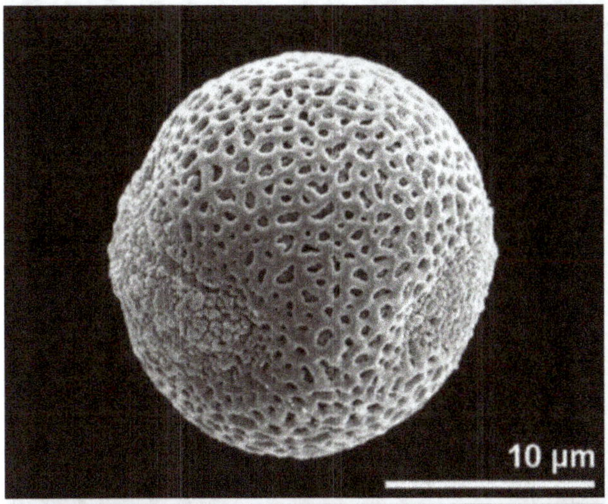

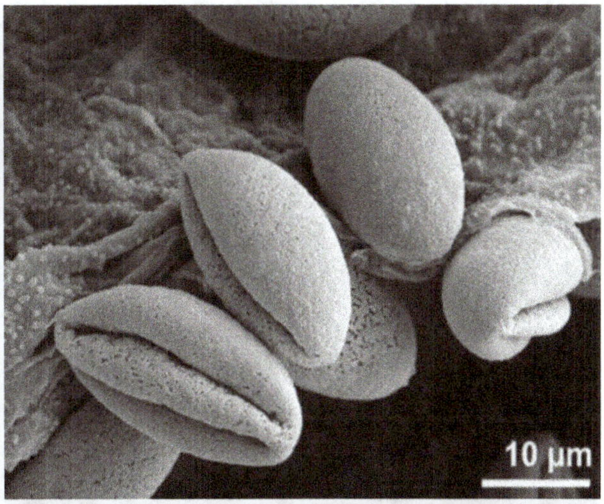

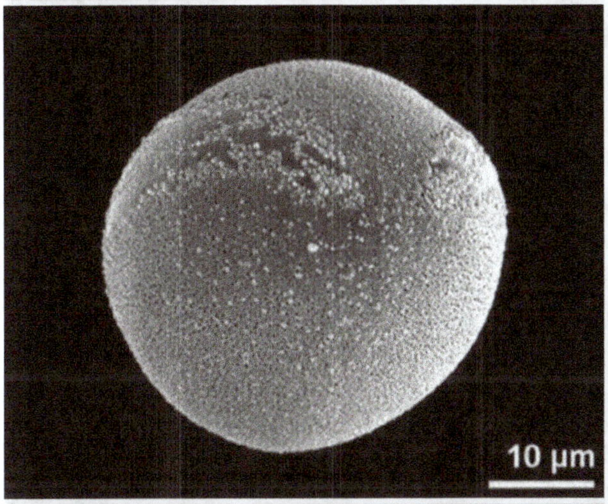

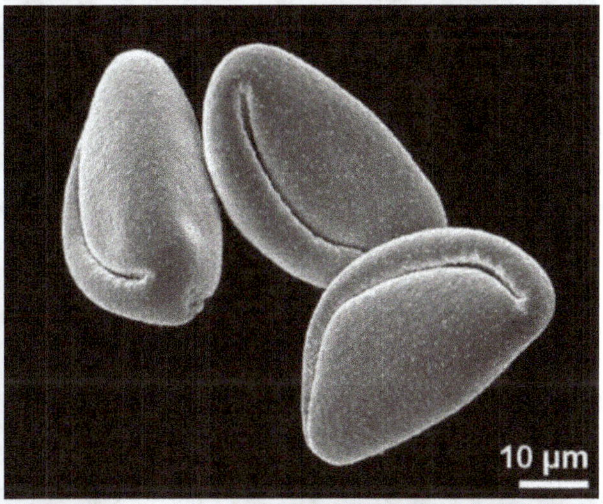

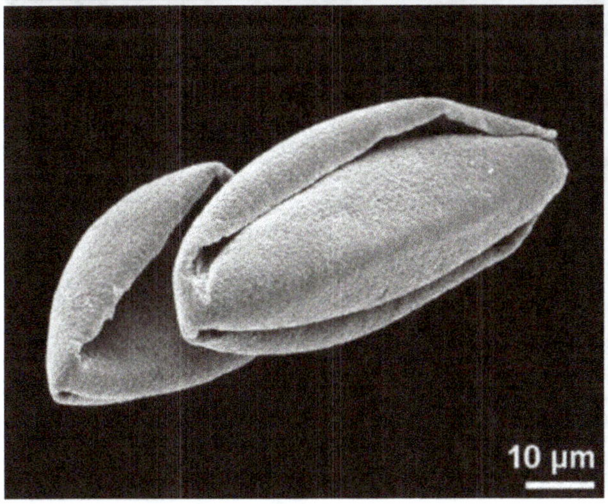

Tofieldia calyculata, Tofieldiaceae
equatorial view

Uvularia grandiflora, Colchicaceae
equatorial view

Eichhornia crassipes, Pontederiaceae
dry pollen

Xerophyta elegans, Velloziaceae
dry pollen

Uvularia grandiflora, Colchicaceae
dry pollen

Crinum x amabile, Amaryllidaceae
dry pollen

sulcus/sulcate, trichotomosulcate

trichotomosulcus/trichotomosulcate: 3-radiate sulcus

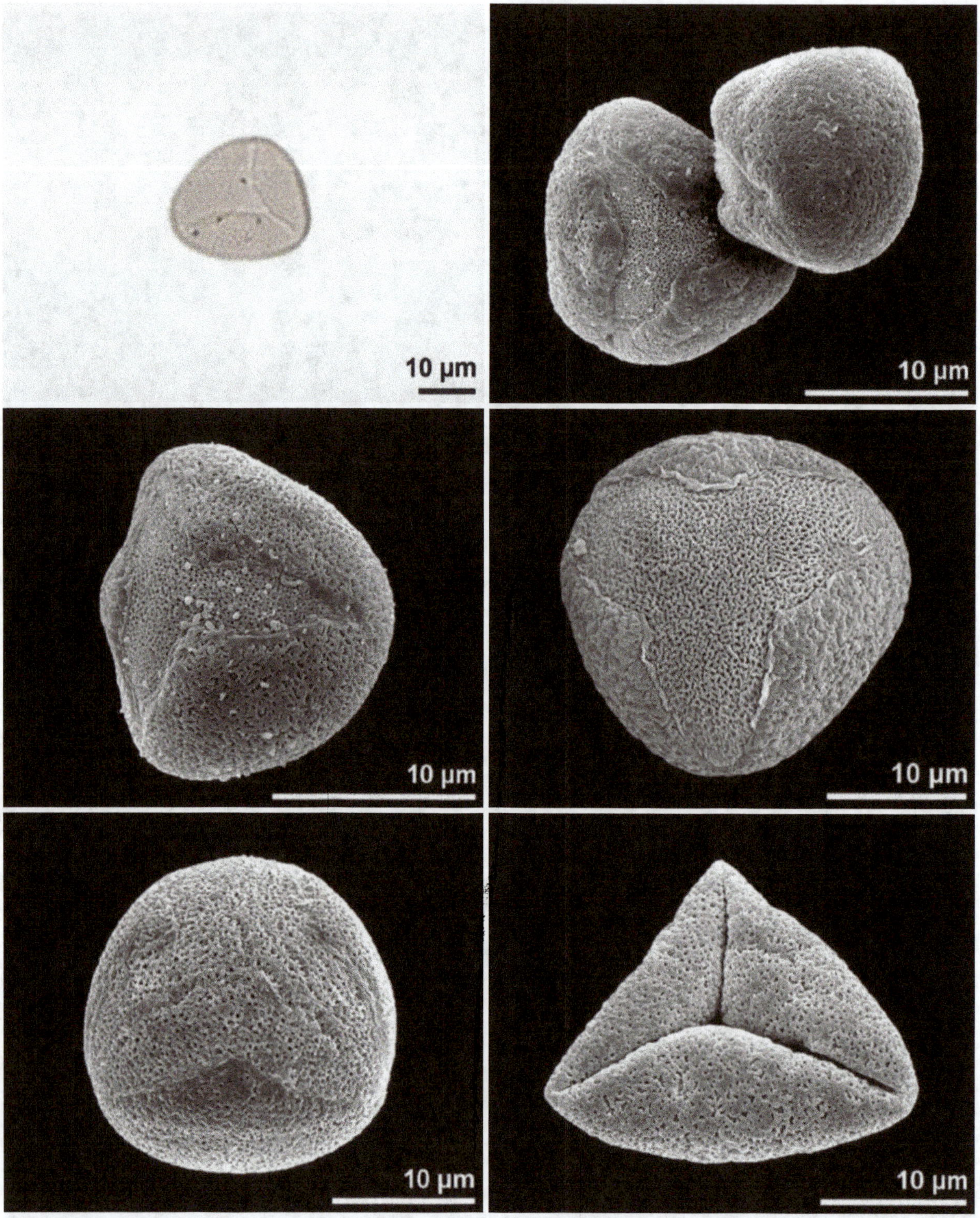

Arecaceae
fossil, early Miocene, South Africa

Dianella intermedia, Xanthorrhoeaceae

Dianella intermedia, Xanthorrhoeaceae
oblique distal polar view

Dianella caerulea, Xanthorrhoeaceae
distal polar view

Dianella tasmanica, Xanthorrhoeaceae
distal polar view

Dianella tasmanica, Xanthorrhoeaceae
distal polar view, dry pollen

sulcus/sulcate, polychotomosulcate

polychotomosulcus/polychotomosulcate: sulcus with more than three arms

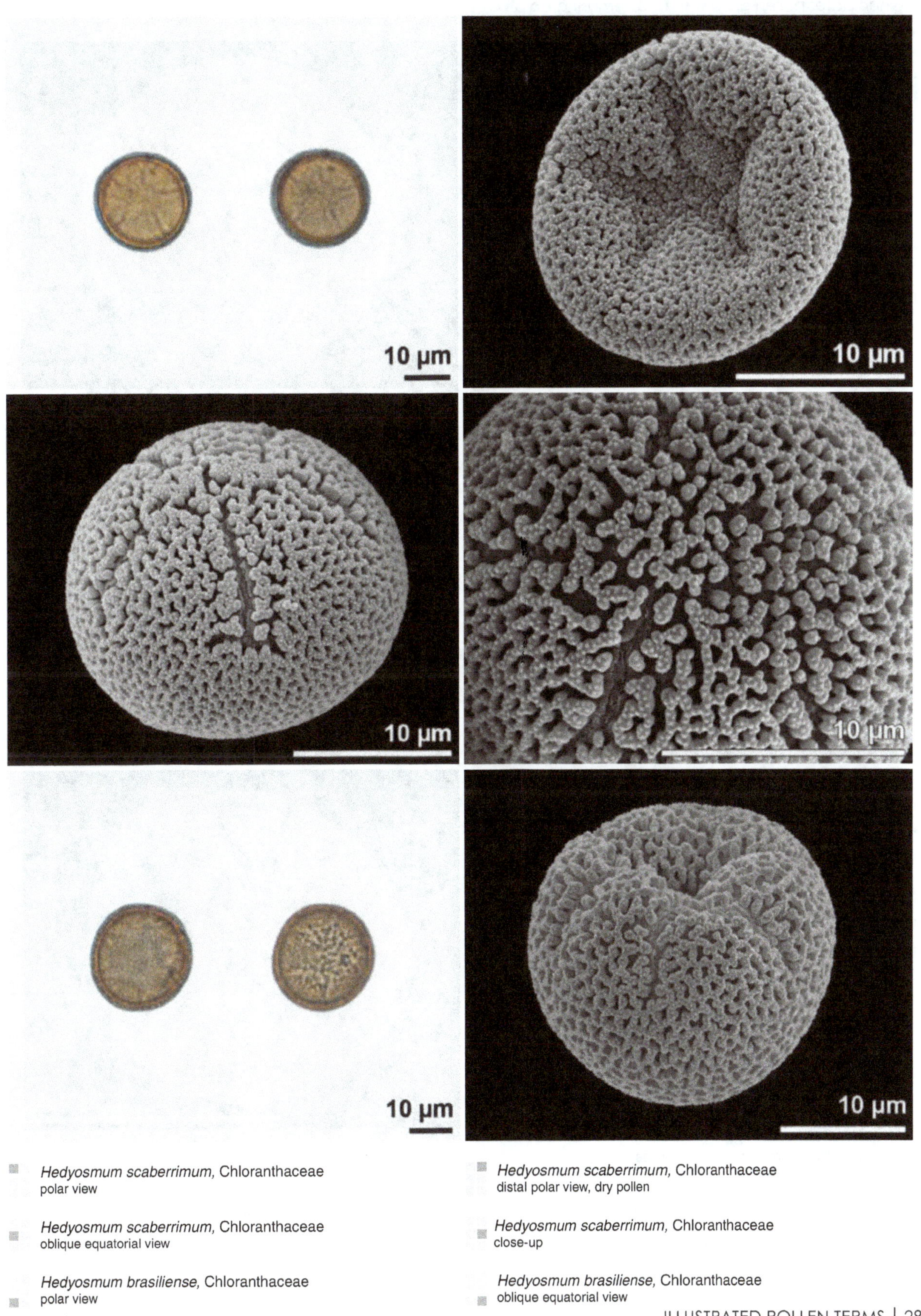

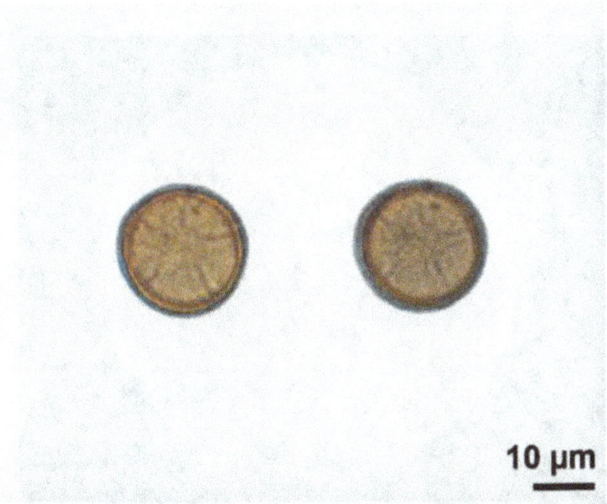

Hedyosmum scaberrimum, Chloranthaceae
polar view

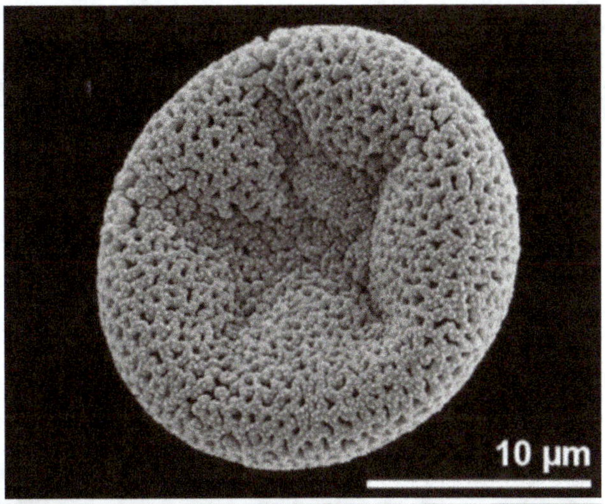

Hedyosmum scaberrimum, Chloranthaceae
distal polar view, dry pollen

Hedyosmum scaberrimum, Chloranthaceae
oblique equatorial view

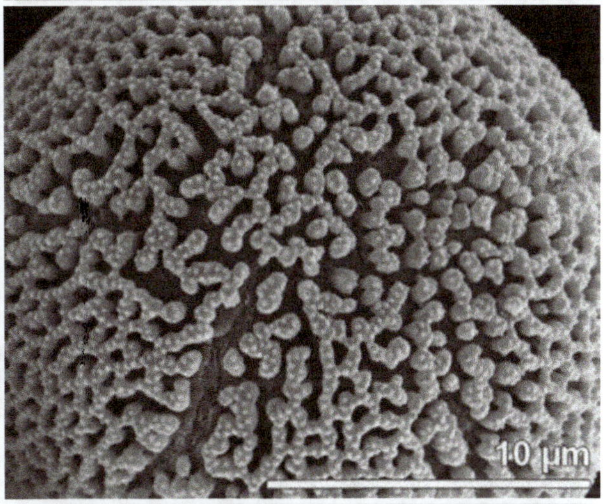

Hedyosmum scaberrimum, Chloranthaceae
close-up

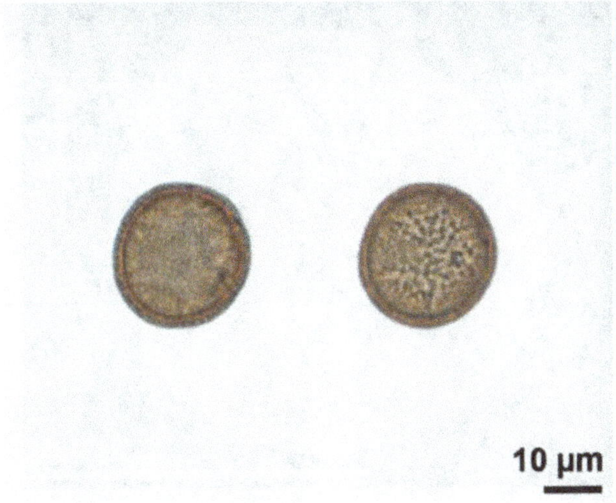

Hedyosmum brasiliense, Chloranthaceae
polar view

Hedyosmum brasiliense, Chloranthaceae
oblique equatorial view

synaperturate, syncolpate, syncolporate

synaperturate: pollen grain with anastomosing apertures
syncolpate: pollen grain with anastomosing colpi

Aetanthus macranthus, Loranthaceae
polar and equatorial view

Pedicularis verticillata, Orobanchaceae
dry pollen

Berberis nervosa, Berberidaceae
dry pollen

Pseudofumaria lutea, Papaveraceae

Barringtonia asiatica, Lecythidaceae
dry pollen

Dodecatheon meadia, Primulaceae
oblique equatorial view

synaperturate, syncolporate

syncolporate: pollen grain with anastomosing colpori

Callistemon coccineus, Myrtaceae
equatorial view

Cassia pulcherrima, Fabaceae

Ardisia crenata, Primulaceae
polar view

Cuphea procumbens, Lythraceae
polar view

Myrtus communis, Myrtaceae
polar view

Onosma visianii, Boraginaceae
dry pollen

ulcus/ulcerate

more or less circular aperture located distally

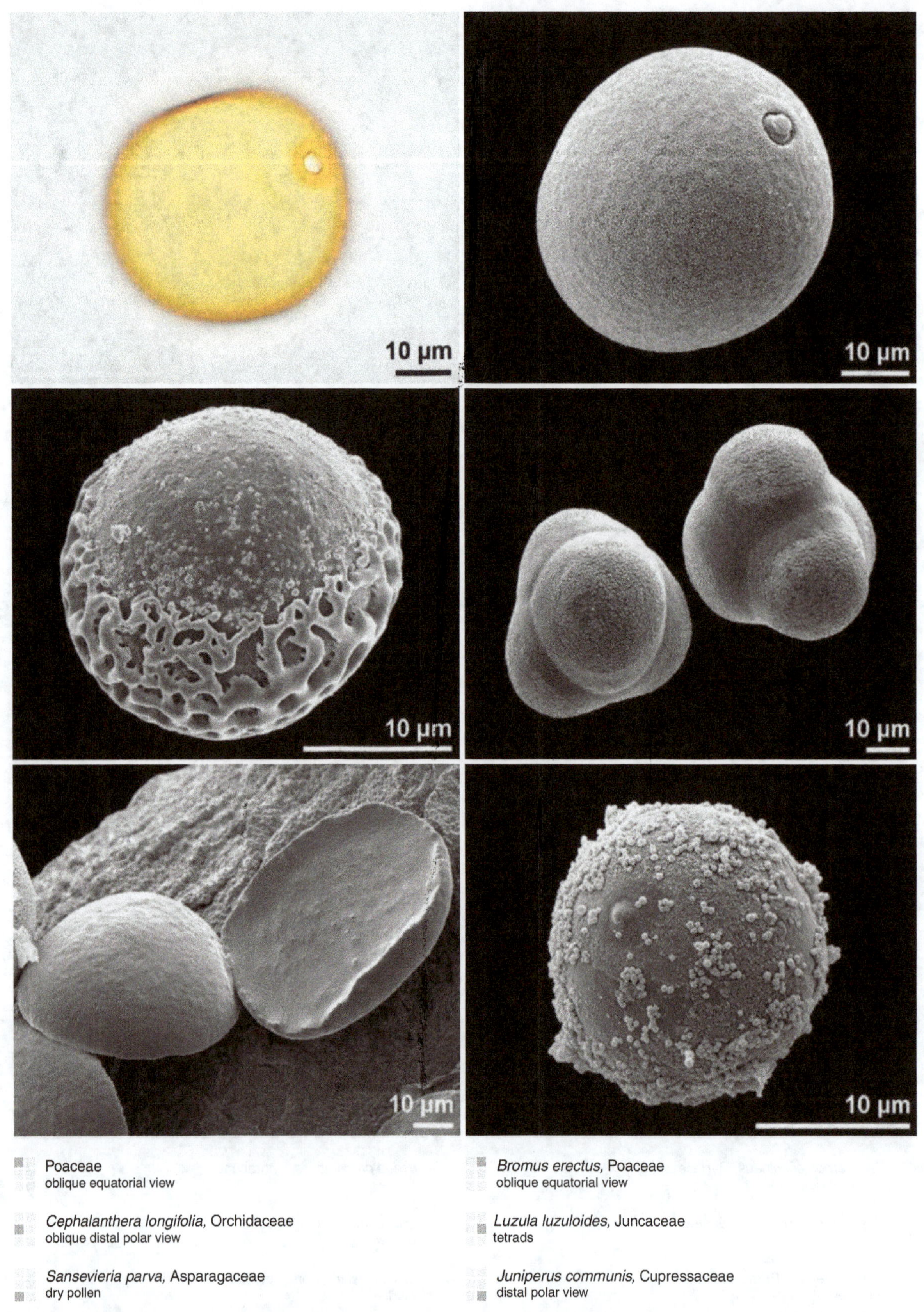

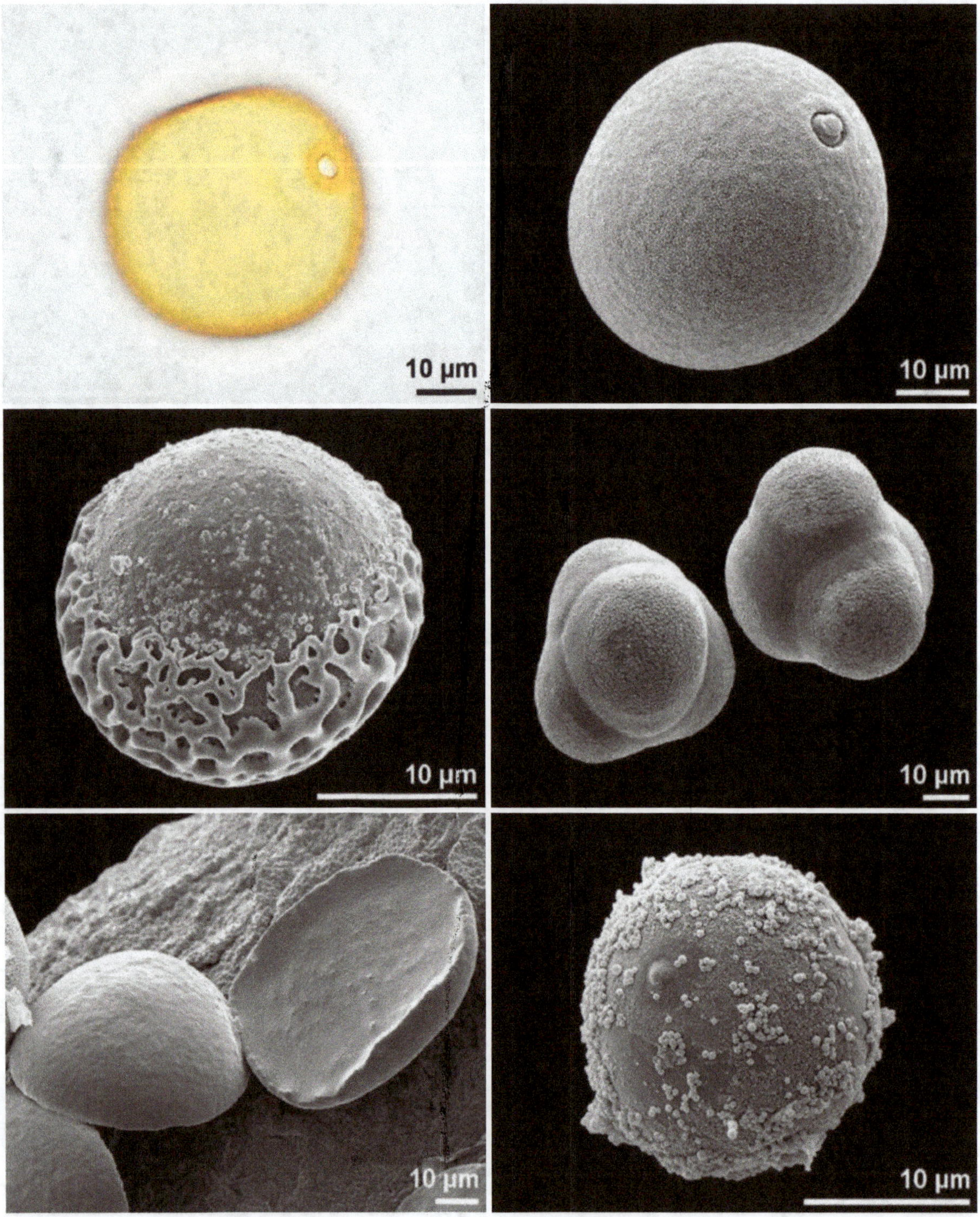

Poaceae
oblique equatorial view

Bromus erectus, Poaceae
oblique equatorial view

Cephalanthera longifolia, Orchidaceae
oblique distal polar view

Luzula luzuloides, Juncaceae
tetrads

Sansevieria parva, Asparagaceae
dry pollen

Juniperus communis, Cupressaceae
distal polar view

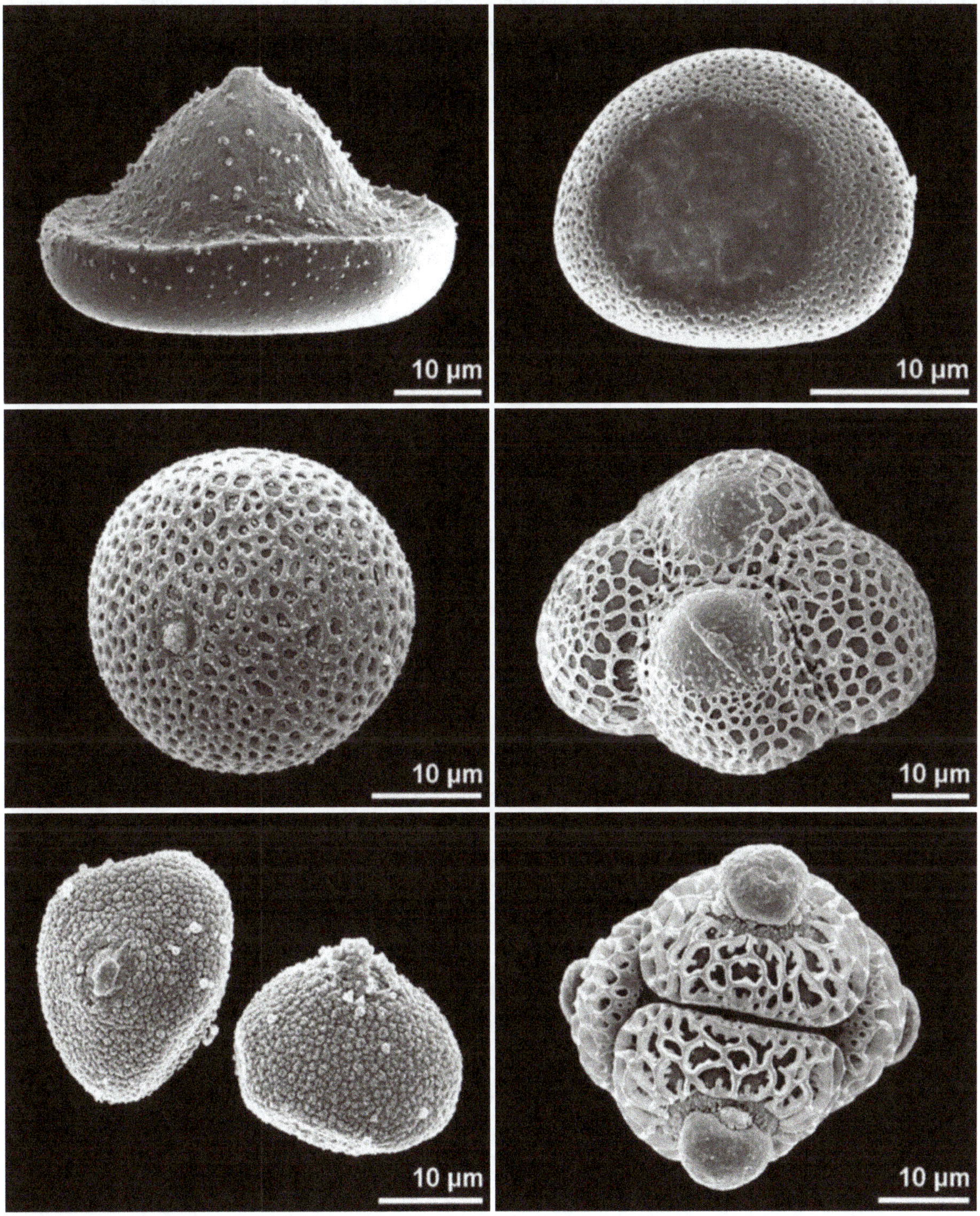

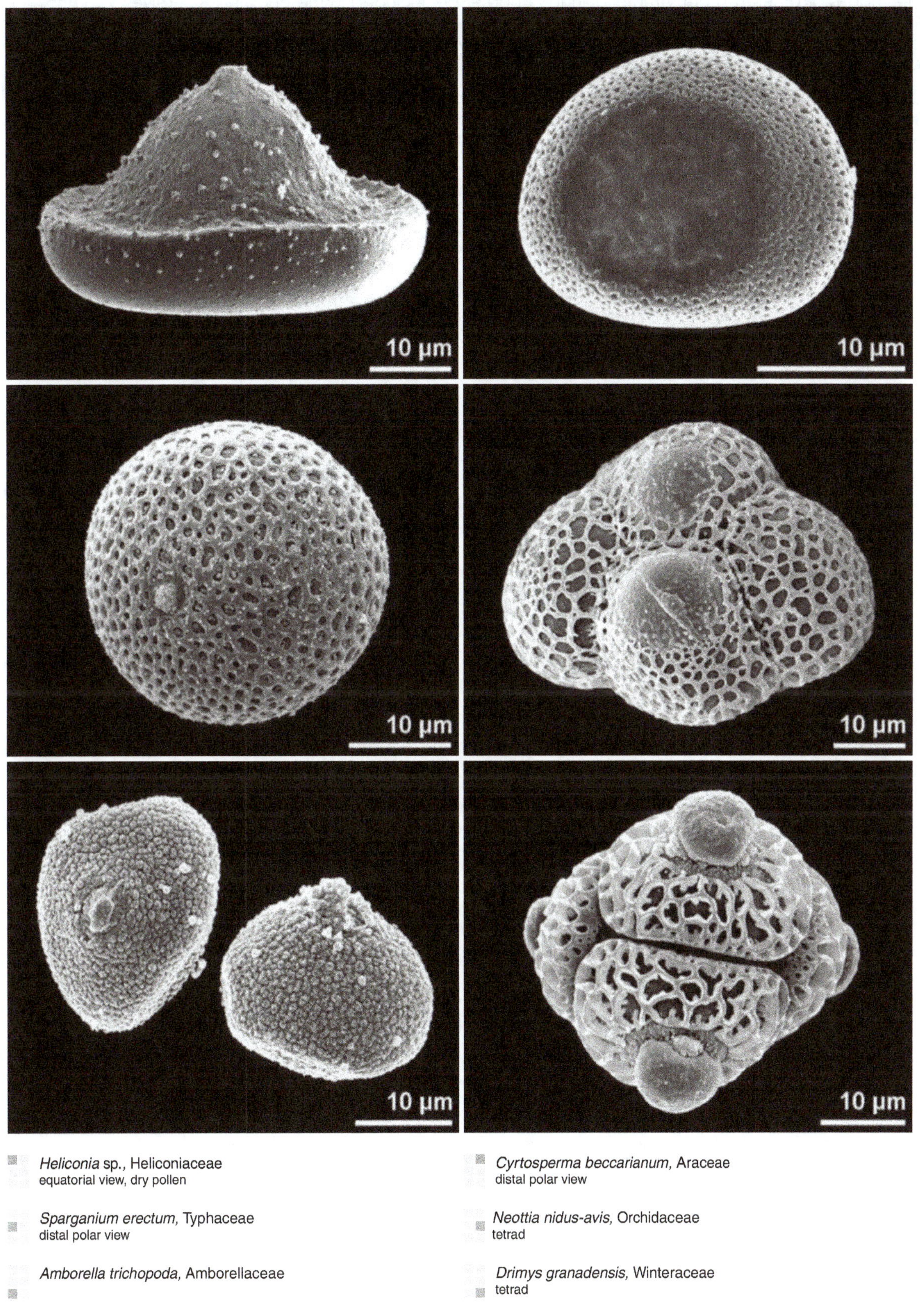

Heliconia sp., Heliconiaceae
equatorial view, dry pollen

Cyrtosperma beccarianum, Araceae
distal polar view

Sparganium erectum, Typhaceae
distal polar view

Neottia nidus-avis, Orchidaceae
tetrad

Amborella trichopoda, Amborellaceae

Drimys granadensis, Winteraceae
tetrad

Ornamentation

H. Halbritter et al., *Illustrated Pollen Terminology*, https://doi.org/10.1007/978-3-319-71365-6_10

areola/areolate

insular ornamentation element

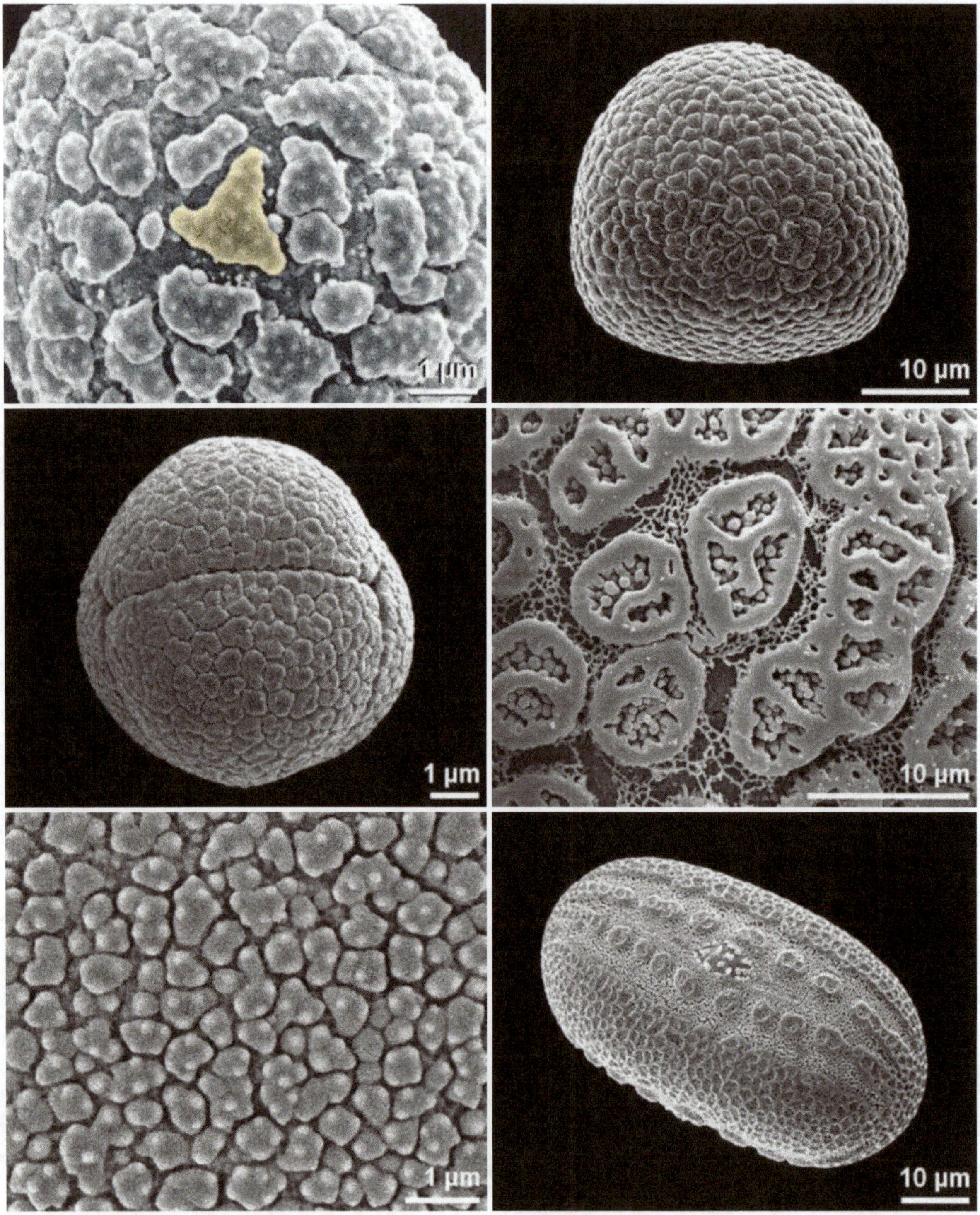

Peperomia rubella, Piperaceae
areola colored

Mimosa pudica, Mimosaceae
tetrad

Cynodon dactylon, Poaceae

Dracunculus vulgaris, Araceae
inaperturate

Poikilacanthus macranthus, Acanthaceae
areolae reticulate

Justicia brandegeeana, Acanthaceae
areolae in aperture area, oblique equatorial view

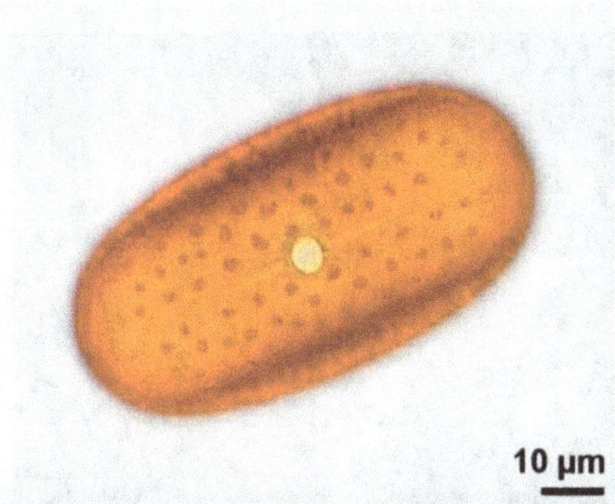

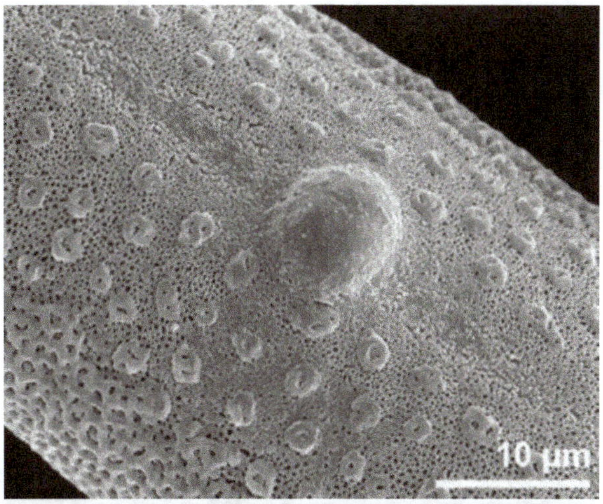

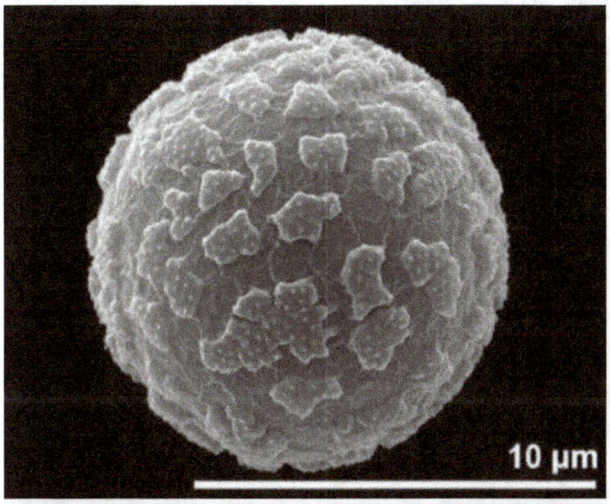

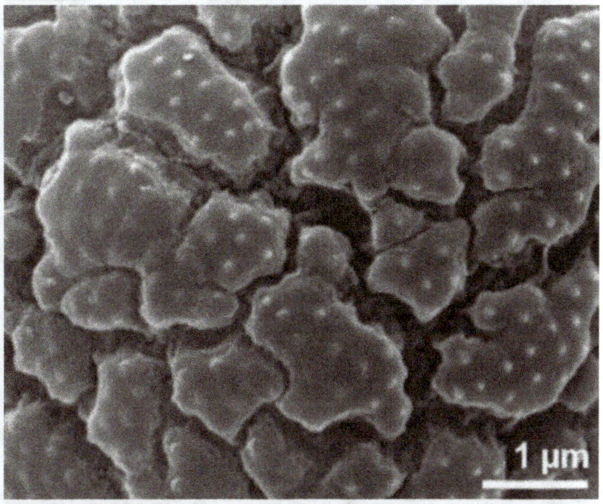

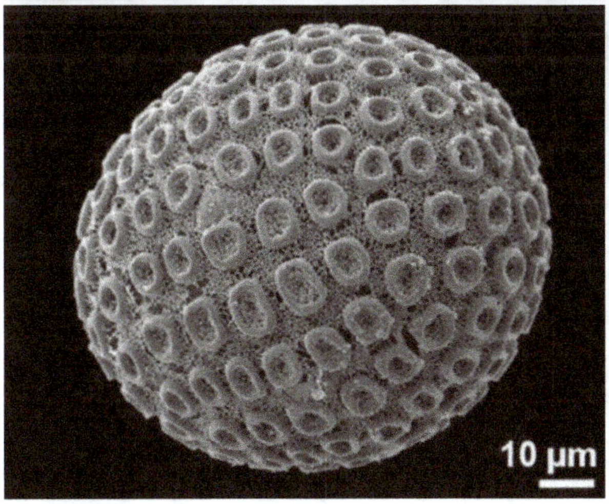

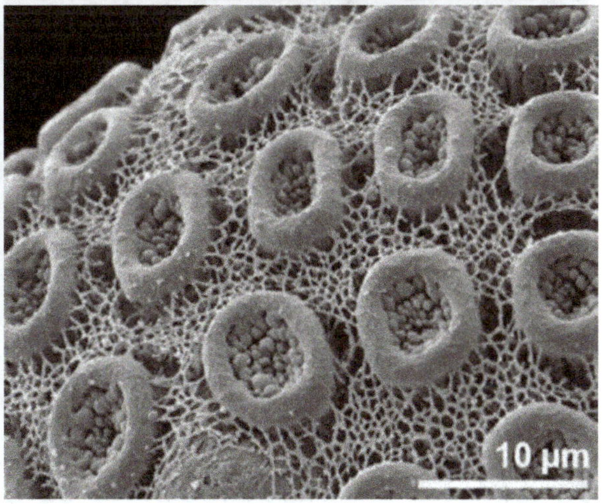

Justicia carnea, Acanthaceae

Justicia carnea, Acanthaceae
aperture area

Peperomia polybotrya, Piperaceae

Peperomia polybotrya, Piperaceae

Megaskepasma erythrochlamys, Acanthaceae
equatorial view

Megaskepasma erythrochlamys, Acanthaceae

baculum/baculate

rod-like, free standing element (never pointed)

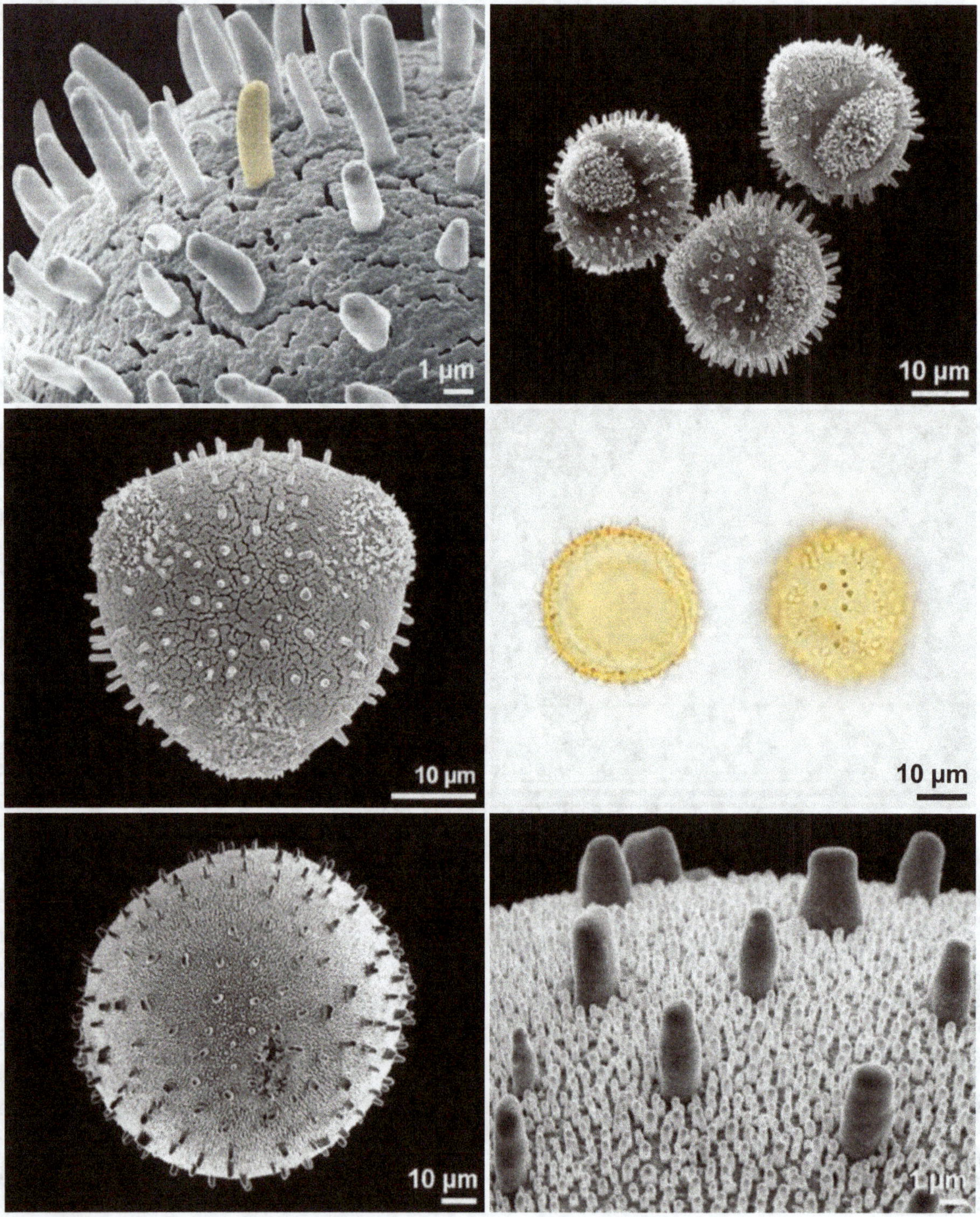

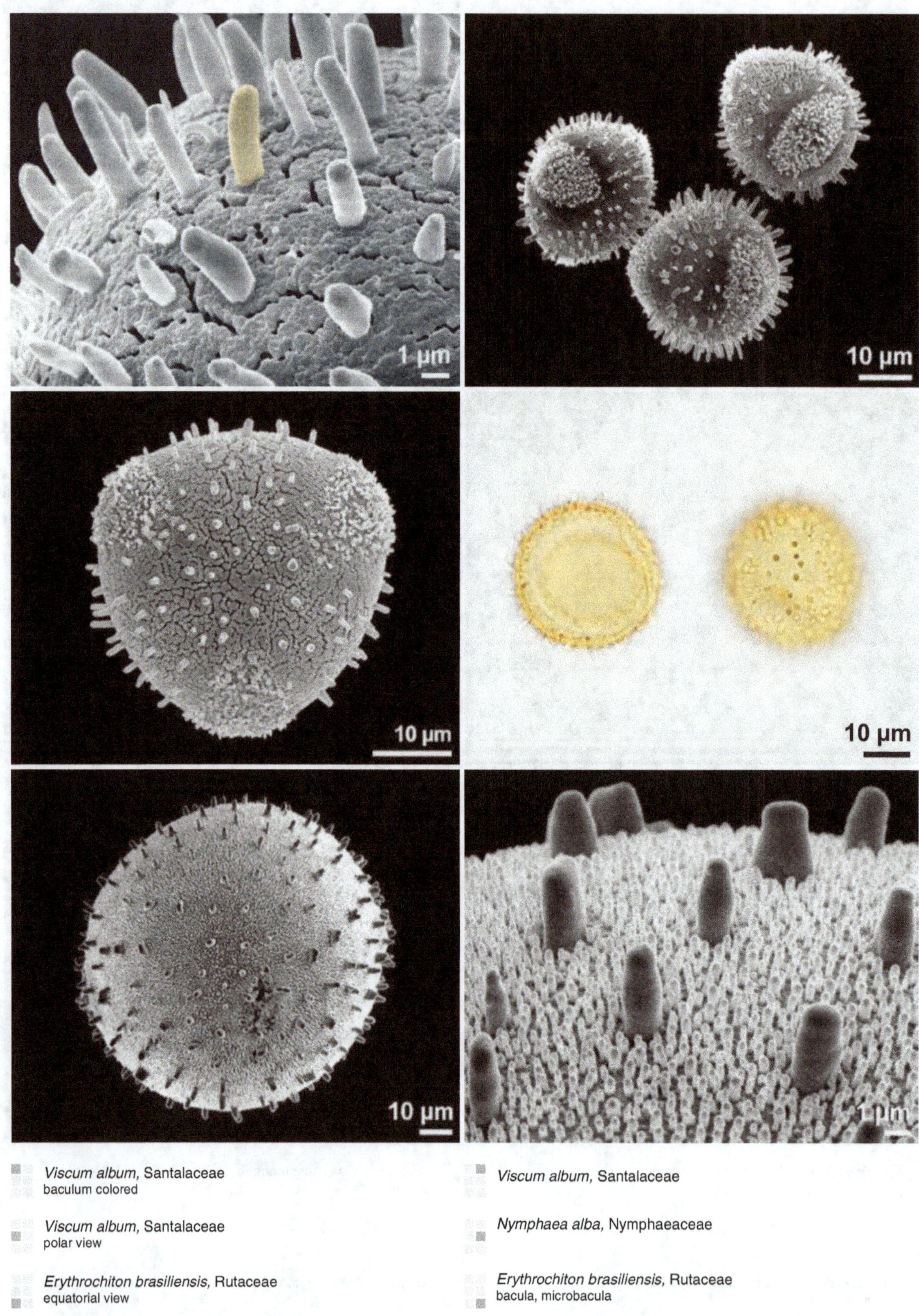

Viscum album, Santalaceae
baculum colored

Viscum album, Santalaceae
polar view

Erythrochiton brasiliensis, Rutaceae
equatorial view

Viscum album, Santalaceae

Nymphaea alba, Nymphaeaceae

Erythrochiton brasiliensis, Rutaceae
bacula, microbacula

bireticulate

reticulate ornamentation, where the lumina of the coarse-meshed reticulum are filled by a fine-meshed reticulum

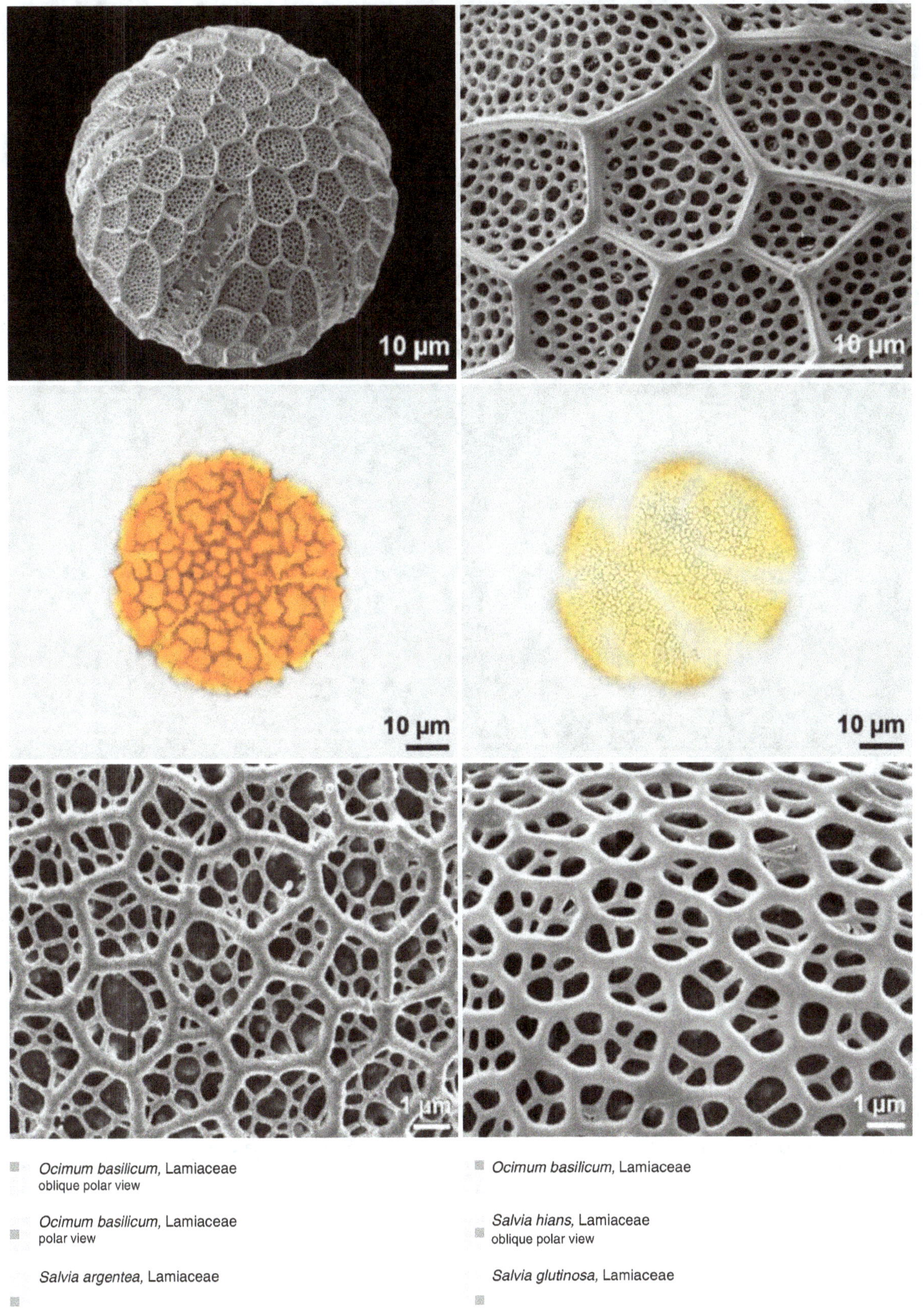

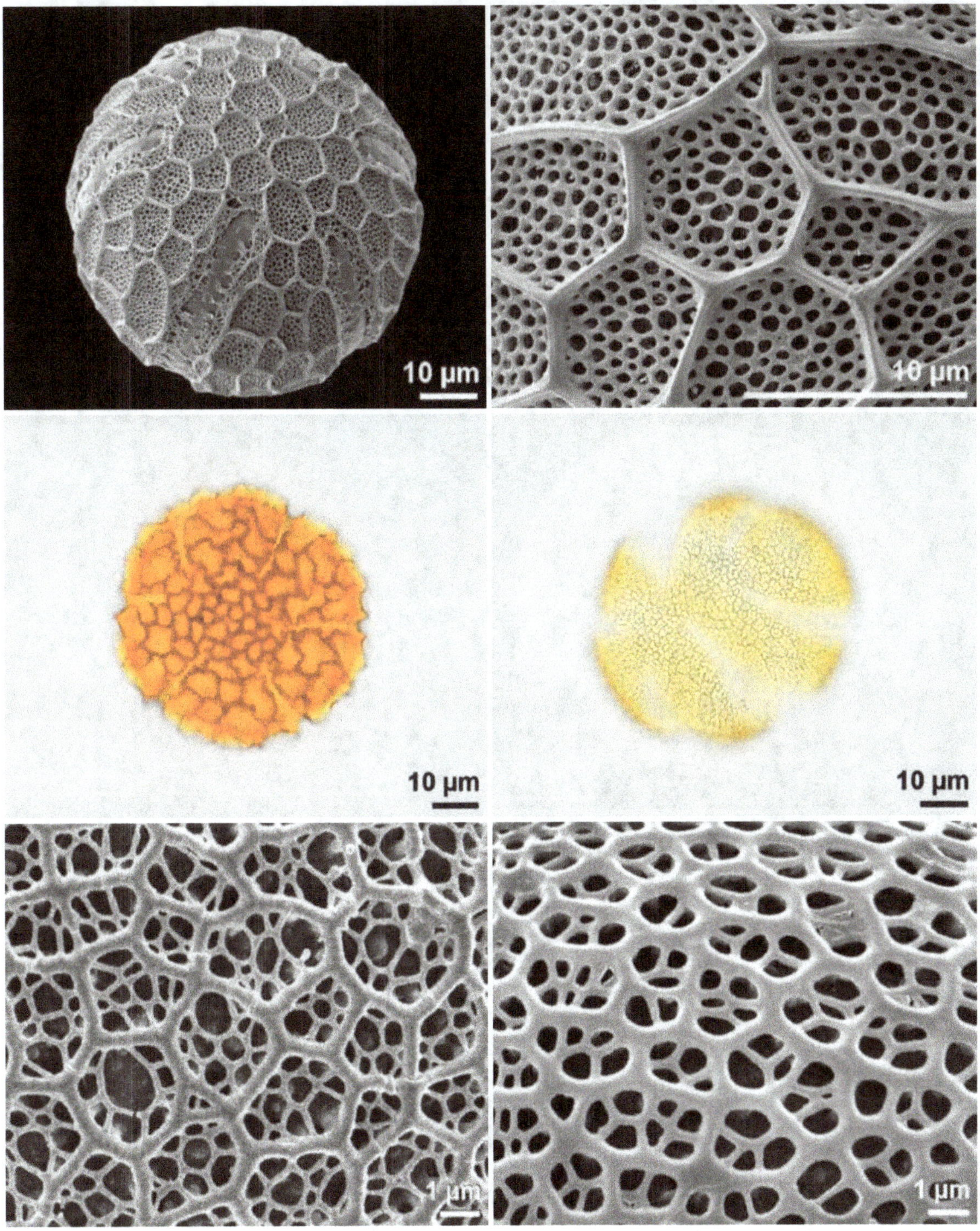

Ocimum basilicum, Lamiaceae
oblique polar view

Ocimum basilicum, Lamiaceae

Ocimum basilicum, Lamiaceae
polar view

Salvia hians, Lamiaceae
oblique polar view

Salvia argentea, Lamiaceae

Salvia glutinosa, Lamiaceae

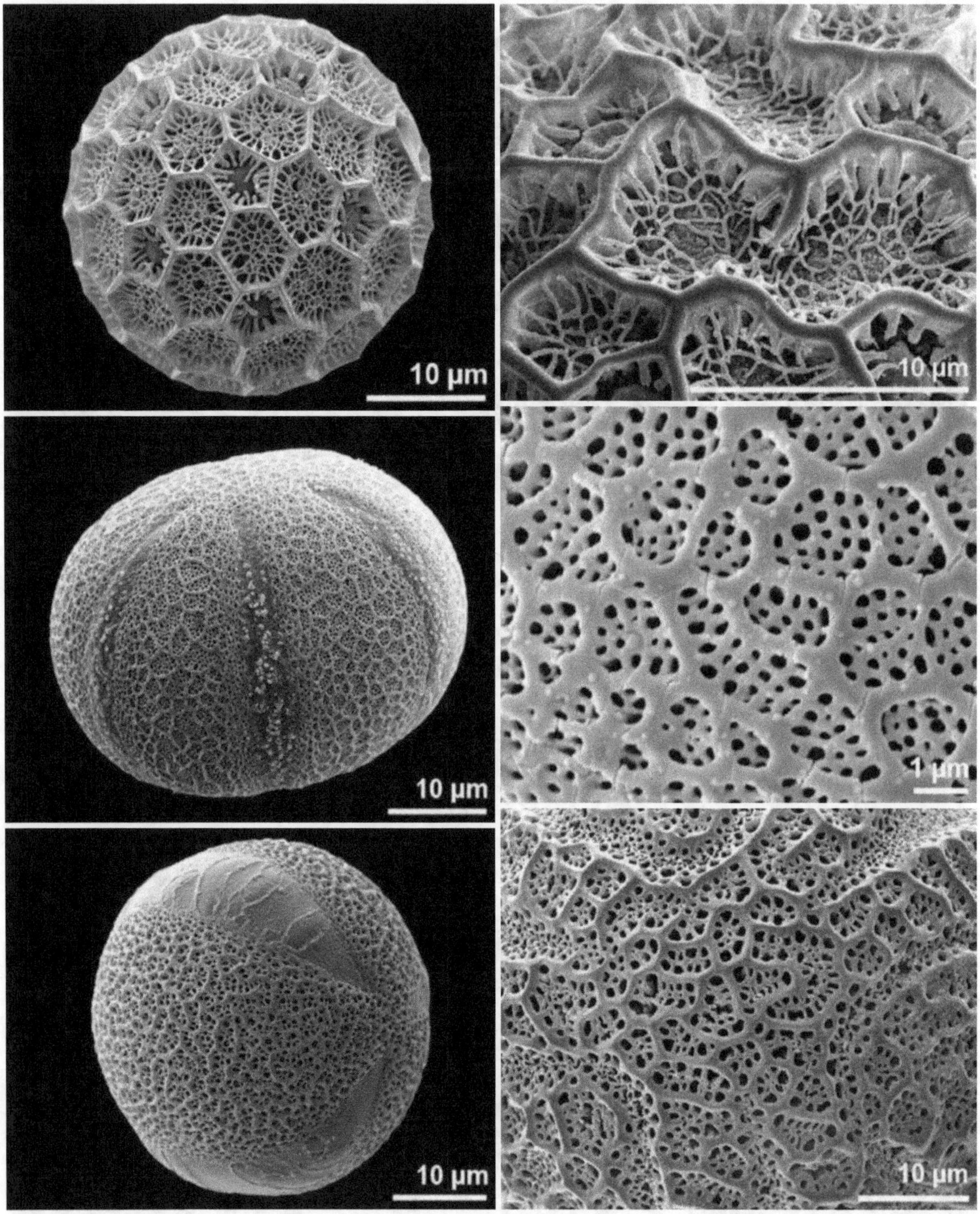

Phlox drummondii, Polemoniaceae
pantoporate

Agastache mexicana, Lamiaceae
oblique equatorial view

Vitex trifolia, Lamiaceae
oblique view

Phlox paniculata, Polemoniaceae

Agastache mexicana, Lamiaceae

Pachystachys lutea, Acanthaceae
polar area

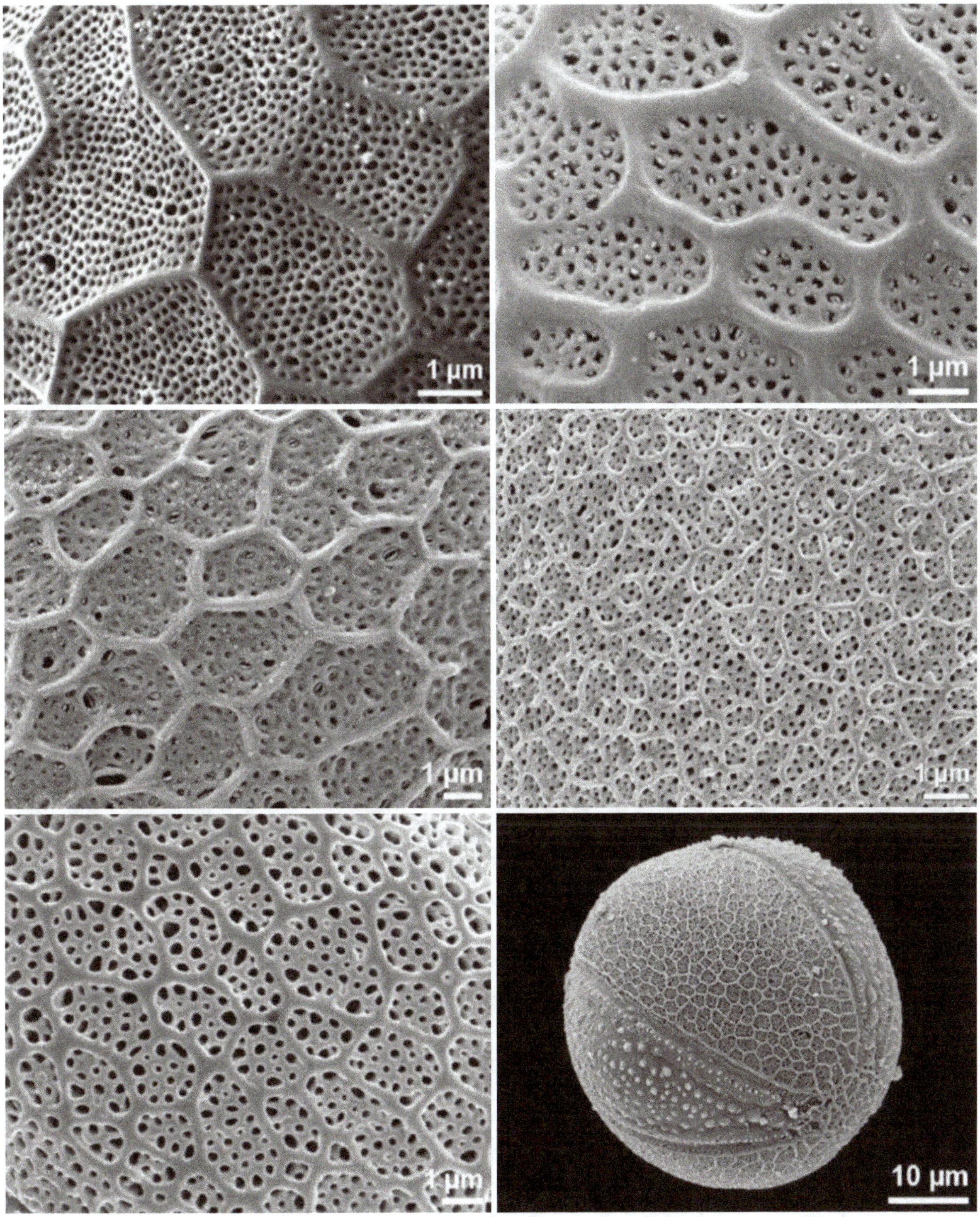

- *Plectranthus ornatus*, Lamiaceae
- *Sparmannia africana*, Malvaceae
- *Prunella grandiflora*, Lamiaceae
- *Hyptis suaveolens*, Lamiaceae
- *Melittis melissophyllum*, Lamiaceae
- *Andrographis paniculata*, Acanthaceae
oblique view

clava/clavate, caput

clava: club-shaped element
caput: distal part of clava

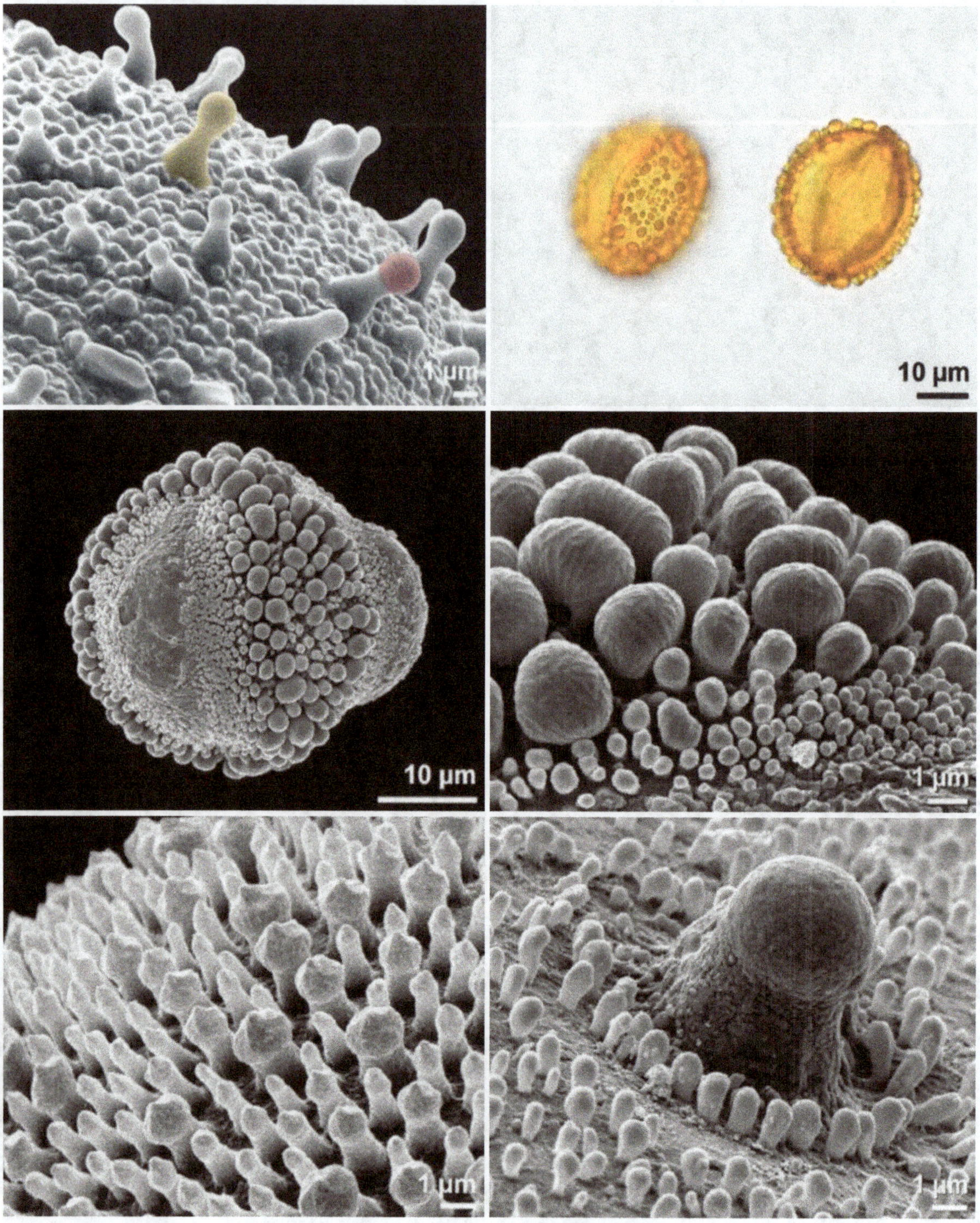

Iris planifolia, Iridaceae
clava (yellow), caput (red)

Ilex aquifolium, Aquifoliaceae
equatorial view

Linum flavum, Linaceae

Ilex sp., Aquifoliaceae
equatorial view

Ilex aquifolium, Aquifoliaceae

Aratitiyopea lopezii, Xyridaceae
clavae of different size

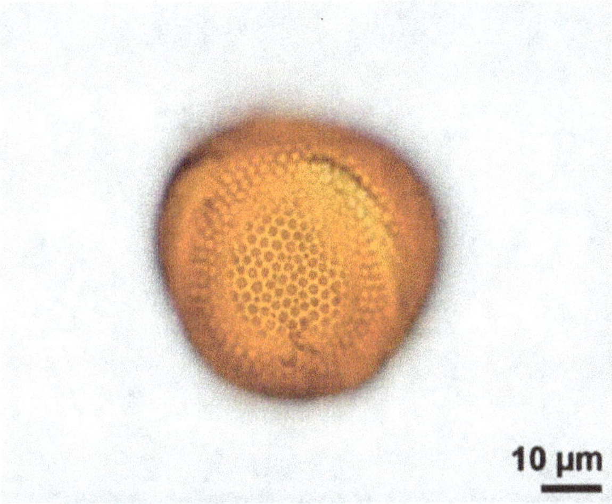

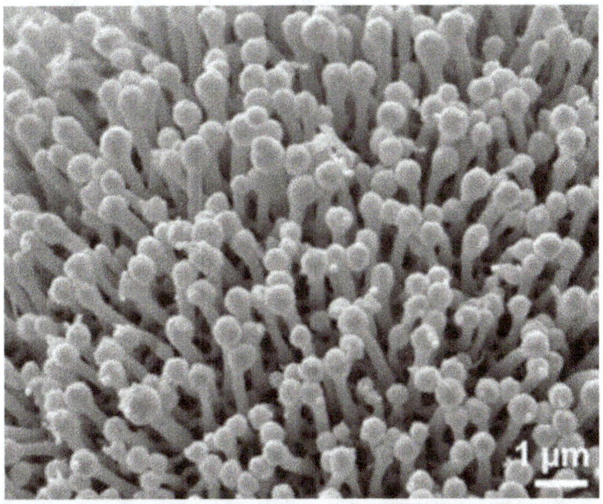

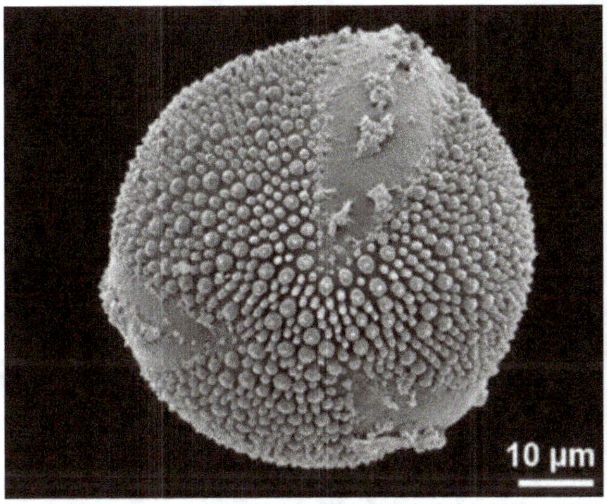

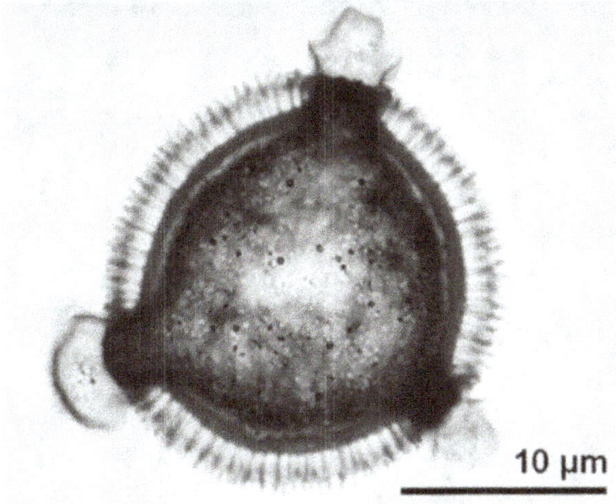

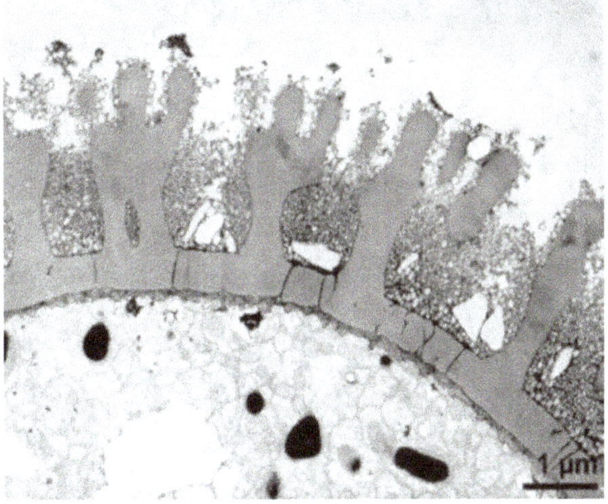

Linum sp., Linaceae

Linum capitatum, Linaceae
polar view

Geranium robertianum, Geraniaceae

Hymenocallis tubiflora, Amaryllidaceae

Linum capitatum, Linaceae

Geranium robertianum, Geraniaceae
cross section of pollen wall

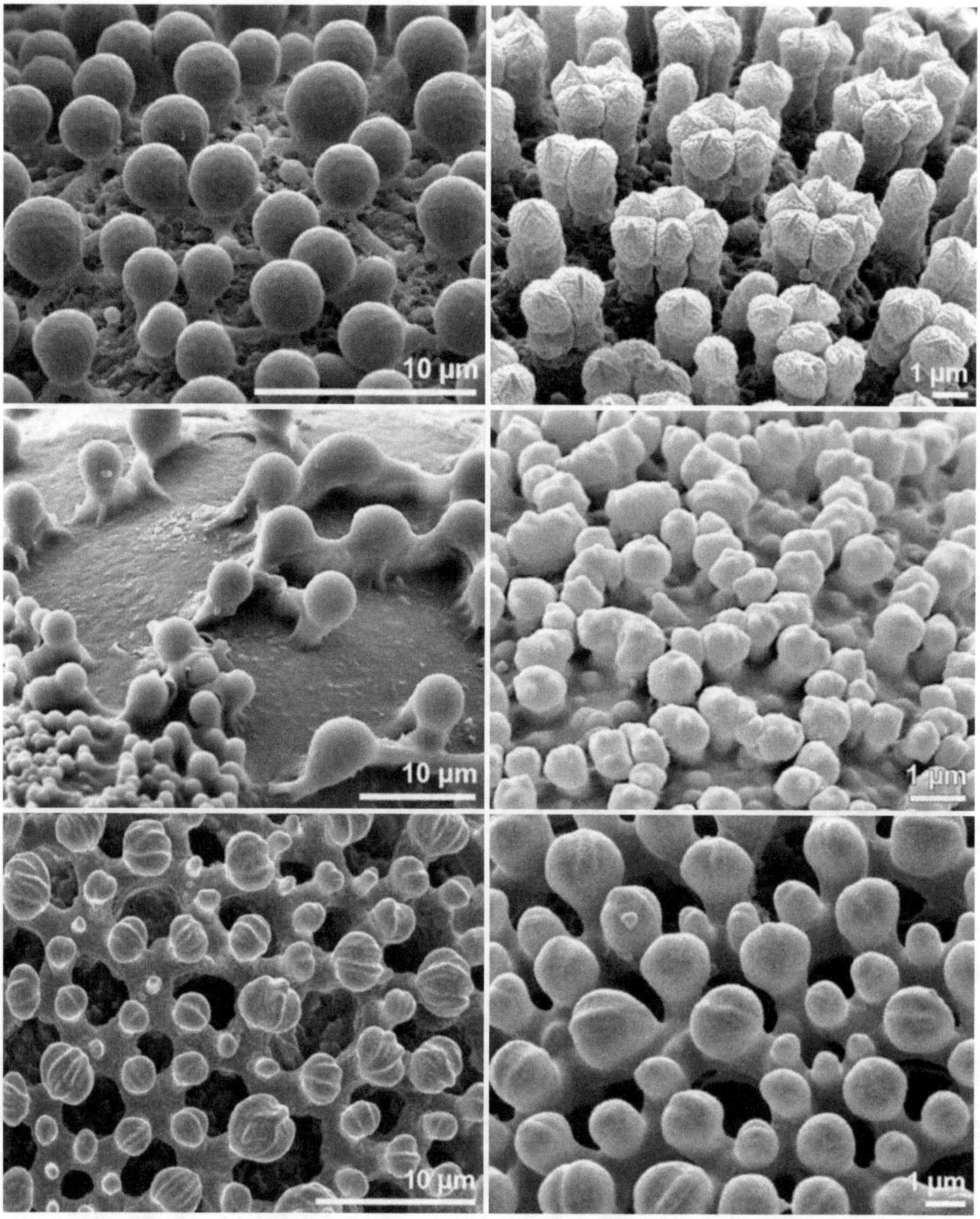

Jatropha multifida, Euphorbiaceae

Plumbago auriculata, Plumbaginaceae

Hymenocallis littoralis, Amaryllidaceae

Viburnum lantana, Adoxaceae

Geranium pratense, Geraniaceae
reticulum cristatum with clavae

Geranium sibiricum, Geraniaceae
reticulum cristatum with clavae

clypeate

pollen with exine subdivided into shields

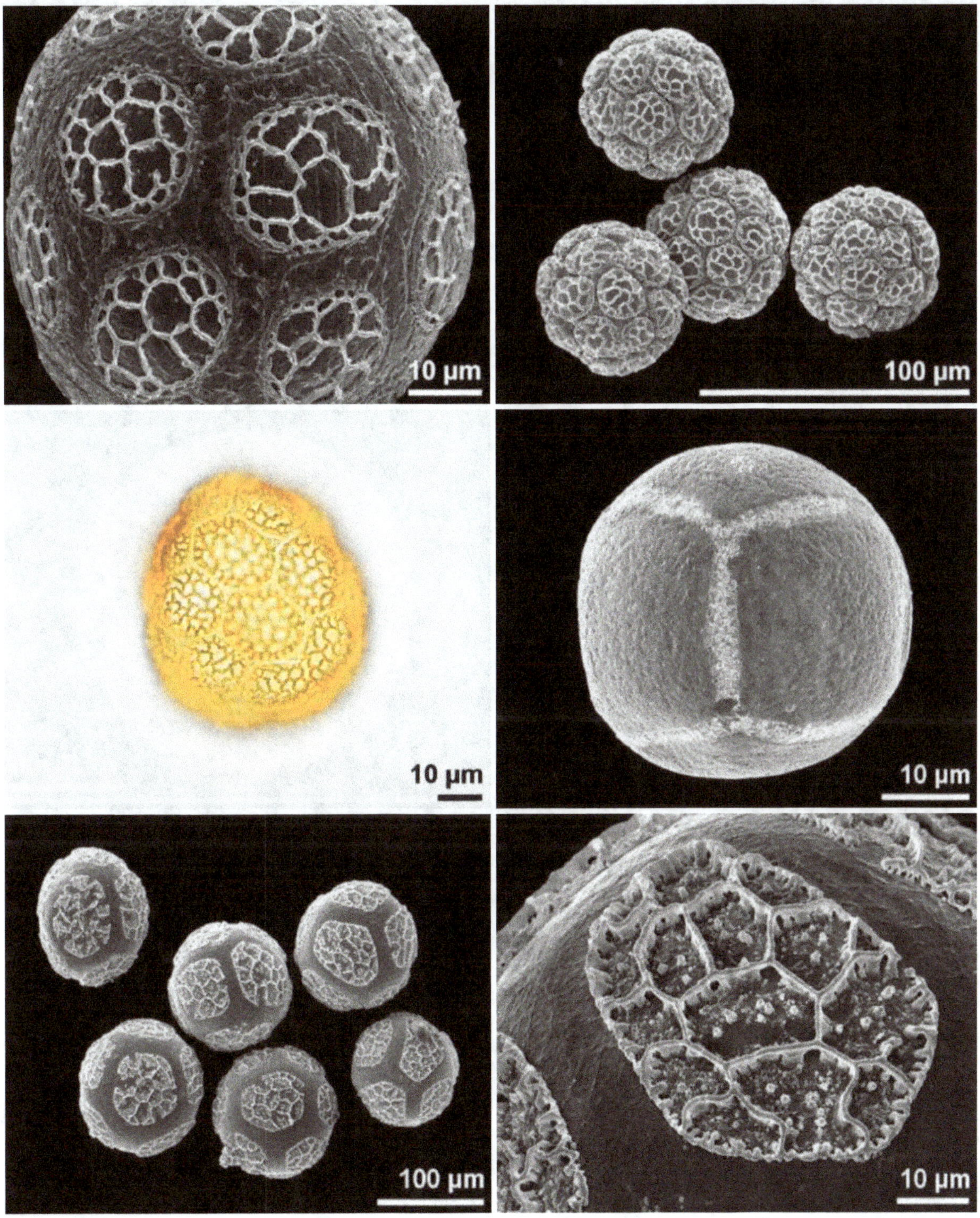

Ibicella lutea, Martyniaceae
inaperturate

Ibicella lutea, Martyniaceae
inaperturate, dry pollen

Proboscidea fragrans, Martyniaceae
inaperturate

Pseudofumaria lutea, Papaveraceae
pantocolpate, syncolpate

Iris bucharica, Iridaceae
inaperturate

Iris bucharica, Iridaceae
inaperturate

Catalpa bignonioides, Bignoniaceae
tetrads, inaperturate, dry pollen

Lophophora williamsii, Cactaceae
pantocolpate, dry pollen

Irlbachia pedunculata, Gentianaceae
tetrads, porate

Iris graeberiana, Iridaceae
inaperturate

Phyllanthus sp., Euphorbiaceae
pantoporate

Banisteria muricata, Malpighiaceae
pantocolporate

croton pattern

special type of reticulum cristatum formed by regularly arranged suprasculpture elements on muri

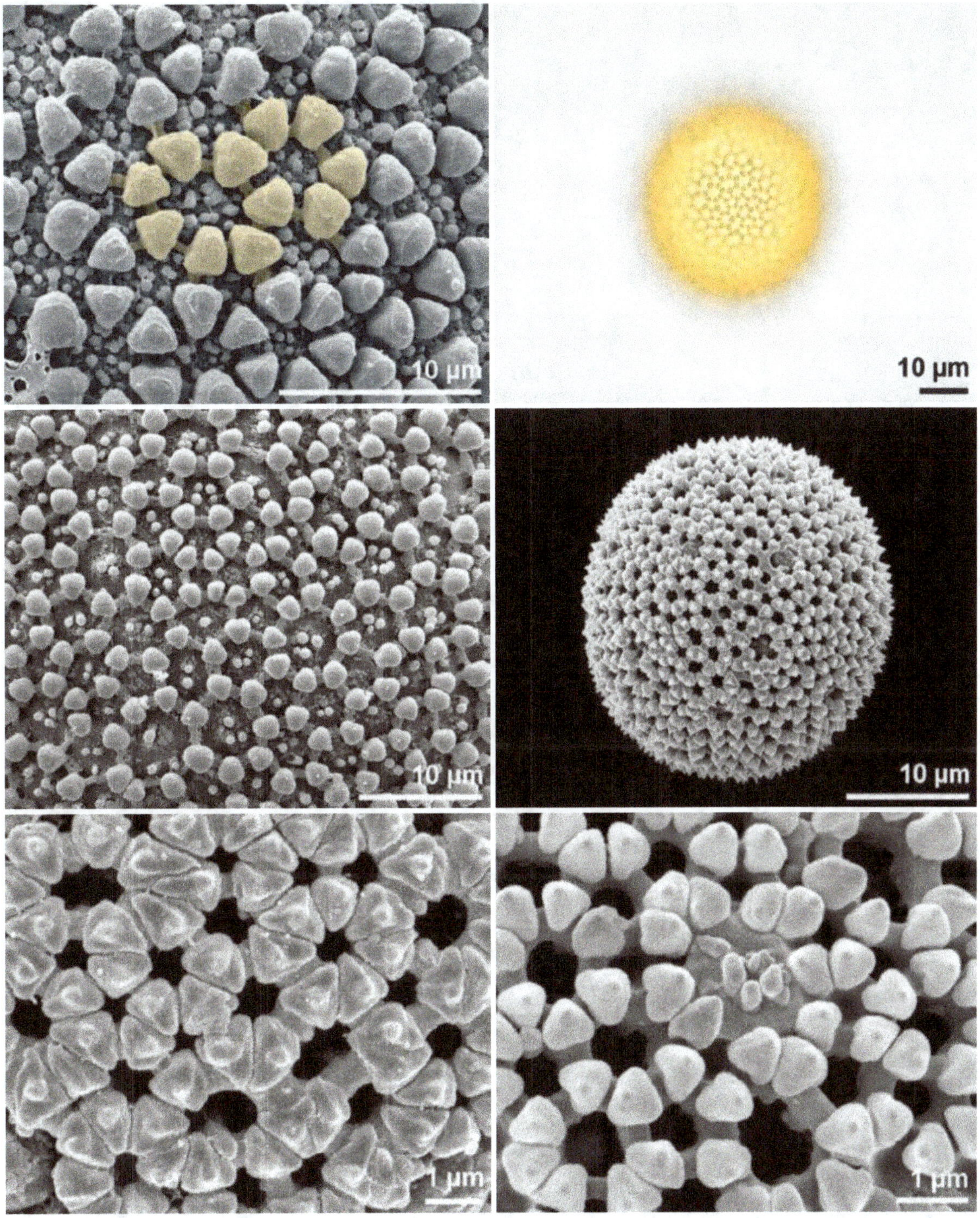

Croton triqueter, Euphorbiaceae
pattern colored

Croton triqueter, Euphorbiaceae

Jatropha podagrica, Euphorbiaceae

Daphne laureola, Thymelaeaceae

Daphne cneorum, Thymelaeaceae

Thymelaea passerina, Thymelaeaceae
surface detail with porus

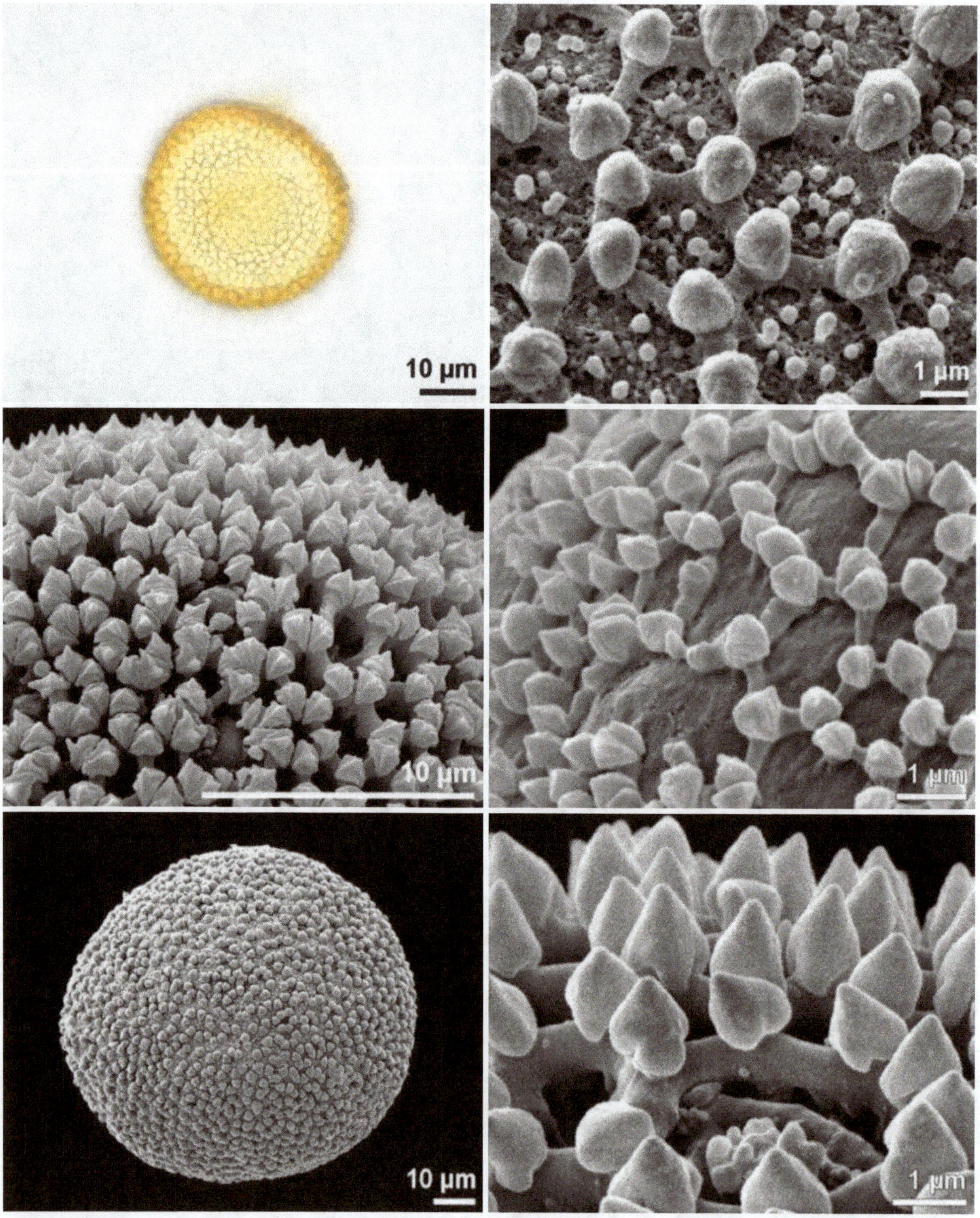

Garcia nutans, Euphorbiaceae

Daphne tangutica, Thymelaeaceae

Croton triqueter, Euphorbiaceae

Garcia nutans, Euphorbiaceae

Callitriche stagnalis, Plantaginaceae

Thymelaea passerina, Thymelaeaceae

echinus/echinate

pointed ornamentation element

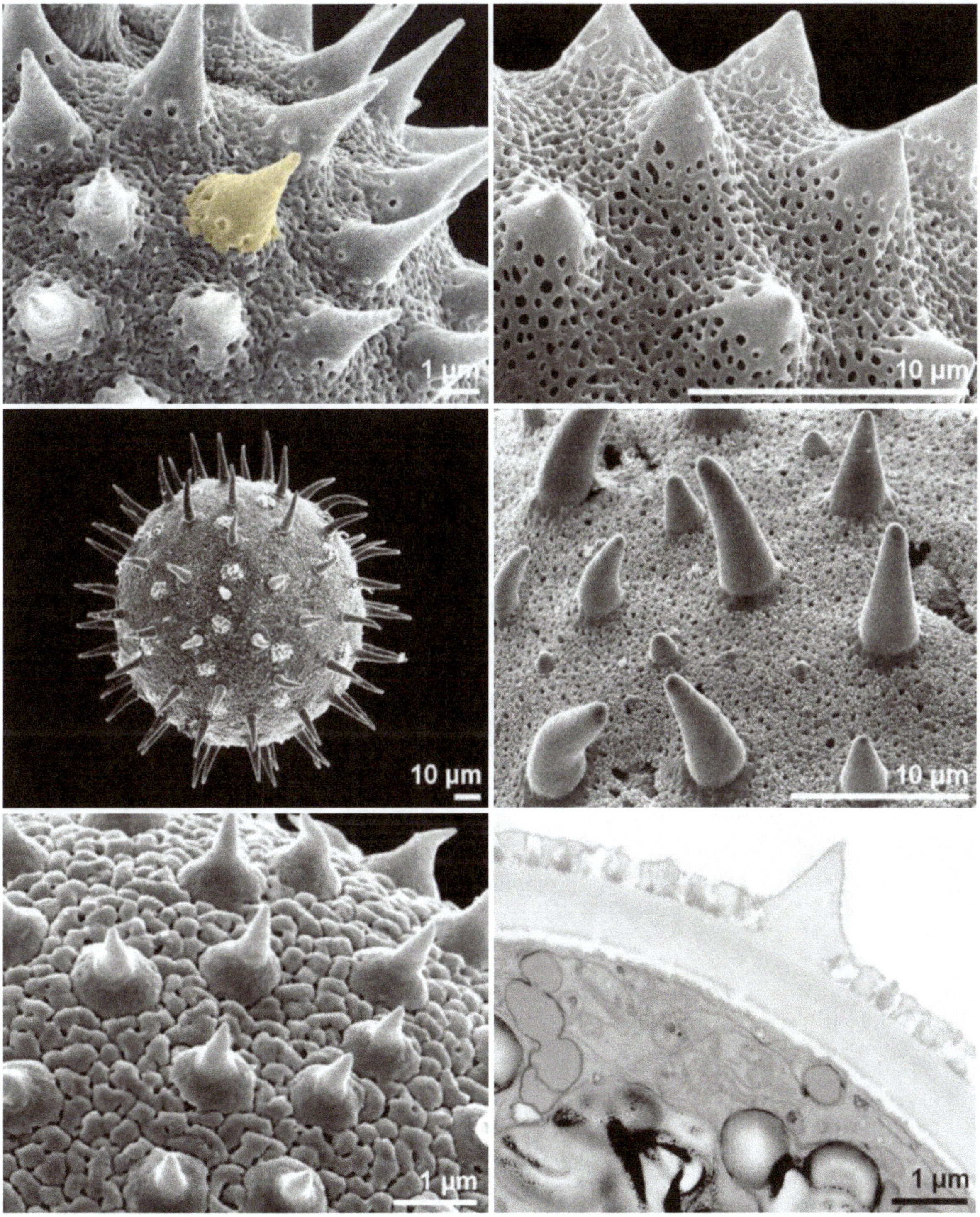

Galinsoga ciliata, Asteraceae
echinus colored

Carduus acanthoides, Asteraceae

Hibiscus trionum, Malvaceae
pantoporate

Lavatera thuringiaca, Malvaceae
pantoporate

Pinellia ternata, Araceae

Pinellia ternata, Araceae
cross section of pollen wall

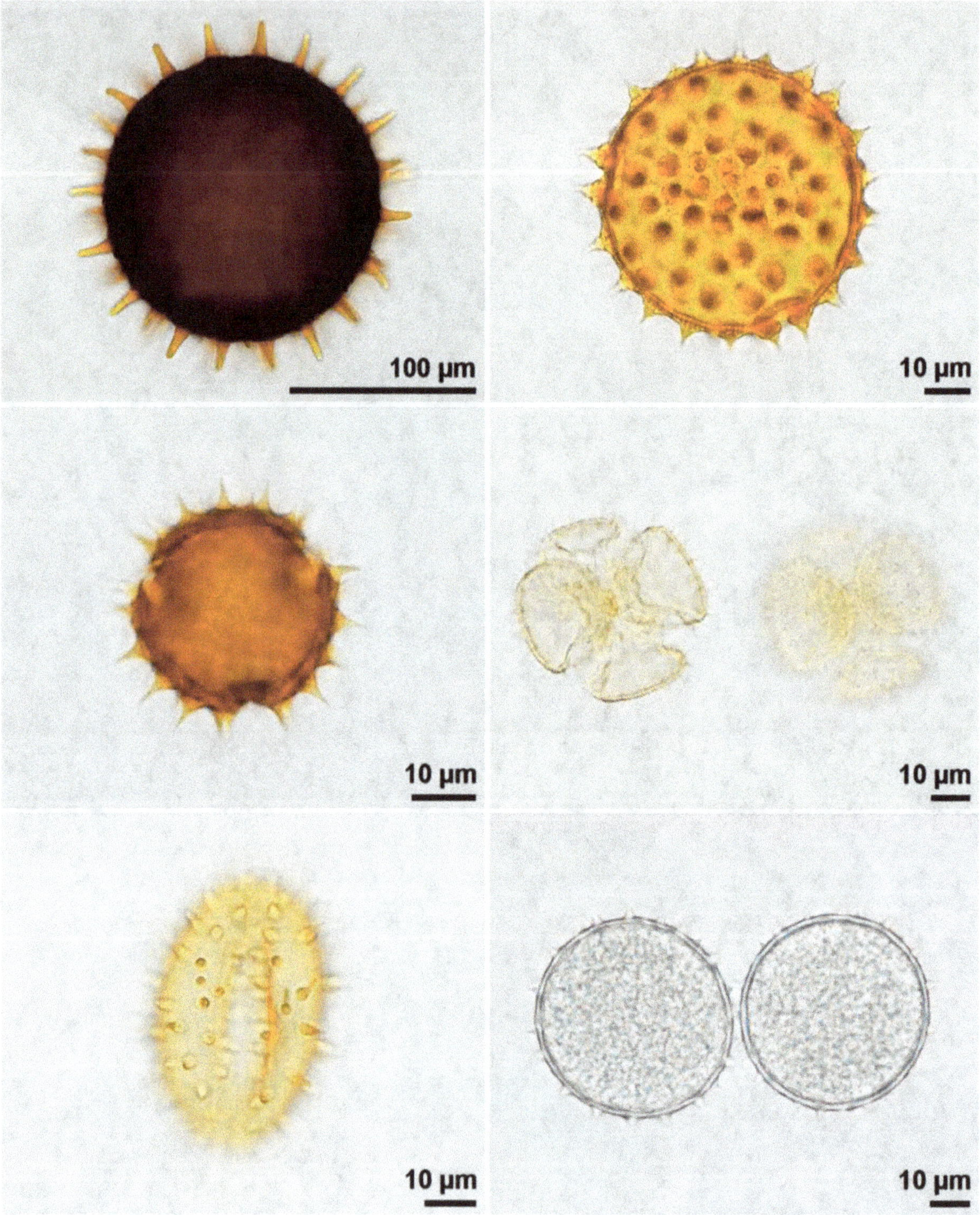

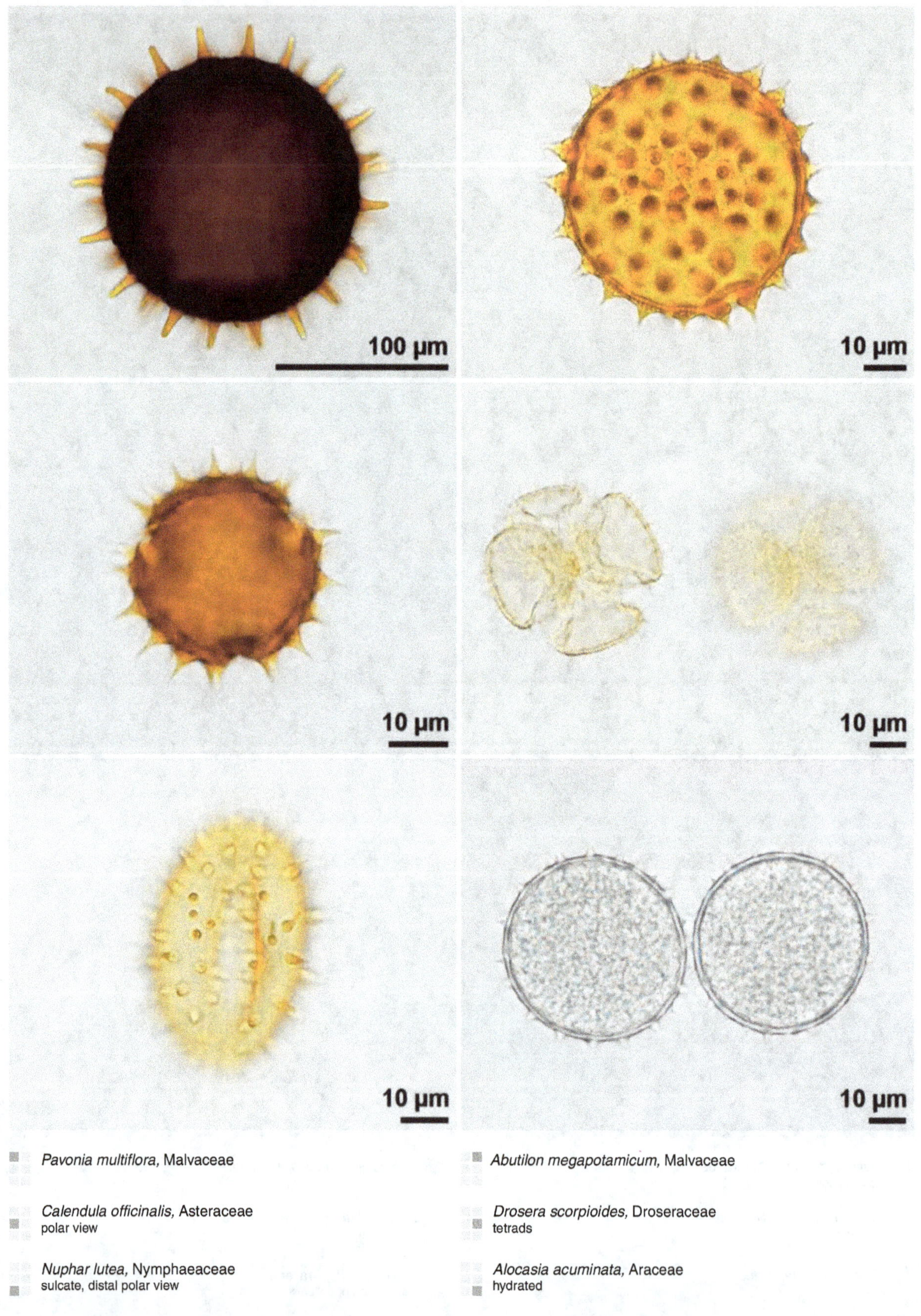

Pavonia multiflora, Malvaceae

Abutilon megapotamicum, Malvaceae

Calendula officinalis, Asteraceae
polar view

Drosera scorpioides, Droseraceae
tetrads

Nuphar lutea, Nymphaeaceae
sulcate, distal polar view

Alocasia acuminata, Araceae
hydrated

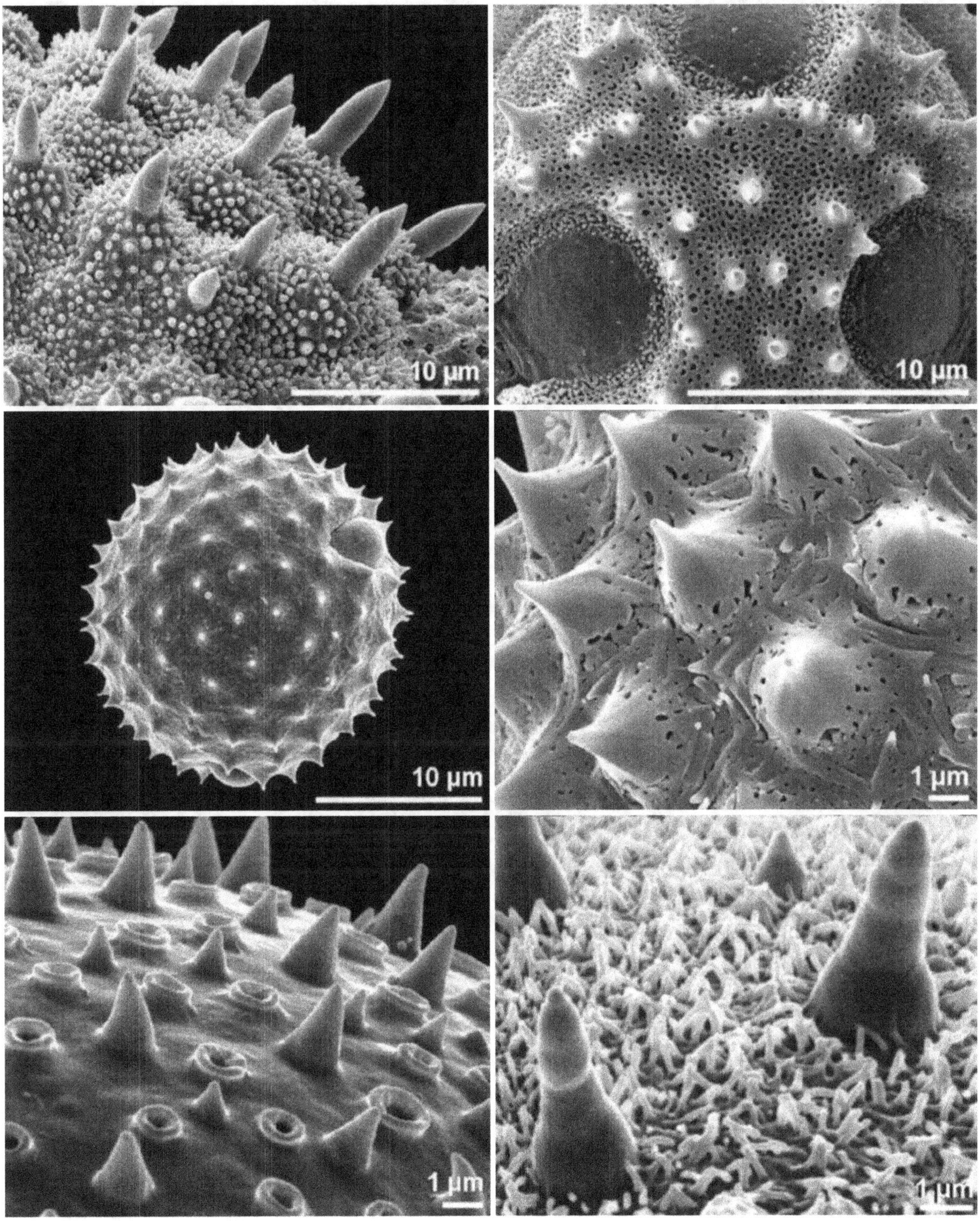

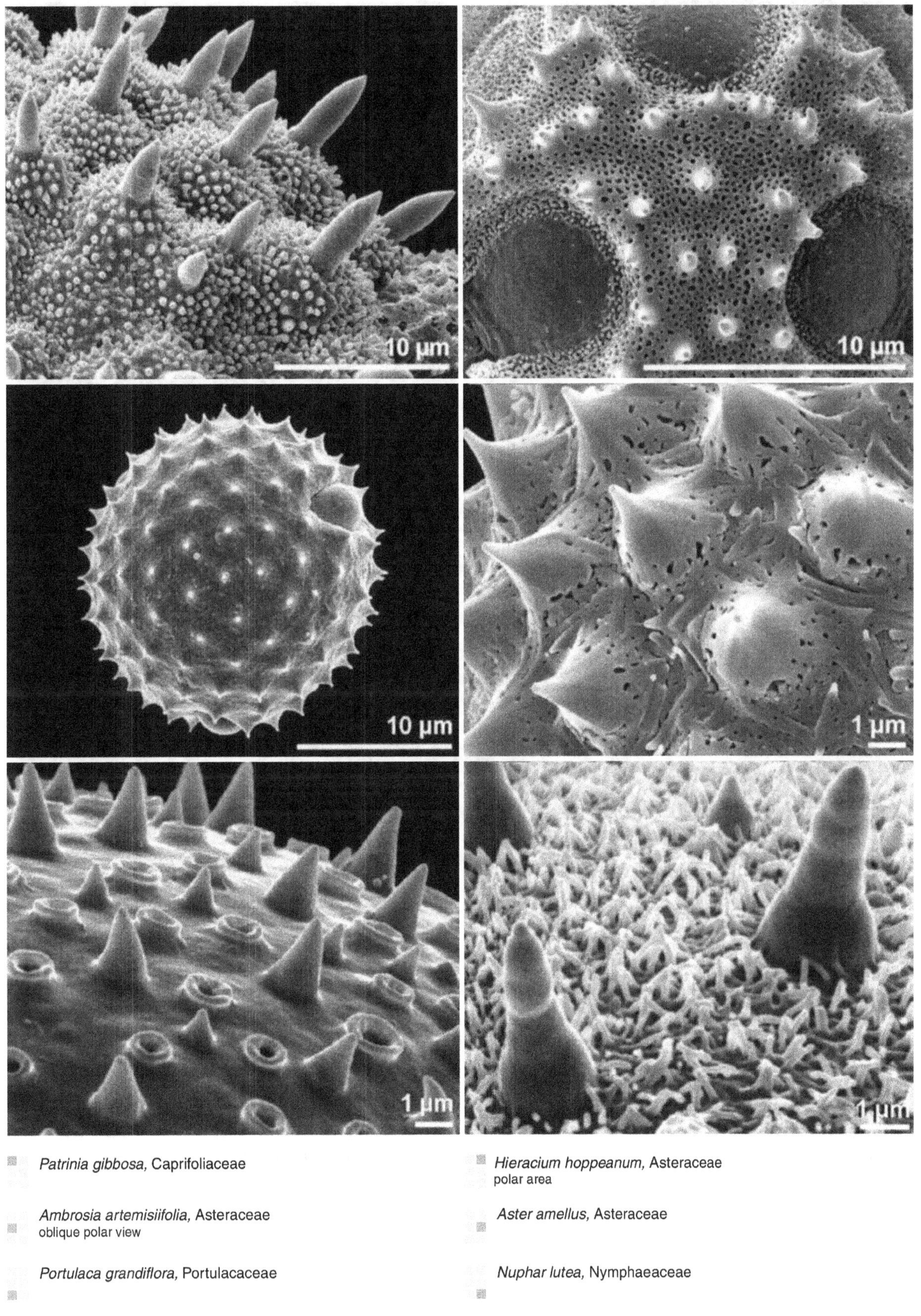

Patrinia gibbosa, Caprifoliaceae

Ambrosia artemisiifolia, Asteraceae
oblique polar view

Portulaca grandiflora, Portulacaceae

Hieracium hoppeanum, Asteraceae
polar area

Aster amellus, Asteraceae

Nuphar lutea, Nymphaeaceae

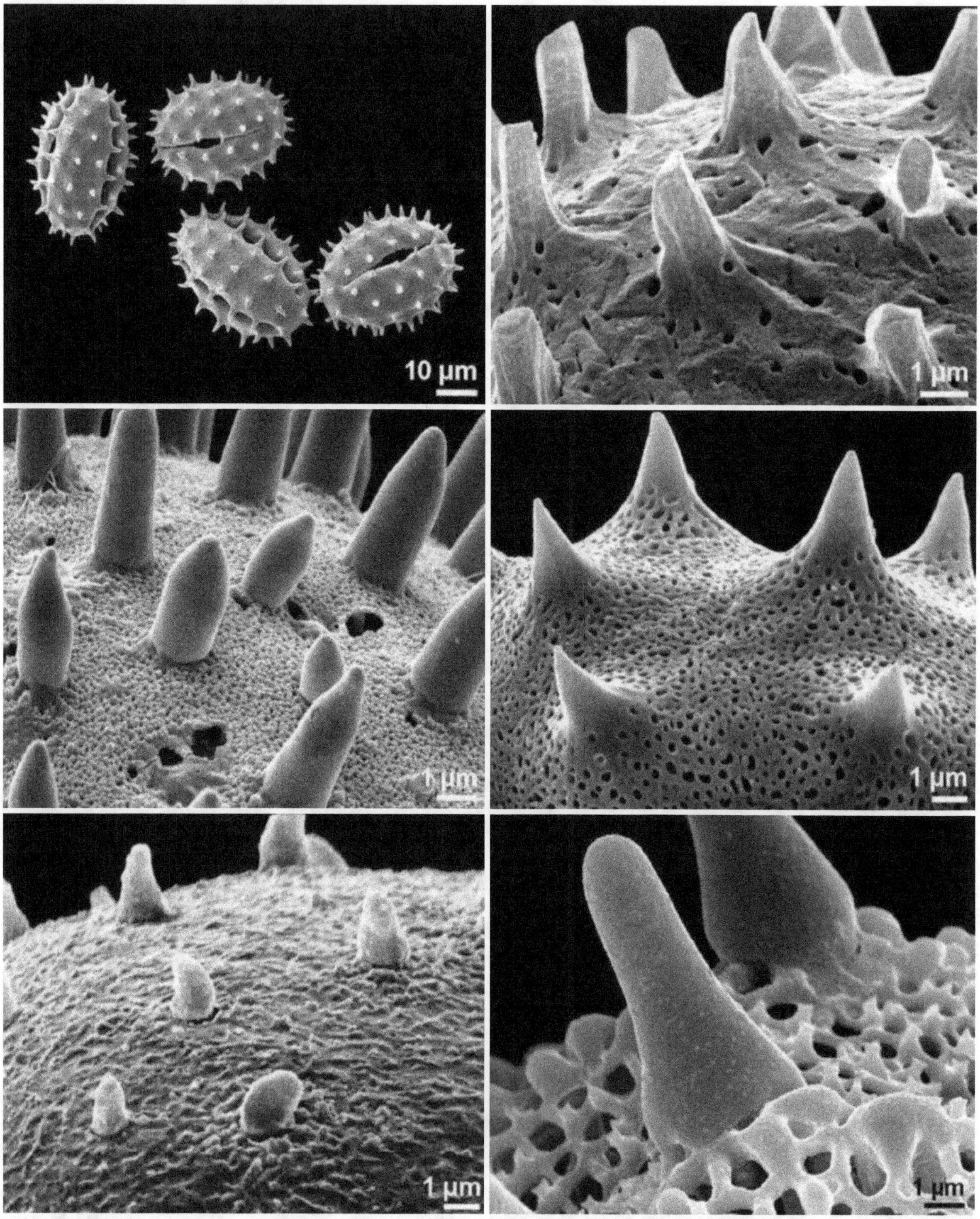

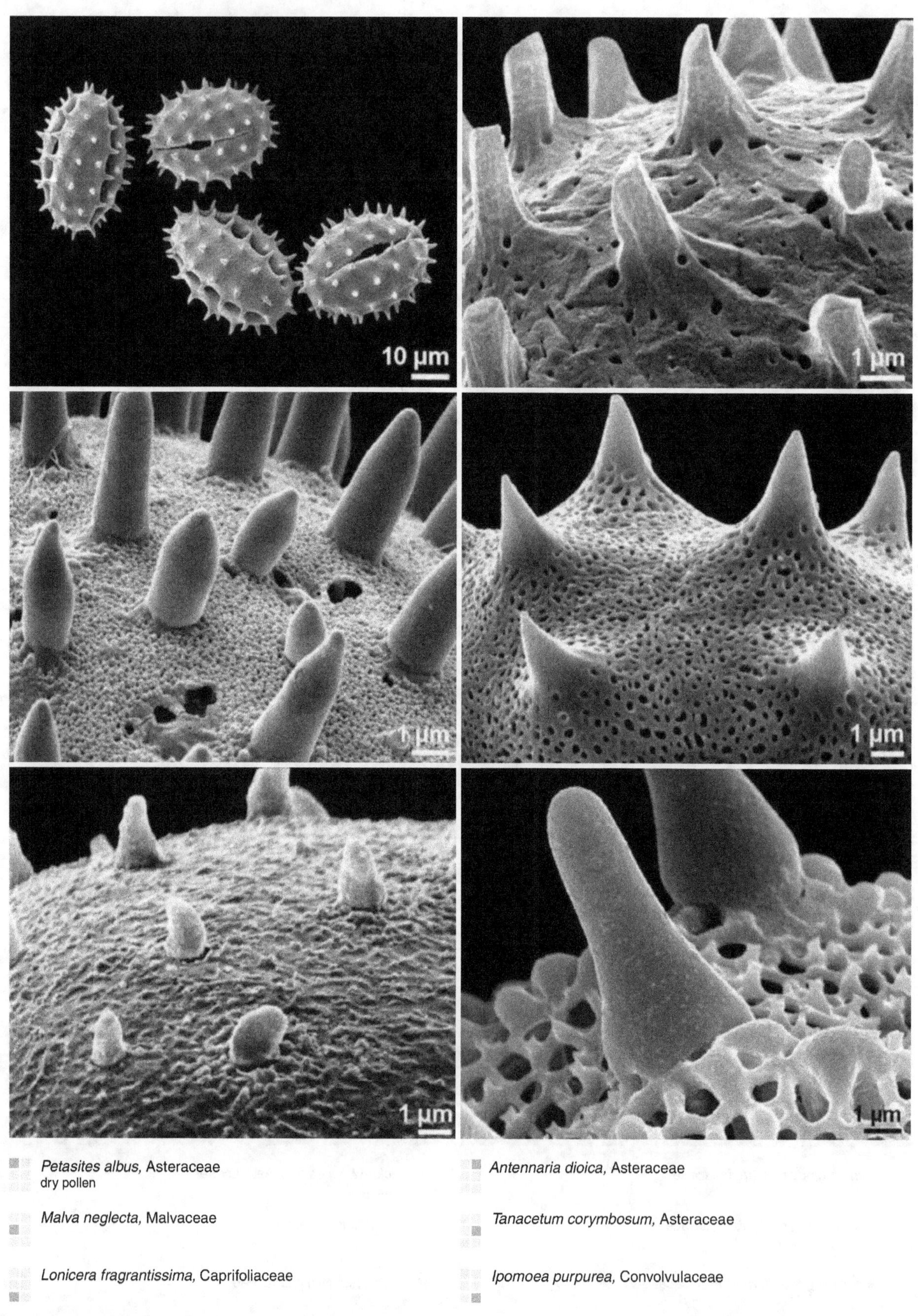

Petasites albus, Asteraceae
dry pollen

Antennaria dioica, Asteraceae

Malva neglecta, Malvaceae

Tanacetum corymbosum, Asteraceae

Lonicera fragrantissima, Caprifoliaceae

Ipomoea purpurea, Convolvulaceae

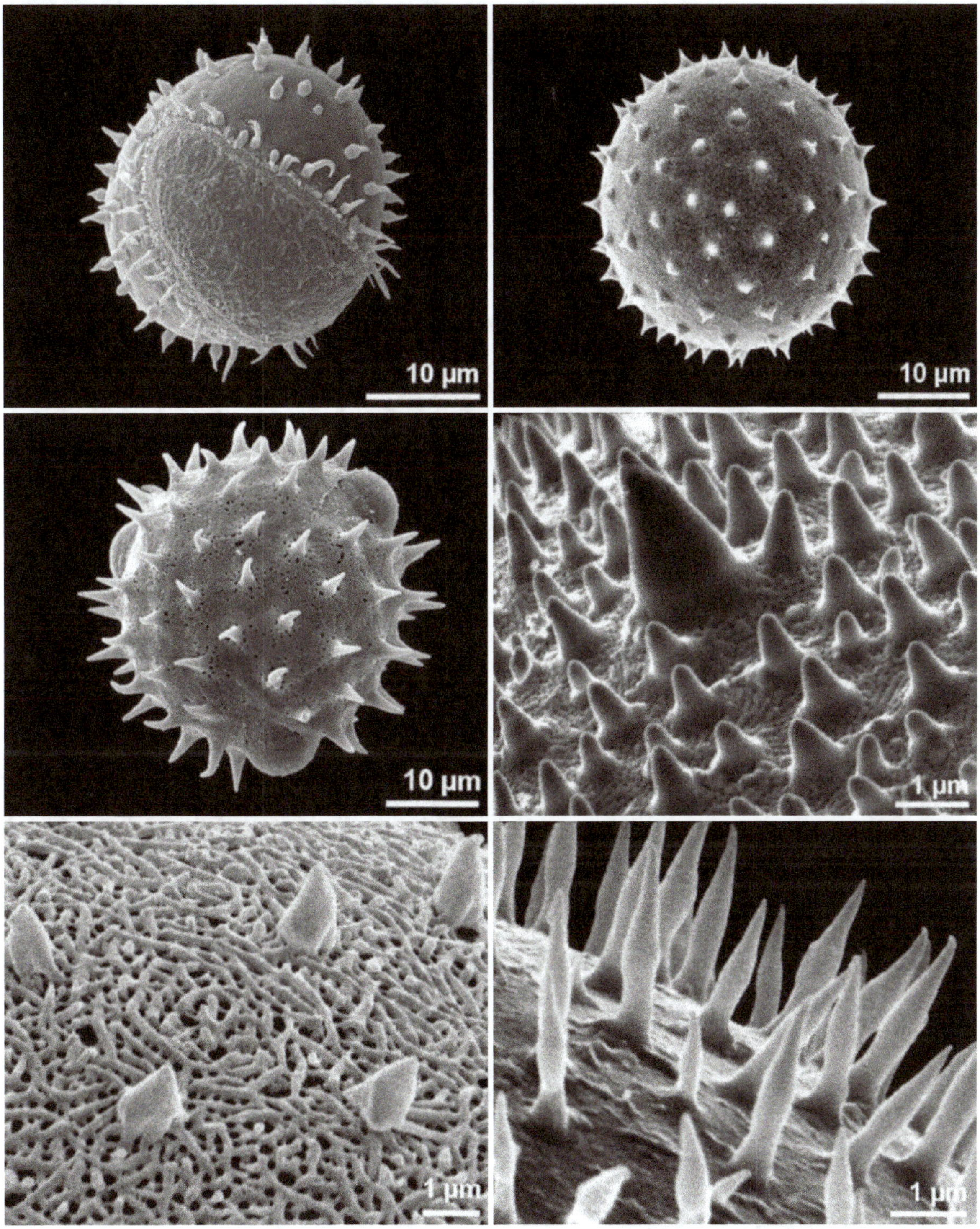

Arenga pinnata, Arecaceae
sulcate, distal polar view

Helianthus annuus, Asteraceae
tricolporate, polar view

Campanula alpina, Campanulaceae

Zomicarpa riedeliana, Araceae
inaperturate

Knautia drymeia, Caprifoliaceae

Ulearum sagittatum, Araceae

Scorzonera austriaca, Asteraceae
tricolporate

Tetrapollinia caerulescens, Gentianacea

Stratiotes aloides, Hydrocharitaceae

Ipomoea batatas, Convolvulaceae

Gynura scandens, Asteraceae
tricolporate, oblique polar view

Hibiscus schizopetalus, Malvaceae
pantoporate

fossula/fossulate

irregular shaped groove

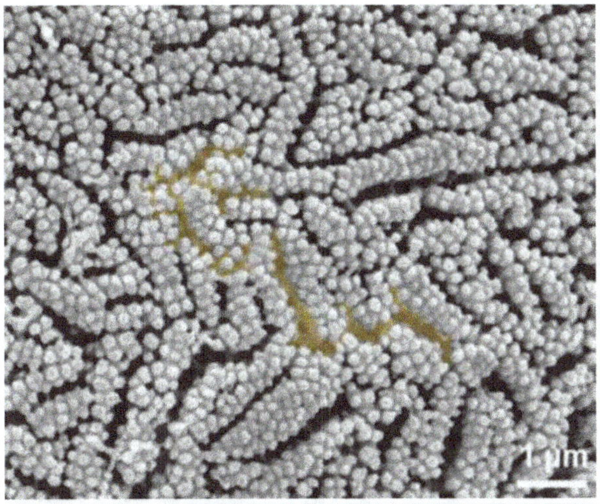

Mendoncia albida, Acanthaceae
fossulae colored

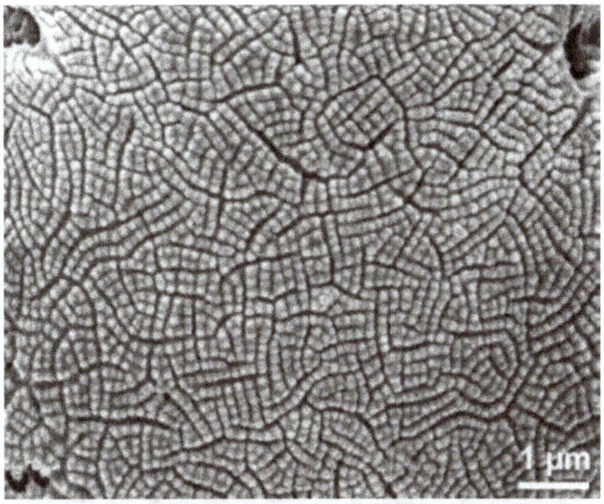

Erica herbacea, Ericaceae

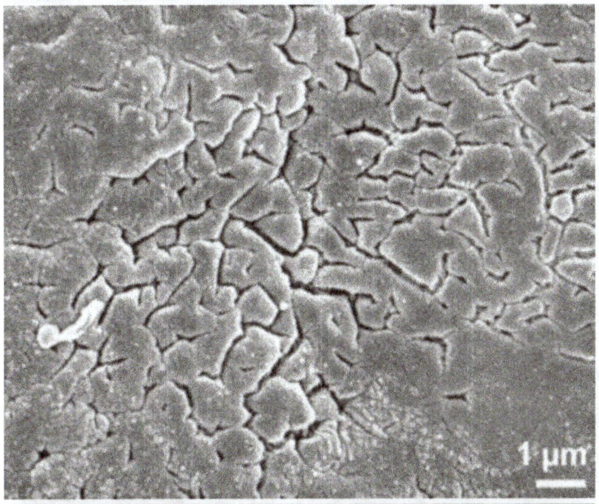

Rhododendron hirsutum, Ericaceae

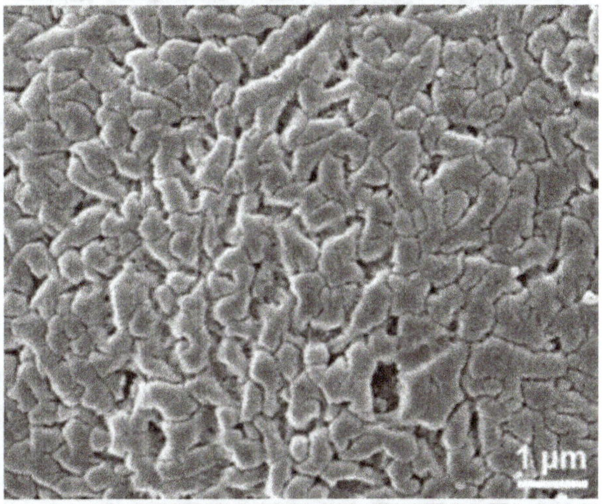

Ledum palustre, Ericaceae

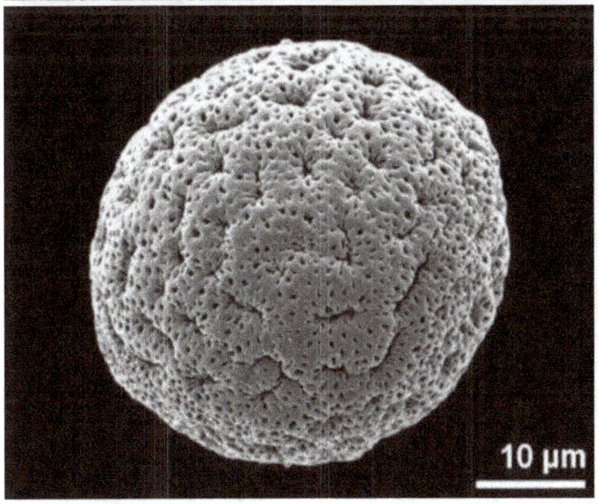

Aristolochia manshuriensis, Aristolochiaceae
fossulate, perforate, inaperturate

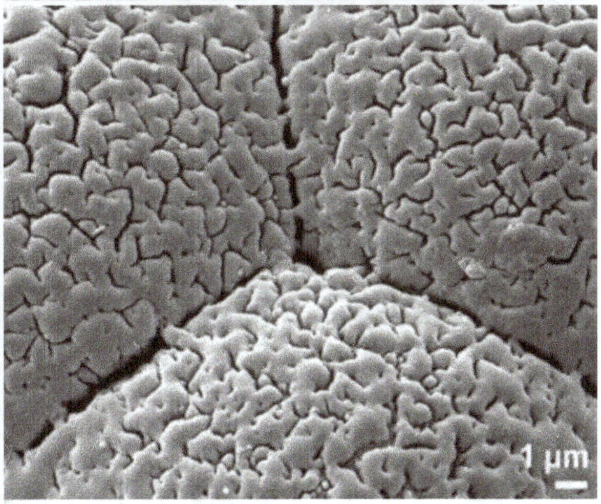

Moneses uniflora, Ericaceae
tetrad, y-shaped fissure between the monads

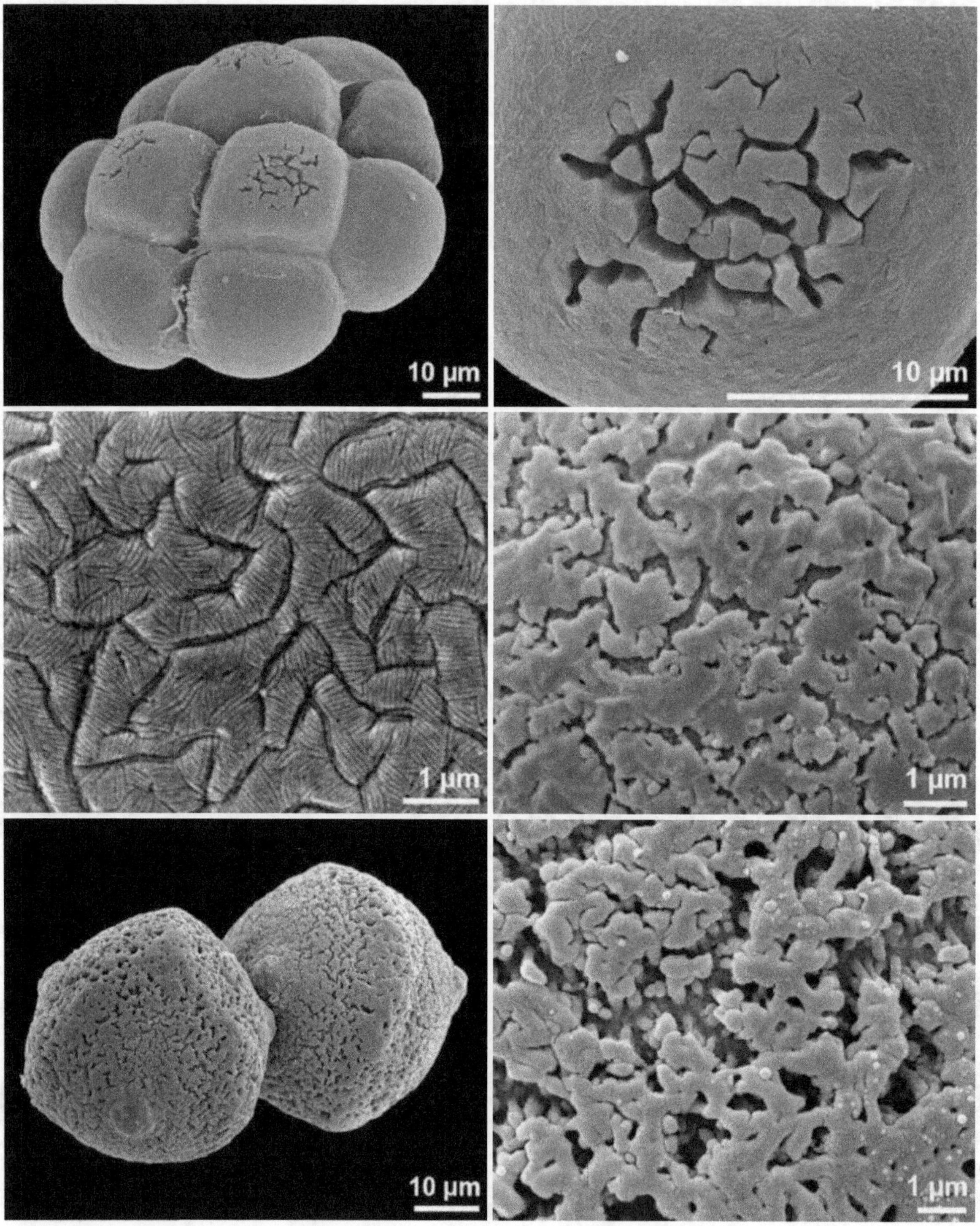

Pithecellobium dulce, Fabaceae
polyad

Pithecellobium dulce, Fabaceae
surface of monad

Gaultheria myrsinoides, Ericaceae

Besleria hirsuta, Gesneriaceae

Lagerstroemia indica, Lythraceae
tricolporate

Lagerstroemia indica, Lythraceae

foveola/foveolate

roundish lumen more than 1 μm in diameter; distance between two adjacent lumina larger than their diameter

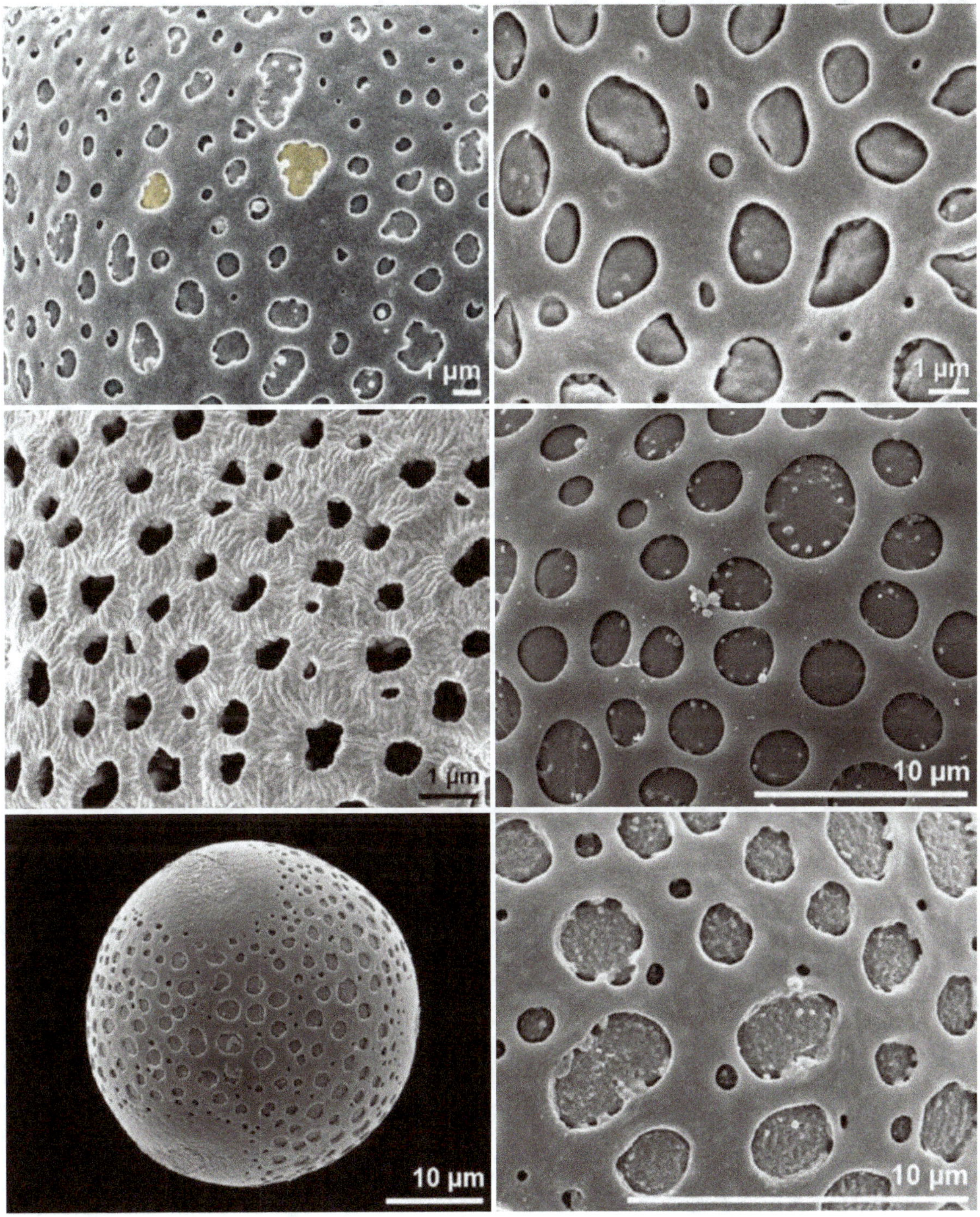

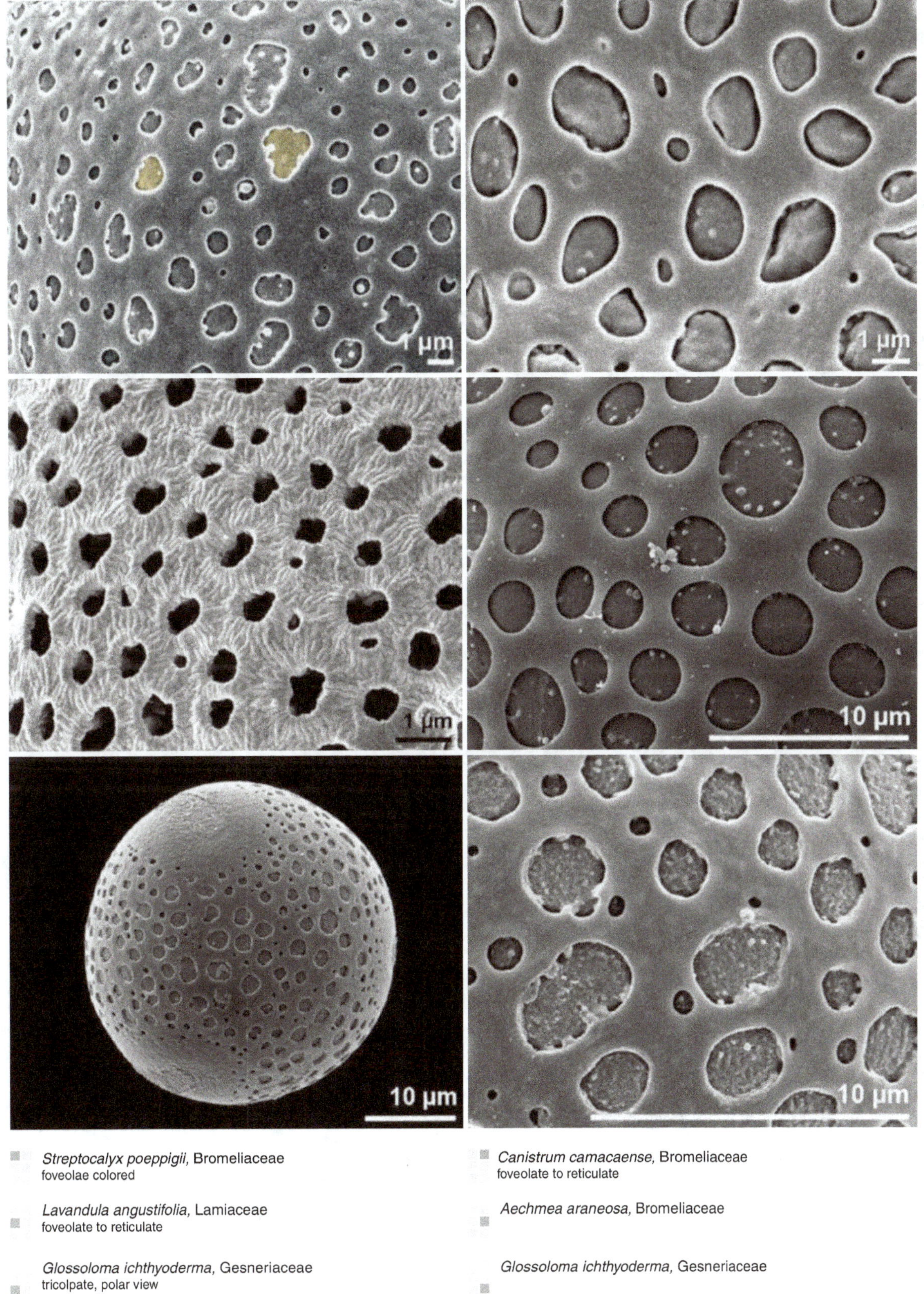

Streptocalyx poeppigii, Bromeliaceae
foveolae colored

Canistrum camacaense, Bromeliaceae
foveolate to reticulate

Lavandula angustifolia, Lamiaceae
foveolate to reticulate

Aechmea araneosa, Bromeliaceae

Glossoloma ichthyoderma, Gesneriaceae
tricolpate, polar view

Glossoloma ichthyoderma, Gesneriaceae

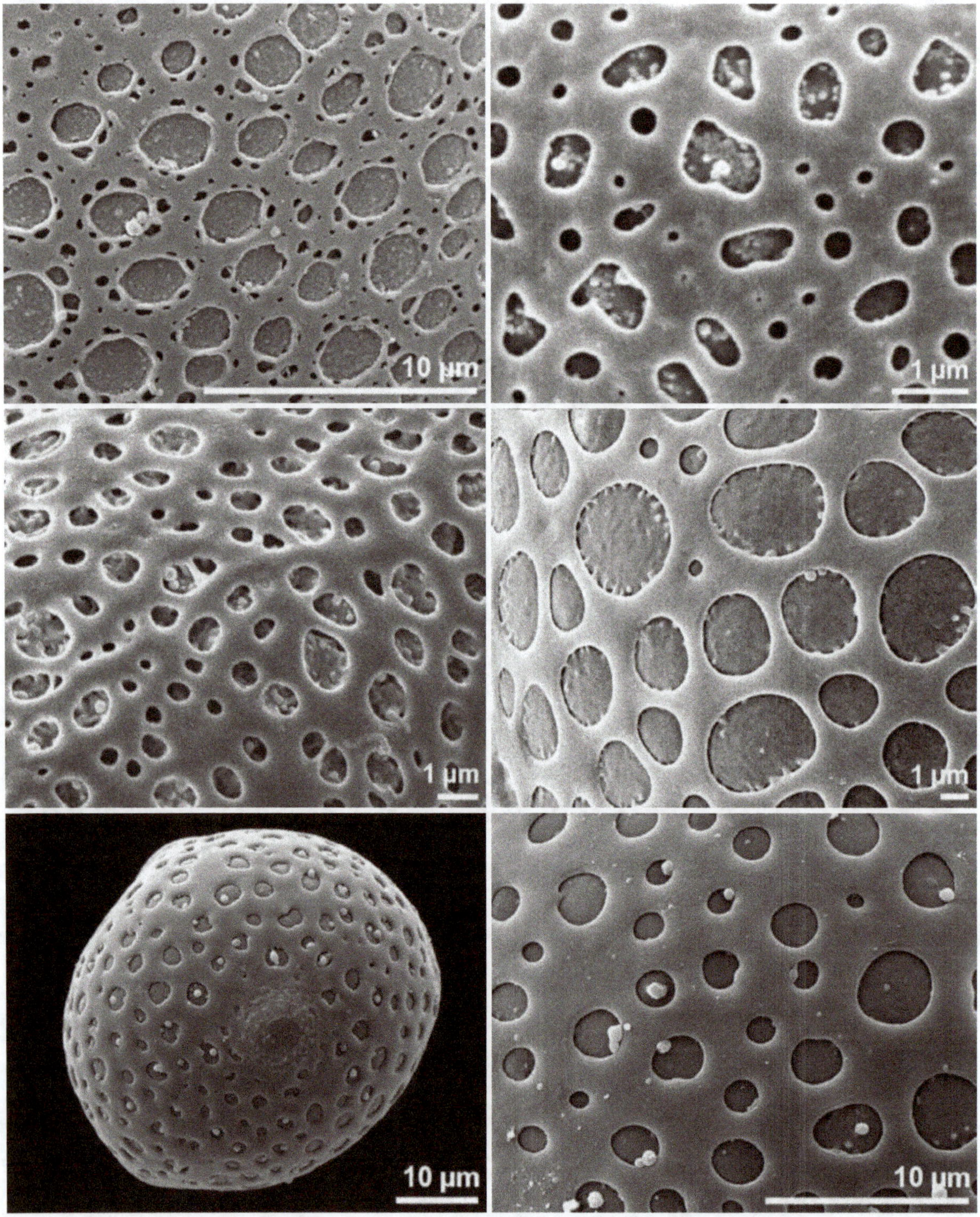

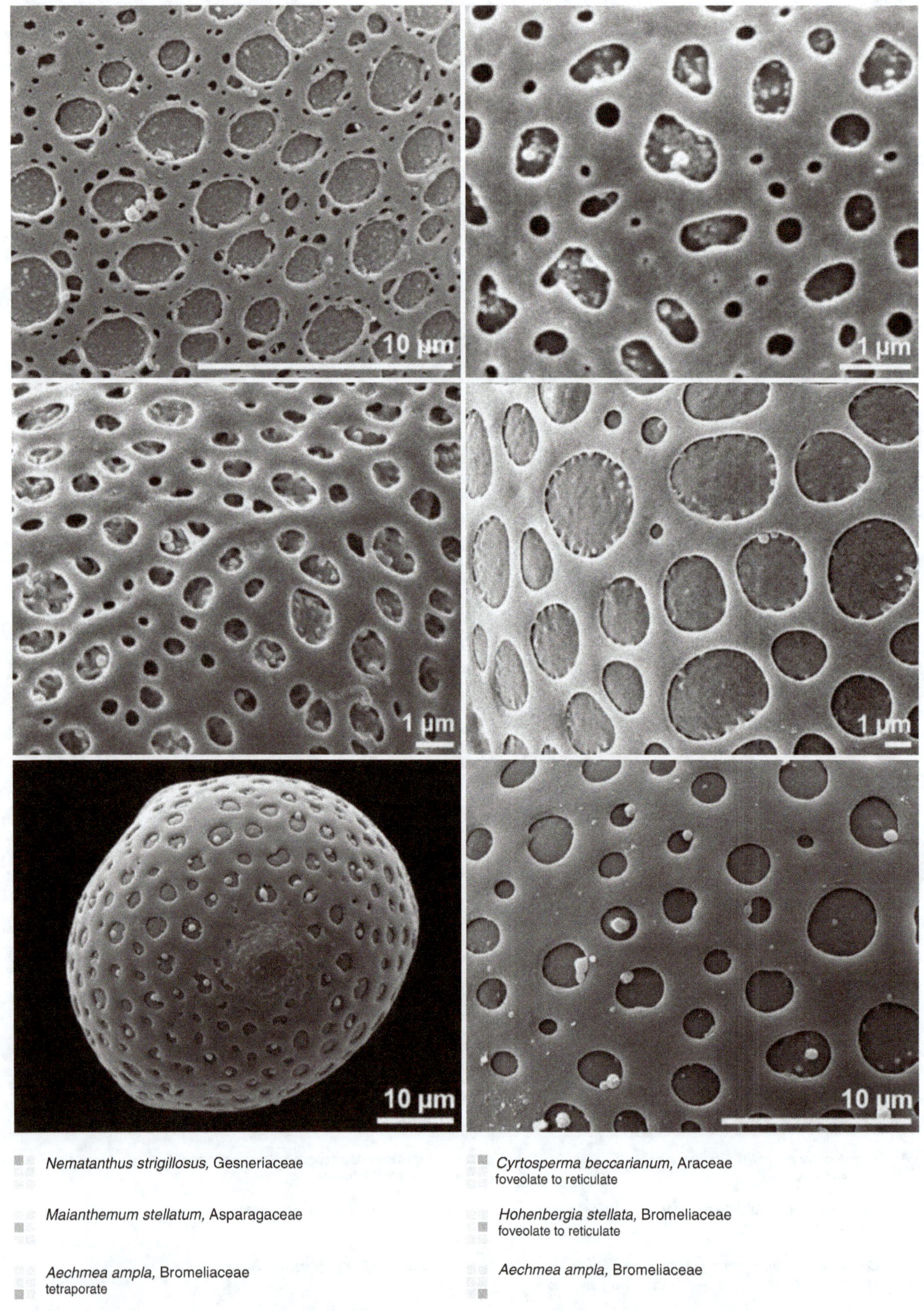

Nematanthus strigillosus, Gesneriaceae

Cyrtosperma beccarianum, Araceae
foveolate to reticulate

Maianthemum stellatum, Asparagaceae

Hohenbergia stellata, Bromeliaceae
foveolate to reticulate

Aechmea ampla, Bromeliaceae
tetraporate

Aechmea ampla, Bromeliaceae

free-standing columellae

columellae not covered by a tectum in semitectate pollen grains

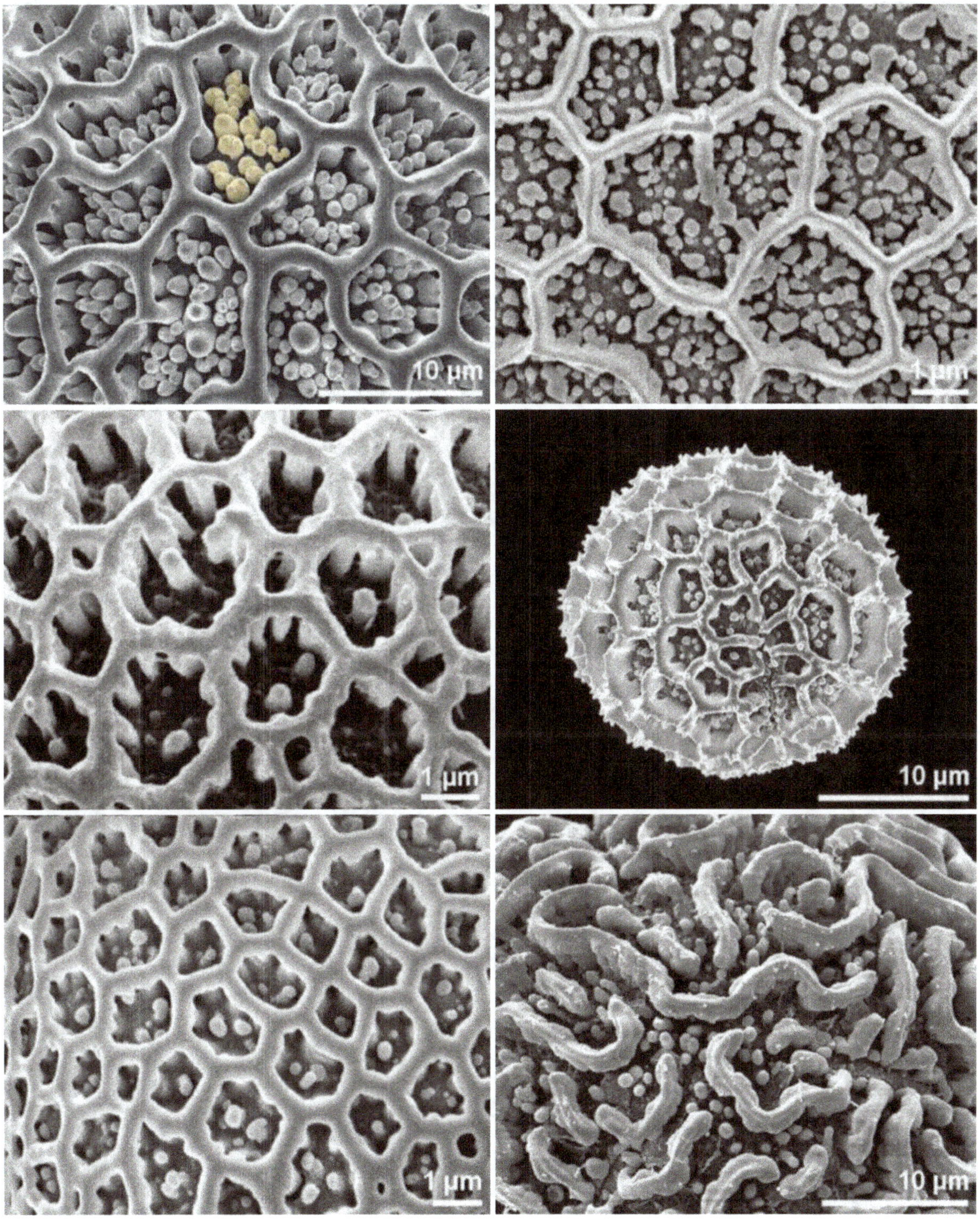

Ruellia makoyana, Acanthaceae
free-standing columellae colored

Erophila verna, Brassicaceae

Viburnum opulus, Adoxaceae

Impatiens parviflora, Balsaminaceae

Bougainvillea sp., Nyctaginaceae
tricolpate, oblique view

Thladiantha hookeri, Cucurbitaceae

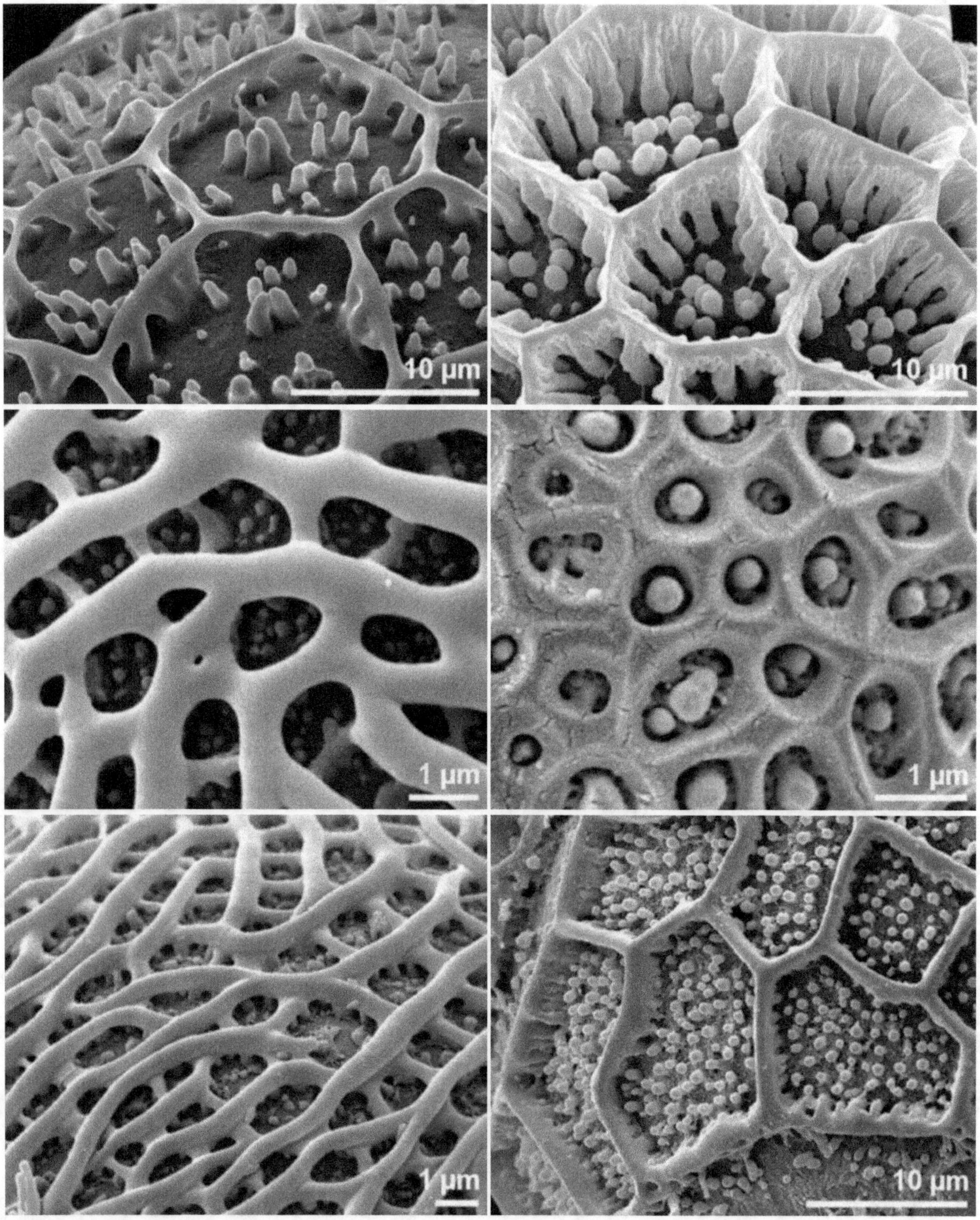

Ruellia devosiana, Acanthaceae

Persicaria chinensis, Polygonaceae

Fouquieria macdougalii, Fouquieriaceae

Ardisia elliptica, Primulaceae

Schotia brachypetala, Fabaceae

Passiflora amethystina, Passifloraceae

gemma/gemmate

globular ornamentation element

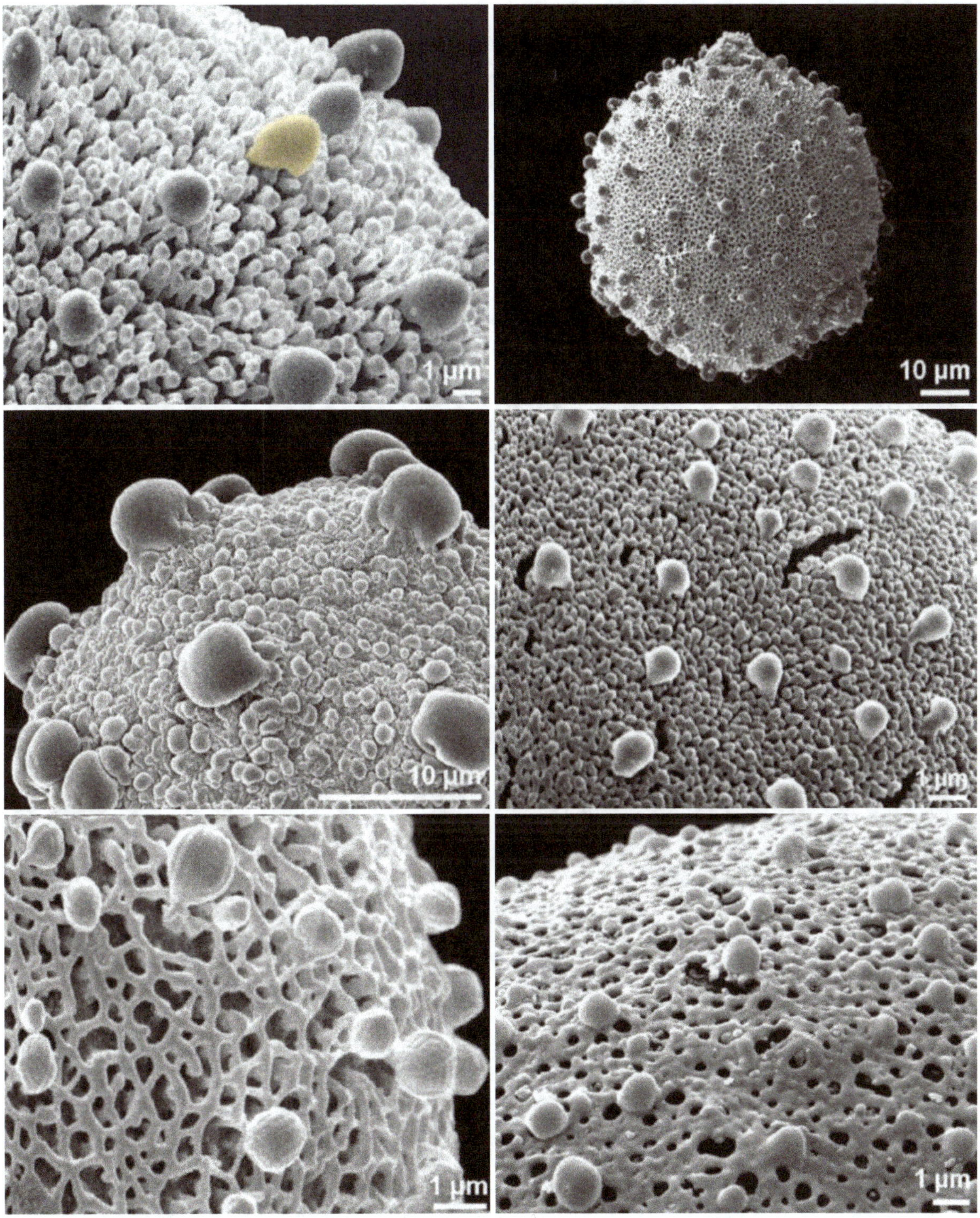

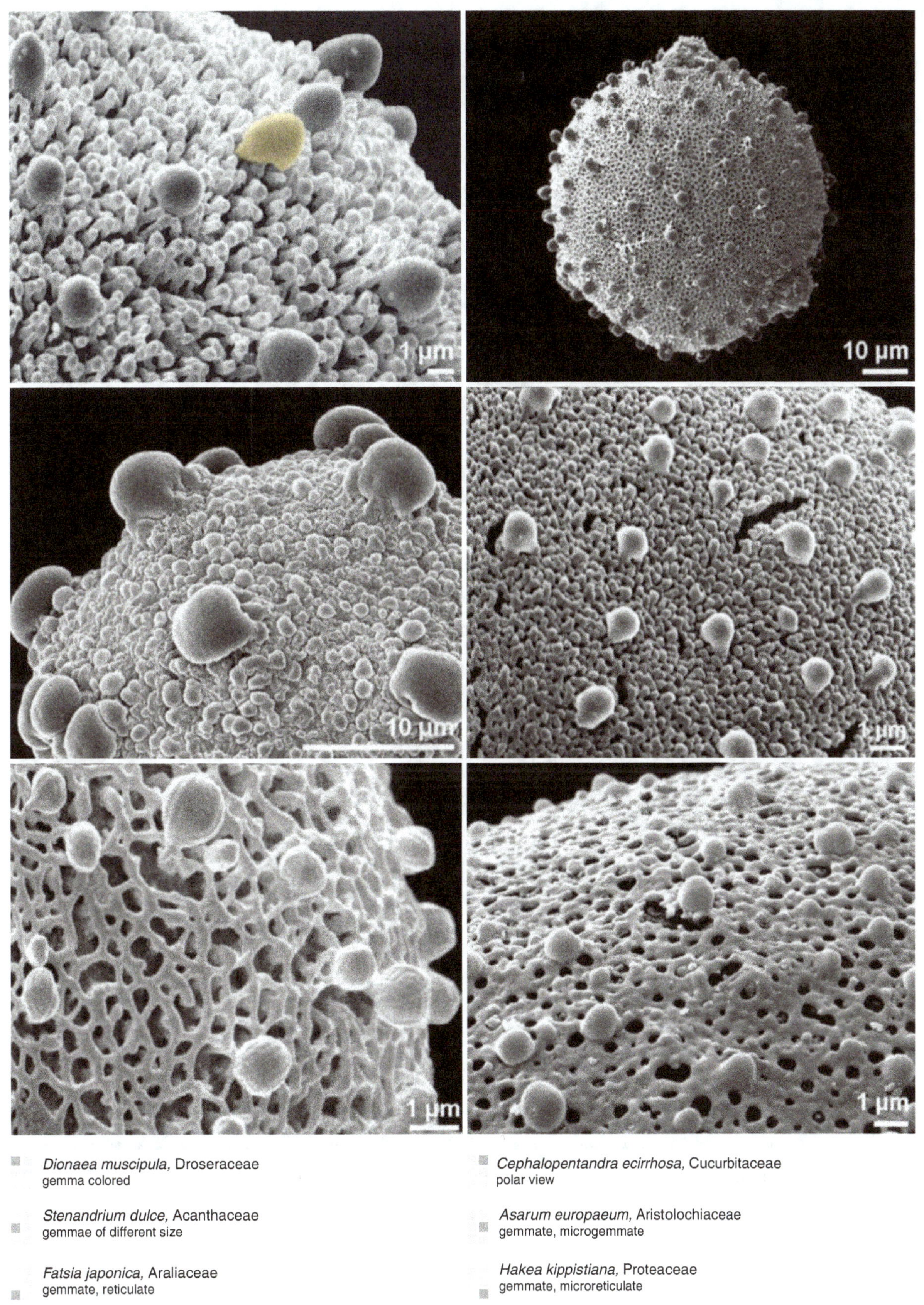

Dionaea muscipula, Droseraceae
gemma colored

Cephalopentandra ecirrhosa, Cucurbitaceae
polar view

Stenandrium dulce, Acanthaceae
gemmae of different size

Asarum europaeum, Aristolochiaceae
gemmate, microgemmate

Fatsia japonica, Araliaceae
gemmate, reticulate

Hakea kippistiana, Proteaceae
gemmate, microreticulate

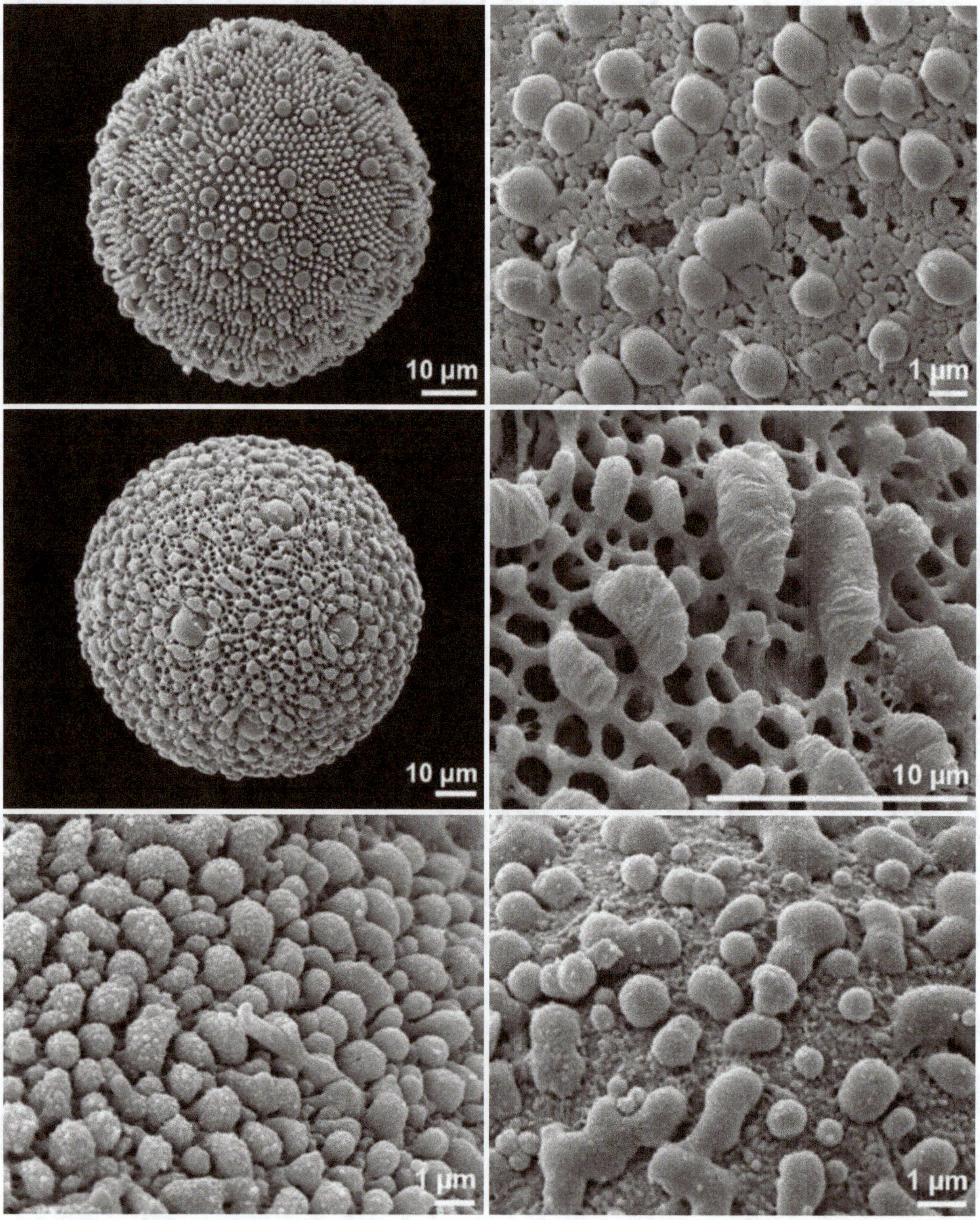

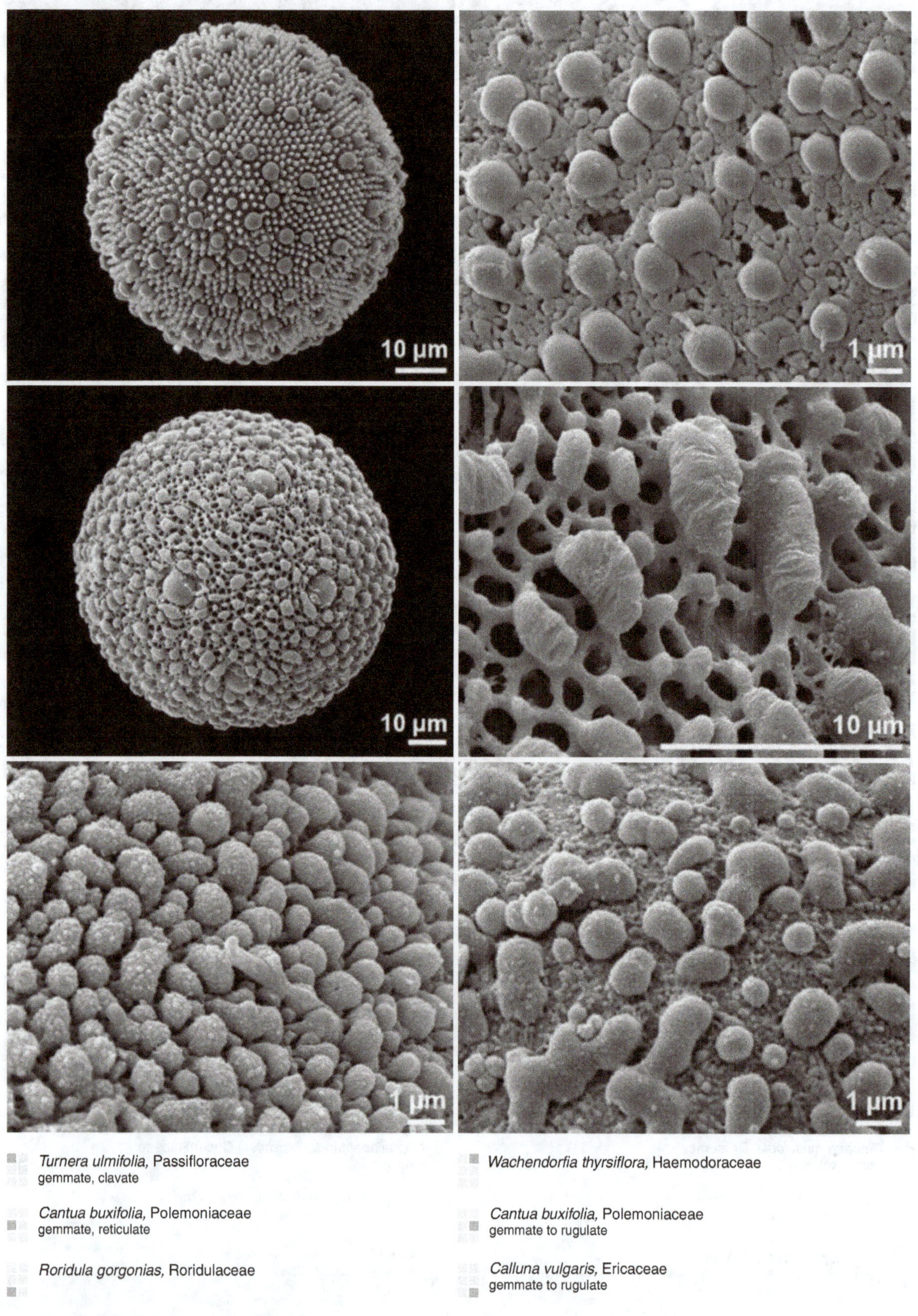

Turnera ulmifolia, Passifloraceae
gemmate, clavate

Wachendorfia thyrsiflora, Haemodoraceae

Cantua buxifolia, Polemoniaceae
gemmate, reticulate

Cantua buxifolia, Polemoniaceae
gemmate to rugulate

Roridula gorgonias, Roridulaceae

Calluna vulgaris, Ericaceae
gemmate to rugulate

granulum/granulate

sculpture element of different/indefinable shape, equal or smaller than 0.1 μm in diameter (hard to outline)

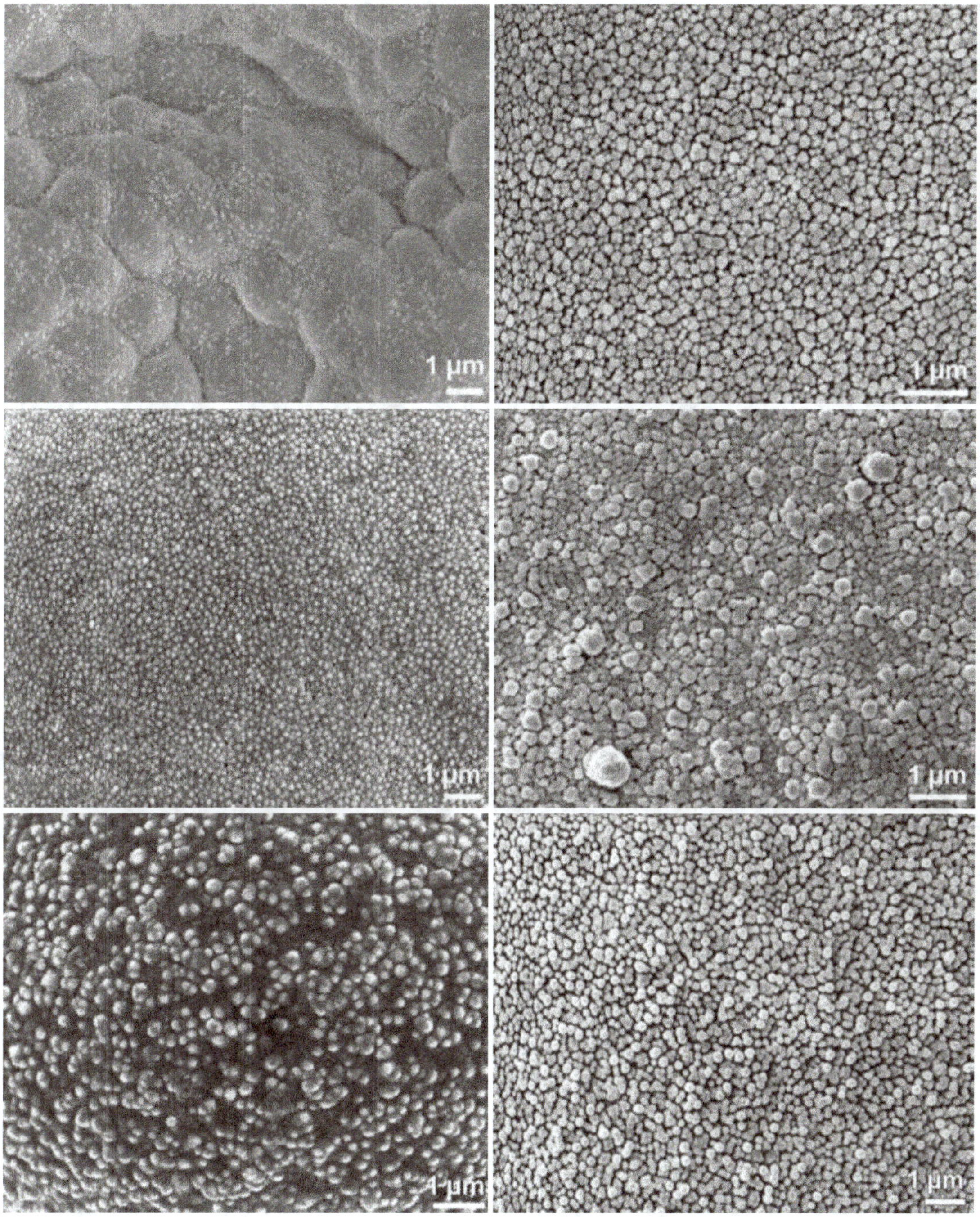

Hemiptelia sp., Ulmaceae
fossil, Miocene, China

Luzula campestris, Juncaceae

Humulus lupulus, Cannabaceae

Clarkia pulchella, Onagraceae

Larix decidua, Pinaceae

Clarkia purpurea, Onagraceae

heterobrochate

reticulate pollen wall with lumina of different sizes

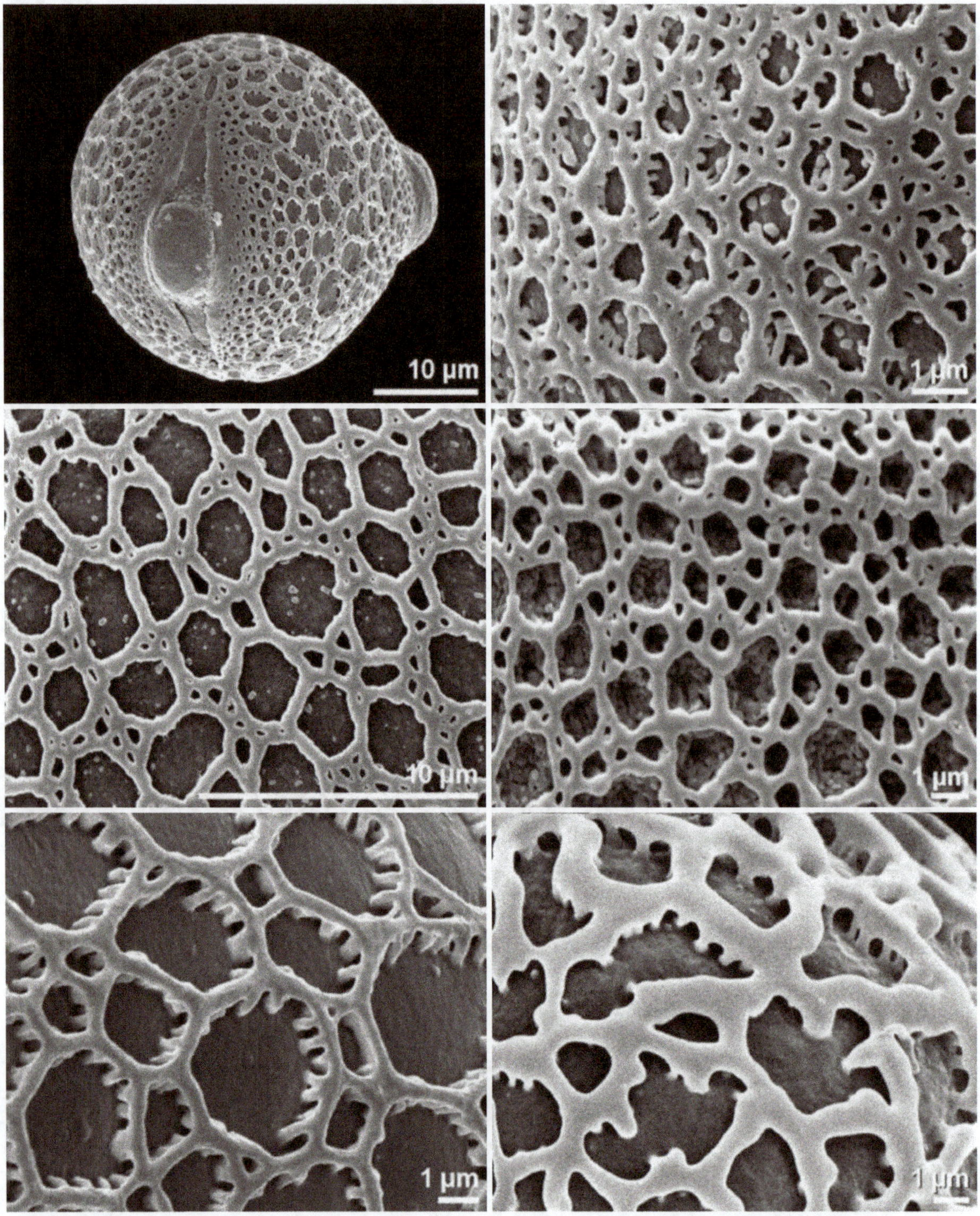

Hedera helix, Araliaceae
equatorial view

Lachenalia aloides, Asparagaceae

Fritillaria meleagris, Liliaceae

Anthericum ramosum, Asparagaceae

Billbergia seidelii, Bromeliaceae

Limodorum abortivum, Orchidaceae

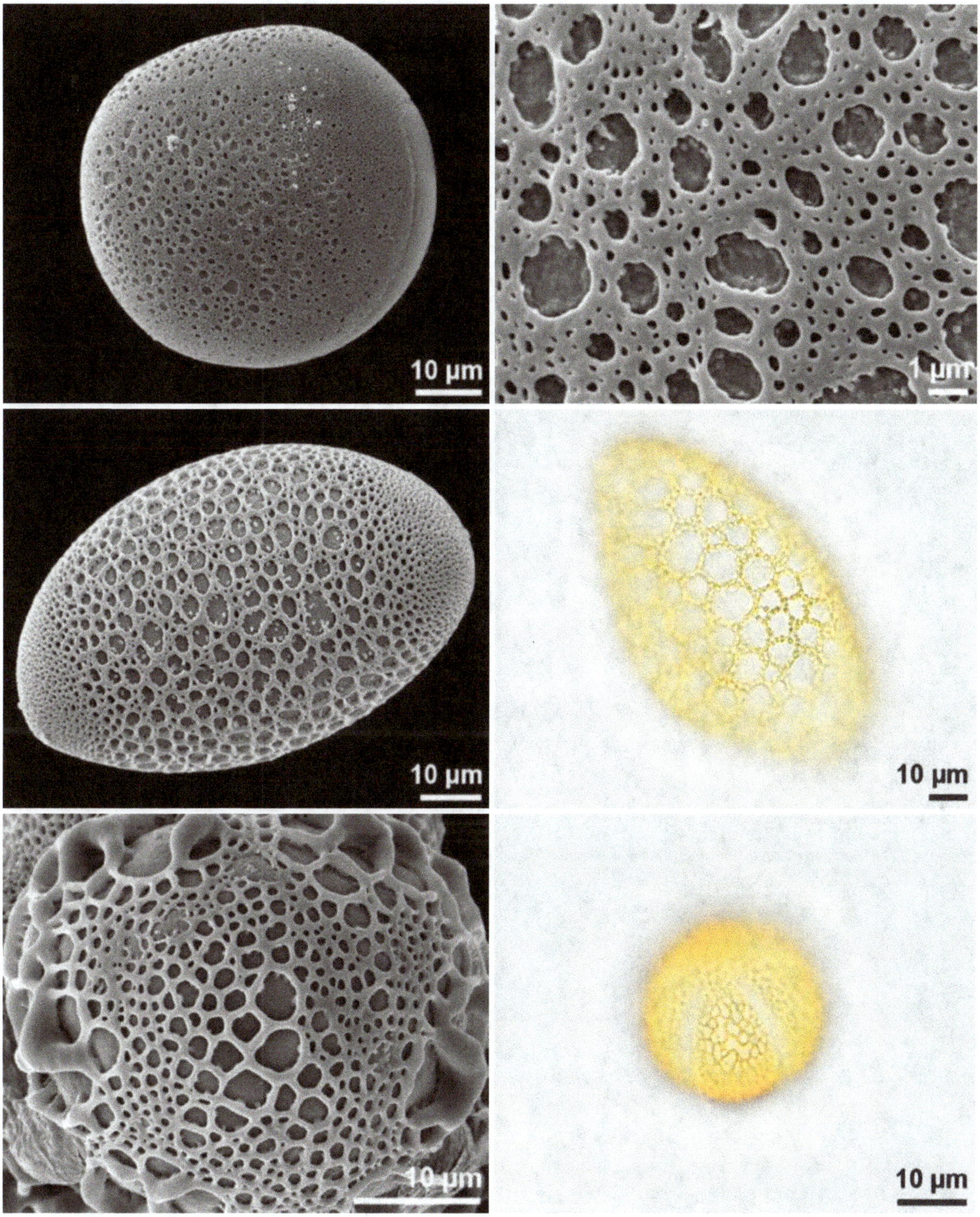

Ornithogalum narbonense, Asparagaceae
equatorial view

Bomarea hirsuta, Alstroemeriaceae
proximal polar view

Chelonanthus alatus, Gentianaceae
polar area

Ornithogalum narbonense, Asparagaceae

Bessera elegans, Asparagaceae

Plectranthus esculentus, Lamiaceae

homobrochate

reticulate pollen wall with lumina of uniform size

Acantholimon glumaceum, Plumbaginaceae

Abeliophyllum distichum, Oleaceae

Eranthemum wattii, Acanthaceae
polar view

Strobilanthes roseus, Acanthaceae

Ruellia sp., Acanthaceae

Thlaspi montanum, Brassicaceae

Armeria pinifolia, Plumbaginaceae
equatorial view

Impatiens parviflora, Balsaminaceae
oblique equatorial view

Kallstroemia maxima, Zygophyllaceae

Armeria pinifolia, Plumbaginaceae

Persicaria chinensis, Polygonaceae
polar view

Ruellia tuberosa, Acanthaceae
triporate, oblique view

lophae/lophate, lacunae

lophae: massive exine ridges
lacunae: depressed areas surrounded by lophae

Leontodon saxatilis, Asteraceae
lophae colored, equatorial view

Cichorium intybus, Asteraceae
lacunae colored, polar view

Opuntia basilaris, Cactaceae

Pfaffia tuberosa, Amaranthaceae

Gazania sp., Asteraceae
tricolporate, polar view

Hieracium hoppeanum, Asteraceae
dry pollen

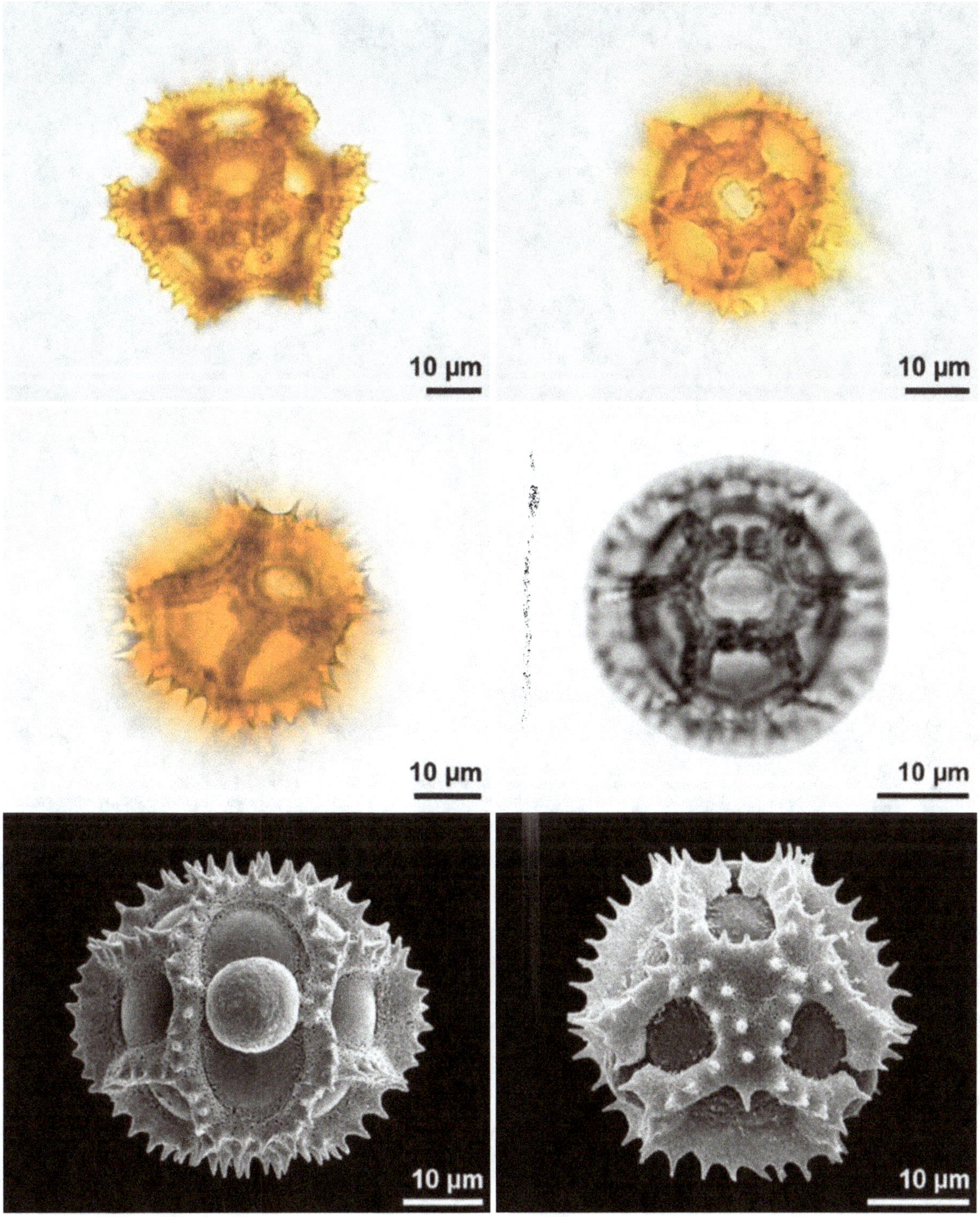

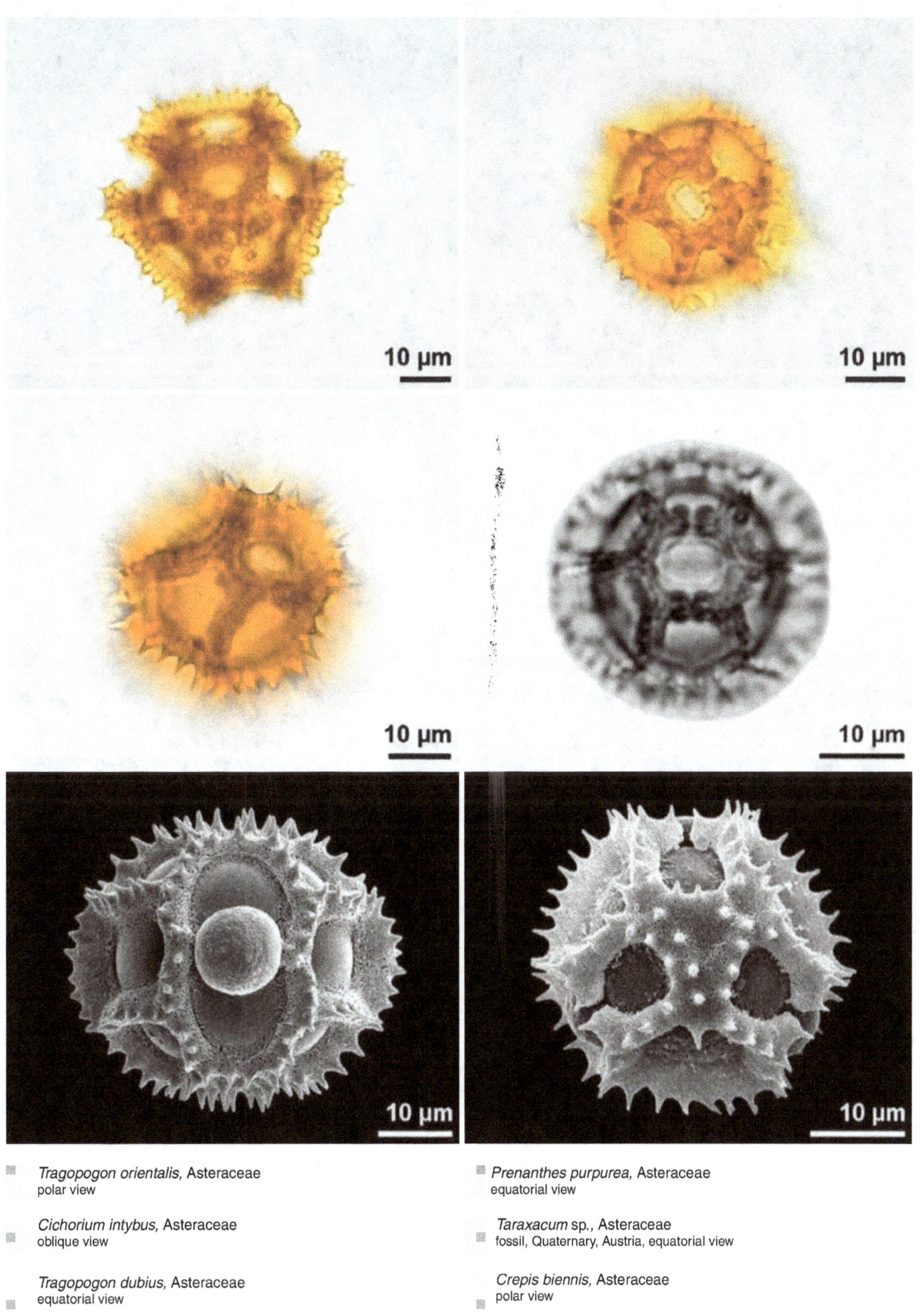

Tragopogon orientalis, Asteraceae
polar view

Cichorium intybus, Asteraceae
oblique view

Tragopogon dubius, Asteraceae
equatorial view

Prenanthes purpurea, Asteraceae
equatorial view

Taraxacum sp., Asteraceae
fossil, Quaternary, Austria, equatorial view

Crepis biennis, Asteraceae
polar view

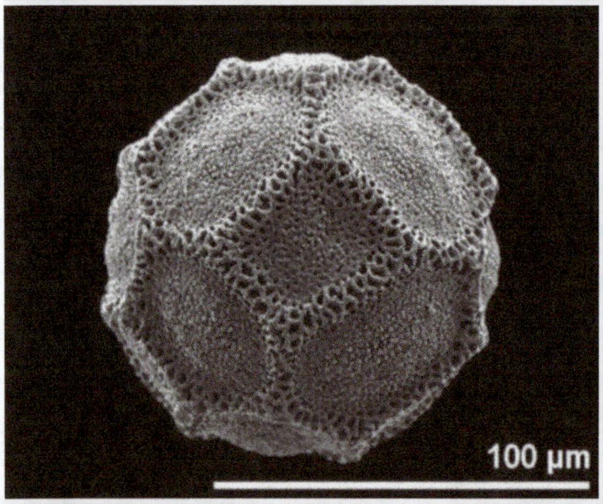

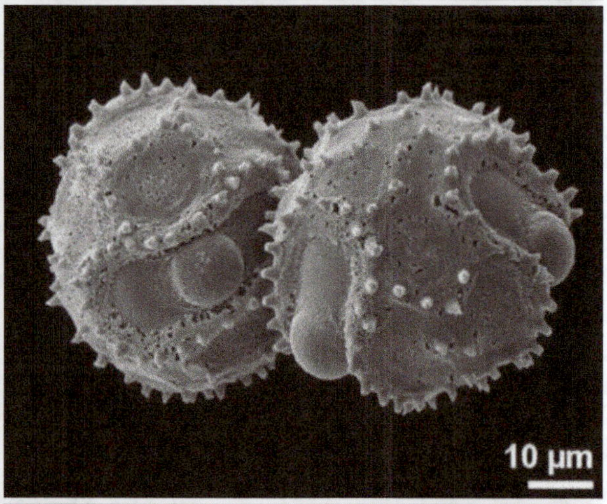

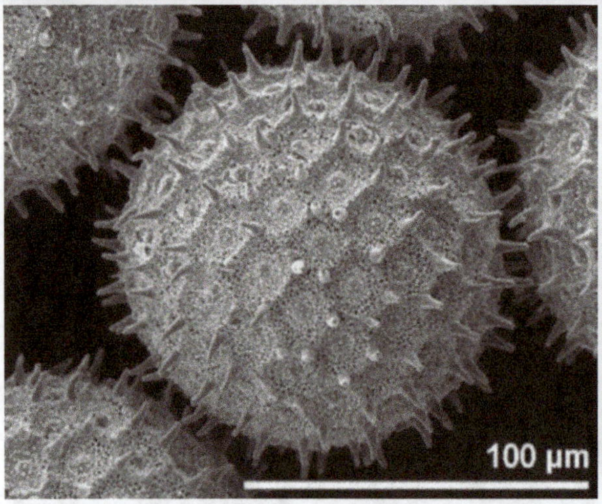

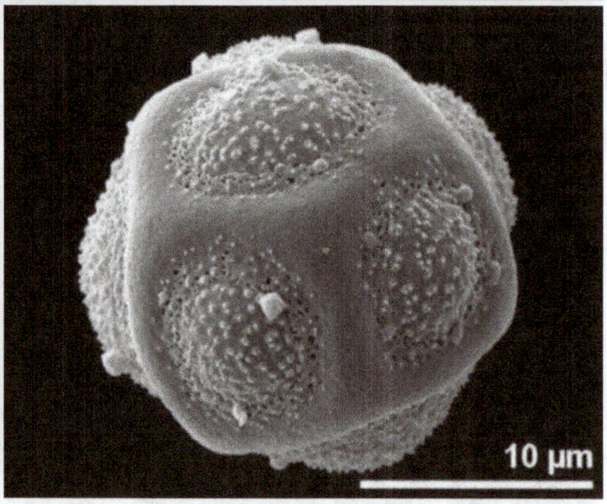

Cyanthillium cinereum, Asteraceae
oblique equatorial view

Gomphrena celosioides, Amaranthaceae

Opuntia polyacantha, Cactaceae

Scorzonera aristata, Asteraceae

Ipomoea caerulea, Convolvulaceae

Herniaria alpina, Caryophyllaceae

micro-

prefix for small; features between 1 and 0.5 µm

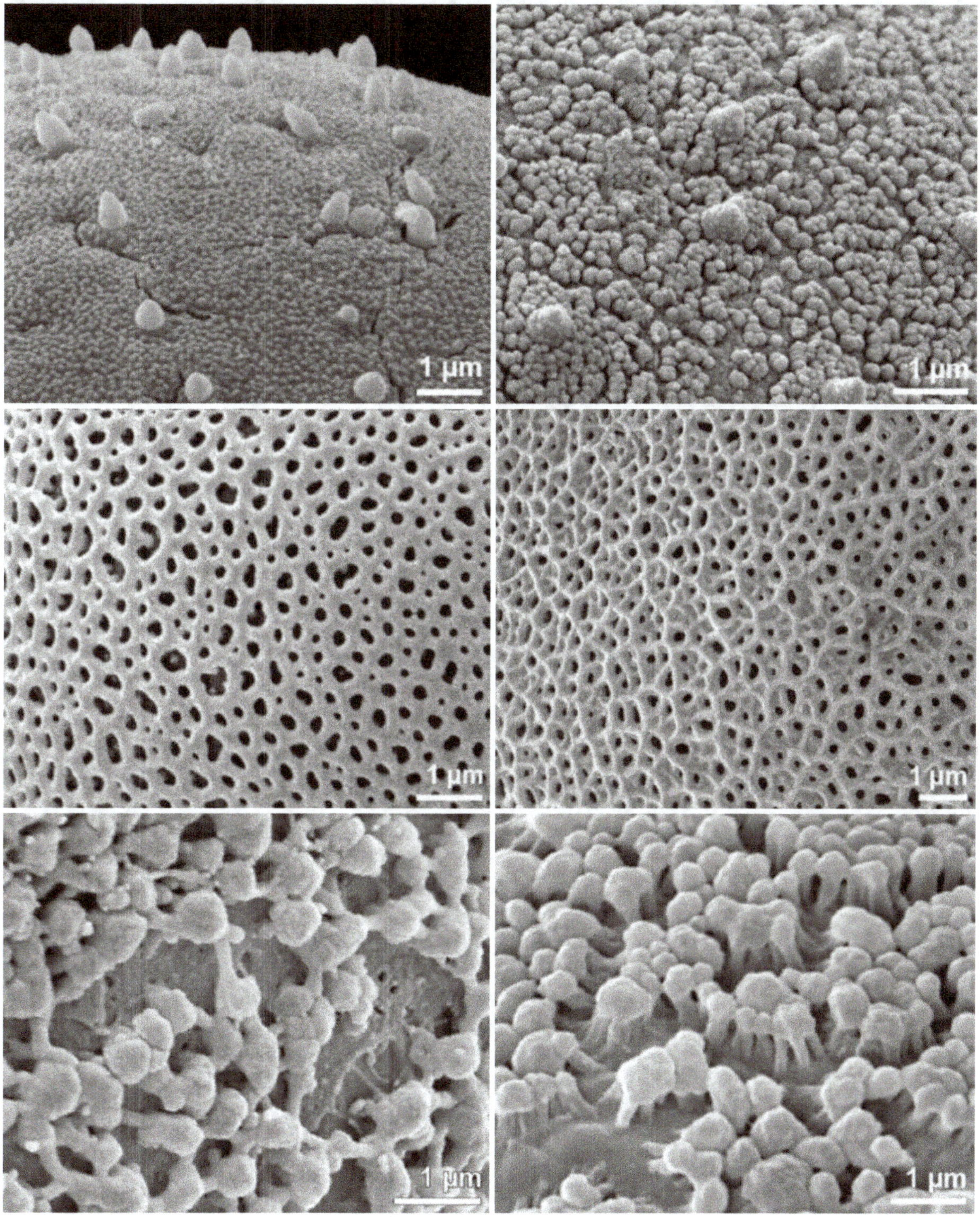

Heloniopsis kawanoi, Melanthiaceae
microechinate, granulate

Kickxia spuria, Plantaginaceae
micro- to nanoreticulate

Aspidistra elatior, Asparagaceae
microgemmate

Trillium grandiflorum, Melanthiaceae
microechinate, nanogemmate to granulate

Lamium purpureum, Lamiaceae
micro- to nanoreticulate

Callisia fragrans, Commelinaceae
microclavate

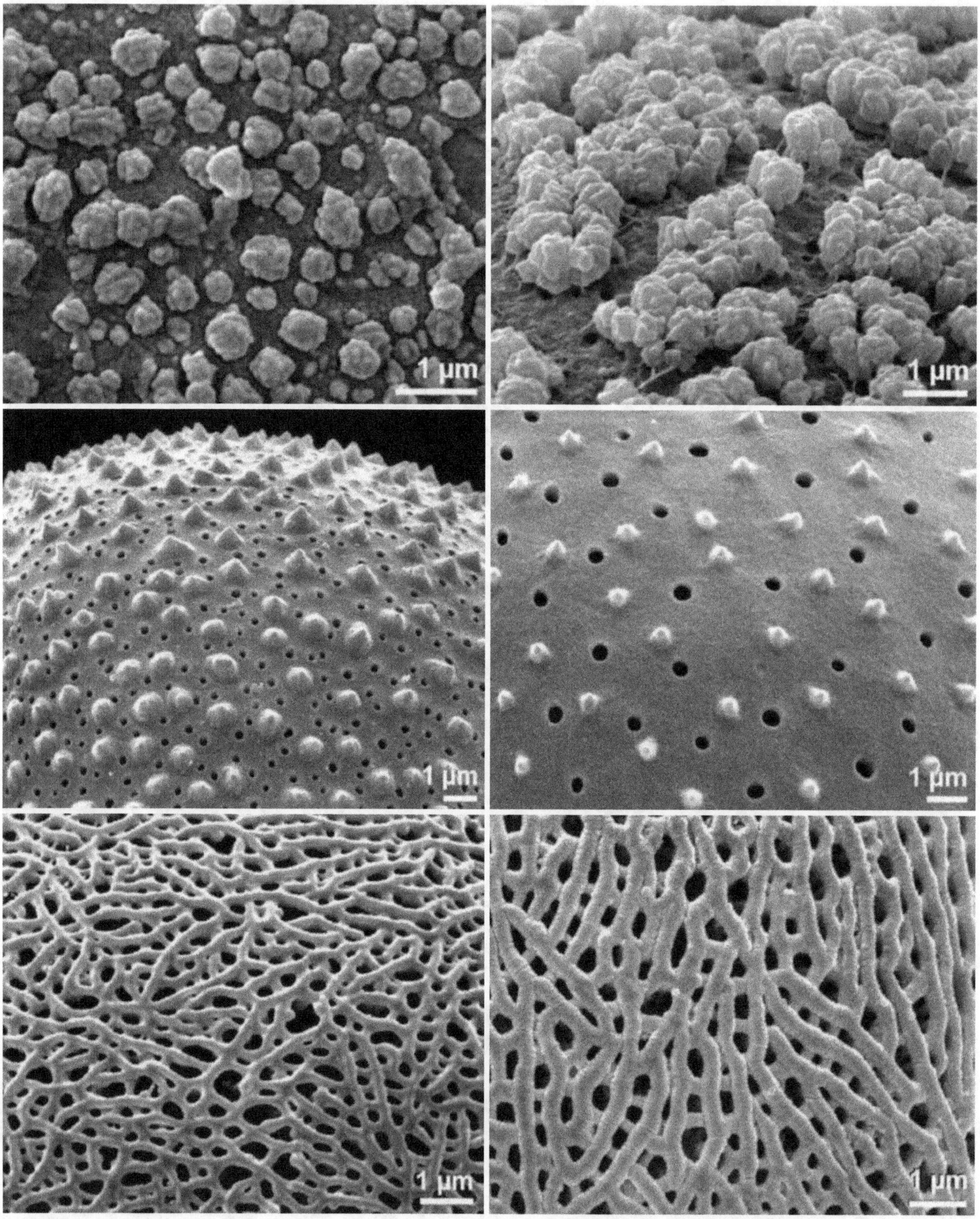

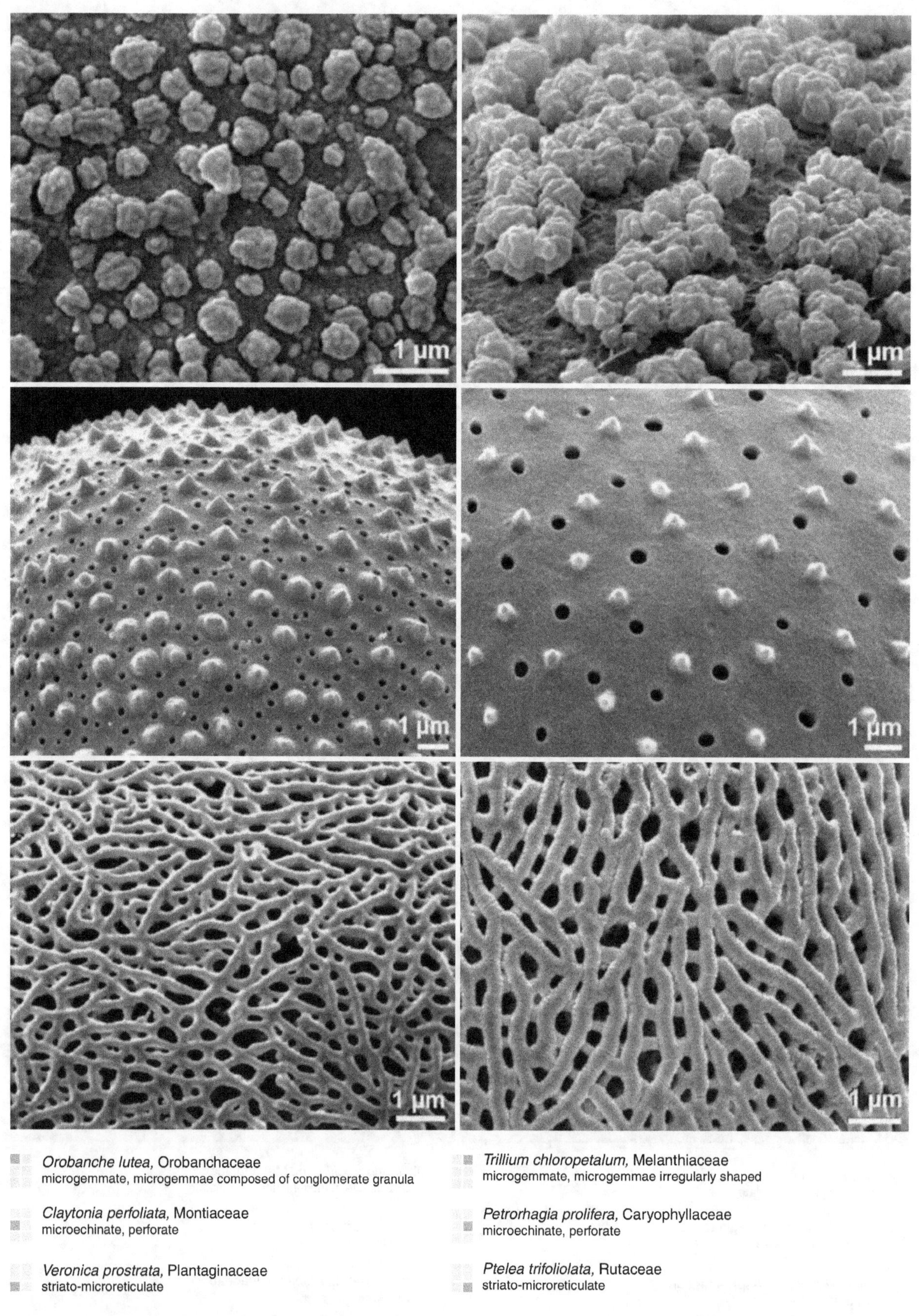

Orobanche lutea, Orobanchaceae
microgemmate, microgemmae composed of conglomerate granula

Trillium chloropetalum, Melanthiaceae
microgemmate, microgemmae irregularly shaped

Claytonia perfoliata, Montiaceae
microechinate, perforate

Petrorhagia prolifera, Caryophyllaceae
microechinate, perforate

Veronica prostrata, Plantaginaceae
striato-microreticulate

Ptelea trifoliolata, Rutaceae
striato-microreticulate

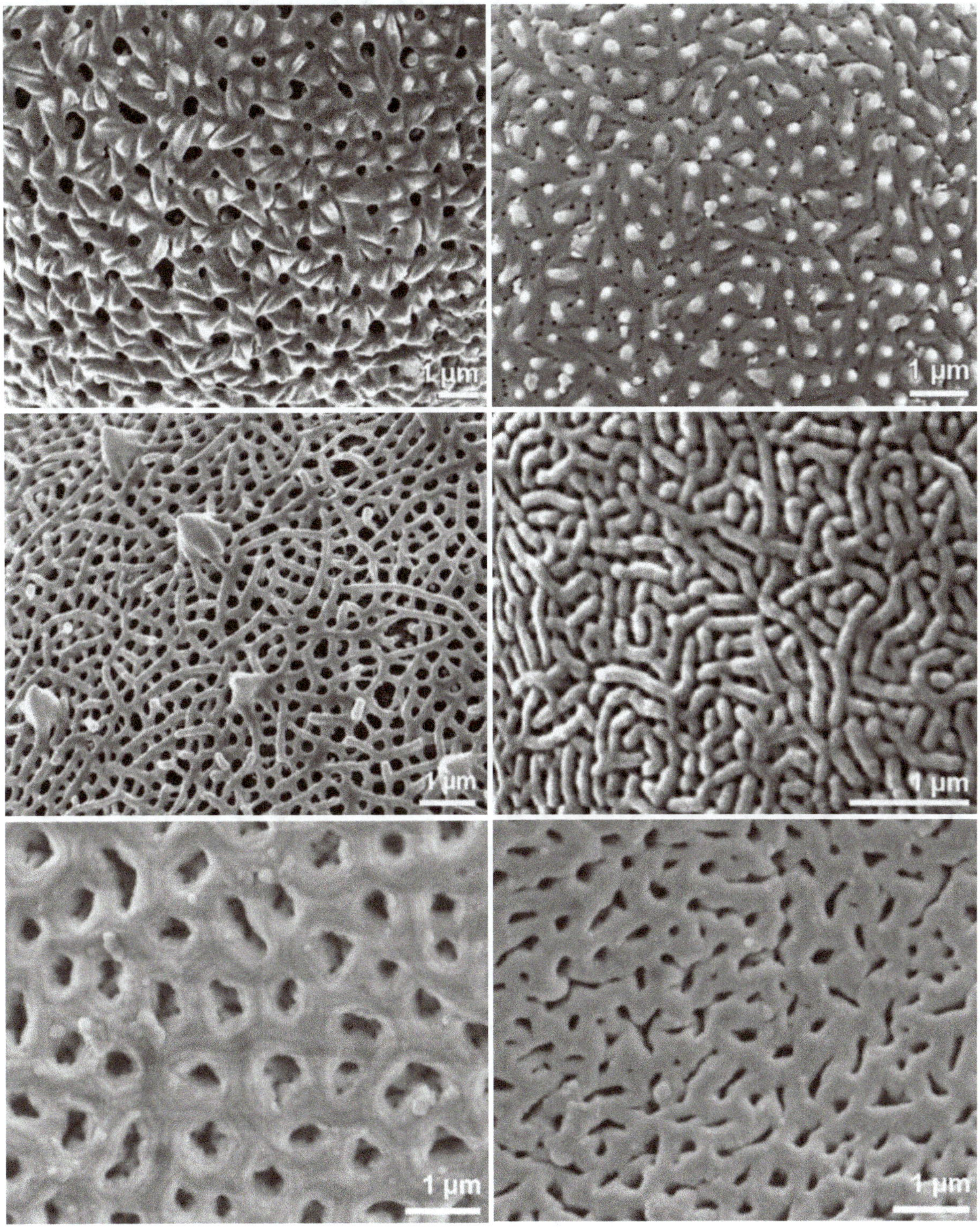

Podophyllum peltatum, Berberidaceae
microechinate, perforate; microechini depressed

Campanula persicifolia, Campanulaceae
microechinate, striato-microreticulate

Tilia sp., Malvaceae
fossil, Miocene, China, microreticulate

Solanum torvum, Solanaceae
microrugulate, microechinate, perforate

Melampyrum pratense, Orobanchaceae
microrugulate

Tilia sp., Malvaceae
fossil, Miocene, China, microreticulate

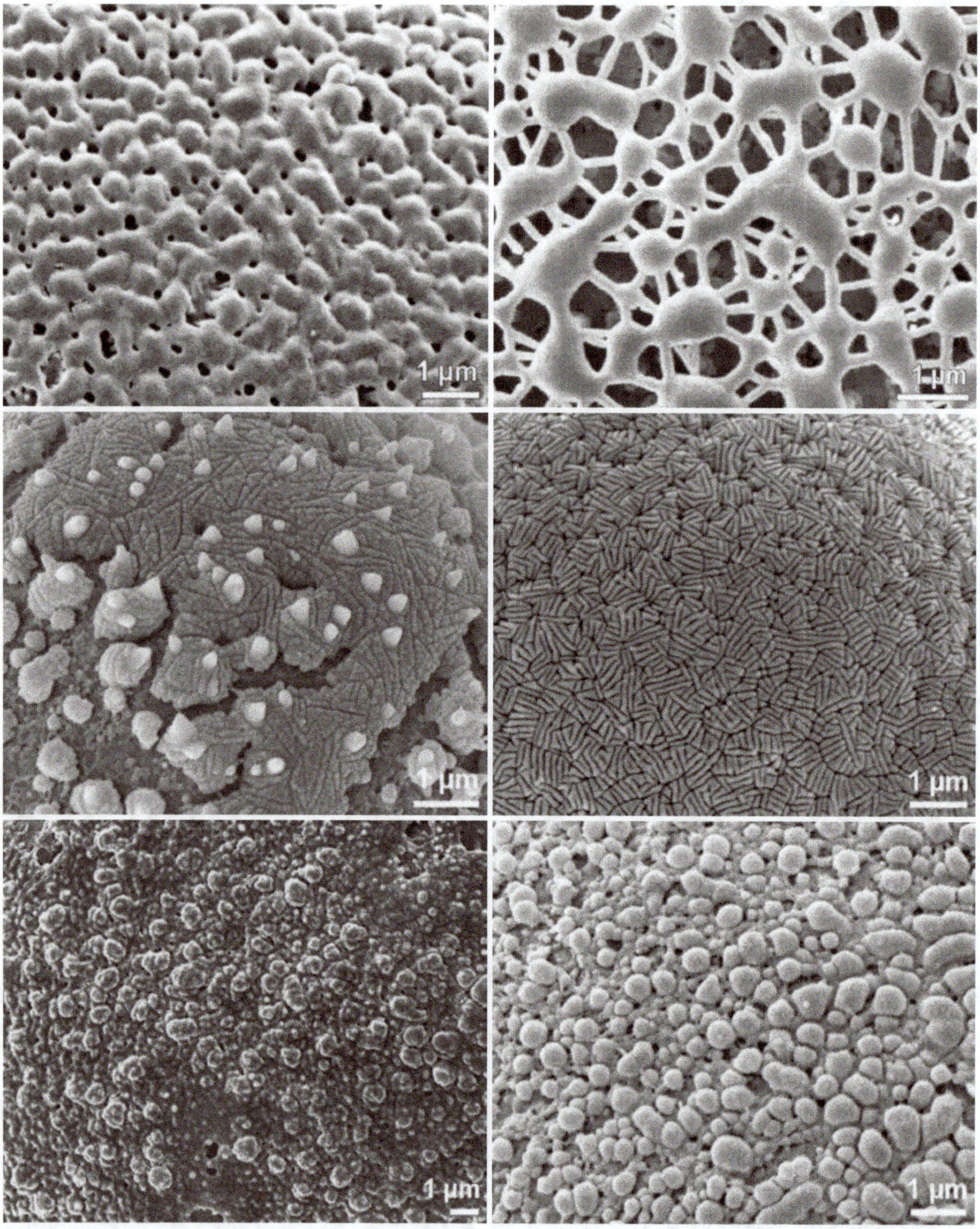

Tradescantia spathacea, Commelinaceae
microverrucate, perforate

Aspidistra locii, Asparagaceae
microverrucate to verrucate, microreticulate

Smilax spinosa, Smilacaceae
microechinate, rugulate to microrugulate; (micro)rugulae segmented

Clethra arborea, Clethraceae
microrugulate

Quercus robur, Fagaceae
microgemmate to granulate

Erica pageana, Ericaceae
micro- to nanogemmate

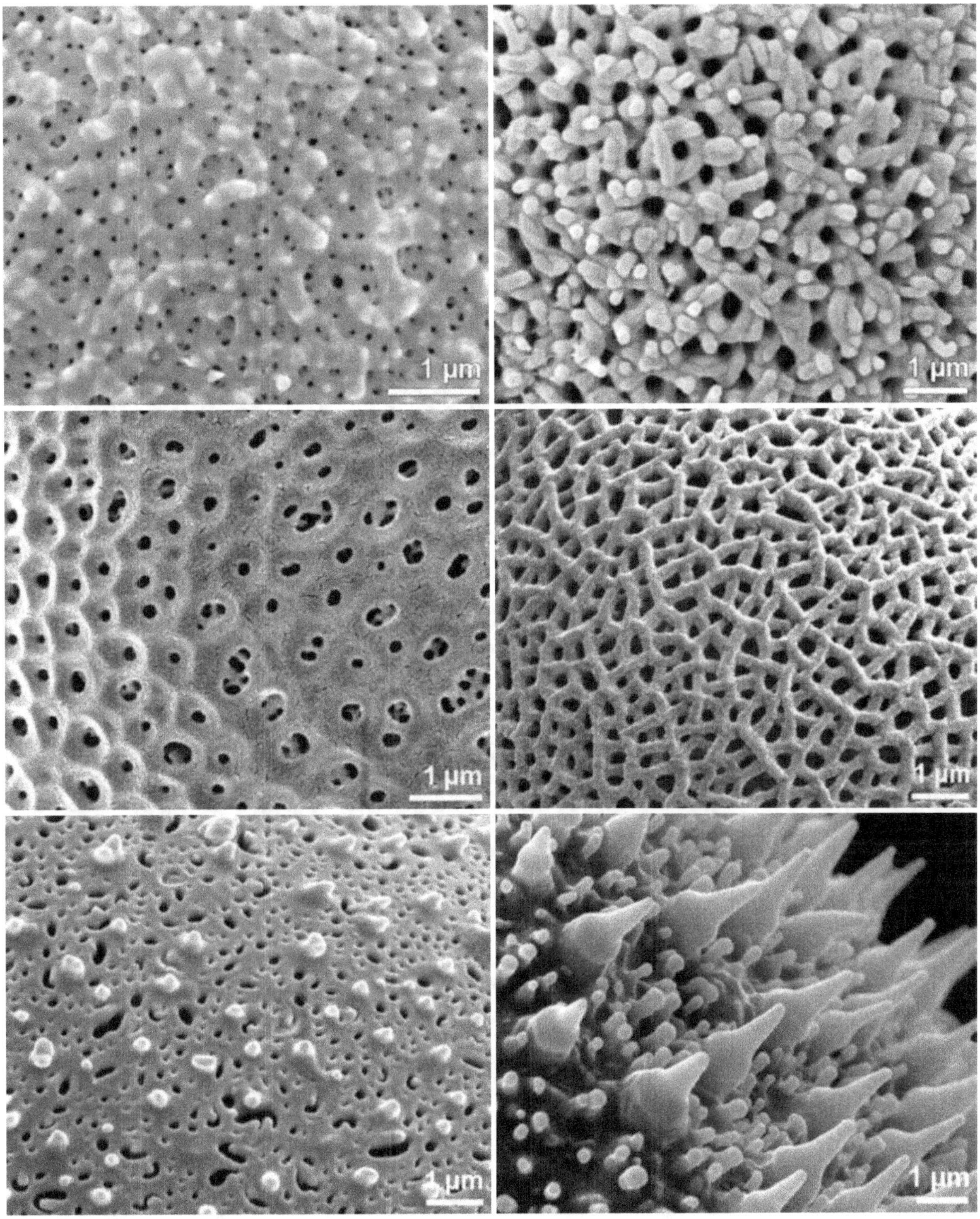

Elaeagnus rhamnoides, Elaeagnaceae
microrugulate, nanoechinate, perforate

Cytisus nigricans, Fabaceae
microreticulate, perforate

Anemone pratensis, Ranunculaceae
microechinate, perforate

Polyscias filicifolia, Araliaceae
microreticulate, microrugulate

Reseda luteola, Resedaceae
microreticulate

Drosera kansaiensis, Droseraceae
echinate, microclavate

nano-

prefix for very small, features between 0.5 and 0.1 µm

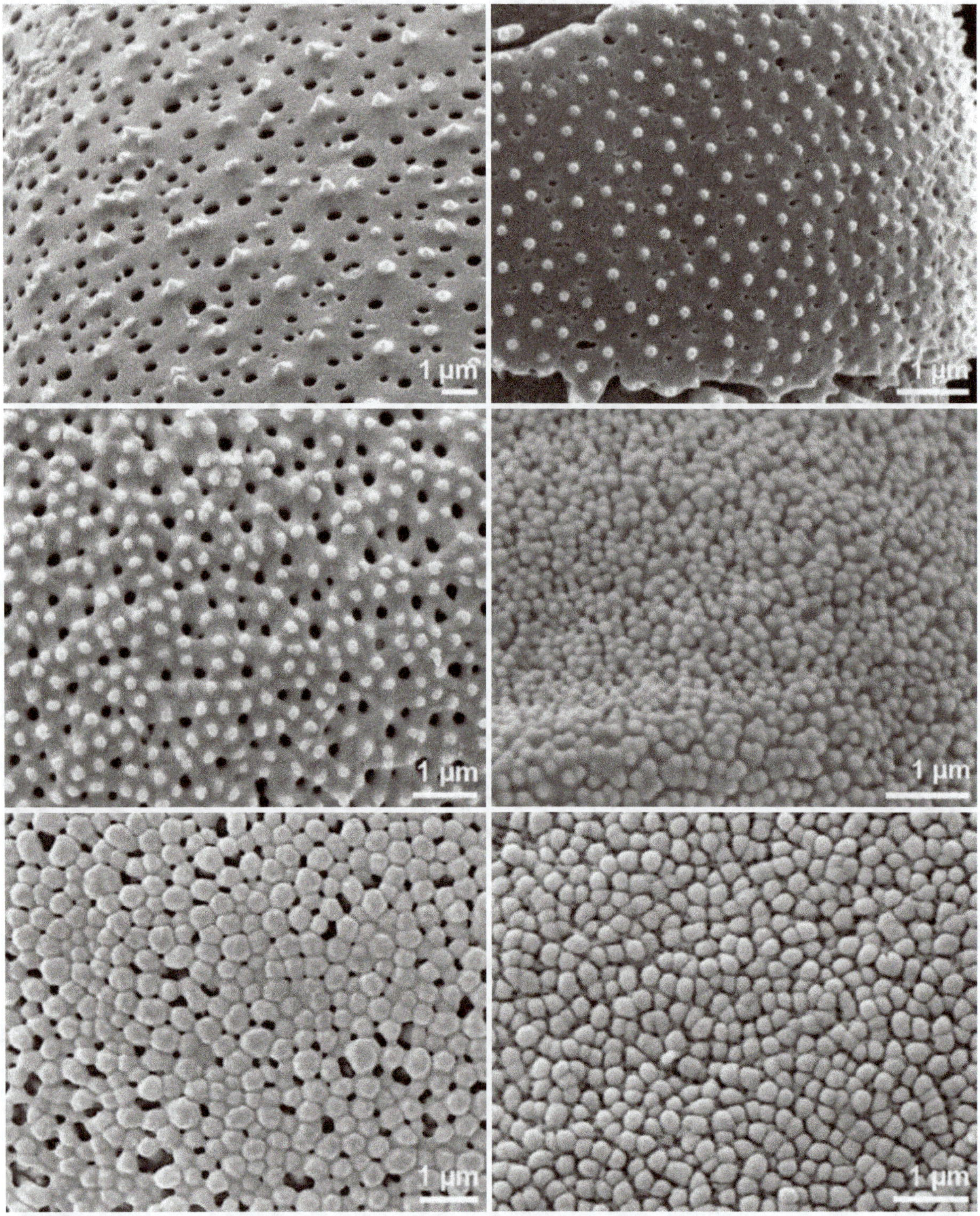

Babiana ecklonii, Iridaceae
nanoechinate

Galium lucidum, Rubiaceae
nanoechinate

Scirpoides holoschoenus, Cyperaceae
nanoechinate to nanogemmate

Poaceae
fossil, Miocene, China, nanoechinate

Callisia fragrans, Commelinaceae
nano- to microgemmate

Oenothera fruticosa, Onagraceae
nanogemmate

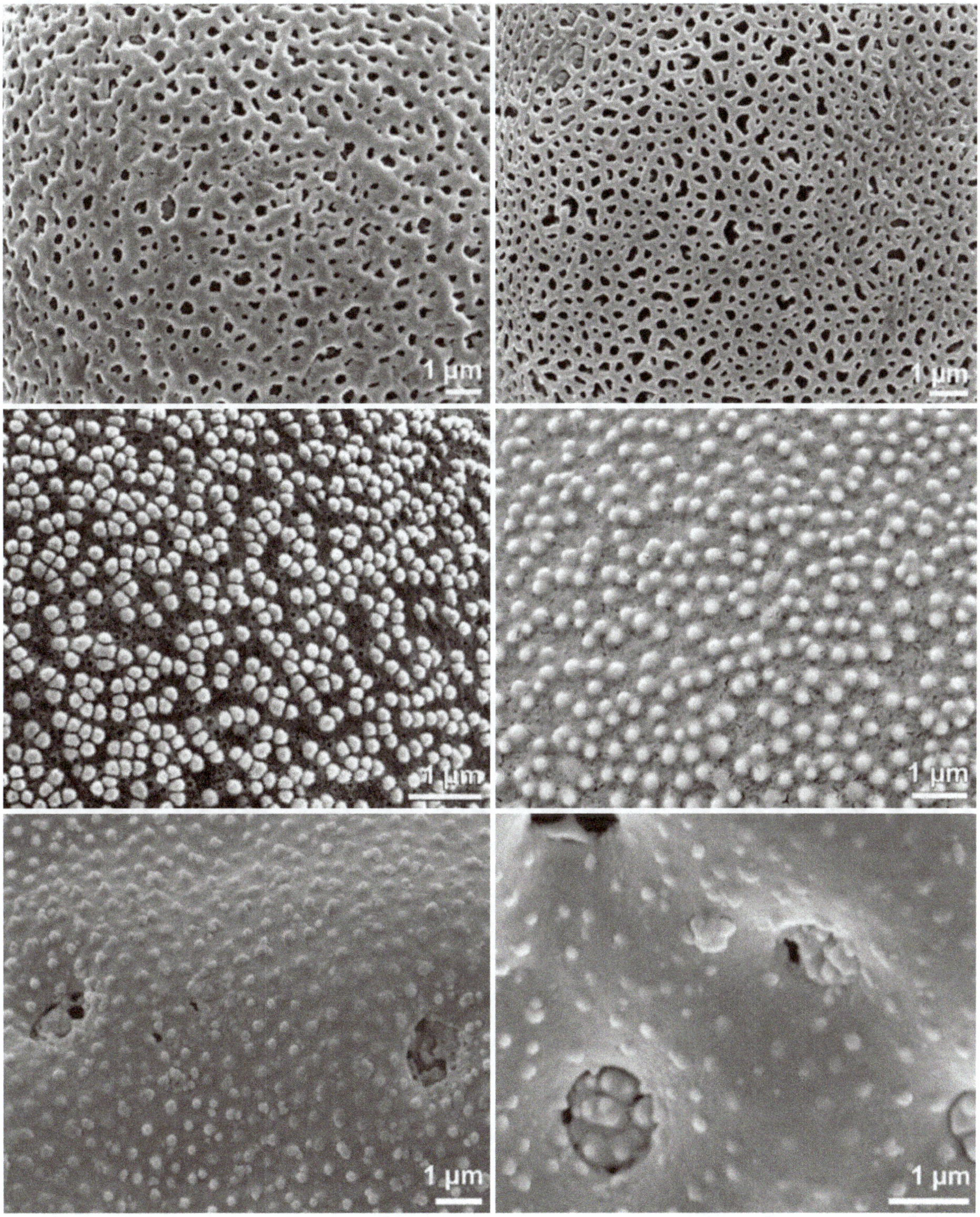

Dianella tasmanica, Xanthorrhoeaceae
nanoreticulate

Symphytum caucasicum, Boraginaceae
nanogemmate

Juglans sp. Juglandaceae
fossil, Miocene, China, nanoechinate

Veronica longifolia, Plantaginaceae
nanoreticulate

Hordeum bulbosum, Poaceae
nanogemmate to nanoverrucate

Amaranthaceae
fossil, Miocene, China, nanoechinate

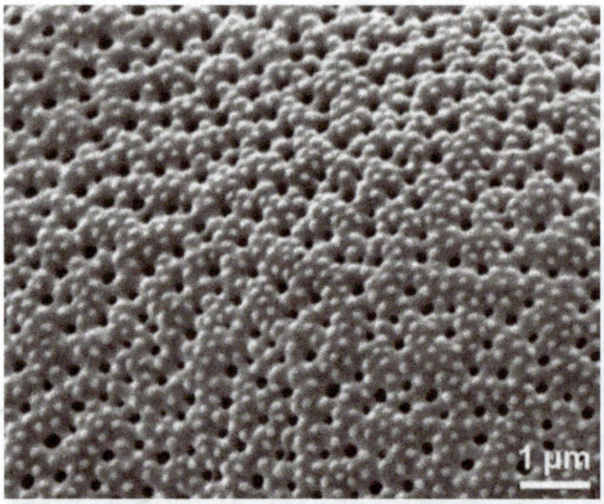

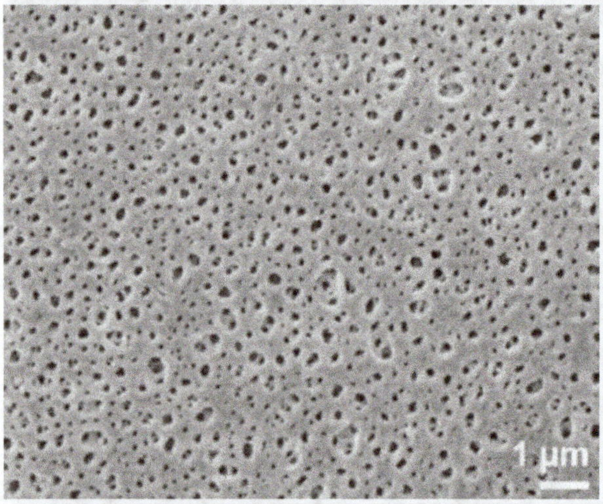

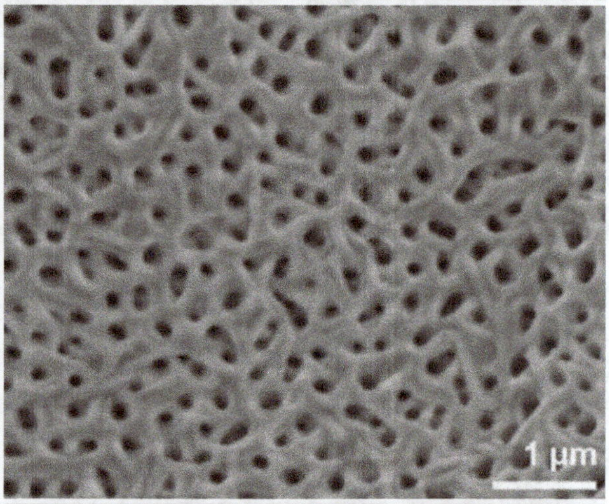

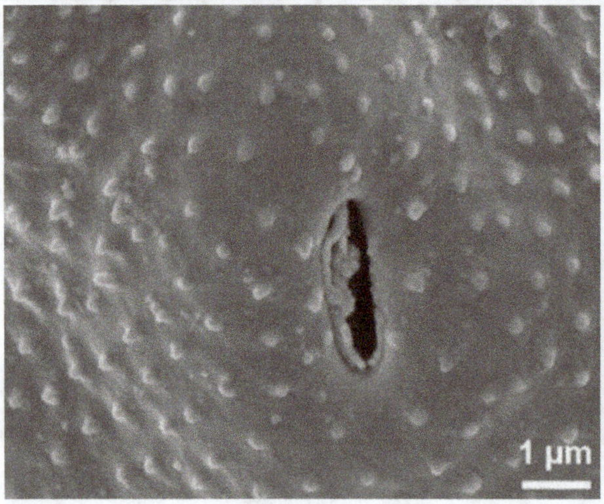

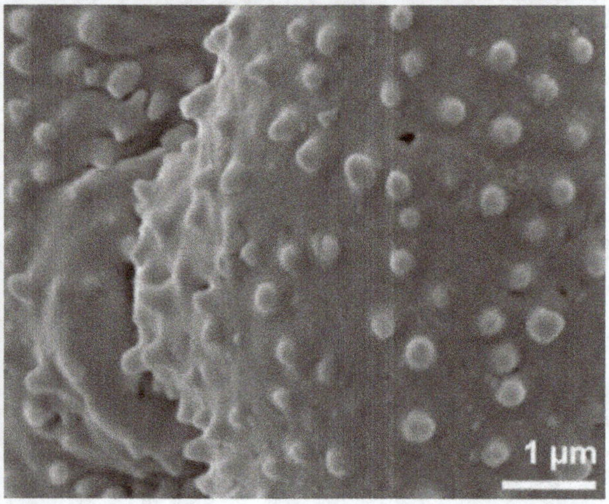

Carex filiformis, Cyperaceae
nanogemmate to nanoverrucate

Viola calcarata, Violaceae
nanoreticulate

Pterocarya sp., Juglandaceae
fossil, Miocene, China, nanoechinate

Cyperaceae
fossil, Miocene, China, nanoverrucate

Tilia sp., Malvaceae
fossil, Miocene, China, nanoreticulate

Rosaceae
fossil, Miocene, China, nano- to microechinate

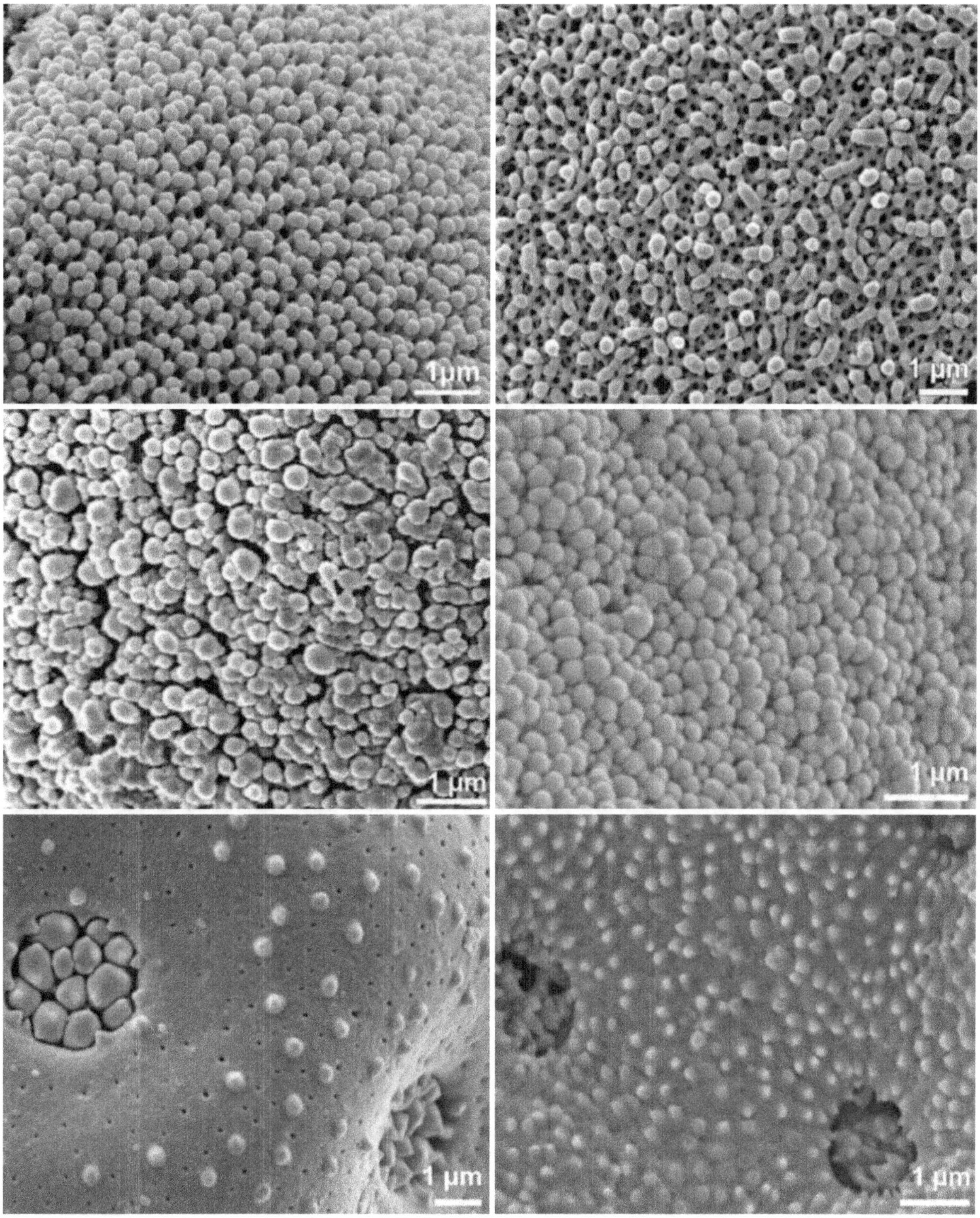

Borago pygmaea, Boraginaceae
nanogemmate

Rhaphidophora africana, Araceae
nanogemmate to nanoverrucate

Caryophyllaceae
fossil, Miocene, China, nanoechinate

Epilobium palustre, Onagraceae
nanogemmate to microrugulate

Trapa sp., Trapaceae
fossil, Miocene, China, nanogemmate

Amaranthaceae
fossil, Miocene, China, nanoechinate

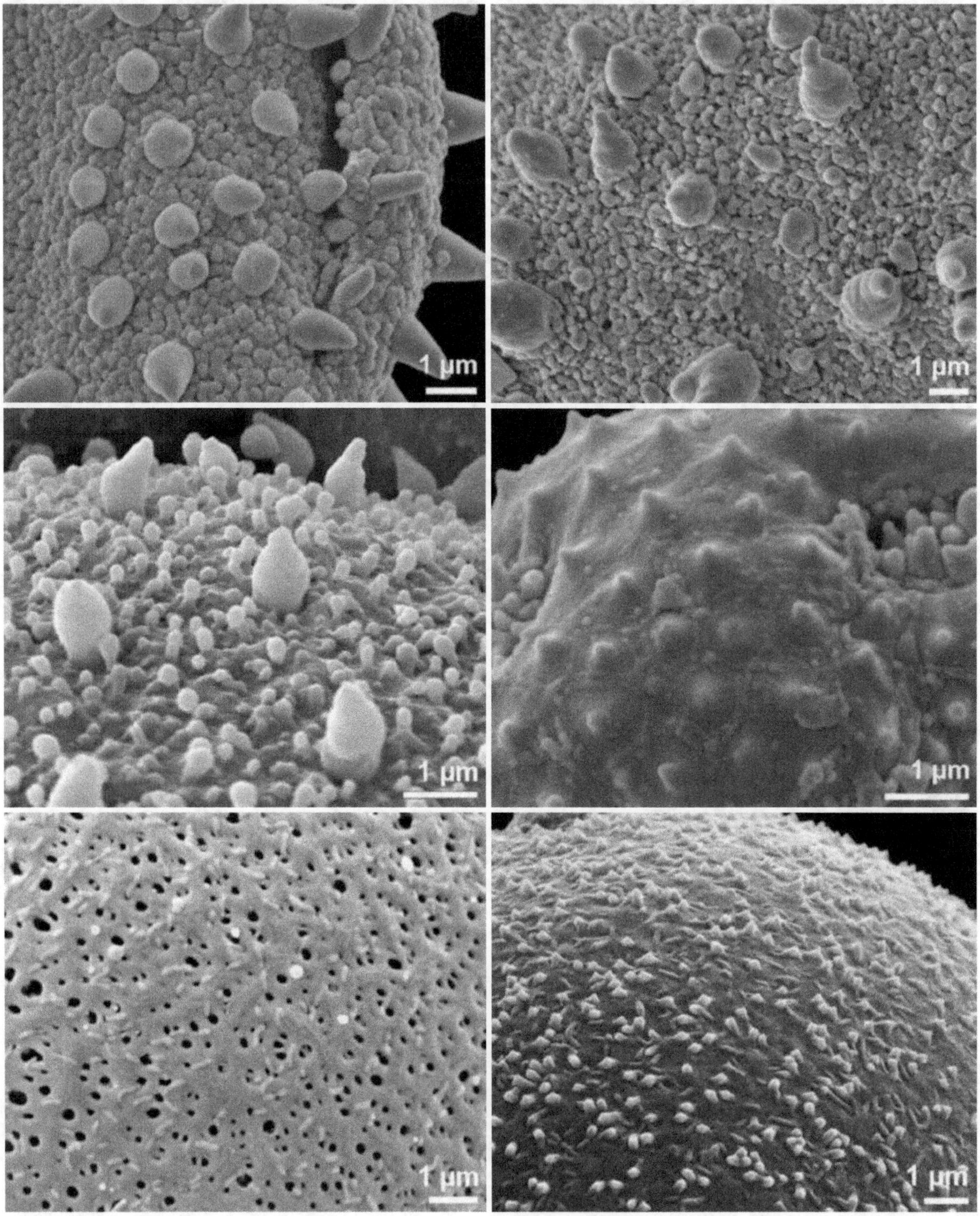

Arceuthobium sp., Santalaceae
fossil, Miocene, China, nanoverrucate, echinate

Caprifoliaceae
fossil, Miocene, China, granulate, nanoverrucate, echinate

Drosera binata, Droseraceae
nanoclavate, echinate

Alisma sp., Alismataceae
fossil, Miocene, China, nanoechinate

Heliohebe raoulii, Plantaginaceae
nanorugulate

Maclura pomifera, Moraceae
nanoechinate

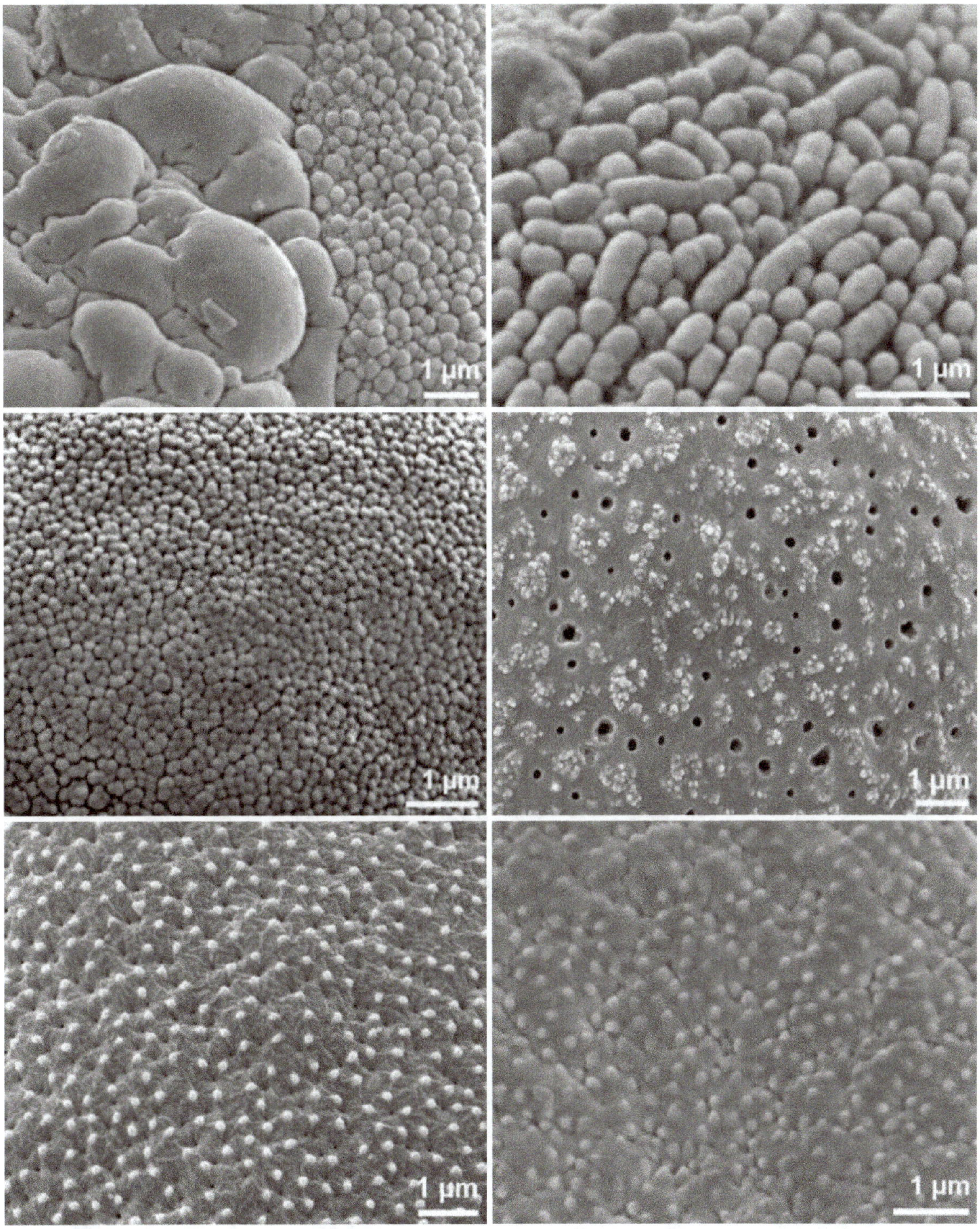

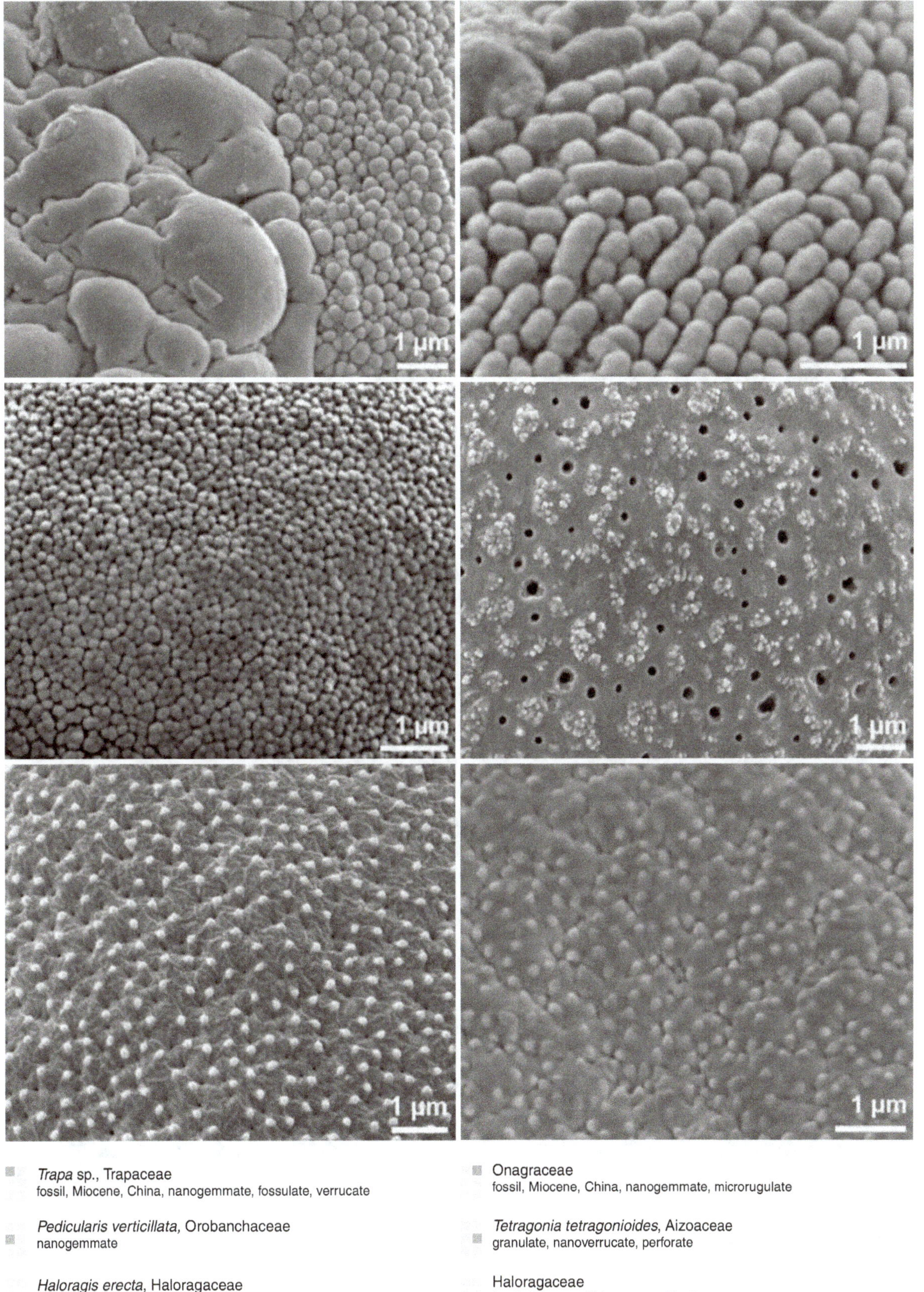

Trapa sp., Trapaceae
fossil, Miocene, China, nanogemmate, fossulate, verrucate

Onagraceae
fossil, Miocene, China, nanogemmate, microrugulate

Pedicularis verticillata, Orobanchaceae
nanogemmate

Tetragonia tetragonioides, Aizoaceae
granulate, nanoverrucate, perforate

Haloragis erecta, Haloragaceae
nanoechinate, nanorugulate

Haloragaceae
fossil, Miocene, China, nanoechinate

perforate

pollen wall with holes less than 1 μm in diameter

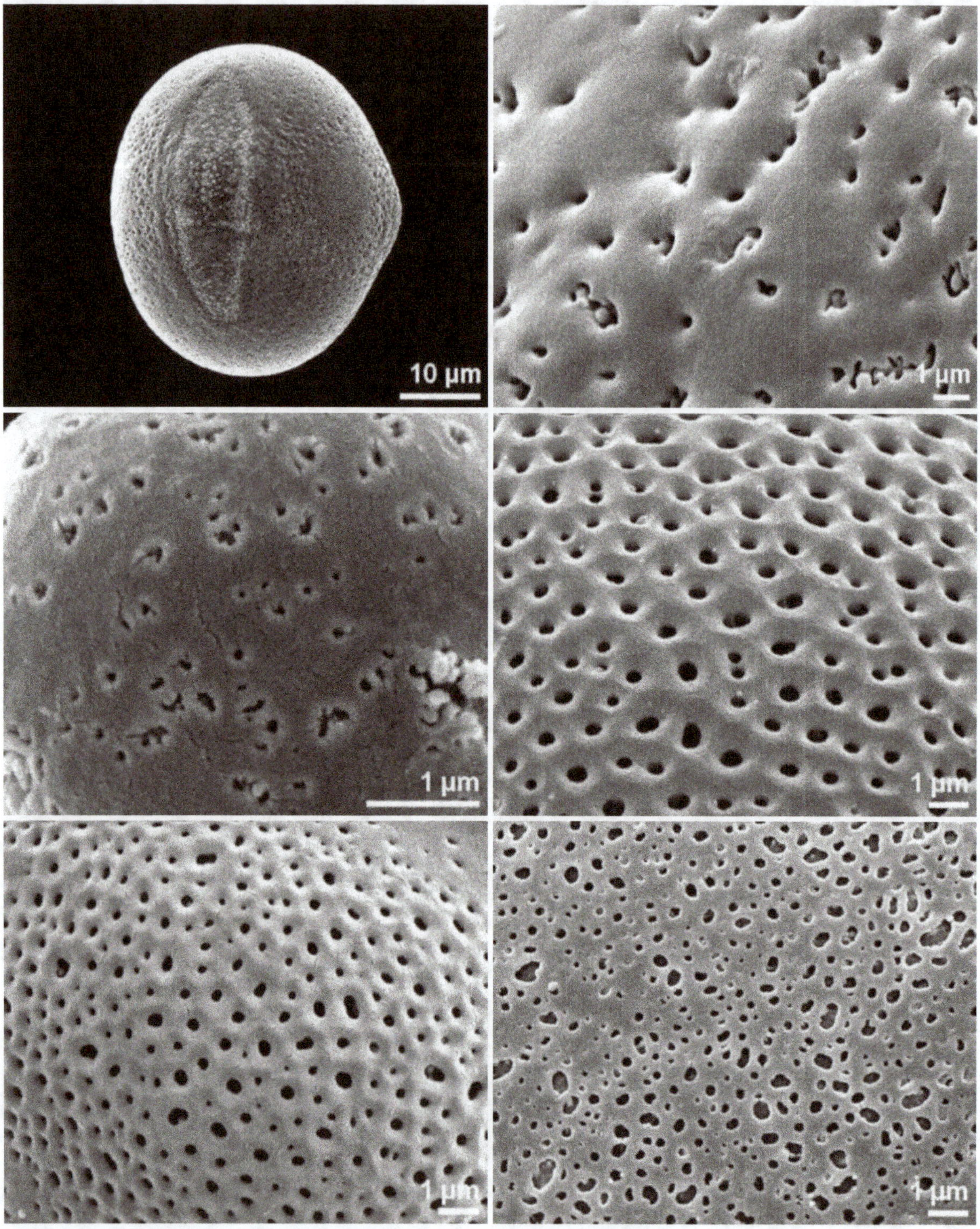

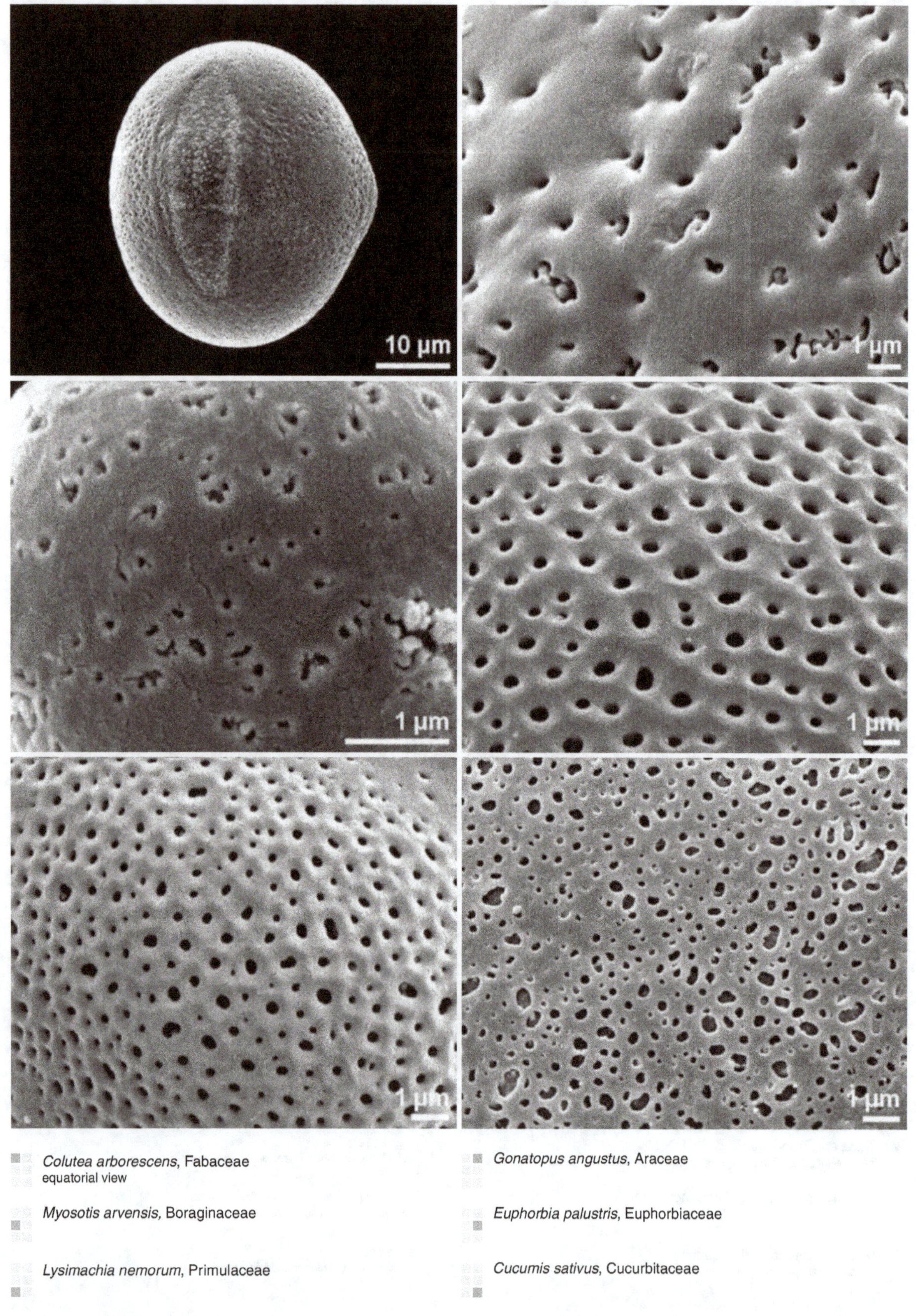

Colutea arborescens, Fabaceae
equatorial view

Myosotis arvensis, Boraginaceae

Lysimachia nemorum, Primulaceae

Gonatopus angustus, Araceae

Euphorbia palustris, Euphorbiaceae

Cucumis sativus, Cucurbitaceae

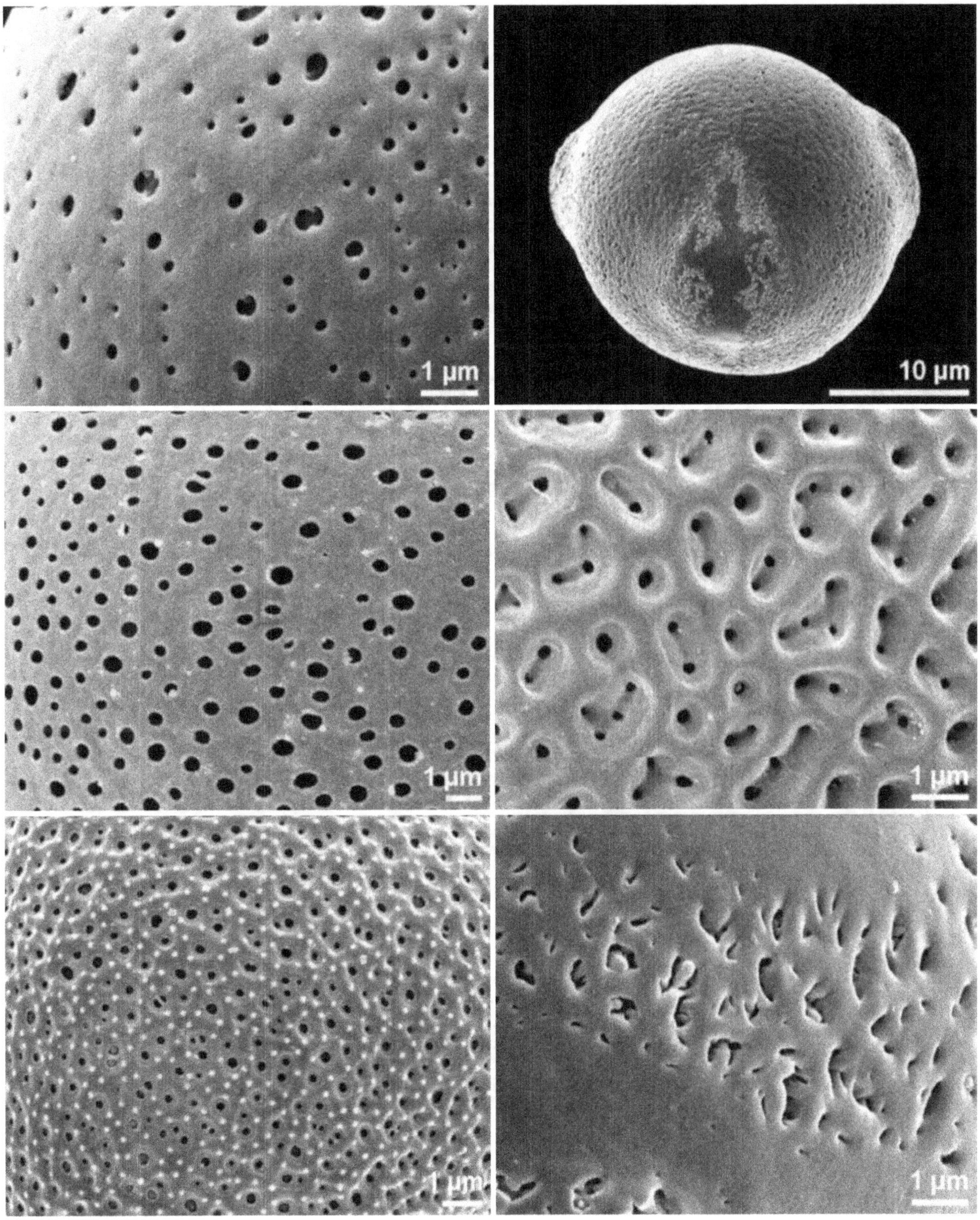

Pulmonaria officinalis, Boraginaceae

Hippocrepis emerus, Fabaceae
equatorial view

Napoleonaea imperialis, Lecythidaceae

Euphorbia helioscopia, Euphorbiaceae

Rumex acetosa, Polygonaceae
perforate, granulate

Trifolium montanum, Fabaceae

plicae/plicate

coarse parallel ridges

Ephedra distachya, Ephedraceae
plica colored

Ephedra sp., Ephedraceae
fossil, Miocene, China, equatorial view

Ephedra sp., Ephedraceae
fossil, Miocene, Austria

Hemigraphis primulaefolia, Acanthaceae
polar (left) and equatorial view (right)

Pistia stratiotes, Araceae
equatorial view

Pistia stratiotes, Araceae
cross section of pollen

Welwitschia mirabilis, Welwitschiaceae

Amorphophallus serrulatus, Araceae
hydrated

Spathiphyllum minor, Araceae
hydrated

Amorphophallus lacourii, Araceae

Brillantaisia owariensis, Acanthaceae
oblique equatorial view

Spathiphyllum cannifolium, Araceae

psilate

pollen wall with smooth surface

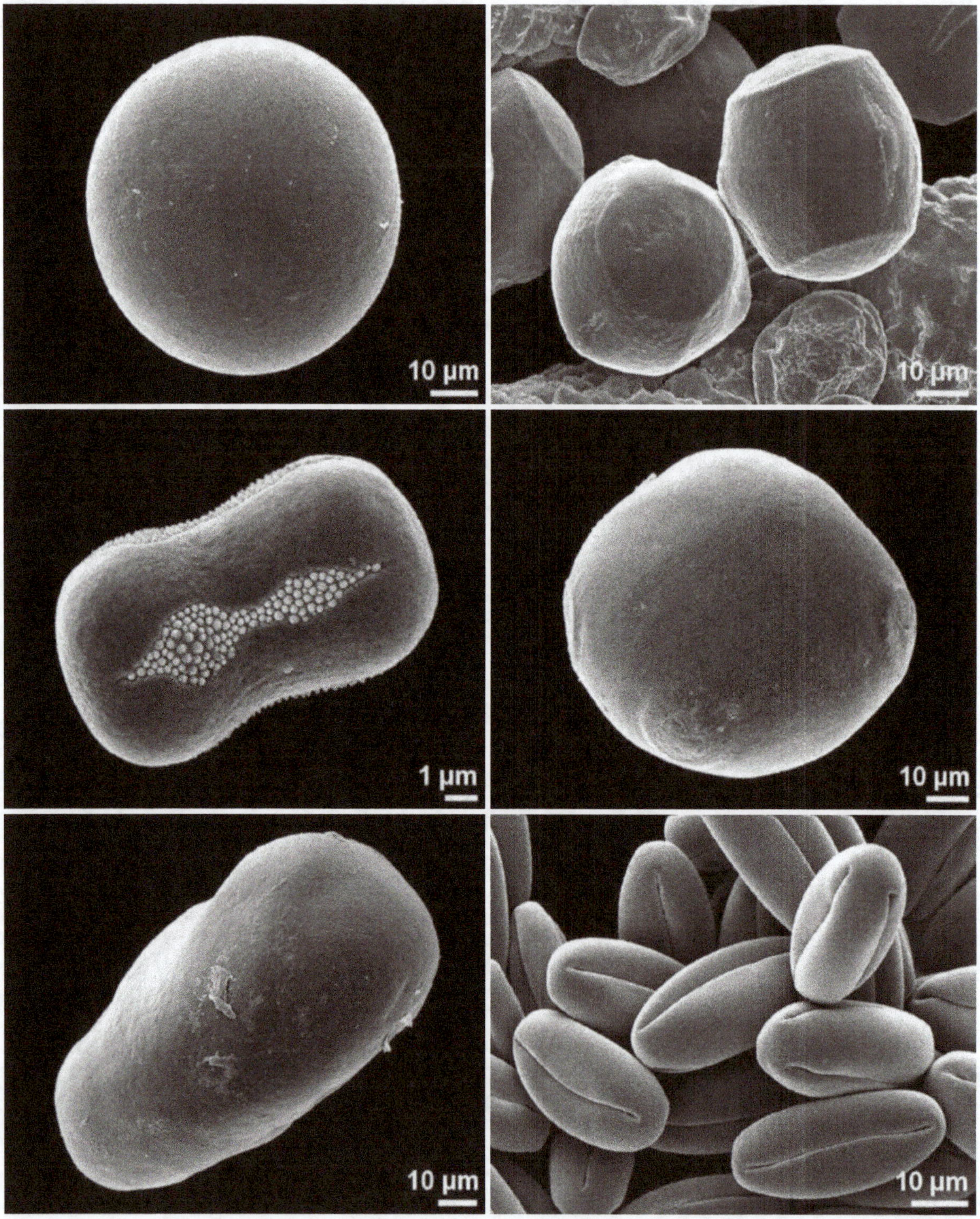

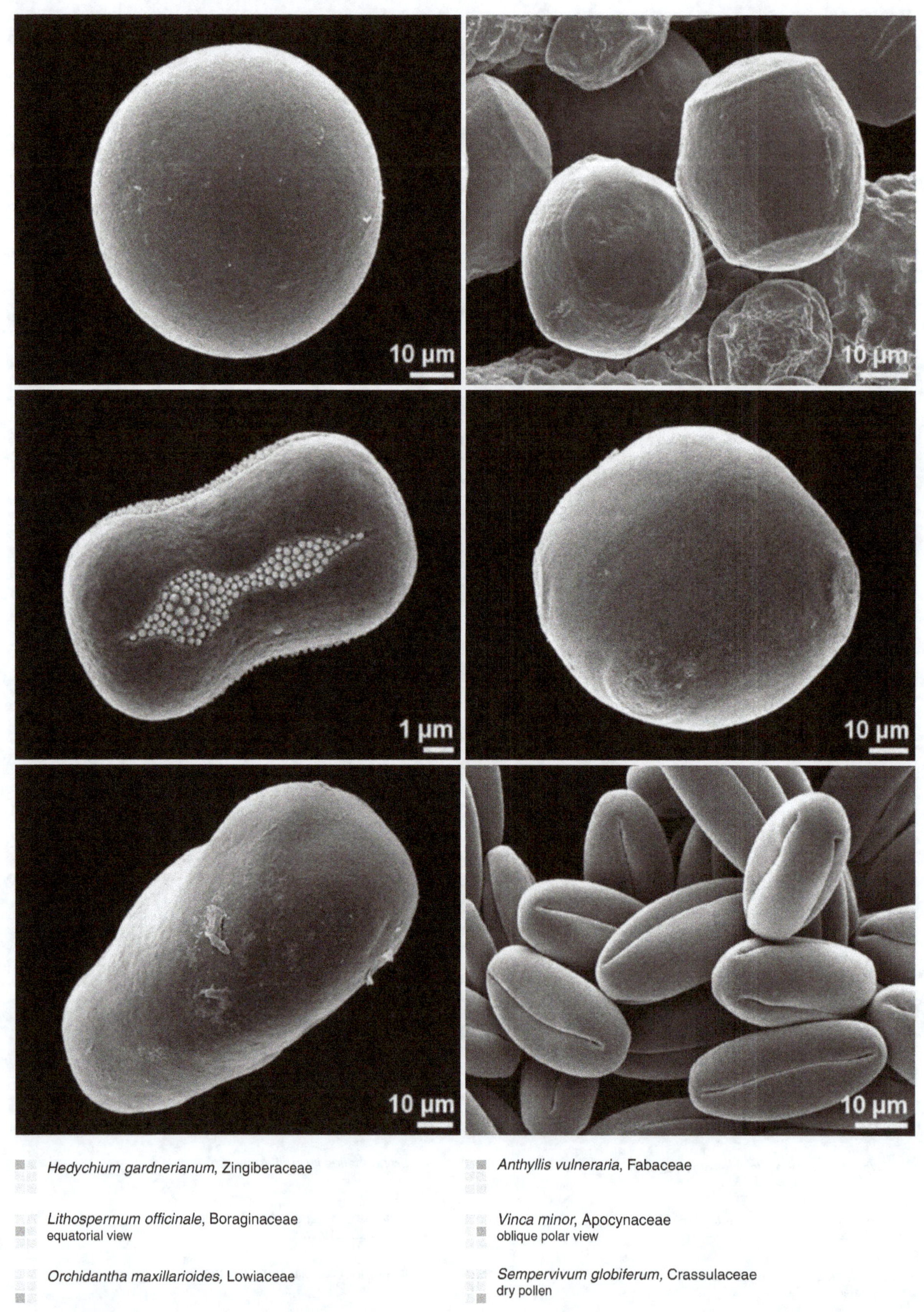

Hedychium gardnerianum, Zingiberaceae

Anthyllis vulneraria, Fabaceae

Lithospermum officinale, Boraginaceae
equatorial view

Vinca minor, Apocynaceae
oblique polar view

Orchidantha maxillarioides, Lowiaceae

Sempervivum globiferum, Crassulaceae
dry pollen

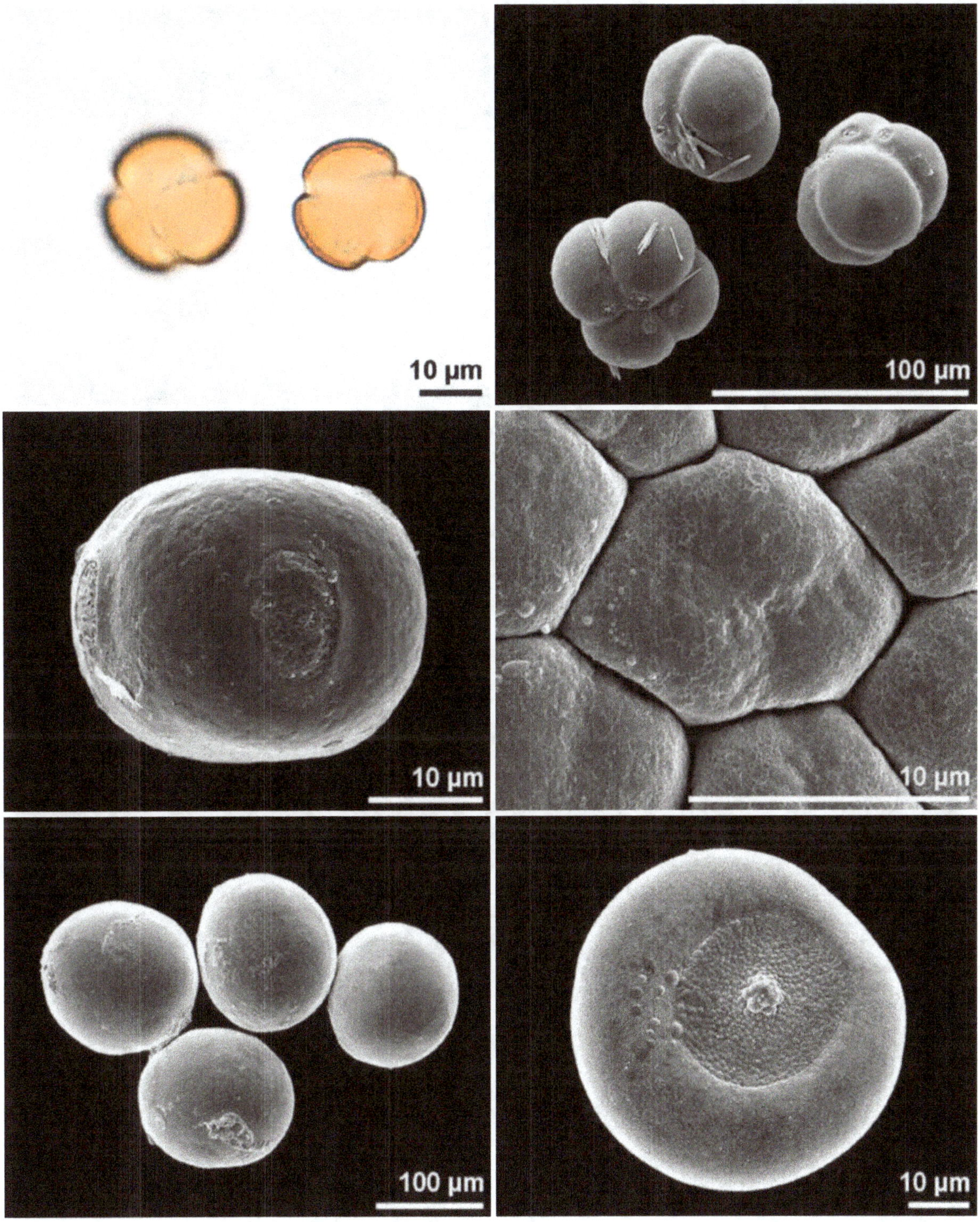

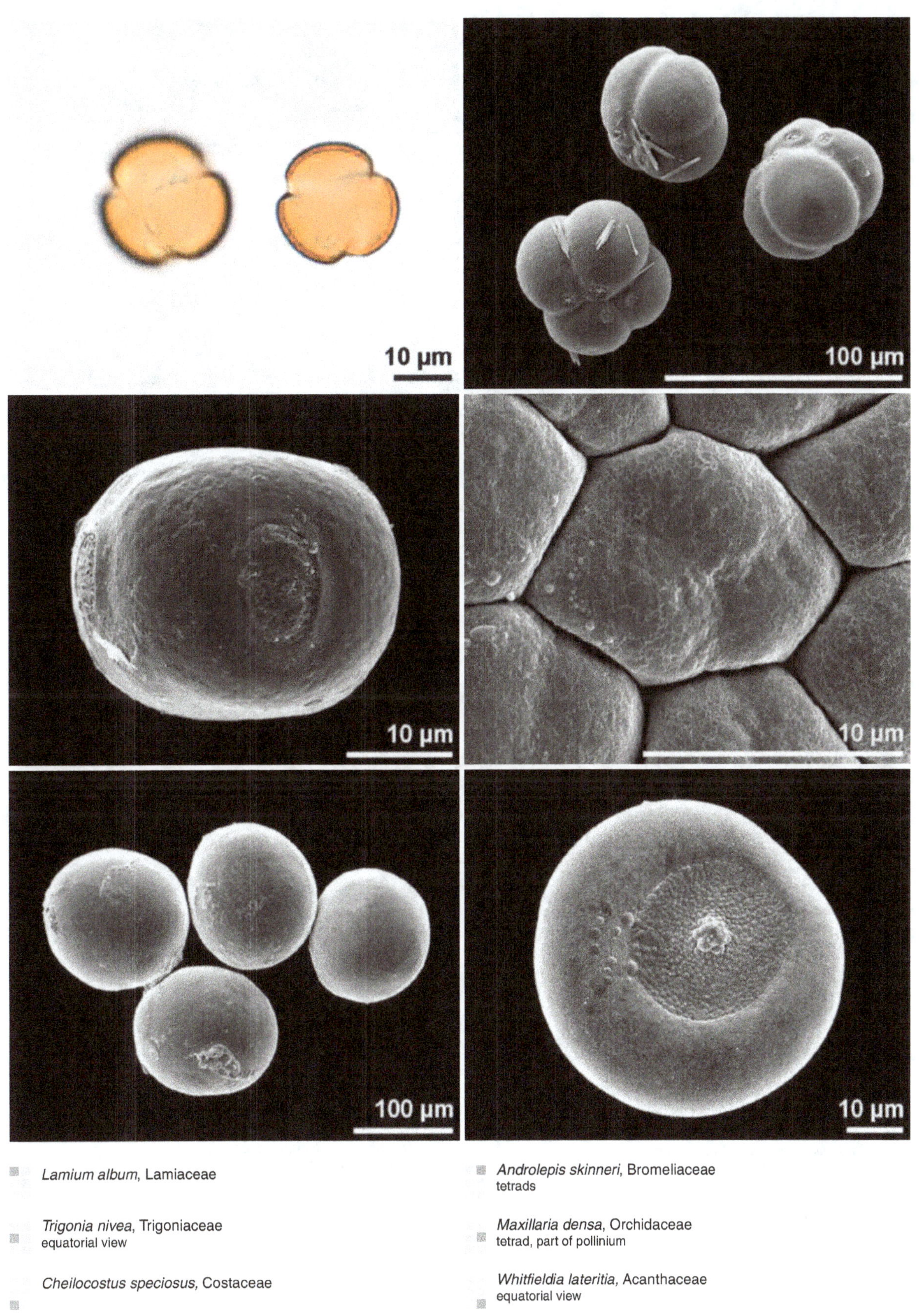

Lamium album, Lamiaceae

Androlepis skinneri, Bromeliaceae
tetrads

Trigonia nivea, Trigoniaceae
equatorial view

Maxillaria densa, Orchidaceae
tetrad, part of pollinium

Cheilocostus speciosus, Costaceae

Whitfieldia lateritia, Acanthaceae
equatorial view

reticulum/reticulate

reticulum: network like pattern consisting of muri and lumina

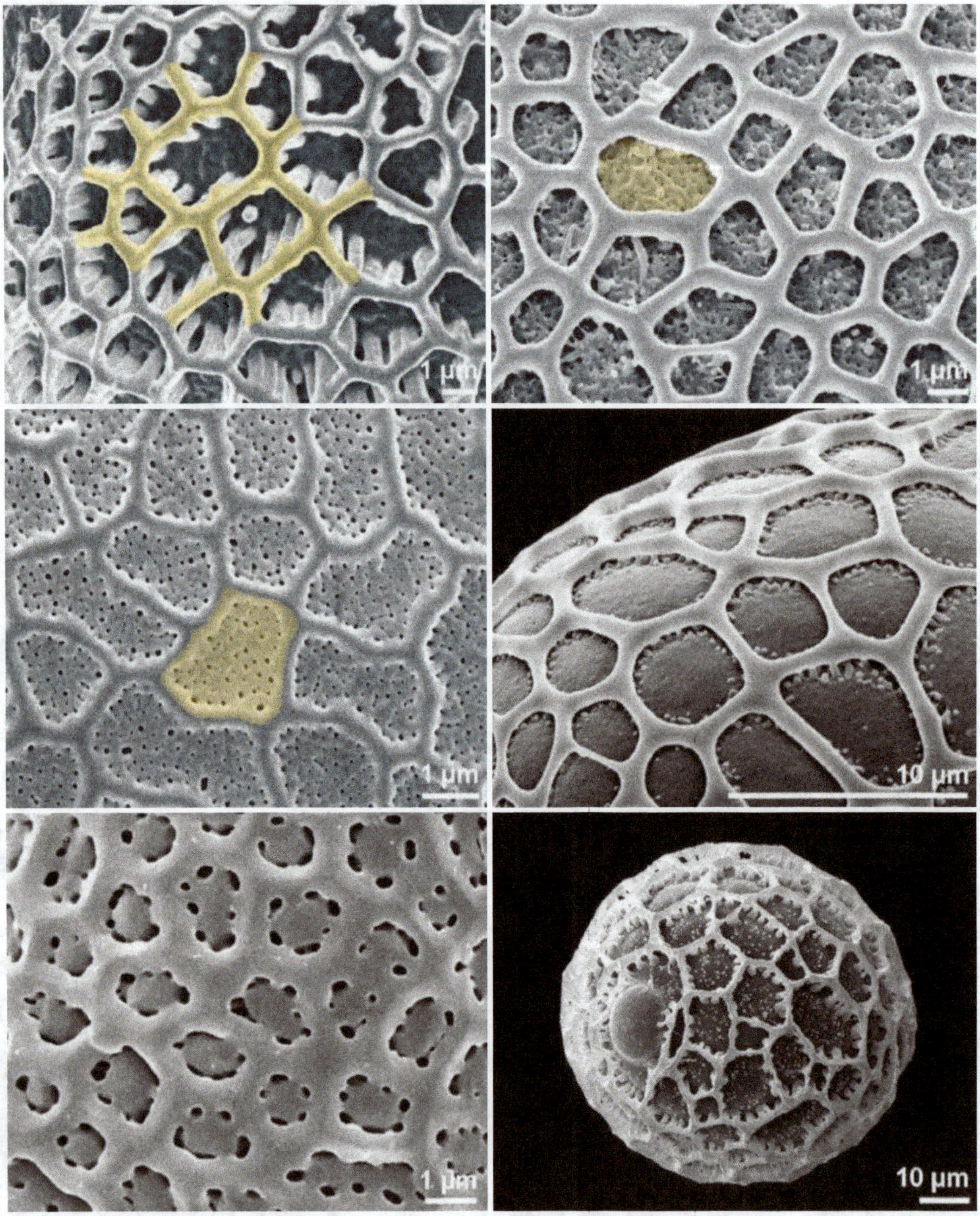

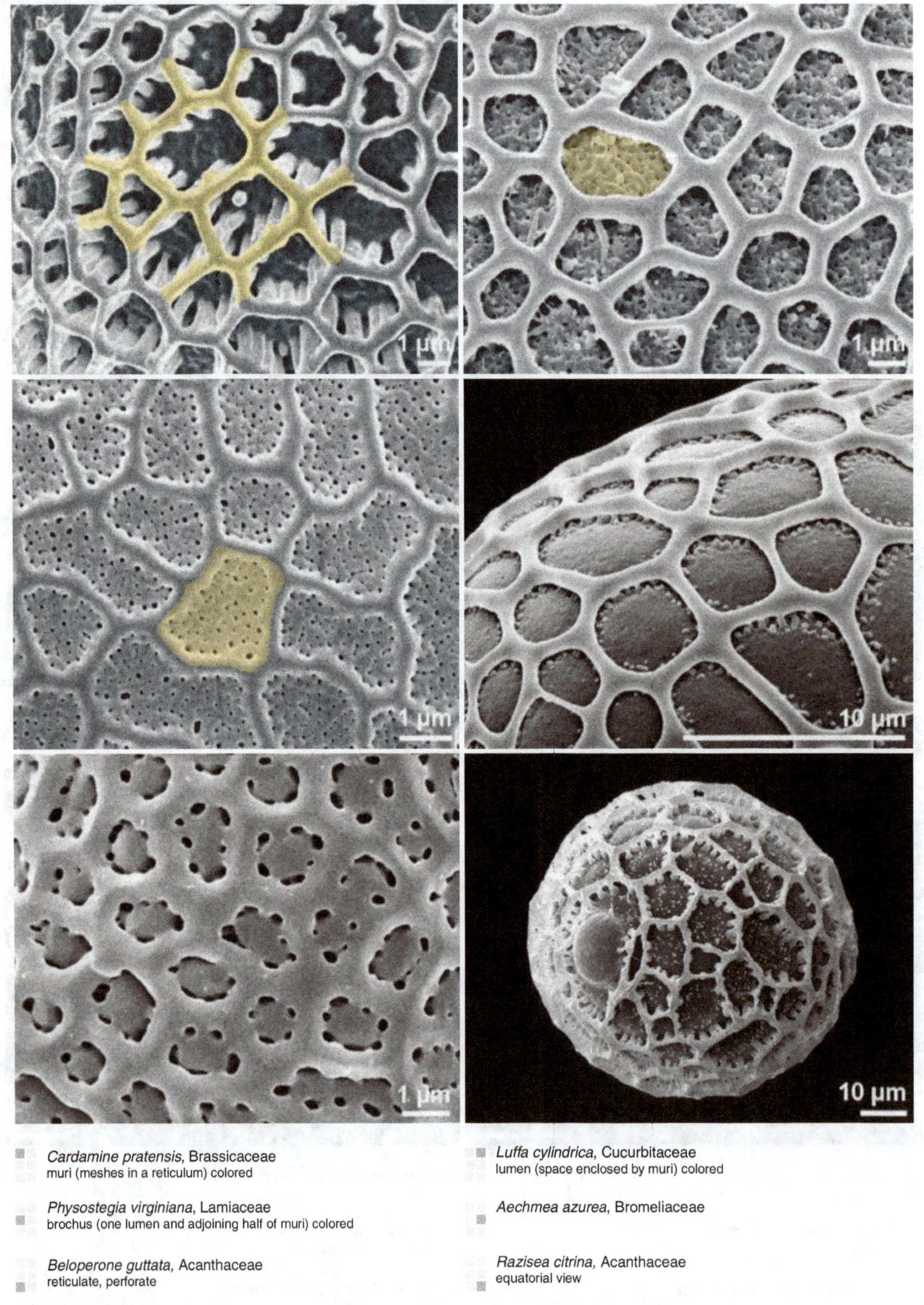

Cardamine pratensis, Brassicaceae
muri (meshes in a reticulum) colored

Physostegia virginiana, Lamiaceae
brochus (one lumen and adjoining half of muri) colored

Beloperone guttata, Acanthaceae
reticulate, perforate

Luffa cylindrica, Cucurbitaceae
lumen (space enclosed by muri) colored

Aechmea azurea, Bromeliaceae

Razisea citrina, Acanthaceae
equatorial view

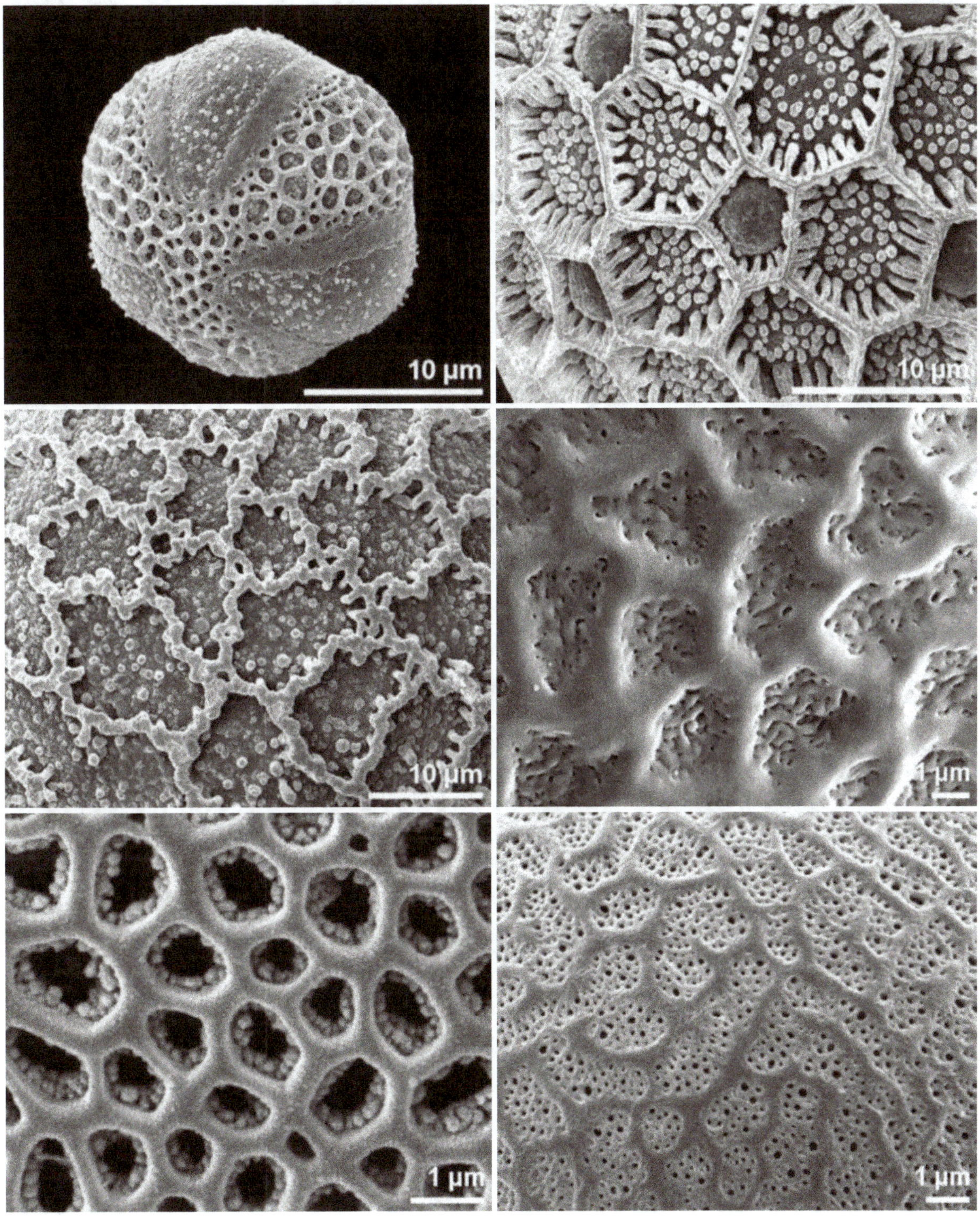

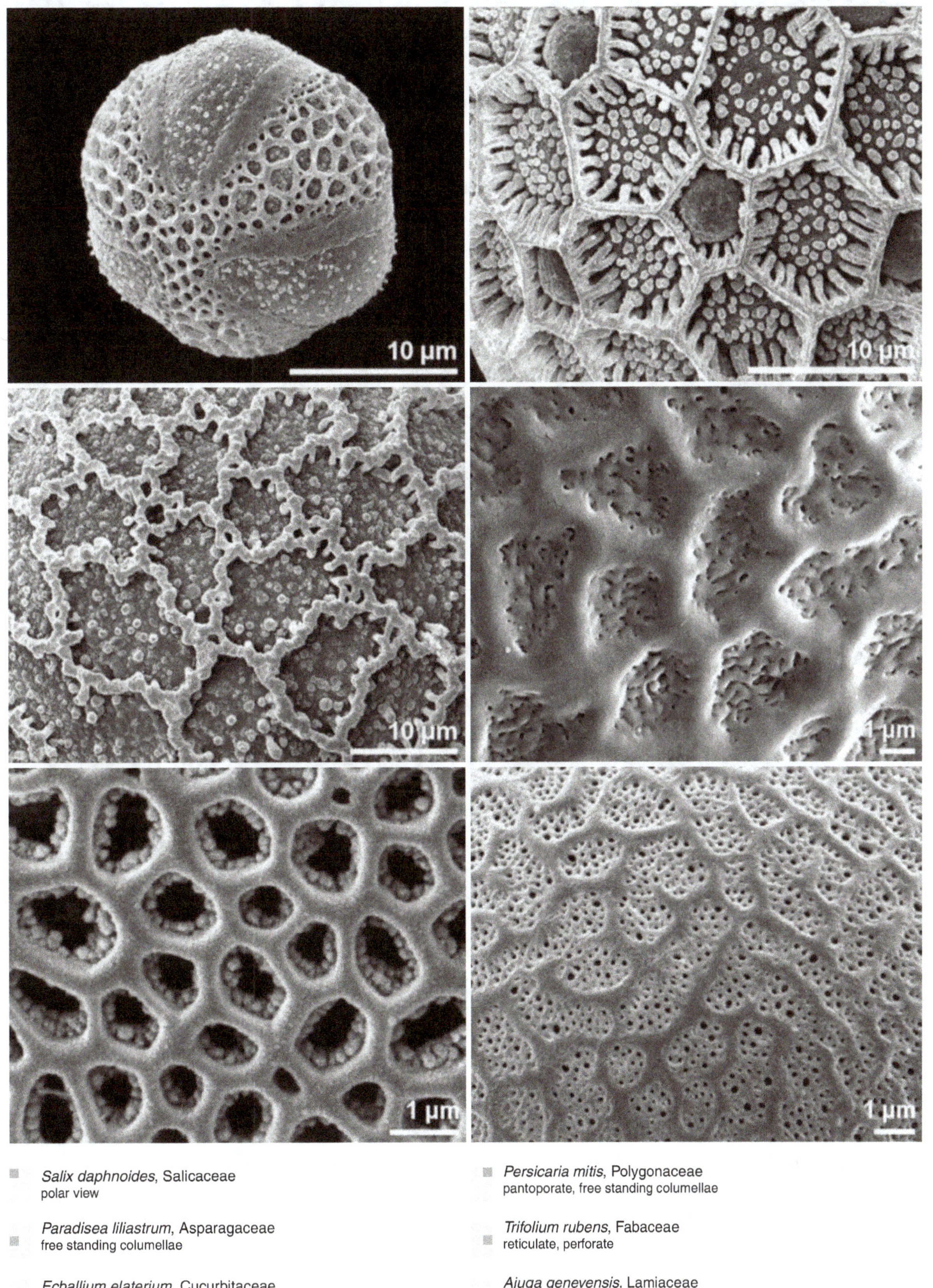

- *Salix daphnoides*, Salicaceae
  polar view
- *Paradisea liliastrum*, Asparagaceae
  free standing columellae
- *Ecballium elaterium*, Cucurbitaceae
- *Persicaria mitis*, Polygonaceae
  pantoporate, free standing columellae
- *Trifolium rubens*, Fabaceae
  reticulate, perforate
- *Ajuga genevensis*, Lamiaceae
  reticulate, perforate

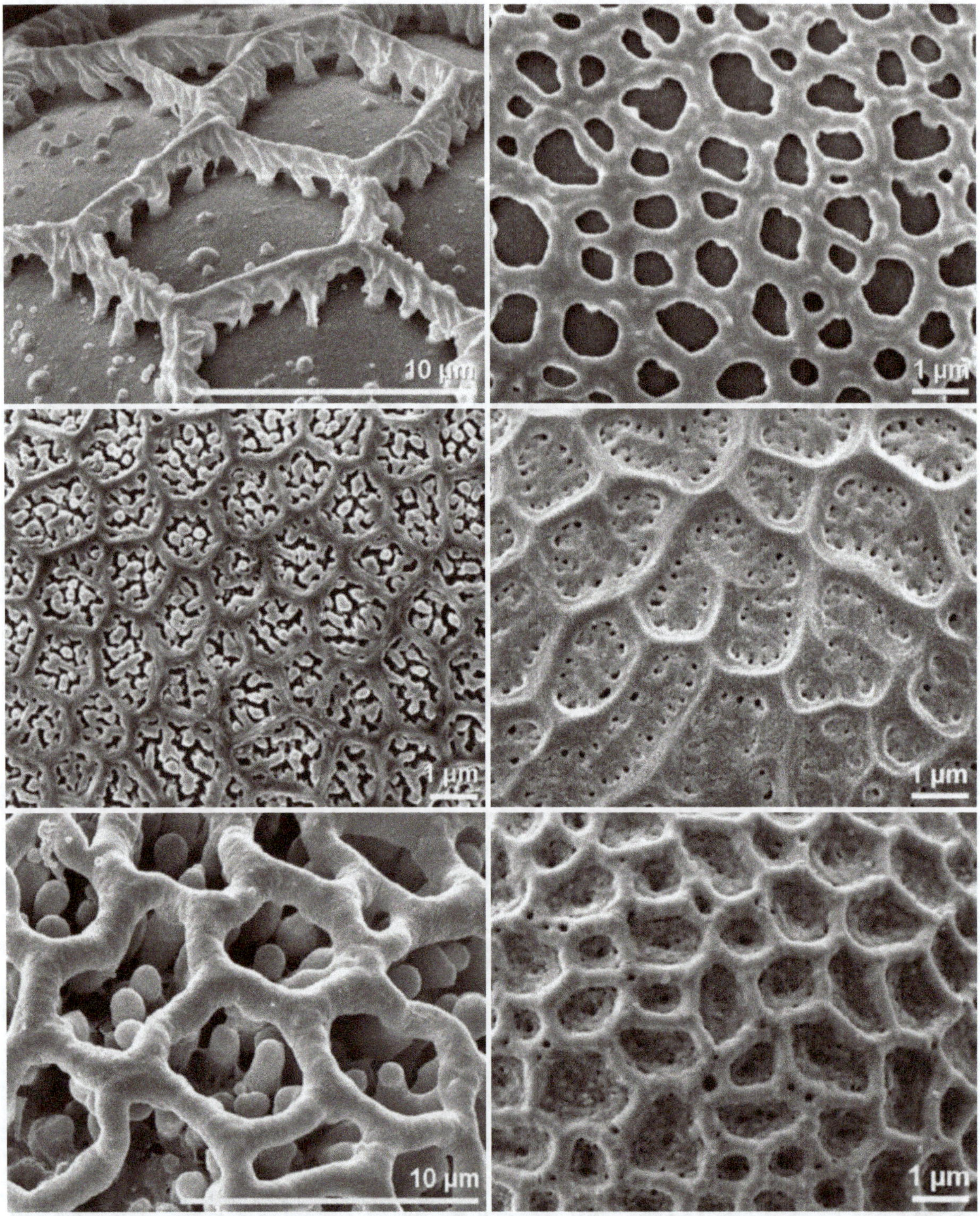

Ruellia brevifolia, Acanthaceae

Impatiens glandulifera, Balsaminaceae

Opuntia paraguayensis, Cactaceae
free standing columellae

Buxus sempervirens, Buxaceae

Lupinus polyphyllus, Fabaceae
reticulate, perforate

Theobroma cacao, Malvaceae
reticulate, perforate

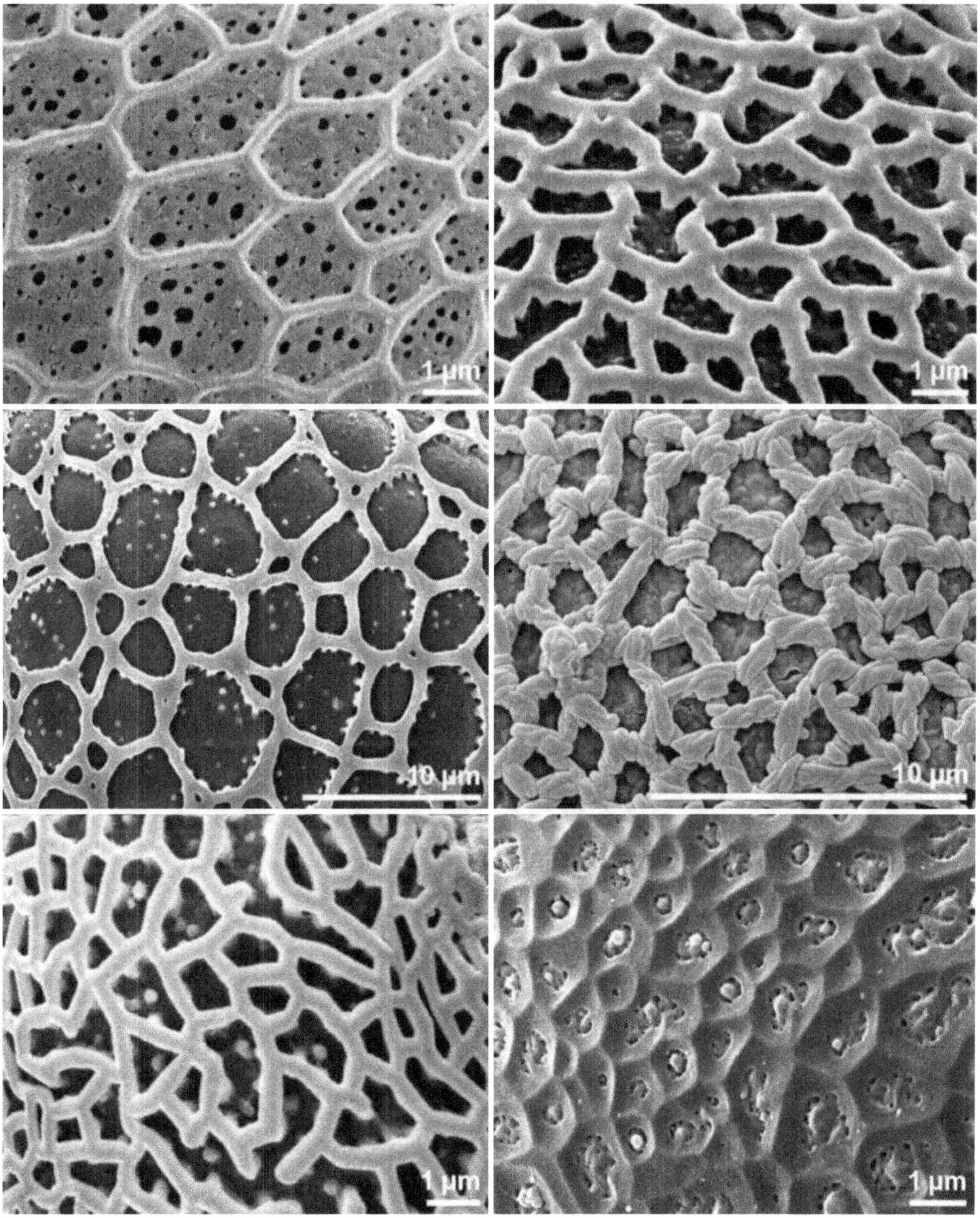

Tropaeolum majus, Tropaeolaceae
reticulate, perforate

Werauhia tarmaensis, Bromeliaceae

Poncirus trifoliata, Rutaceae

Veratrum album, Melanthiaceae

Erythronium dens-canis, Liliaceae
reticulate with suprasculpture

Melilotus officinalis, Fabaceae
reticulate, perforate

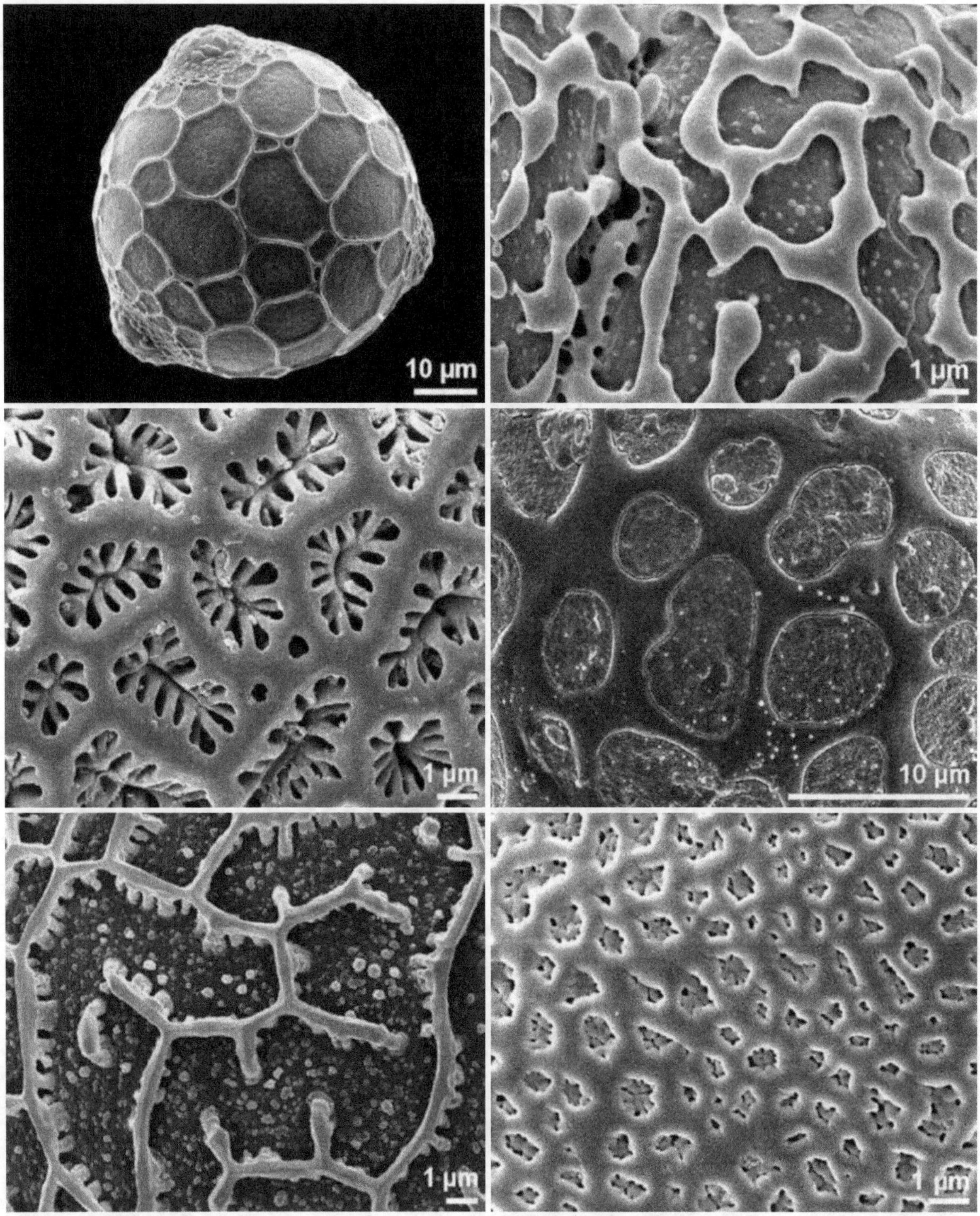

Vigna speciosa, Fabaceae
polar view

Harpochilus neesianus, Acanthaceae
reticulate, brochi with inclined columellae

Adenia fruticosa, Passifloraceae
incomplete reticulum

Epipactis helleborine, Orchidaceae
incomplete reticulum

Polygala major, Polygalaceae
reticulate to foveolate

Pisum sativum, Fabaceae
reticulate, perforate

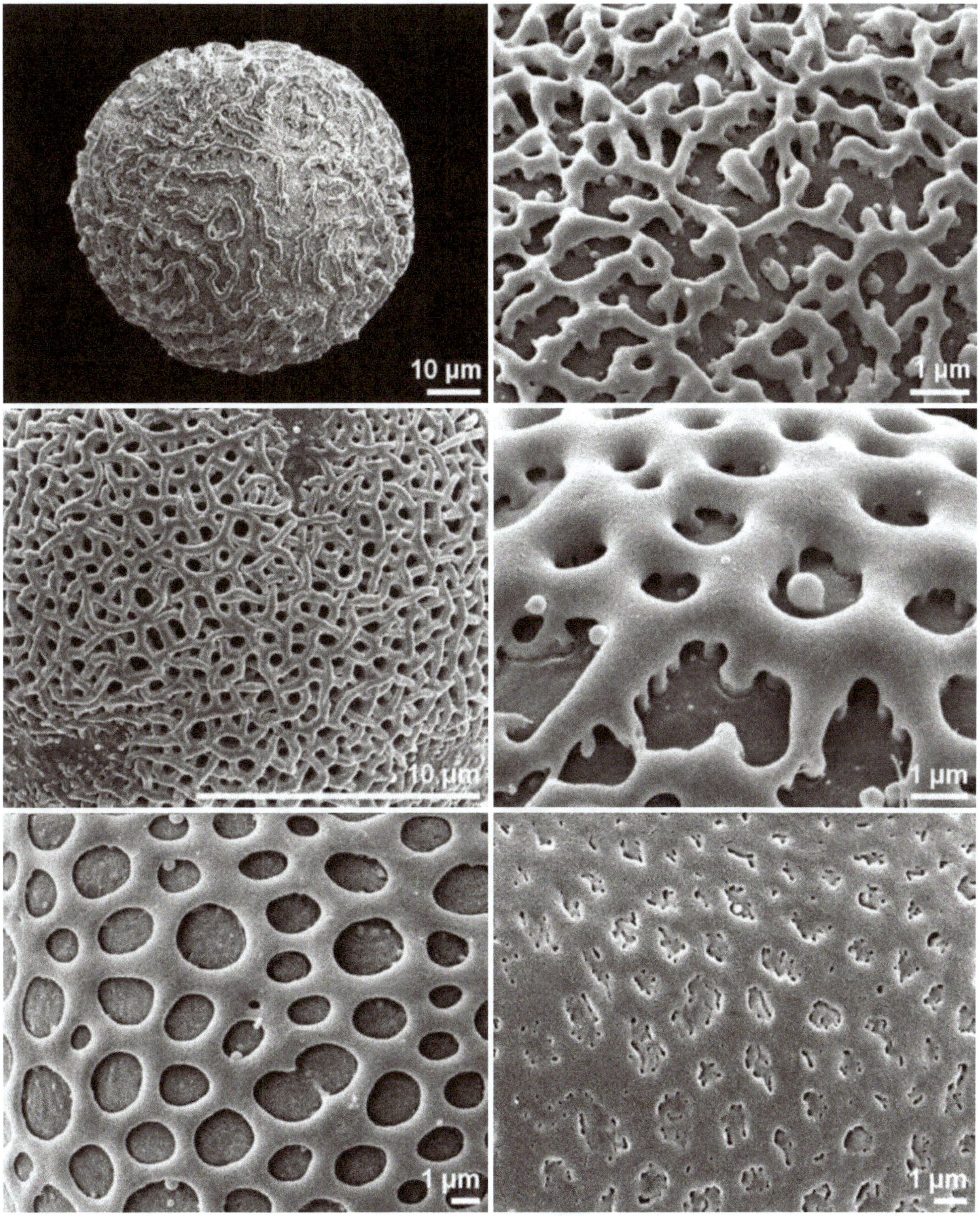

Thladiantha hookeri, Cucurbitaceae
oblique equatorial view, incomplete reticulum

Pinguicula alpina, Lentibulariaceae
incomplete reticulum

Billardiera heterophylla, Pittosporaceae
reticulate to rugulate

Cephalanthera longifolia, Orchidaceae

Aechmea allenii, Bromeliaceae
reticulate to foveolate

Lathyrus vernus, Fabaceae
reticulate, perforate

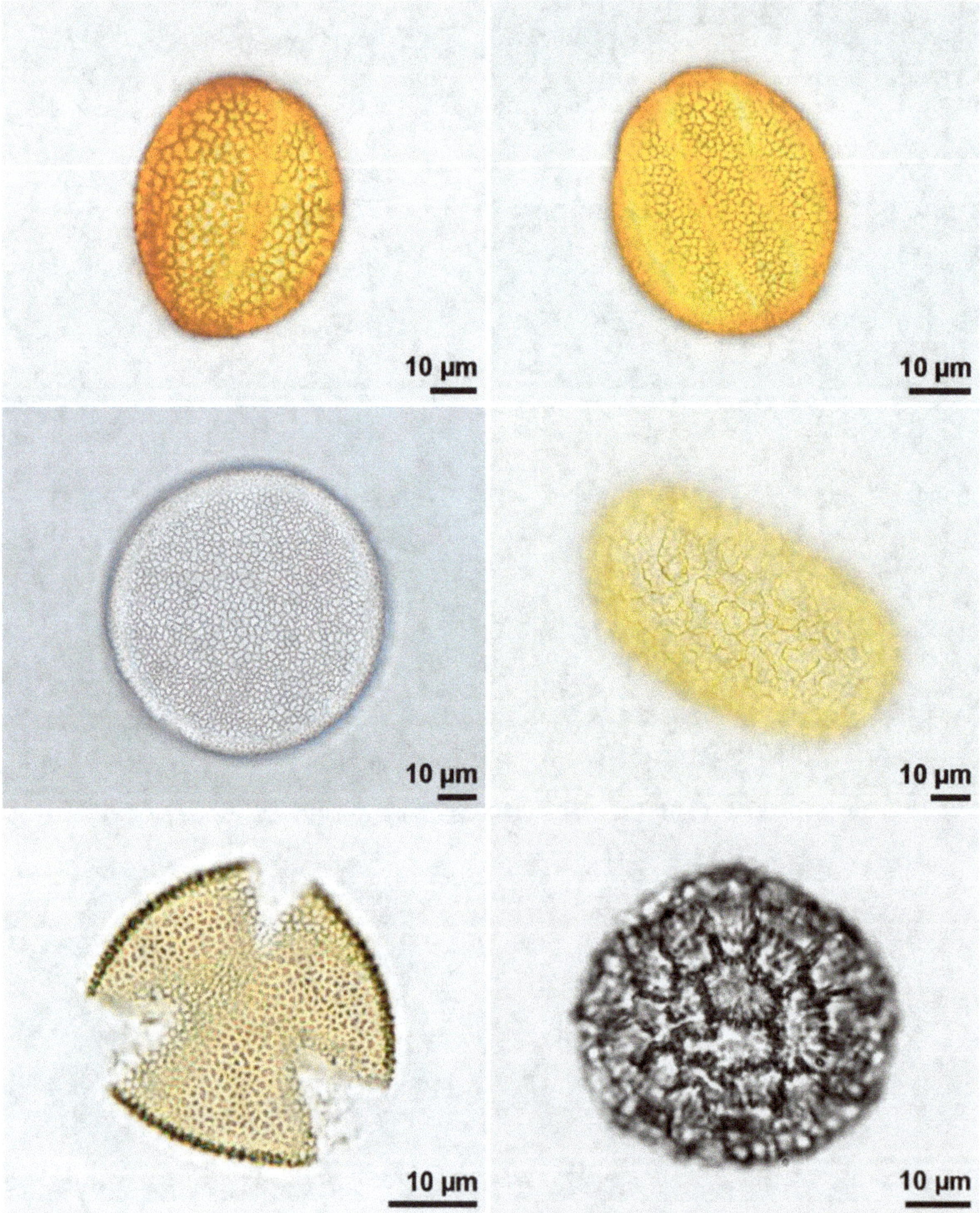

Jasminum nudiflorum, Oleaceae
equatorial view

Lamiaceae
equatorial view

Iris pseudacorus, Iridaceae
proximal polar view

Iris domestica, Iridaceae
proximal polar view

Fraxinus excelsior, Oleaceae
polar view

Persicaria sp., Polygonaceae
fossil, Miocene, Austria

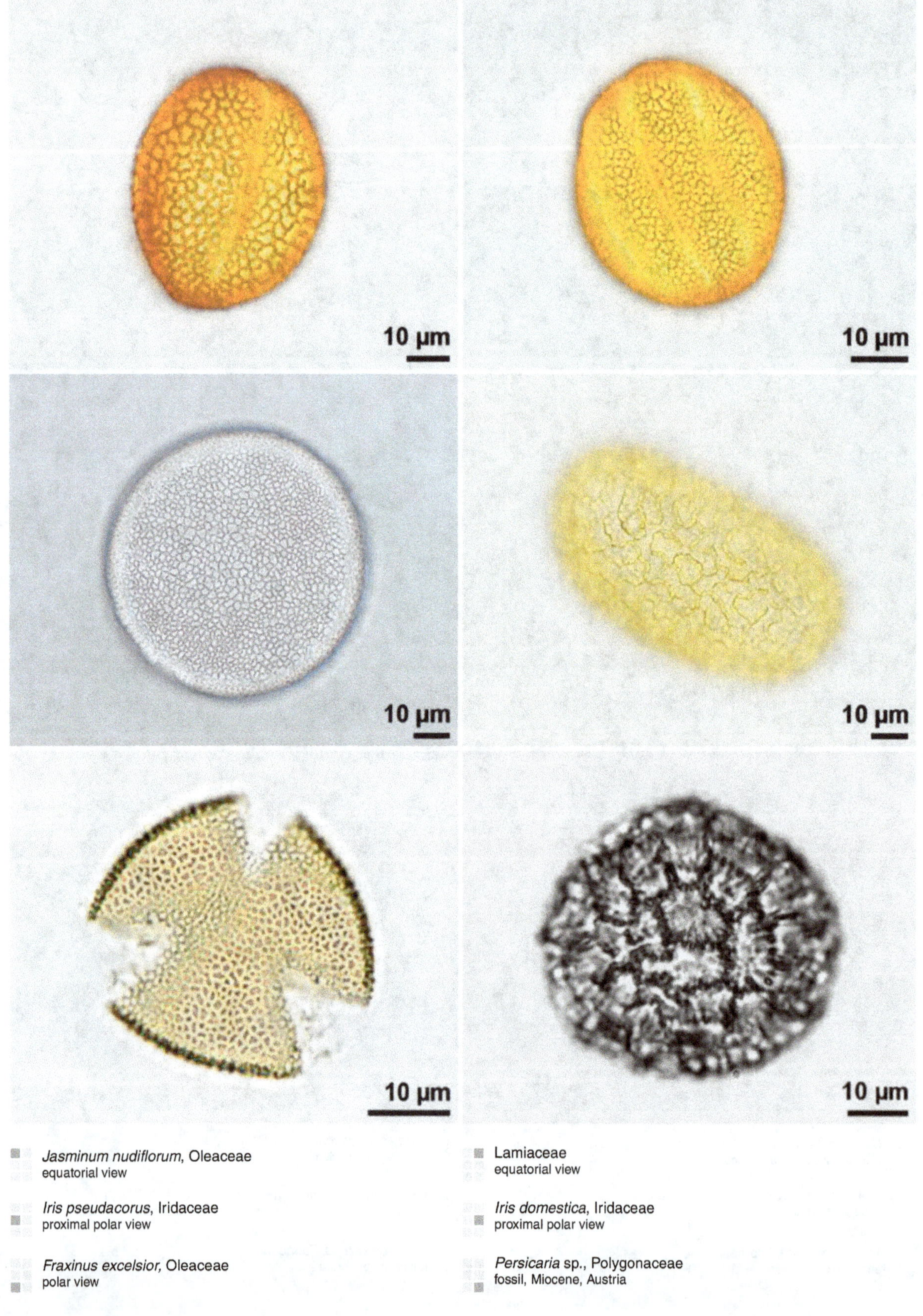

reticulum cristatum

special type of reticulum; muri with prominent suprasculpture

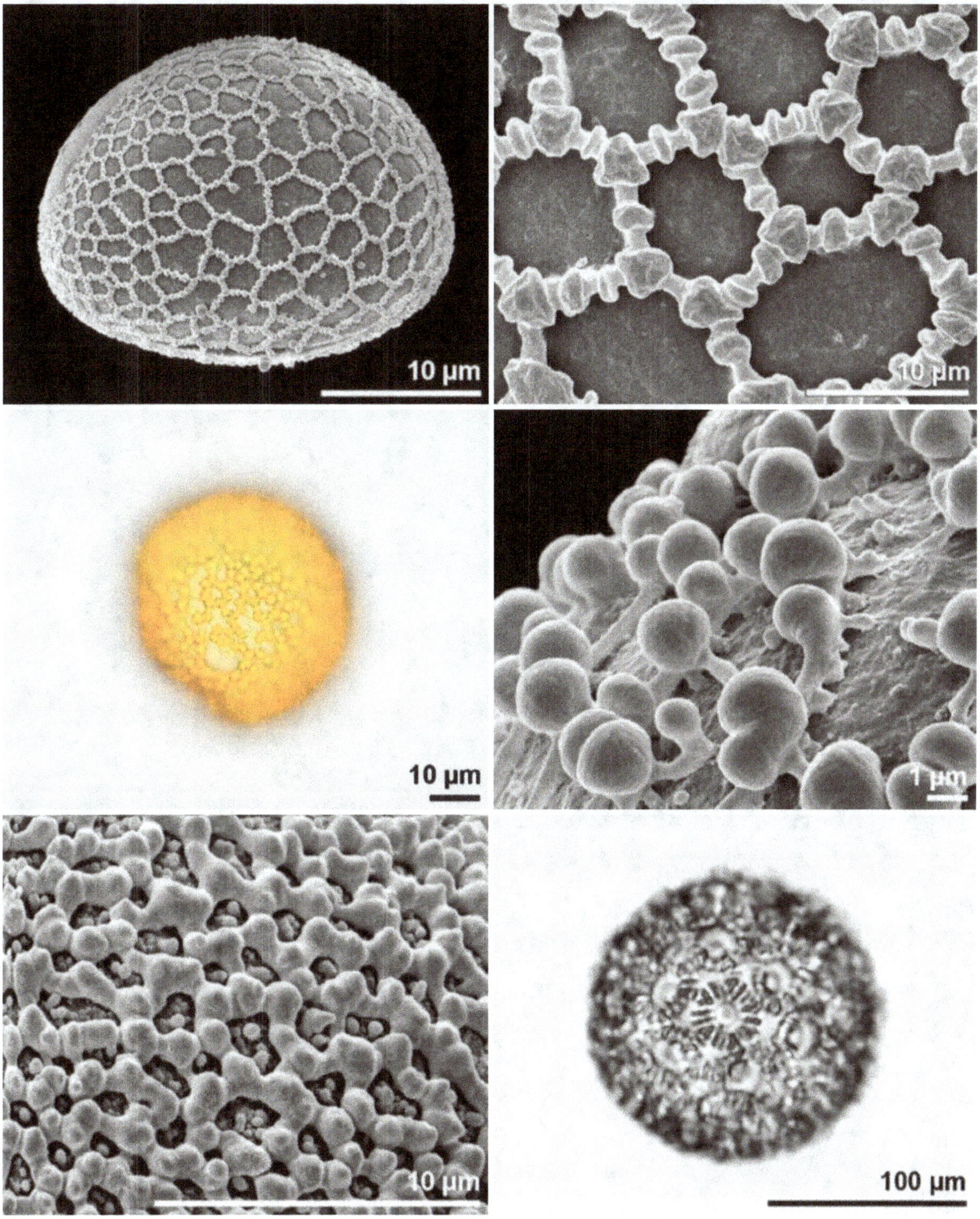

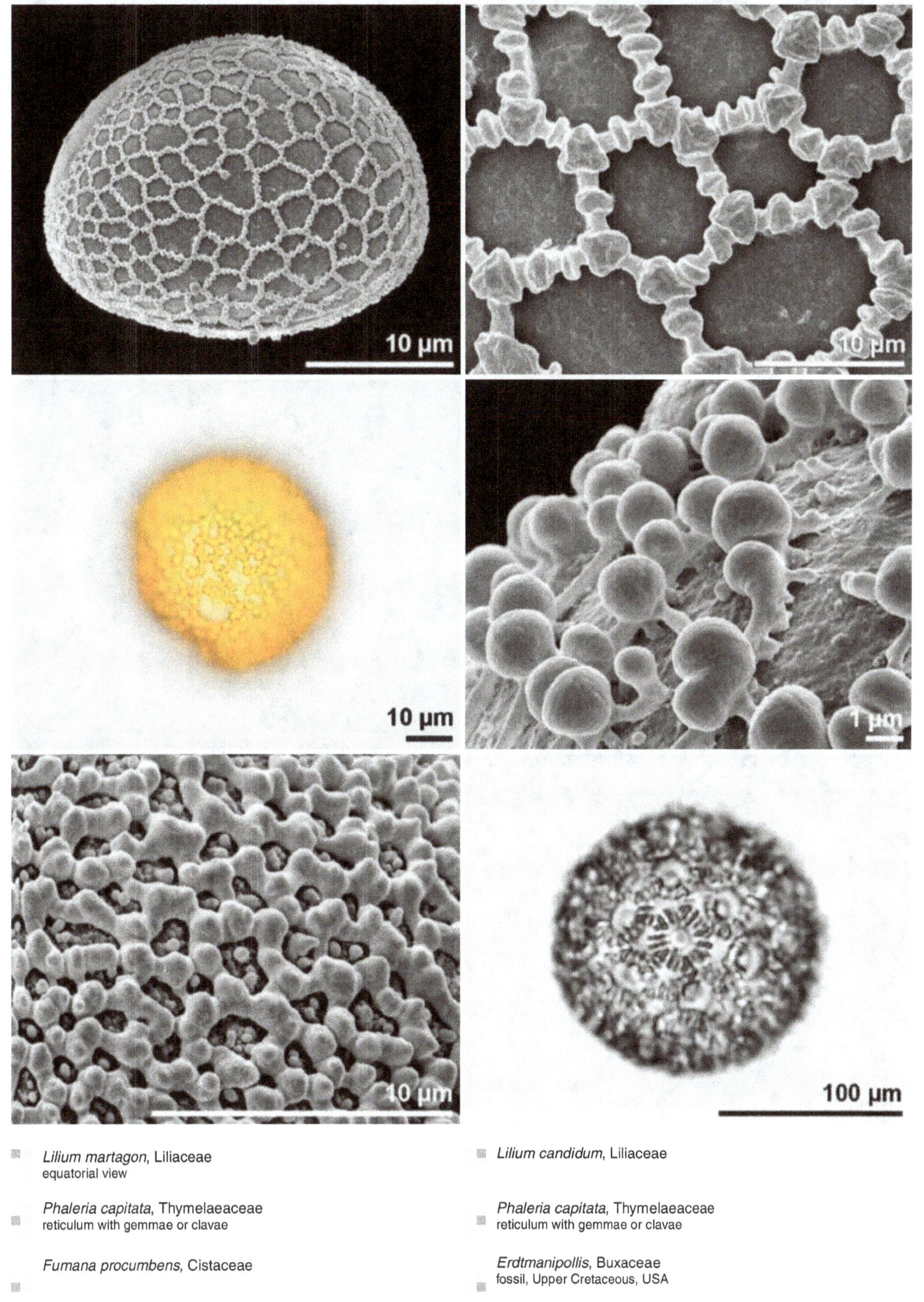

- *Lilium martagon*, Liliaceae
  equatorial view
- *Lilium candidum*, Liliaceae
- *Phaleria capitata*, Thymelaeaceae
  reticulum with gemmae or clavae
- *Phaleria capitata*, Thymelaeaceae
  reticulum with gemmae or clavae
- *Fumana procumbens*, Cistaceae
- *Erdtmanipollis*, Buxaceae
  fossil, Upper Cretaceous, USA

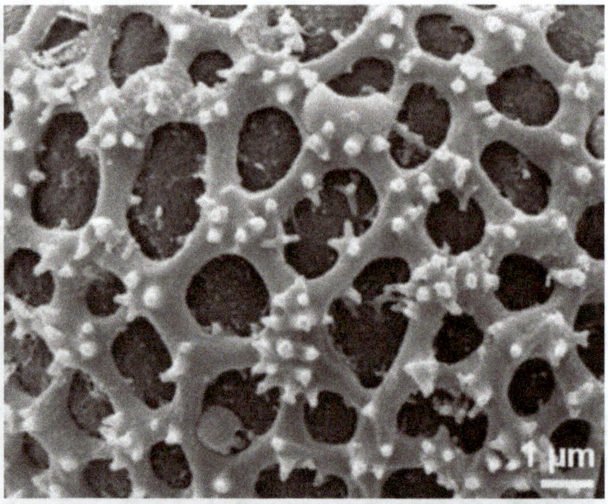

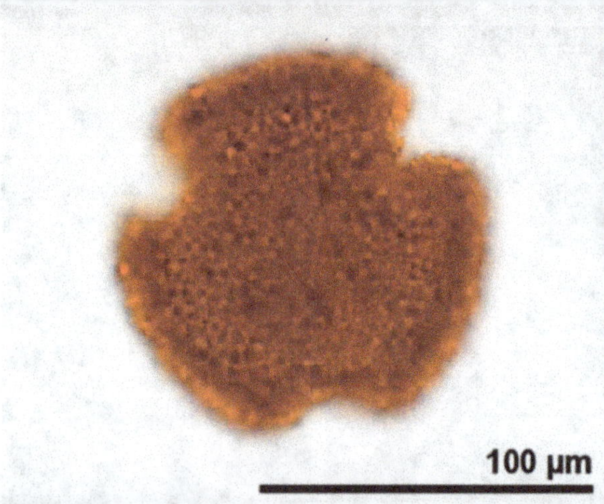

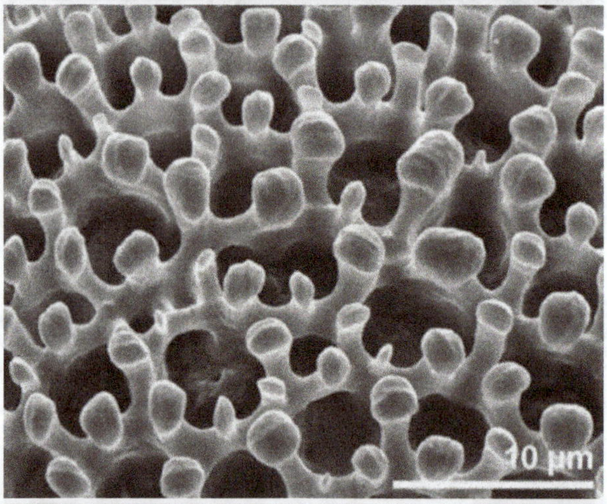

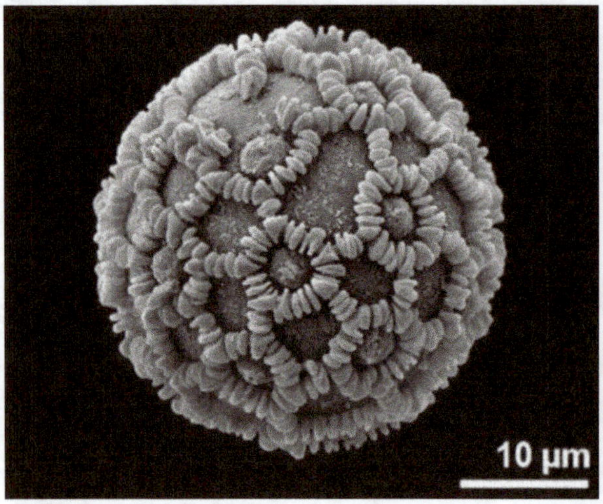

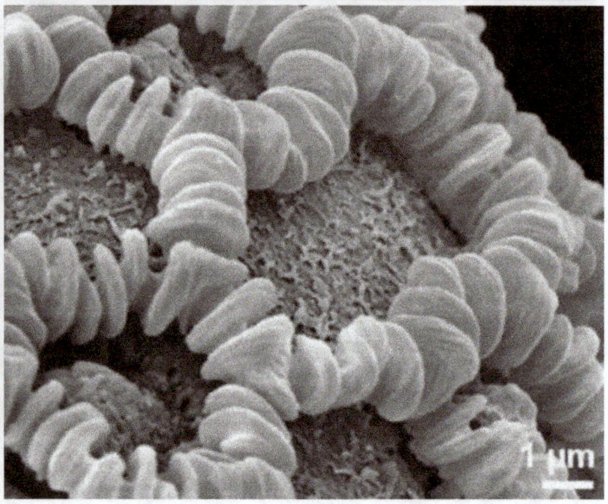

Anthurium gracile, Araceae

Anthurium gracile, Araceae
reticulum with microechini

Geranium pratense, Geraniaceae

Geranium reuteri, Geraniaceae
reticulum with clavae

Sarcococca pruniformis, Buxaceae
pantoporate

Sarcococca pruniformis, Buxaceae

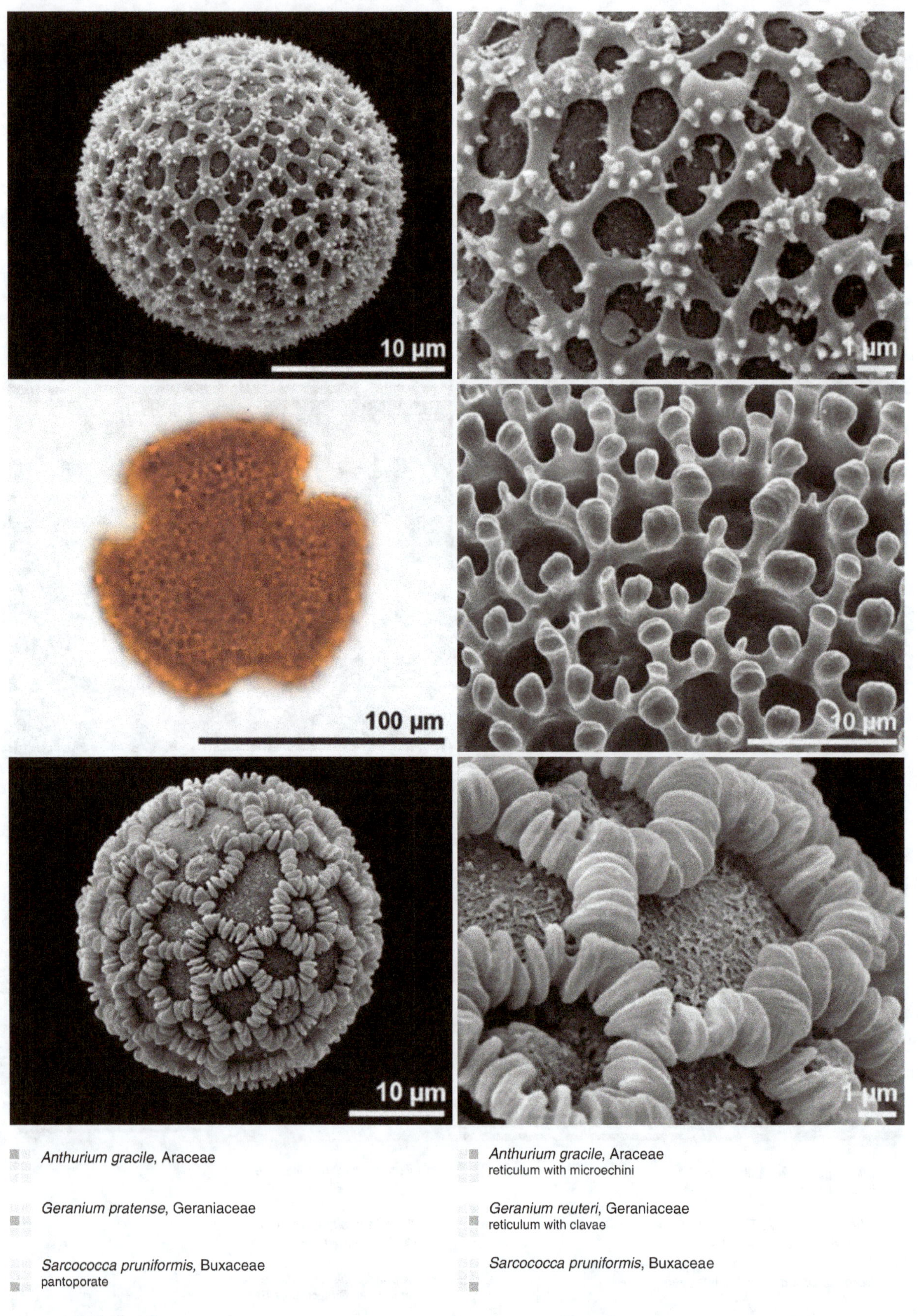

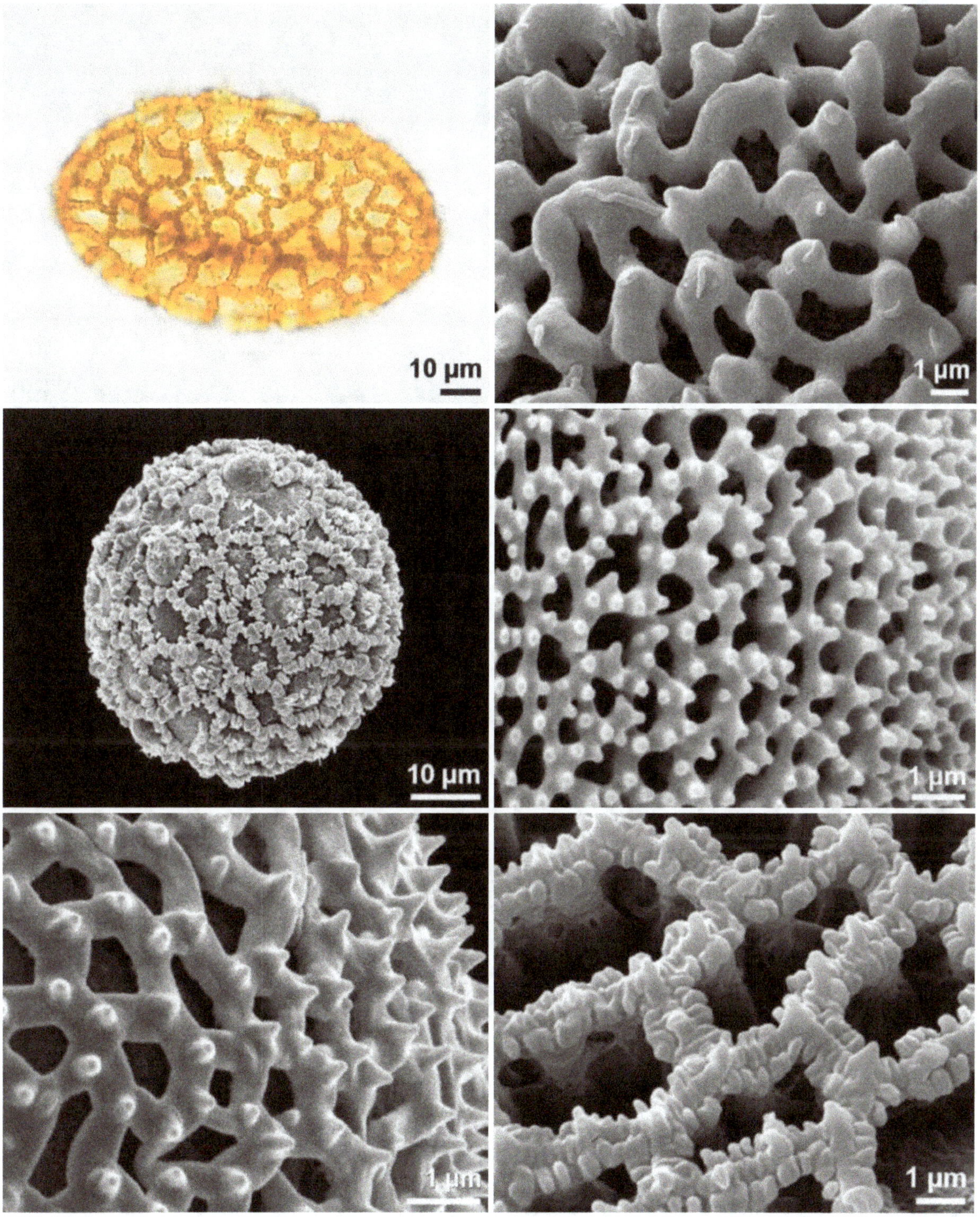

Liliaceae
proximal polar view

Pachira aquatica, Malvaceae

Pachysandra terminalis, Buxaceae

Mercurialis perennis, Euphorbiaceae
reticulum with microechini

Aponogeton masoalaensis, Aponogetonaceae
reticulum with microechini

Armeria maritima, Plumbaginaceae

rugulae/rugulate

elongated ornamentation elements irregularly arranged

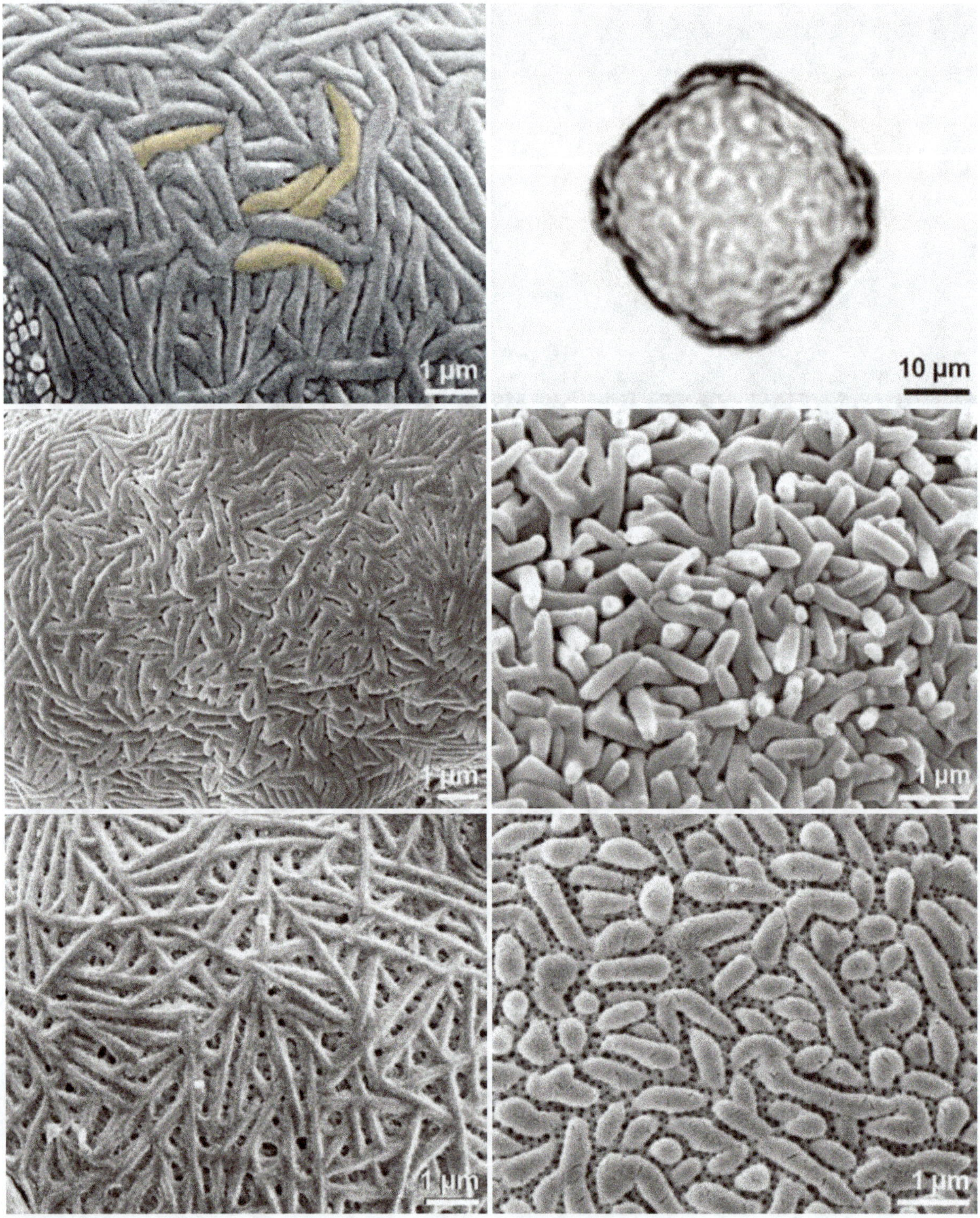

Securigera varia, Fabaceae
rugulae colored

Zelkova sp., Ulmaceae
fossil, Miocene, Austria, polar view

Peucedanum cervaria, Apiaceae

Fagus sp., Fagaceae
fossil, Miocene, Austria, polar view

Sedum acre, Crassulaceae
rugulate, perforate

Circaea lutetiana, Onagraceae
rugulate, perforate

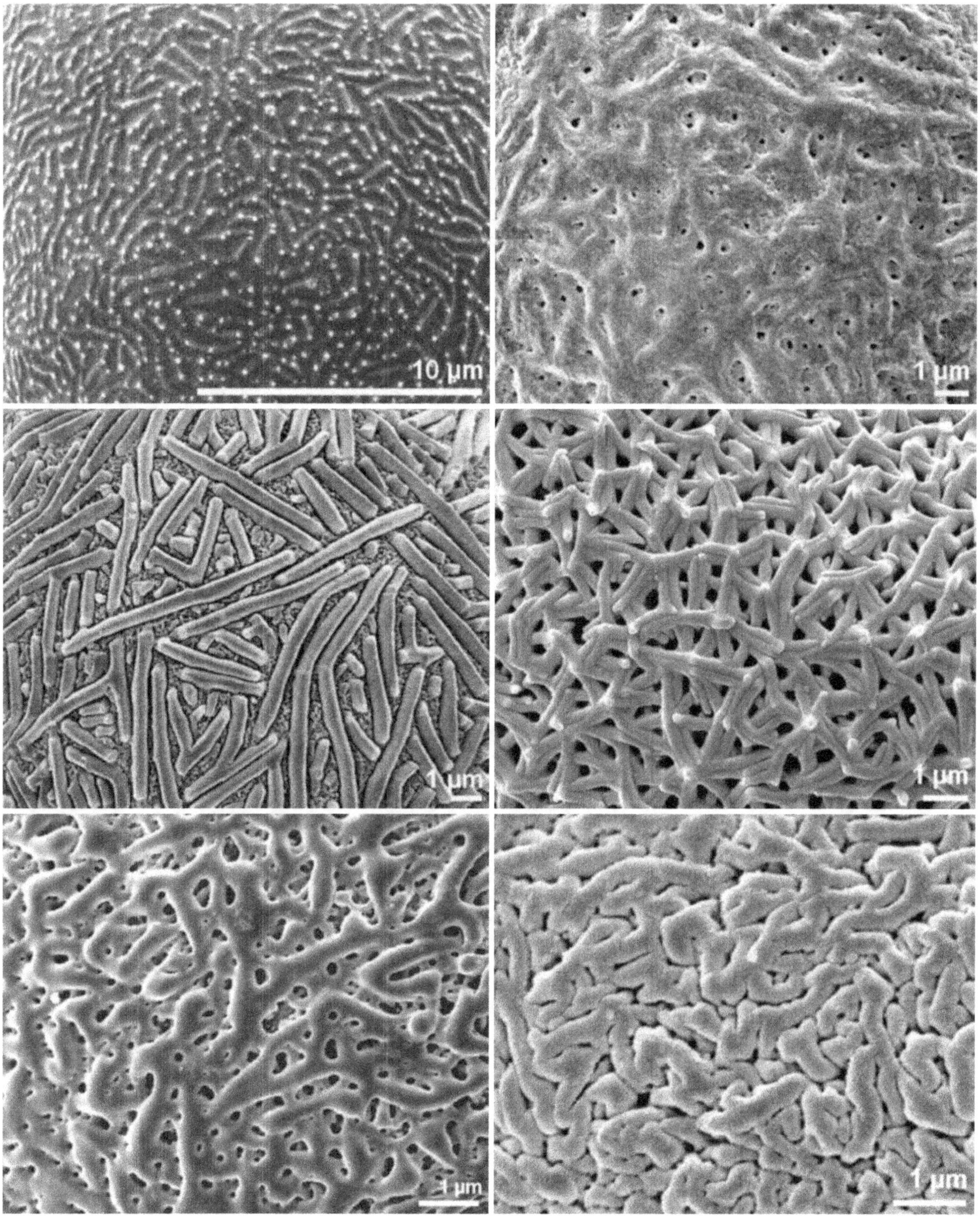

Carpinus betulus, Betulaceae
rugulate, granulate

Nymphoides peltata, Menyanthaceae

Nicotiana tabacum, Solanaceae
rugulate, perforate

Acer negundo, Sapindaceae
rugulate, perforate

Leucadendron discolor, Proteaceae
rugulate, perforate or microreticulate

Myrrhis odorata, Apiaceae

scabrate

term used for light microscopy only, describing minute sculpture elements of undefined shape and of a size close to the resolution limit of the light microscope

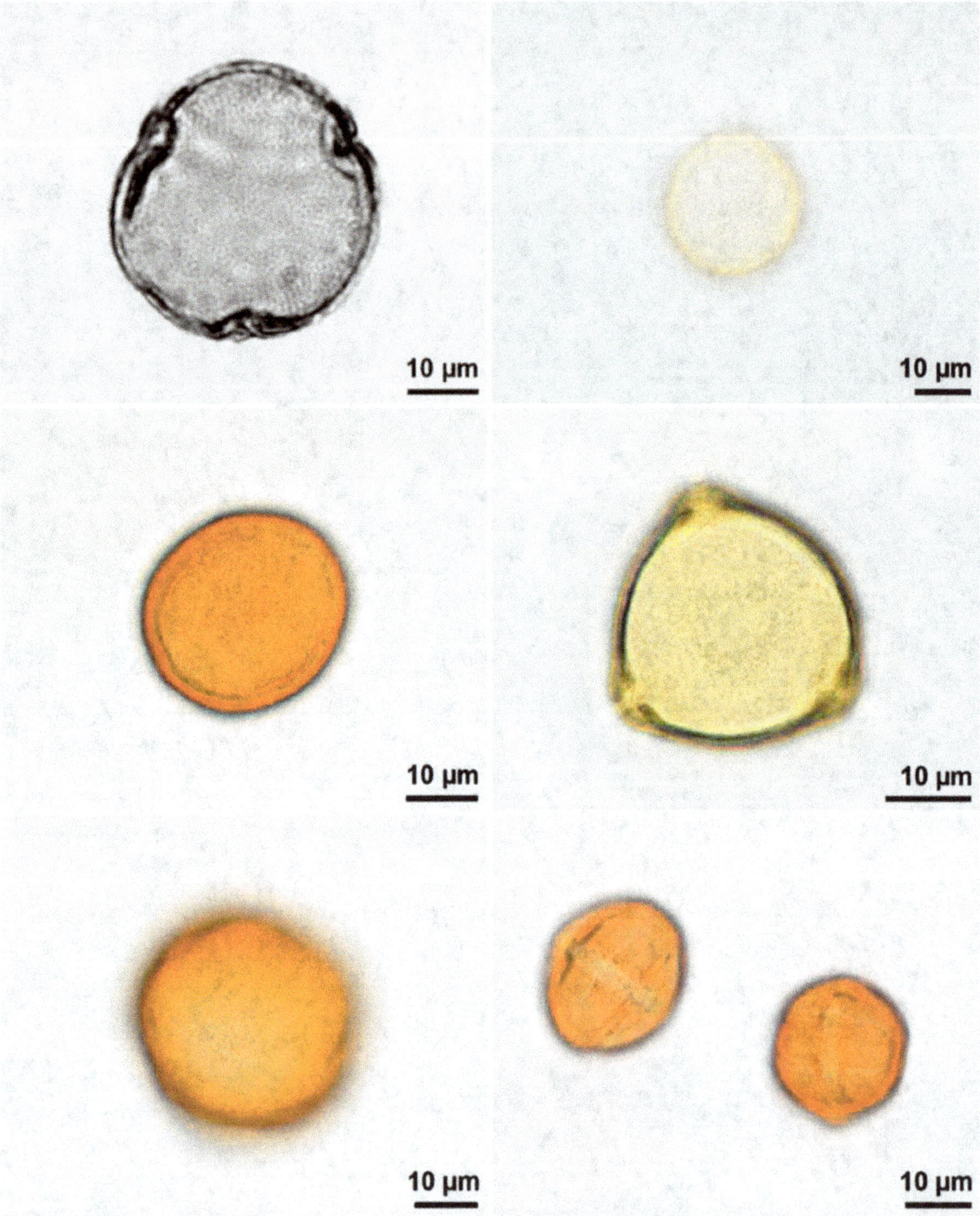

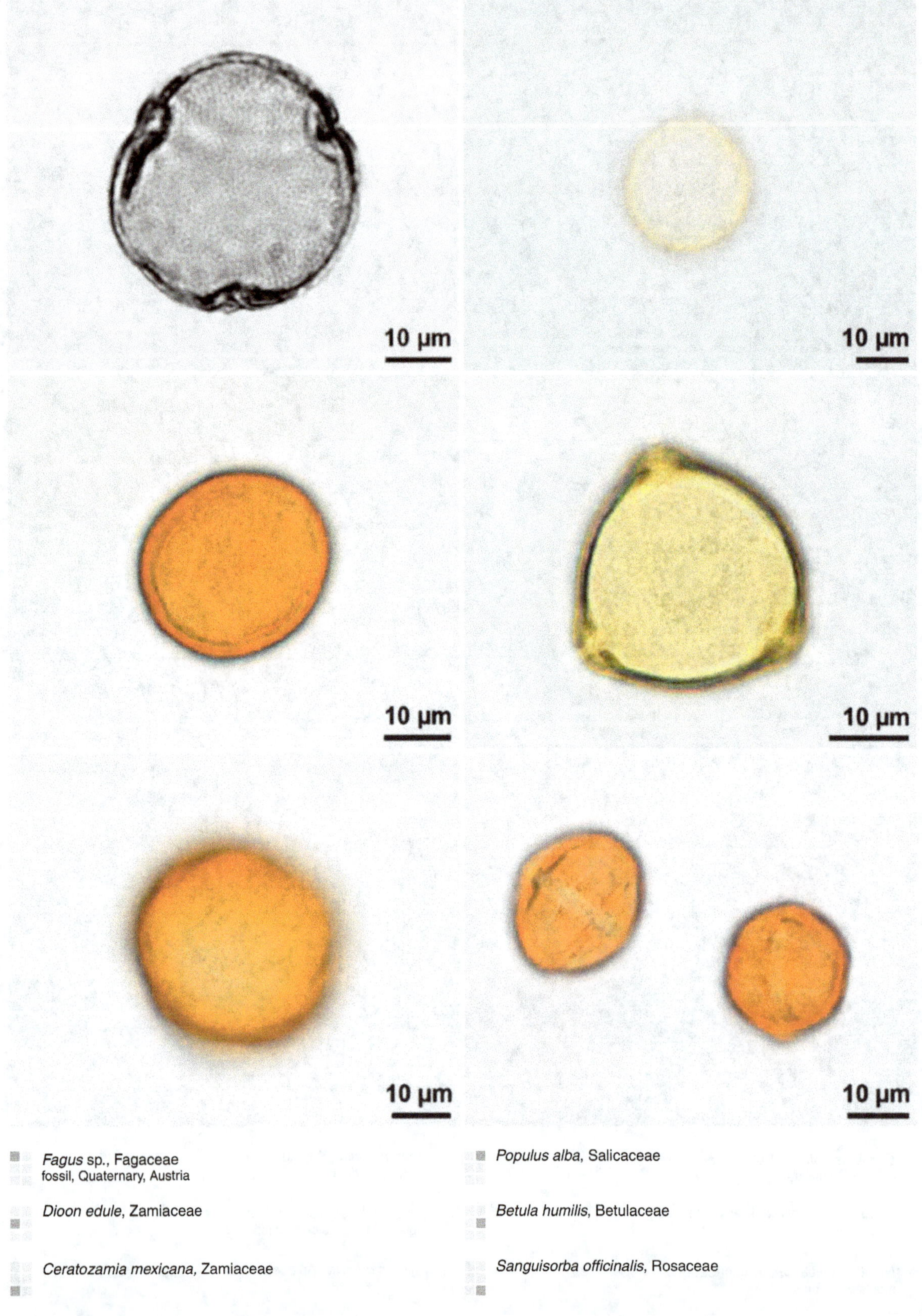

Fagus sp., Fagaceae
fossil, Quaternary, Austria

Populus alba, Salicaceae

Dioon edule, Zamiaceae

Betula humilis, Betulaceae

Ceratozamia mexicana, Zamiaceae

Sanguisorba officinalis, Rosaceae

striae/striate

elongated ornamentation elements separated by grooves parallelly arranged

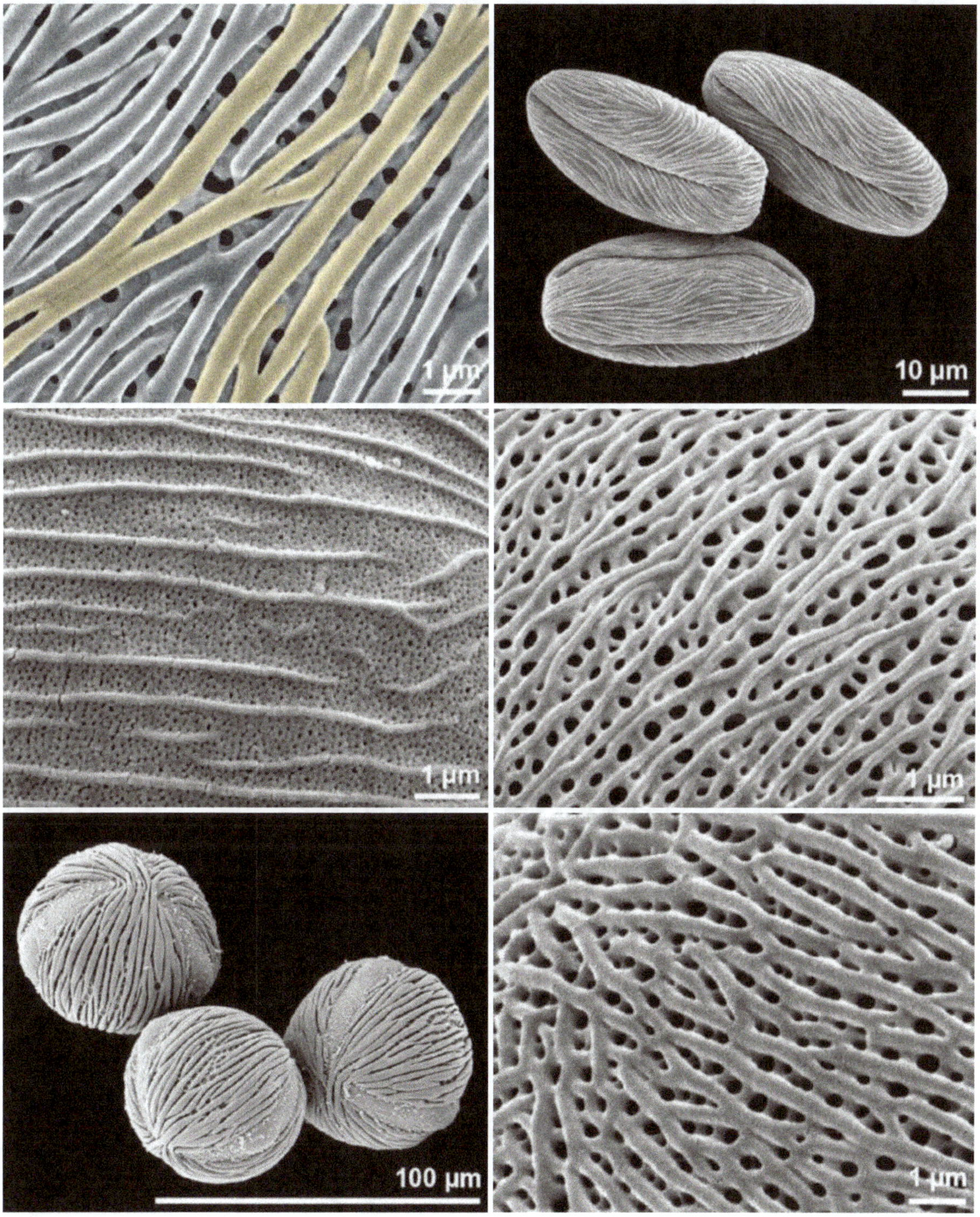

Acer pseudoplatanus, Sapindaceae
striae colored

Potentilla inclinata, Rosaceae
striate, perforate

Amherstia nobilis, Fabaceae

Prunus avium, Rosaceae
dry pollen

Veronica cinerea, Plantaginaceae
striate, perforate

Gentiana lutea, Gentianaceae
striate, perforate

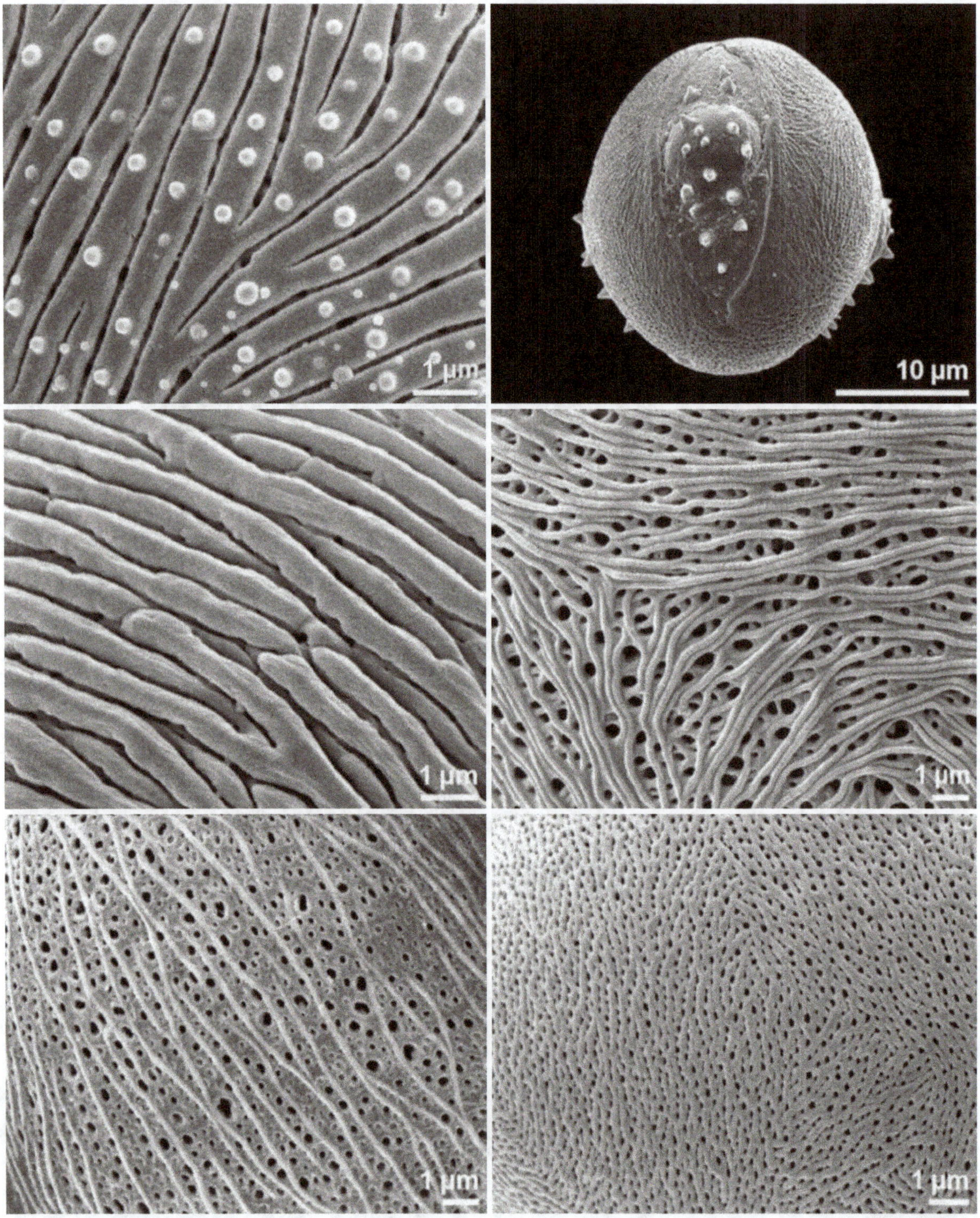

Saxifraga rotundifolia, Saxifragaceae
striate, microgemmate

Lycium barbarum, Solanaceae

Rubus caesius, Rosaceae
striate, perforate

Aesculus hippocastanum, Sapindaceae
equatorial view, striate, perforate

Chaenomeles sinensis, Rosaceae
striate, perforate

Allium flavum, Amaryllidaceae
striate, perforate

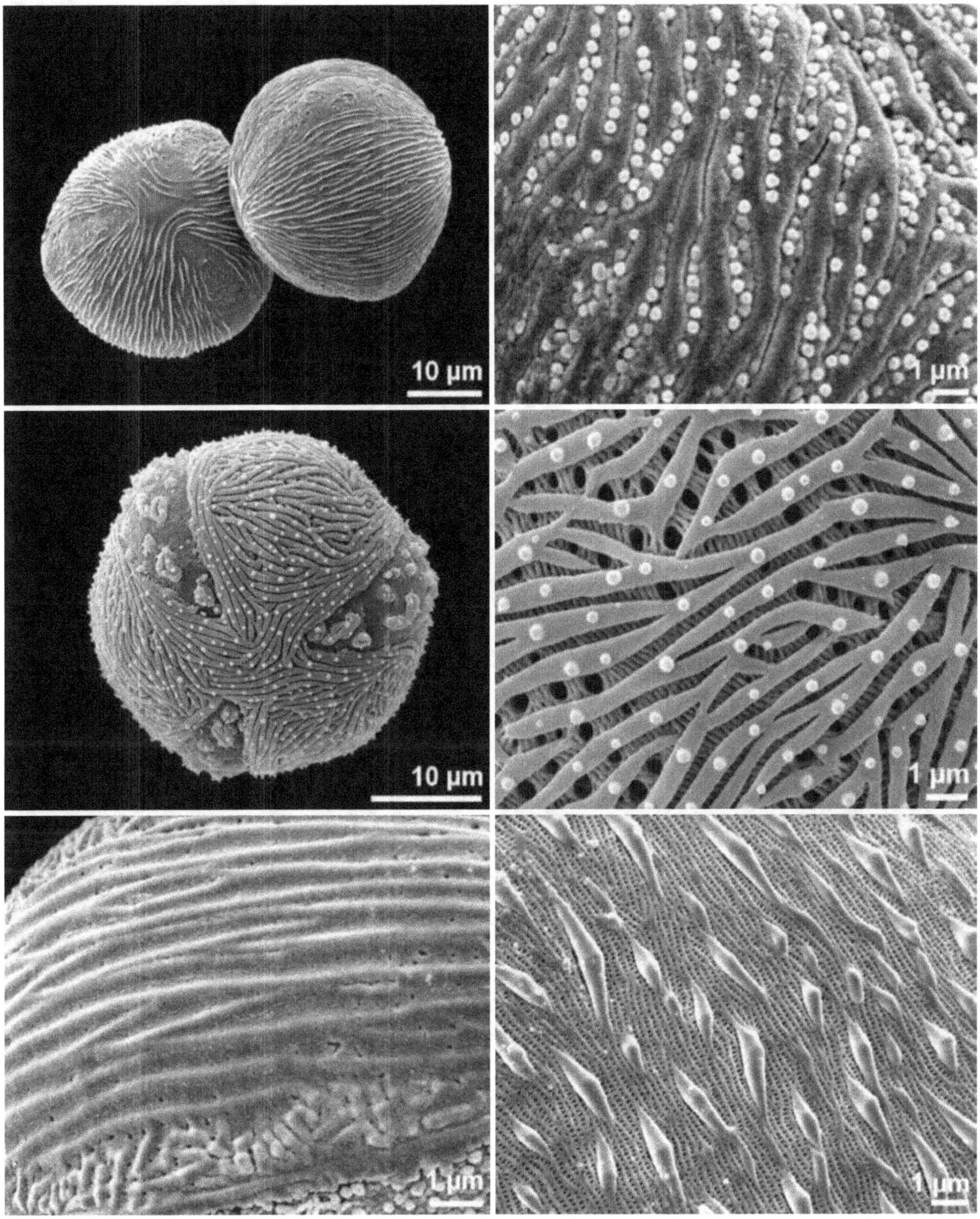

Geum reptans, Rosaceae
polar (left) and equatorial (right) view

Sanguisorba minor, Rosaceae
striate, nanogemmate

Saxifraga taygetea, Saxifragaceae
polar view

Saxifraga taygetea, Saxifragaceae
striate, microgemmate, perforate

Begonia heracleifolia, Begoniaceae

Cabomba palaeformis, Cabombaceae
striate, perforate

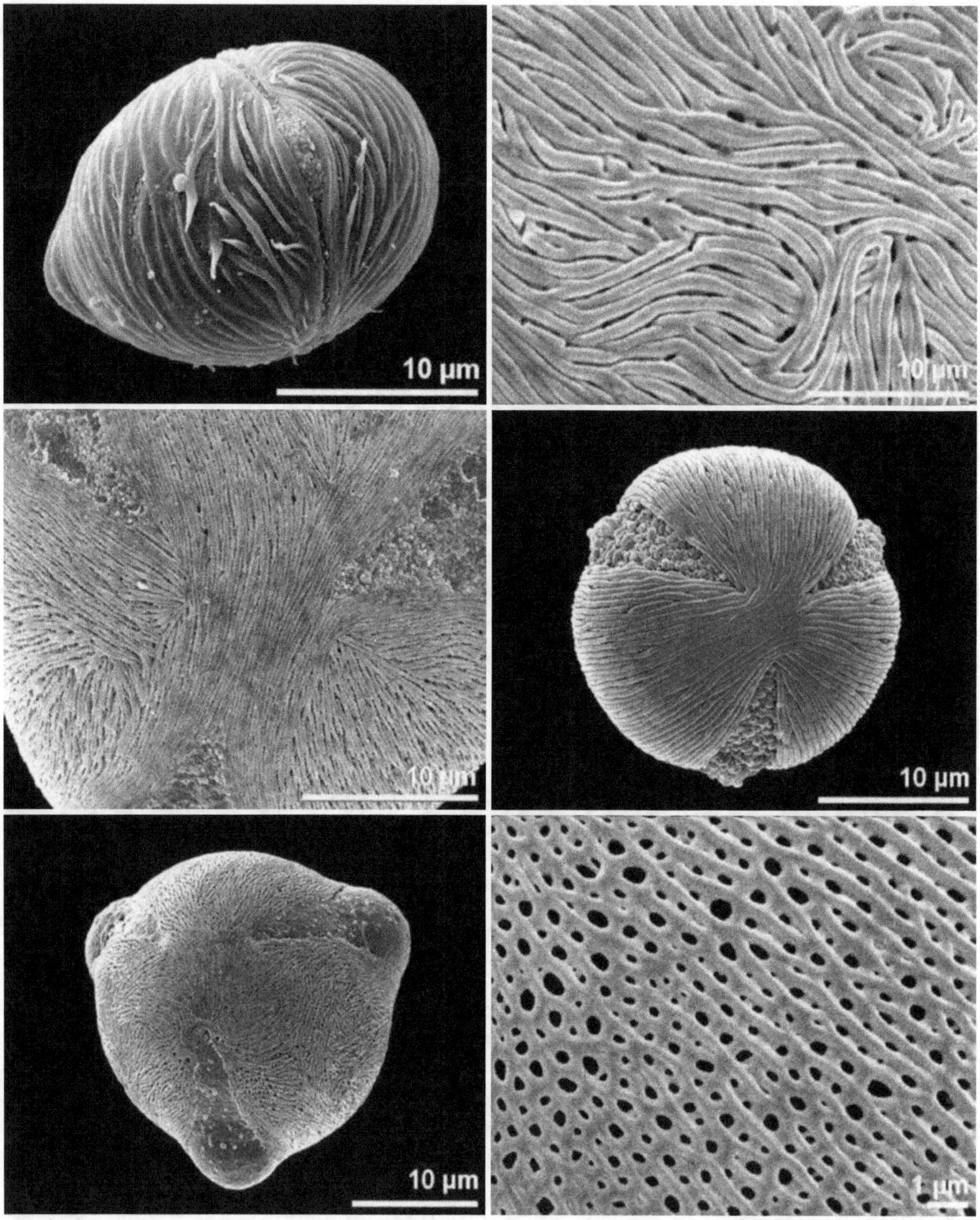

Cuphea llavea, Lythraceae
equatorial view

Malus sylvestris, Rosaceae
striate, perforate, polar area

Ruta graveolens, Rutaceae
polar view

Crataegus laevigata, Rosaceae
striate, perforate

Neoalsomitra sarcophylla, Cucurbitaceae
polar view

Helianthemum nummularium, Cistaceae
striate, perforate

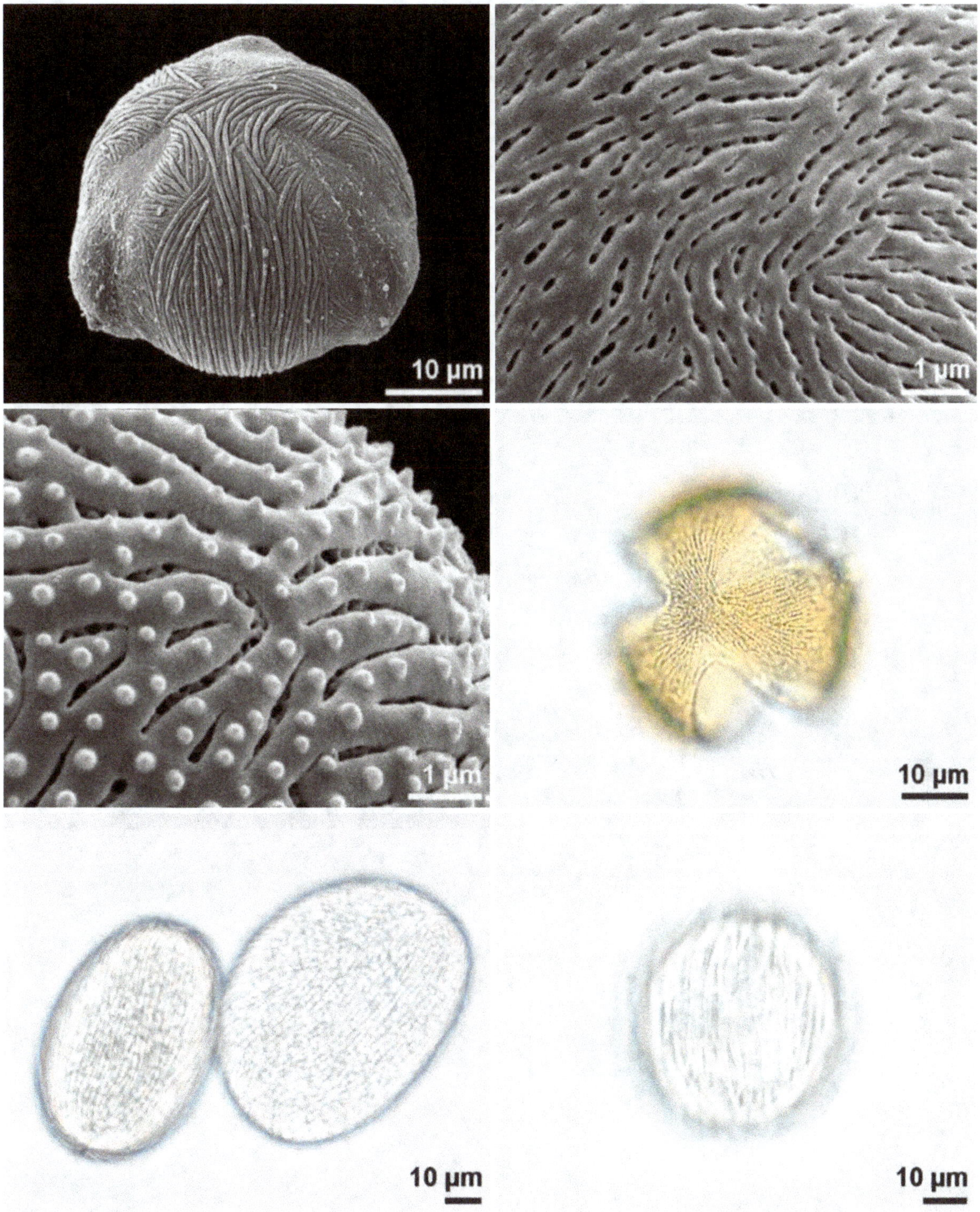

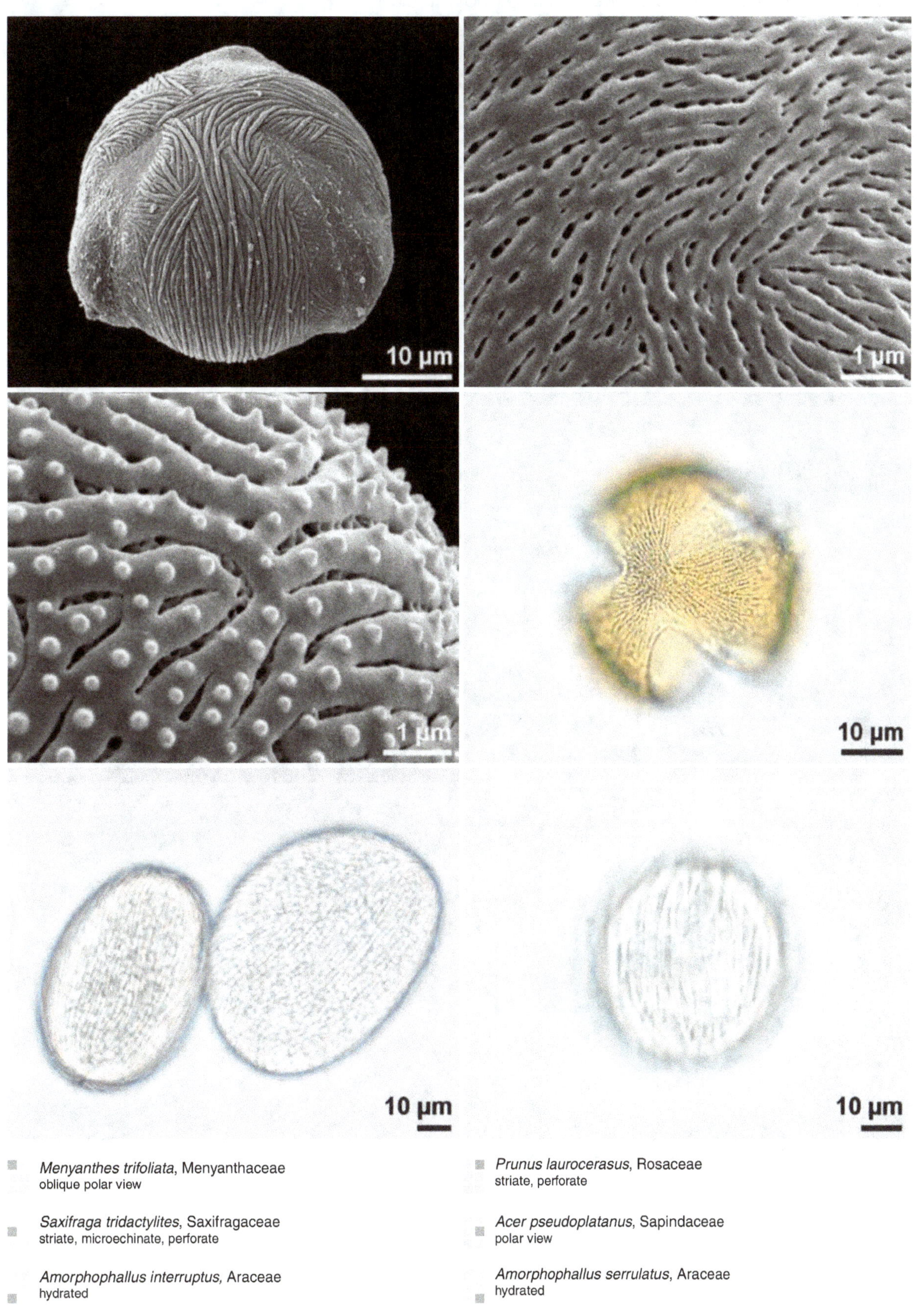

Menyanthes trifoliata, Menyanthaceae
oblique polar view

Prunus laurocerasus, Rosaceae
striate, perforate

Saxifraga tridactylites, Saxifragaceae
striate, microechinate, perforate

Acer pseudoplatanus, Sapindaceae
polar view

Amorphophallus interruptus, Araceae
hydrated

Amorphophallus serrulatus, Araceae
hydrated

striato-reticulate

ornamentation intermediate between striate and reticulate

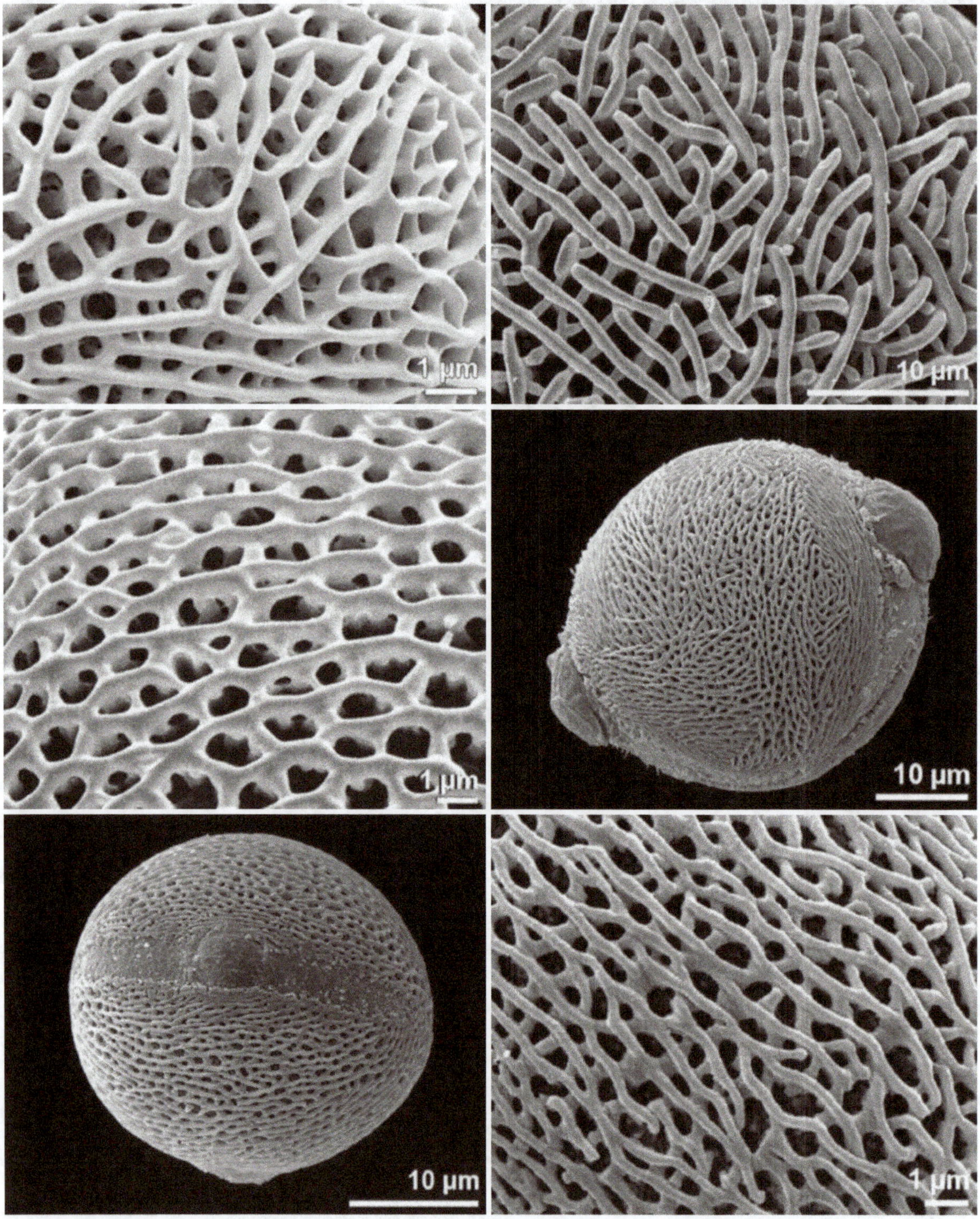

Blackstonia perfoliata, Gentianaceae

Gentianella austriaca, Gentianaceae

Cotinus coggygria, Anacardiaceae
equatorial view

Erodium cicutarium, Geraniaceae

Gelsemium sempervirens, Gelsemiaceae
equatorial view

Cotinus coggygria, Anacardiaceae

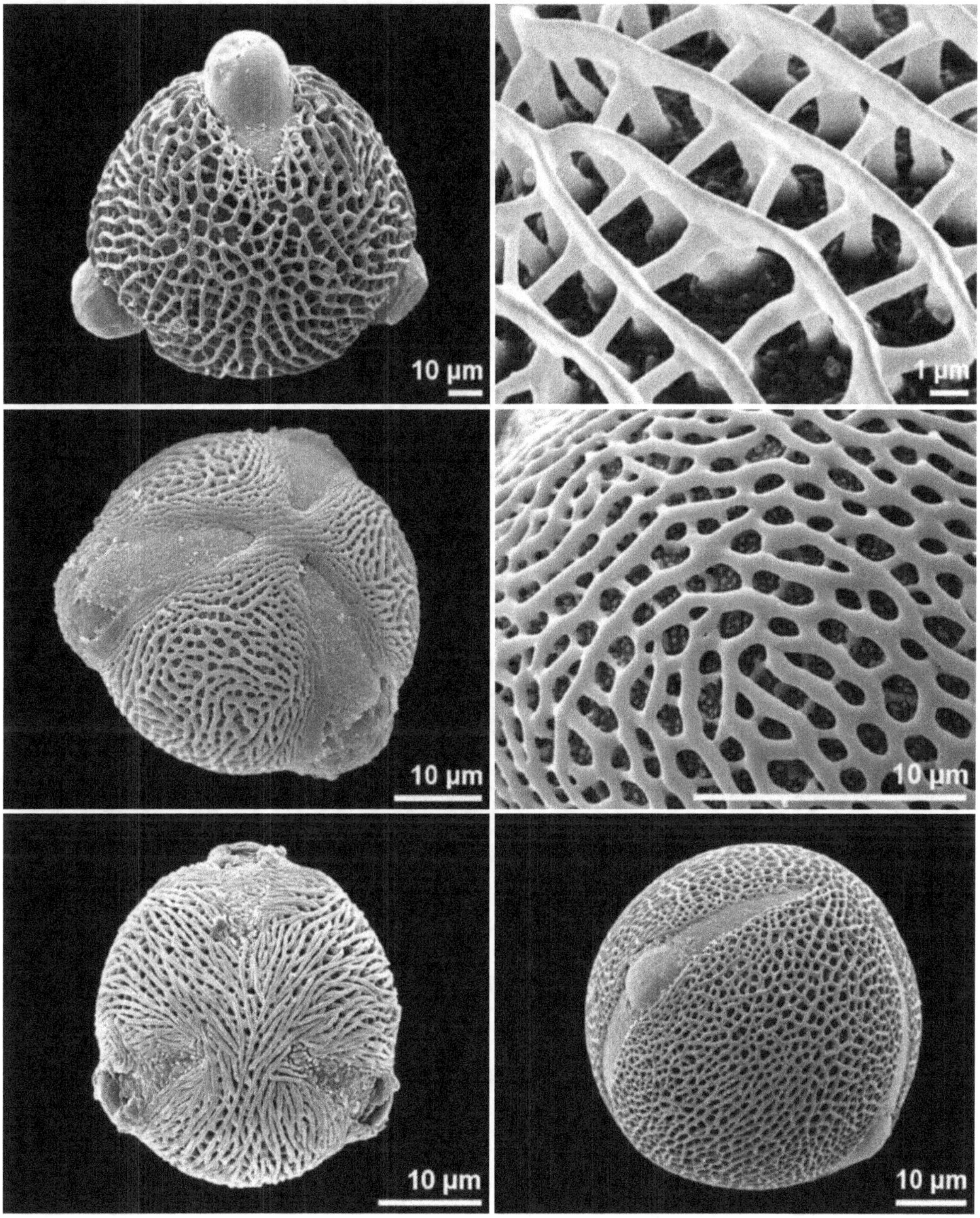

Pelargonium tetragonum, Geraniaceae
polar view

Pelargonium carnosum, Geraniaceae

Fouquieria macdougalii, Fouquieriaceae
polar view

Fouquieria macdougalii, Fouquieriaceae

Ailanthus altissima, Simaroubaceae
polar view

Cistus clusii, Cistaceae
equatorial view

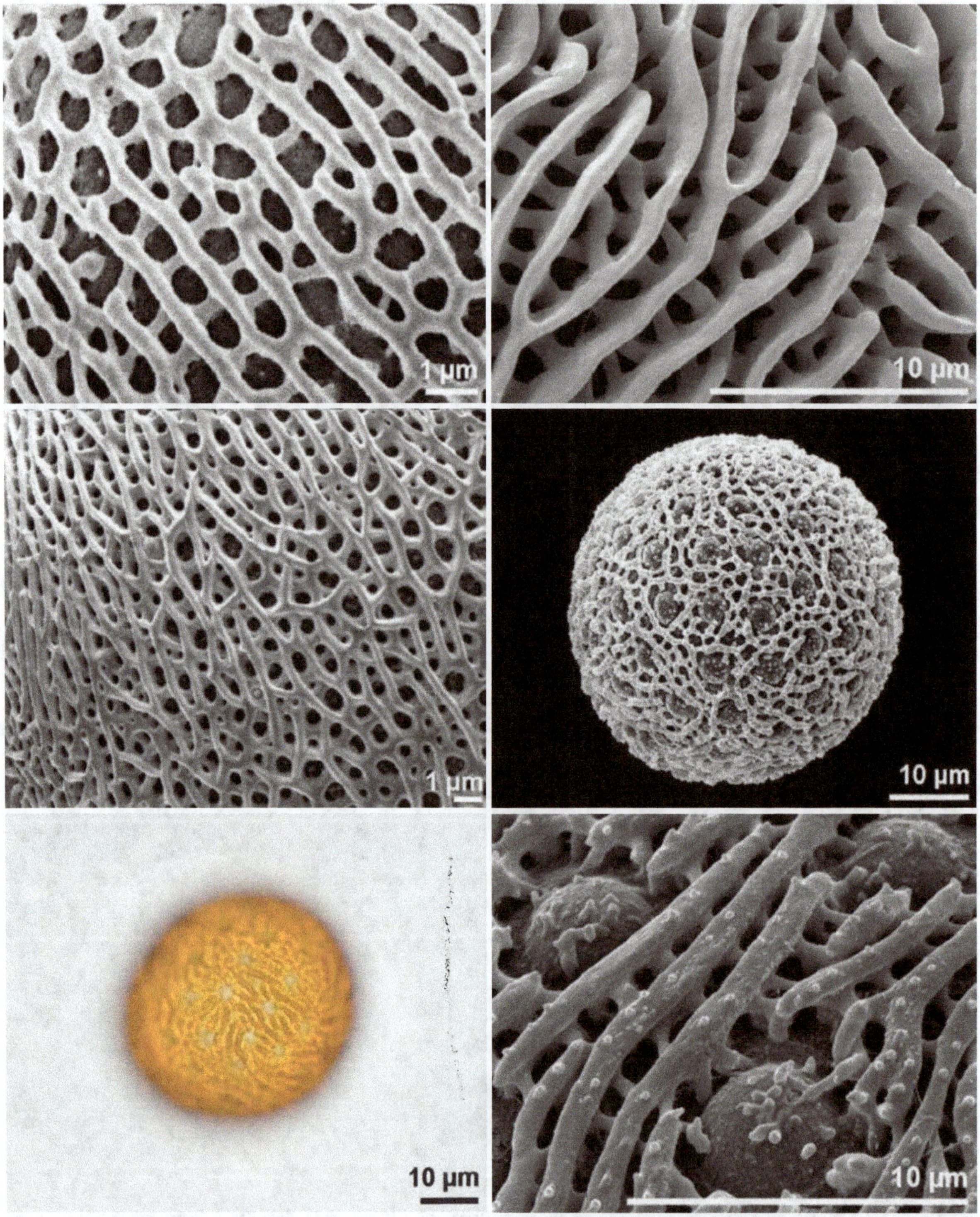

Solandra longiflora, Solanaceae

Pelargonium punctatum, Geraniaceae

Gentiana acaulis, Gentianaceae

Polemonium caeruleum, Polemoniaceae
pantoporate

Polemonium caeruleum, Polemoniaceae
pantoporate

Polemonium pauciflorum, Polemoniaceae

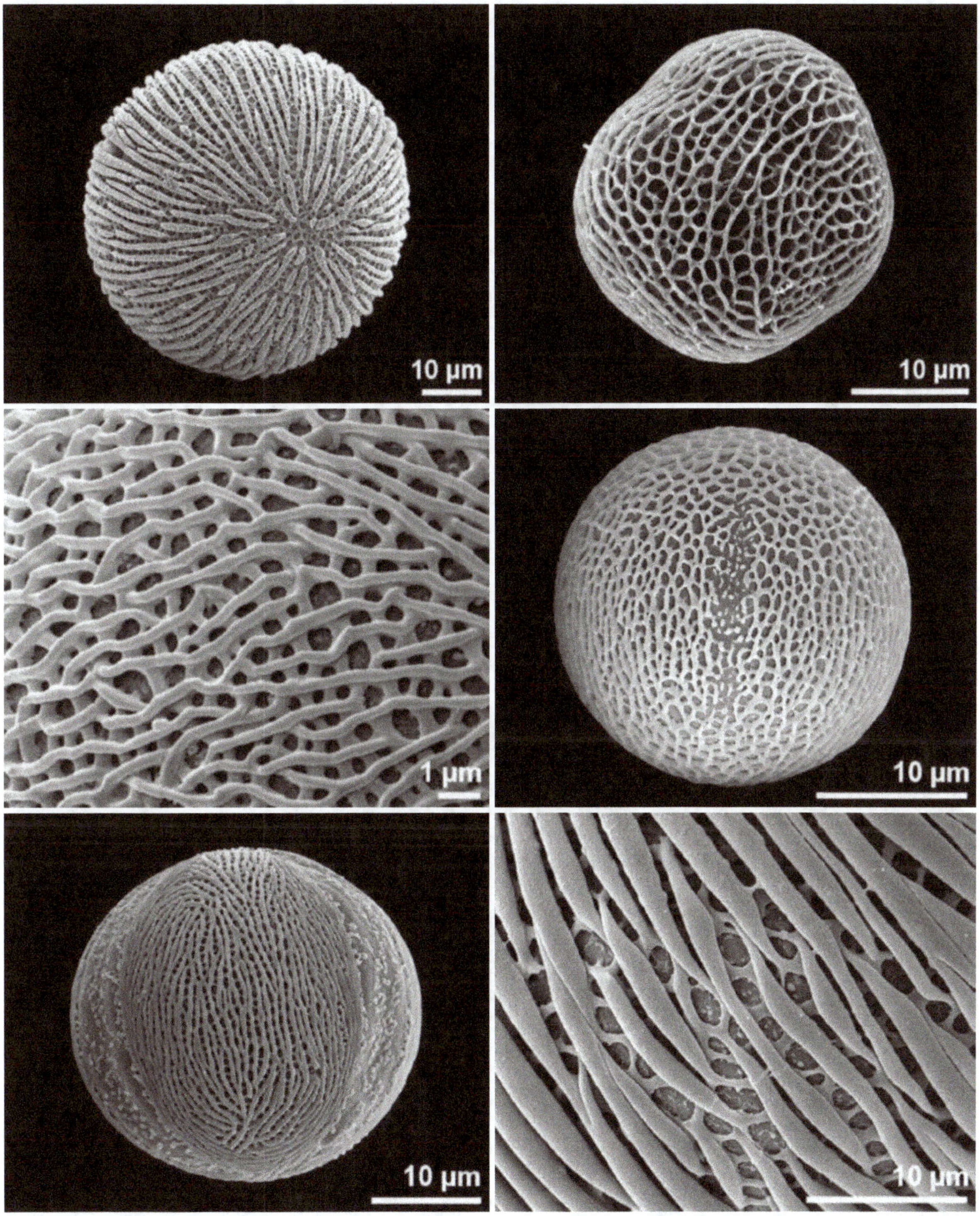

Brugmansia suaveolens, Solanaceae
polar view

Chlorospatha pubescens, Araceae
tetrad

Schinus molle, Anacardiaceae

Solandra maxima, Solanaceae
equatorial view

Plagiorhegma dubium, Berberidaceae
equatorial view

Amherstia nobilis, Fabaceae

suprasculpture

secondary sculpture elements positioned on the primary sculpture of the pollen surface

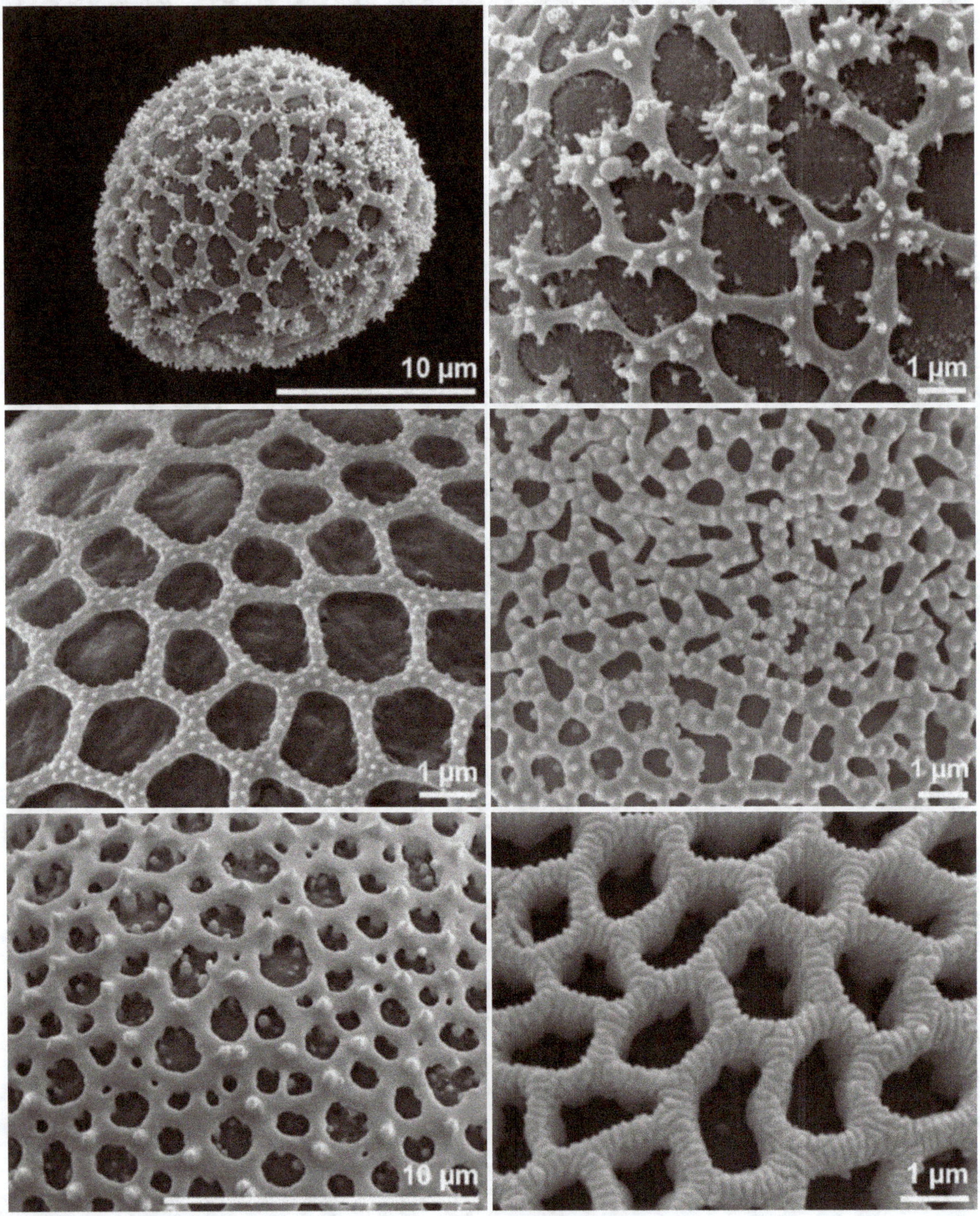

Anthurium gracile, Araceae
nano- to microechinate

Anthurium gracile, Araceae
nano- to microechinate

Triglochin maritima, Juncaginaceae
nanoechinate

Ascarina lucida, Chloranthaceae
nanogemmate

Lophophora williamsii, Cactaceae
microechinate

Phillyrea angustifolia, Oleaceae
microrugulate

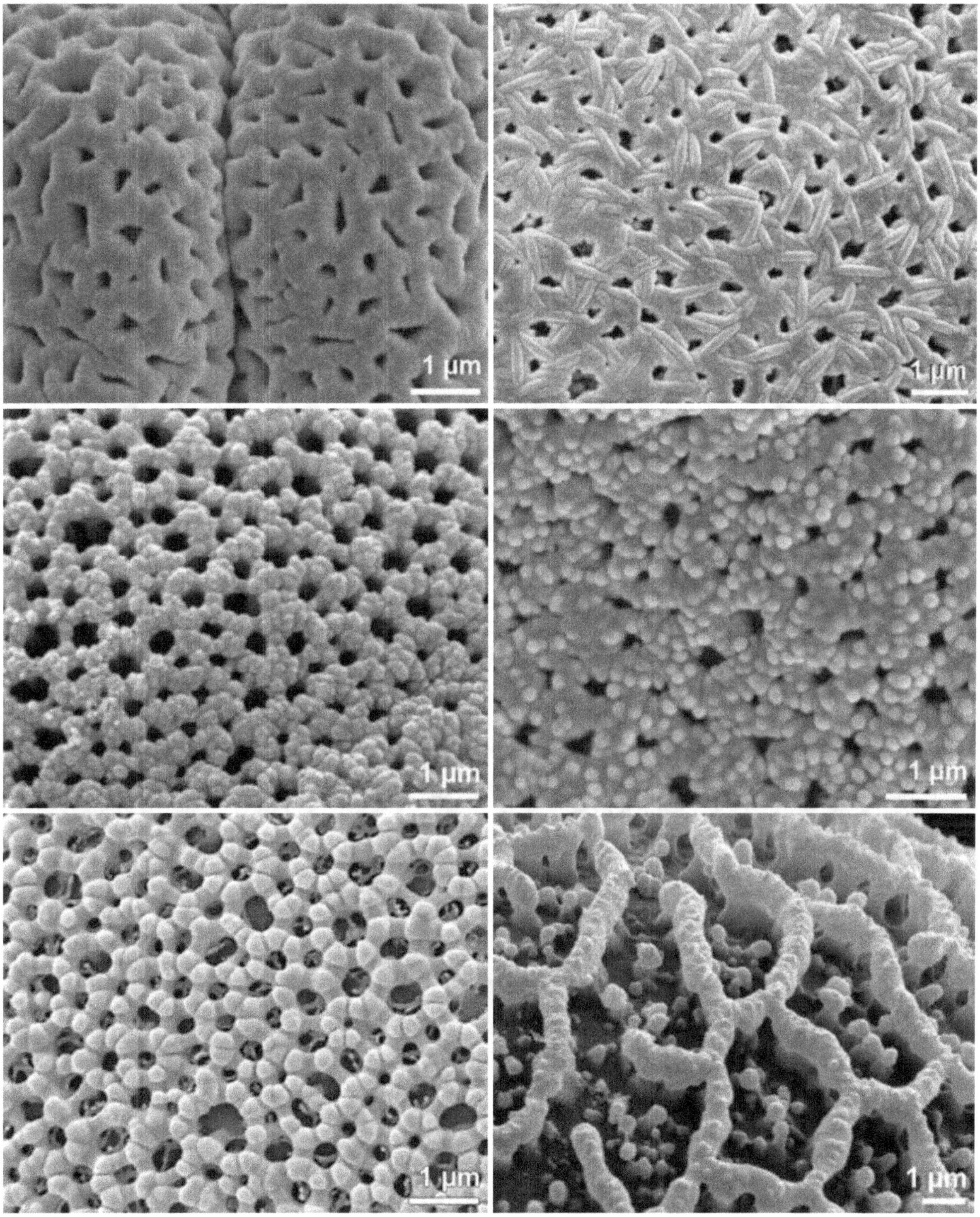

Fraxinus sp., Oleaceae
fossil, Miocene, China, microrugulate

Tulipa turkestanica, Liliaceae
microrugulate

Helleborus foetidus, Ranunculaceae
nanoverrucate to nanogemmate

Cercidiphyllum sp., Cercidiphyllaceae
fossil, Miocene, China, nanogemmate

Parentucellia latifolia, Orobanchaceae
microgemmate

Viburnum utile, Adoxaceae
micro- to nanoechinate

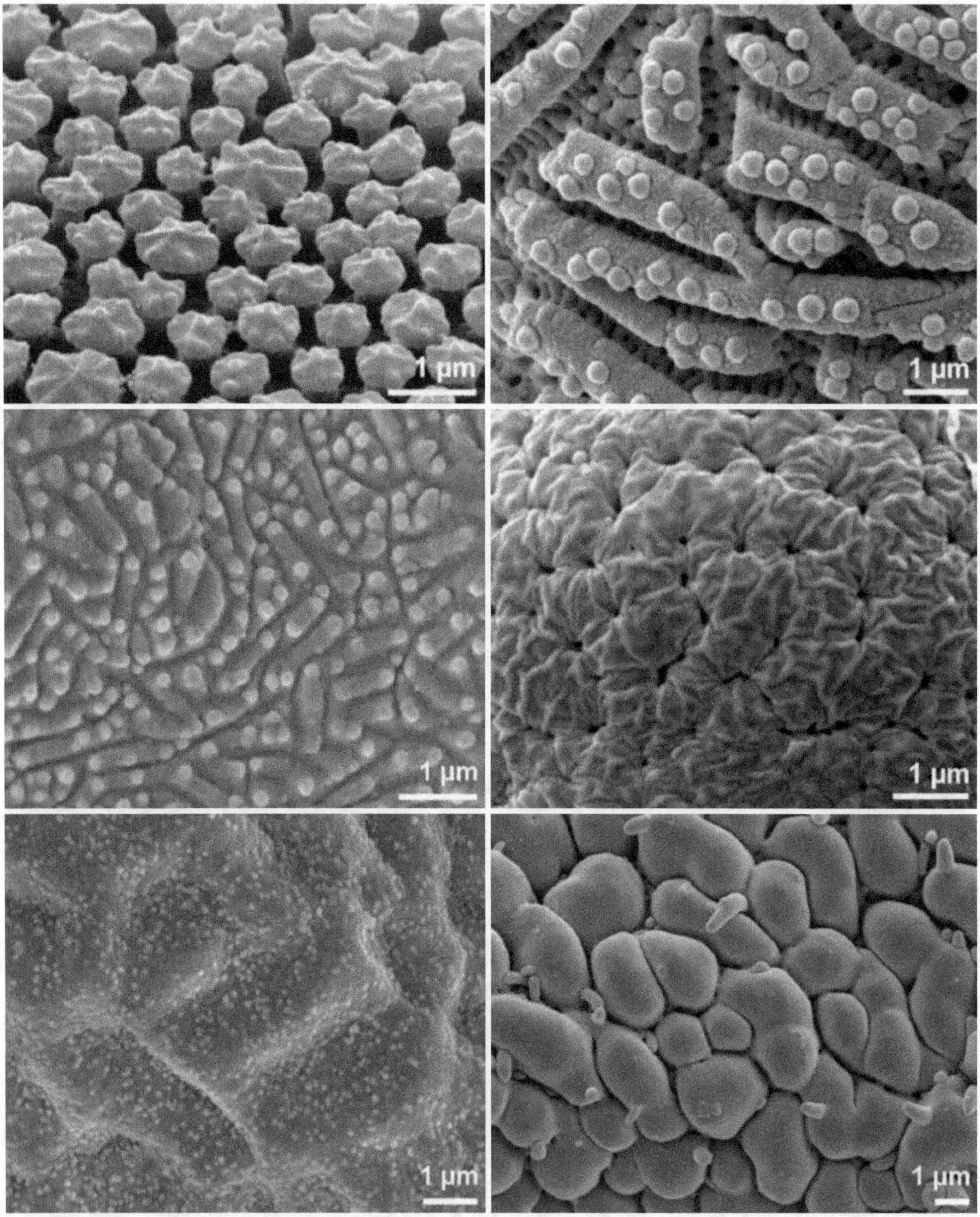

Linum bienne, Linaceae
micro- to nanorugulate, nanoechinate

Saxifraga hostii, Saxifragaceae
microgemmate

Betula sp., Betulaceae
fossil, Miocene, China, nanogemmate to nanoechinate

Piper auritum, Piperaceae
microrugulate

Ulmus sp., Ulmaceae
fossil, Miocene, China, granulate

Tsuga sp., Pinaceae
fossil, Miocene, China, echinate to baculate

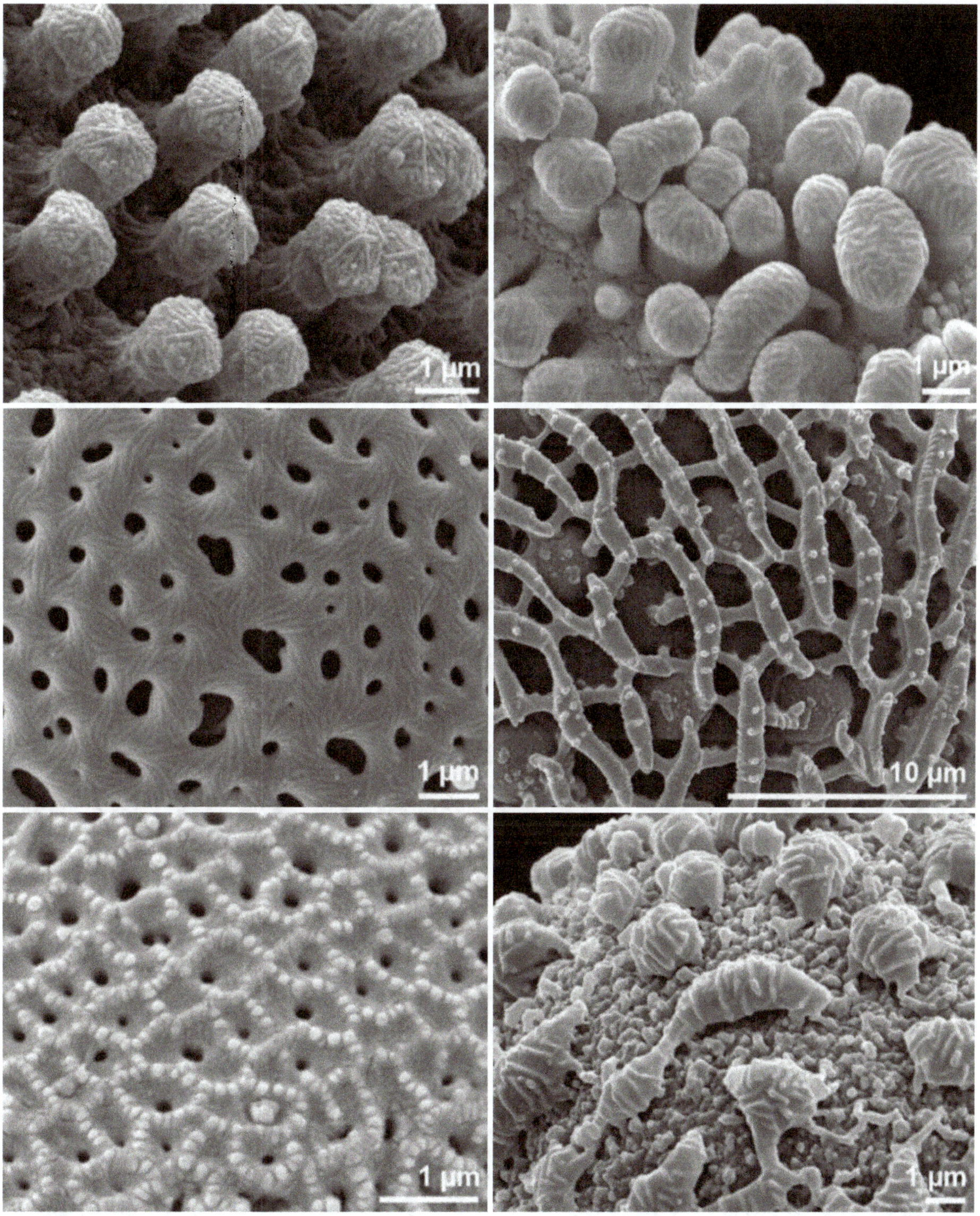

Ceratostigma plumbaginoides, Plumbaginaceae
micro- to nanorugulate, granulate

Akebia quinata, Lardizabalaceae
microrugulate to rugulate

Ouratea hexasperma, Ochnaceae
nanogemmate

Ilex sp., Aquifoliaceae
fossil, Miocene, China, microrugulate

Polemonium caeruleum, Polemoniaceae
microgemmate

Coffea arabica, Rubiaceae
rugulate to microrugulate

supratectal element

sculpture element positioned on top of the tectum

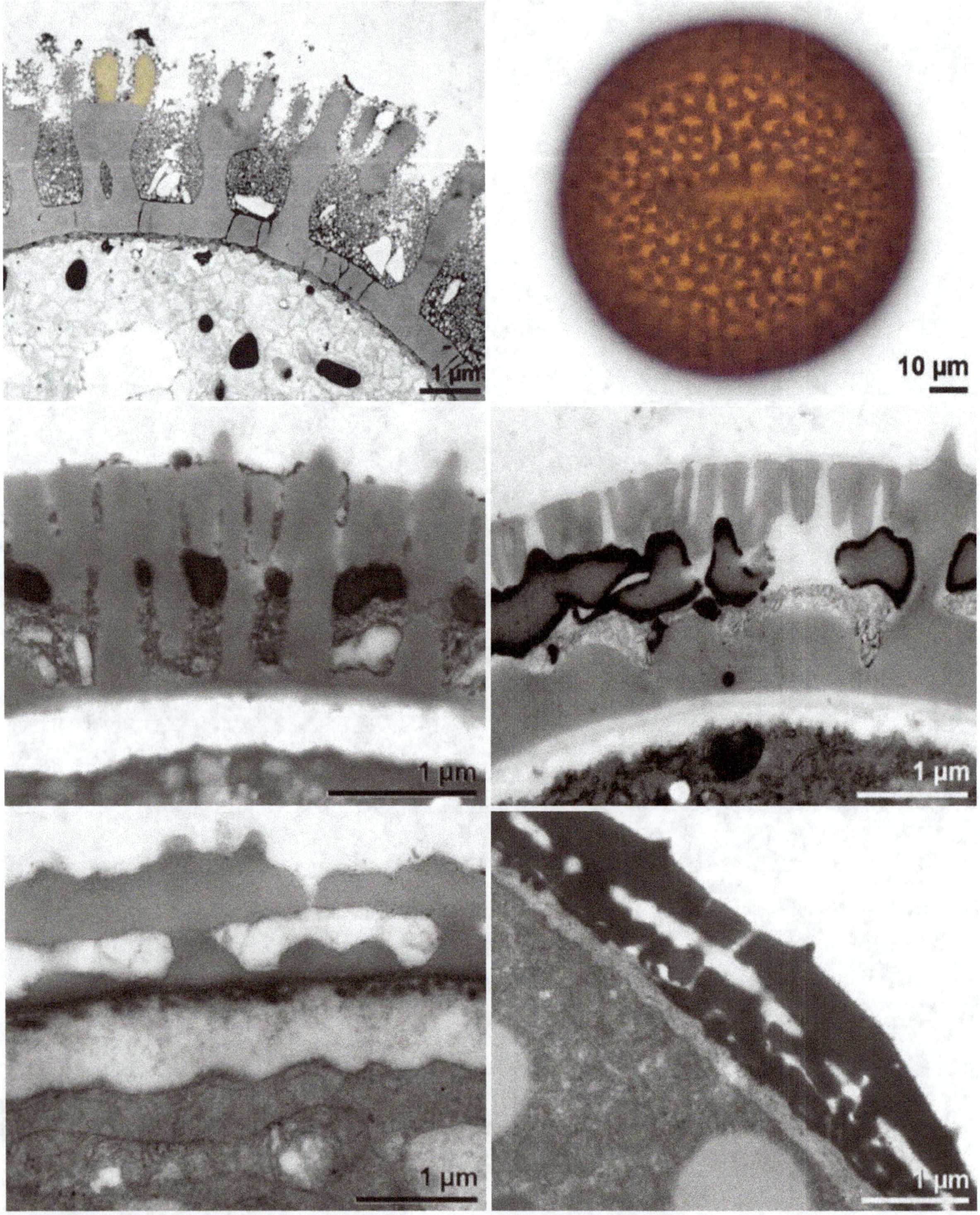

Geranium robertianum, Geraniaceae
clavae on tectum colored

Atriplex sagittata, Amaranthaceae
echini on tectum

Quercus robur, Fagaceae
echini on tectum

Geranium robertianum, Geraniaceae
clavae on reticulum

Pulsatilla grandis, Ranunculaceae
echini on tectum

Plantago maritima, Plantaginaceae
echini on tectum

verruca/verrucate

wart-like element broader than high

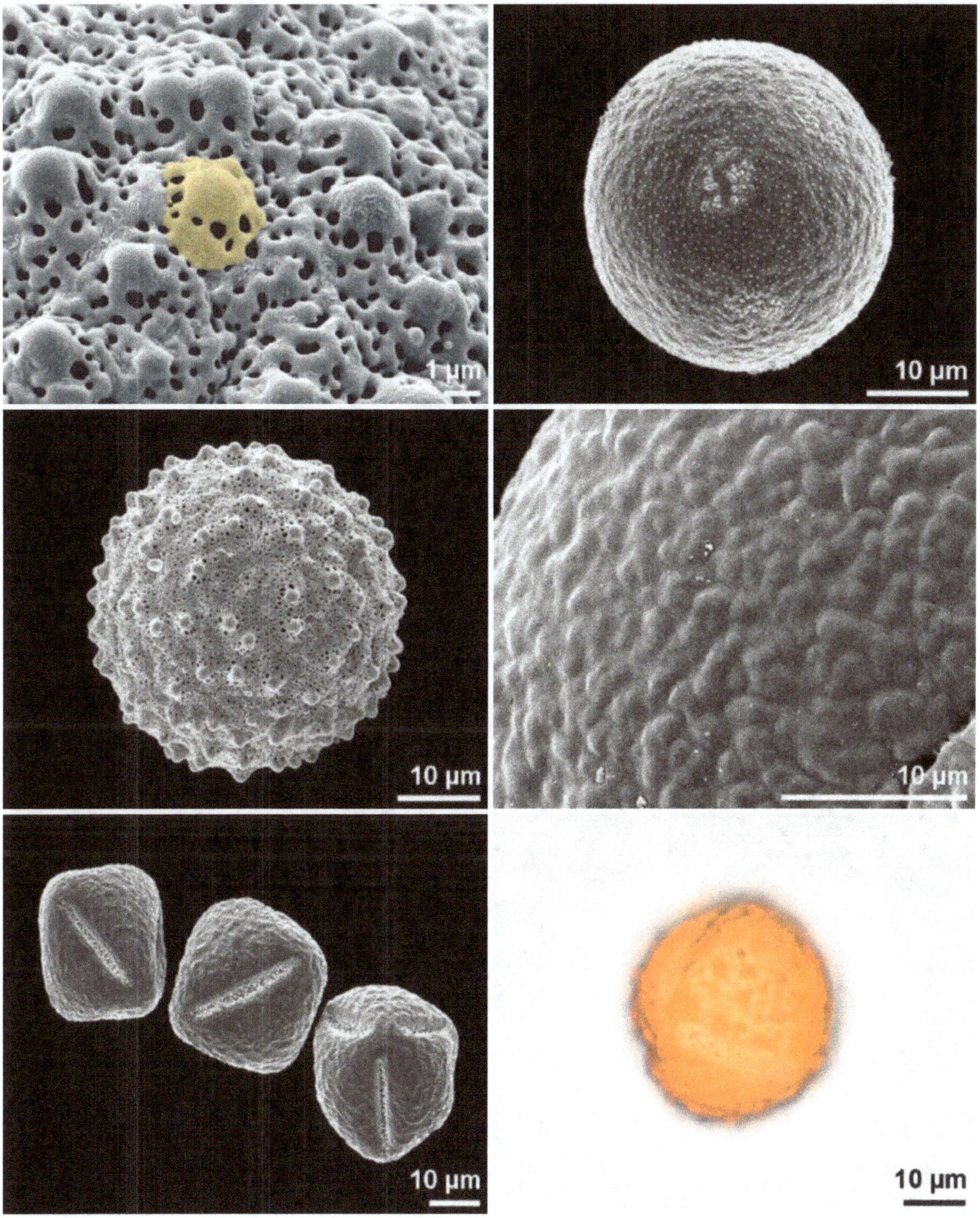

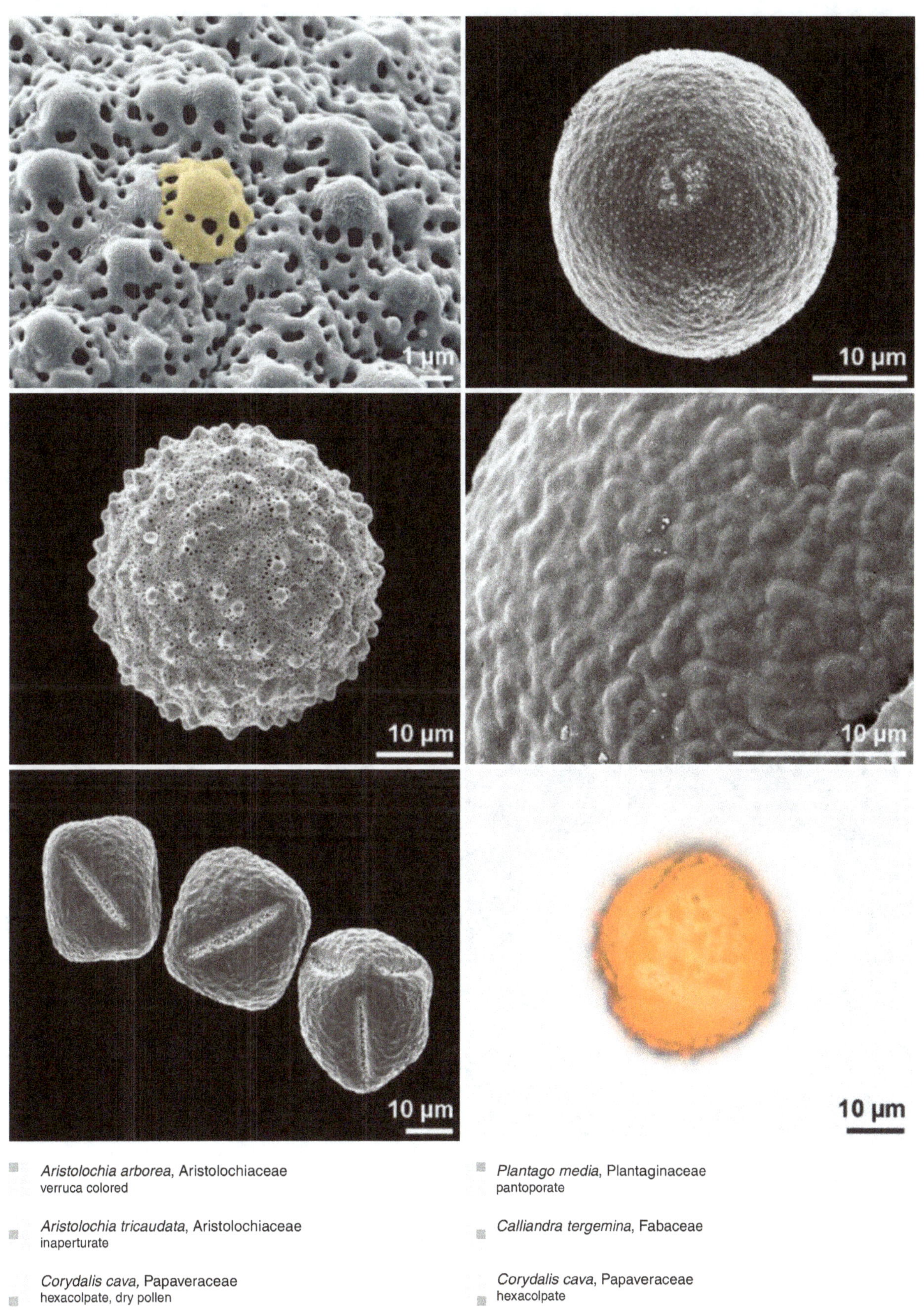

Aristolochia arborea, Aristolochiaceae
verruca colored

Aristolochia tricaudata, Aristolochiaceae
inaperturate

Corydalis cava, Papaveraceae
hexacolpate, dry pollen

Plantago media, Plantaginaceae
pantoporate

Calliandra tergemina, Fabaceae

Corydalis cava, Papaveraceae
hexacolpate

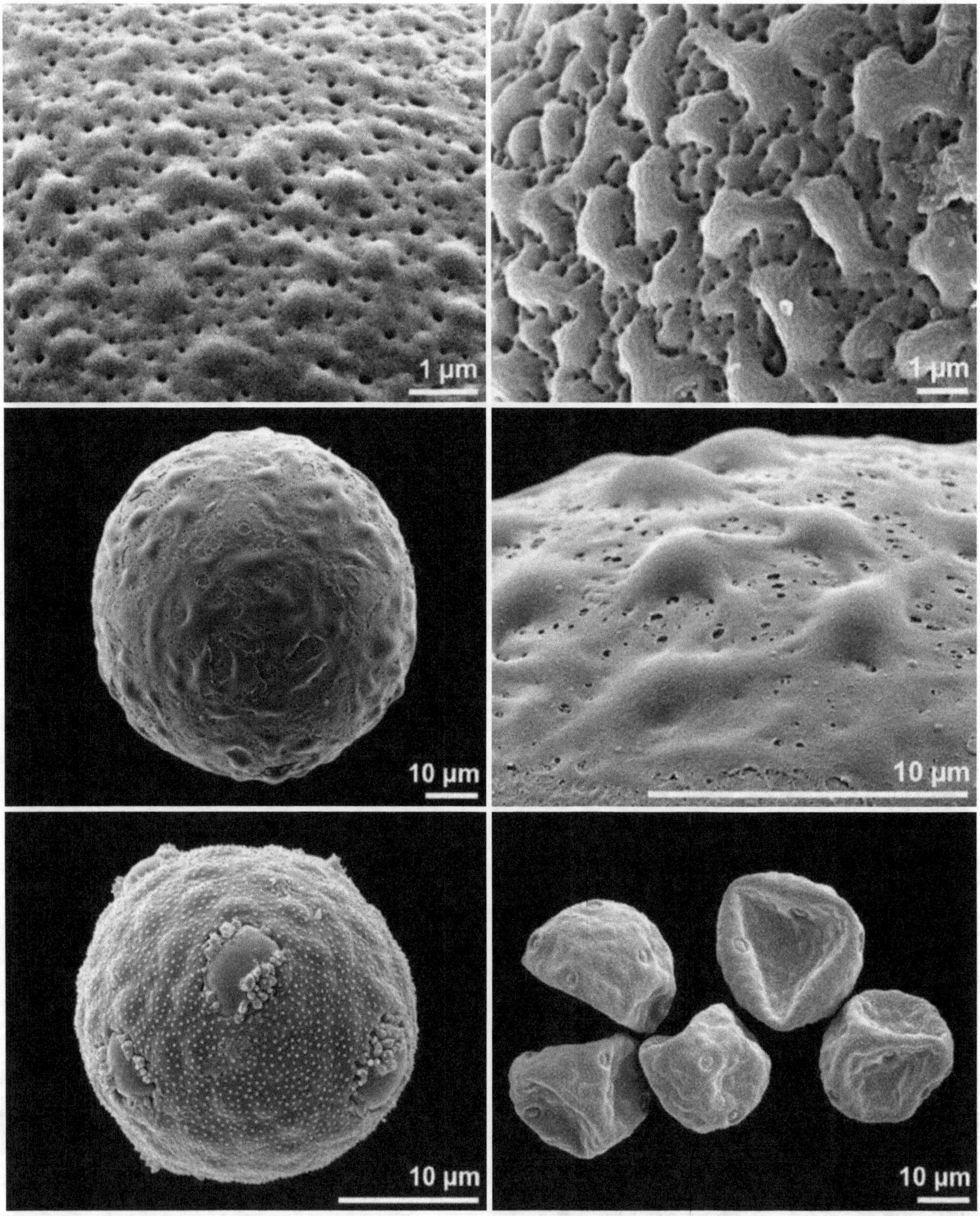

Teucrium chamaedrys, Lamiaceae
microverrucate, perforate

Cyanastrum cordifolium, Tecophilaeaceae
verrucate, perforate

Platycapnos tenuilobus, Papaveraceae

Platycapnos tenuilobus, Papaveraceae
verrucate, perforate

Plantago reniformis, Plantaginaceae
verrucate, microechinate

Plantago lanceolata, Plantaginaceae
verrucate, microechinate, dry pollen

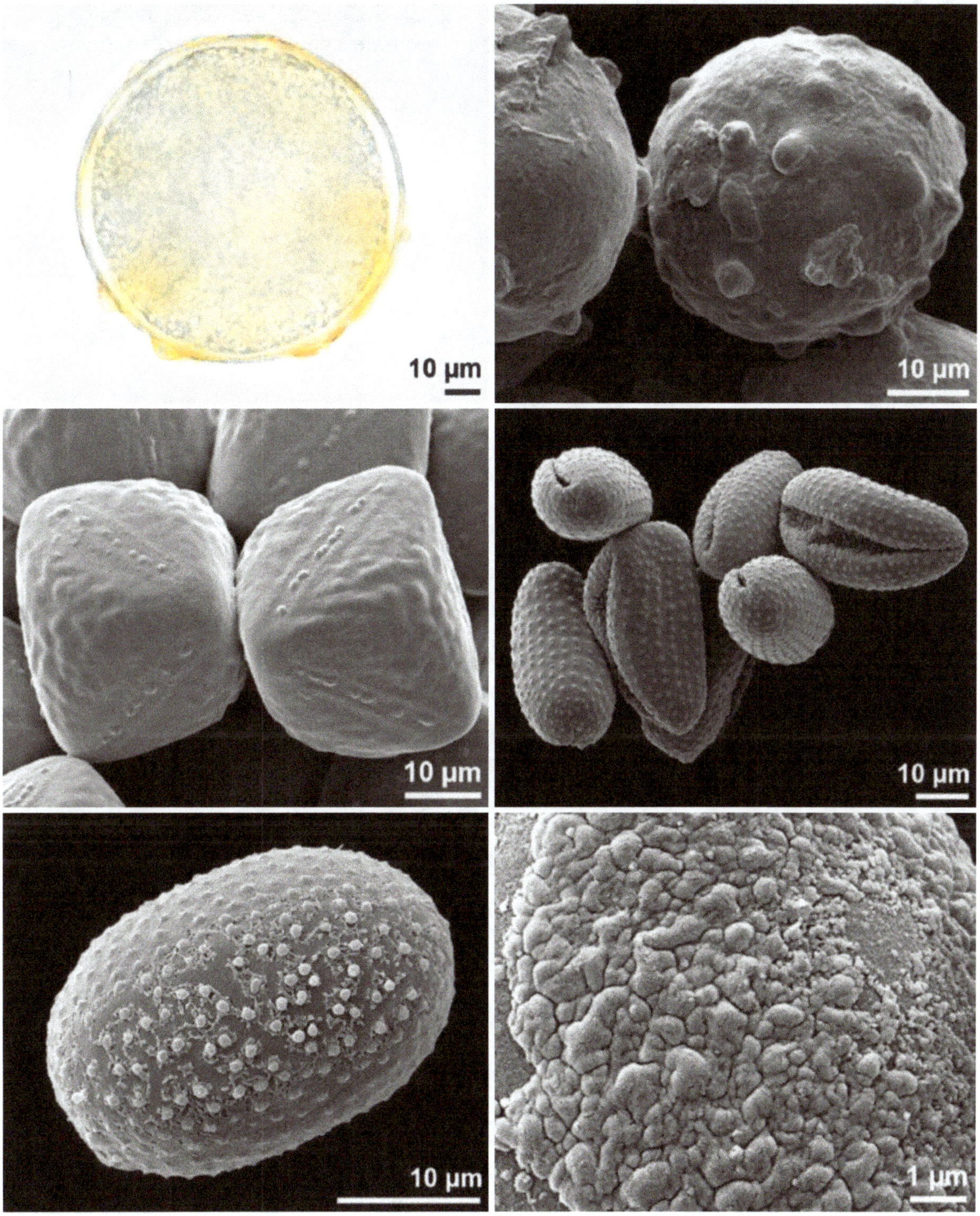

Amorphophallus stuhlmannii, Araceae
hydrated

Amorphophallus stuhlmannii, Araceae
inaperturate

Sarcocapnos enneaphylla, Papaveraceae
hexacolpate, dry pollen

Commelina erecta, Commelinaceae
sulcate, dry pollen

Stanfieldiella imperforata, Commelinaceae
sulcate, distal polar view

Callistemon comboynensis, Myrtaceae
verrucate to microverrucate

Pollen Wall

H. Halbritter et al., *Illustrated Pollen Terminology*, https://doi.org/10.1007/978-3-319-71365-6_11

tectum/tectate

outer more or less continuous ektexine layer; tectum condition can be eutectate or semitectate

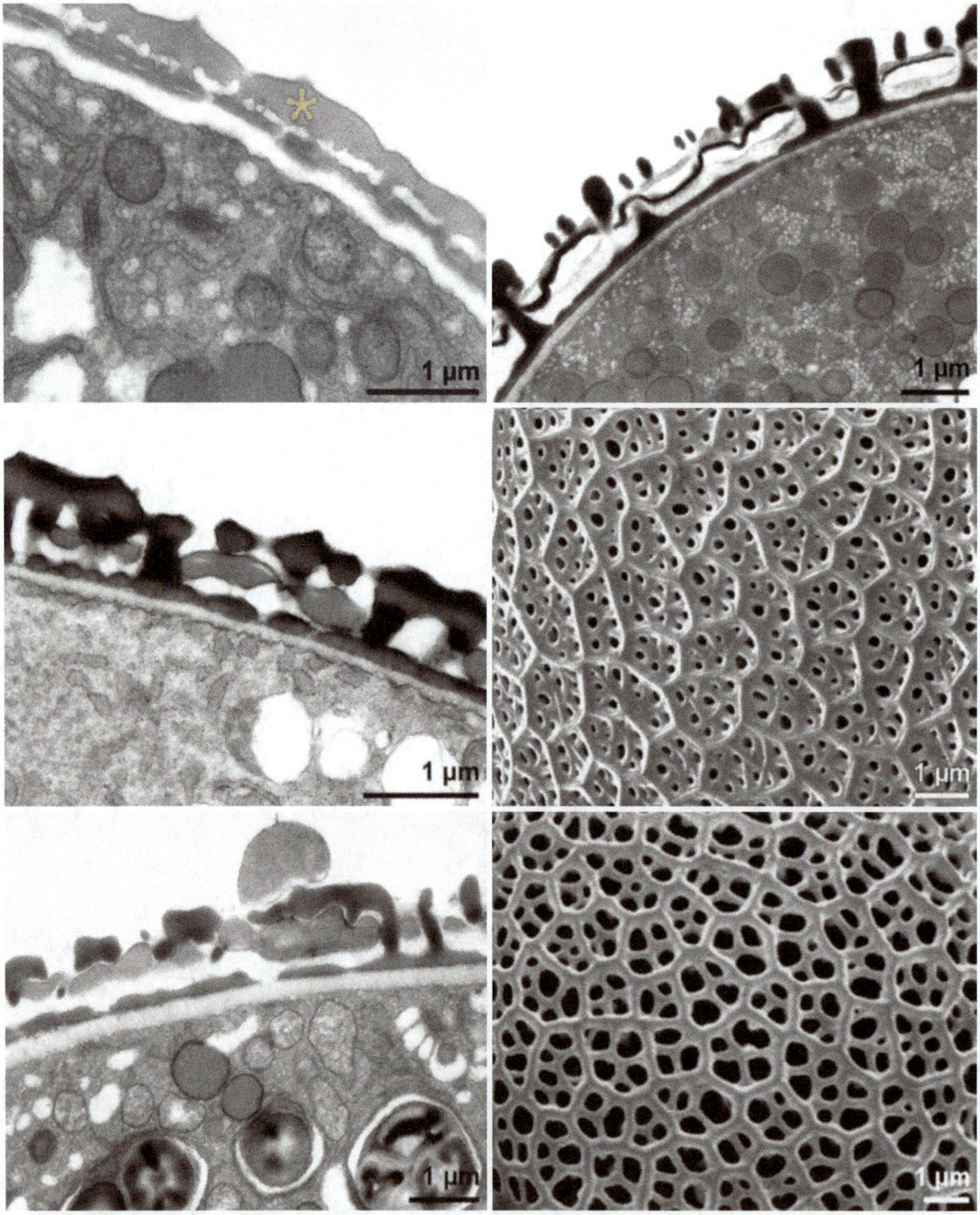

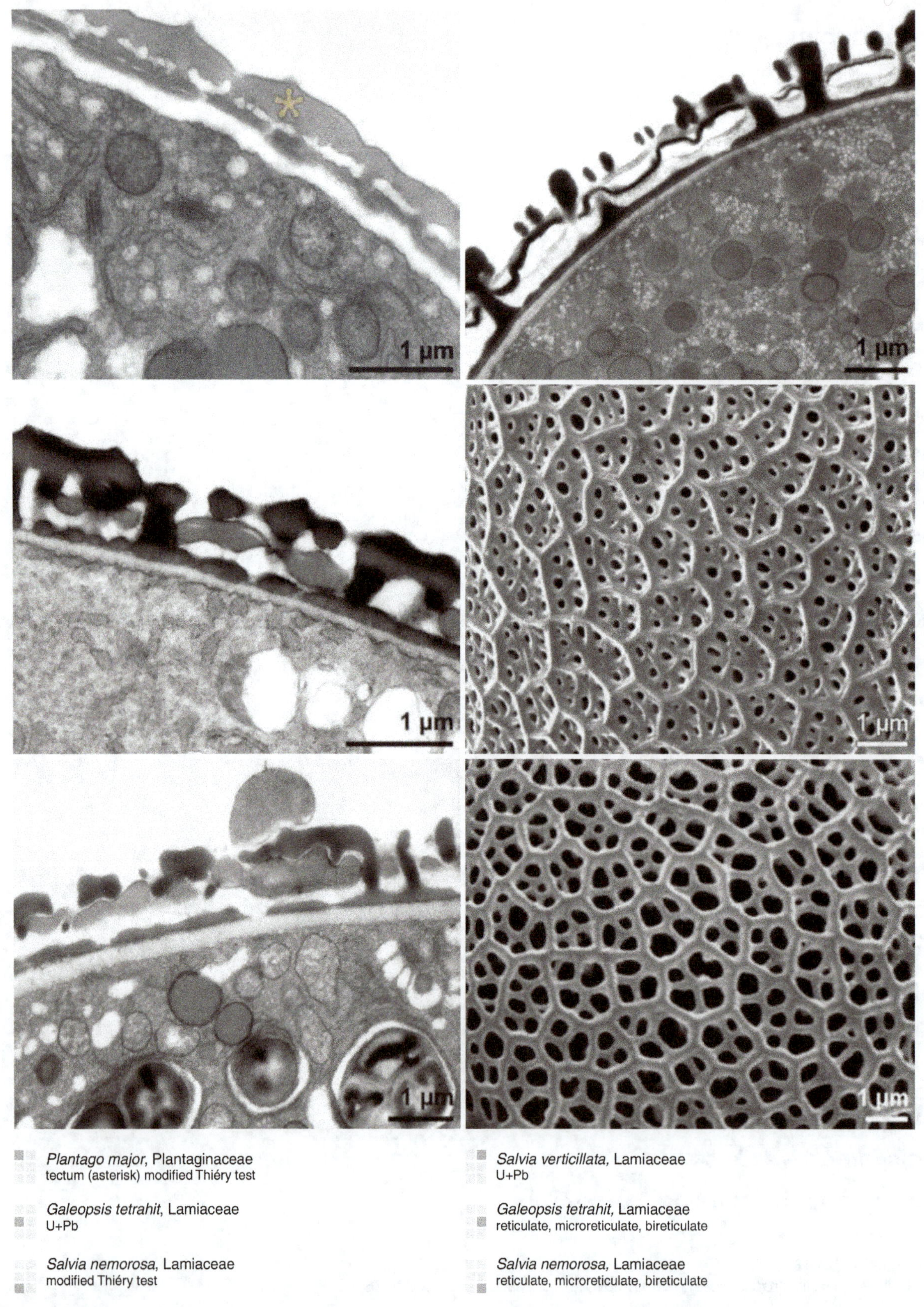

Plantago major, Plantaginaceae
tectum (asterisk) modified Thiéry test

Salvia verticillata, Lamiaceae
U+Pb

Galeopsis tetrahit, Lamiaceae
U+Pb

Galeopsis tetrahit, Lamiaceae
reticulate, microreticulate, bireticulate

Salvia nemorosa, Lamiaceae
modified Thiéry test

Salvia nemorosa, Lamiaceae
reticulate, microreticulate, bireticulate

eutectate

pollen grain with a predominantly continuous tectum

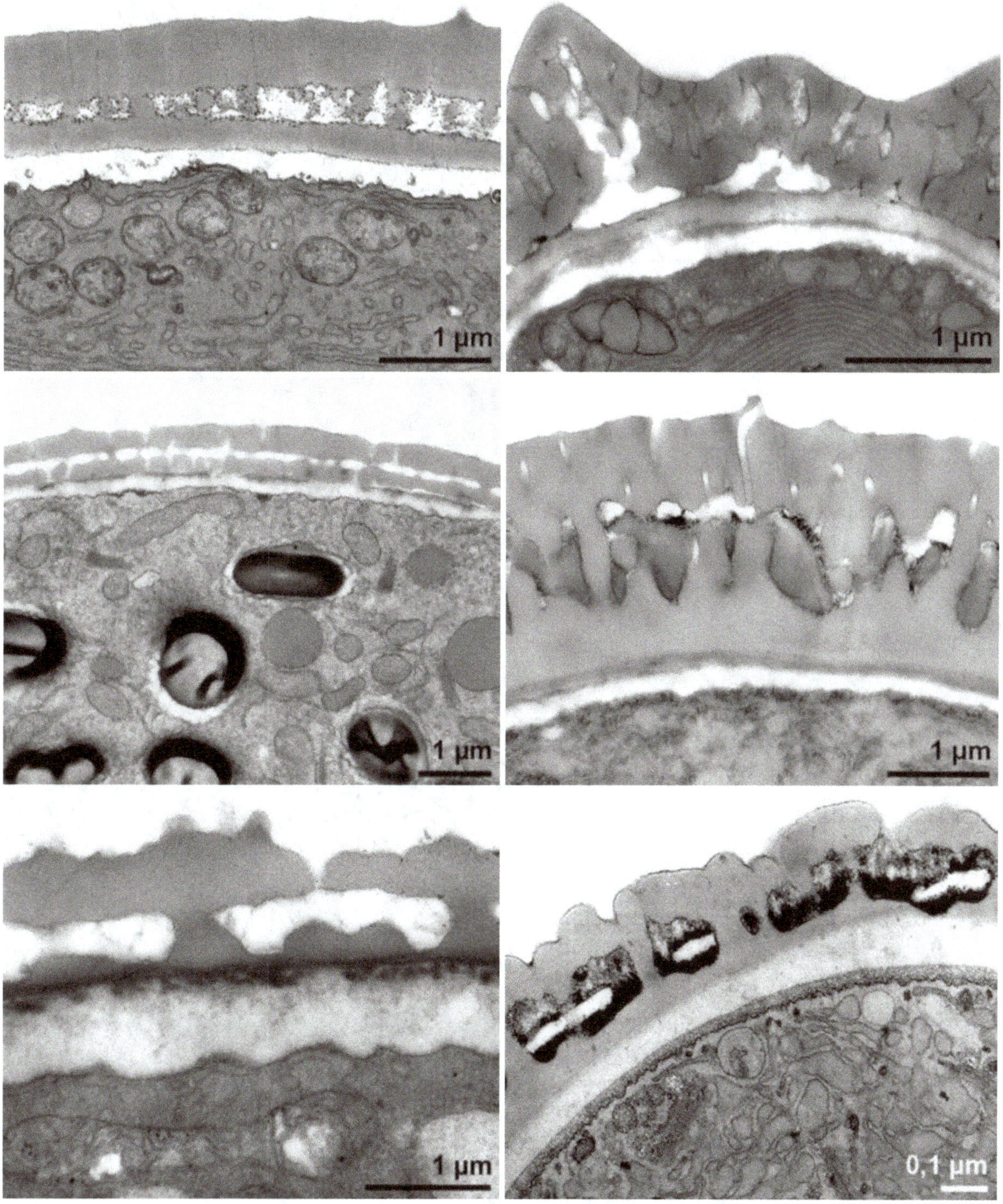

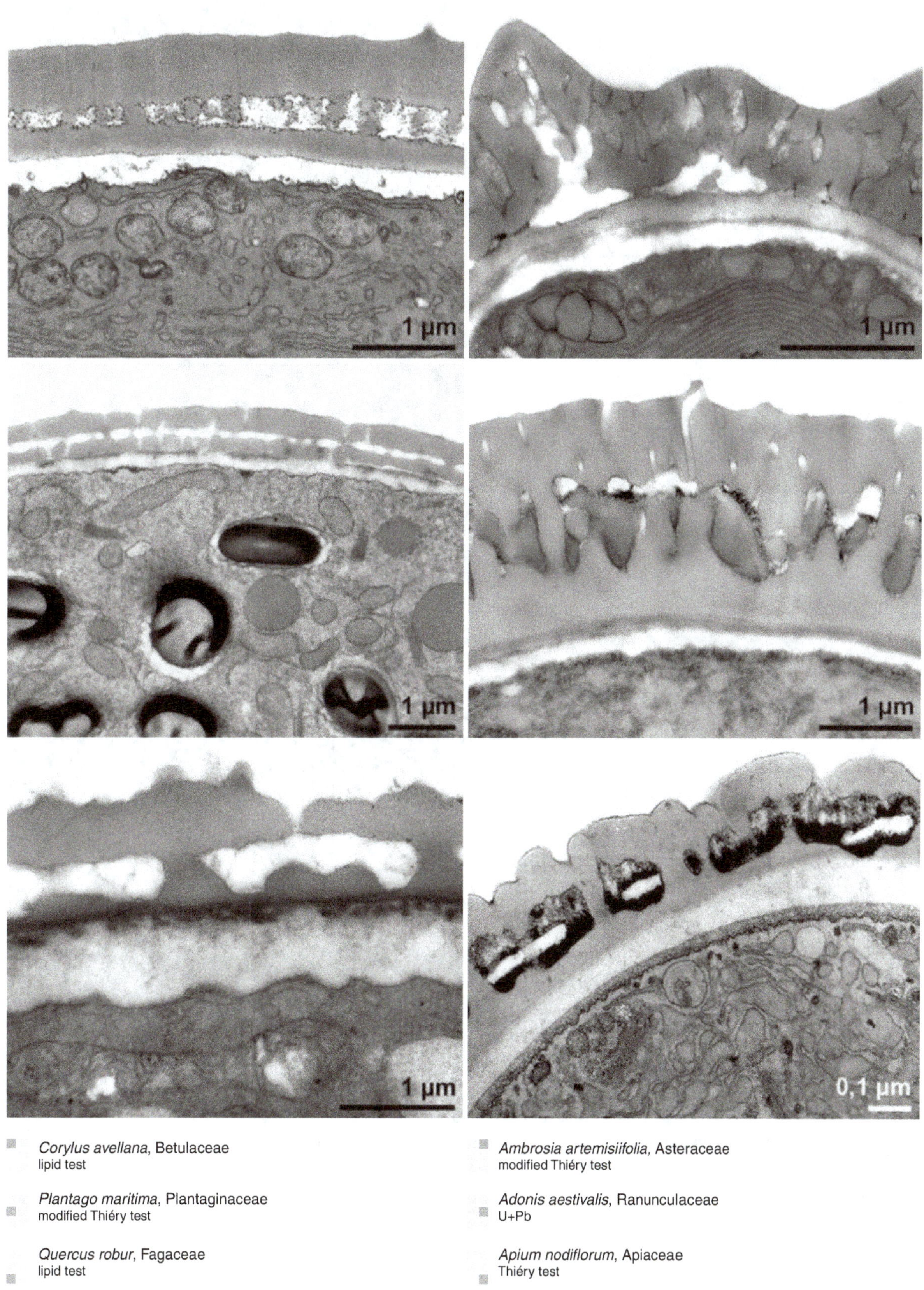

Corylus avellana, Betulaceae
lipid test

Plantago maritima, Plantaginaceae
modified Thiéry test

Quercus robur, Fagaceae
lipid test

Ambrosia artemisiifolia, Asteraceae
modified Thiéry test

Adonis aestivalis, Ranunculaceae
U+Pb

Apium nodiflorum, Apiaceae
Thiéry test

semitectate

pollen grain with a discontinuous tectum

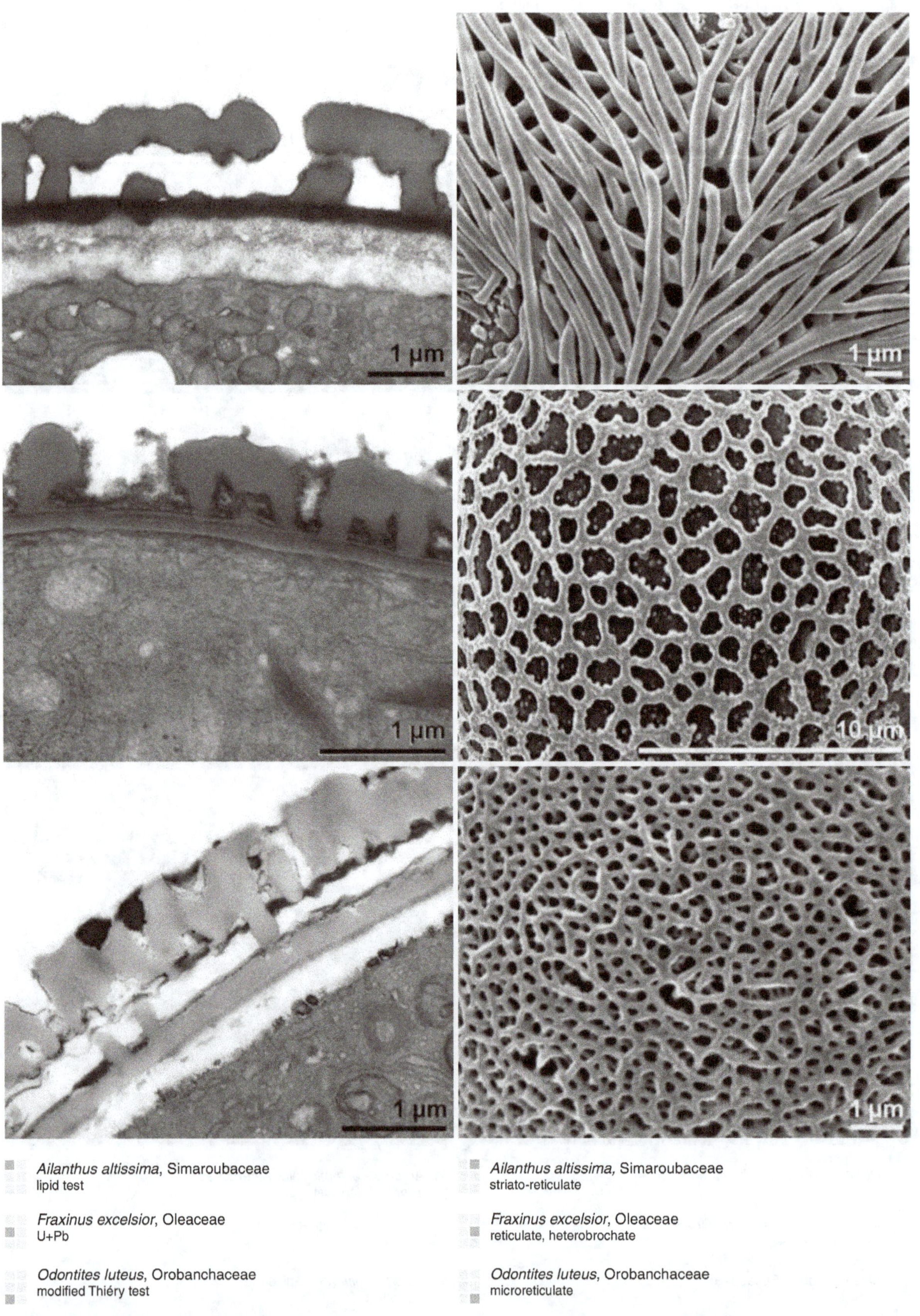

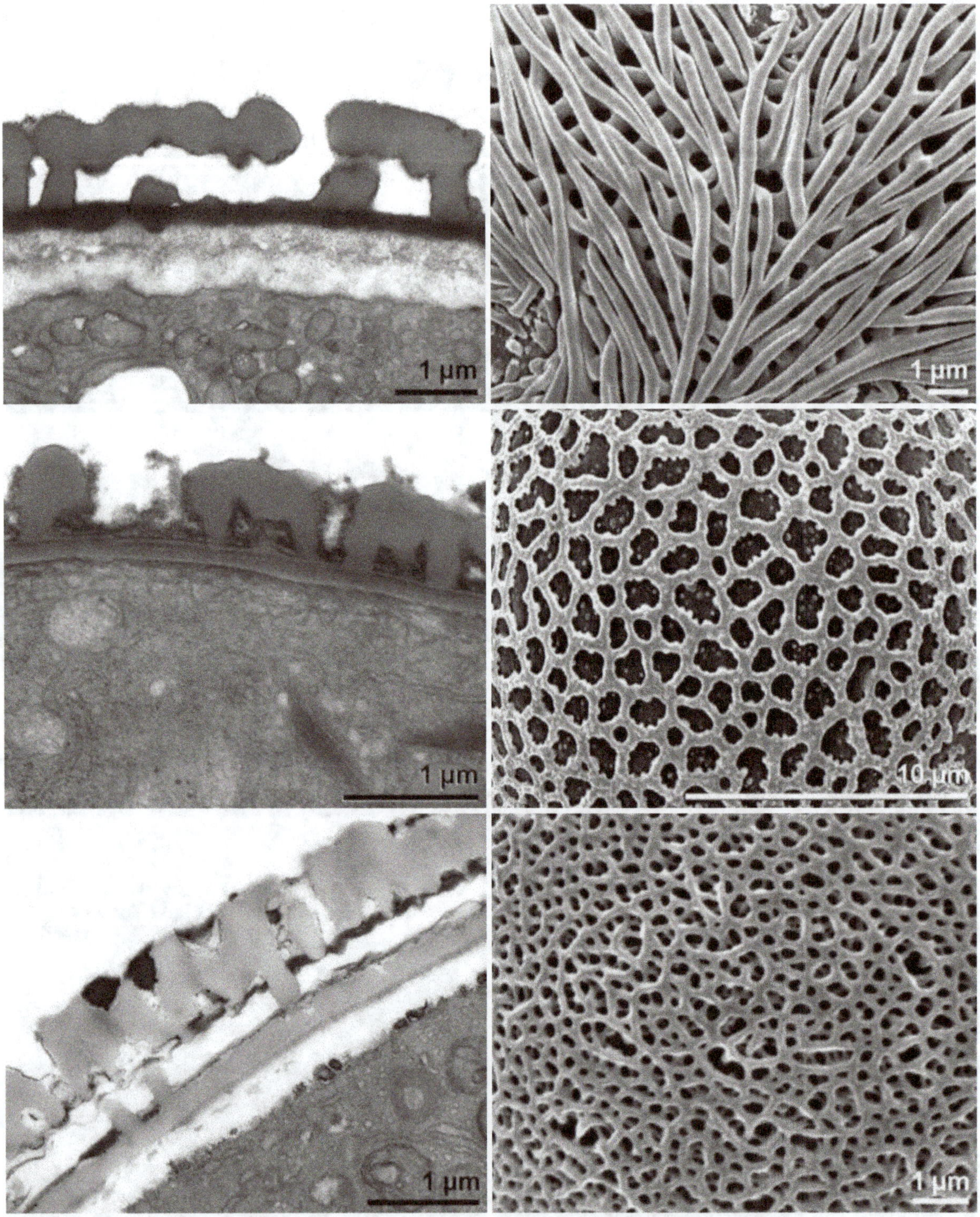

Ailanthus altissima, Simaroubaceae
lipid test

Fraxinus excelsior, Oleaceae
U+Pb

Odontites luteus, Orobanchaceae
modified Thiéry test

Ailanthus altissima, Simaroubaceae
striato-reticulate

Fraxinus excelsior, Oleaceae
reticulate, heterobrochate

Odontites luteus, Orobanchaceae
microreticulate

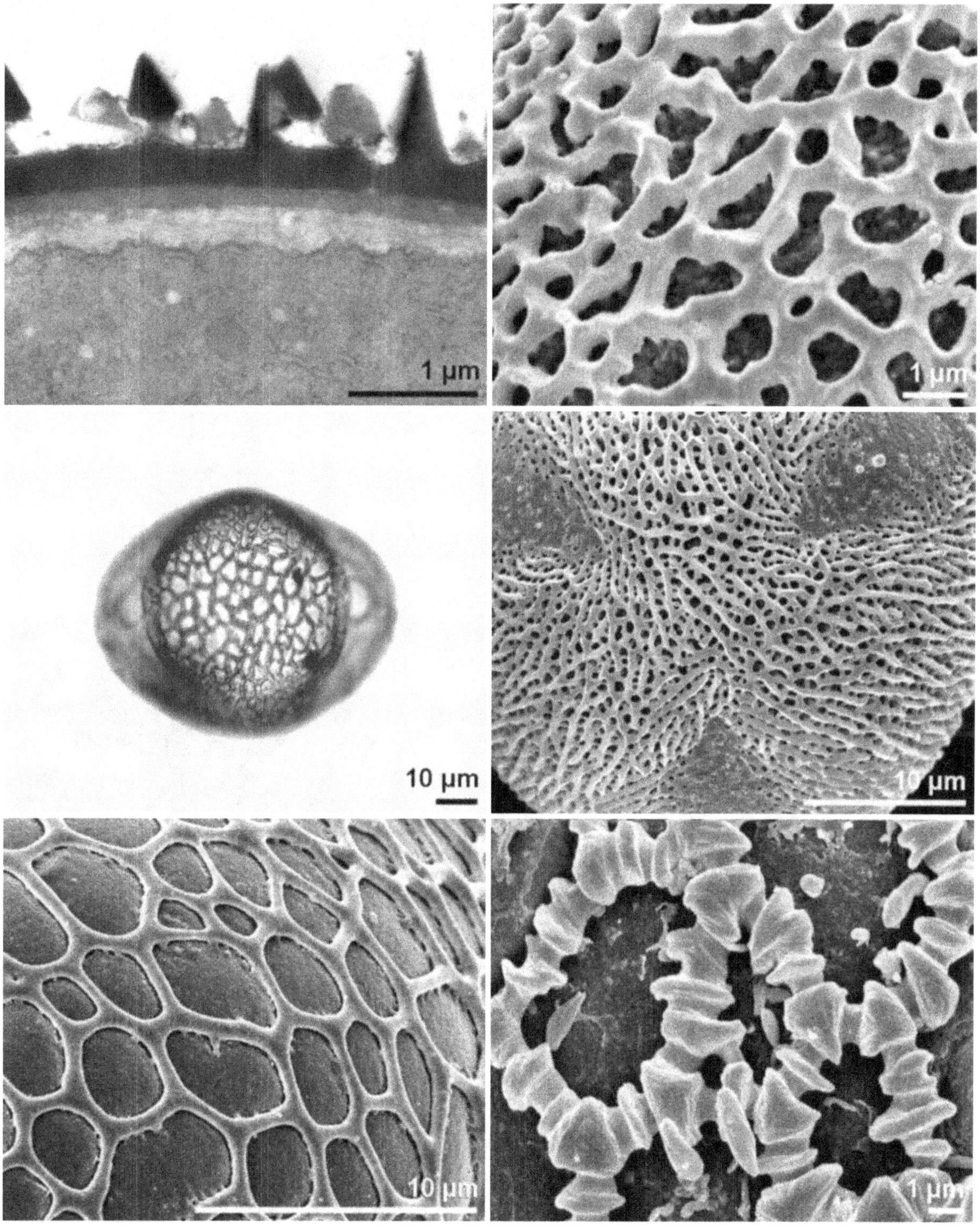

Salix fragilis, Salicaceae
U+Pb

Alangium sp., Cornaceae,
fossil reticulate, equatorial view

Quesnelia lateralis, Bromeliaceae
reticulate

Salix fragilis, Salicaceae
reticulate

Lomatogonium carinthiacum, Gentianaceae
striato-microreticulate, polar area

Pachysandra terminalis, Buxaceae
reticulum cristatum

atectate

pollen grain lacking a tectum

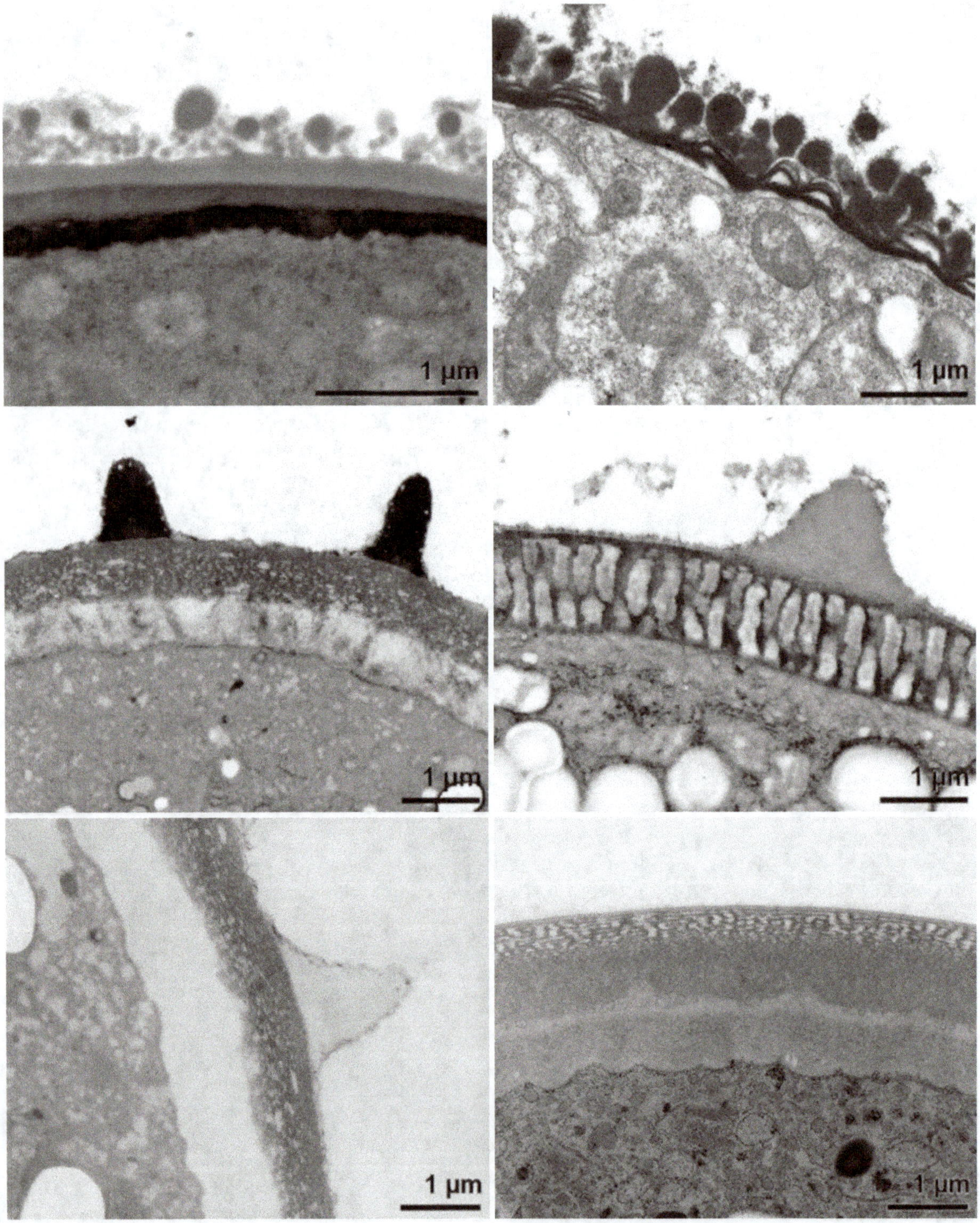

Rhaphidophora africana, Araceae
U+Pb

Sauromatum venosum, Araceae
Thiéry test

Sauromatum venosum, Araceae
lipid test

Orobanche hederae, Orobanchaceae
potassium permanganate

Globba schomburgkii, Zingiberaceae
U+Pb

Spathicarpa sp., Araceae
Thiéry test

infratectum alveolate

infratectum with compartments of irregular size and shape

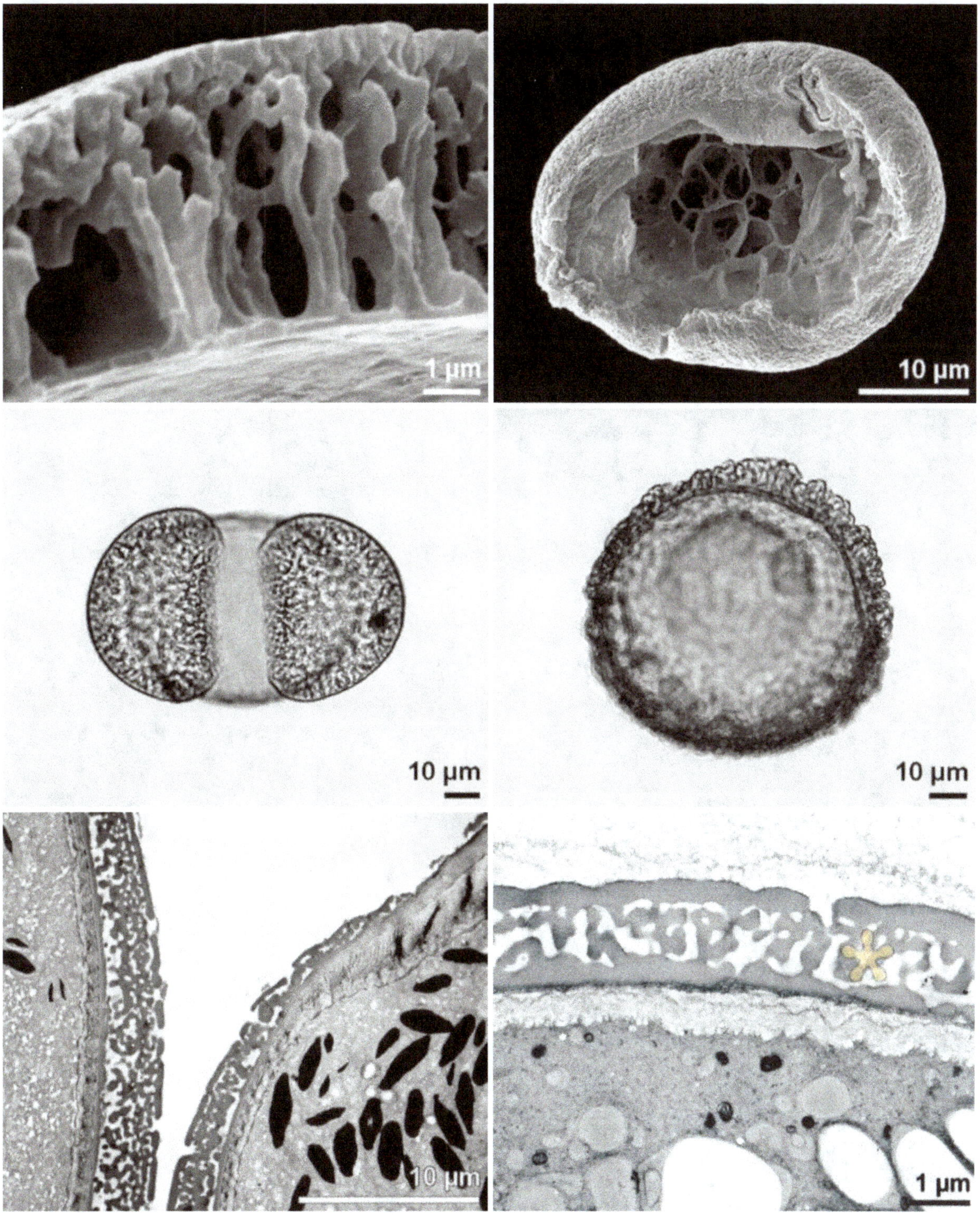

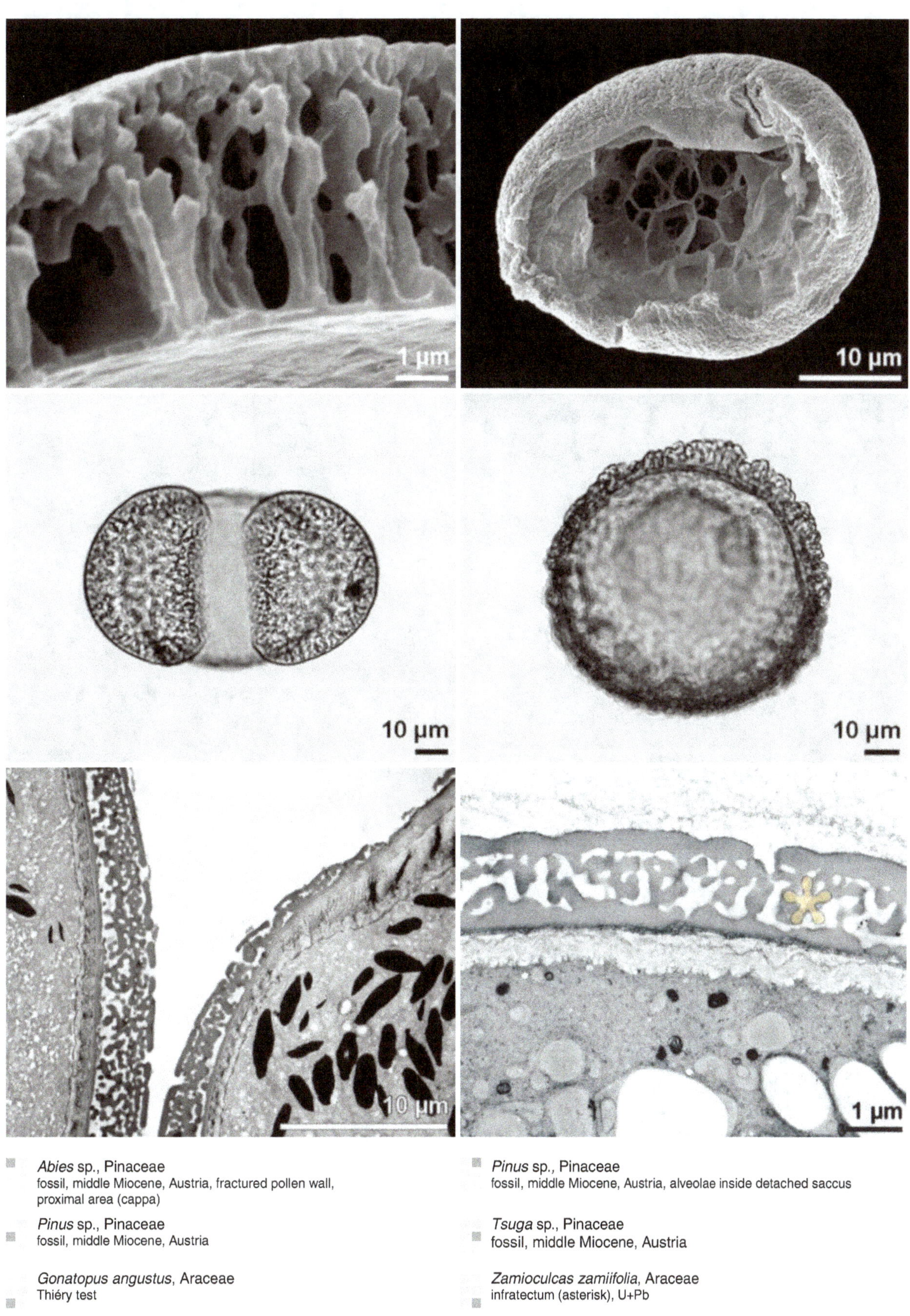

Abies sp., Pinaceae
fossil, middle Miocene, Austria, fractured pollen wall, proximal area (cappa)

Pinus sp., Pinaceae
fossil, middle Miocene, Austria

Gonatopus angustus, Araceae
Thiéry test

Pinus sp., Pinaceae
fossil, middle Miocene, Austria, alveolae inside detached saccus

Tsuga sp., Pinaceae
fossil, middle Miocene, Austria

Zamioculcas zamiifolia, Araceae
infratectum (asterisk), U+Pb

infratectum columellate

infratectum with columellae

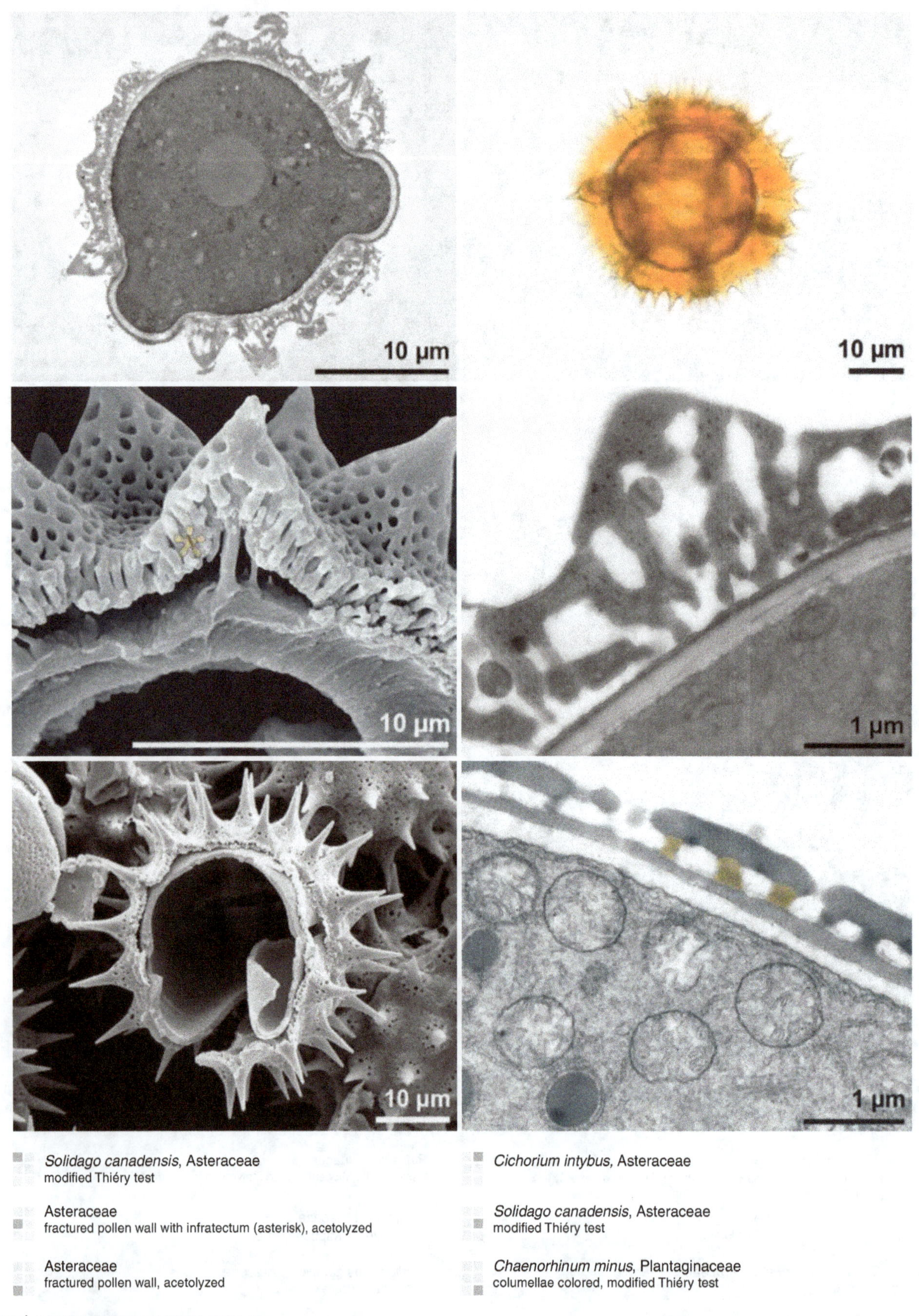

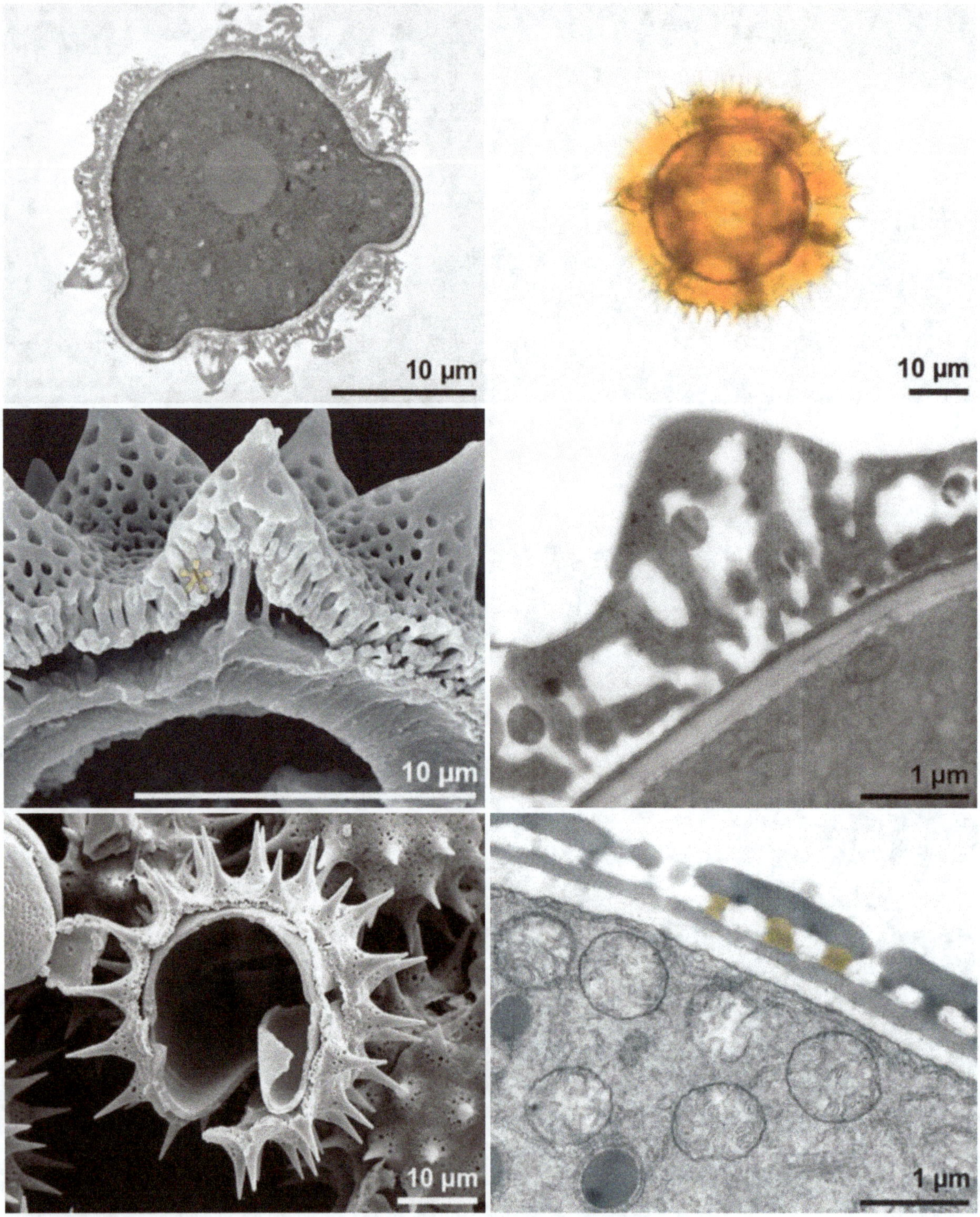

Solidago canadensis, Asteraceae
modified Thiéry test

Asteraceae
fractured pollen wall with infratectum (asterisk), acetolyzed

Asteraceae
fractured pollen wall, acetolyzed

Cichorium intybus, Asteraceae

Solidago canadensis, Asteraceae
modified Thiéry test

Chaenorhinum minus, Plantaginaceae
columellae colored, modified Thiéry test

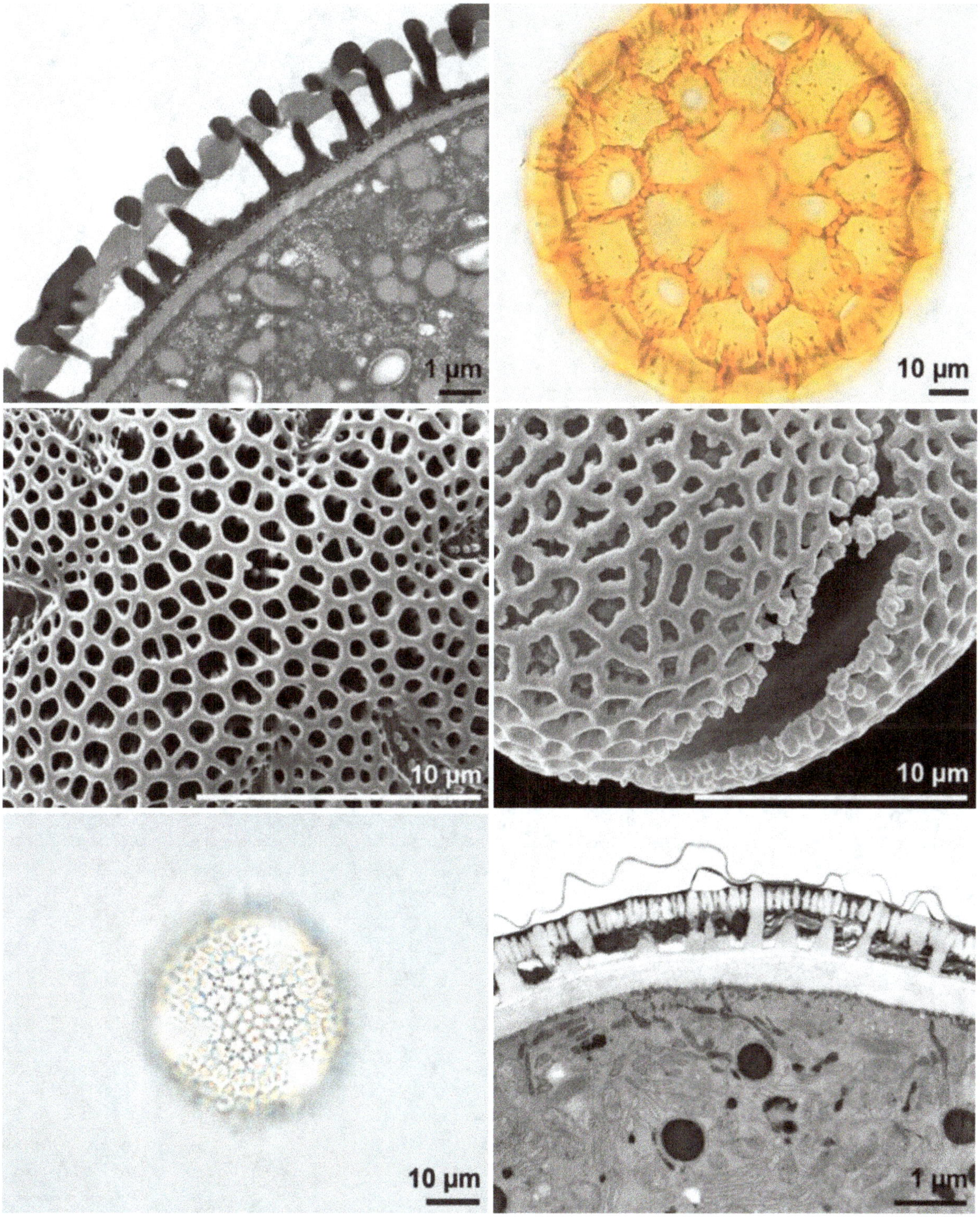

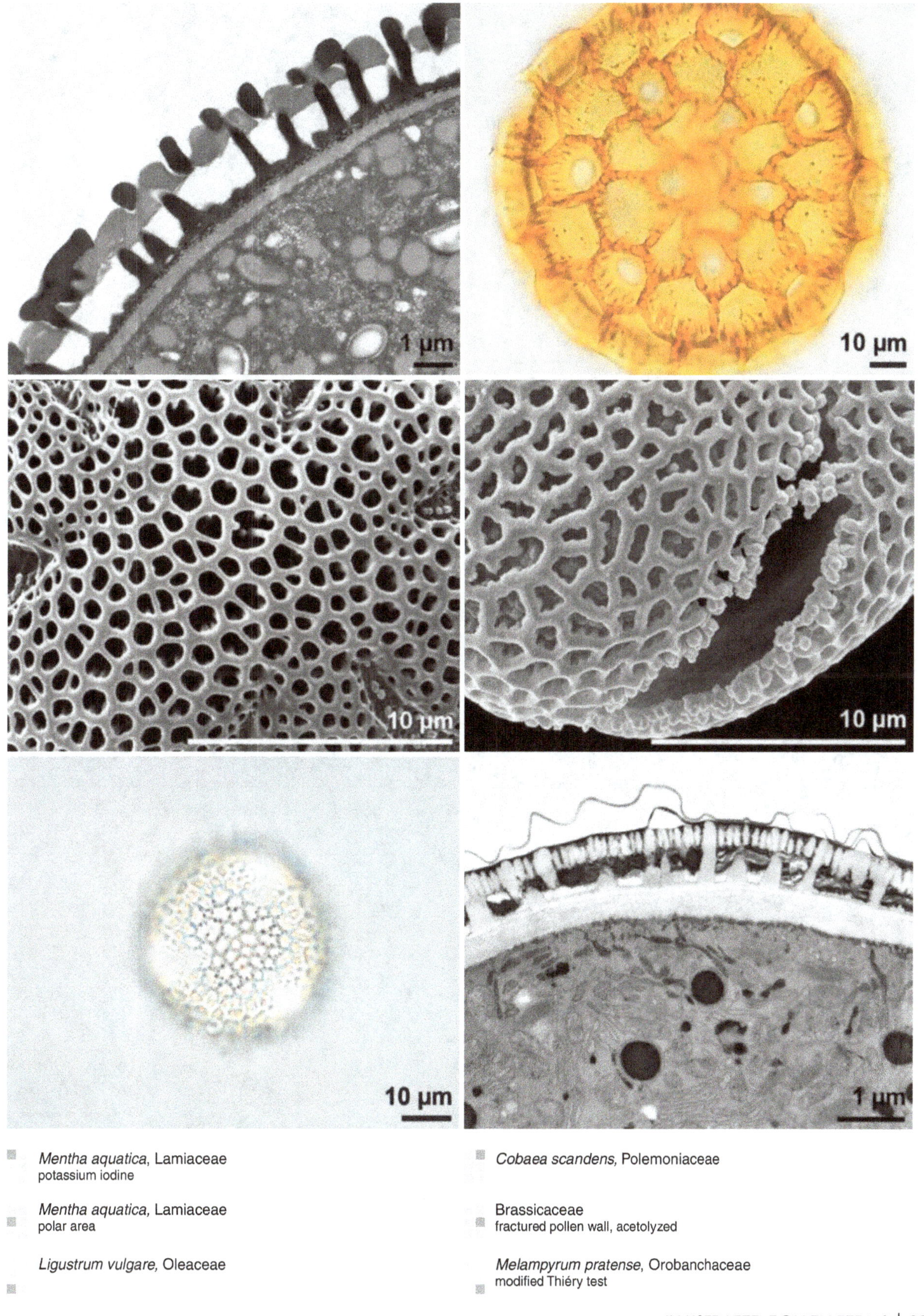

Mentha aquatica, Lamiaceae
potassium iodine

Mentha aquatica, Lamiaceae
polar area

Ligustrum vulgare, Oleaceae

Cobaea scandens, Polemoniaceae

Brassicaceae
fractured pollen wall, acetolyzed

Melampyrum pratense, Orobanchaceae
modified Thiéry test

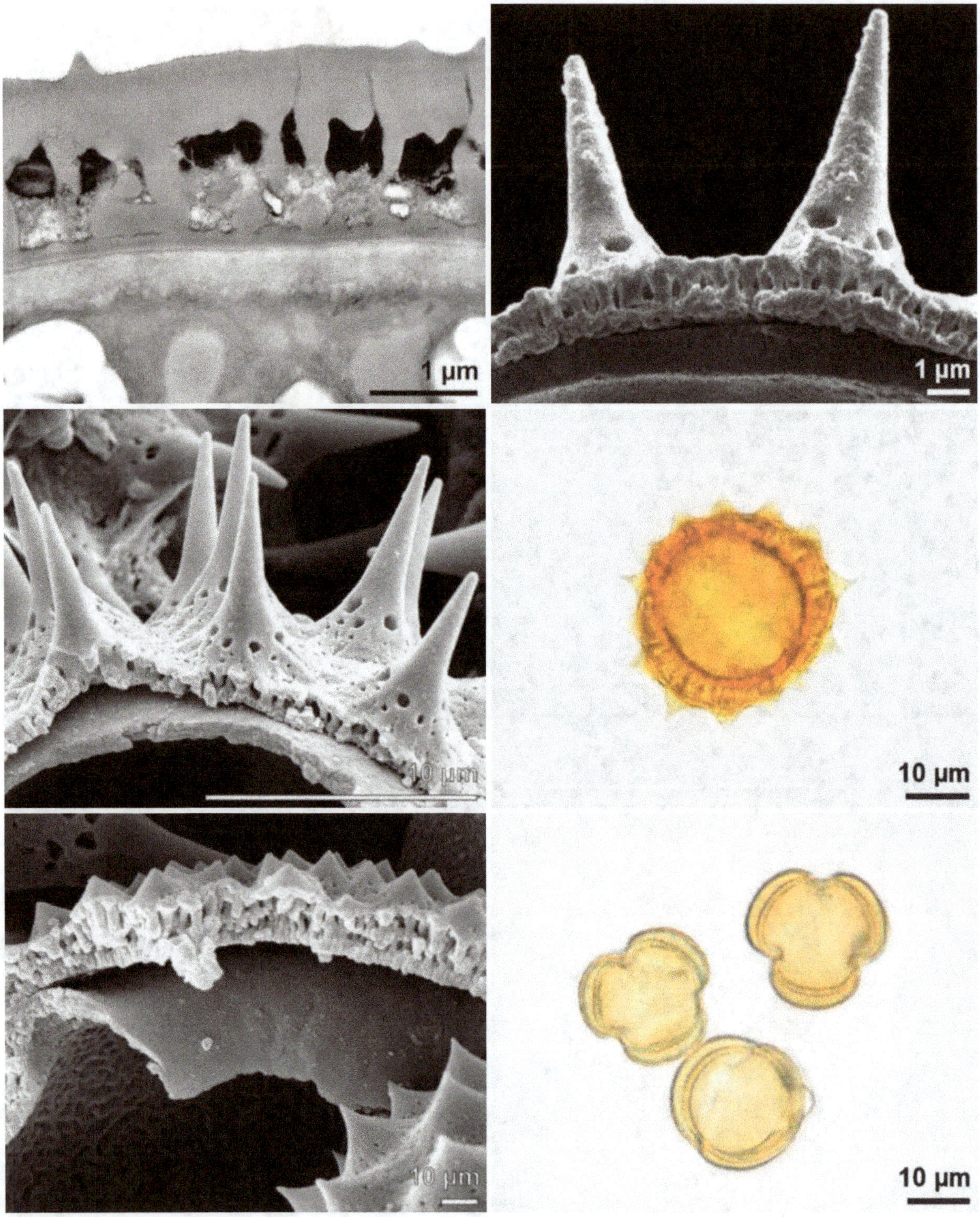

Bassia scoparia, Amaranthaceae
U+Pb

Asteraceae
fractured pollen wall, acetolyzed

Asteraceae
fractured pollen wall, acetolyzed

Asteraceae
fractured pollen wall, acetolyzed

Tanacetum corymbosum, Asteraceae

Artemisia vulgaris, Asteraceae

infratectum granular

infratectum composed of granula, cluster of granula or elements of different size and shape

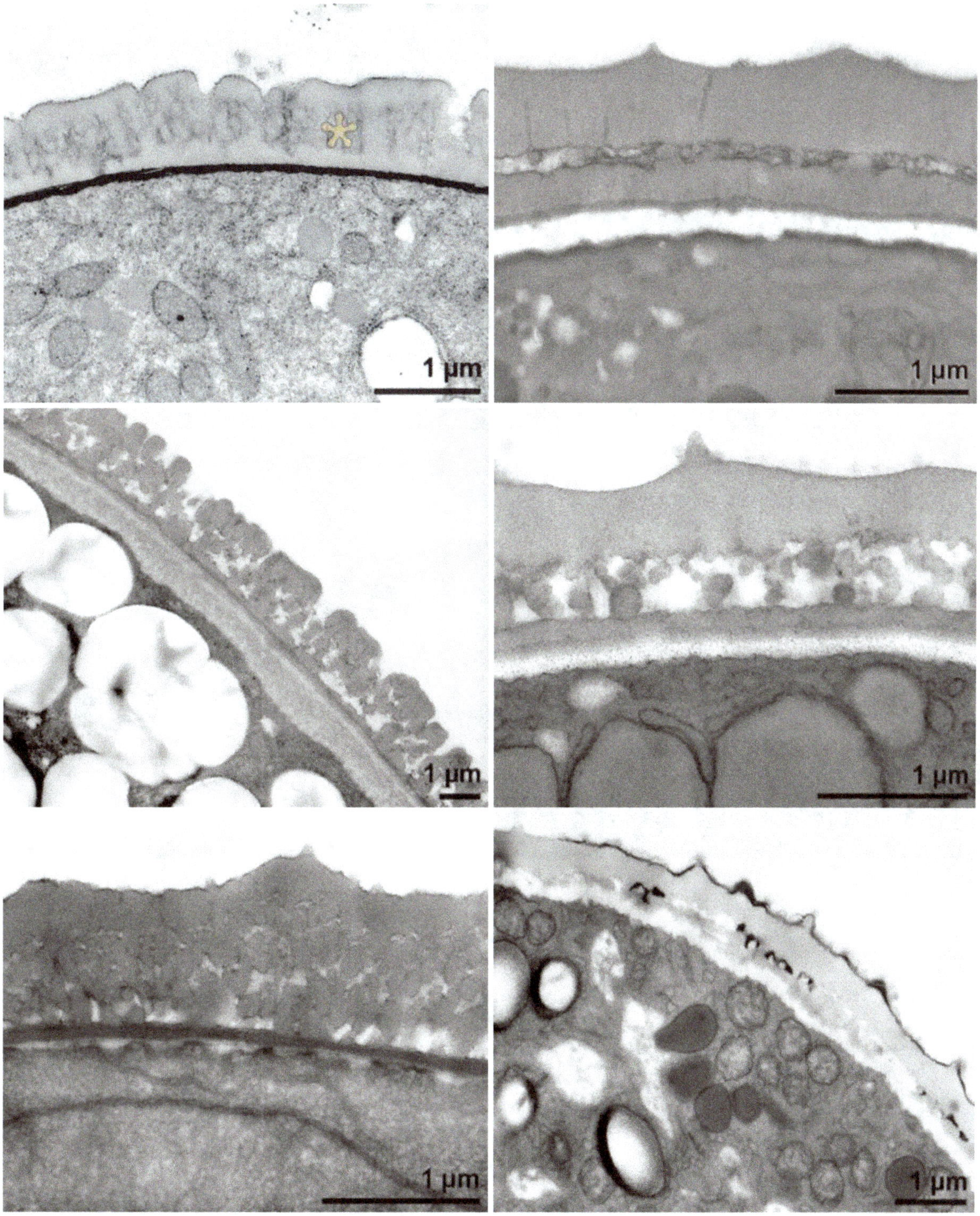

Amydrium medium, Araceae
infratectum (asterisk), U+Pb

Viola tricolor, Violaceae
U+Pb

Ulmus laevis, Ulmaceae
U+Pb

Corylus colurna, Betulaceae
modified Thiéry test

Juglans regia, Juglandaceae
modified Thiéry test

Plantago lanceolata, Plantaginaceae
modified Thiéry test

infratectum absent

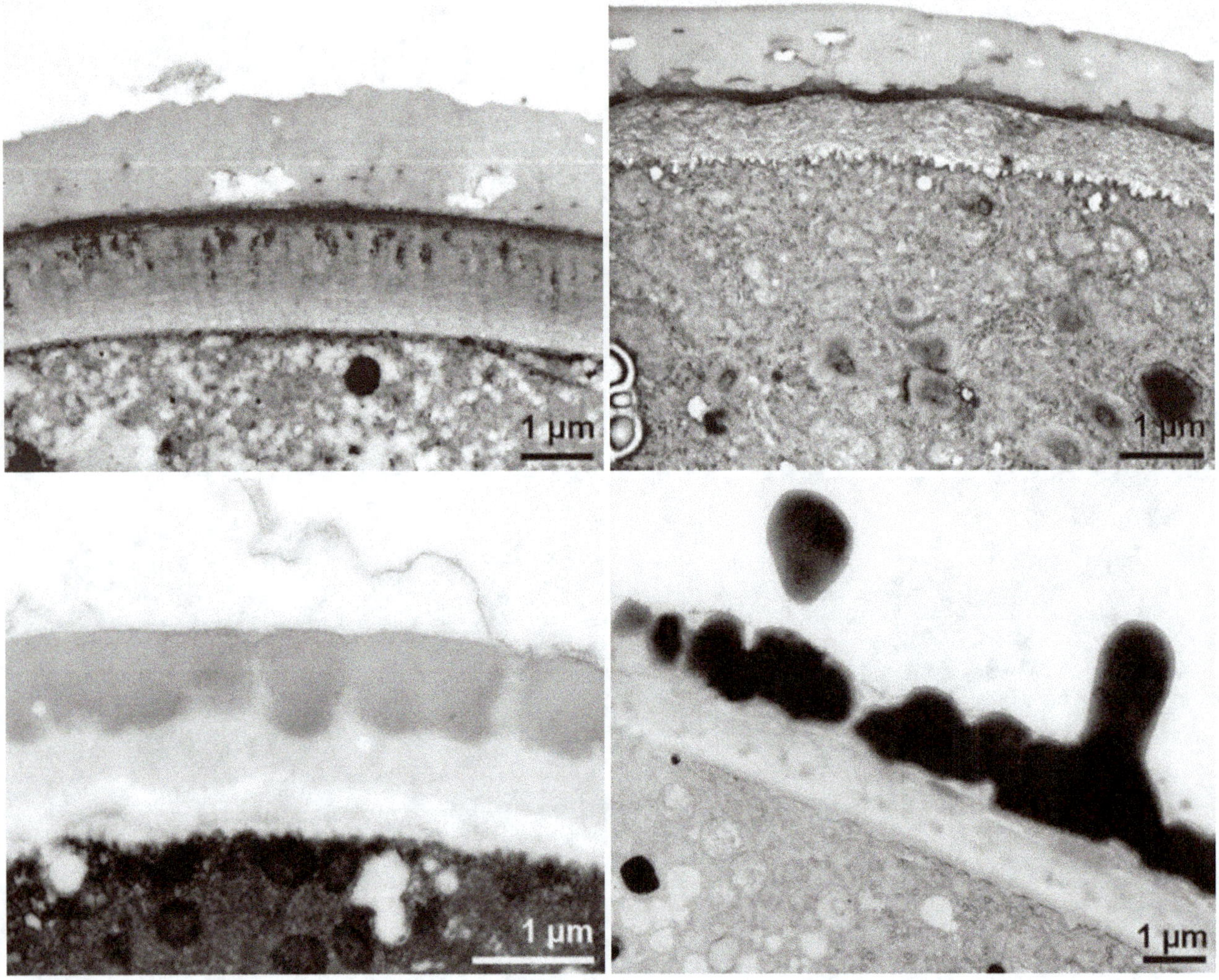

Dieffenbachia humilis, Araceae
U+Pb

Mahonia aquifolium, Berberidaceae
U+Pb

Berberis vulgaris, Berberidaceae
modified Thiéry test

Iris pumila, Iridaceae
modified Thiéry test

internal tectum

additional more or less continuous layer within the infratectum

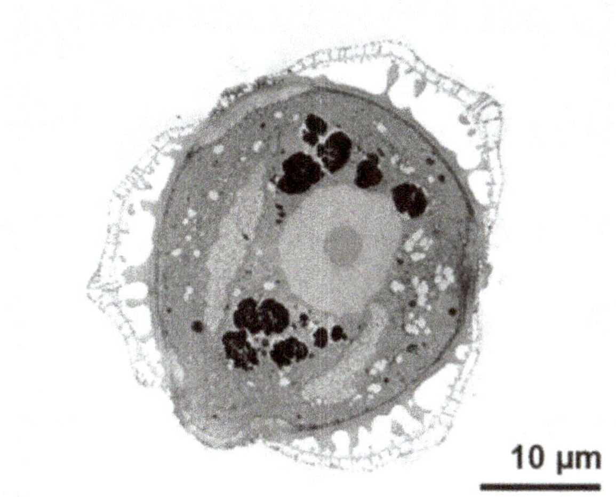

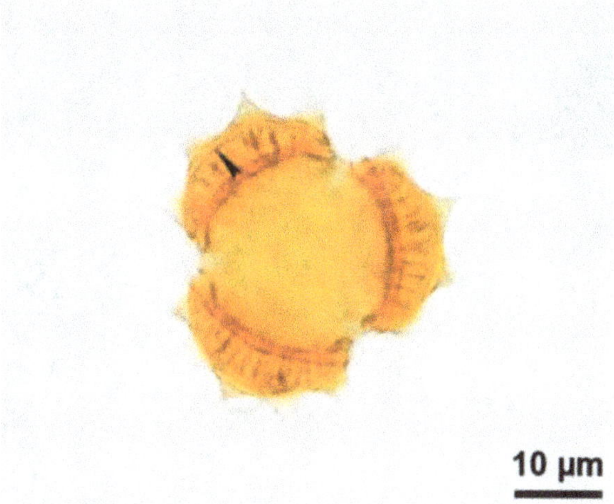

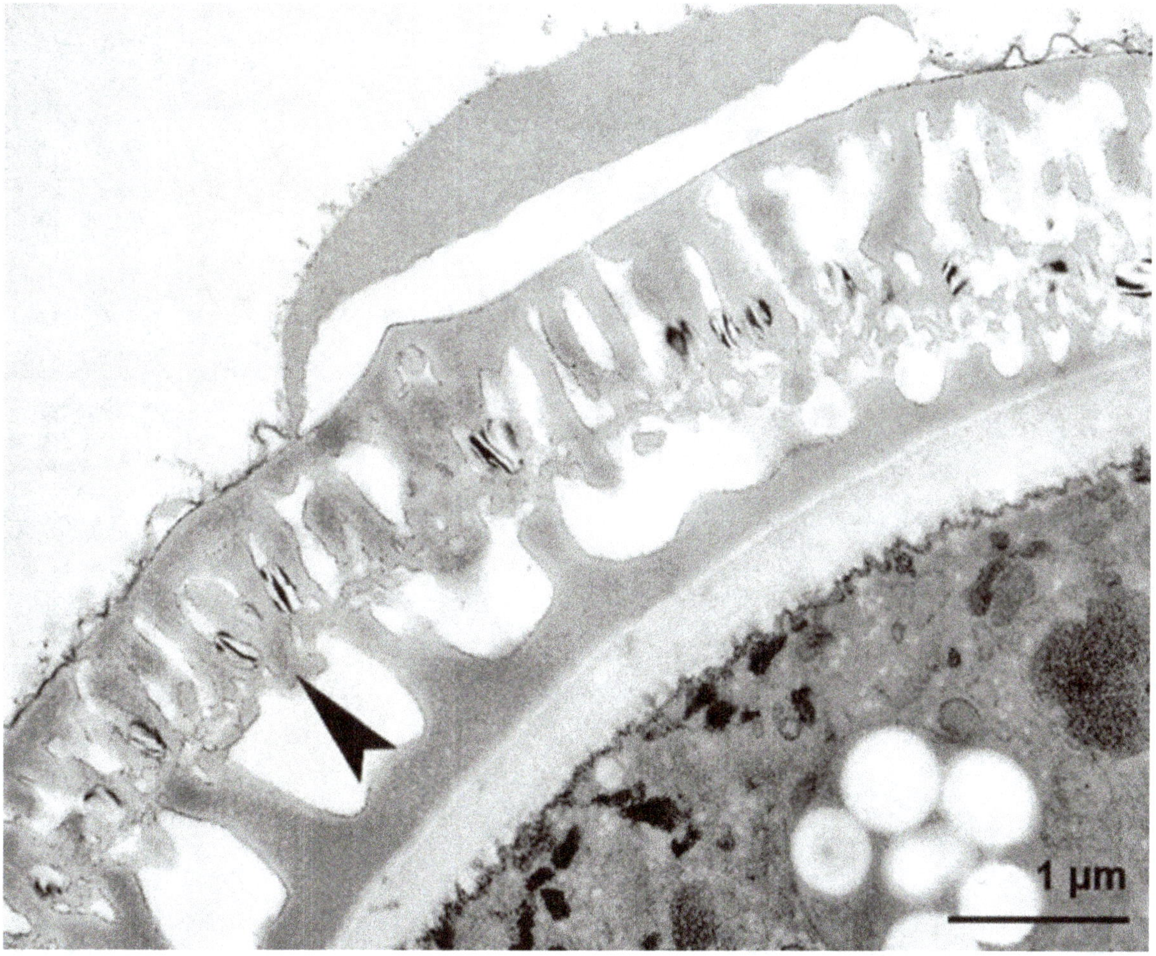

Argyranthemum sp., Asteraceae
internal tectum colored, U+Pb

Tanacetum corymbosum, Asteraceae
internal tectum (arrowhead)

Cyanus segetum, Asteraceae
internal tectum (arrowhead), modified Thiéry test

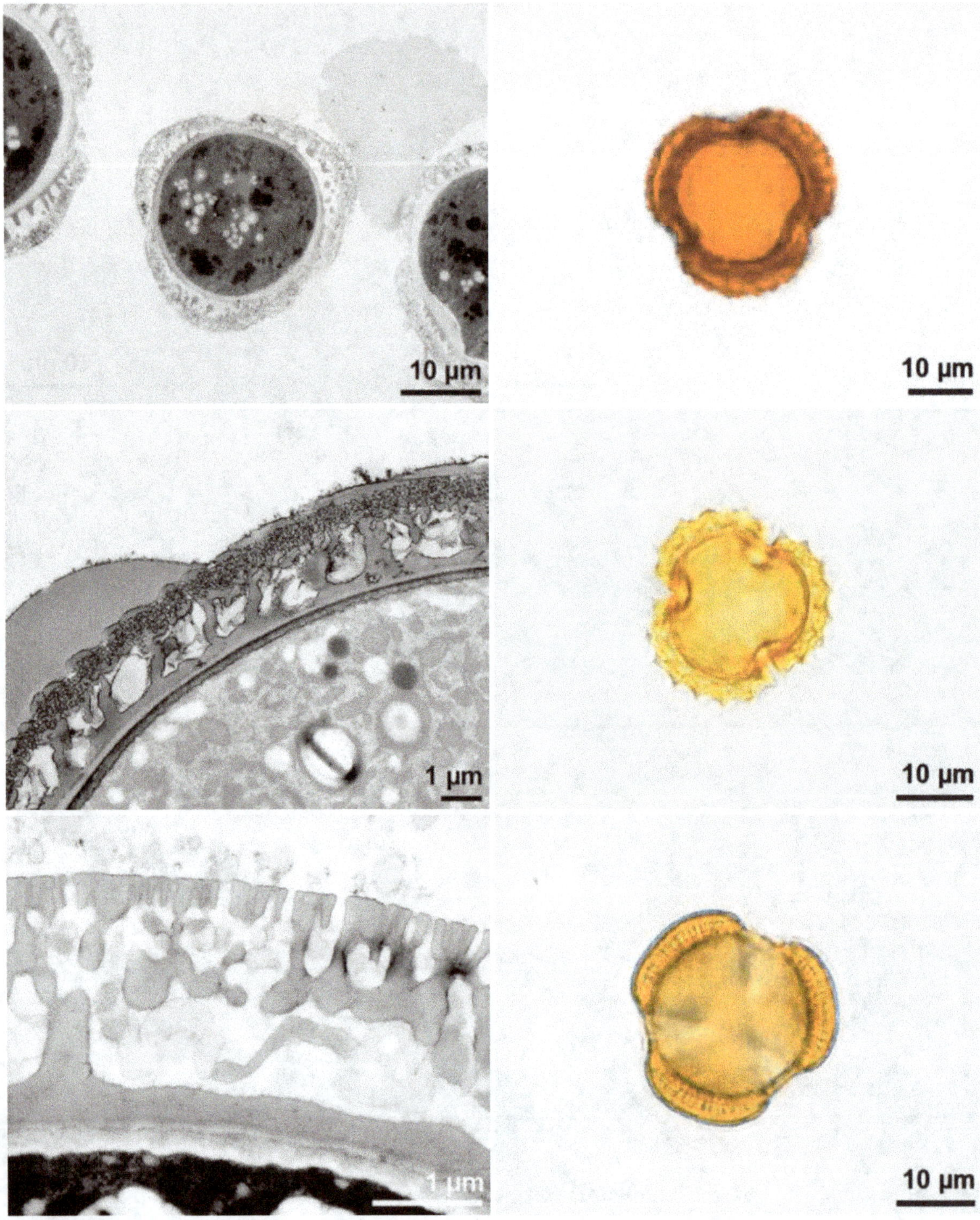

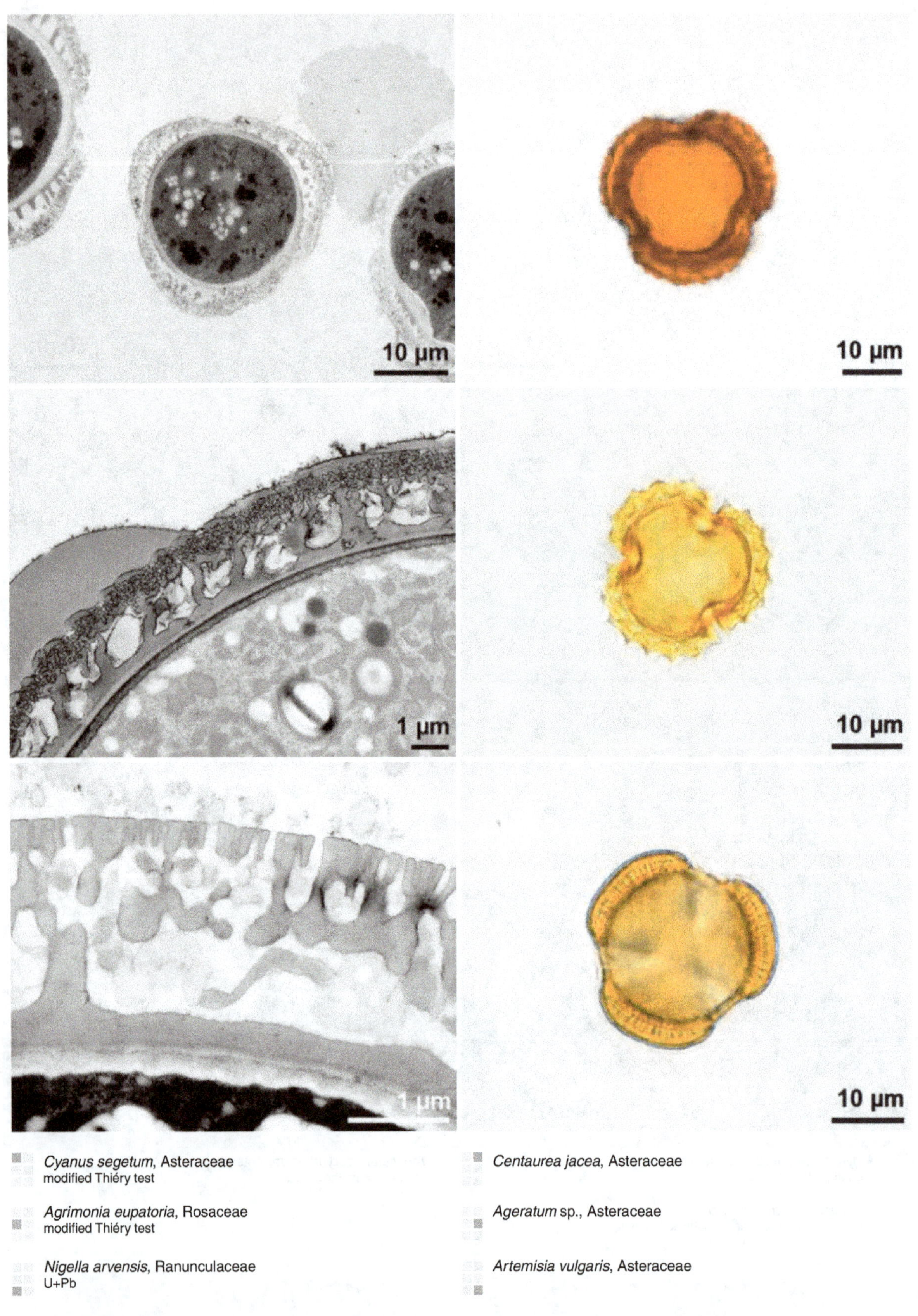

Cyanus segetum, Asteraceae
modified Thiéry test

Agrimonia eupatoria, Rosaceae
modified Thiéry test

Nigella arvensis, Ranunculaceae
U+Pb

Centaurea jacea, Asteraceae

Ageratum sp., Asteraceae

Artemisia vulgaris, Asteraceae

foot layer

inner layer of an ektexine that can be continuous, discontinuous, perforated or absent

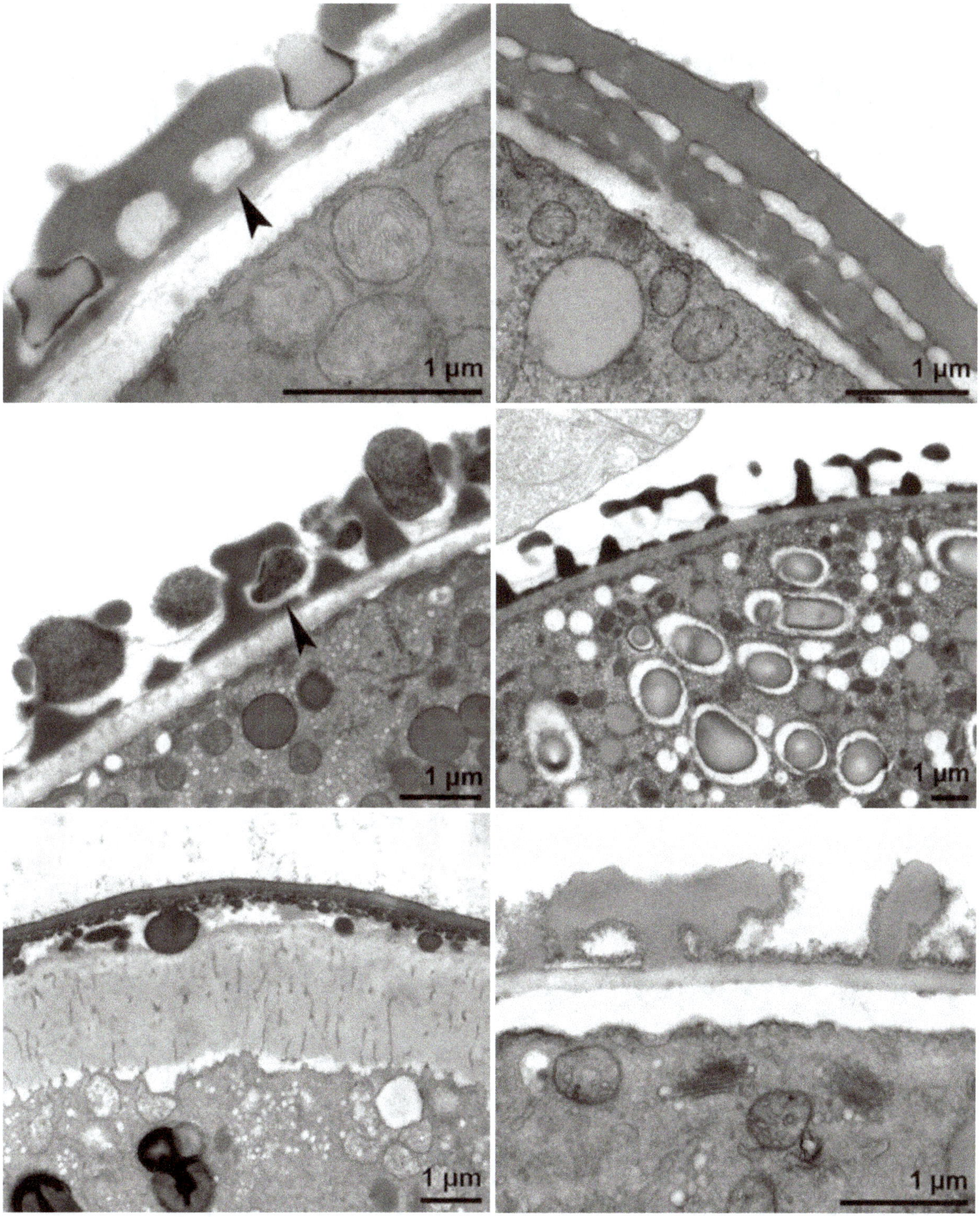

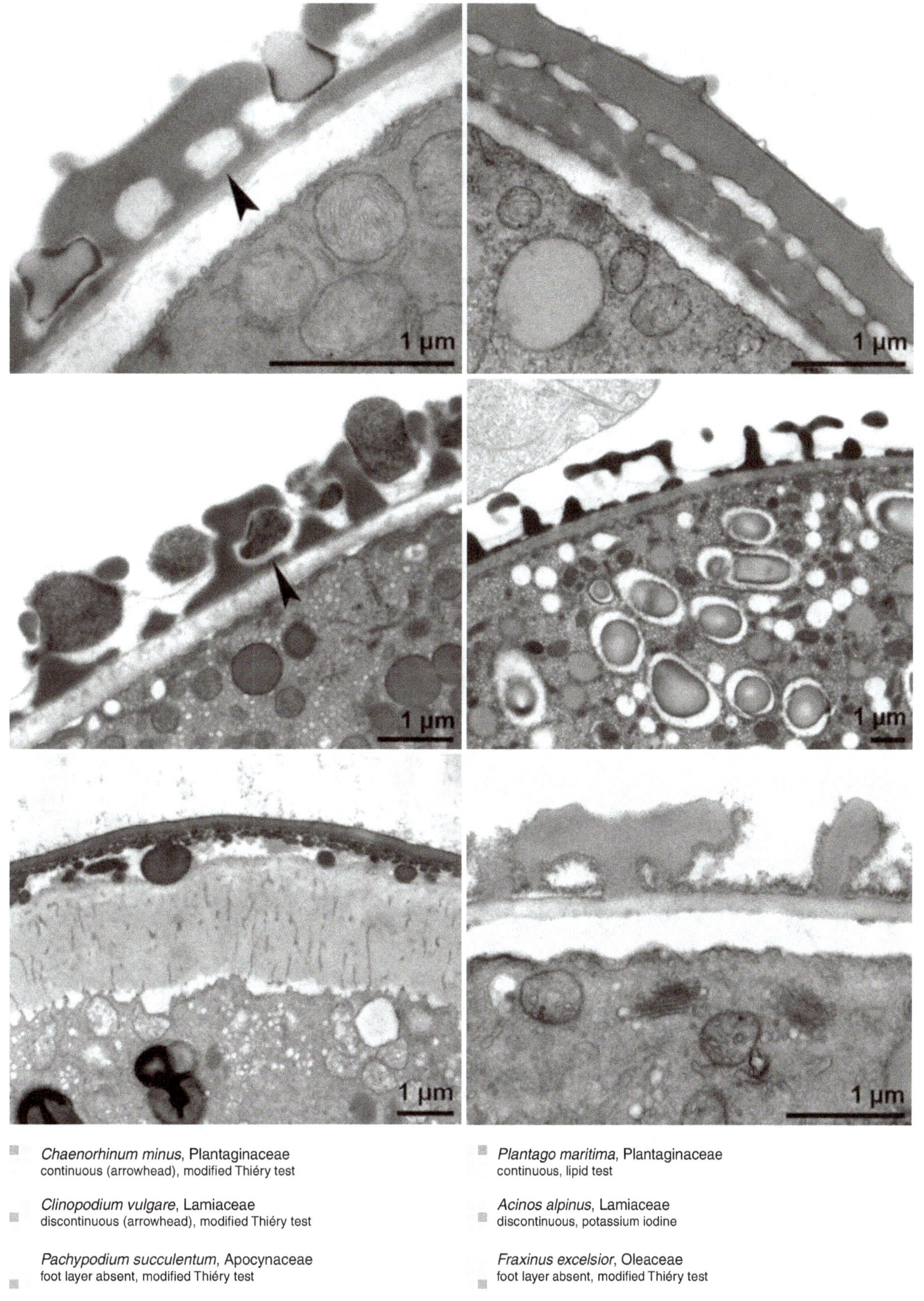

Chaenorhinum minus, Plantaginaceae
continuous (arrowhead), modified Thiéry test

Clinopodium vulgare, Lamiaceae
discontinuous (arrowhead), modified Thiéry test

Pachypodium succulentum, Apocynaceae
foot layer absent, modified Thiéry test

Plantago maritima, Plantaginaceae
continuous, lipid test

Acinos alpinus, Lamiaceae
discontinuous, potassium iodine

Fraxinus excelsior, Oleaceae
foot layer absent, modified Thiéry test

endexine compact-continuous

distinct exine layer between ektexine and intine

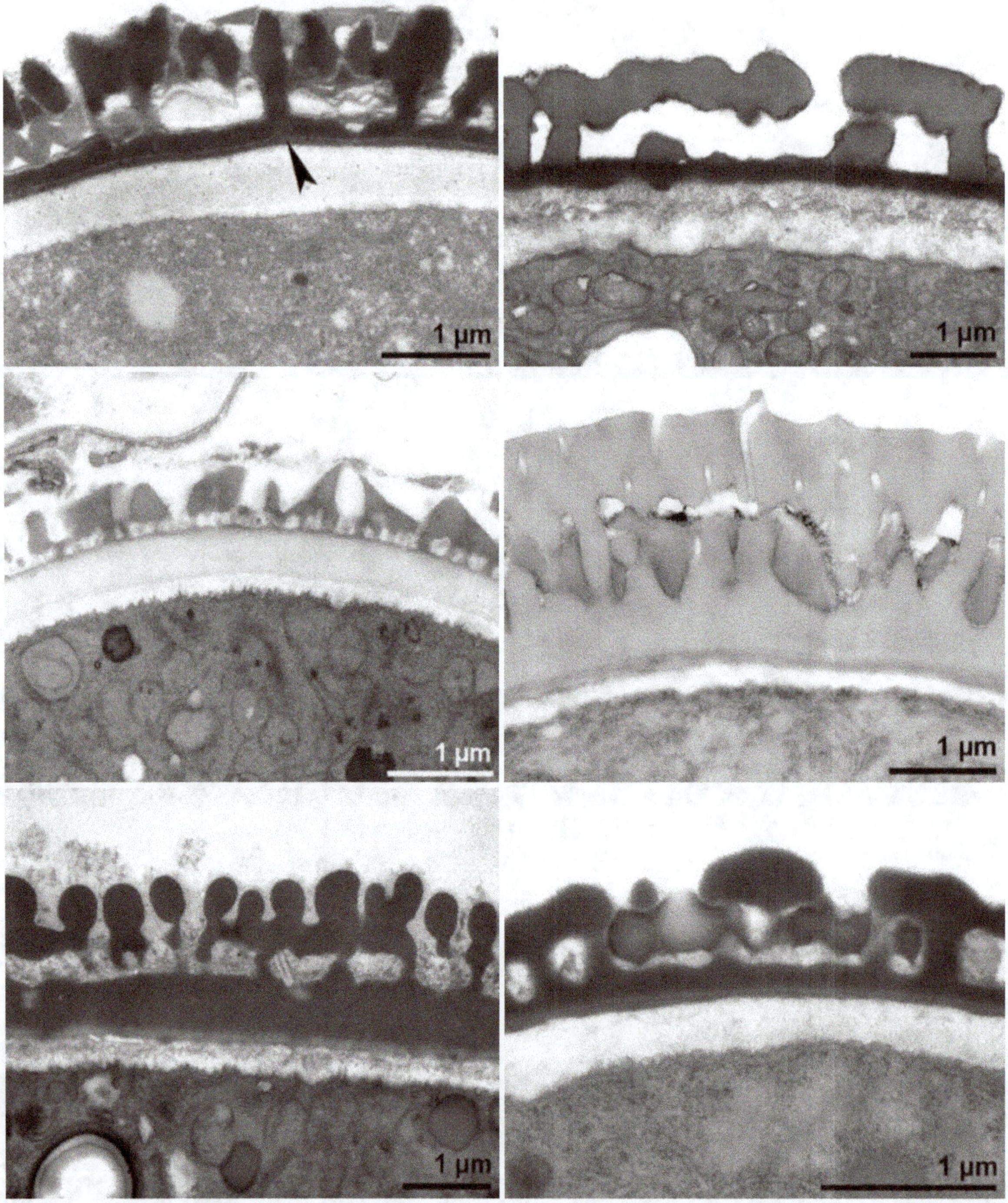

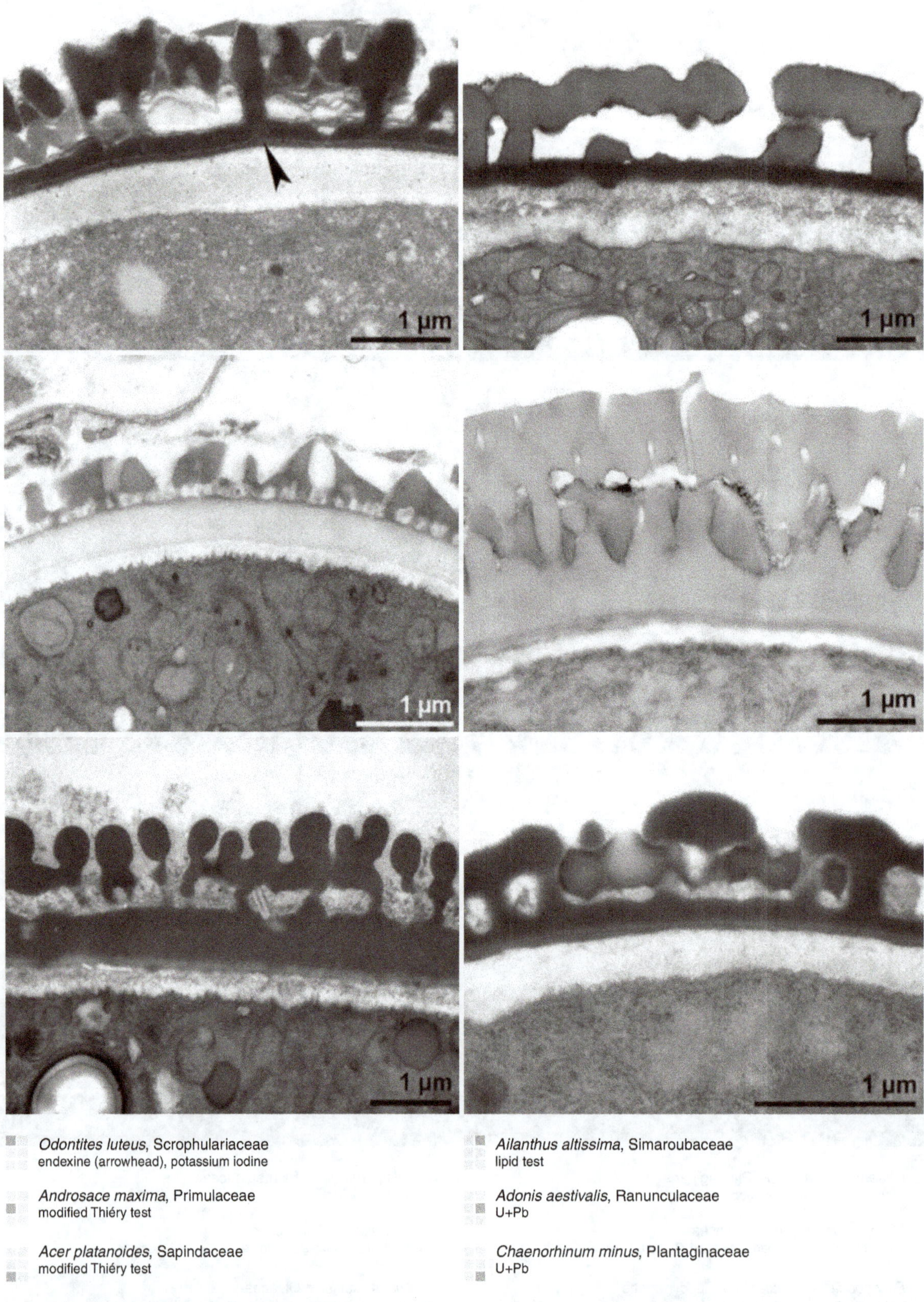

Odontites luteus, Scrophulariaceae
endexine (arrowhead), potassium iodine

Androsace maxima, Primulaceae
modified Thiéry test

Acer platanoides, Sapindaceae
modified Thiéry test

Ailanthus altissima, Simaroubaceae
lipid test

Adonis aestivalis, Ranunculaceae
U+Pb

Chaenorhinum minus, Plantaginaceae
U+Pb

endexine compact-discontinuous

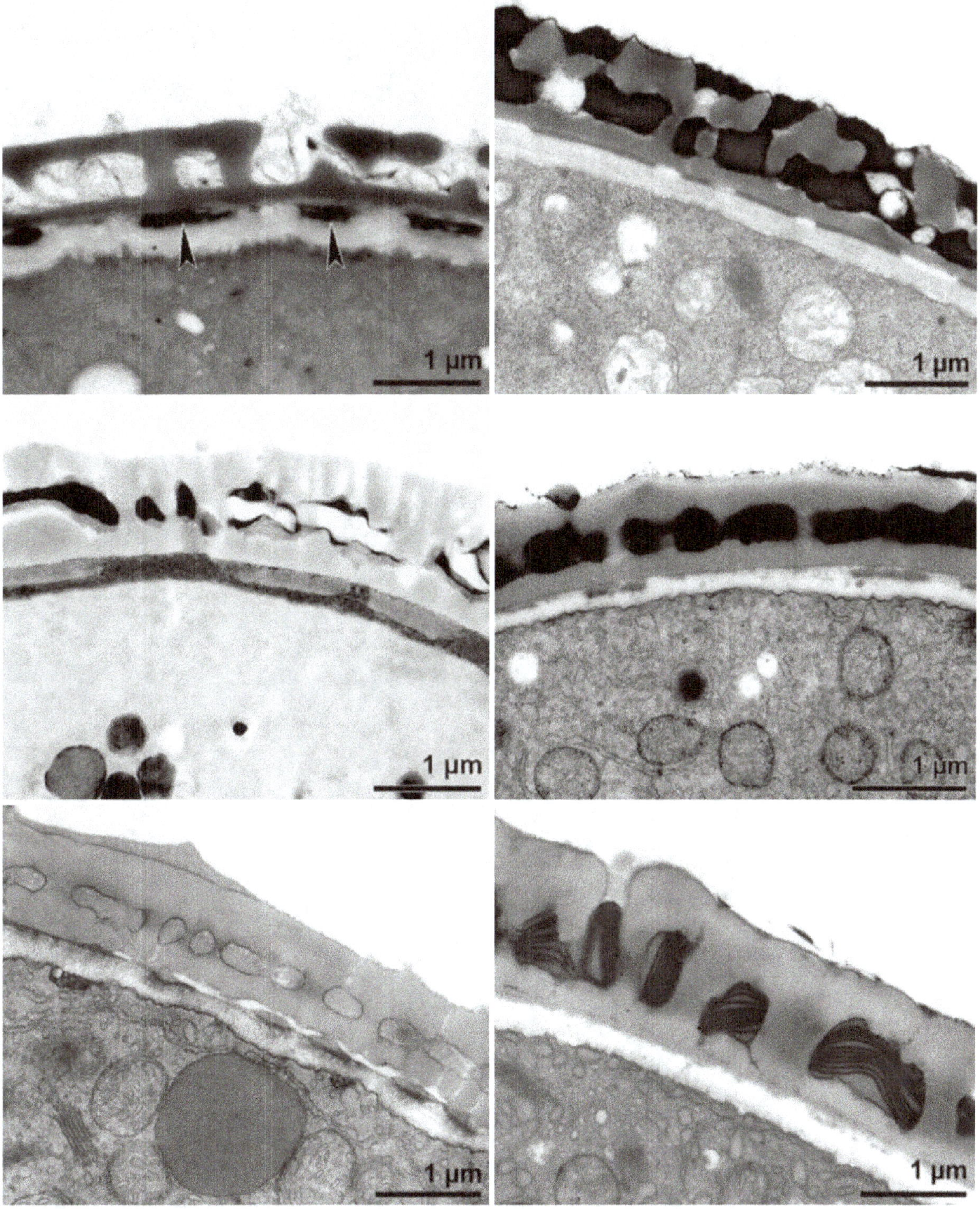

Glaucium flavum, Papaveraceae
endexine (arrowheads), U+Pb

Ranunculus trichophyllus, Ranunculaceae
Thiéry test

Plantago maritima, Plantaginaceae
modified Thiéry test

Odontites vulgaris, Orobanchaceae
modified Thiéry test

Delphinium elatum, Ranunculaceae
modified Thiéry test

Stachys officinalis, Lamiaceae
modified Thiéry test

endexine spongy-continuous

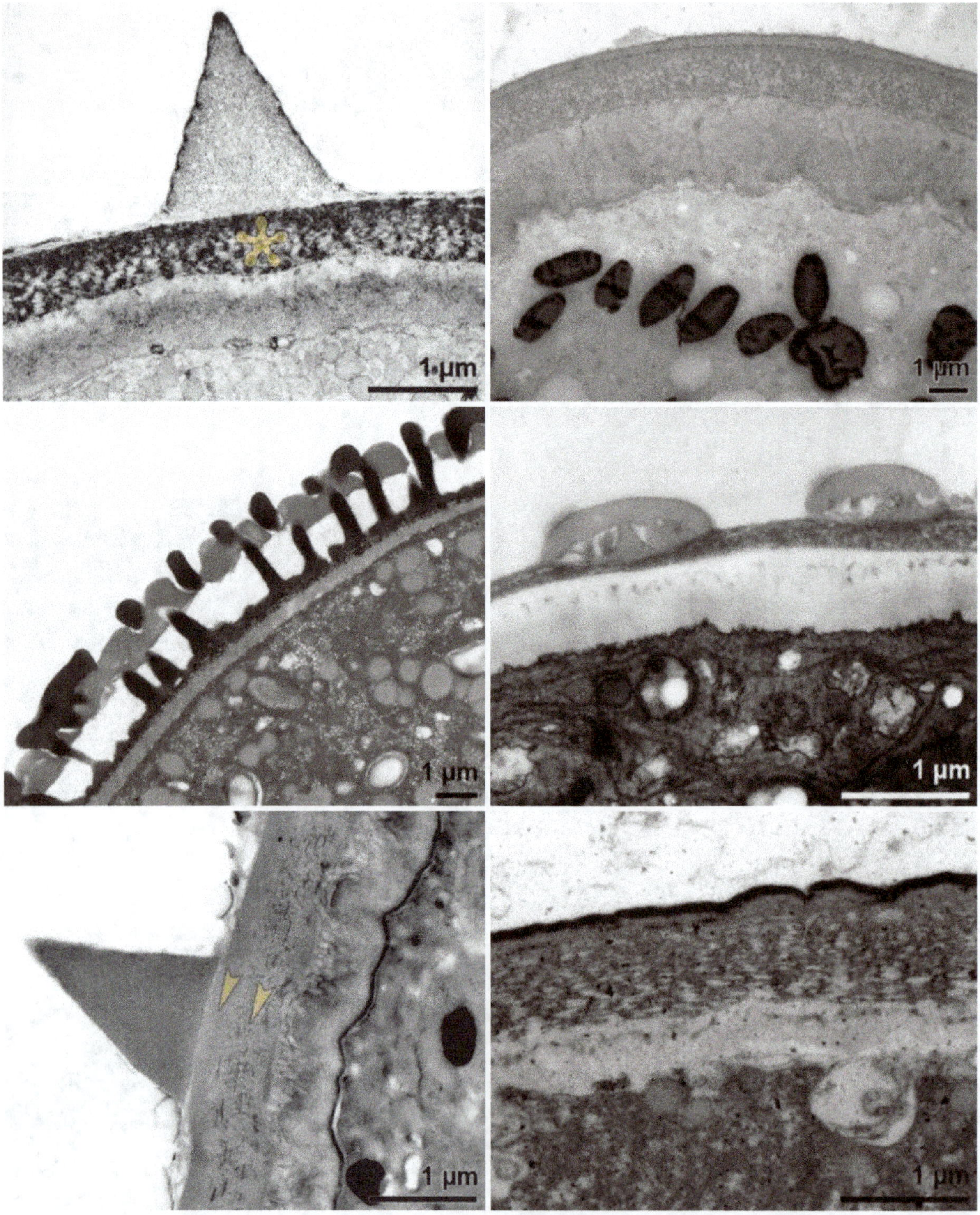

Pistia stratiotes, Araceae
endexine (asterisk), Thiéry test

Mentha aquatica, Lamiaceae
potassium iodine

Arophyton buchetii, Araceae
endexine bilayered (arrowheads), U+Pb

Amorphophallus mossambicensis, Araceae
Thiéry test

Spathiphyllum blandum, Araceae
lipid test

Homalomena wallisii, Araceae
bilayered endexine, potassium permanganate

endexine lamellar-continuous

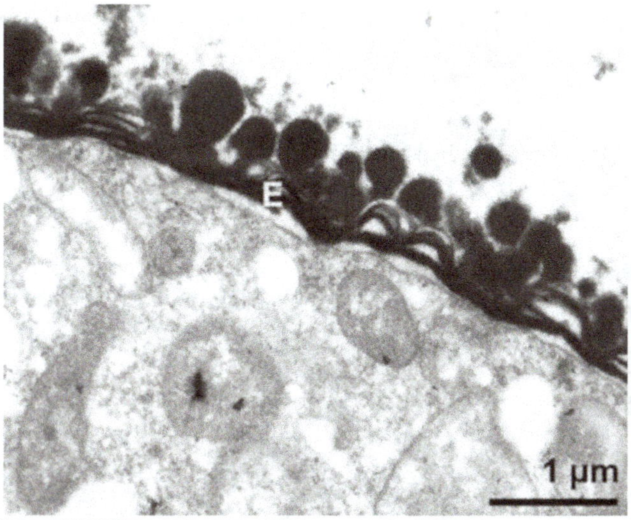

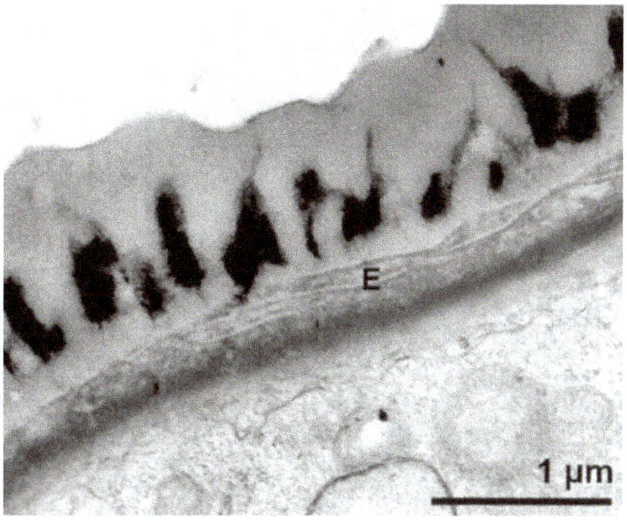

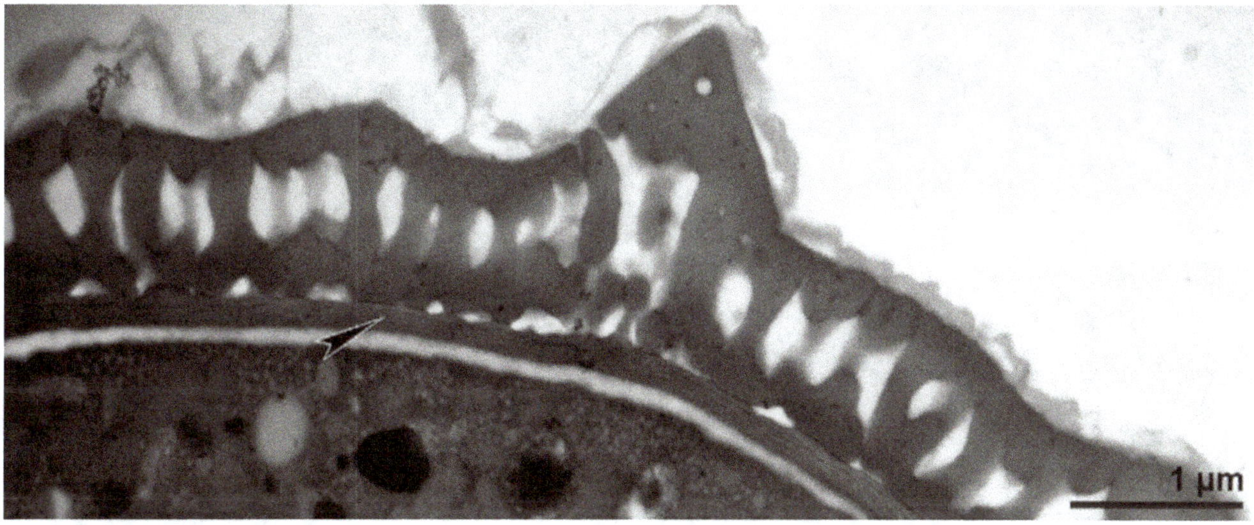

Orobanche hederae, Orobanchaceae
endexine (E), potassium iodine

Thalictrum flavum, Ranunculaceae
endexine (E), modified Thiéry test

Ambrosia artemisiifolia, Asteraceae
endexine (arrowhead), potassium permanganate

endexine in aperture only

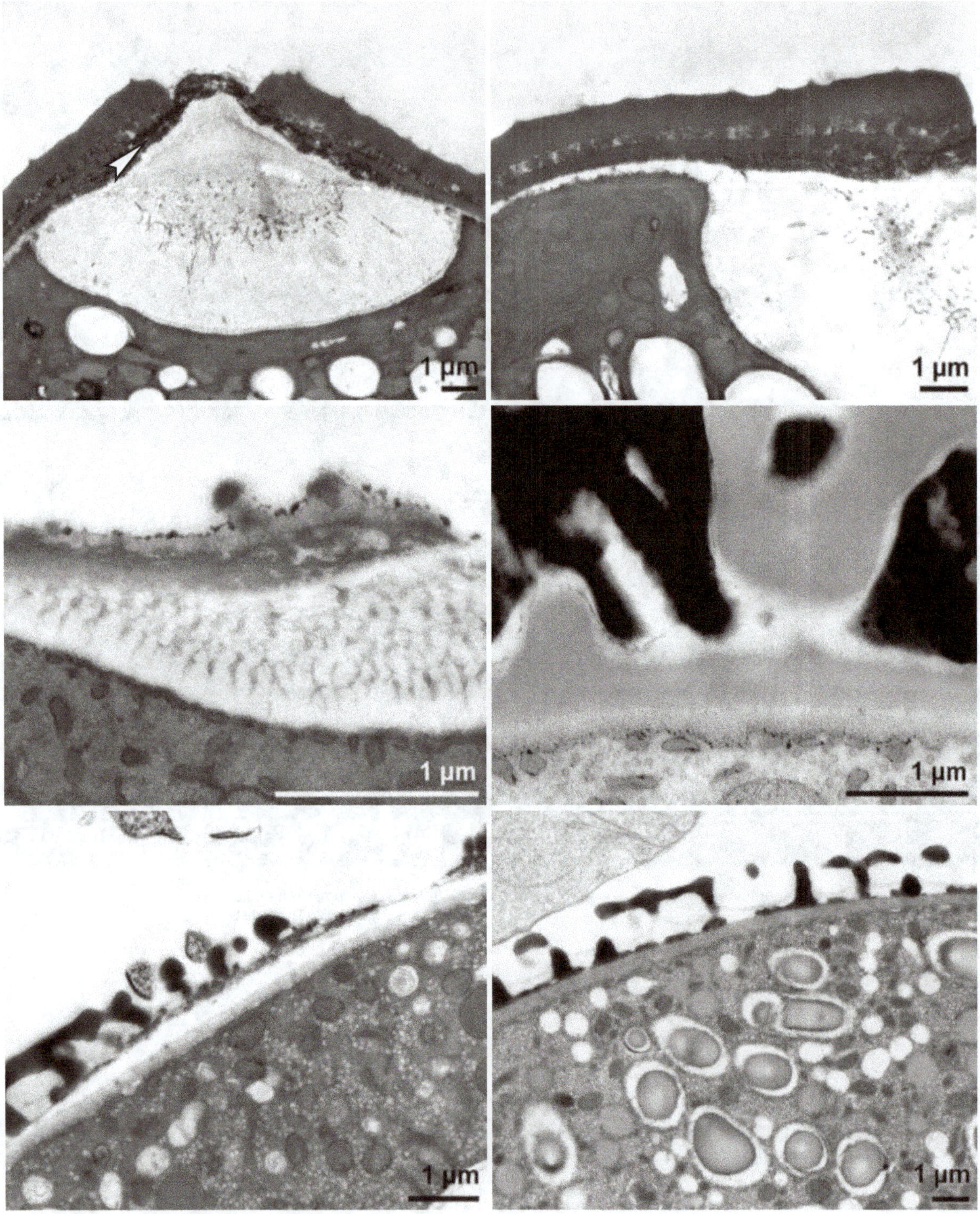

Corylus avellana, Betulaceae
endexine in aperture only (arrowhead), U+Pb

Corylus avellana, Betulaceae
lipid test

Syringa vulgaris, Oleaceae
modified Thiéry test

Syringa vulgaris, Oleaceae
endexine absent in interapertural area, lipid test

Acinos alpinus, Lamiaceae
U+Pb

Acinos alpinus, Lamiaceae
endexine absent in interapertural area, potassium permanganate

endexine absent

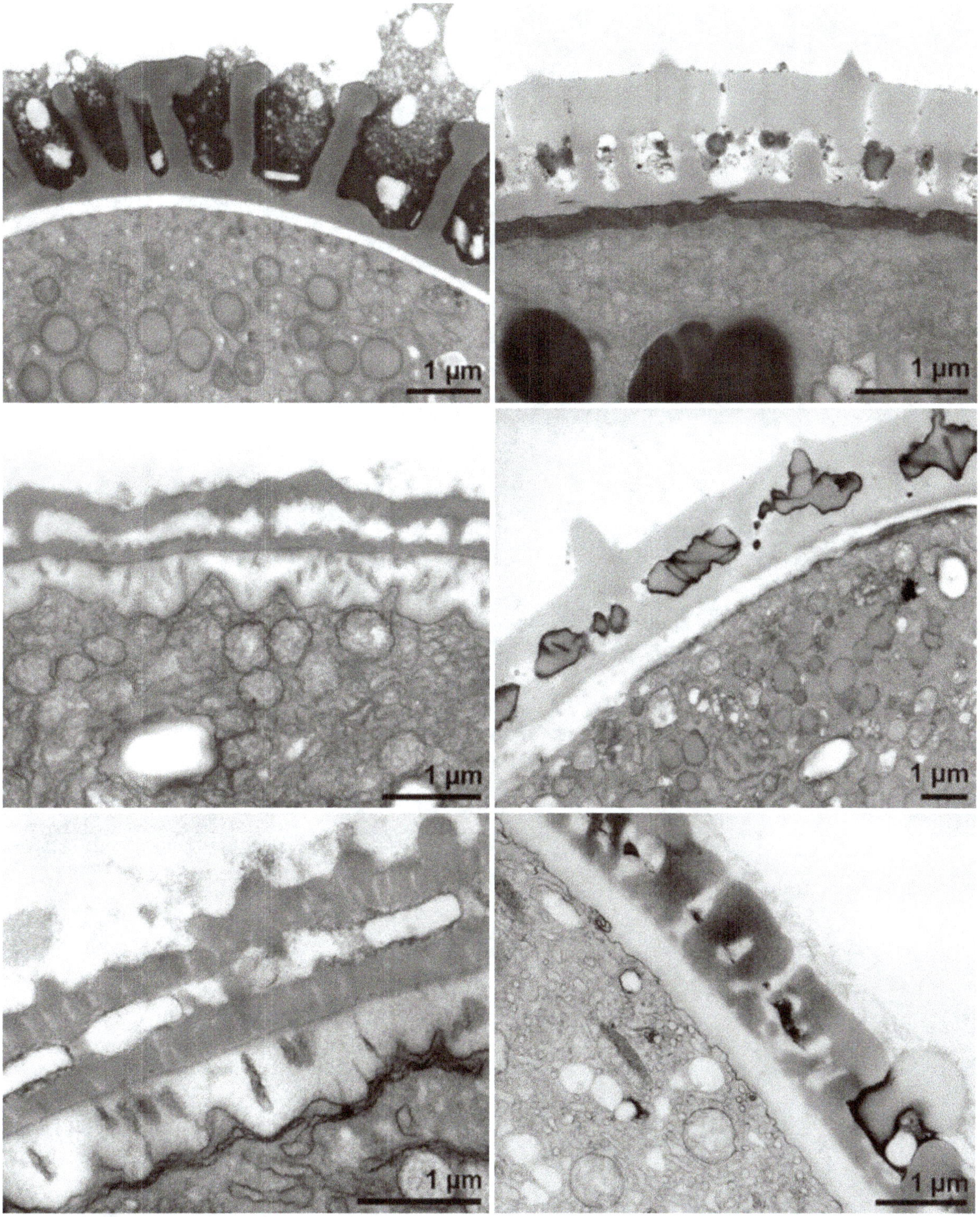

Brassica napus, Brassicaceae
modified Thiéry test

Trisetum flavescens, Poaceae
modified Thiéry test

Avena sativa, Poaceae
modified Thiéry test

Chenopodium album, Amaranthaceae
Thiéry test

Cereus sp., Cactaceae
modified Thiéry test

Ornithogalum nutans, Asparagaceae
modified Thiéry test

intine, ektintine, endintine

intine: part of the pollen wall next to the cytoplasm, can be monolayered or bilayered (ektintine and endintine)

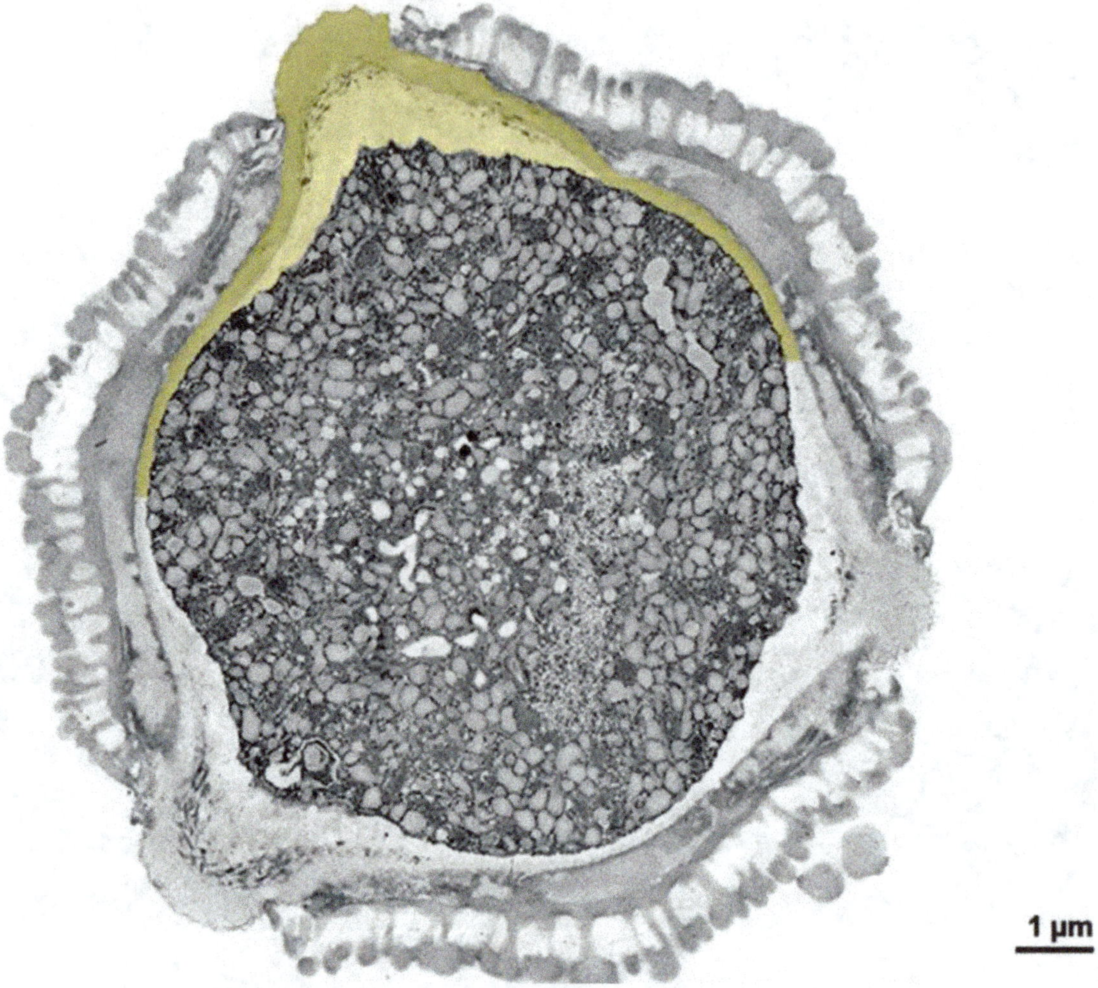

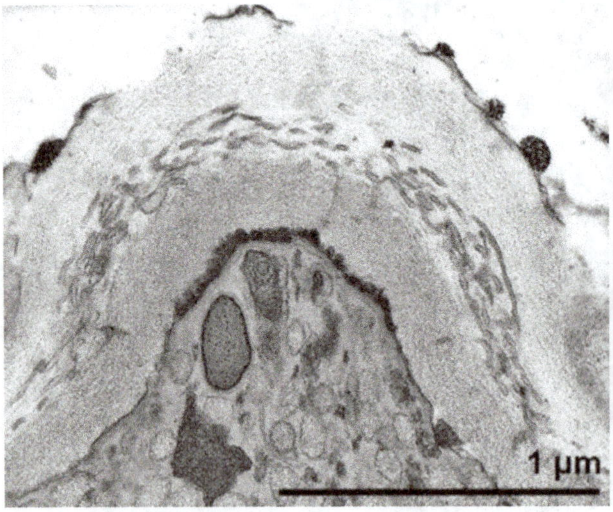

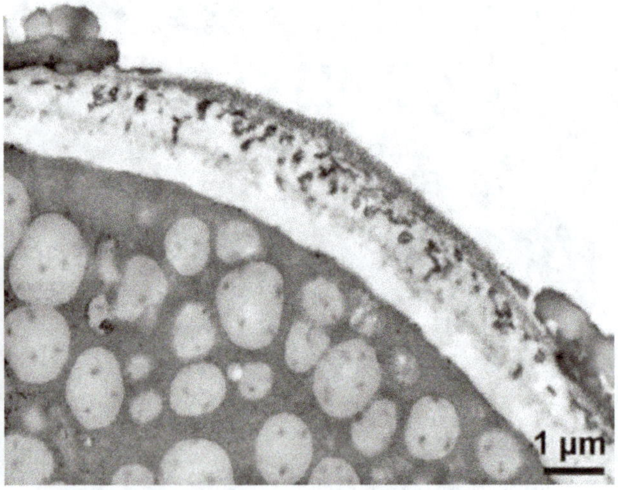

Apium nodiflorum, Apiaceae
intine bilayered in aperture area (colored) U+Pb

Apium nodiflorum, Apiaceae
ektintine (e-transparent), endintine (e-dense), Thiéry test

Quercus robur, Fagaceae
ektintine channelled, lipid test

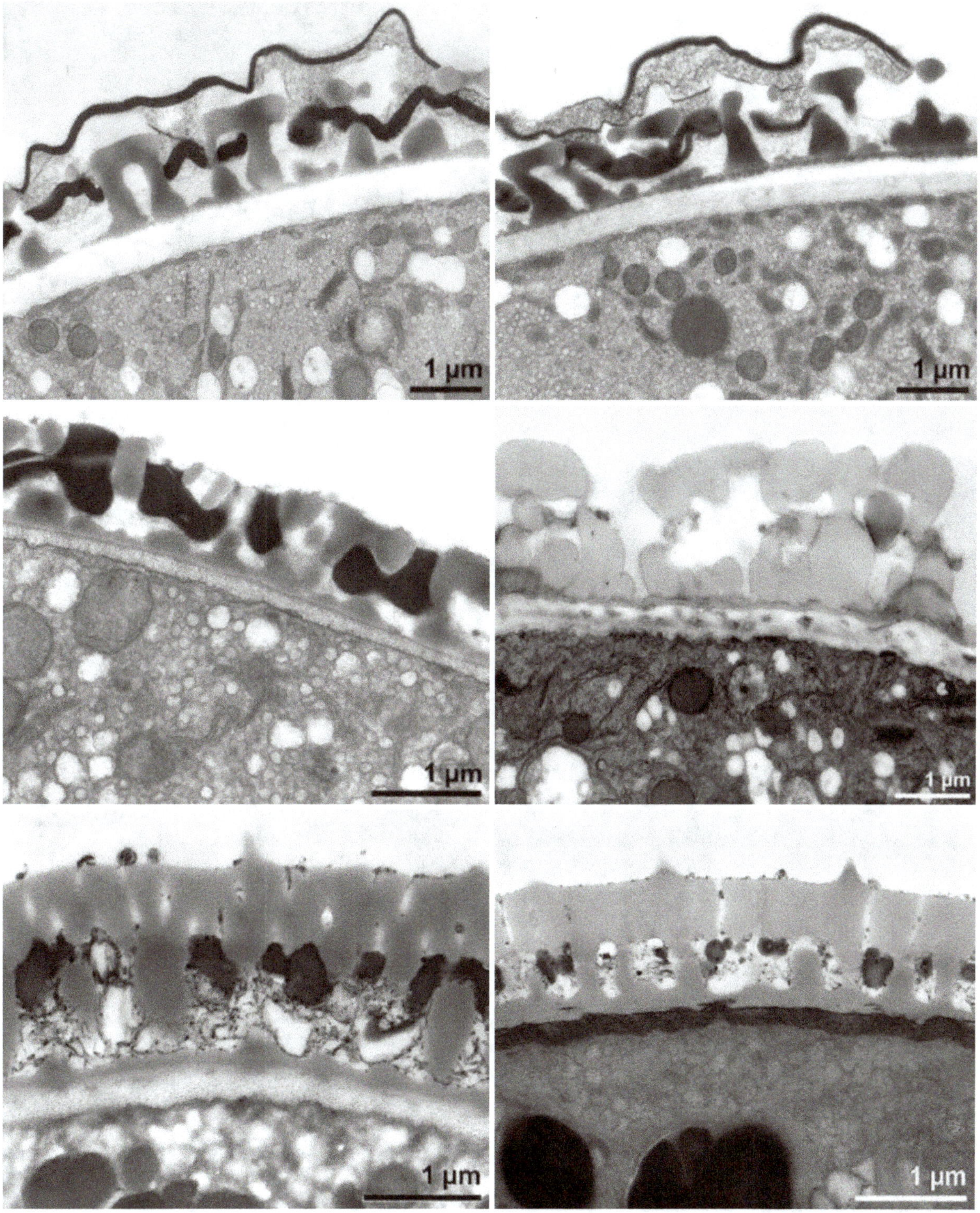

Acinos alpinus, Lamiaceae
modified Thiéry test

Acinos alpinus, Lamiaceae
U+Pb

Veronica anagallis-aquatica, Plantaginaceae
modified Thiéry test

Anthericum liliago, Asparagaceae
modified Thiéry test

Atriplex tatarica, Amaranthaceae
modified Thiéry test

Chenopodium album, Amaranthaceae
intine (polysaccharides) stain electron dense with Thiéry test

ektintine (outer layer) and endintine (inner layer) of a bilayered intine

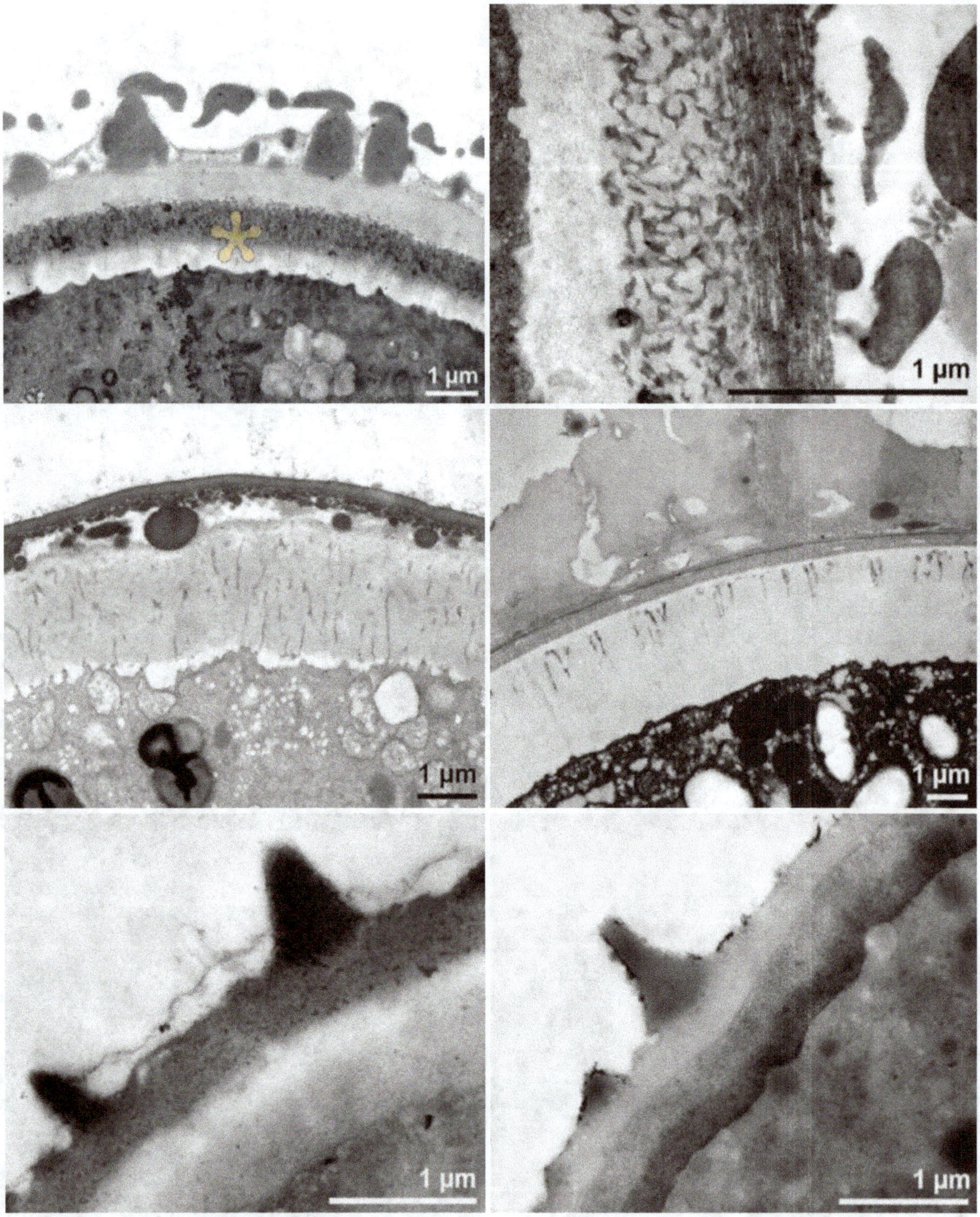

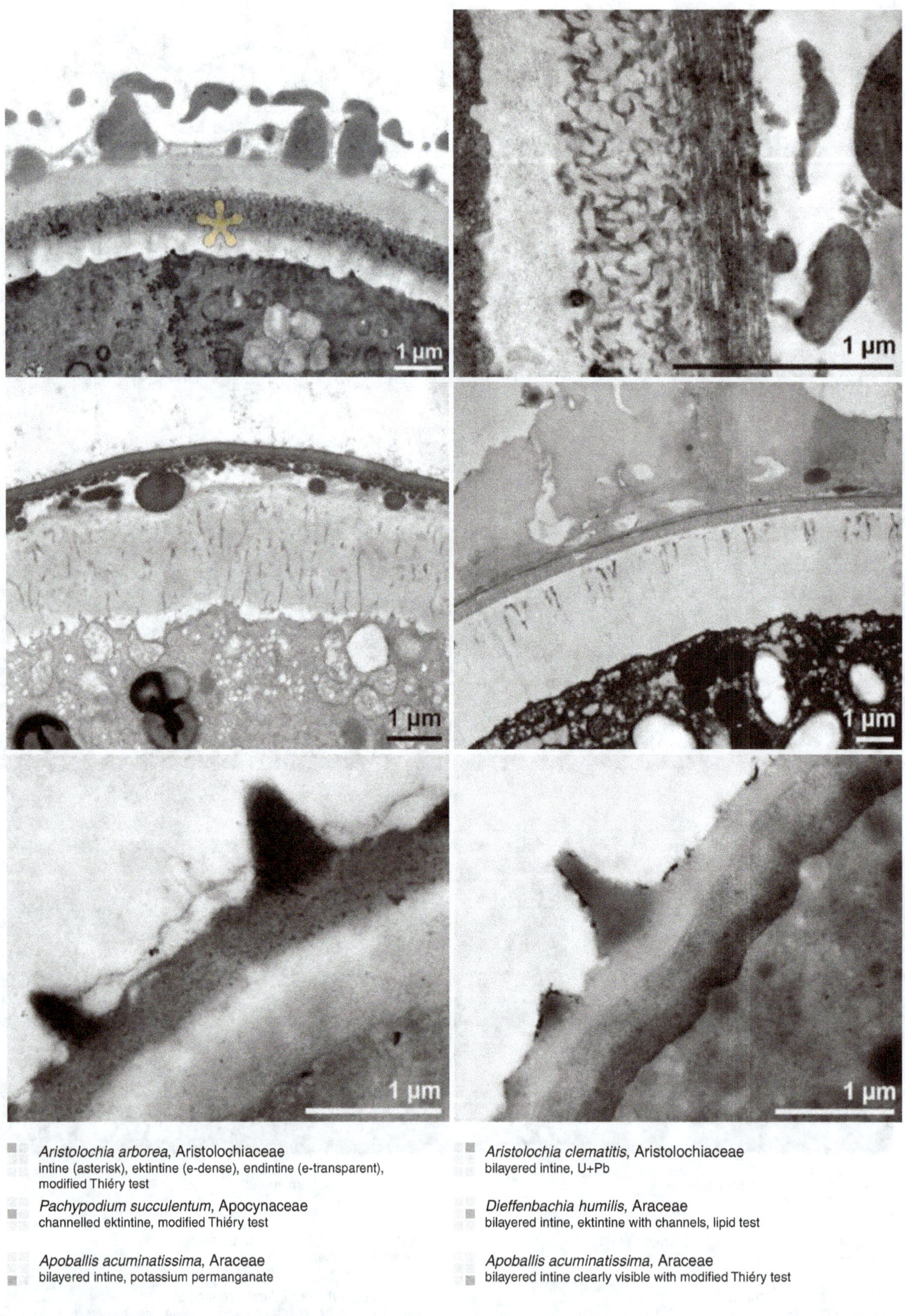

Aristolochia arborea, Aristolochiaceae
intine (asterisk), ektintine (e-dense), endintine (e-transparent), modified Thiéry test

Aristolochia clematitis, Aristolochiaceae
bilayered intine, U+Pb

Pachypodium succulentum, Apocynaceae
channelled ektintine, modified Thiéry test

Dieffenbachia humilis, Araceae
bilayered intine, ektintine with channels, lipid test

Apoballis acuminatissima, Araceae
bilayered intine, potassium permanganate

Apoballis acuminatissima, Araceae
bilayered intine clearly visible with modified Thiéry test

primexine

polysaccharidic layer formed during early developmental stage wherein the later exine structures are preformed

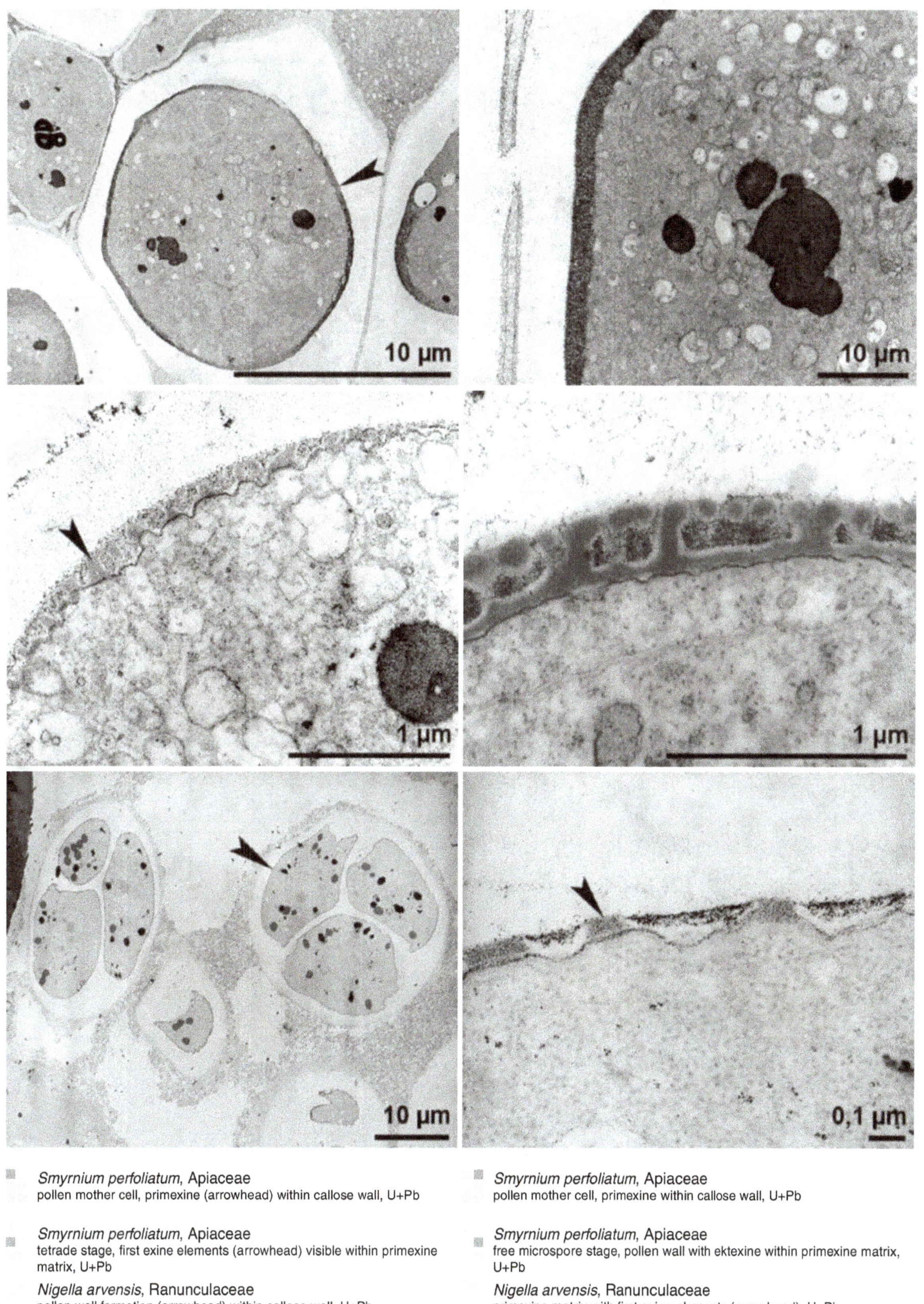

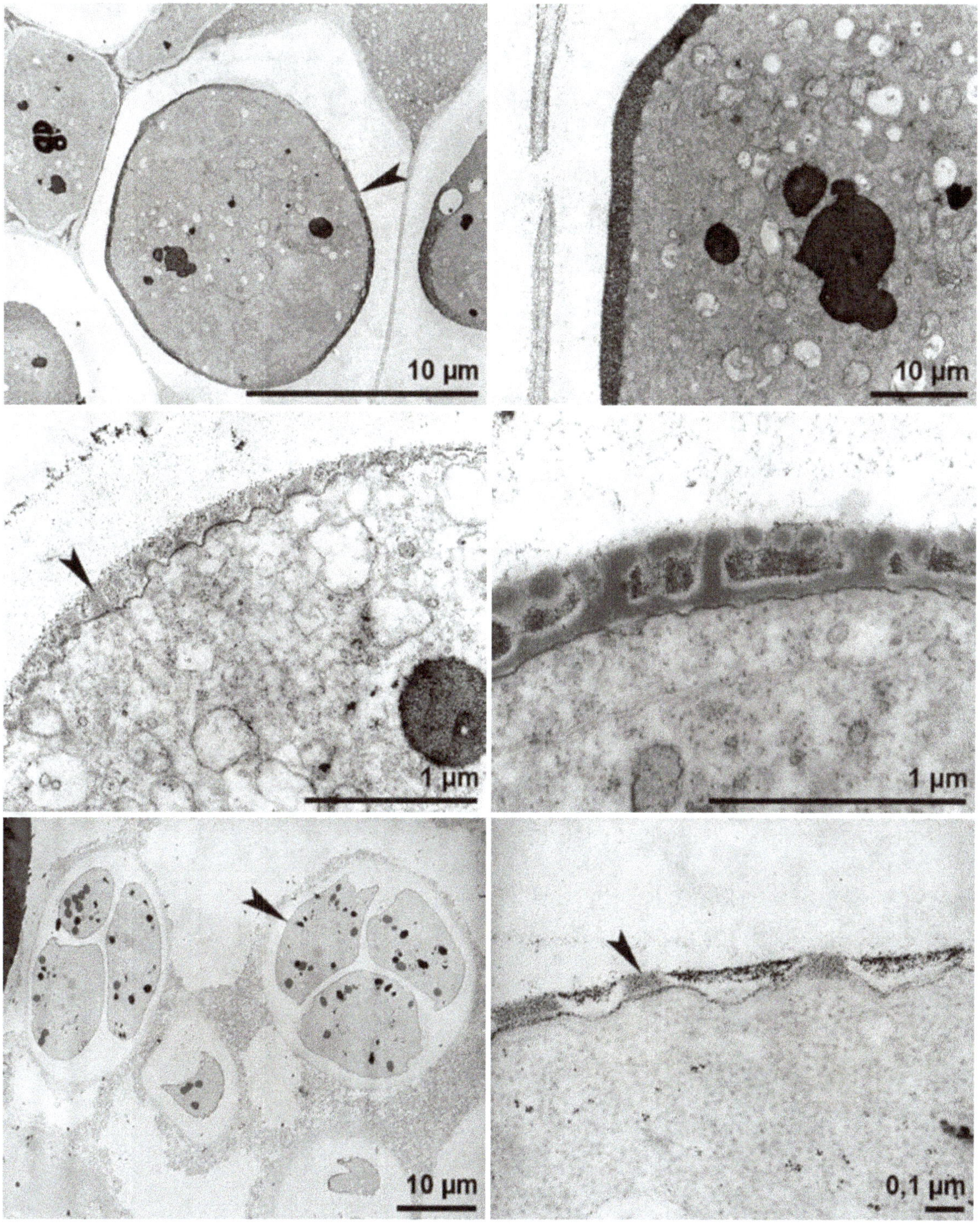

Smyrnium perfoliatum, Apiaceae
pollen mother cell, primexine (arrowhead) within callose wall, U+Pb

Smyrnium perfoliatum, Apiaceae
pollen mother cell, primexine within callose wall, U+Pb

Smyrnium perfoliatum, Apiaceae
tetrade stage, first exine elements (arrowhead) visible within primexine matrix, U+Pb

Smyrnium perfoliatum, Apiaceae
free microspore stage, pollen wall with ektexine within primexine matrix, U+Pb

Nigella arvensis, Ranunculaceae
pollen wall formation (arrowhead) within callose wall, U+Pb

Nigella arvensis, Ranunculaceae
primexine matrix with first exine elements (arrowhead), U+Pb

sexine, nexine

terms used for light microscopy, describing the outer and inner layers of the exine

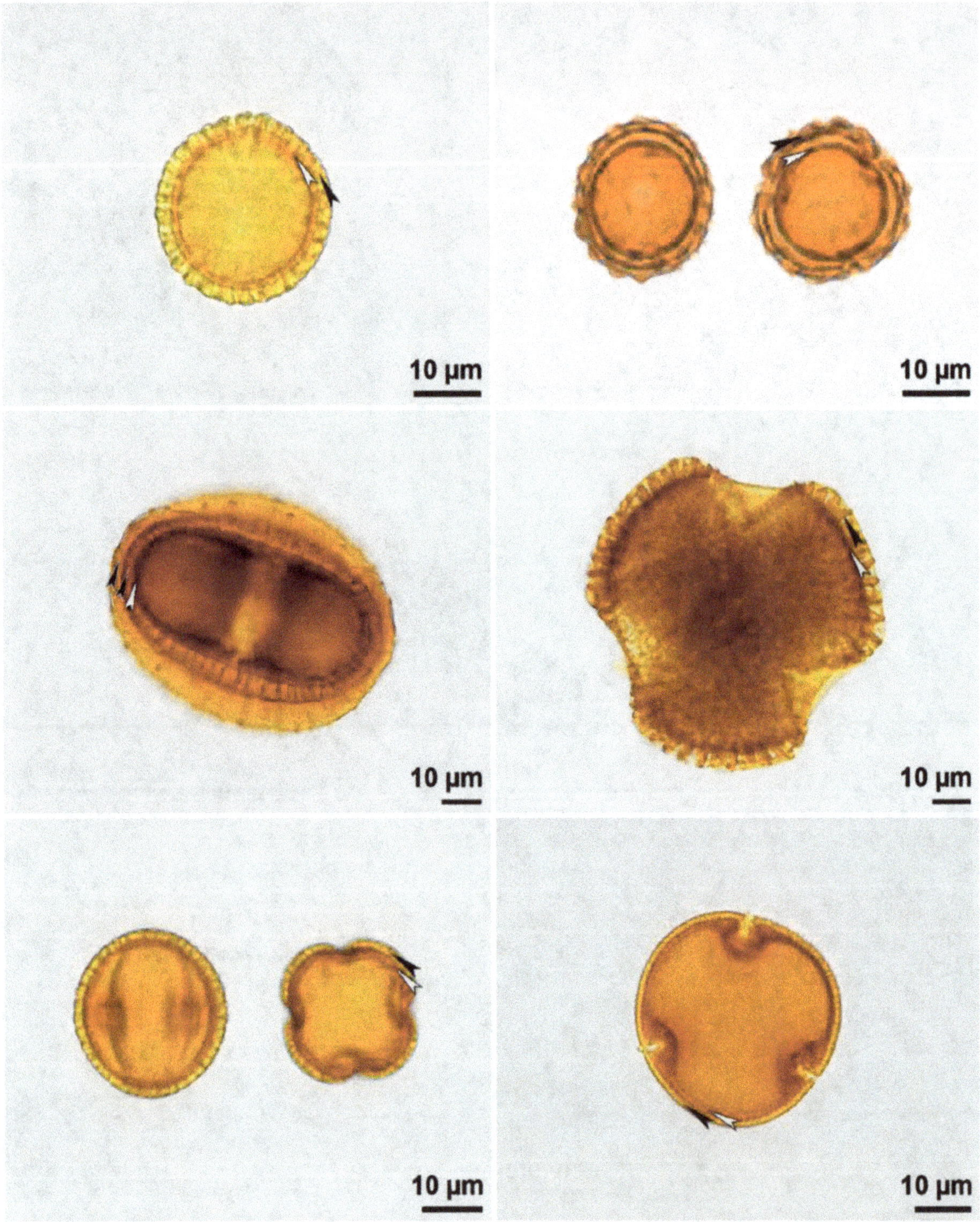

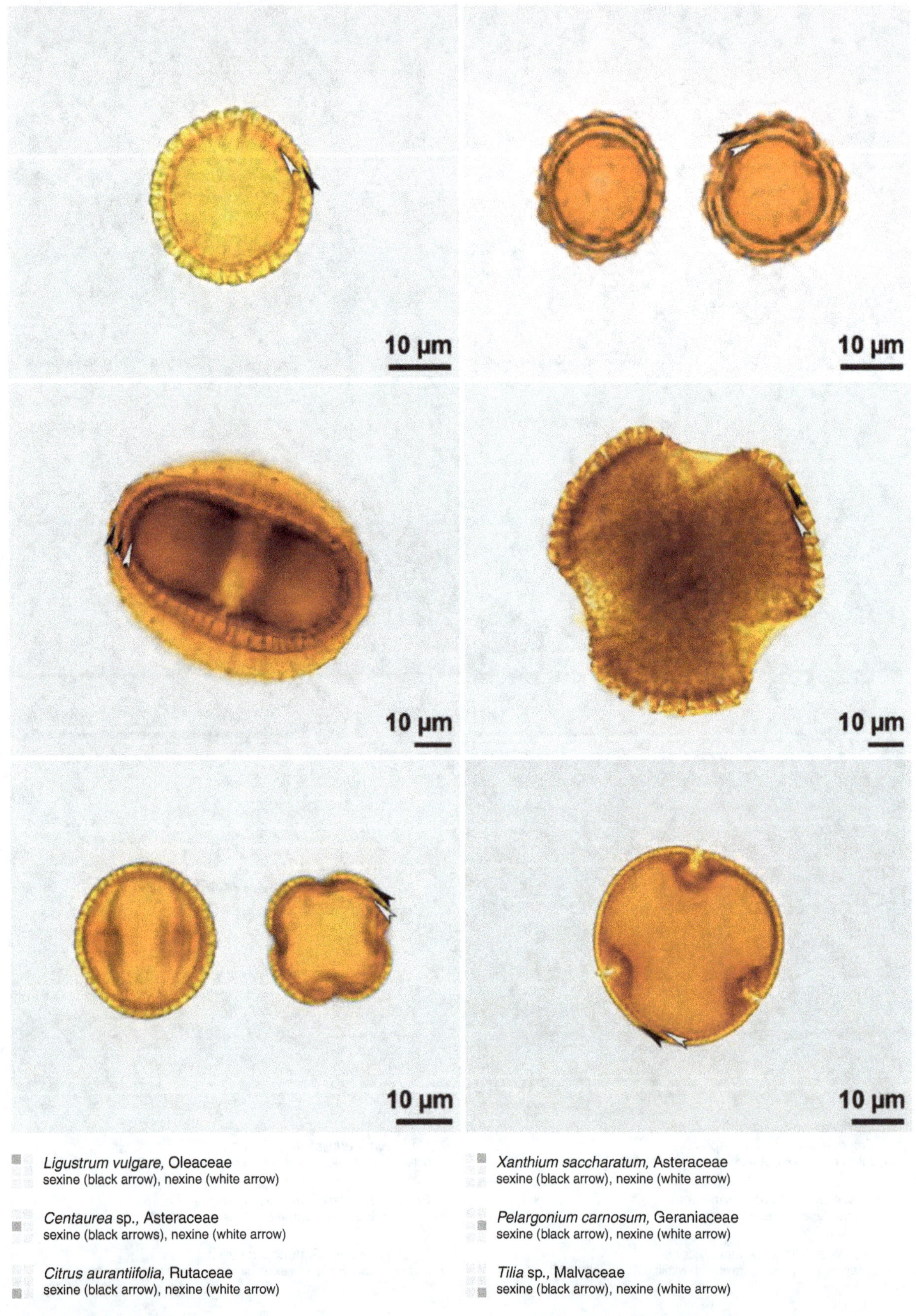

Ligustrum vulgare, Oleaceae
sexine (black arrow), nexine (white arrow)

Xanthium saccharatum, Asteraceae
sexine (black arrow), nexine (white arrow)

Centaurea sp., Asteraceae
sexine (black arrows), nexine (white arrow)

Pelargonium carnosum, Geraniaceae
sexine (black arrow), nexine (white arrow)

Citrus aurantiifolia, Rutaceae
sexine (black arrow), nexine (white arrow)

Tilia sp., Malvaceae
sexine (black arrow), nexine (white arrow)

calymmate

units covered by a continuous exine envelope

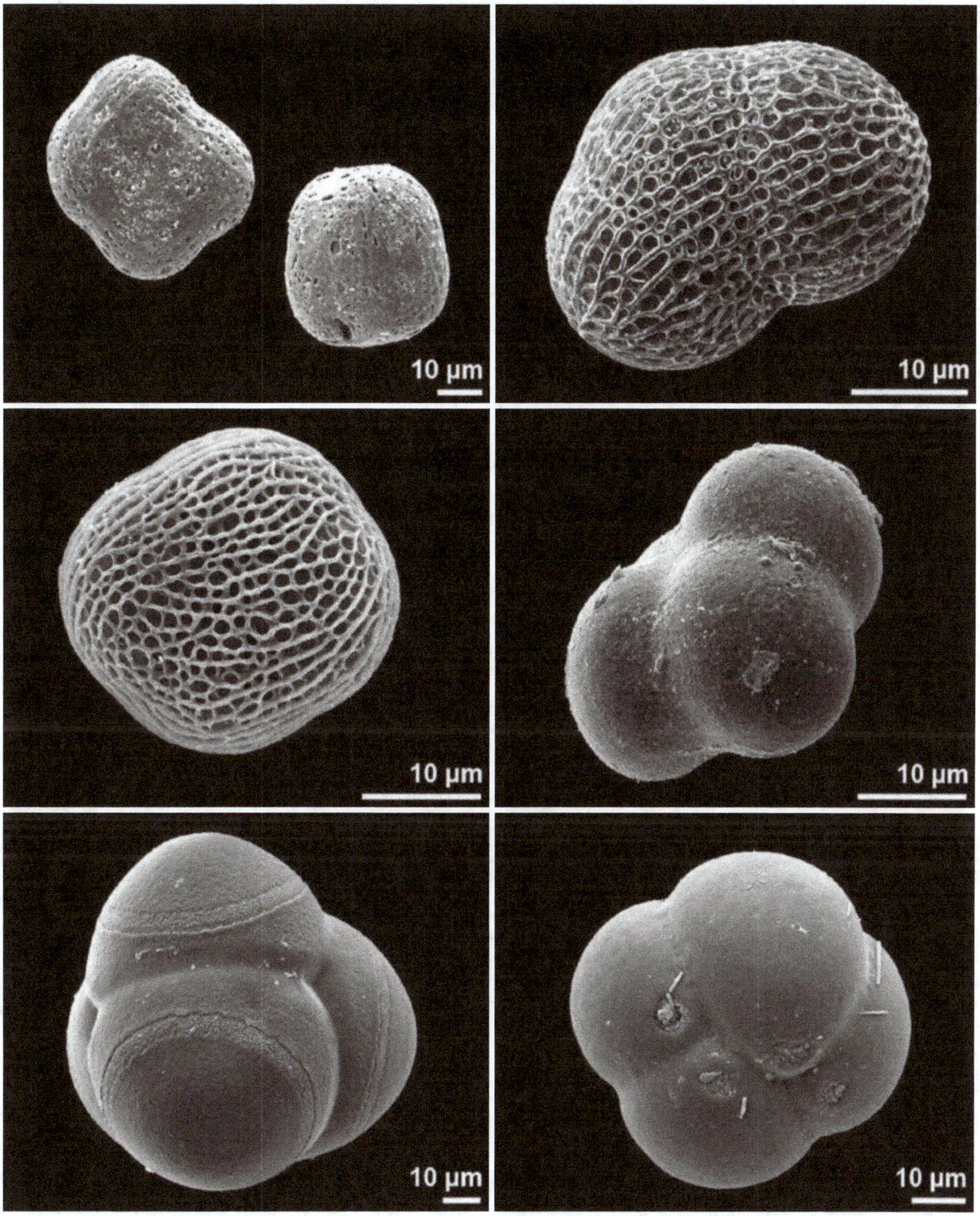

Chlorospatha kolbii, Araceae
tetrads

Chlorospatha pubescens, Araceae
tetrad

Victoria regia, Nymphaeaceae
tetrad

Chlorospatha hannoniae, Araceae
tetrad

Chlorospatha oblongifolia, Araceae
tetrad

Androlepis skinneri, Bromeliaceae
tetrad

acalymmate

units covered by an exine envelope which is discontinuous at the junctions between monads

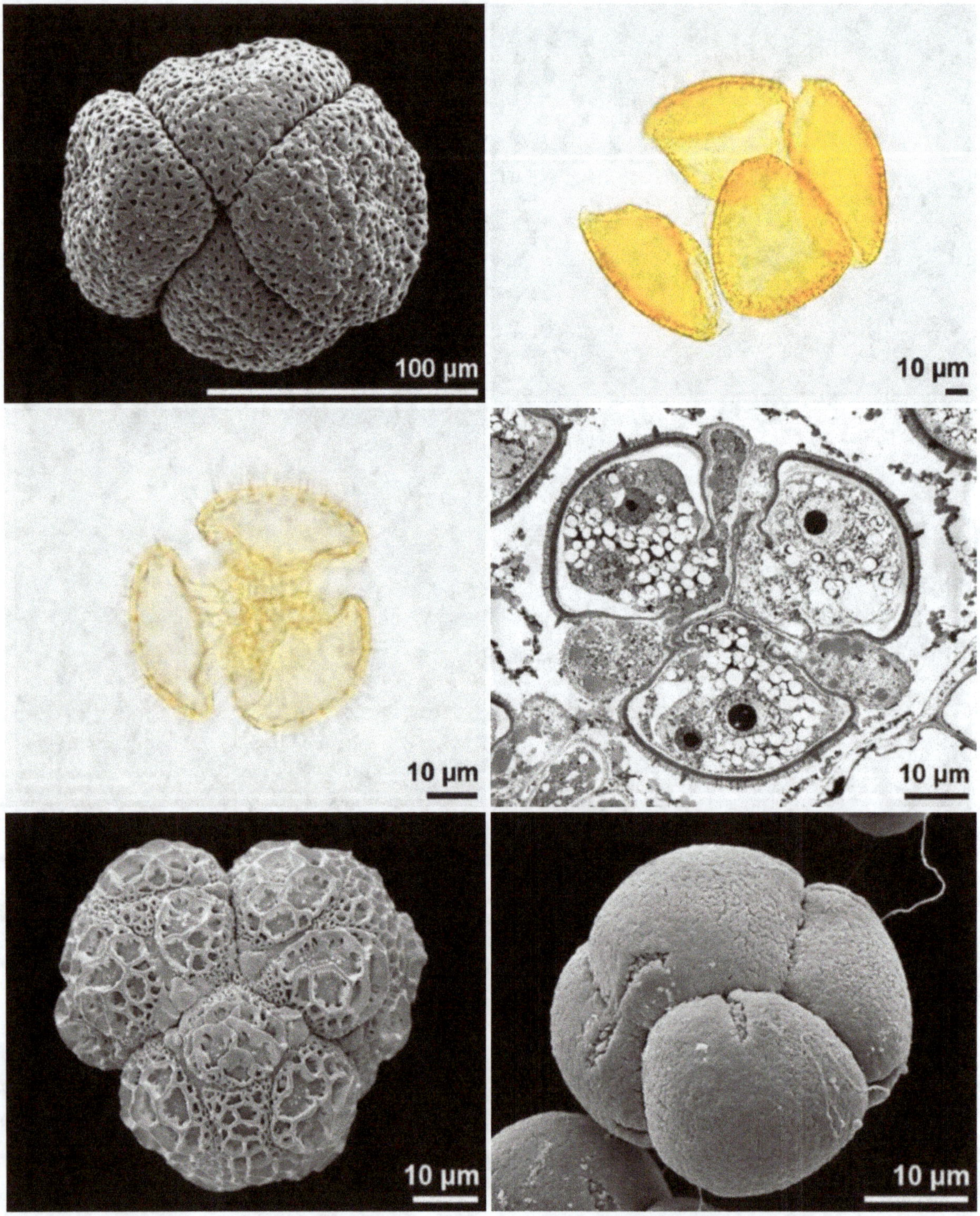

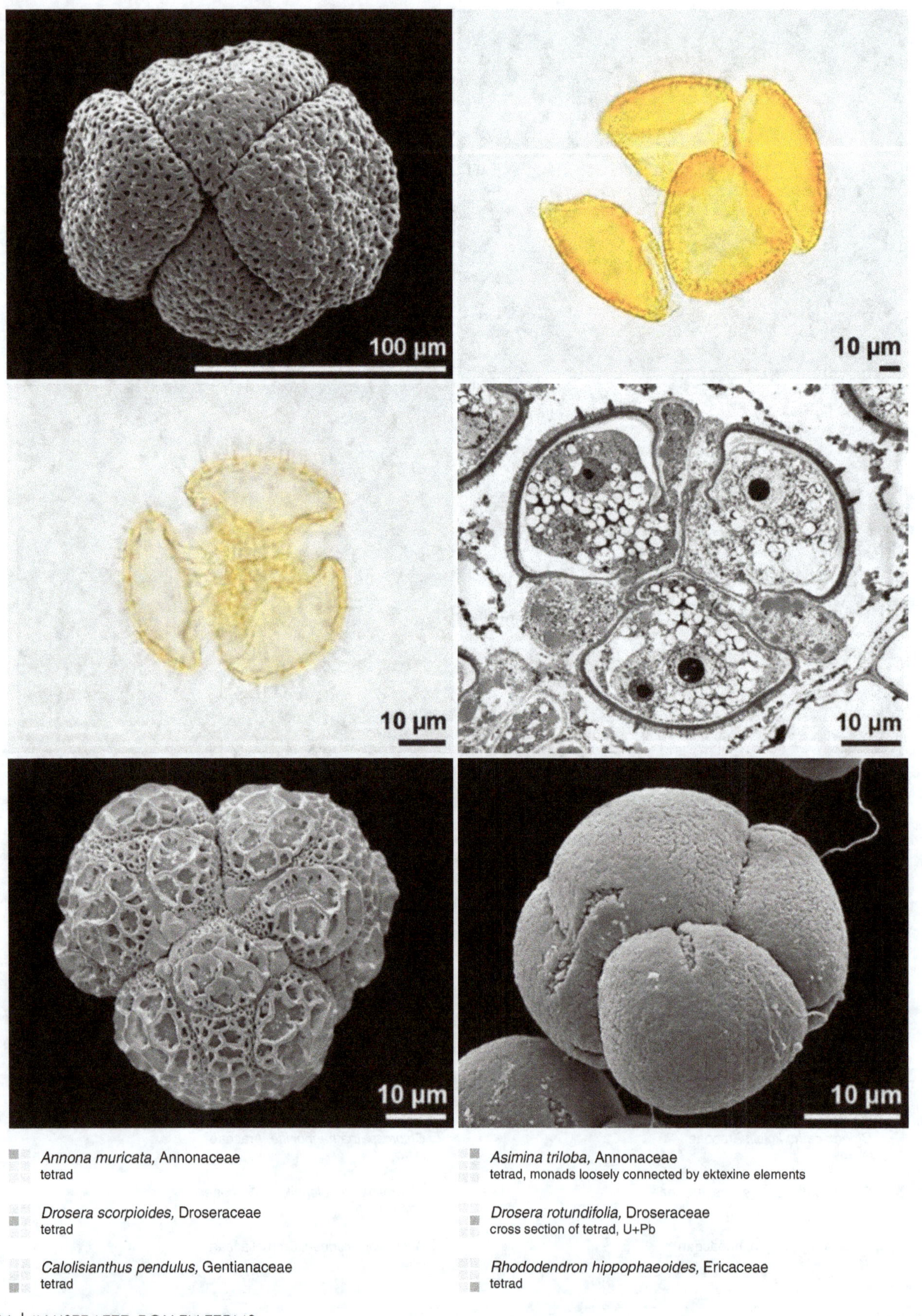

Annona muricata, Annonaceae
tetrad

Asimina triloba, Annonaceae
tetrad, monads loosely connected by ektexine elements

Drosera scorpioides, Droseraceae
tetrad

Drosera rotundifolia, Droseraceae
cross section of tetrad, U+Pb

Calolisianthus pendulus, Gentianaceae
tetrad

Rhododendron hippophaeoides, Ericaceae
tetrad

arcus/arcuate

a curved wall thickening interconnecting apertures

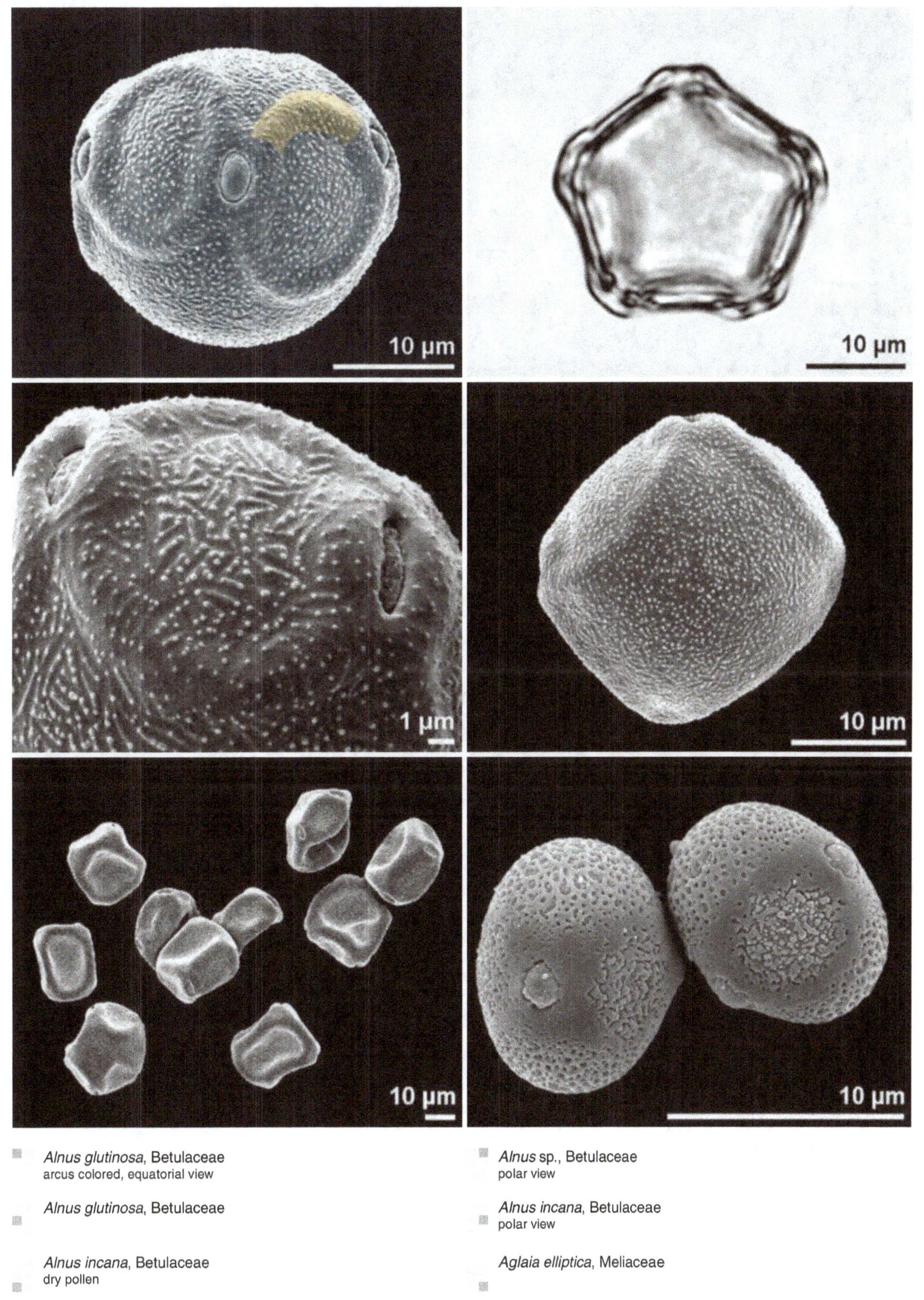

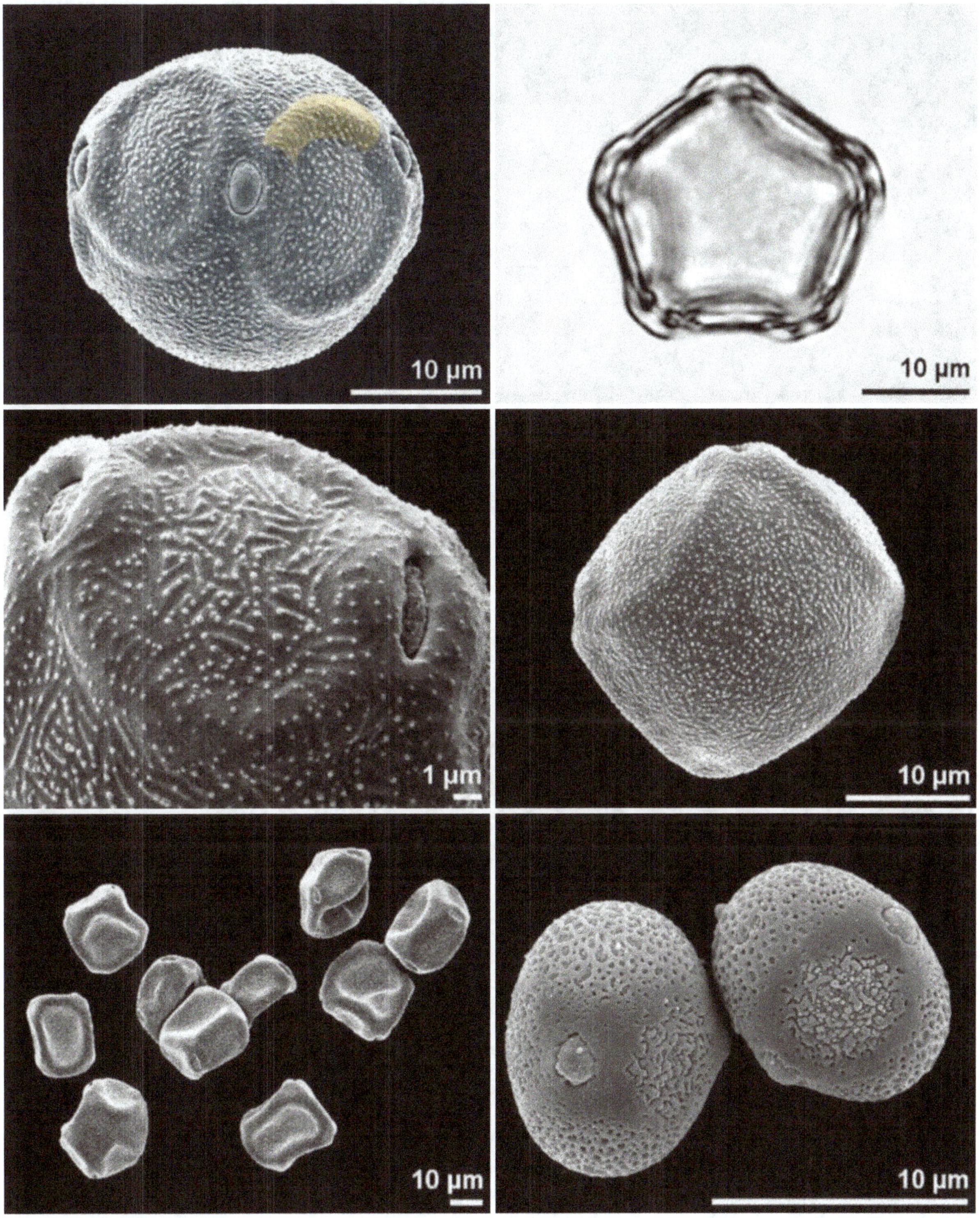

Alnus glutinosa, Betulaceae
arcus colored, equatorial view

Alnus glutinosa, Betulaceae

Alnus incana, Betulaceae
dry pollen

Alnus sp., Betulaceae
polar view

Alnus incana, Betulaceae
polar view

Aglaia elliptica, Meliaceae

tenuitas

general term for a thinning of the pollen wall

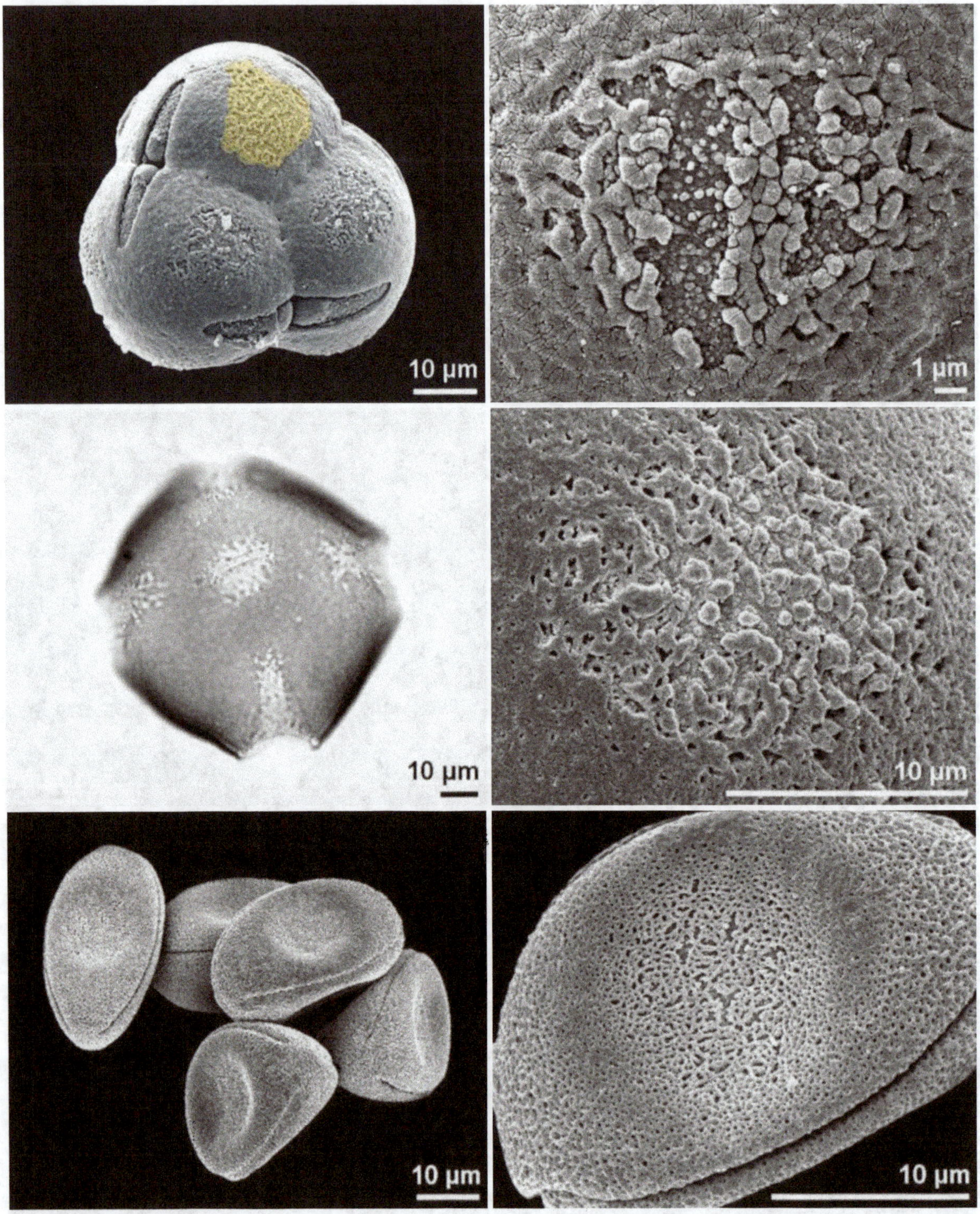

Agapetes variegata, Ericaceae
tenuitas colored, tetrad

Agapetes variegata, Ericaceae

Viola tricolor, Violaceae
polar view

Viola tricolor, Violaceae

Odontites vulgaris, Orobanchaceae
dry pollen

Odontites vulgaris, Orobanchaceae
dry pollen

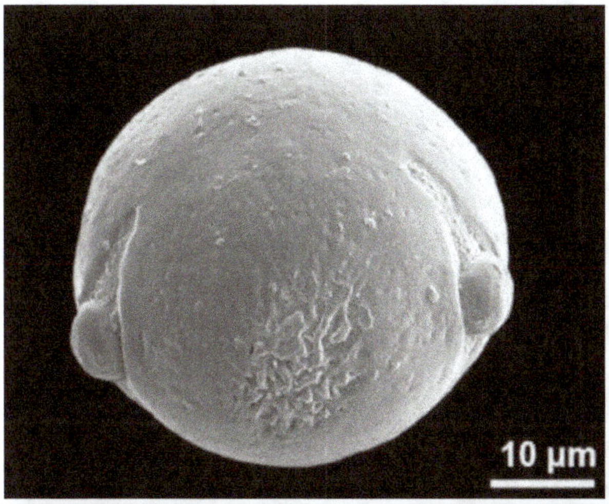

Acokanthera oblongifolia, Apocynaceae
oblique equatorial view

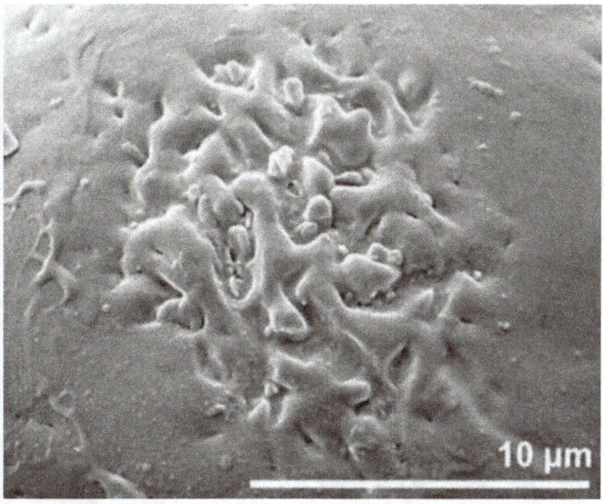

Acokanthera oblongifolia, Apocynaceae

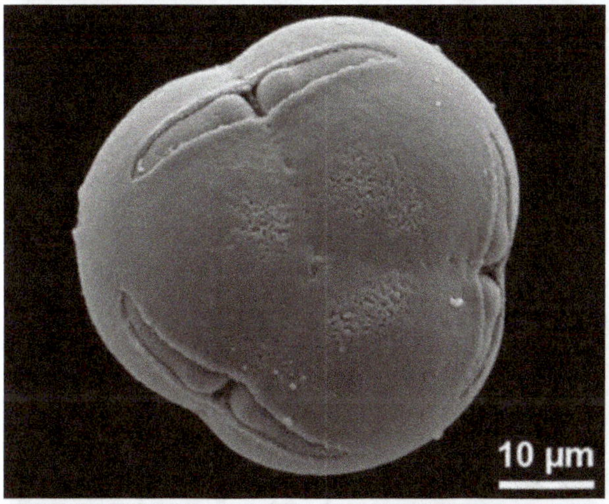

Satyria warszewiczii, Ericaceae
tetrad

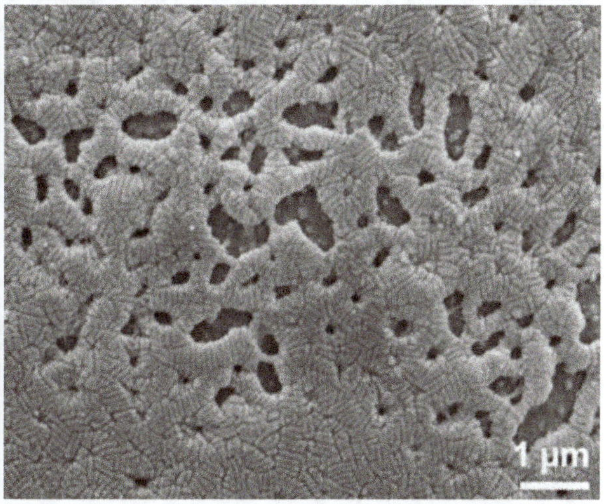

Satyria warszewiczii, Ericaceae

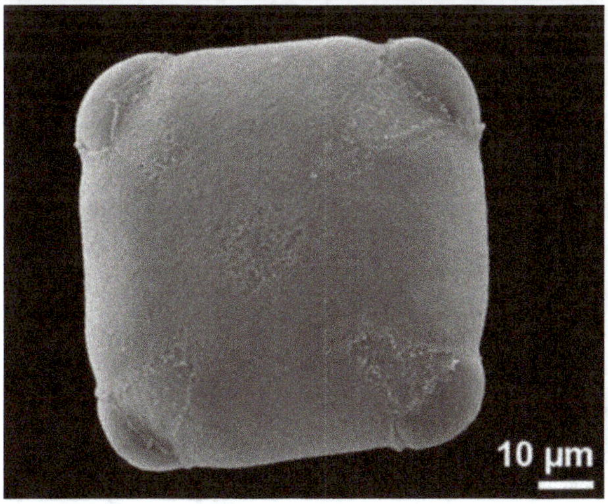

Viola calcarata, Violaceae
polar view

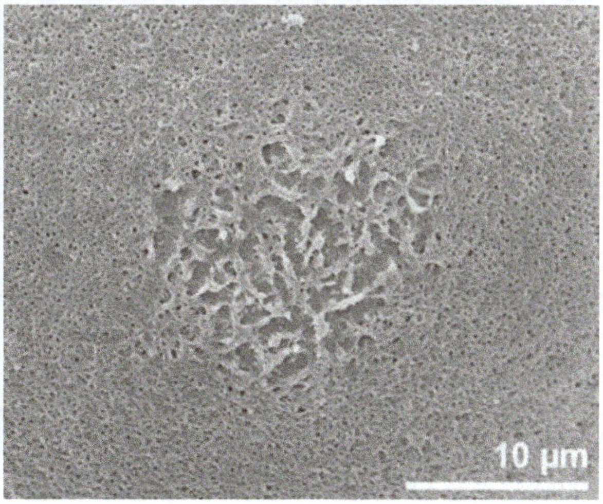

Viola calcarata, Violaceae
polar area

Axinaea lehmannii, Melastomataceae
polar view

Axinaea lehmannii, Melastomataceae
debatable: tenuitas or pseudocolpus

Tulipa linifolia, Liliaceae
sulcate with 2 additional proximal tenuitates, proximal polar view

Tulipa linifolia, Liliaceae
sulcus with operculum, 2 tenuitates proximally, dry pollen, equatorial view

Tulipa kaufmanniana, Liliaceae
sulcate with 2 tenuitates proximally, debatable trisulcate

Tulipa kaufmanniana, Liliaceae
equatorial view

pollen coating, pollenkitt

pollen coating consisting of sticky substances, mainly lipids

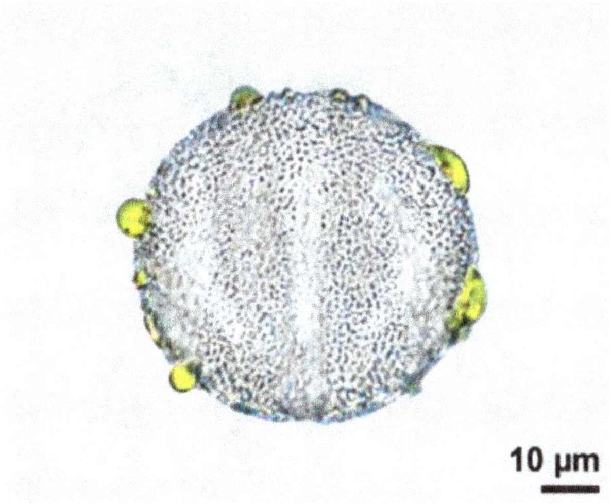

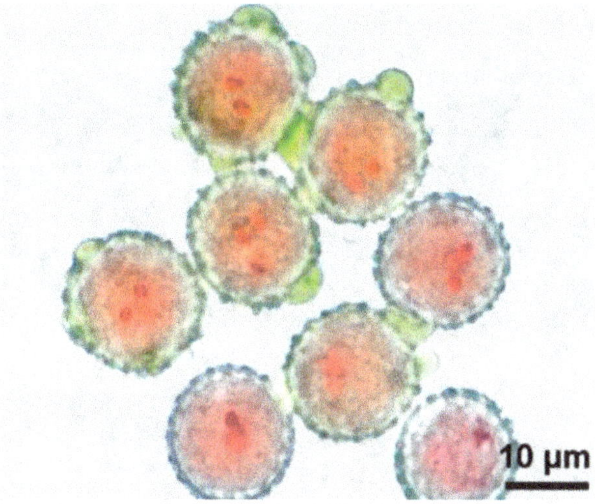

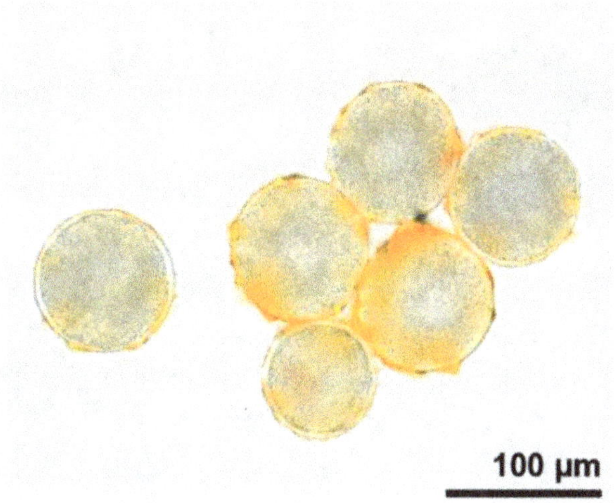

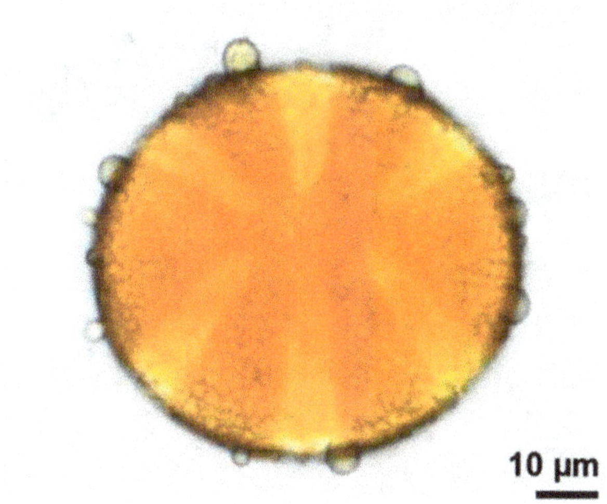

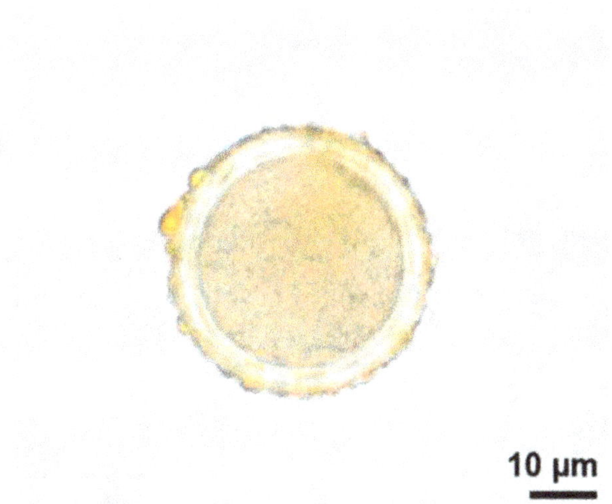

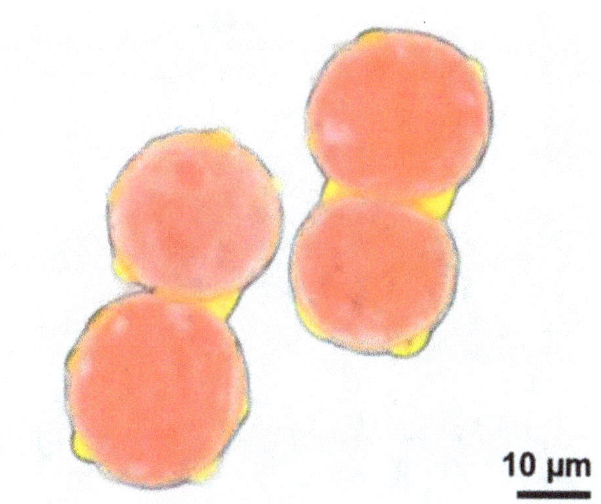

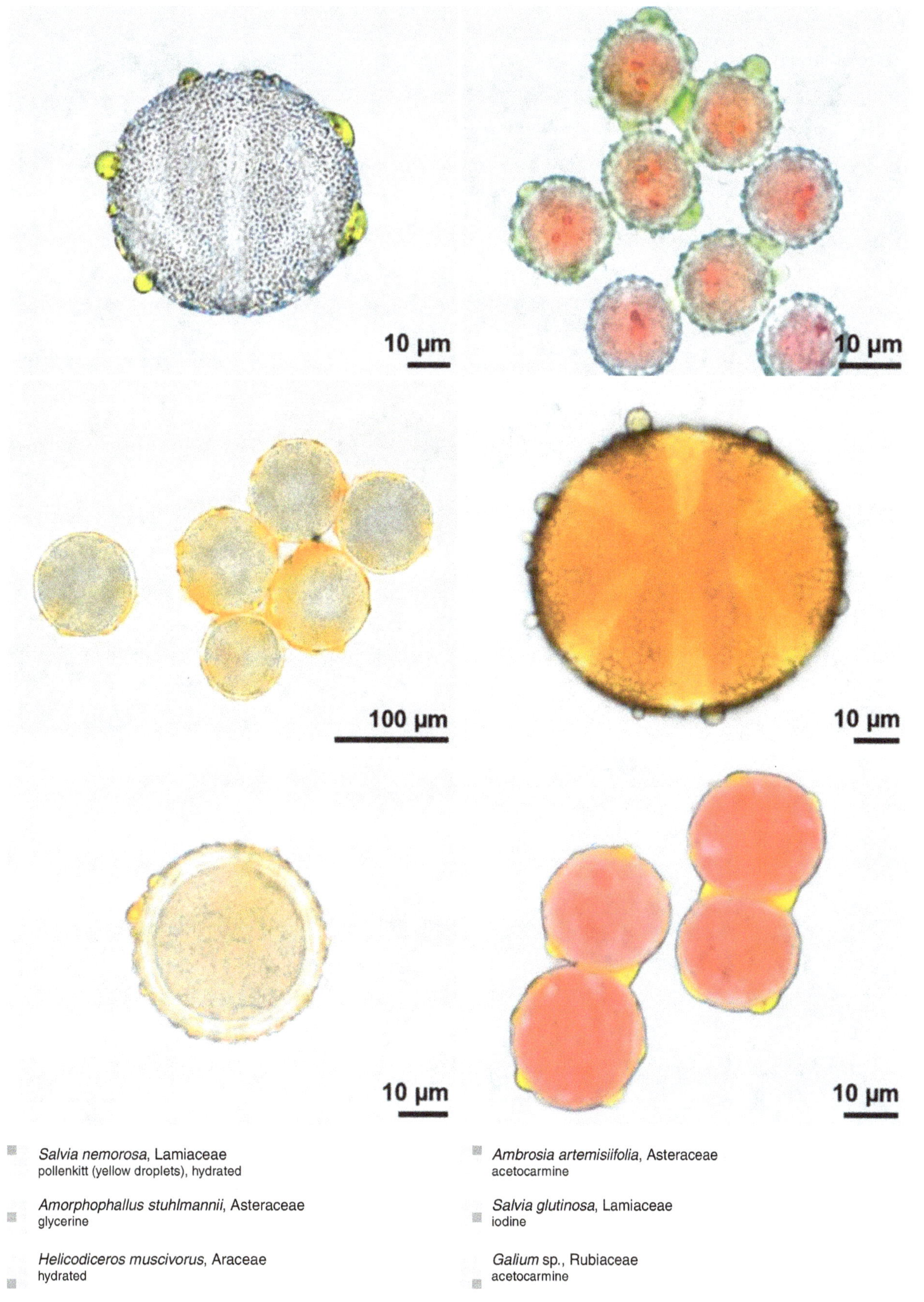

Salvia nemorosa, Lamiaceae
pollenkitt (yellow droplets), hydrated

Ambrosia artemisiifolia, Asteraceae
acetocarmine

Amorphophallus stuhlmannii, Asteraceae
glycerine

Salvia glutinosa, Lamiaceae
iodine

Helicodiceros muscivorus, Araceae
hydrated

Galium sp., Rubiaceae
acetocarmine

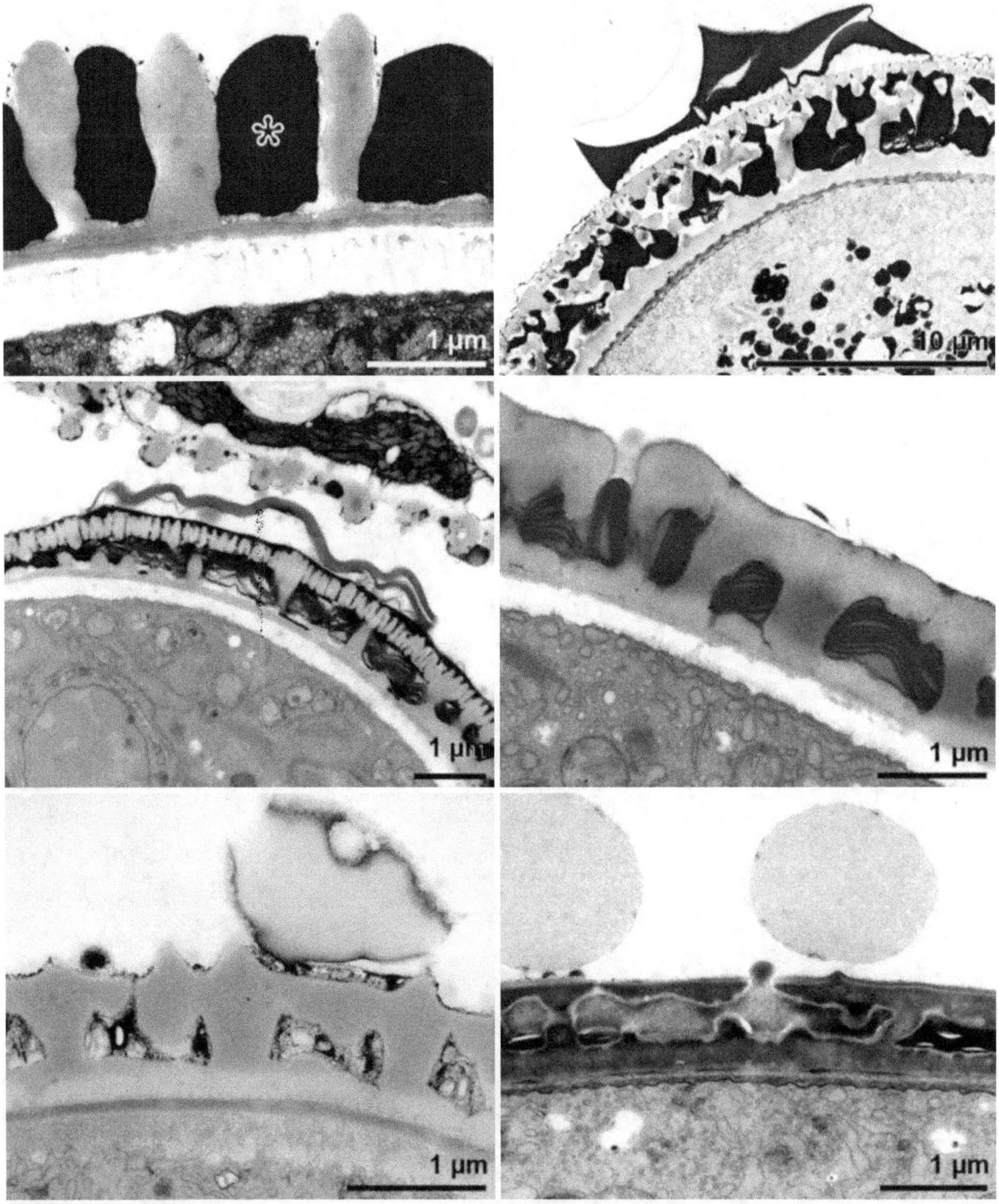

Jasminum nudiflorum, Oleaceae
pollenkitt (asterisk), lipid test

Nigella arvensis, Ranunculaceae
Thiéry test

Melampyrum nemorosum, Orobanchaceae
modified Thiéry test

Betonica officinalis, Lamiaceae
modified Thiéry test

Anemone ranunculoides, Ranunculaceae
modified Thiéry test

Consolida regalis, Ranunculaceae
U+Pb

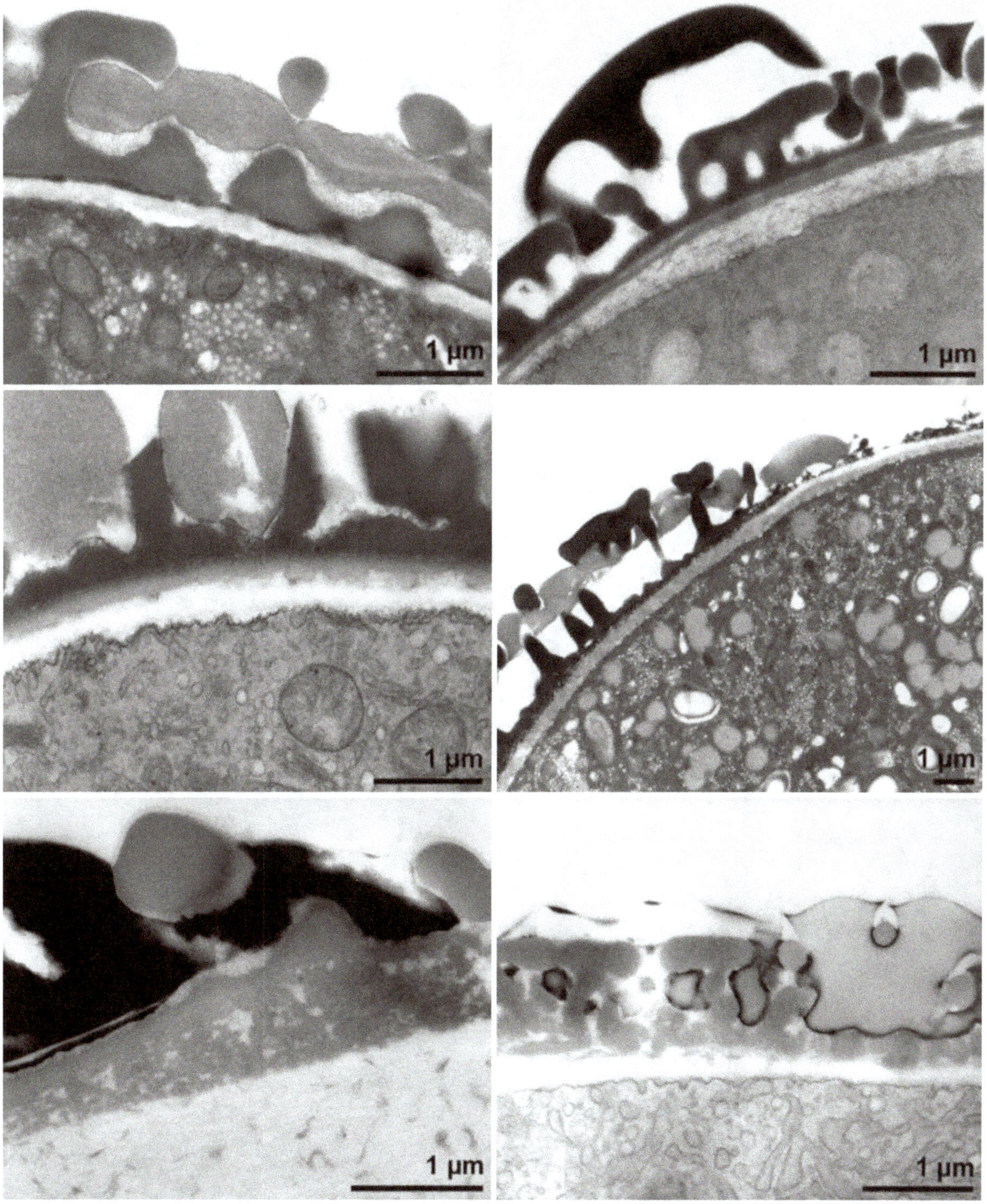

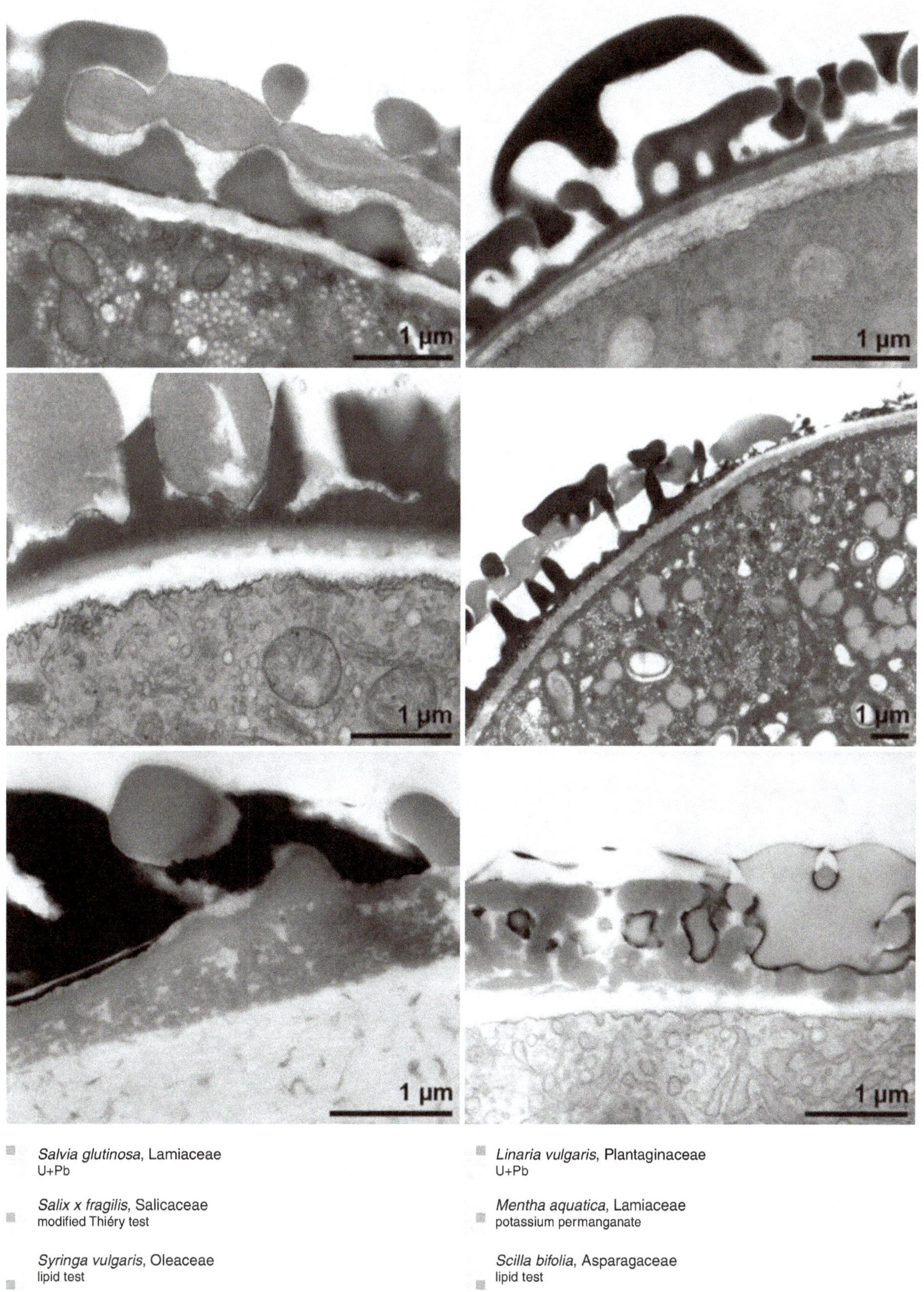

Salvia glutinosa, Lamiaceae
U+Pb

Linaria vulgaris, Plantaginaceae
U+Pb

Salix x fragilis, Salicaceae
modified Thiéry test

Mentha aquatica, Lamiaceae
potassium permanganate

Syringa vulgaris, Oleaceae
lipid test

Scilla bifolia, Asparagaceae
lipid test

pollen coating, primexine matrix

pollen coating consisting of primexine remnants in mature pollen grains

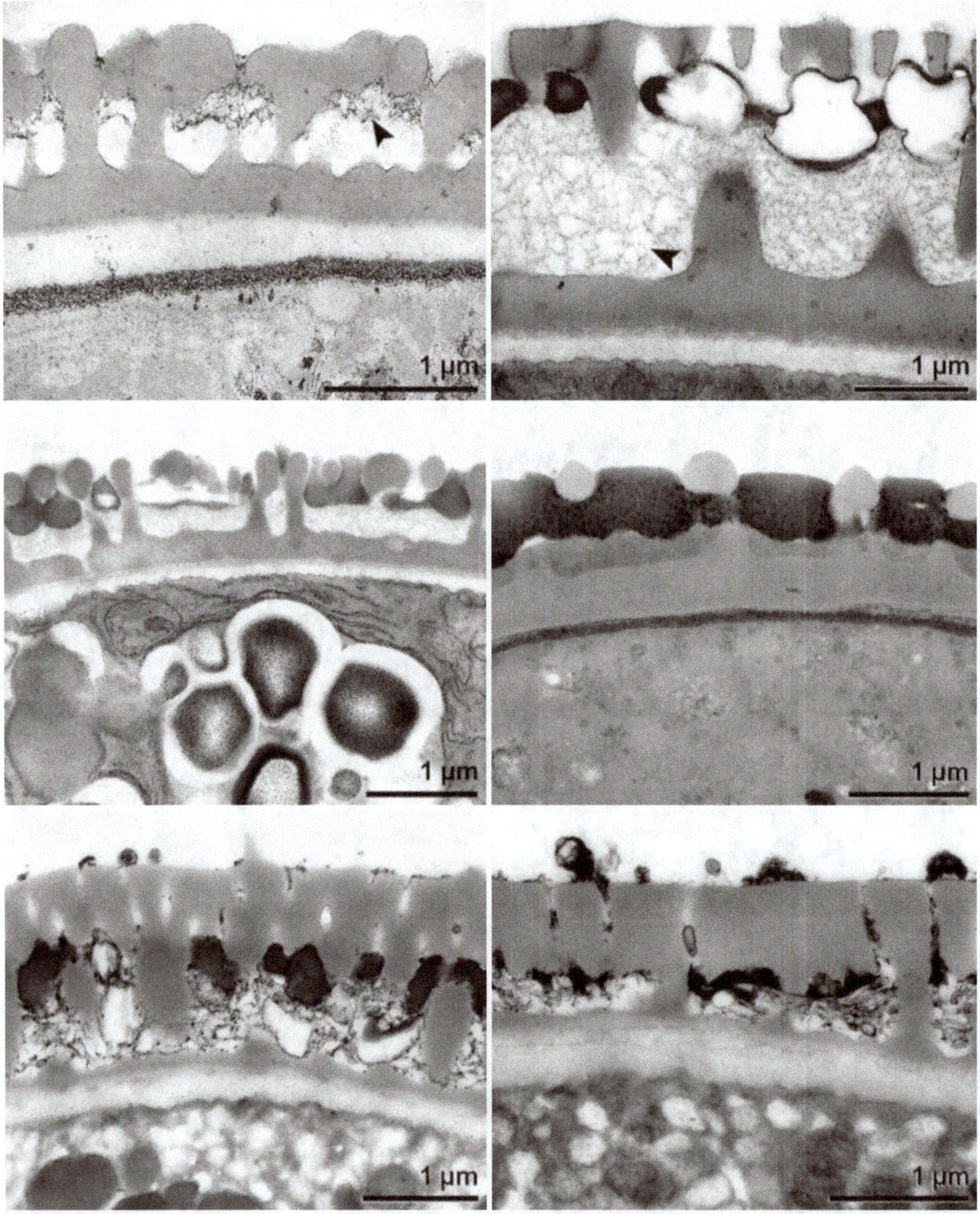

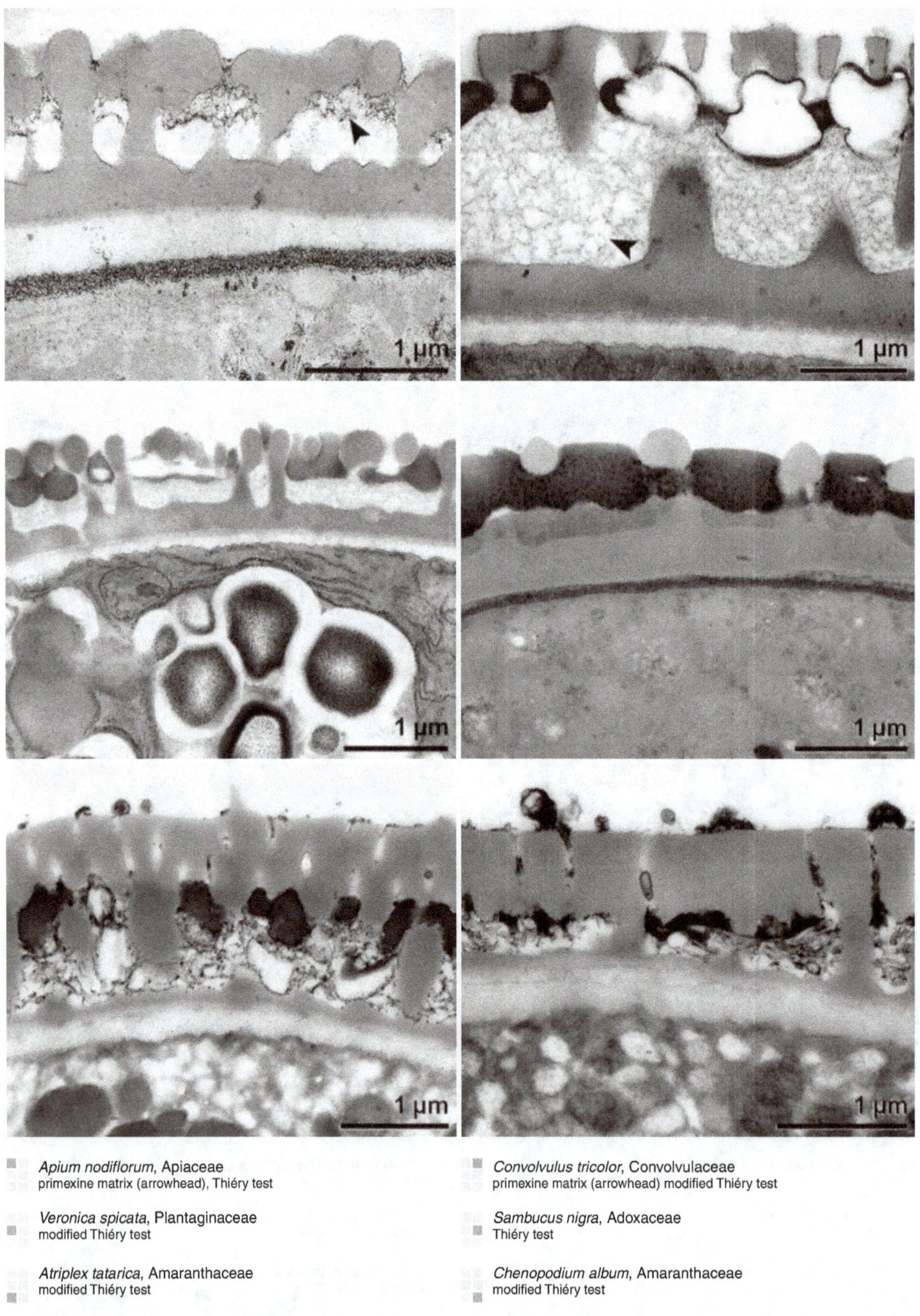

Apium nodiflorum, Apiaceae
primexine matrix (arrowhead), Thiéry test

Convolvulus tricolor, Convolvulaceae
primexine matrix (arrowhead) modified Thiéry test

Veronica spicata, Plantaginaceae
modified Thiéry test

Sambucus nigra, Adoxaceae
Thiéry test

Atriplex tatarica, Amaranthaceae
modified Thiéry test

Chenopodium album, Amaranthaceae
modified Thiéry test

pollen coating, tryphine

pollen coating consisting mainly of lipids mixed with membrane remnants

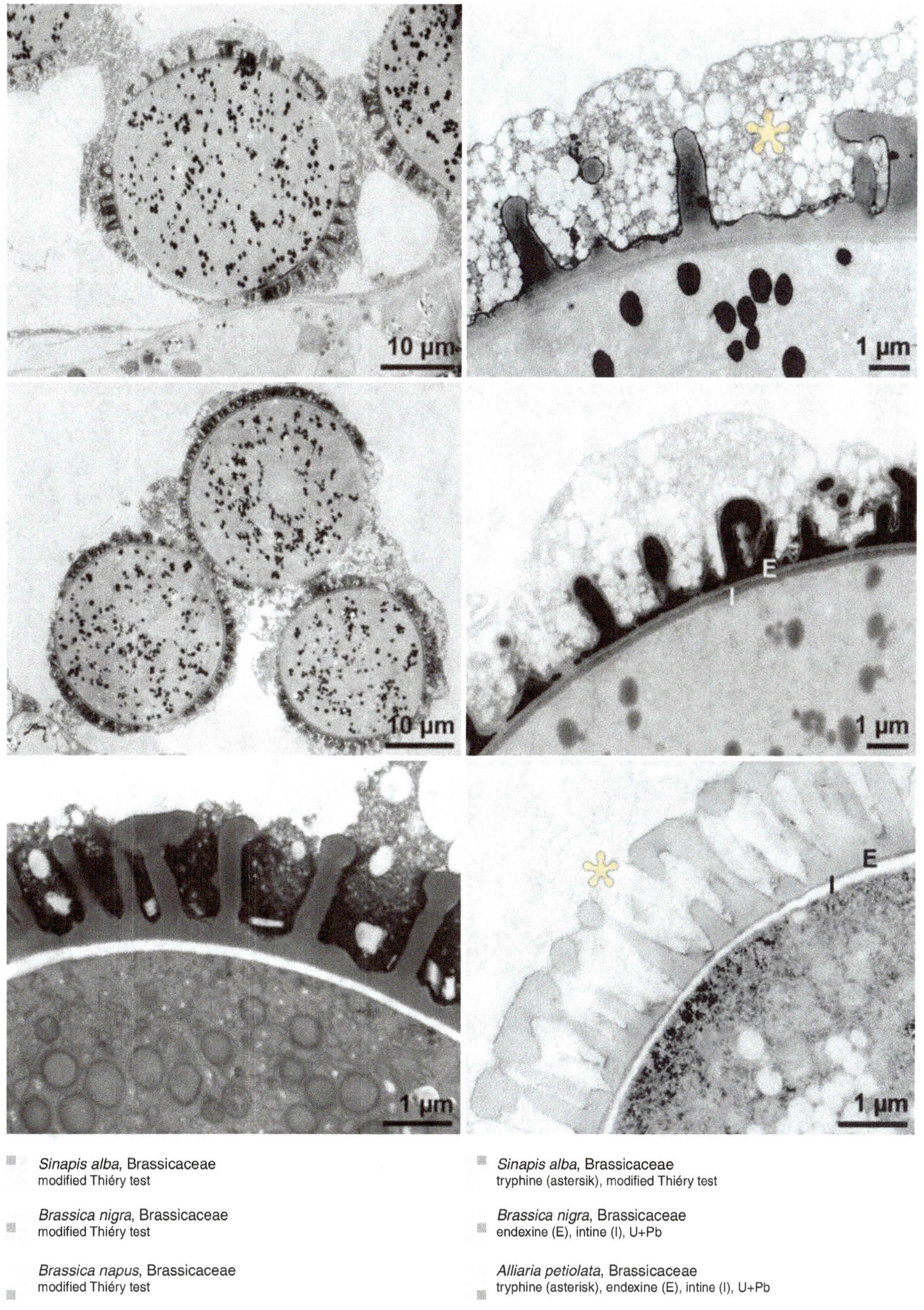

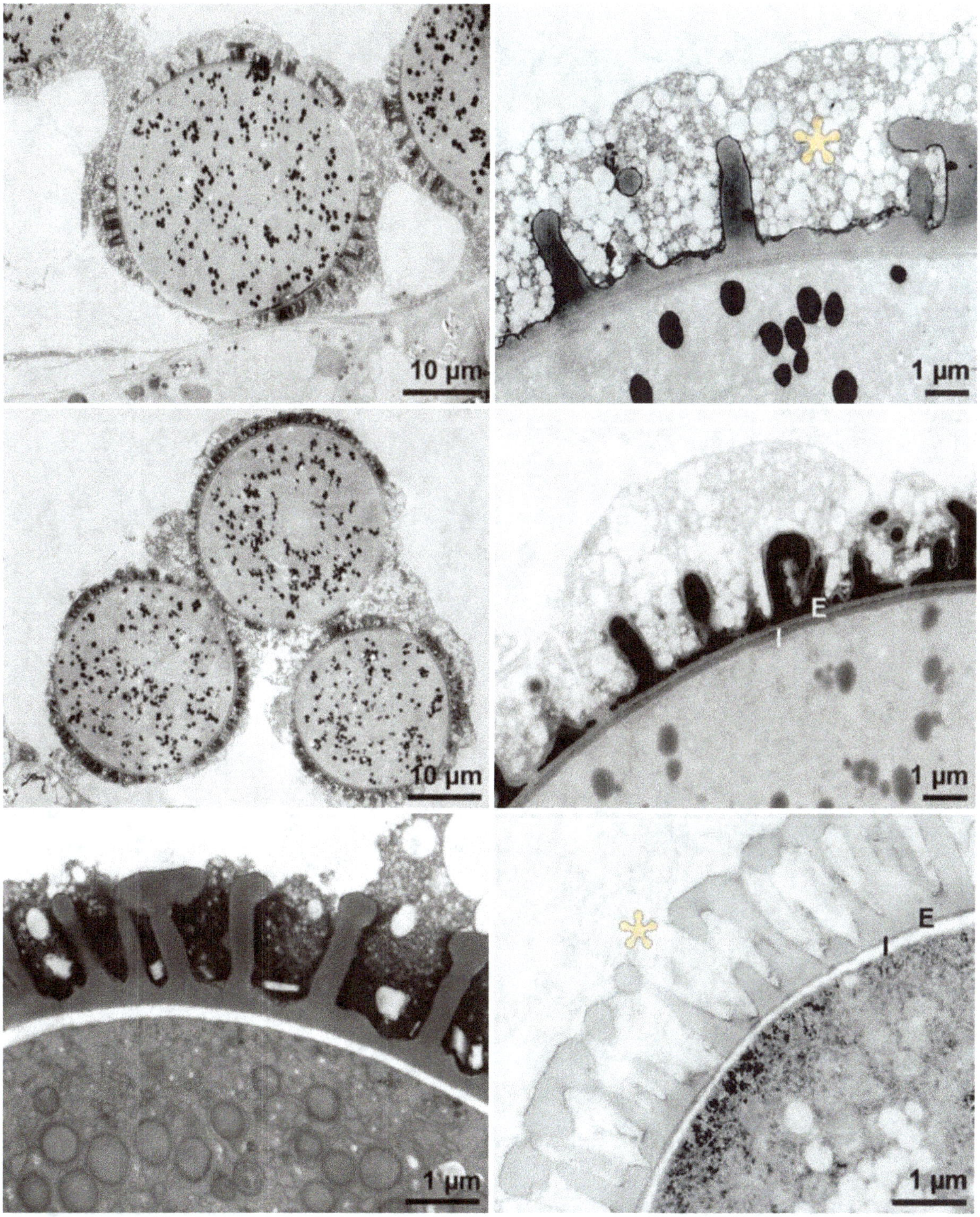

Sinapis alba, Brassicaceae
modified Thiéry test

Sinapis alba, Brassicaceae
tryphine (astersik), modified Thiéry test

Brassica nigra, Brassicaceae
modified Thiéry test

Brassica nigra, Brassicaceae
endexine (E), intine (I), U+Pb

Brassica napus, Brassicaceae
modified Thiéry test

Alliaria petiolata, Brassicaceae
tryphine (asterisk), endexine (E), intine (I), U+Pb

elastoviscin

highly elastic, not acetolysis resistant substance in Orchidaceae, which interconnects the subunits (monads, tetrads or massulae) of a pollinium and builds up the caudicles

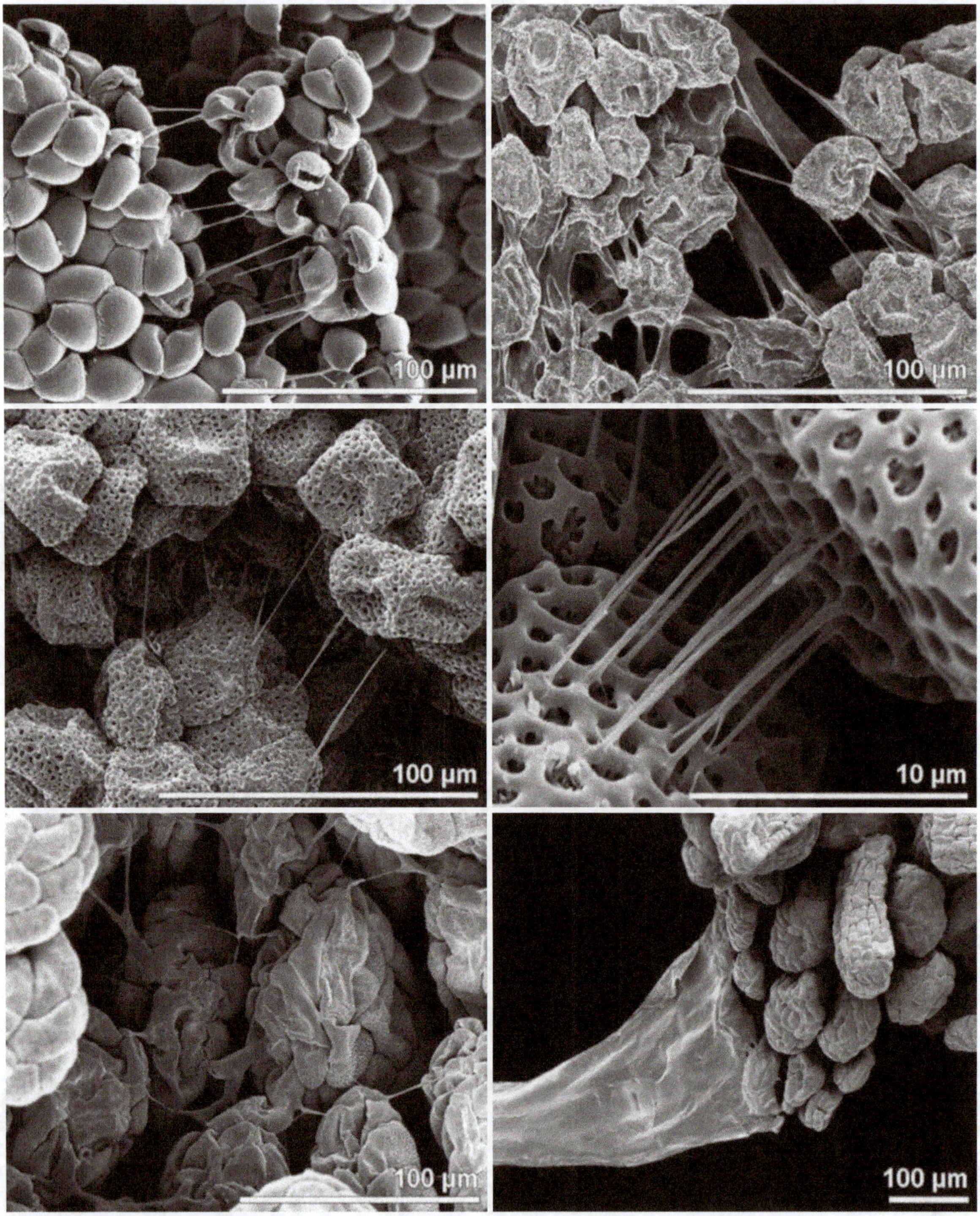

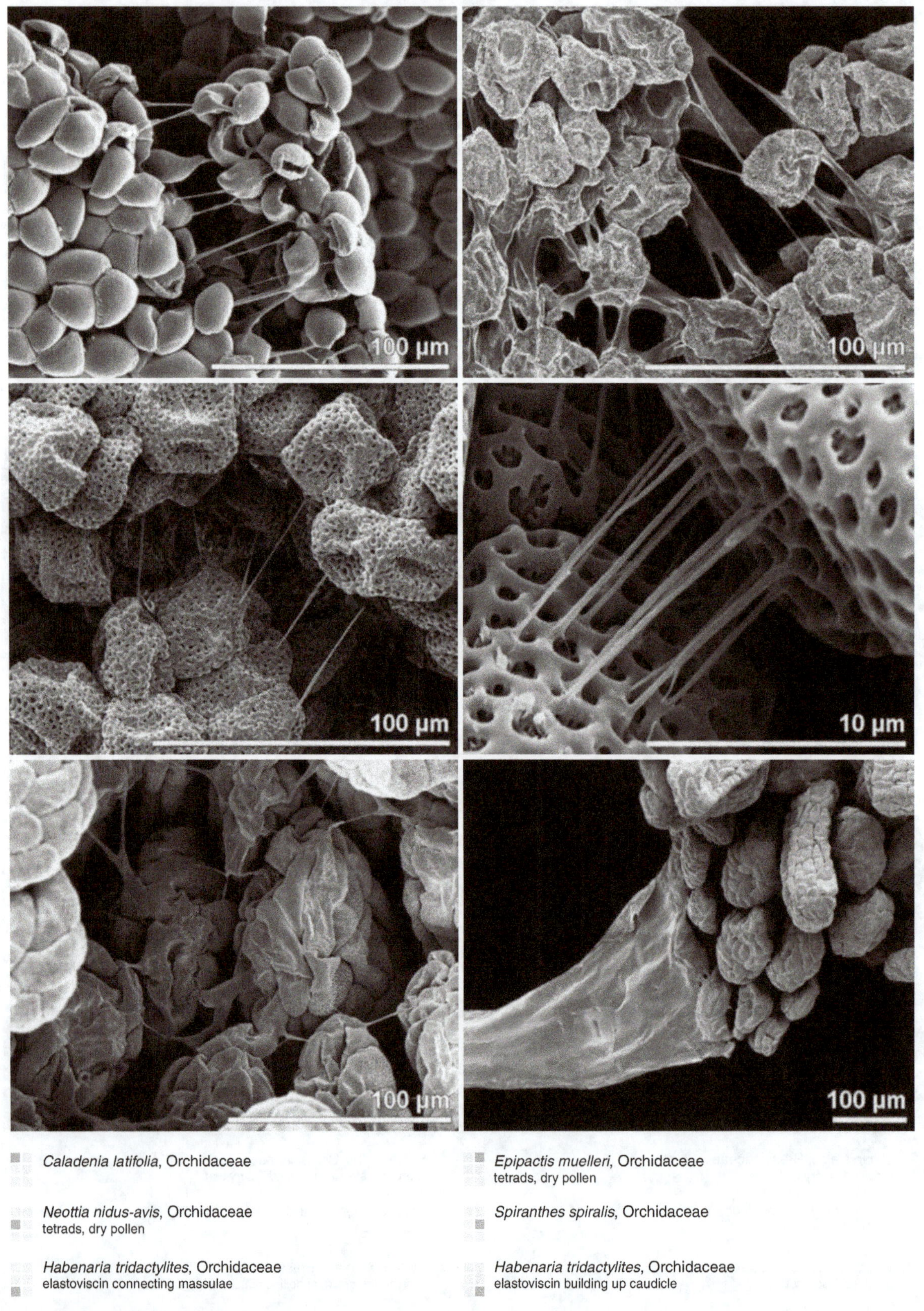

Caladenia latifolia, Orchidaceae

Epipactis muelleri, Orchidaceae
tetrads, dry pollen

Neottia nidus-avis, Orchidaceae
tetrads, dry pollen

Spiranthes spiralis, Orchidaceae

Habenaria tridactylites, Orchidaceae
elastoviscin connecting massulae

Habenaria tridactylites, Orchidaceae
elastoviscin building up caudicle

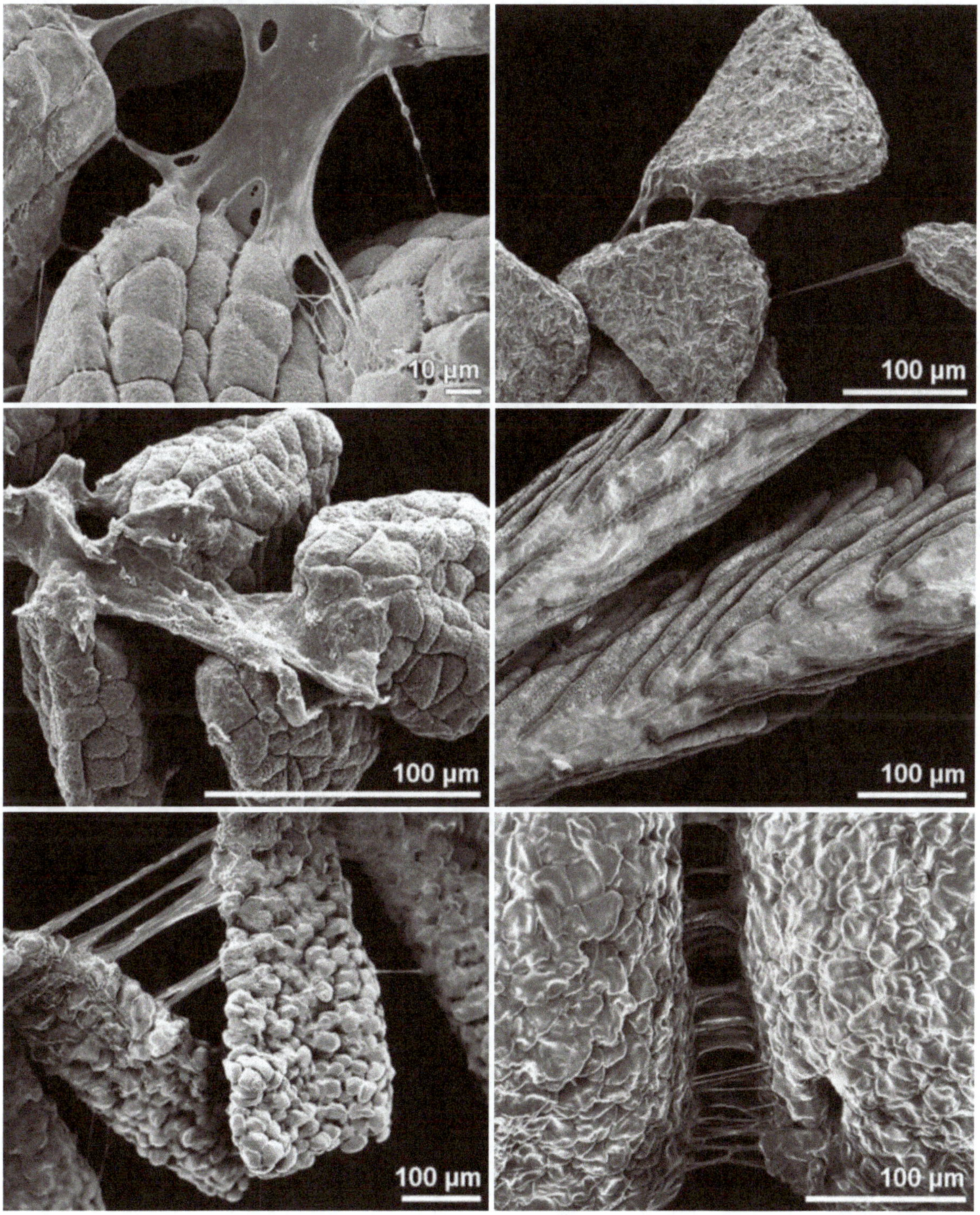

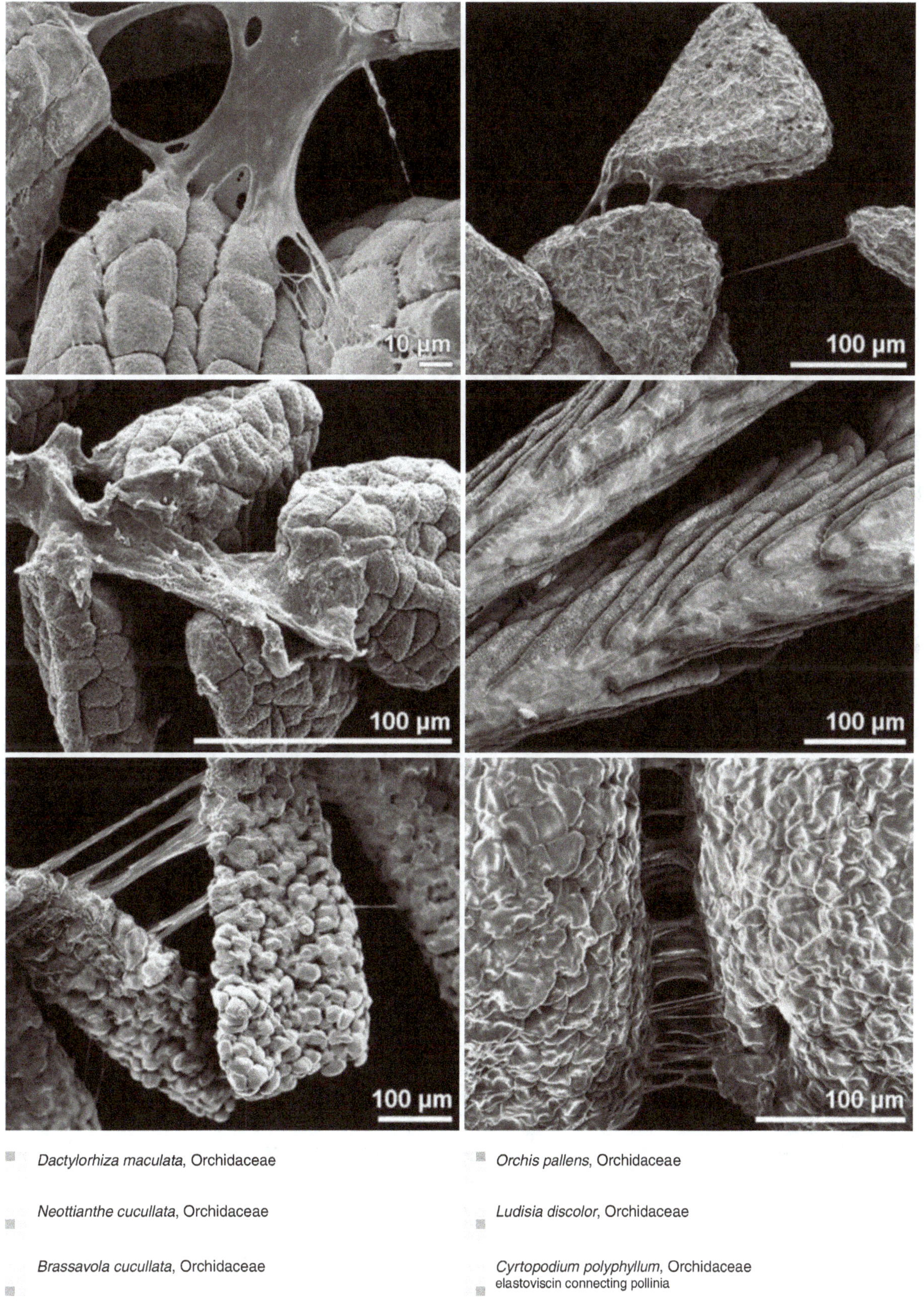

Dactylorhiza maculata, Orchidaceae

Orchis pallens, Orchidaceae

Neottianthe cucullata, Orchidaceae

Ludisia discolor, Orchidaceae

Brassavola cucullata, Orchidaceae

Cyrtopodium polyphyllum, Orchidaceae
elastoviscin connecting pollinia

viscin thread

acetolysis resistant thread arising from the exine

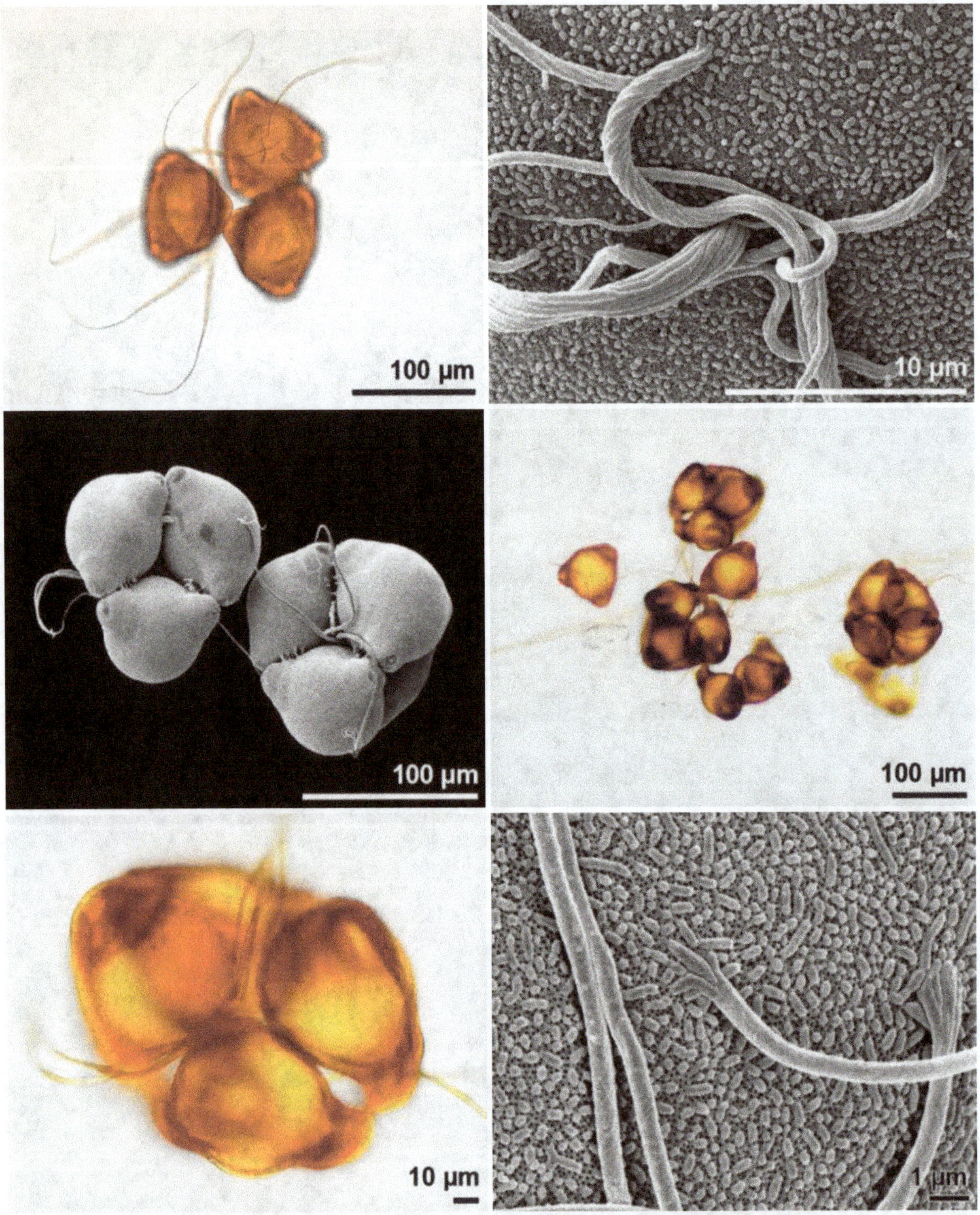

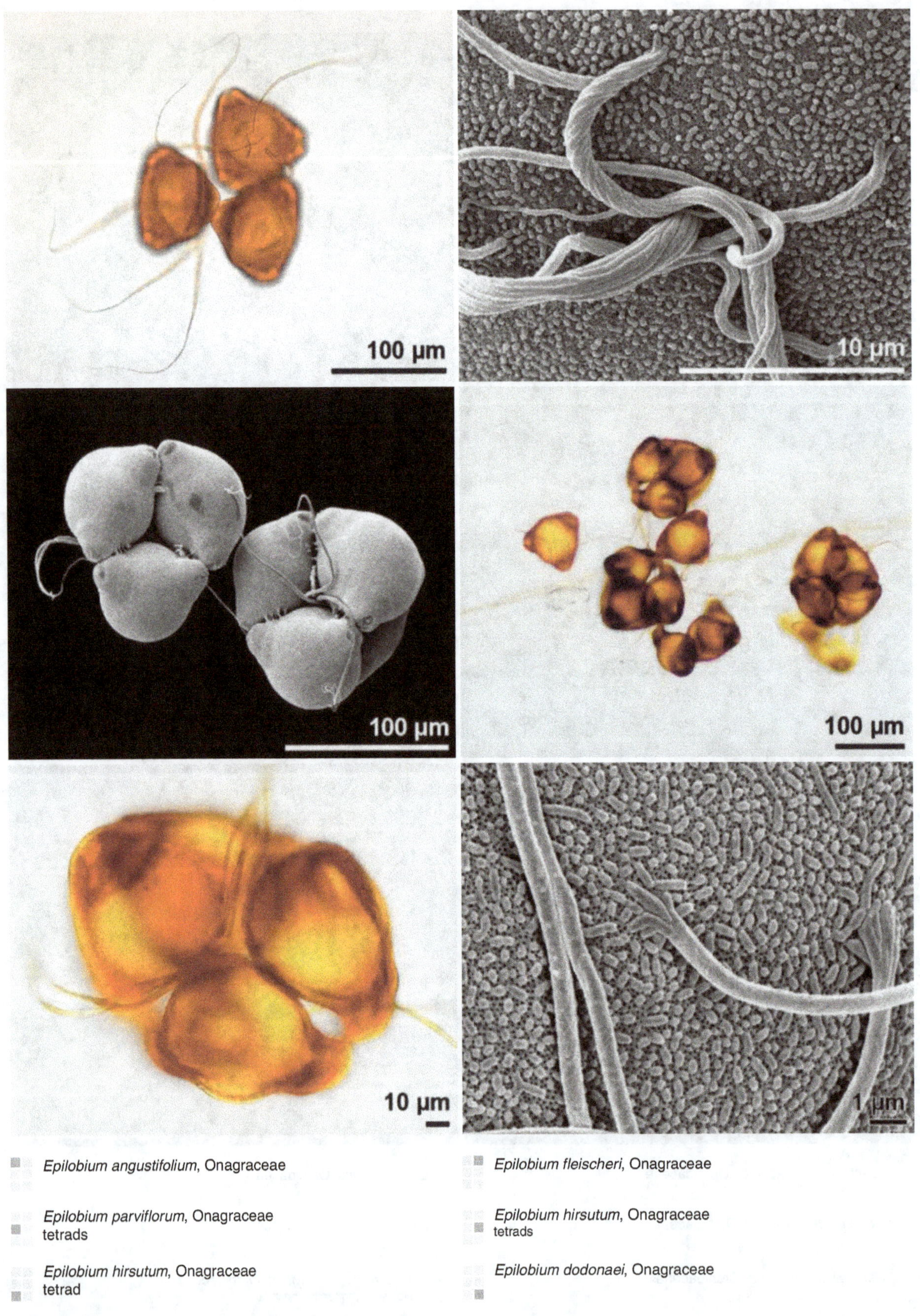

Epilobium angustifolium, Onagraceae

Epilobium fleischeri, Onagraceae

Epilobium parviflorum, Onagraceae
tetrads

Epilobium hirsutum, Onagraceae
tetrads

Epilobium hirsutum, Onagraceae
tetrad

Epilobium dodonaei, Onagraceae

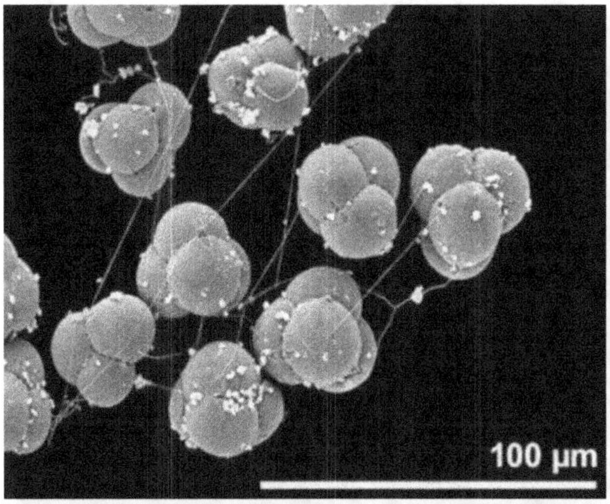

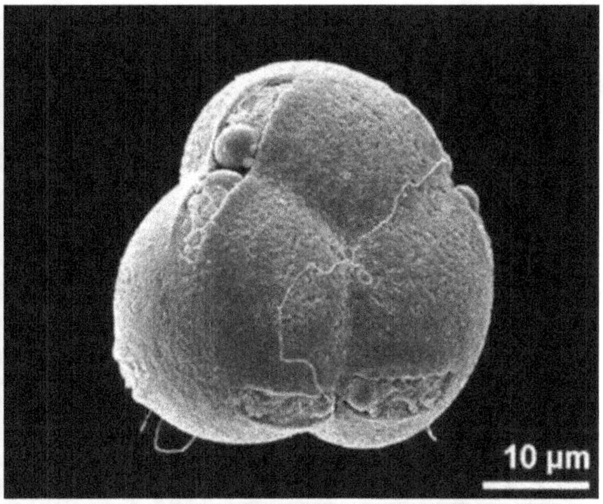

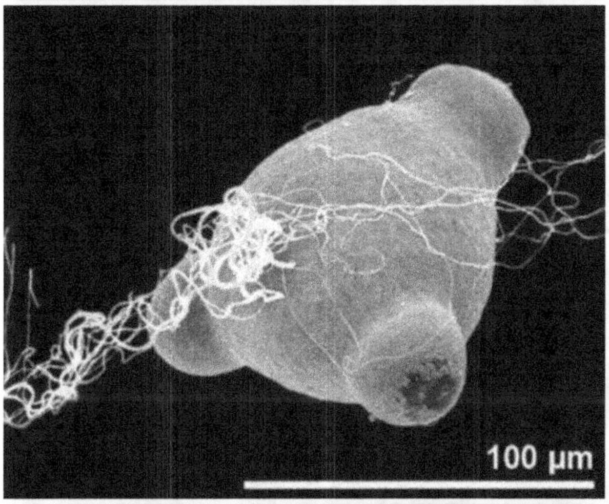

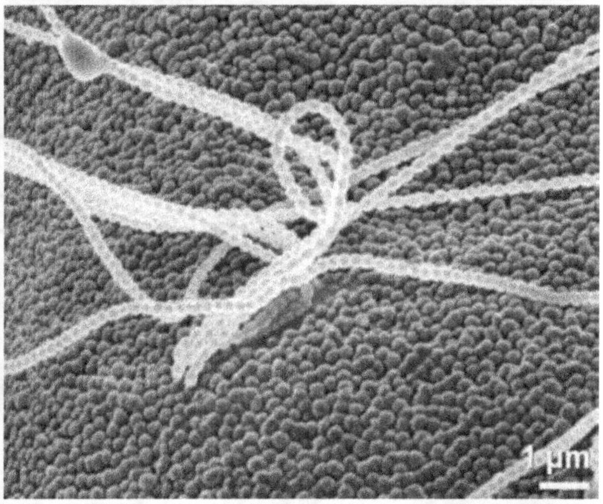

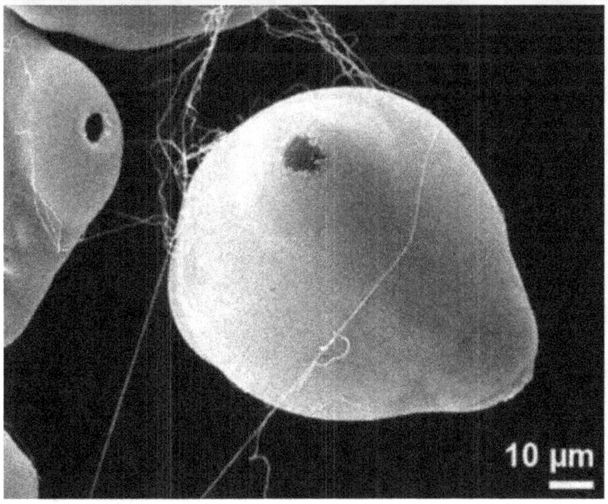

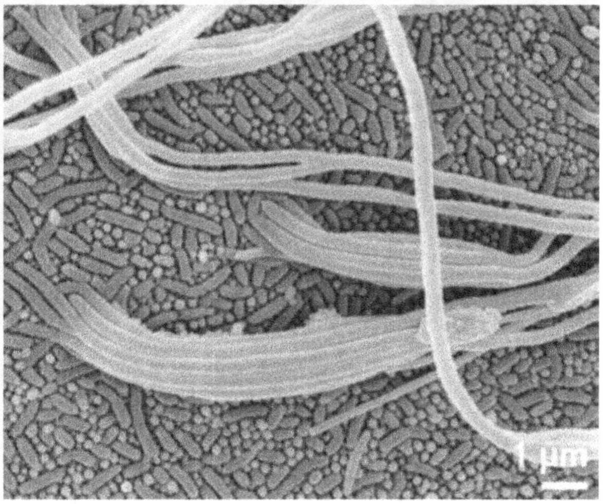

Kalmia latifolia, Ericaceae
tetrads

Oenothera biennis, Onagraceae
oblique view

Clarkia pulchella, Onagraceae
equatorial view

Ledum palustre, Ericaceae
tetrad

Oenothera biennis, Onagraceae

Clarkia unguiculata, Onagraceae

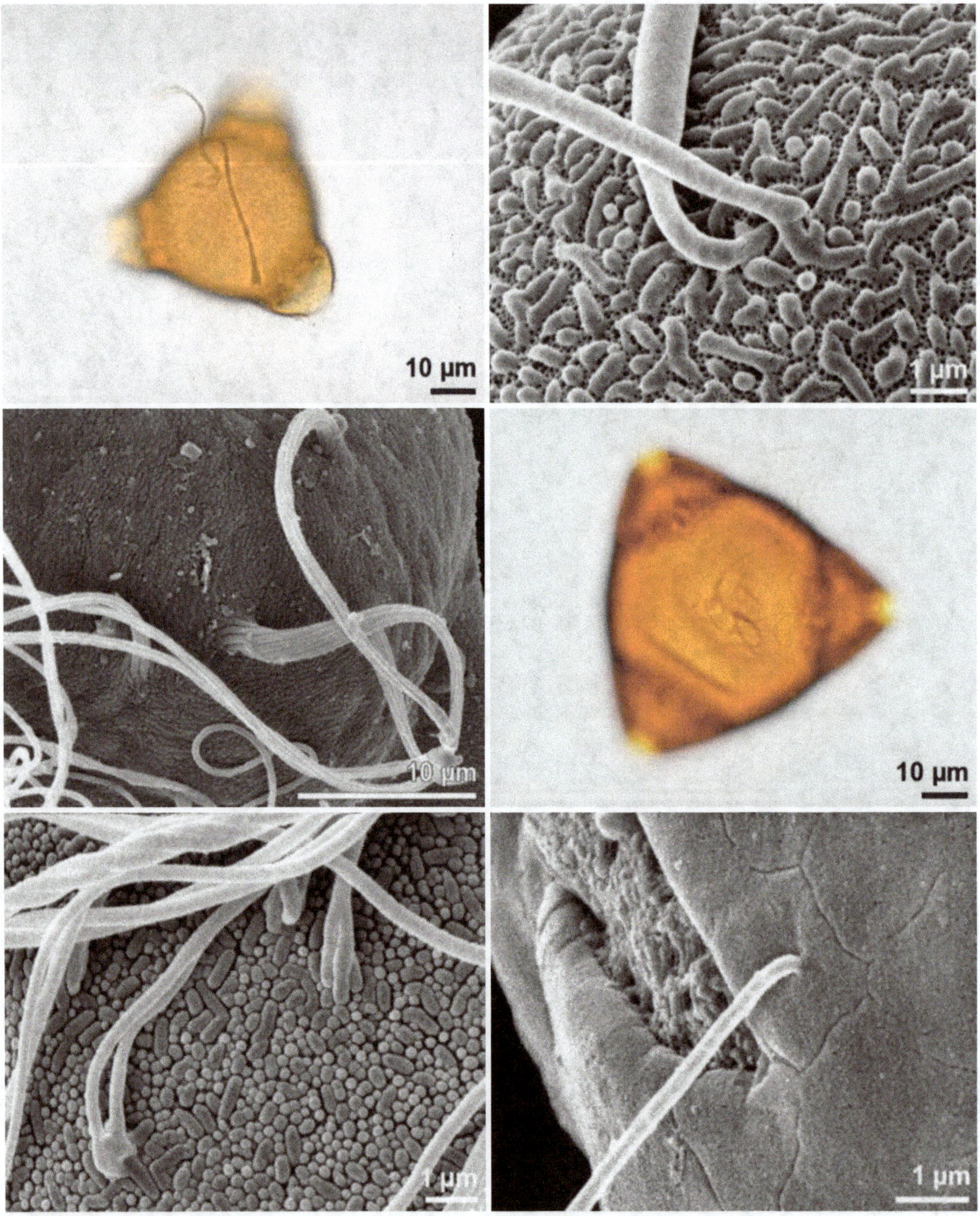

Circaea lutetiana, Onagraceae

Circaea lutetiana, Onagraceae

Fuchsia magellanica, Onagraceae

Lopezia racemosa, Onagraceae

Clarkia purpurea, Onagraceae

Rhododendron hirsutum, Ericaceae

Ubisch body (Ubisch bodies)

polymorphic sporopollenin-element produced by the tapetum

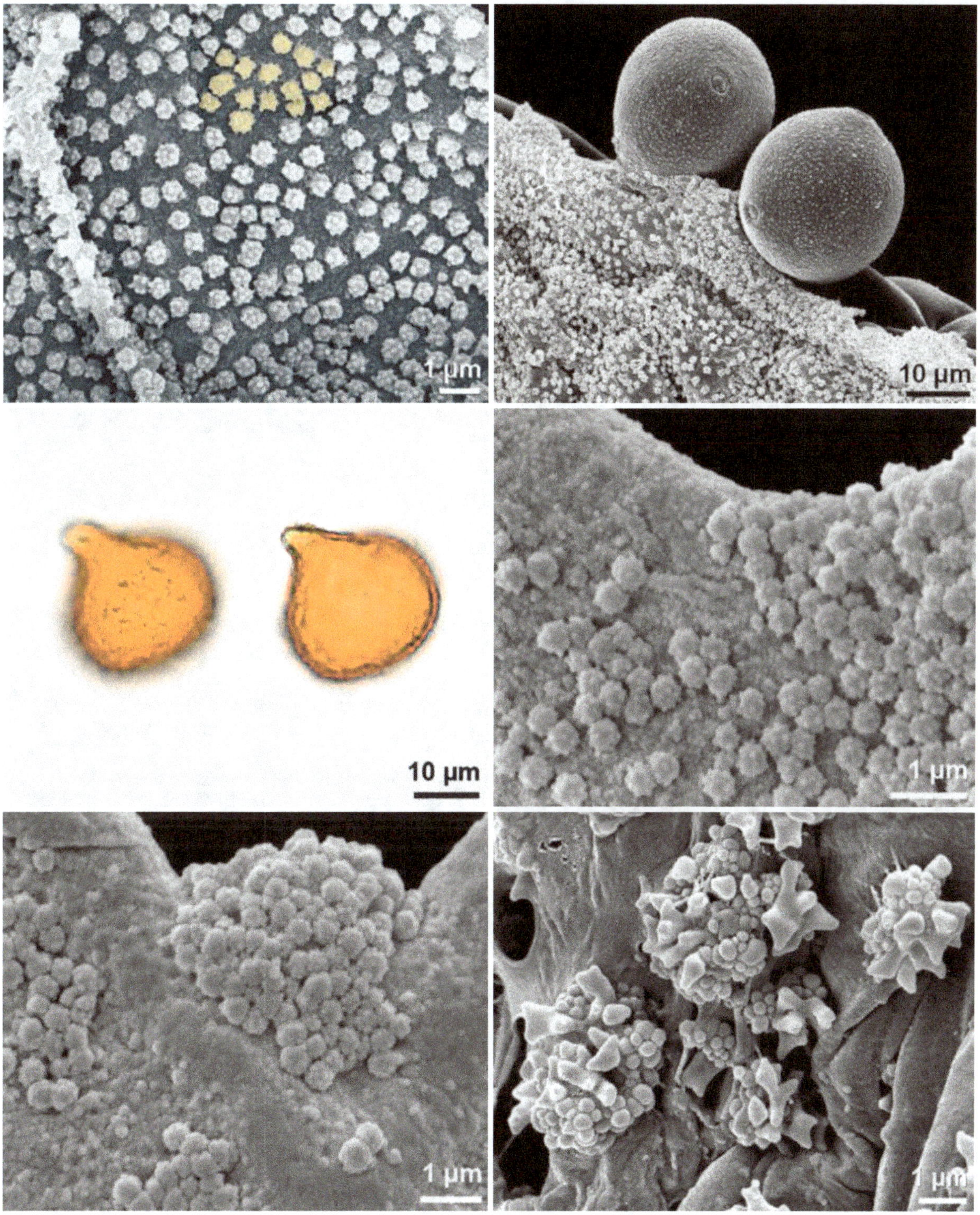

Trisetum flavescens, Poaceae
Ubisch bodies (colored) on locular wall

Corylus avellana, Betulaceae
two pollen grains attached to locular wall

Metasequoia glyptostroboides, Cupressaceae
Ubisch bodies attached to pollen surface (left)

Cupressaceae
fossil, Miocene, China

Cupressaceae
fossil, Miocene, China

Stellaria graminea, Caryophyllaceae

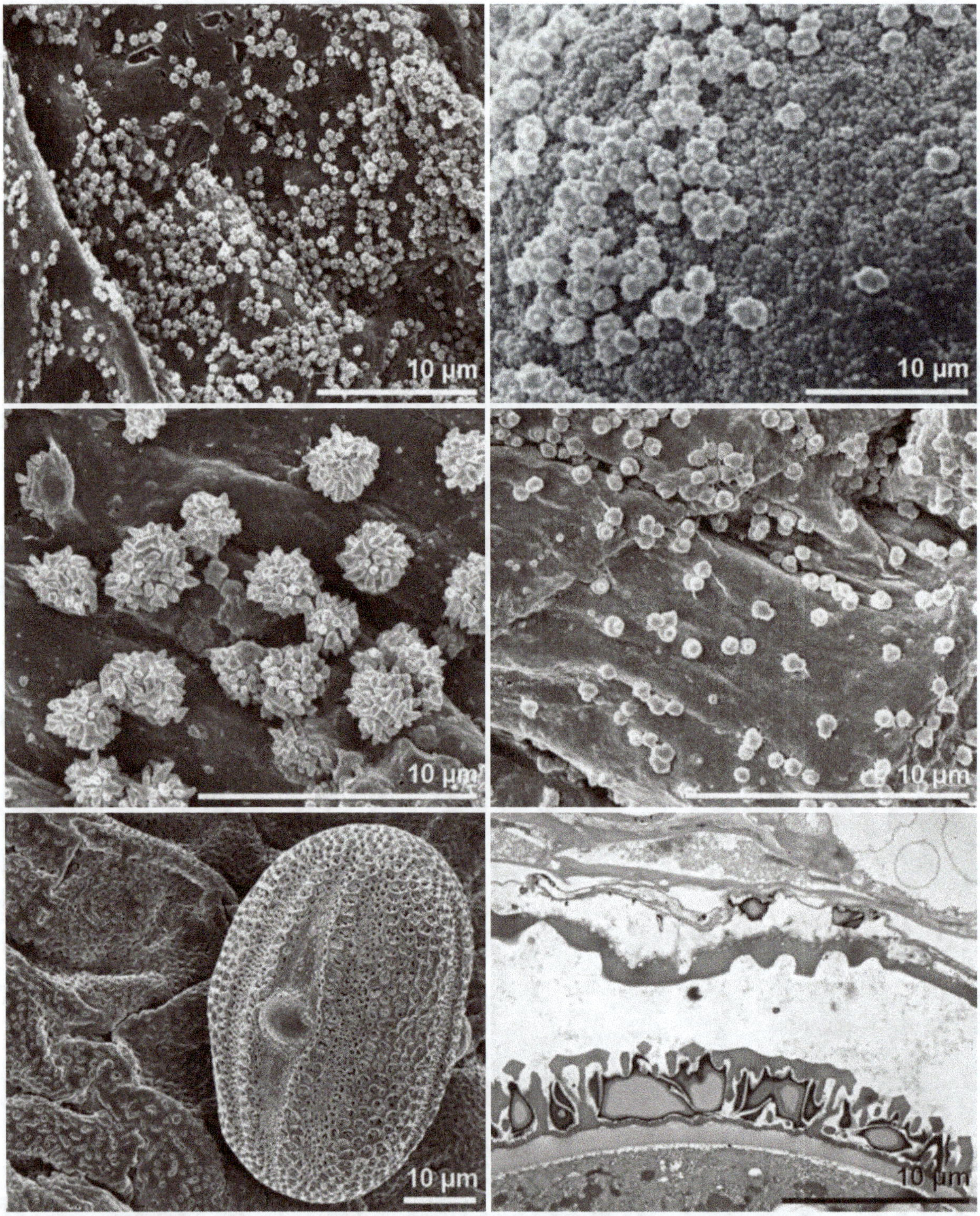

Chamaecyparis lawsoniana, Cupressaceae
Ubisch bodies on locular wall

Chamaecyparis lawsoniana, Cupressaceae
Ubisch bodies attached to pollen surface

Gladiolus illyricus, Iridaceae

Quercus robur, Fagaceae

Ruspolia seticalyx, Acanthaceae
Ubisch bodies forming a reticulum

Ruspolia seticalyx, Acanthaceae
U+Pb

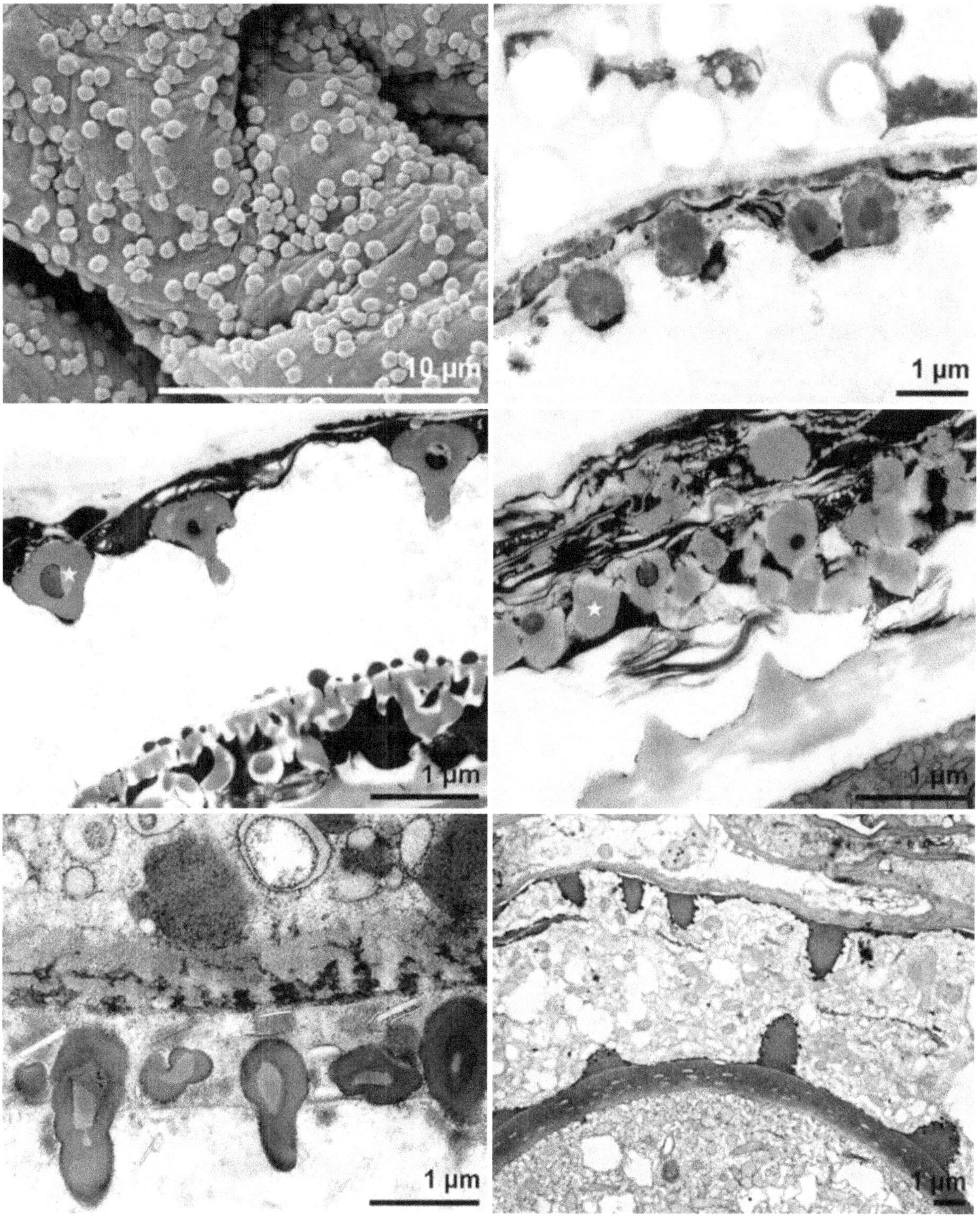

Ephedra foeminea, Ephedraceae

Odontites luteus, Orobanchaceae
U+Pb

Nigella arvensis, Ranunculaceae
modified Thiéry test

Ficaria verna, Ranunculaceae
modified Thiéry test

Tilia platyphyllos, Malvaceae
U+Pb

Sauromatum venosum, Araceae
U+Pb

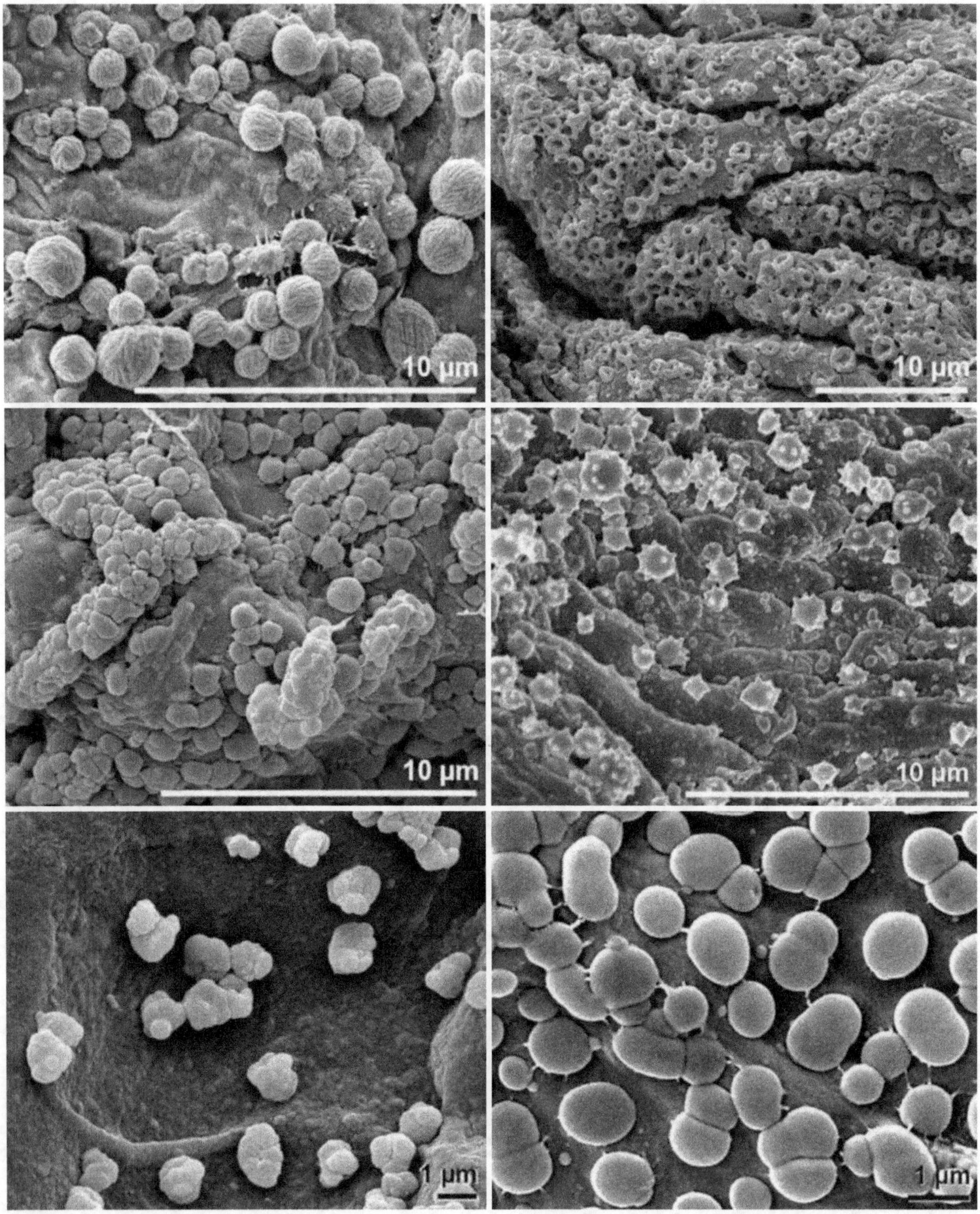

Akebia quinata, Lardizabalaceae

Acacia binervia, Mimosaceae

Calluna vulgaris, Ericaceae

Atriplex sagittata, Amaranthaceae

Justicia brandegeeana, Acanthaceae

Acacia binervia, Fabaceae

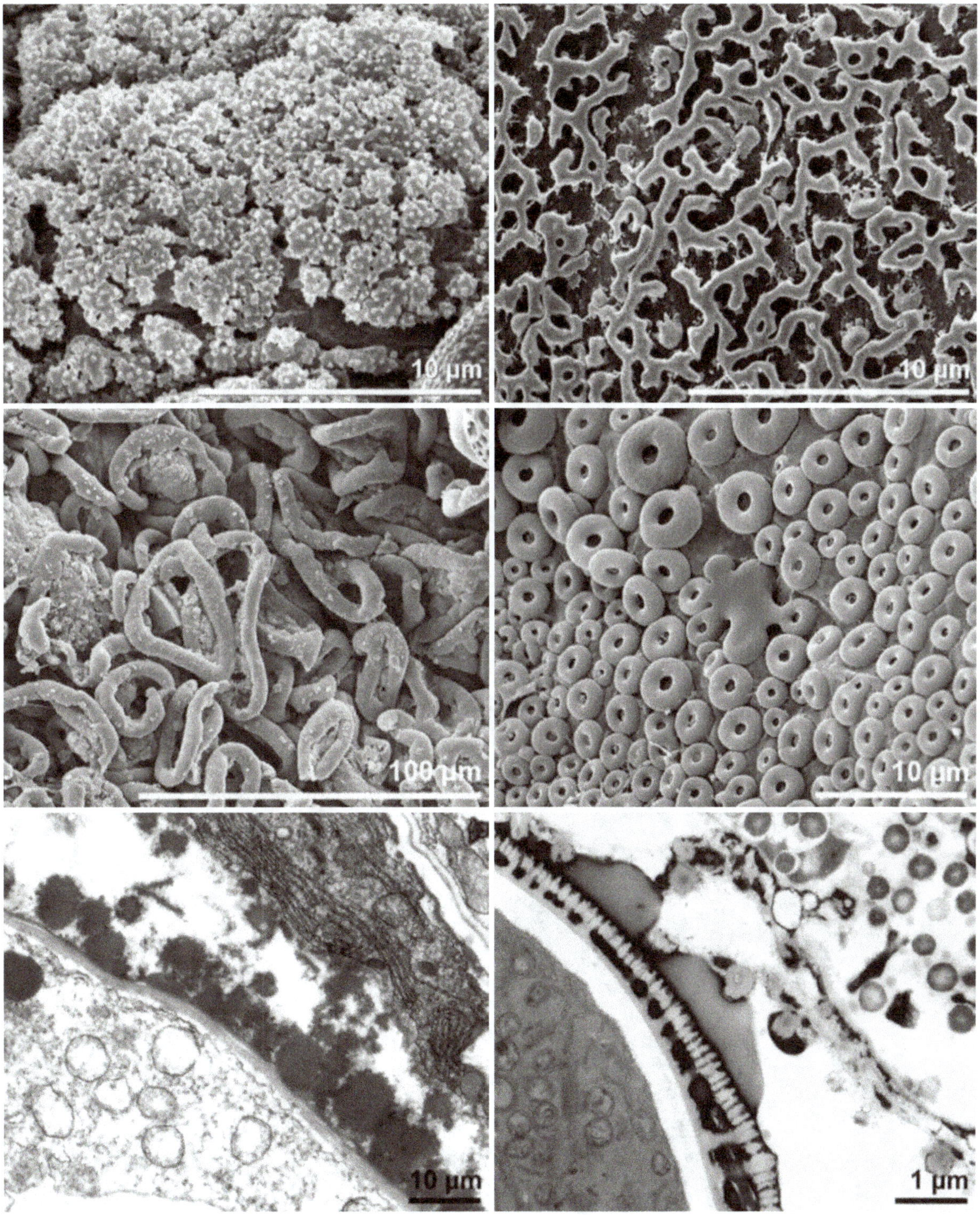

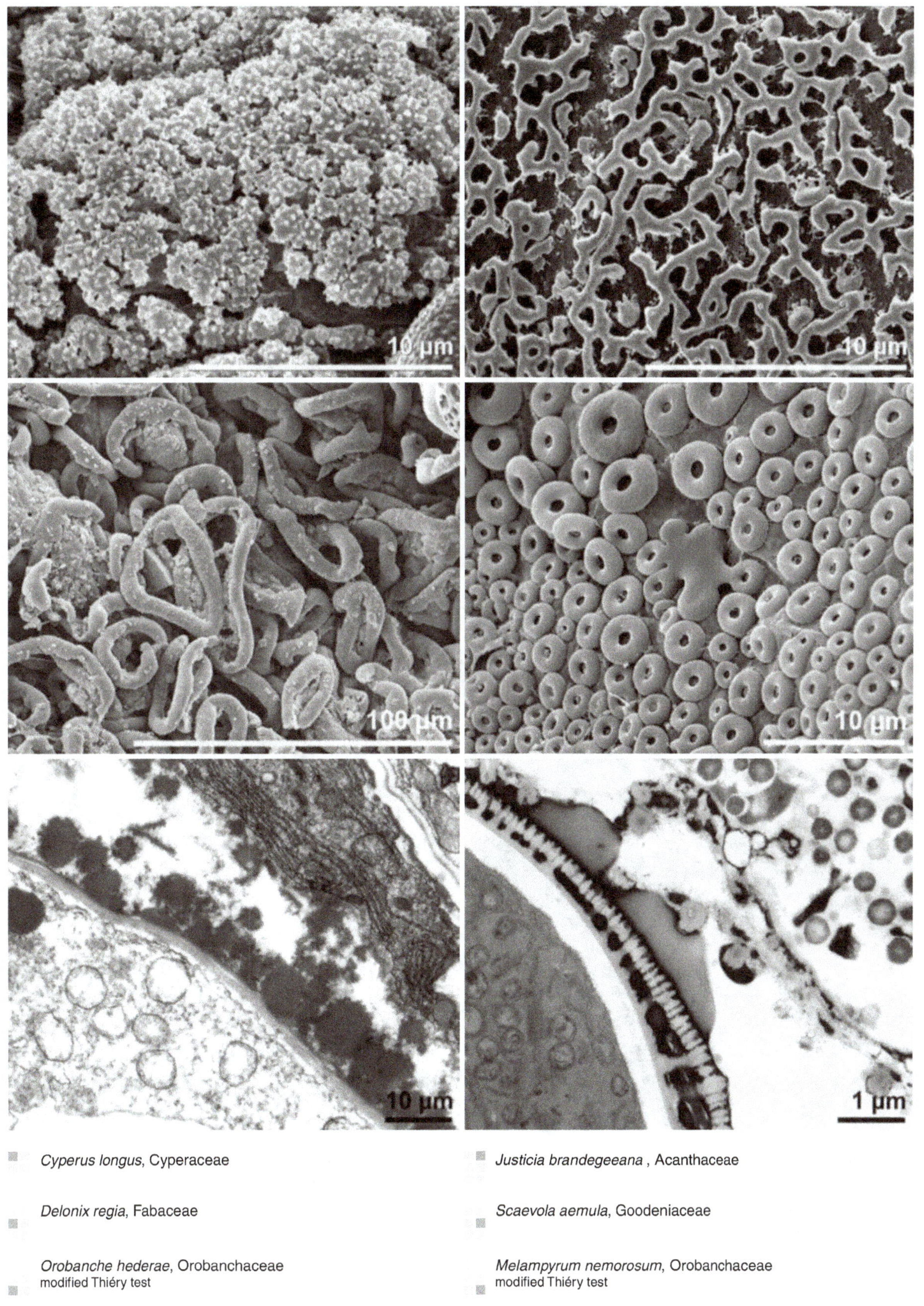

Cyperus longus, Cyperaceae

Justicia brandegeeana , Acanthaceae

Delonix regia, Fabaceae

Scaevola aemula, Goodeniaceae

Orobanche hederae, Orobanchaceae
modified Thiéry test

Melampyrum nemorosum, Orobanchaceae
modified Thiéry test

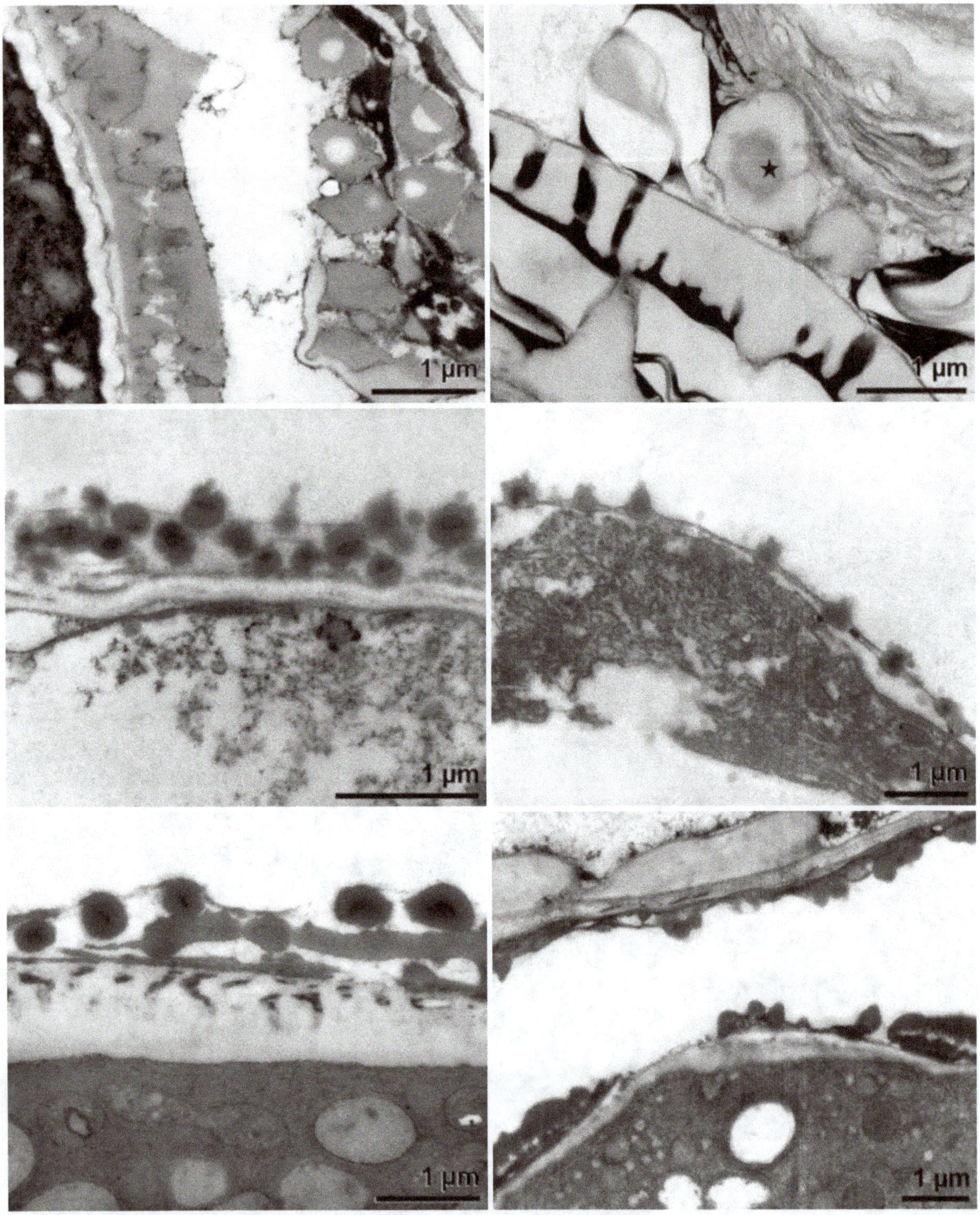

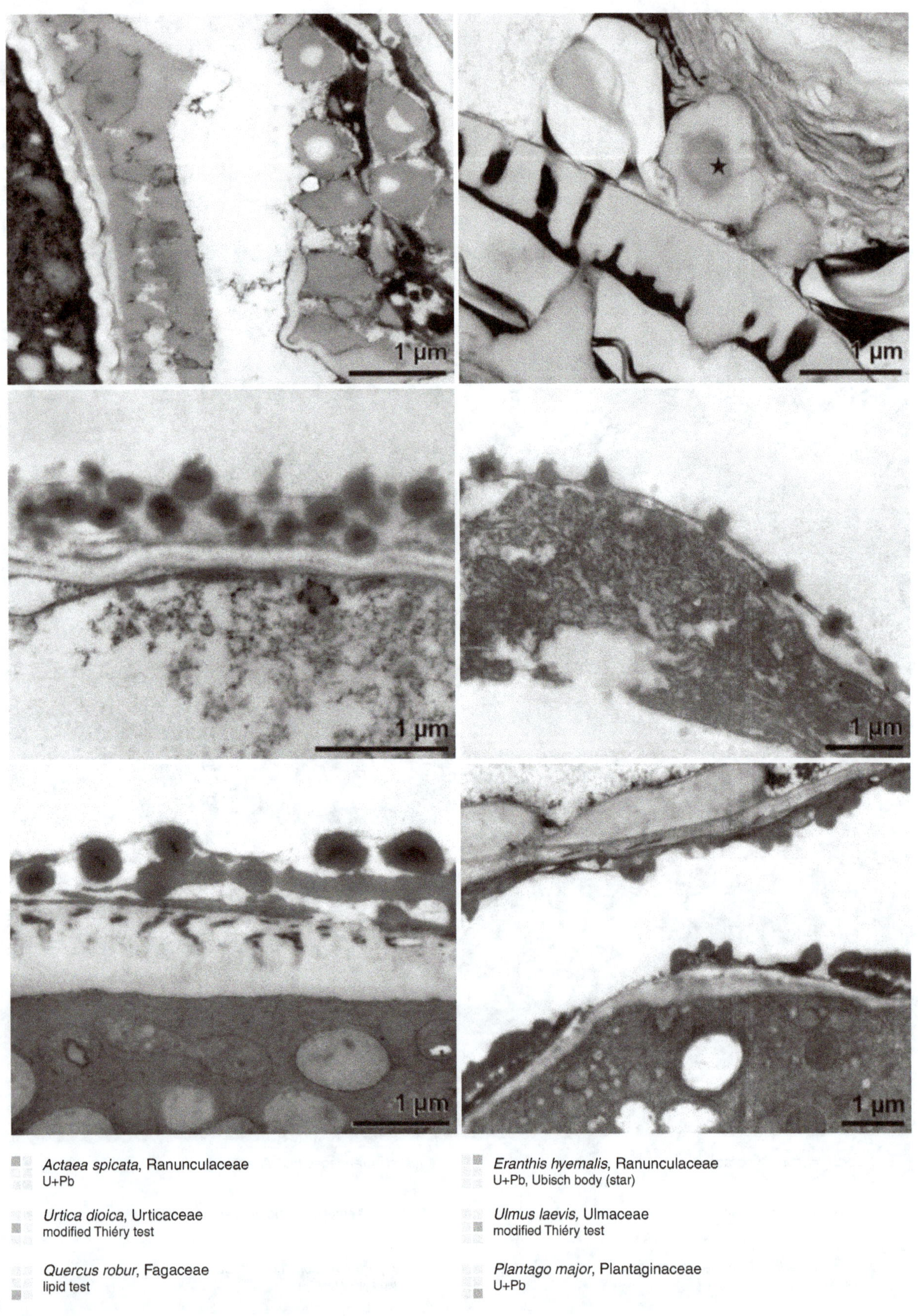

Actaea spicata, Ranunculaceae
U+Pb

Eranthis hyemalis, Ranunculaceae
U+Pb, Ubisch body (star)

Urtica dioica, Urticaceae
modified Thiéry test

Ulmus laevis, Ulmaceae
modified Thiéry test

Quercus robur, Fagaceae
lipid test

Plantago major, Plantaginaceae
U+Pb

Pollen Class

H. Halbritter et al., *Illustrated Pollen Terminology*, https://doi.org/10.1007/978-3-319-71365-6_12

pollen class

artificial grouping of pollen grains that share one or more distinctive characters

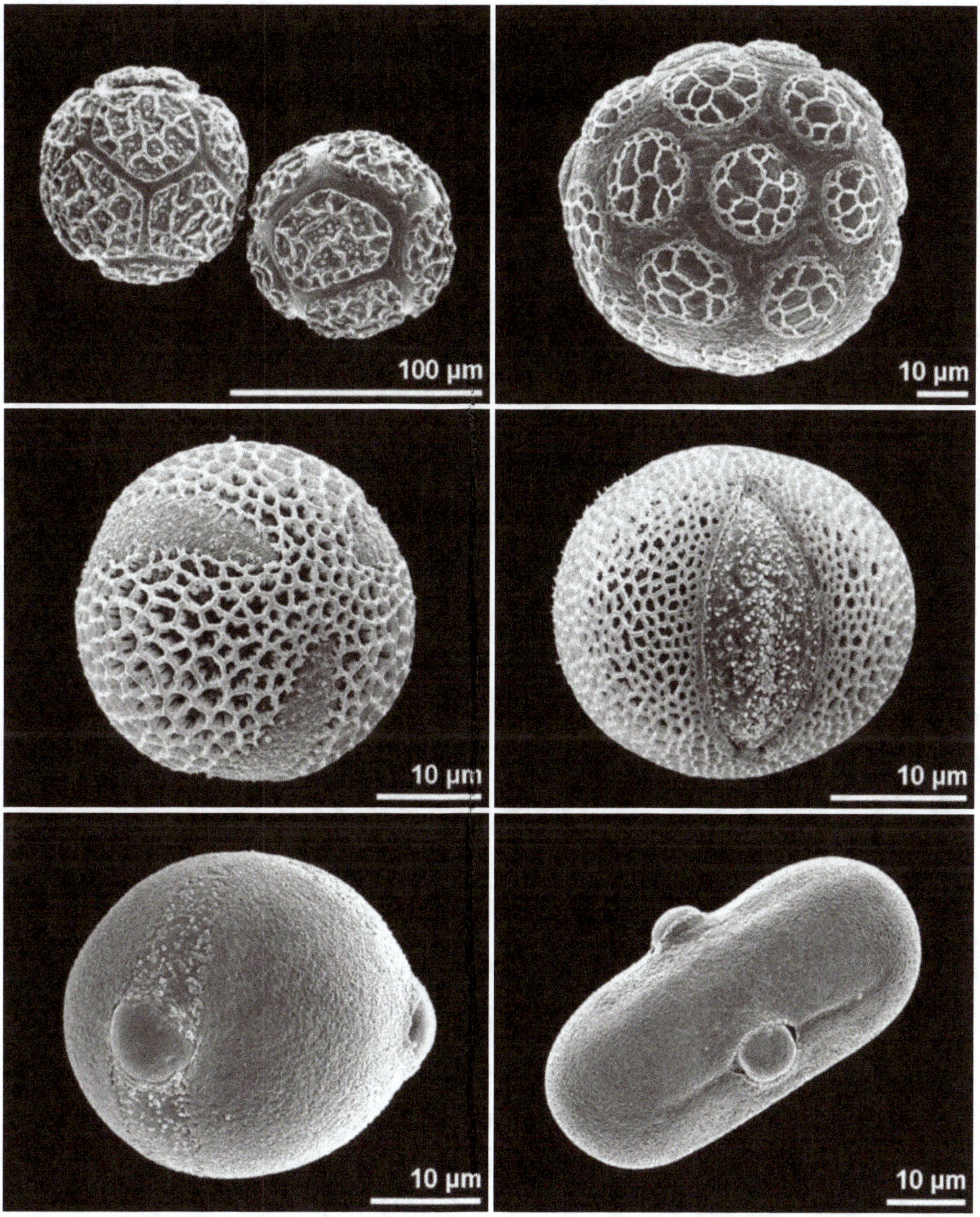

Iris bucharica, Iridaceae
clypeate

Ibicella lutea, Martyniaceae
clypeate

Bunias orientalis, Brassicaceae
colpate

Corylopsis glabrescens, Hamamelidaceae
colpate

Viola alba, Violaceae
colporate

Orlaya grandiflora, Apiaceae
colporate

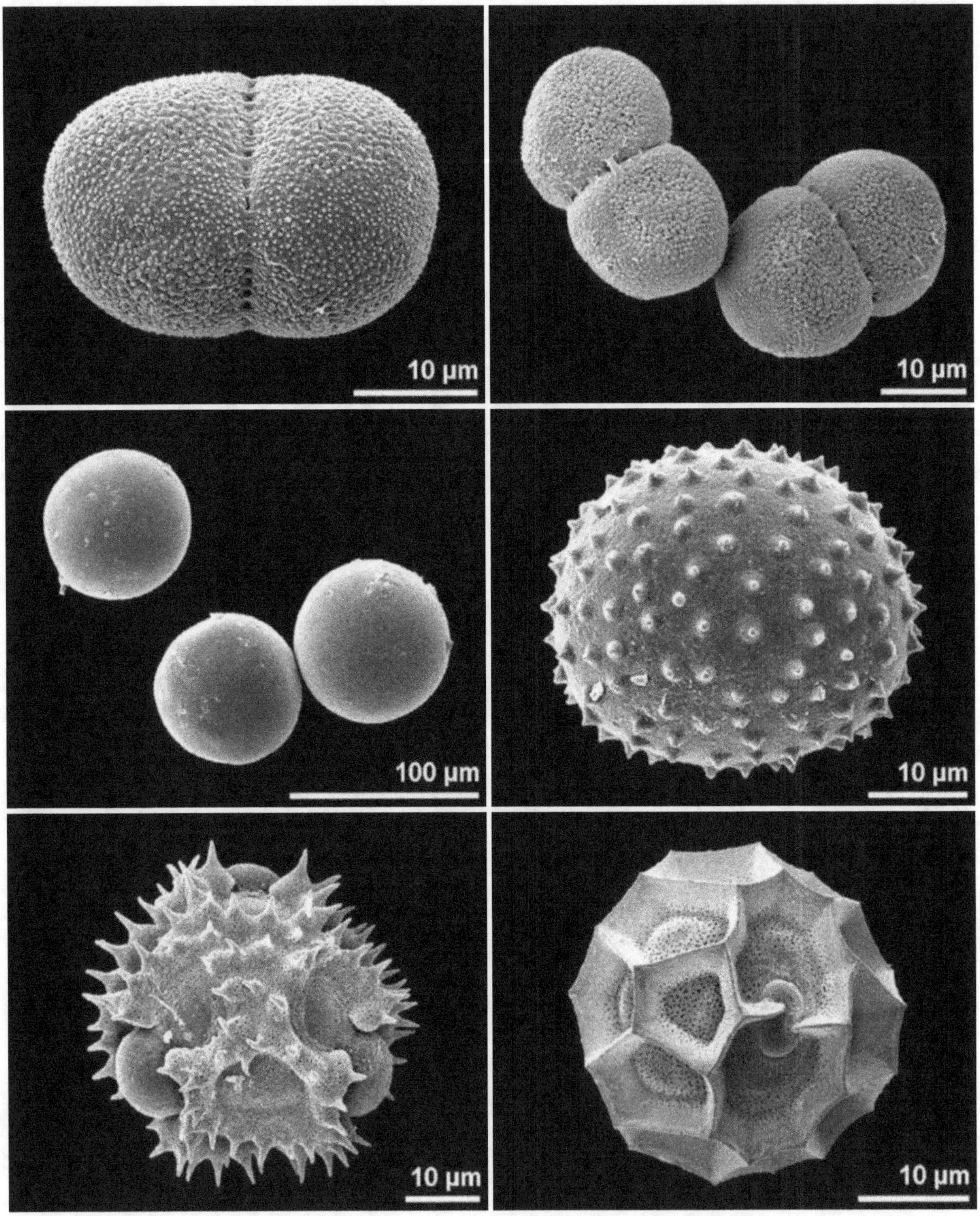

Zeylanidium subulatum, Podostemaceae
dyad

Podostemum munnarense, Podostemaceae
dyad

Hedychium gardnerianum, Zingiberaceae
inaperturate

Sauromatum venosum, Araceae
inaperturate

Prenanthes purpurea, Asteraceae
lophate

Gazania sp., Asteraceae
lophate

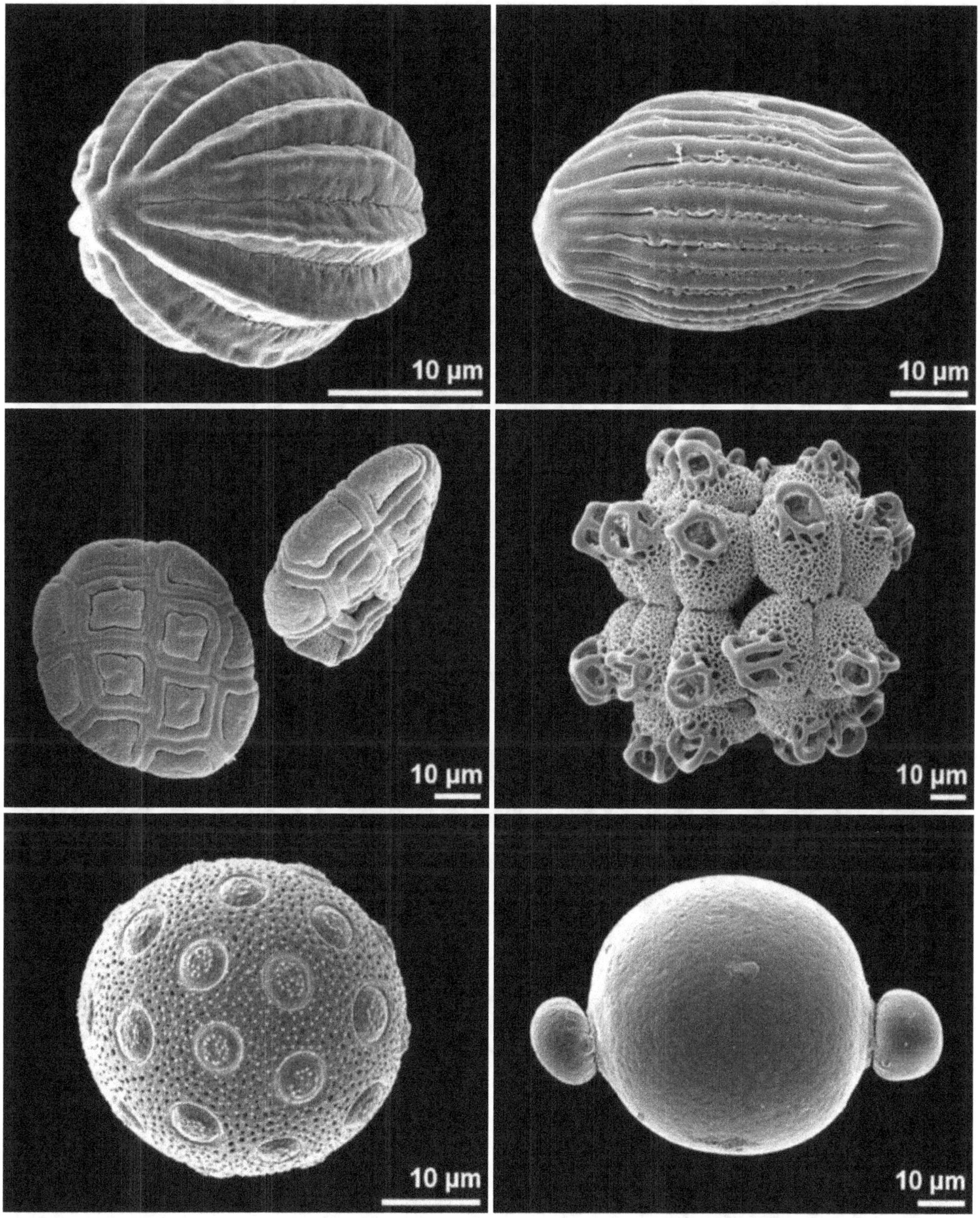

Ephedra distachya, Ephedraceae
plicate

Welwitschia mirabilis, Welwitschiaceae
plicate

Acacia dealbata, Mimosaceae
polyad

Chelonanthus purpurascens, Gentianaceae
polyad

Silene flos-cuculi, Caryophyllaceae
porate

Pachypodium saundersii, Apocynaceae
porate

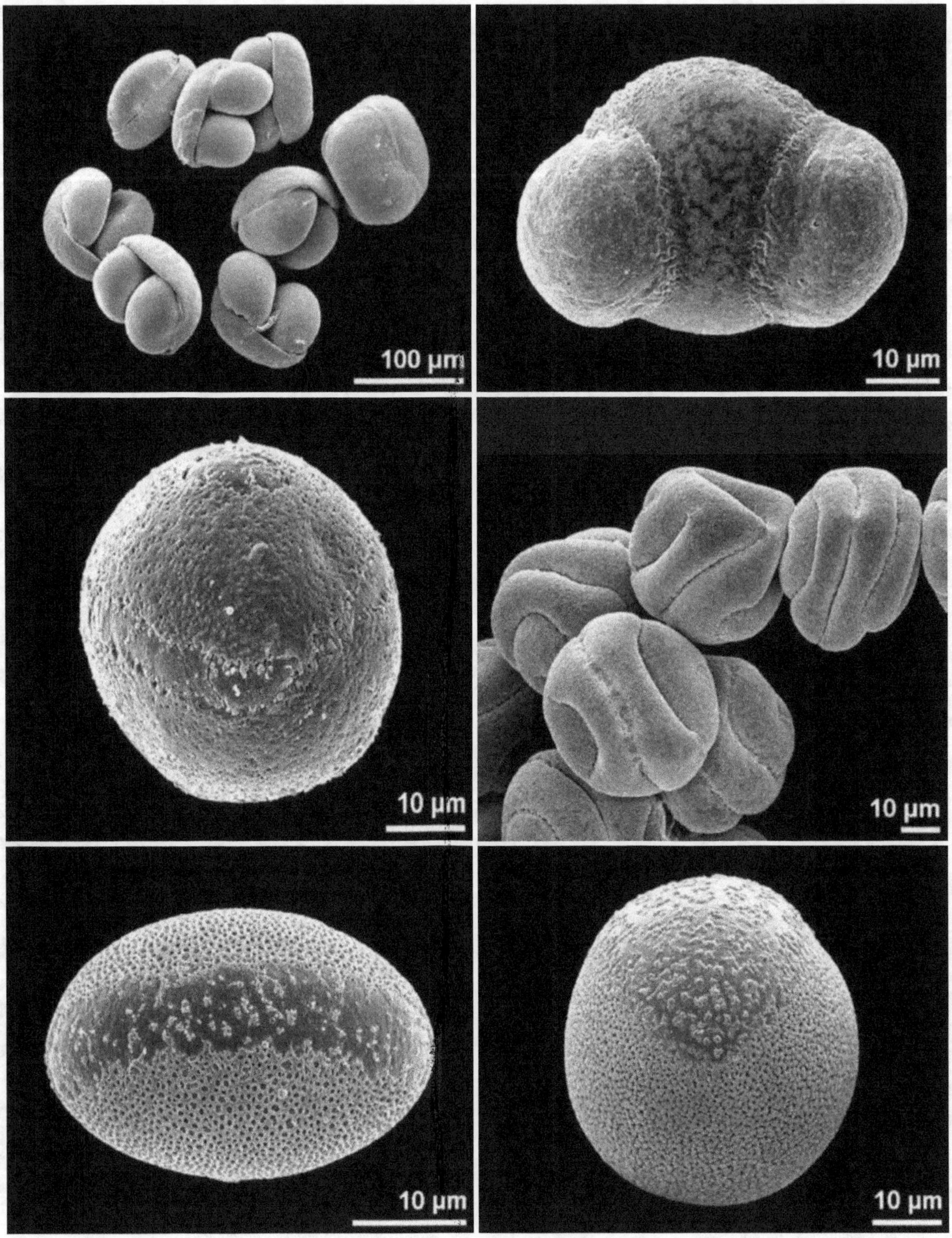

Abies cephalonica, Pinaceae
saccate

Pinus mugo, Pinaceae
saccate

Berberis thunbergii, Berberidaceae
spiraperturate

Thunbergia alata, Acanthaceae
spiraperturate

Sandersonia vaurantiaca, Colchicaceae
sulcate

Gagea villosa, Liliaceae
sulcate

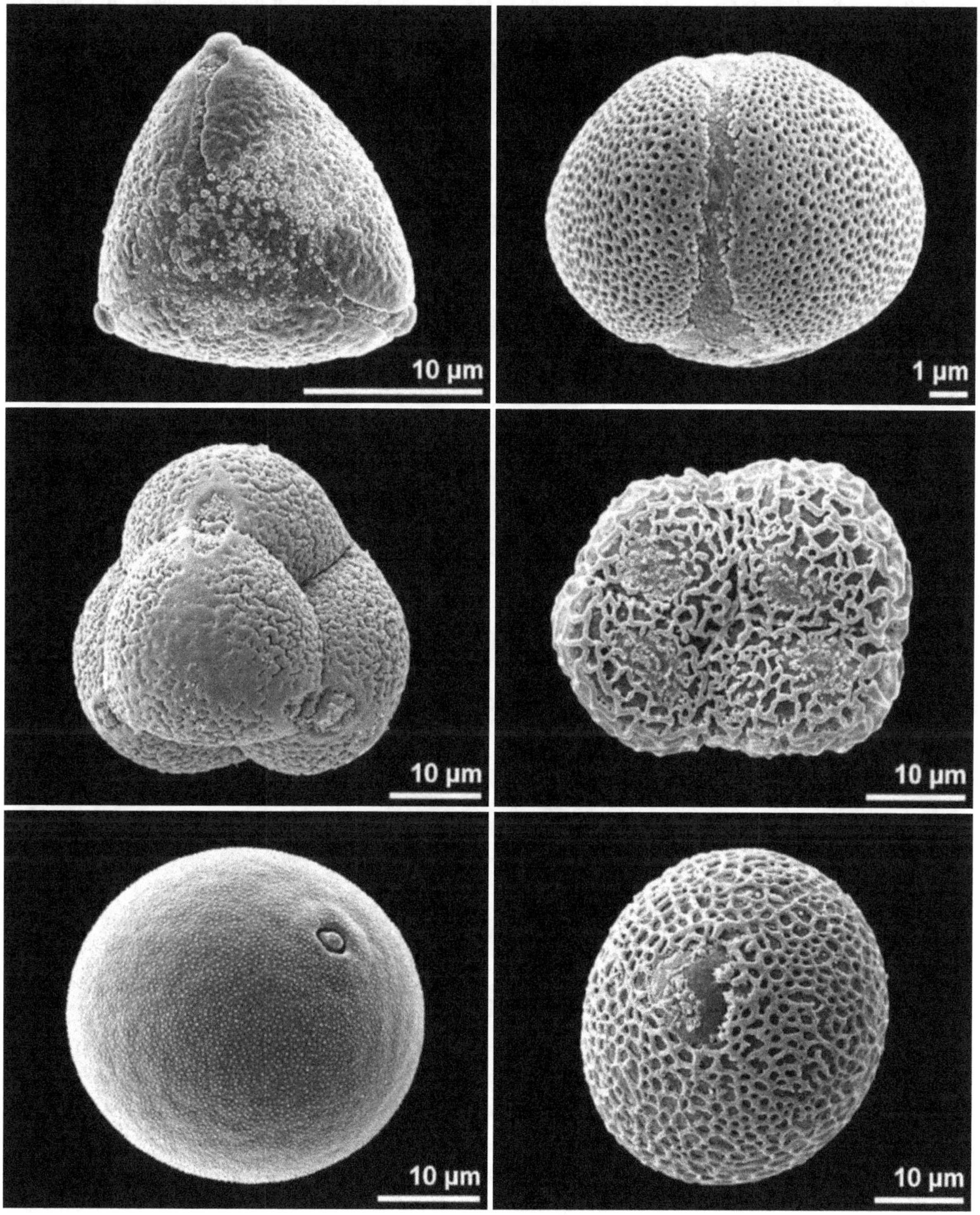

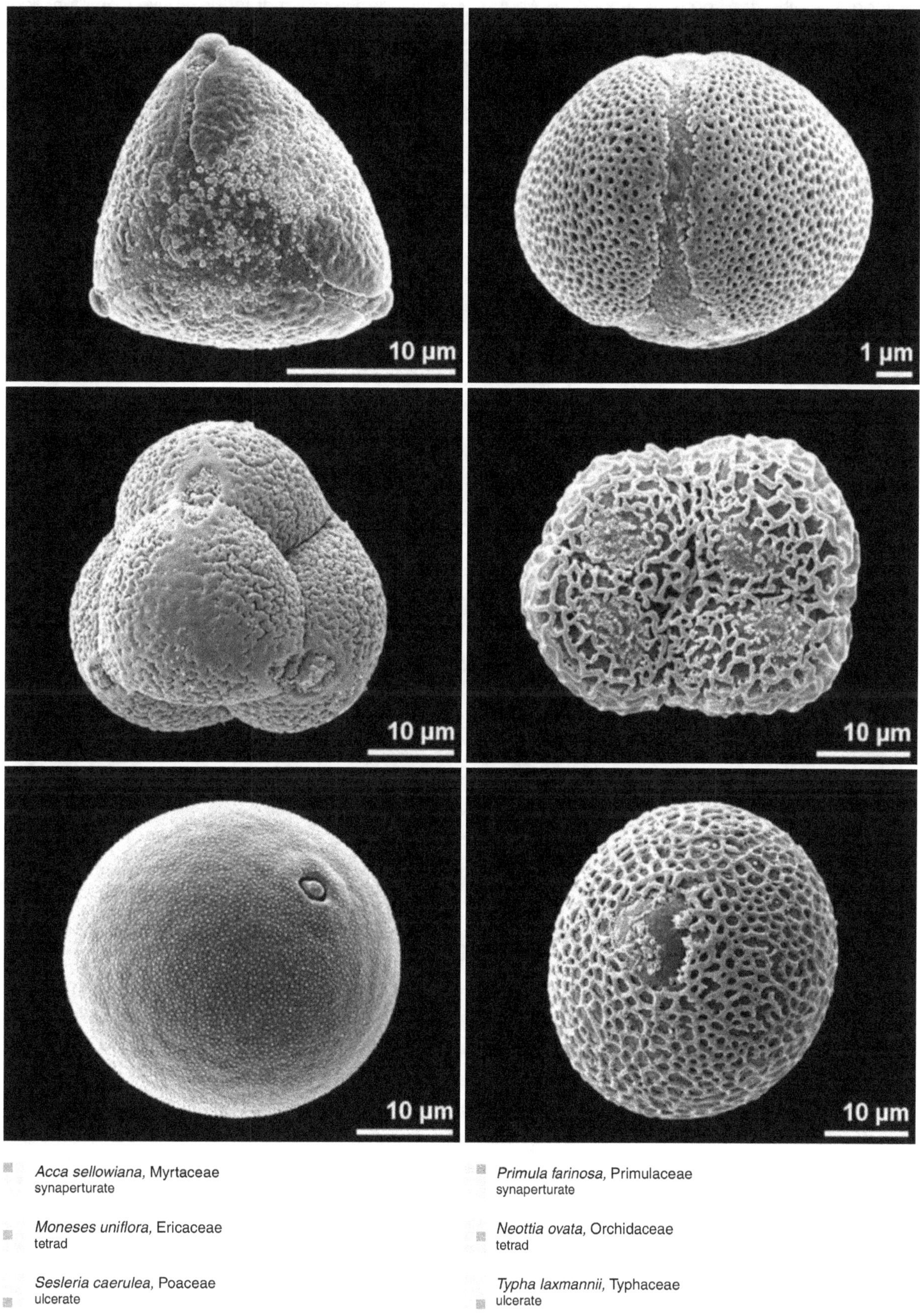

Acca sellowiana, Myrtaceae
synaperturate

Moneses uniflora, Ericaceae
tetrad

Sesleria caerulea, Poaceae
ulcerate

Primula farinosa, Primulaceae
synaperturate

Neottia ovata, Orchidaceae
tetrad

Typha laxmannii, Typhaceae
ulcerate

Palynological Terms

Contents

Glossary of Palynological Terms

All **important terms** in palynology are listed here and explained. Terms figured in the chapters of the "Illustrated Pollen Terms" part are indicated by bold page numbers. **Non-recommended terms** are only provided with an explanatory comment. For consistency, phrases are standardized, for example, features of ornamentation are stereotypically defined as "**pollen wall with** …", and pollen wall features (or pollen shape and size) as "**pollen grain with** …".

Both the singular and the plural are given for Latin terms. The English spelling of the Latin term is added (porus, pl. pori, engl. pore) if it is preferable. Cross-references are given to terms that are **synonyms** (the preferable one is printed in bold) or that indicate the opposite condition (**antonyms**), e.g., homo- and heterobrochate. If both a Greek and a corresponding Latin form exist for a prefix, then the Greek form is used consistently: panto- (not peri-), ekto- (not ecto-), or the Greek di- (dis-), and not the Latin bi- (bis-). There are few exceptions from this rule. If the Latin form is more widely accepted, then the term is used as *nomen conservandum*, for example, bisaccate is found exclusively in the literature and not the Greek form disaccate. Sometimes two adjectival variants (-ate, -ar) are used, but in two different meanings. For example, from the noun granulum derive two adjectival forms: granular and granulate (both meaning "with granules"). These are corresponding terms used in two quite different contexts: **granular** describes a distinct type of infratectum hence a structural feature whereas **granulate** refers to an ornamentation feature—a sculpture element.

Terms not listed in the glossary belong to spores, or are considered as redundant (e.g., multiplanar tetrad), superfluous (e.g., polyplicate, because plicate pollen grains are always equipped with several to many plicae), or may be a permanent source of confusion (zon-, zona-, zoni-, zono-).

H. Halbritter et al., *Illustrated Pollen Terminology*, https://doi.org/10.1007/978-3-319-71365-6_13

a
prefix meaning absent

acalymmate 406
dyads, tetrads, and polyads covered by an exine envelope which is discontinuous at the junctions between the monads
Antonym: calymmate

acetolysis 77, 100, 103, 123
widely used technique for preparing pollen and spores especially for light microscopy

actuopalynology
the study of pollen and spores of extant plants

aeropalynology
the study of palynomorphs found in the atmosphere

alveolate 50, 385
infratectum with compartments of irregular size and shape

alveolus (lat., pl. **alveoli**) 50, 385
compartment(s) of irregular size and shape

angulaperturate 208
pollen grain with an angular outline where the apertures are located at the angles
Antonym: planaperturate

annulate 209
pollen grain with an annulus or annuli

annulus (lat. pl. **annuli**) 209
ring like wall thickening surrounding a porus or ulcus
Comment: "anulus" orthographical variant of "annulus"

aperturate 211
pollen grain with one or more apertures
Antonym: inaperturate

aperture 42, 73
a region of the pollen wall that differs significantly from its surroundings in morphology and/or anatomy, presumed to function usually as germination site and to play a role in harmomegathy

apertures sunken 57, 194
characteristic shape of pollen grains in dry condition as a consequence of harmomegathy

aperture membrane 42, 214
exine layer covering an aperture; aperture membrane (colpus or porus membrane) can be psilate or ornamented

apex (lat., pl. **apices**)
applied to outermost tips in the equatorial plane of pollen

apocolpium (lat., pl. **apocolpia**)
see: polar area
Comment: "polar area" is the more general term independent of the aperture type

apoporium (lat., pl. **apoporia**)
see: polar area
Comment: "polar area" is the more general term independent of the aperture type

arcuate 407
pollen grain with arcus

arcus (lat., pl. **arcus**) 407
curved wall thickening interconnecting apertures

areola (lat., pl. **areolae**) 78, 296
insular ornamentation element

areolate 78, 296
pollen wall with areolae

atectate 47, 384
pollen grain lacking a tectum
Antonym: "tectate"

atrium (lat., pl. **atria**) 220
space between diverging exine layers within the aperture

baculate 298
pollen wall with bacula longer and/or wider than 1 µm

baculum (lat., pl. **bacula**) 298
rod-like, free standing element (never pointed)

bi-
prefix for two

biporate
see: diporate
Comment: "diporate" is the more common term

bireticulate 299
reticulate ornamentation, where the lumina of the coarse-meshed reticulum are filled by a fine-meshed reticulum

bisaccate 50, 190
pollen grain with two sacci
Comment: nomen conservandum

bisulcate
see: disulcate
Comment: "disulcate" is the more common term

boat-shaped 196
characteristic shape of sulcate pollen grains in dry condition as a consequence of harmomegathy

brevi-
prefix meaning short

brevicolpate 221
pollen grain with brevicolpi

brevicolporate 222
pollen grain with brevicolpori

brevicolporus (lat., pl. **brevicolpori**) 222
short colpus in a compound aperture situated equatorially

brevicolpus (lat., pl. **brevicolpi**) 221
short colpus situated equatorially

bridge 223
exine connection(s) between the margins of an aperture
Comment: the term is often used in a more general context, e.g., for exine connections within tetrads

brochus (lat., pl. **brochi**) 348
mesh of a reticulum consisting of one lumen and the adjoining half of the muri

calymmate 57, 405
dyads, tetrads, and polyads covered by a continuous exine envelope
Antonym: acalymmate

cappa (lat., pl. **cappae**) 50, 385
the thick-walled proximal face of the corpus in a saccate pollen grain

cappula (lat., pl. **cappulae**)
see: leptoma
Comment: may be confused with "cappa" which points to the proximal side, while "cappula" refers to distal

caput (lat., pl. **capita**) **302**
distal part of a clava

cavea (lat., pl. **caveae**) 47
cavity between the sexine and nexine in the interapertural area

caveate 47
pollen wall with caveae

circular **156**
see: outline
Comment: a general term, used in palynology describing, e.g., "outline"

clava (lat., pl. **clavae**) 51, **302**
club-shaped element

clavate 51, **302**
pollen wall with clavae longer and/or wider than 1 µm

clypeate **305, 430**
pollen with exine subdivided into shields

colpate 42, 72, **225**
pollen grain with colpi

colporate 42, **233**
pollen grain with colpori

colporoidate 42
used for light microscopy only, describing compound apertures composed of a colpus (ektoaperture) with an indistinct endoaperture

colporus (lat., pl. **colpori**) 42, **233**
compound aperture composed of a colpus (ektoaperture) combined with an endoaperture of variable size and shape

colpus (lat., pl. **colpi**) 42, 72, **225**
elongated aperture (length/width ratio > 2) situated at the equator or globally distributed

columella (lat., pl. **columellae**) 45, 50, **319, 386**
rod-like structure element, supporting a tectum

columellate 13, 50, **386**
infratectum with columellae

compound aperture 42, **222, 233, 240**
aperture with two or more components that are situated in more than one wall layer, e.g., colporus

copropalynology 9
the study of palynomorphs in coprolites or feces

corpus (lat., pl. **corpora**) 50, **188**
body of a saccate pollen grain

costa (lat., pl. **costae**) 50, **347**
thickening of the nexine/endexine bordering an endoaperture

costate 50, **347**
pollen grain with costae

croton pattern 10, **307**
special type of reticulum cristatum formed by regularly arranged suprasculpture elements on muri

cryopalynology
the study of palynomorphs found in ice

cup-shaped 42, **199**
characteristic shape of pollen grains in dry condition as a consequence of harmomegathy

di-
prefix meaning two

diaperturate 42
pollen grain with two apertures: disulcate, dicolpate, dicolporate, diporate

dicolpate 74, **225**
pollen grain with two colpi

dicolporate **233**
pollen grain with two colpori

diploxylon-type 50
see: *Pinus* subgenus *Pinus* type

diporate **263**
pollen grain with two pori

dispersal unit 38, 76, **131**
unit in which pollen is dispersed (e.g., monad, tetrad, pollinarium)

distal 38
directing away from the center of a tetrad (deduced from tetrad stage)

disulcate 74, **287**
pollen grain with two sulci

dyad 38, **134, 431**
unit of two pollen grains

echinate **309**
pollen wall with echini longer and/or wider than 1 µm

echinus (lat., pl. **echini**) **309**
pointed ornamentation element
Comment: the plural "echinae" is linguistically incorrect

ektexine 45, 50, **380, 393**
outer layer of an exine

ektintine **400**
the outer layer of the intine which is adjacent to the exine

ekto-
prefix meaning outer

ektoaperture 42, **240**
outer part of a compound aperture

elastoviscin **416**
highly elastic, not acetolysis resistant substance in Orchidaceae, which interconnects the subunits (monads, tetrads, or massulae) of a pollinium and builds up the caudicles

elliptic **158**
see: outline
Comment: a general term, used in palynology describing, e.g., "outline"

endexine 13, 45, 50, 80, 117, **394**
distinct exine layer between ektexine and intine; endexine can be compact, spongy or lamellar as well as continuous, discontinuous, absent, or in aperture only
Comment: the endexine can be monolayered or bilayered; characteristic for the endexine is the increasing thickness close to the aperture

endintine 400
inner layer of the intine which is adjacent to the cytoplasm

endo-
prefix meaning inner

endoaperture 42, 98, **240**
inner part of a compound aperture

endoplica
fold of the inner exine layer

equator 38
imaginary line encircling a pollen grain between the proximal and distal poles

equatorial 38
preposition indicating a region on the pollen surface

equatorial diameter 39, **168**
diameter of a pollen grain at the equator

equatorial plane 38
imaginary plane at the equator, perpendicular to the polar axis

equatorial view 39
view of a pollen grain where the equator is directed towards the observer

eu-
prefix meaning true

eurypalynous 13
plant taxa characterized by a significant variation in pollen morphology
Antonym: stenopalynous

eutectate 45, **381**
pollen grain with a predominantly continuous tectum
Antonym: semitectate

exine 6, 45
outer layer of the pollen wall, usually resistant to acetolysis

fenestrate
see: lophate
Comment: as there is no corresponding substantive to "fenestrate", we prefer the terms "lophate" and "lophae"

Fischer's law/rule 42
refers to the most frequent aperture arrangement where a pair of apertures occur at six points in a tetrad

foot layer 45, 50, **393**
inner layer of an ektexine that can be continuous, discontinuous, perforated or absent

forensic palynology 9, 16
the study of palynomorphs found in crime related samples

fossula (pl. **fossulae**) 78, **315**
irregular shaped groove

fossulate 78, **315**
pollen wall with fossulae

foveola (pl. **foveolae**) **317**
roundish lumen more than 1 µm in diameter; distance between two adjacent lumina larger than their diameter

foveolate **317**
pollen wall with foveolae

free-standing columellae 45, **319**
columellae not covered by a tectum in semitectate pollen grains

frustrate
special mental condition of palynologists discussing terminology of pollen; see *PalDat* (www.paldat.org)

Garside's law/rule 42
refers to the unusual arrangement of apertures where a group of three apertures occur at four points in a tetrad

gemma (lat., pl. **gemmae**) **321**
globular ornamentation element

gemmate **321**
pollen wall with gemmae larger than 1 µm in diameter

generative cell 24
progenitor cell of the sperm cells

geniculum (pl. **genicula**) 42
colpus buckled in the equatorial region

granular 50, **389, 439**
infratectum composed of granula, cluster of granula or elements of different size and shape (never solid and rod-like)
Comment: not to be confused with "granulate", which is a type of ornamentation

granulate 54, **323, 439**
pollen wall with granula
Comment: not to be confused with "granular", which is a feature of the pollen wall structure

granulum (lat., pl. **granula**) **323, 439**
sculpture element of different/indefinable shape, equal or smaller than 0,1 µm in diameter (hard to outline)
Comment: only applicable if sculpture element cannot be defined more precisely by improved microscopic resolution

haploxylon-type 50
see: *Pinus* subgenus *Strobus* type

harmomegathy 43, 57, **194**
mechanism permitting changes in shape and size of the pollen grain due to varying hydration status

hetero-
prefix meaning different

heteroaperturate 42, **242, 273**
pollen grain with different types of apertures; only one type presumed to function as germination site; see: pseudocolpus

heterobrochate **324**
reticulate pollen wall with lumina of different sizes
Antonym: homobrochate

heterocolpate
see: heteroaperturate
Comment: unfortunately the term "heterocolpate" is commonly incorrectly used for pollen grains with alternating colpi and colpori, but "heterocolpate" means two different types of colpi; therefore we prefer the more general term "heteroaperturate"

heteropolar 39, **178**
pollen grain with different proximal and distal faces
Antonym: isopolar

hexa- 50
prefix meaning six

hexaaperturate 42, **165, 230, 238**
pollen grain with 6 apertures: hexacolpate, hexacolporate, hexaporate

homo-
prefix meaning equal

homobrochate **326**
reticulate pollen wall with lumina of uniform size
Antonym: heterobrochate

iatropalynology 9
the study of palynomorphs causing allergies

impression mark 68
a linear or Y-shaped mark on the proximal polar area of a pollen grain retained from the tetrad stage

in-
prefix meaning absent

inaperturate 42, **244**
pollen grain without distinct apertures
Antonym: aperturate

infoldings 57, 73, **194**
consequence of harmomegathy in dry condition, e.g., boat-shaped, cup-shaped, irregularly infolded

infra-
prefix meaning beneath

infratectum 45, 50, **385**
layer between tectum and foot layer or endexine (if foot layer is absent); infratectum can be alveolate, columellate, granular or absent

intectate
see: atectate
Comment: "atectate" is the more common term

inter-
prefix for in between

interapertural area 47, 57, **160, 201**
region between apertures

interapertural area sunken **201**
characteristic shape of pollen grains in dry condition as a consequence of harmomegathy

intercolpium
see: interapertural area
Comment: "interapertural area" is the more general term independent from the aperture type

internal tectum **391**
additional more or less continuous layer within the infratectum

interporium
see: interapertural area
Comment: "interapertural area" is the more general term independently from the aperture type

interstitium
see: infratectum
Comment: "infratectum" is the more common term

intine 6, 45, 117, **400**
part of the pollen wall next to the cytoplasm, mainly consisting of polysaccharides; the intine can be monolayered or bilayered; see also "ektintine" and "endintine"
Comment: characteristic for a monolayered intine is the increasing thickness in the aperture region and that it becomes bilayered (ektintine and endintine)

intra-
prefix for within

irregular **203**
Comment: a general term, used in palynology describing, e.g., "outline", "shape"

iso-
prefix meaning identical

isodiametric 39, **171**
pollen grain with a polar axis equal to the equatorial diameter
Comment: a general term, used in palynology describing, e.g., "shape"

isopolar 39, **176**
pollen grain with identical proximal and distal faces
Antonym: heteropolar

lacuna (lat., pl. **lacunae**) **328**
depressed area surrounded by ridges (lophae) in lophate pollen grains

laesura (lat., pl. **laesurae**) 43
a single arm of a tetrad mark

lalongate **240**
endoaperture elongated equatorially

leptoma (gr., pl. **leptomata**) 50, **247**
thinning of the pollen wall on the distal face in conifers, presumed to function as germination area

LO-analysis 7
light microscopic method for analyzing pollen surfaces

lobate **160**
outline in polar view of a pollen grain with bulged interapertural areas (mainly in dry pollen grains)

lolongate **240**
endoaperture elongated meridionally

LO-pattern
ornamentation identified by LO-analysis

lophae (lat., sing. **lopha**) **328**
massive exine ridges

lophate **328**
pollen wall with coarse meshed pattern formed by lophae and lacunae

lumen (lat., pl. **lumina**) **299, 317, 324, 326, 348**
space enclosed by muri in reticulate pollen grains

margo (lat., pl. **margines**) **248**
exine area with different ornamentation bordering a colpus/colporus/sulcus

massula (lat., pl. **massulae**) 76, **147**
unit of more than 4 pollen grains but less than the locular content of a theca
Comment: in angiosperms only used for Orchidaceae with sectile "pollinia"

megaspore
the larger spore in vascular plants

melissopalynology 9, 16
the study of palynomorphs found in honey

melittopalynology
see: melissopalynology
Comment: the term "melittopalynology" is the Greek variant of the Latin "melissopalynology"

meridian
imaginary line on the pollen surface connecting proximal and distal poles

meridional
preposition indicating a direction on the pollen surface

meso-
prefix meaning middle

mesocolpium (lat., pl. **mesocolpia**) 94
see: interapertural area
Comment: "interapertural area" is the more general term independent of the aperture type

micro- **331**
prefix for small; features between 1-0,5 μm: microbaculate, microclavate, microechinate, microgemmate, microrugulate, microreticulate, microverrucate; not used in combination with striate and foveolate

microspore 24
the smaller spore of heterosporous vascular plants

microspore mother cell 24
see: pollen mother cell

monad 38, **132**
unit consisting of a single pollen grain

mono-
prefix meaning one

monocolpate
see: sulcate
Comment: superfluous term; as far as known, there is no example of a pollen grain with a single colpus (situated equatorially); in all pollen grains with a single elongated aperture the latter is situated distally (sulcus)

monoporate
see: ulcerate
Comment: superfluous term; as far as known, there is no example of a pollen grain with a single porus (situated equatorially); in all pollen grains with a single porus the latter is situated distally (ulcus)

monosaccate **189**
pollen grain with a single saccus

monosulcate
see: sulcate
Comment: superfluous term, because "sulcate" implies a single elongated aperture (sulcus)

muri (lat. sing. **murus**) **348**
ornamentation elements forming the meshes in a reticulum

nano- **336**
prefix for very small, features between 0.5 - 0.1 μm: nanobaculate, nanoclavate, nanoechinate, nanogemmate, nanorugulate, nanoreticulate, nanoverrucate; not used in combination with striate and foveolate

nexine 47, **404**
term used for light microscopy, describing the inner, unstructured layer of the exine

nodulum (lat., pl. **nodula**) 50
small body located on the nexine of the central saccus area

Normapolles 11
group of Cretaceous and lower Paleogene pollen, usually triaperturate, with a complex pore apparatus

oblate 39, **168**
pollen grain with a polar axis shorter than the equatorial diameter
Antonym: prolate

oblique view
view of a pollen grain neither in polar nor in equatorial view

omniaperturate
see: inaperturate
Comment: the term refers to the functional aspect only, therefore we prefer "inaperturate"

oncus (lat., pl. **onci**) **220**
lens-shaped body located beneath the aperture, not resistant to acetolysis

operculate **251**
aperture with an operculum

operculum (lat., pl. **opercula**) 42, **251**
distinctly delimited exine structure covering an aperture

orbicule
see: Ubisch body
Comment: "orbicule" implies a globular element; therefore, we recommend the term "Ubisch body", as these are polymorphic

ornamentation 51, 295
applied in palynology to surface features

outline **156, 158, 160, 163, 166, 167**
describes the contour of pollen grains in polar and/or equatorial view, e.g., circular, elliptic, triangular, quadrangular, polygonal, irregular, lobate
see also: amb

P/E-ratio 39, **168, 171, 173**
refers to the length of the polar axis between the two poles compared to the equatorial diameter

paleo(palaeo-)palynology 6, 11, 89, 119
the study of fossil palynomorphs

palynogram
diagram summarizing the main morphological features of a palynomorph

palynology 4
the study of palynomorphs

palynomorph 4, 6, 118
general term for all biological entities found in palynological samples, e.g., pollen, spores, cysts, diatoms

panto-
prefix for global

pantoaperturate 42, **255, 256, 257**
pollen grain with apertures distributed more or less regularly over the surface: pantocolpate, pantocolporate, pantoporate

pantocolpate 255
see: pantoaperturate

pantoporate 256, 257
see: pantoaperturate

pantocolporate 256
see: pantoaperturate

papilla (lat., pl. **papillae**) 16, 259
small protuberance typical for Taxodioideae pollen located distally

pedium
see: foot layer
Comment: outdated term

penta-
prefix meaning five

pentaaperturate 229, 238, 267
pollen grain with 5 apertures: pentacolpate, pentacolporate, pentaporate

pentacolpate 229
see: pentaaperturate

pentacolporate 238
see: pentaaperturate

pentaporate 267
see: pentaaperturate

perforate 342
pollen wall with holes less than 1 µm in diameter

peri-
see: panto-

pharmacopalynology
the study of palynomorphs in drugs

pilate
see: clavate

pilum (pl. pila)
see: clava
Comment: the term "pilum" does not refer to the palynological feature; "pilum" means "dart" or "javelin"

***Pinus* subgenus *Strobus* type** 16, 50
bisaccate pollen grain with hemispherical sacci
Synonym: haploxylon type

***Pinus* subgenus *Pinus* type** 16, 50
bisaccate pollen grain with balloon-like sacci
Synonym: diploxylon-type

planaperturate 70, **260**
pollen grain with an angular outline, where the apertures are situated between the angles
Antonym: angulaperturate

plicae (lat., sing. **plica**) **344**
coarse parallel ridges

plicate 51, **344**
pollen wall with plicae

pluricolumellate
reticulate pollen wall with more than one row of columellae beneath the muri

polar area
region at and around the pole(s)

polar axis 38
imaginary line between the proximal and the distal pole of a pollen grain

polar view
view of a pollen grain in which the pole is directed towards the observer

polarity 38
orientation of the proximal and distal pole of a pollen grain resulting from tetrad stage

pole 38
outermost proximal and/or distal point of a pollen grain

pollen 24
see: pollen grain

pollen analysis
study of assemblages of dispersed palynomorphs
Comment: does not mean the morphological description of a pollen grain
see: palynogram

pollen class 10, **429**
artificial grouping of pollen grains that share one or more distinctive characters

pollen coating 24, 54, 57, **411, 414, 415**
general term applied to organic compounds usually produced by the tapetum, located on the exine and/or in exine cavities; see also: pollenkitt, tryphine, primexine matrix

pollen coating vesicles
tapetum derived vesicles covering the perforated footlayer in mature pollen grains in, e.g., *Geranium* and *Pelargonium*

pollen grain (engl., pl. **pollen grains or pollen**) 4, 24
the male gametophyte of seed plants; source and carrier for the male gametes (spermatozoids or sperm cells)

pollen mother cell 24
diploid cell forming a tetrad of four haploid microspores by meiosis

pollen tube 6, 16, 24, 44
tube produced by the vegetative cell as part of the fertilization process (transports the male gametes)

pollen type 6, 9, 57, 61
term categorizing pollen grains; often used in connection with a distinct taxon

pollen unit 10, 24, 38, 76, **131**
morphological unit of mature pollen grains (e.g., monad, tetrad)

pollen wall 6, 24, 45, 50, 105, **379**
layer(s) enclosing the cytoplasm of a pollen grain

pollenkitt 24, 47, 54, **411**
pollen coating consisting of sticky substances, mainly lipids

pollinarium (lat., pl. **pollinaria**) **150**
dispersal unit of pollinium (or pollinia) plus secretions and/or tissues that aid in the removal of the structure from the flower

pollination 4, 6, 13, 54, 57
transfer of pollen from the male to the female reproductive organs in seed plants

pollinium (lat., pl. **pollinia**) 76, **148**
unit of a more or less interconnected loculiform pollen mass

Comment: loculi may be subdivided by septae, thus resulting in more than 2 "pollinia"

poly-
prefix for many

polyad 10, 38, 57, 72, 76, **145**
unit of more than 4 pollen grains (multiple of 4)

polychotomosulcate **289**
pollen grain with a polychotomosulcus

polychotomosulcus **289**
sulcus with more than three arms

polygonal 10, **167, 171, 182**
Comment: a general term, used in palynology describing, e.g., "outline" and "shape"

polyplicate
see: plicate
Comment: a "plicate" pollen grain has always more than one plica, therefore the term "polyplicate" is superfluous

pontoperculate **262**
aperture with a pontoperculum

pontoperculum (lat., pl. **pontopercula**) **262**
elongated operculum linked to the ends of the aperture

porate 10, 42, **263, 264, 266, 267, 268, 269**
pollen grain with pori

poroid 42, 73, **272**
indistinct circular or elliptic aperture

poroidate 73, **272**
pollen grain with poroids

pororate 42
pollen grain with compound apertures composed of a circular "ektoporus" and "endoporus"

porus (lat., pl. **pori;** engl. **pore,** engl., pl. **pores**) 10, 42, **263, 264, 266, 267, 268, 269**
more or less circular aperture; pori located at the equator or regularly spread over the pollen grain

prae-
prefix for before

pre(prae)-pollen 43
microspores of certain extinct basal seed plants occurring from the Late Devonian until the Cretaceous, characterized by proximal and sometimes additional distal apertures, and presumed proximal germination

primexine 24, **403**
polysaccharidic layer formed during early developmental stage wherein the later exine structures are preformed

primexine matrix **414**
pollen coatings consisting of primexine remnants in mature pollen grains

prolate 39, **173**
pollen grain with a polar axis longer than the equatorial diameter
Antonym: oblate

proximal 38
pollen face/pole/side directing towards the center of the tetrad (deduced from tetrad stage)
Antonym: distal

pseudocolpus 43, **273**
colpus in a heteroaperturate pollen grain, presumed not to function as germination site

pseudomonad **135**
unit of a permanent tetrad with 3 rudimentary pollen grains

psilate 42, 54, 77, **346**
pollen wall with smooth surface

punctate
see: perforate
Comment: "punctum" does not describe the three-dimensional character of a perforation

quadrangular **166**
Comment: a general term, used in palynology describing, e.g., "outline"

reticulate 51, **348**
pollen wall with reticulum

reticulum (lat. pl. **reticula**) 51, **348**
network like pattern consisting of muri and lumina

reticulum cristatum 76, **355**
special type of reticulum; muri with prominent suprasculpture

retipilate 76
reticulum formed by rows of pila instead of muri
Comment: To the best of our knowledge there is no example of a reticulum formed by rows of pila instead of muri. Earlier observations were based on light microscopy. SEM-investigations reveal that the given example of Cuscuta and Callitriche do not fit the definition.

ring-like aperture 16, 73, 74, 75, **274**
circumferential aperture (situated more or less equatorially or, rarely, meridionally)

rugulae (lat., sing. **rugula**) **358**
elongated ornamentation elements irregularly arranged

rugulate **358**
pollen wall with rugulae

saccate **188, 189, 190, 193**
pollen grain with one or more air sacs

saccus (lat., pl. **sacci**) **188, 189, 190, 193**
exinous expansion forming an air sac

scabrate 16, 50
term used for light microscopy only, describing minute sculpture elements of undefined shape and of a size close to the resolution limit of the light microscope

sculpture 45, 51
ornamentation elements on the pollen surface, e.g., echini, bacula, clavae, verrucae

semi-
prefix for half

semitectate **319, 380, 382**
pollen grain with a discontinuous tectum

sexine 47, **404**
term used for light microscopy, describing the structured/sculptured outer layer of the exine

shape 10, 16, 38, **155, 181**
3-dimensional form of a pollen grain in relation to the P/E-ratio

size 9, 57
Comment: we recommend the size categories, very small (<10 μm), small (10–25 μm), medium (26–50 μm), large (51–100 μm), and very large (>100 μm)

sperm cell 6, 24
male gamete

spheroidal **171, 181**
Comment: a general term, used in palynology describing, e.g., "shape"

spine
see: echinus
Comment: the terms "spine", "spinulate", "spinus", "spinous", and "spinose" are linguistically inconsequent

spinose
see: echinate
see: spine

spiral aperture **277**
elongated, coiled aperture

spiraperturate 10, **277, 433**
pollen grain with one or more spiral aperture(s)

spore 4, 6, 11, 12, 44
general term for a reproductive unit (sexual, asexual) in plants and fungi

sporoderm 45
general term for the pollen/spore wall

sporopollenin 6, 7, 24, 45, 47, 49, 54, 56, **421**
main component of the exine, consisting of acetolysis-resistant biopolymers

stenopalynous 13
plant taxa characterized by only slight variation in pollen morphology
Antonym: eurypalynous

stephano-
prefix meaning equatorially situated

stephanoaperturate 42, 75, **279, 231, 238, 268**
apertures situated at the equator (term usually used for 6 or more apertures); see also: stephanocolpate, stephanocolporate, stephanoporate

stephanocolpate **231**
see: stephanoaperturate

stephanocolporate **238, 239**
see: stephanoaperturate

stephanoporate **268**
see: stephanoaperturate

striae (lat., sing. **stria**) 78, **361**
elongated ornamentation elements separated by grooves parallelly arranged
Comment: the term "striae" is used inconsistently in the literature; we use the term for the elevated elements and not for the grooves

striate 78, **361, 332**
pollen wall with striae

striato-microreticulate **332, 333, 383**
see: micro-

striato-reticulate 78, **366**
ornamentation intermediate between striate and reticulate

structure 45
the stratification of a pollen wall

sub-
prefix for less than

sulcate 10, 13, 70, 74, **281, 287, 288, 433**
pollen grain with a sulcus (sulci)

sulcus (lat., pl. **sulci**) 10, 13, 70, 74, **281, 287, 288, 433**
elongated aperture located distally

supra-
prefix for above

suprasculpture 61, 76, 92, **370**
secondary sculpture elements positioned on the primary sculpture of the pollen surface

supratectal element 47, 50, **374**
sculpture element positioned on top of the tectum
Comment: can only be defined in cross section of a pollen wall, see also "suprasculpture"

syn-
prefix for together

synaperturate 10, 69, **290, 291**
pollen grain with anastomosing apertures: syncolpate, syncolporate

syncolpate 69, **290**
pollen grain with anastomosing colpi

syncolporate 69, **290, 291**
pollen grain with anastomosing colpori

tapetum 24, 49, 52, **421**
specialized layer of cells lining the locule and participating in the nourishment of pollen grains, wall formation, and synthesis of pollen coatings

tectate 45, 47, 51, **380**
pollen grain with a tectum
Antonym: atectate

tectum (lat., pl. **tecta**) 45, 47, 51, **380**
outer more or less continuous ektexine layer; tectum condition can be eutectate or semitectate

tenuitas (lat., pl. **tenuitates**) 50, 68, 70, 73, **408**
general term for a thinning of the pollen wall

tetra-
prefix meaning four

tetraaperturate **229, 237, 266**
pollen grain with 4 apertures: tetracolpate, tetracolporate, tetraporate

tetracolpate **229**
see: tetraaperturate

tetracolporate **237**
see: tetraaperturate

tetraporate **266**
see: tetraaperturate

tetrad **137, 138, 142, 143**
unit of 4 pollen grains

tetrad decussate **142**
unit of 4 pollen grains arranged in 2 pairs in 2 different plains
Comment: these tetrads are a consequence of successive cytokinesis; without systematic relevance

tetrad linear 143
special case of tetrad planar

tetrad mark 43
mark on the proximal face of a spore retained from the postmeiotic stage functioning as germination area (linear = monolete, y-shaped = trilete)

tetrad planar 26, **137**, **143**
unit of 4 pollen grains arranged in one plane; can be: tetragonal, T-shaped, linear
Comment: these tetrads are a consequence of successive cytokinesis; without systematic relevance

tetrad stage 24, 38, 99
step in pollen development

tetrad tetrahedral 24, **138**
unit of 4 pollen grains in which the centers of the grains define a tetrahedron

tetrad T-shaped **143**, **144**
special case of tetrad planar

tetrad tetragonal **143**, **144**
special case of tetrad planar

treme 6, 12
suffix for a germination site (aperture, laesura); for kata-, ana-, zono-, and pantotreme

tri-
prefix meaning three

triangular 42, 68, **163**
Comment: a general term, used in palynology describing, e.g., "outline"

triaperturate **226**, **234**, **264**
pollen grain with 3 apertures: tricolpate, tricolporate, triporate

trichotomosulcate 70, **288**
pollen grain with a trichotomosulcus

trichotomosulcus 70, **288**
3-radiate sulcus

tricolpate **226**
pollen grain with 3 colpi

tricolporate **234**
pollen grain with 3 colpori

trilete 44
see: laesura and tetrad mark

triporate **264**
pollen grain with 3 pori

trisaccate **188**, **193**
pollen grain with three sacci

tryphine 24, **415**
pollen coating consisting mainly of lipids mixed with membrane remnants

Ubisch body (engl., pl. **Ubisch bodies**) 24, 52, 94, **421**
polymorphic sporopollenin-element produced by the tapetum
Comment: the "Ubisch body" is named after Gerta von Ubisch, who described these bodies for the first time

ulcerate 10, 74, **292**, **434**
pollen grain with an ulcus

ulcus (lat., pl. **ulci**)
more or less circular aperture located distally

vegetative cell 24
the larger of the two cells formed after first pollen mitosis

vegetative nucleus 24
the nucleus of the vegetative cell

verruca (lat., pl. **verrucae**) 78, **375**
wart-like element broader than high

verrucate 78, **375**
pollen wall with verrucae

vesiculate
see: saccate
Comment: "saccate" is the more common term

vestibulum (lat., pl. **vestibula**)
see: atrium
Comment: "atrium" is the more common term

viscin thread 24, 94, **428**
acetolysis resistant thread arising from the exine

zona- 75, **439**
prefix for ring-like
Comment: we prefer the term "ring-like"

zono- 12, 75, **439**
prefix for equatorially located features
Comment: we prefer the term "stephano-"

zonaaperturate
see: ring-like aperture

zonoaperturate
see: stephanoaperturate

Zwischenkörper 6
see: oncus
Comment: "oncus" is the more common term

Annex

Picture Copyrights

Part 1 General Chapters

Chapter "Palynology: History and Systematic Aspects"

Figure 1: Hyde and Williams (1944)
Figure 2: Grew (1682)
Figures 3–4: Fritzsche (1837)
Figure 5: *PalDat*
Figure 6, all pictures: Zetter, Reinhard
Figures 7–10: *PalDat*

Chapter "Pollen Development"

Figure 1: Illustration by Buchner, Ralf & Halbritter, Heidemarie
Figure 2: *PalDat*
Figure 3: Weber, Martina
Figures 4–10: *PalDat*
Figure 11: Ulrich Silvia (pictures A–D), Weber, Martina (pictures E–F)

Chapter "Pollen Morphology and Ultrastructure"

Figures 1–2: Illustration by Buchner, Ralf & Halbritter, Heidemarie
Figure 3, pictures A–B: Illustration by Buchner, Ralf & Halbritter, Heidemarie;
pictures C–D: *PalDat*
Figure 4, pictures A, C, E, G: Illustration by Buchner, Ralf & Halbritter, Heidemarie; pictures B, D, F, H: *PalDat*
Figure 5: Illustration by Buchner, Ralf & Halbritter, Heidemarie
Figure 6: Zetter, Reinhard
Figure 7: Illustration by Buchner, Ralf & Halbritter, Heidemarie
Figure 8: Illustration by Buchner, Ralf & Halbritter, Heidemarie
Figure 9, pictures A–C: Zetter, Reinhard; picture D: Weber, Martina
Figure 10: Zetter, Reinhard
Figure 11, pictures A–C: Zetter, Reinhard, picture D: *PalDat*
Figure 12: *PalDat*
Figure 13, pictures A–B: Weber, Martina picutres C–D: Ulrich, Silvia
Figure 14: *PalDat*
Figure 15, picture A: Illustration by Buchner, Ralf & Halbritter, Heidemarie, pictures B–C: *PalDat*
Figure 16, pictures A, D: Illustration by Buchner, Ralf & Halbritter, Heidemarie, pictures B, C, E, F: *PalDat*
Figure 17: Weber, Martina
Figure 18: Zetter, Reinhard
Figures 19–20: Zetter, Reinhard
Figures 21–24: *PalDat*
Figure 25, A, C, F: Weber, Martina; pictures B, D, E, G: *PalDat*
Figures 26–30: *PalDat*

Chapter "Misinterpretations in Palynology"

Figure 1, picture A–B: *PalDat* C: Zetter, Reinhard
Figures 2–4: *PalDat*
Figure 5: Zetter, Reinhard
Figures 6–7: *PalDat*
Figure 8, pictures A–B: Zetter, Reinhard, picuture C: *PalDat*
Figure 9: *PalDat*
Figure 10, pictures A–C: Grímsson, Friðgeir, D: *PalDat*, E: Weber, Martina
Figure 11: Zetter, Reinhard
Figures 12–14: *PalDat*
Figure 15, picture A: Weber, Martina, pictures B–C: *PalDat*
Figure 16: *PalDat*
Figure 17, picture A. Erdtman (1952), picture B. Punt et al. 2007,
picture C. Weber, Martina, pictures D–E: *PalDat*
Figure 18: Svojtka, Matthias
Figure 19: Ulrich, Silvia
Figure 20: *PalDat*
Figure 21: Ulrich, Silvia
Figure 22: *PalDat*

Chapter "How to Describe and Illustrate Pollen Grains"

Figures 1–2: Weber Martina & Ulrich Silvia
Figures 3–9 Table 1: Grímsson, Friðgeir & Zetter, Reinhard

Chapter "Methods in Palynology"

Figure 1: Grímsson, Friðgeir & Zetter, Reinhard
Figures 2–5: Ulrich, Silvia

Figures 6–8: Weber, Martina
Figure 9: Grímsson, Friðgeir & Zetter, Reinhard
Figure 10: Ulrich, Silvia & Halbritter, Heidemarie
Figures 11–12: Weber Martina & Ulrich Silvia
Figure 13, pictures A–J, L–M: Weber Martina & Ulrich Silvia, picture K: Böhmdorfer, Gudrun
Figures 14–15: Weber Martina & Ulrich Silvia
Figure 16: Weber, Martina
Figures 17–22: Weber Martina & Ulrich Silvia
Figures 23–27: Grímsson, Friðgeir & Zetter, Reinhard
Figure 28: Weber, Martina & Ulrich, Silvia

Part 2 Illustrated Pollen Terms

Chapter: "Pollen- and Dispersal Units"

Picture: *Alocasia odora*, page 132: Ulrich, Silvia
Picture: *Tilia tomentosa*, page 132: *PalDat*
Picture: *Betula humilis*, page 132: *PalDat*
Picture: *Lathyrus pratensis*, page 132: *PalDat*
Picture: *Tragopogon orientalis*, page 132: Weber, Martina
Picture: *Gonatopus bovinii*, page 132: Ulrich, Silvia
Picture: *Haworthia herbacea*, page 133: *PalDat*
Picture: *Pinus strobus*, page 133: *PalDat*
Picture: *Tragopogon orientalis*, page 133: *PalDat*
Picture: *Bituminaria bituminosa*, page 133: *PalDat*
Picture: *Lamium galeobdolon*, page 133: *PalDat*
Picture: *Dactylis glomerata*, page 133: *PalDat*
Picture: *Polypleurum stylosum*, page 134: *PalDat*
Picture: *Polypleurum stylosum*, page 134: *PalDat*
Picture: *Zeylanidium olivaceum*, page 134: *PalDat*
Picture: *Zeylanidium subulatum*, page 134: *PalDat*
Picture: *Thelethylax minutiflora*, page 134: *PalDat*
Picture: *Scheuchzeria palustris*, page 134: *PalDat*
Picture: *Carex* sp., page 135: *PalDat*
Picture: *Carex atrata*, page 135: *PalDat*
Picture: *Schoenoplectus lacustris*, page 135: *PalDat*
Picture: *Scirpus sylvaticus*, page 135: *PalDat*
Picture: *Carex distans*, page 135: *PalDat*
Picture: *Cyperus conglomeratus*, page 135: *PalDat*
Picture: *Carex digitata*, page 136: Weber, Martina
Picture: *Trichophorum cespitosum*, page 136: *PalDat*
Picture: *Cyperus diffusus*, page 136: *PalDat*
Picture: *Scirpoides holoschoenus*, page 136: *PalDat*
Picture: *Eleocharis palustris*, page 136: *PalDat*
Picture: *Cyperus papyrus*, page 136: *PalDat*
Picture: *Erica herbacea*, page 137: *PalDat*
Picture: *Chlorospatha kolbii*, page 137: *PalDat*
Picture: *Epipactis helleborine*, page 137: *PalDat*
Picture: *Cyprinia gracilis*, page 137: *PalDat*
Picture: *Epidendrum centropetalum*, page 137: *PalDat*
Picture: *Chlorospatha ceronii*, page 137: *PalDat*
Picture: Ericaceae, page 138: Grímsson, Friðgeir
Picture: *Epilobium hirsutum*, page 138: Weber, Martina
Picture: *Drosera scorpioides*, page 138: Weber, Martina
Picture: *Erica tetralix*, page 138: Weber, Martina
Picture: *Vaccinium myrtillus*, page 138: Weber, Martina
Picture: *Andromeda polifolia*, page 138: Weber, Martina
Picture: *Rhododendron hirsutum*, page 139: *PalDat*
Picture: *Victoria regia*, page 139: *PalDat*
Picture: *Luzula campestris*, page 139: *PalDat*
Picture: *Drimys granatensis*, page 139: *PalDat*
Picture: *Epilobium montanum*, page 139: *PalDat*
Picture: *Oxyanthus subpunctatus*, page 139: *PalDat*
Picture: *Epilobium parviflorum*, page 140: *PalDat*
Picture: *Arbutus unedo*, page 140: *PalDat*
Picture: *Chelonanthus alatus*, page 140: *PalDat*
Picture: *Mimosa pudica*, page 140: *PalDat*
Picture: *Ludwigia octovalvis*, page 140: *PalDat*
Picture: *Agapetes macrantha*, page 140: *PalDat*
Picture: *Dionaea muscipula*, page 141: *PalDat*
Picture: *Juncus effusus*, page 141: *PalDat*
Picture: *Moneses uniflora*, page 141: *PalDat*
Picture: *Drosera binata*, page 141: *PalDat*
Picture: *Vaccinium vitis-idaea*, page 141: *PalDat*
Picture: *Calolisianthus pendulus*, page 141: *PalDat*
Picture: *Calla palustris*, page 142: Ulrich, Silvia
Picture: *Androlepis skinneri*, page 142: *PalDat*
Picture: *Calluna vulgaris*, page 142: *PalDat*
Picture: *Chlorospatha dodsonii*, page 142: *PalDat*
Picture: *Rhodothamnus chamaecistus*, page 142: *PalDat*
Picture: *Catalpa bungei*, page 142: *PalDat*
Picture: *Typha latifolia*, page 143: Zetter, Reinhard
Picture: *Typha latifolia*, page 143: *PalDat*
Picture: *Typha latifolia*, page 143: Zetter, Reinhard
Picture: *Typha latifolia*, page 143: *PalDat*
Picture: *Typha latifolia*, page 143: Zetter, Reinhard
Picture: *Typha latifolia*, page 143: *PalDat*
Picture: *Asimina triloba*, page 144: Weber, Martina
Picture: *Calluna vulgaris*, page 144: Weber, Martina
Picture: *Chlorospatha kolbii*, page 144: *PalDat*
Picture: *Chlorospatha dodsonii*, page 144: *PalDat*
Picture: *Cyprinia gracilis*, page 144: *PalDat*
Picture: *Drosera peltata*, page 144: *PalDat*
Picture: *Acacia myrtifolia*, page 145: *PalDat*
Picture: *Calliandra tergemina*, page 145: *PalDat*
Picture: *Pithecellobium dulce*, page 145: *PalDat*
Picture: *Acacia* sp.page 145: Weber, Martina
Picture: *Cymbopetalum aequale*, page 145: *PalDat*

Picture: *Chelonanthus purpurascens*, page 145: *PalDat*
Picture: *Acacia* sp.page 146: *PalDat*
Picture: *Acacia dealbata*, page 146: *PalDat*
Picture: *Acacia karroo*, page 146: *PalDat*
Picture: *Acacia karroo*, page 146: *PalDat*
Picture: *Albizia julibrissin*, page 146: *PalDat*
Picture: *Albizia saman*, page 146: *PalDat*
Picture: *Traunsteinera globosa*, page 147: *PalDat*
Picture: *Epipogium aphyllum*, page 147: *PalDat*
Picture: *Herminium monorchis*, page 147: *PalDat*
Picture: *Gennaria diphylla*, page 147: Svojtka, Matthias
Picture: *Orchis italica*, page 147: Svojtka, Matthias
Picture: *Orchis purpurea*, page 147: Svojtka, Matthias
Picture: *Cephalanthera longifolia*, page 148: Svojtka, Matthias
Picture: *Cephalanthera longifolia*, page 148: Svojtka, Matthias
Picture: *Dendrobium farmeri*, page 148: *PalDat*
Picture: *Dendrobium farmeri*, page 148: *PalDat*
Picture: *Steveniella satyrioides*, page 148: *PalDat*
Picture: *Steveniella satyrioides*, page 148: *PalDat*
Picture: *Hammarbya paludosa*, page 149: *PalDat*
Picture: *Plectrophora cultrifolia*, page 149: *PalDat*
Picture: *Restrepia muscifera*, page 149: *PalDat*
Picture: *Malaxis monophyllos*, page 149: *PalDat*
Picture: *Stephanotis floribunda*, page 149: *PalDat*
Picture: *Hoodia flava*, page 149: *PalDat*
Picture: *Spiranthes spiralis*, page 150: *PalDat*
Picture: *Caladenia latifolia*, page 150: *PalDat*
Picture: *Ornithocephalus myrticola*, page 150: *PalDat*
Picture: *Aerides multiflora*, page 150: *PalDat*
Picture: *Maxillaria densa*, page 150: *PalDat*
Picture: *Brassavola cucullata*, page 150: Svojtka, Matthias
Picture: *Coelogyne fimbriata*, page 151: *PalDat*
Picture: *Pleurothallis loranthophylla*, page 151: *PalDat*
Picture: *Oncidium maizaefolium*, page 151: *PalDat*
Picture: *Schoenorchis fragrans*, page 151: Svojtka, Matthias
Picture: *Haraella odorata*, page 151: Svojtka, Matthias
Picture: *Stanhopea oculata*, page 151: *PalDat*
Picture: *Herminium monorchis*, page 152: *PalDat*
Picture: *Goodyera repens*, page 152: *PalDat*
Picture: *Gennaria diphylla*, page 152: *PalDat*
Picture: *Steveniella satyrioides*, page 152: *PalDat*
Picture: *Ludisia discolor*, page 152: *PalDat*
Picture: *Anacamptis pyramidalis*, page 152: *PalDat*
Picture: *Stephanotis floribunda*, page 153: *PalDat*
Picture: *Ceropegia sandersonii*, page 153: *PalDat*
Picture: *Hoya carnosa*, page 153: *PalDat*
Picture: *Hoya multiflora*, page 153: *PalDat*
Picture: *Frerea indica*, page 153: *PalDat*
Picture: *Orbeanthus hardyi*, page 153: *PalDat*

Chapter: "Shape and Polarity"

Picture: *Ligustrum* sp., page: 156: Zetter, Reinhard
Picture: Chenopodiaceae, page: 156: Weber, Martina
Picture: *Hedyosmum scaberrimum*, page: 156: *PalDat*
Picture: *Anthurium ovatifolium*, page: 156: *PalDat*
Picture: *Abutilon theophrasti*, page: 156: *PalDat*
Picture: *Corydalis ophiocarpa*, page: 156: *PalDat*
Picture: *Mayna odorata*, page: 157: *PalDat*
Picture: *Saruma henryi*, page: 157: *PalDat*
Picture: *Phleum pratense*, page: 157: *PalDat*
Picture: *Fraxinus ornus*, page: 157: *PalDat*
Picture: *Galium lucidum*, page: 157: *PalDat*
Picture: *Ginkgo biloba*, page: 157: *PalDat*
Picture: *Impatiens parviflora*, page: 158: Weber, Martina
Picture: *Impatiens parviflora*, page: 158: *PalDat*
Picture: *Amorphophallus interruptus*, page: 158: Ulrich, Silvia
Picture: *Allium oleraceum*, page: 158: *PalDat*
Picture: *Ambrosina bassi*, page: 158: Ulrich, Silvia
Picture: *Aechmea dealbata*, page: 158: *PalDat*
Picture: *Salvia coccinea*, page: 159: *PalDat*
Picture: *Commelina erecta*, page: 159: *PalDat*
Picture: *Billbergia porteana*, page: 159: *PalDat*
Picture: *Zamia loddigesii*, page: 159: *PalDat*
Picture: *Galeopsis tetrahit*, page: 159: *PalDat*
Picture: *Physostegia virginiana*, page: 159: *PalDat*
Picture: *Acer pseudoplatanus*, page: 160: *PalDat*
Picture: *Artemisia pontica*, page: 160: *PalDat*
Picture: *Sanguisorba officinalis*, page: 160: *PalDat*
Picture: *Orthilia secunda*, page: 160: *PalDat*
Picture: *Gunnera tinctoria*, page: 160: *PalDat*
Picture: *Gunnera tinctoria*, page: 160: *PalDat*
Picture: *Sedum rupestre*, page: 161: *PalDat*
Picture: *Viola alba*, page: 161: *PalDat*
Picture: *Clematis heracleifolia*, page: 161: *PalDat*
Picture: *Sanicula europaea*, page: 161: *PalDat*
Picture: *Pinguicula ehlersiae*, page: 161: *PalDat*
Picture: *Bellis perennis*, page: 161: *PalDat*
Picture: *Artemisia* sp.page: 162: Weber, Martina
Picture: *Nicotiana tabacum*, page: 162: Weber, Martina
Picture: *Hypecoum imberbe*, page: 162: *PalDat*
Picture: *Barringtonia asiatica*, page: 162: *PalDat*
Picture: *Pelargonium punctatum*, page: 162: *PalDat*
Picture: *Viola riviniana*, page: 162: *PalDat*
Picture: *Lopezia racemosa*, page: 163: Weber, Martina
Picture: *Loranthus europaeus*, page: 163: *PalDat*

Picture: *Macadamia ternifolia*, page: 163: Weber, Martina
Picture: *Tropaeolum emarginatum*, page: 163: *PalDat*
Picture: *Kolkwitzia amabilis*, page: 163: Weber, Martina
Picture: *Acicarpha tribuloides*, page: 163: *PalDat*
Picture: *Callistemon coccineus*, page: 164: *PalDat*
Picture: *Hypoestes phyllostachya*, page: 164: *PalDat*
Picture: *Echinops ritro*, page: 164: *PalDat*
Picture: *Bupleurum rotundifolium*, page: 164: *PalDat*
Picture: *Paullinia tomentosa*, page: 164: *PalDat*
Picture: *Primula denticulata*, page: 164: *PalDat*
Picture: *Orlaya grandiflora*, page: 165: *PalDat*
Picture: *Circaea lutetiana*, page: 165: *PalDat*
Picture: *Sempervivum globiferum*, page: 165: *PalDat*
Picture: *Cunonia capensis*, page: 165: *PalDat*
Picture: *Dipsacus fullonum*, page: 165: *PalDat*
Picture: *Potentilla inclinata*, page: 165: *PalDat*
Picture: *Anchusa officinalis*, page: 166: *PalDat*
Picture: *Nonea pulla*, page: 166: *PalDat*
Picture: *Viola tricolor*, page: 166: *PalDat*
Picture: *Eremurus robustus*, page: 166: *PalDat*
Picture: *Herniaria glabra*, page: 166: *PalDat*
Picture: *Sideritis romana*, page: 166: *PalDat*
Picture: *Viola arvensis*, page: 167: Weber, Martina
Picture: *Viola arvensis*, page: 167: PalDat
Picture: *Opuntia basilaris*, page: 167: *PalDat*
Picture: *Talinum paniculatum*, page: 167: *PalDat*
Picture: *Sarracenia alata*, page: 167: *PalDat*
Picture: *Stellaria holostea*, page: 167: *PalDat*
Picture: *Carya* sp., page: 168: Zetter, Reinhard
Picture: *Corylus avellana*, page: 168: *PalDat*
Picture: *Plectranthus esculentus*, page: 168: Weber, Martina
Picture: *Salvia argentea*, page: 168: *PalDat*
Picture: *Impatiens* sp., page: 168: Weber, Martina
Picture: *Impatiens glandulifera*, page: 168: *PalDat*
Picture: *Sarracenia alata*, page: 169: *PalDat*
Picture: *Hymenocallis tubiflora*, page: 169: *PalDat*
Picture: *Knautia drymeia*, page: 169: *PalDat*
Picture: *Cuphea procumbens*, page: 169: *PalDat*
Picture: *Hakea kippistiana*, page: 169: *PalDat*
Picture: *Roridula gorgonias*, page: 169: *PalDat*
Picture: *Clarkia unguiculata*, page: 170: *PalDat*
Picture: *Vriesea neoglutinosa*, page: 170: *PalDat*
Picture: *Acicarpha tribuloides*, page: 170: *PalDat*
Picture: *Heliconia* sp., page: 170: *PalDat*
Picture: *Amsonia ciliata*, page: 170: *PalDat*
Picture: *Clarkia purpurea*, page: 170: *PalDat*
Picture: *Parnassia palustris*, page: 171: *PalDat*
Picture: *Campanula fenestrellata*, page: 171: *PalDat*
Picture: *Roemeria hybrida*, page: 171: *PalDat*
Picture: *Silene nutans*, page: 171: *PalDat*
Picture: *Iris pumila*, page: 171: *PalDat*
Picture: *Sarcocapnos enneaphylla*, page: 171: *PalDat*
Picture: *Whitfieldia lateritia*, page: 172: *PalDat*
Picture: *Whitfieldia lateritia*, page: 172: *PalDat*
Picture: *Schoepfia schreberi*, page: 172: *PalDat*
Picture: *Thesium arvense*, page: 172: *PalDat*
Picture: *Pedicularis gyroflexa*, page: 172: *PalDat*
Picture: *Basella alba*, page: 172: *PalDat*
Picture: *Justicia carnea*, page: 173: Weber, Martina
Picture: *Aphelandra arborea*, page: 173: *PalDat*
Picture: *Aesculus* sp., page: 173: Weber, Martina
Picture: *Lysimachia lichiangensis*, page: 173: *PalDat*
Picture: *Colutea arborescens*, page: 173: Weber, Martina
Picture: *Oxytropis jacquinii*, page: 173: *PalDat*
Picture: *Crossandra flava*, page: 174: *PalDat*
Picture: *Jurinea mollis*, page: 174: *PalDat*
Picture: *Torilis arvensis*, page: 174: *PalDat*
Picture: *Peucedanum cervaria*, page: 174: *PalDat*
Picture: *Astragalus onobrychis*, page: 174: *PalDat*
Picture: *Symphytum officinale*, page: 174: *PalDat*
Picture: *Buglossoides purpurocaerulea*, page: 175: *PalDat*
Picture: *Vitaliana primuliflora*, page: 175: *PalDat*
Picture: *Platycodon grandiflorus*, page: 175: *PalDat*
Picture: *Stenandrium guineense*, page: 175: *PalDat*
Picture: *Lathyrus tuberosus*, page: 175: *PalDat*
Picture: *Salvia sclarea*, page: 175: *PalDat*
Picture: Apiaceae, page: 176: Weber, Martina
Picture: *Bifora radians*, page: 176: *PalDat*
Picture: *Columnea magnifica*, page: 176: *PalDat*
Picture: *Asperula tinctoria*, page: 176: *PalDat*
Picture: *Monotropa hypopitys*, page: 176: *PalDat*
Picture: *Myosotis scorpioides*, page: 176: *PalDat*
Picture: *Viburnum tinus*, page: 177: *PalDat*
Picture: *Cerinthe minor*, page: 177: *PalDat*
Picture: *Luffa cylindrica*, page: 177: *PalDat*
Picture: *Crossandra flava*, page: 177: *PalDat*
Picture: *Pulmonaria angustifolia*, page: 177: *PalDat*
Picture: *Aesculus flava*, page: 177: *PalDat*
Picture: *Onosma visianii*, page: 178: *PalDat*
Picture: *Billbergia seidelii*, page: 178: *PalDat*
Picture: *Chaenorhinum minus*, page: 178: *PalDat*
Picture: *Limnanthes douglasii*, page: 178: *PalDat*
Picture: *Sesleria albicans*, page: 178: *PalDat*
Picture: *Elaeagnus angustifolia*, page: 178: *PalDat*
Picture: *Heliconia* sp., page: 179: *PalDat*
Picture: *Quesnelia augusto-coburgii*, page: 179: *PalDat*
Picture: *Erica arborea*, page: 179: *PalDat*
Picture: *Pinus strobus*, page: 179: *PalDat*
Picture: *Nuphar lutea*, page: 179: *PalDat*
Picture: *Sansevieria parva*, page: 179: *PalDat*

Picture: *Amorphophallus yunnanensis*, page: 180: Ulrich, Silvia
Picture: *Austrobaileya scandens*, page: 180: *PalDat*
Picture: *Echium italicum*, page: 180: *PalDat*
Picture: *Adenanthos sericeus*, page: 180: *PalDat*
Picture: *Calluna vulgaris*, page: 180: *PalDat*
Picture: *Alkanna corcyrensis*, page: 180: *PalDat*
Picture: *Adansonia gregorii*, page: 181: *PalDat*
Picture: *Ruellia macrantha*, page: 181: *PalDat*
Picture: *Cirsium oleraceum*, page: 181: *PalDat*
Picture: *Phlox paniculata*, page: 181: *PalDat*
Picture: *Crossandra flava*, page: 181: *PalDat*
Picture: *Justicia carnea*, page: 181: *PalDat*
Picture: *Basella alba*, page: 182: *PalDat*
Picture: *Herniaria glabra*, page: 182: *PalDat*
Picture: *Cerastium dubium*, page: 182: *PalDat*
Picture: *Eremogone procera*, page: 182: *PalDat*
Picture: *Paronychia polygonifolia*, page: 182: *PalDat*
Picture: *Paronychia polygonifolia*, page: 182: *PalDat*
Picture: *Alkanna corcyrensis*, page: 183: *PalDat*
Picture: *Montrichardia arborescens*, page: 183: *PalDat*
Picture: *Heliconia rostrata*, page: 183: *PalDat*
Picture: *Echium plantagineum*, page: 183: *PalDat*
Picture: *Cyperus longus*, page: 183: *PalDat*
Picture: *Ephedra foeminea*, page: 183: *PalDat*
Picture: *Brugmansia suaveolens*, page: 184: *PalDat*
Picture: *Anthyllis vulneraria*, page: 184: *PalDat*
Picture: *Corydalis cheilanthifolia*, page: 184: *PalDat*
Picture: *Schoepfia schreberi*, page: 184: *PalDat*
Picture: *Aechmea drakeana*, page: 184: *PalDat*
Picture: *Cardiospermum halicacabum*, page: 184: *PalDat*
Picture: *Whitfieldia lateritia*, page: 185: *PalDat*
Picture: *Pedicularis portenschlagii*, page: 185: *PalDat*
Picture: *Billbergia pyramidalis*, page: 185: *PalDat*
Picture: *Quesnelia imbricata*, page: 185: *PalDat*
Picture: *Clarkia unguiculata*, page: 185: *PalDat*
Picture: *Fuchsia paniculata*, page: 185: *PalDat*
Picture: *Acicarpha tribuloides*, page: 186: *PalDat*
Picture: *Nicandra physalodes*, page: 186: *PalDat*
Picture: *Myosotis alpestris*, page: 186: *PalDat*
Picture: *Loranthus europaeus*, page: 186: *PalDat*
Picture: *Sansevieria suffruticosa*, page: 186: *PalDat*
Picture: *Juncus jacquinii*, page: 186: *PalDat*
Picture: *Gaura lindheimeri*, page: 187: *PalDat*
Picture: *Eremurus robustus*, page: 187: *PalDat*
Picture: *Hyacinthoides italica*, page: 187: *PalDat*
Picture: *Galanthus nivalis*, page: 187: *PalDat*
Picture: *Limnanthes douglasii*, page: 187: *PalDat*
Picture: *Thesium dollineri*, page: 187: *PalDat*
Picture: *Picea* sp., page: 188: Grímsson, Friðgeir
Picture: *Abies cephalonica*, page: 188: Weber, Martina
Picture: *Pinus cembra*, page: 188: Weber, Martina
Picture: *Picea* sp., page: 188: Weber, Martina
Picture: *Tsuga canadensis*, page: 188: Weber, Martina
Picture: *Dacrycarpus dacrydioides*, page: 188: *PalDat*
Picture: *Tsuga* sp., page: 189: Zetter, Reinhard
Picture: *Tsuga* sp., page: 189: Zetter, Reinhard
Picture: *Tsuga canadensis*, page: 189: Weber, Martina
Picture: *Tsuga canadensis*, page: 189: *PalDat*
Picture: *Tsuga* sp., page: 189: Grímsson, Friðgeir
Picture: *Tsuga* sp., page: 189: Grímsson, Friðgeir
Picture: *Picea* sp.page: 190: Grímsson, Friðgeir
Picture: *Picea* sp.page: 190: Grímsson, Friðgeir
Picture: *Picea* sp.page: 190: Grímsson, Friðgeir
Picture: *Pinus* sp., page: 190: Grímsson, Friðgeir
Picture: *Pinus* sp., page: 190: Grímsson, Friðgeir
Picture: *Abies cephalonica*, page: 190: Weber, Martina
Picture: *Abies cephalonica*, page: 191: *PalDat*
Picture: *Picea abies*, page: 191: *PalDat*
Picture: *Picea abies*, page: 191: *PalDat*
Picture: *Pinus mugo*, page: 191: *PalDat*
Picture: *Abies nordmanniana*, page: 191: *PalDat*
Picture: *Picea pungens*, page: 191: *PalDat*
Picture: *Pinus heldreichii*, page: 192: *PalDat*
Picture: *Pinus nigra*, page: 192: *PalDat*
Picture: *Podocarpus* sp., page: 192: *PalDat*
Picture: *Podocarpus* sp., page: 192: *PalDat*
Picture: *Pinus mugo*, page: 192: Weber, Martina
Picture: *Pinus contorta*, page: 192: *PalDat*
Picture: Podocarpaceae, page: 193: Grímsson, Friðgeir
Picture: *Pherosphaera hookeriana*, page: 193: *PalDat*
Picture: *Pherosphaera hookeriana*, page: 193: *PalDat*
Picture: *Dacrycarpus dacrydioides*, page: 193: *PalDat*
Picture: *Abies concolor*, page: 193: *PalDat*
Picture: *Abies concolor*, page: 193: *PalDat*
Picture: *Fritillaria pontica*, page: 194: Weber, Martina
Picture: *Veratrum album*, page: 194: *PalDat*
Picture: *Galium odoratum*, page: 194: *PalDat*
Picture: *Sparmannia africana*, page: 194: *PalDat*
Picture: *Roemeria hybrida*, page: 194: *PalDat*
Picture: *Bifora radians*, page: 194: *PalDat*
Picture: *Artemisia pontica*, page: 195: *PalDat*
Picture: *Carex alba*, page: 195: *PalDat*
Picture: *Lachenalia aloides*, page: 195: *PalDat*
Picture: *Luzula sylvatica*, page: 195: *PalDat*
Picture: *Moehringia muscosa*, page: 195: *PalDat*
Picture: *Anemone hortensis*, page: 195: *PalDat*
Picture: *Dracontium asperum*, page: 196: Ulrich, Silvia

Picture: *Billbergia seidelii*, page: 196: *PalDat*
Picture: *Lilium candidum*, page: 196: *PalDat*
Picture: *Nuphar lutea*, page: 196: *PalDat*
Picture: *Ginkgo biloba*, page: 196: *PalDat*
Picture: *Asphodeline lutea*, page: 196: *PalDat*
Picture: *Lysichiton americanus*, page: 197: *PalDat*
Picture: *Piper nigrum*, page: 197: *PalDat*
Picture: *Gagea lutea*, page: 197: *PalDat*
Picture: *Sparganium erectum*, page: 197: *PalDat*
Picture: *Dioon edule*, page: 197: *PalDat*
Picture: *Symplocarpus foetidus*, page: 197: *PalDat*
Picture: *Asphodelus fistulosus*, page: 198: *PalDat*
Picture: *Eremurus thiodanthus*, page: 198: *PalDat*
Picture: *Piper auritum*, page: 198: *PalDat*
Picture: *Wachendorfia thyrsiflora*, page: 198: *PalDat*
Picture: *Haemanthus coccineus*, page: 198: *PalDat*
Picture: *Hymenocallis tubiflora*, page: 198: *PalDat*
Picture: *Bougainvillea* sp., page: 199: *PalDat*
Picture: *Heliconia* sp., page: 199: *PalDat*
Picture: *Tilia euchlora*, page: 199: *PalDat*
Picture: *Elaeagnus angustifolia*, page: 199: *PalDat*
Picture: *Luzula campestris*, page: 199: *PalDat*
Picture: *Tsuga canadensis*, page: 199: *PalDat*
Picture: *Adenanthos sericeus*, page: 200: *PalDat*
Picture: *Heliconia stricta*, page: 200: *PalDat*
Picture: *Petrea volubilis*, page: 200: *PalDat*
Picture: *Leucadendron brunioides*, page: 200: *PalDat*
Picture: *Cunninghamia lanceolata*, page: 200: *PalDat*
Picture: *Hibiscus schizopetalus*, page: 200: *PalDat*
Picture: *Alnus glutinosa*, page: 201: *PalDat*
Picture: *Erica arborea*, page: 201: *PalDat*
Picture: *Bupleurum rotundifolium*, page: 201: *PalDat*
Picture: *Melampyrum arvense*, page: 201: *PalDat*
Picture: *Leucadendron discolor*, page: 201: *PalDat*
Picture: *Melastoma sanguineum*, page: 201: *PalDat*
Picture: *Verbena officinalis*, page: 202: *PalDat*
Picture: *Tsusiophyllum tanakae*, page: 202: *PalDat*
Picture: *Ardisia crenata*, page: 202: *PalDat*
Picture: *Thesium arvense*, page: 202: *PalDat*
Picture: *Grevillea banksii*, page: 202: *PalDat*
Picture: *Tropaeolum moritzianum*, page: 202: *PalDat*
Picture: Poaceae, page: 203: Weber, Martina
Picture: *Dorstenia contrajerva*, page: 203: *PalDat*
Picture: *Guzmania elvallensis*, page: 203: *PalDat*
Picture: *Smilax spinosa*, page: 203: *PalDat*
Picture: *Callitriche stagnalis*, page: 203: *PalDat*
Picture: *Vanilla planifolia*, page: 203: *PalDat*
Picture: *Urtica dioica*, page: 204: *PalDat*
Picture: *Populus alba*, page: 204: *PalDat*
Picture: *Sesleria albicans*, page: 204: *PalDat*
Picture: *Anthurium radicans*, page: 204: *PalDat*
Picture: *Coriaria nepalensis*, page: 204: *PalDat*
Picture: *Orobanche hederae*, page: 204: *PalDat*

Chapter: "Aperture"

Picture: *Symplocos* sp., page 208: Zetter, Reinhard
Picture: *Acalypha macrostachya*, page 208: *PalDat*
Picture: *Tropaeolum majus*, page 208: *PalDat*
Picture: *Anthyllis vulneraria*, page 208: *PalDat*
Picture: *Cardiospermum halicacabum*, page 208: *PalDat*
Picture: *Oenothera biennis*, page 208: *PalDat*
Picture: *Betula humilis*, page 209: *PalDat*
Picture: *Triticum aestivum*, page 209: *PalDat*
Picture: Poaceae, page 209: Weber, Martina
Picture: *Secale cereale*, page 209: *PalDat*
Picture: *Circaea lutetiana*, page 209: *PalDat*
Picture: *Fumaria officinalis*, page 209: *PalDat*
Picture: *Myriophyllum spicatum*, page 210: *PalDat*
Picture: *Morina longifolia*, page 210: *PalDat*
Picture: *Epilobium angustifolium*, page 210: *PalDat*
Picture: *Dorstenia contrajerva*, page 210: *PalDat*
Picture: *Trichosanthes cucumerina*, page 210: *PalDat*
Picture: *Zea mays*, page 210: *PalDat*
Picture: *Pinus strobus*, page 211: Weber, Martina
Picture: *Alnus glutinosa*, page 211: *PalDat*
Picture: *Tilia* sp., page 211: Weber, Martina
Picture: *Elaeagnus angustifolia*, page 211: Weber, Martina
Picture: *Rubus caesius*, page 211: Weber, Martina
Picture: *Acinos alpinus*, page 211: *PalDat*
Picture: *Eupatorium cannabinum*, page 212: *PalDat*
Picture: *Clinopodium vulgare*, page 212: *PalDat*
Picture: *Eucharis grandiflora*, page 212: *PalDat*
Picture: *Alisma plantago-aquatica*, page 212: *PalDat*
Picture: *Sesleria sadleriana*, page 212: *PalDat*
Picture: *Carex alba*, page 212: *PalDat*
Picture: *Apium nodiflorum*, page 213: *PalDat*
Picture: *Apium nodiflorum*, page 213: *PalDat*
Picture: *Veronica spicata*, page 213: *PalDat*
Picture: *Mentha aquatica*, page 213: *PalDat*
Picture: *Ophiorrhiza* sp., page 213: *PalDat*
Picture: *Geranium robertianum*, page 213: *PalDat*
Picture: *Scrophularia vernalis*, page 214: *PalDat*
Picture: *Doryanthes palmeri*, page 214: *PalDat*
Picture: *Melampyrum subalpinum*, page 214: *PalDat*
Picture: *Aechmea caesia*, page 214: *PalDat*
Picture: *Kickxia spuria*, page 214: *PalDat*
Picture: *Tragopogon dubius*, page 214: *PalDat*
Picture: *Aesculus x carnea*, page 215: Weber, Martina

Picture: *Calystegia sepium*, page 215: Weber, Martina
Picture: *Lamprocapnos spectabilis*, page 215: Weber, Martina
Picture: *Papaver rhoeas*, page 215: Weber, Martina
Picture: *Ficaria verna*, page 215: Weber, Martina
Picture: *Lupinus* sp., page 215: Weber, Martina
Picture: *Convolvulus tricolor*, page 216: *PalDat*
Picture: *Salix alba*, page 216: *PalDat*
Picture: *Aesculus hippocastanum*, page 216: *PalDat*
Picture: *Moltkia petraea*, page 216: *PalDat*
Picture: *Billbergia macrocalyx*, page 216: *PalDat*
Picture: *Arenaria serpyllifolia*, page 216: *PalDat*
Picture: *Saxifraga vandellii*, page 217: *PalDat*
Picture: *Galeopsis tetrahit*, page 217: *PalDat*
Picture: *Veronica wyomingensis*, page 217: *PalDat*
Picture: *Clarkia pulchella*, page 217: *PalDat*
Picture: *Nuphar lutea*, page 217: *PalDat*
Picture: *Gagea villosa*, page 217: *PalDat*
Picture: *Commelina erecta*, page 218: *PalDat*
Picture: *Sparmannia africana*, page 218: *PalDat*
Picture: *Vigna speciosa*, page 218: *PalDat*
Picture: *Clerodendrum thomsoniae*, page 218: *PalDat*
Picture: *Roridula gorgonias*, page 218: *PalDat*
Picture: *Carthamus lanatus*, page 218: *PalDat*
Picture: *Lamiastrum galeobdolon*, page 219: *PalDat*
Picture: *Aconitum lycoctonum*, page 219: *PalDat*
Picture: *Mercurialis perennis*, page 219: *PalDat*
Picture: *Ulmus minor*, page 219: *PalDat*
Picture: *Chenopodium hybridum*, page 219: *PalDat*
Picture: *Aesculus carnea*, page 219: *PalDat*
Picture: *Betula* sp., page 220: Grímsson, Friðgeir
Picture: *Betula pendula*, page 220: *PalDat*
Picture: *Betula humilis*, page 220: *PalDat*
Picture: *Betula humilis*, page 220: *PalDat*
Picture: *Corylus avellana*, page 220: *PalDat*
Picture: *Tilia platyphyllos*, page 220: *PalDat*
Picture: *Pistacia* sp., page 221: Weber, Martina
Picture: *Impatiens columbaria*, page 221: *PalDat*
Picture: *Mendoncia albida*, page 221: *PalDat*
Picture: *Scabiosa ochroleuca*, page 221: *PalDat*
Picture: *Petrea volubilis*, page 221: *PalDat*
Picture: *Succisa pratensis*, page 221: *PalDat*
Picture: *Tilia platyphyllos*, page 222: Weber, Martina
Picture: *Coriaria myrtifolia*, page 222: *PalDat*
Picture: *Dalechampia spathulata*, page 222: *PalDat*
Picture: *Symphytum orientale*, page 222: *PalDat*
Picture: *Borago pygmaea*, page 222: *PalDat*
Picture: *Tabernaemontana simulans*, page 222: *PalDat*
Picture: *Elaeagnus angustifolia*, page 223: *PalDat*
Picture: *Malus baccata*, page 223: Weber, Martina
Picture: *Bifora radians*, page 223: *PalDat*
Picture: *Gazania rigens*, page 223: *PalDat*
Picture: *Cunonia capensis*, page 223: *PalDat*
Picture: *Colutea arborescens*, page 223: Weber, Martina
Picture: *Typha latifolia*, page 224: *PalDat*
Picture: *Epilobium hirsutum*, page 224: *PalDat*
Picture: *Ludwigia octovalvis*, page 224: *PalDat*
Picture: *Thelethylax minutiflora*, page 224: *PalDat*
Picture: *Aechmea subintegerrima*, page 224: *PalDat*
Picture: *Aechmea subintegerrima*, page 224: *PalDat*
Picture: *Hypecoum imberbe*, page 225: *PalDat*
Picture: *Hypecoum procumbens*, page 225: *PalDat*
Picture: *Mayna odorata*, page 225: *PalDat*
Picture: *Mayna odorata*, page 225: *PalDat*
Picture: *Pedicularis elongata*, page 225: *PalDat*
Picture: *Pedicularis elongata*, page 225: *PalDat*
Picture: *Lamium maculatum*, page 226: *PalDat*
Picture: *Erysimum odoratum*, page 226: *PalDat*
Picture: *Nelumbo nucifera*, page 226: *PalDat*
Picture: *Lonicera fragrantissima*, page 226: *PalDat*
Picture: *Stachys palustris*, page 226: *PalDat*
Picture: *Ceratostigma plumbaginoides*, page 226: *PalDat*
Picture: *Fraxinus excelsior*, page 227: *PalDat*
Picture: *Odontites luteus*, page 227: *PalDat*
Picture: *Nandina domestica*, page 227: *PalDat*
Picture: *Corylopsis platypetala*, page 227: *PalDat*
Picture: *Trollius europaeus*, page 227: *PalDat*
Picture: *Veronica serpyllifolia*, page 227: *PalDat*
Picture: *Convolvulus arvensis*, page 228: Weber, Martina
Picture: *Petrea volubilis*, page 228: *PalDat*
Picture: *Fraxinus excelsior*, page 228: Weber, Martina
Picture: *Melampyrum arvense*, page 228: *PalDat*
Picture: *Acer* sp., page 228: Zetter, Reinhard
Picture: *Cleistocactus straussii*, page 228: *PalDat*
Picture: *Impatiens parviflora*, page 229: Weber, Martina
Picture: *Impatiens glandulifera*, page 229: *PalDat*
Picture: *Impatiens parviflora*, page 229: *PalDat*
Picture: *Sideritis romana*, page 229: *PalDat*
Picture: *Didymaea mexicana*, page 229: *PalDat*
Picture: *Mendoncia albida*, page 229: *PalDat*
Picture: *Cruciata laevipes*, page 230: *PalDat*
Picture: *Merremia umbellata*, page 230: *PalDat*
Picture: *Platycodon grandiflorus*, page 230: *PalDat*
Picture: *Primula veris*, page 230: *PalDat*
Picture: *Origanum vulgare*, page 230: Weber, Martina
Picture: *Clinopodium vulgare*, page 230: *PalDat*
Picture: *Galium glaucum*, page 231: *PalDat*

Picture: *Galium lucidum*, page 231: *PalDat*
Picture: *Sherardia arvensis*, page 231: *PalDat*
Picture: *Galium odoratum*, page 231: *PalDat*
Picture: *Codonopsis pilosula*, page 231: *PalDat*
Picture: *Sechium edule*, page 231: *PalDat*
Picture: *Sarcocapnos enneaphylla*, page 232: *PalDat*
Picture: *Talinum paniculatum*, page 232: *PalDat*
Picture: *Pseudofumaria lutea*, page 232: *PalDat*
Picture: *Maripa nicaraguensis*, page 232: *PalDat*
Picture: *Mollugo verticillata*, page 232: *PalDat*
Picture: *Turbinicarpus pseudomacrochele*, page 232: *PalDat*
Picture: *Justicia procumbens*, page 233: *PalDat*
Picture: *Justicia procumbens*, page 233: *PalDat*
Picture: *Adhatoda schimperiana*, page 233: *PalDat*
Picture: *Justicia macrantha*, page 233: *PalDat*
Picture: *Justicia carnea*, page 233: Weber, Martina
Picture: *Justicia xylosteoides*, page 233: *PalDat*
Picture: *Lathyrus vernus*, page 234: *PalDat*
Picture: *Kraussia floribunda*, page 234: *PalDat*
Picture: *Hieracium hoppeanum*, page 234: *PalDat*
Picture: *Erica herbacea*, page 234: *PalDat*
Picture: *Rumex acetosa*, page 234: *PalDat*
Picture: *Aruncus dioicus*, page 234: *PalDat*
Picture: *Tricolporopollenites wackersdorfensis*, page 235: Zetter, Reinhard
Picture: Fabaceae, page 235: Weber, Martina
Picture: *Fagopyrum* sp., page 235: Weber, Martina
Picture: *Lathyrus sylvestris*, page 235: Weber, Martina
Picture: *Rhus* sp., page 235: Weber, Martina
Picture: *Euphorbia peplus*, page 235: Weber, Martina
Picture: *Centaurea scabiosa*, page 236: *PalDat*
Picture: *Cirsium oleraceum*, page 236: *PalDat*
Picture: *Echium vulgare*, page 236: *PalDat*
Picture: *Gardenia thunbergia*, page 236: *PalDat*
Picture: *Myrrhis odorata*, page 236: *PalDat*
Picture: *Fatsia japonica*, page 236: *PalDat*
Picture: *Citrus swinglei*, page 237: *PalDat*
Picture: *Pulmonaria mollis*, page 237: *PalDat*
Picture: *Nicotiana tabacum*, page 237: *PalDat*
Picture: *Genlisea violacea*, page 237: *PalDat*
Picture: *Poncirus trifoliata*, page 237: *PalDat*
Picture: *Tridax procumbens*, page 237: *PalDat*
Picture: *Viola arvensis*, page 238: *PalDat*
Picture: *Justicia menesii*, page 238: *PalDat*
Picture: *Sanguisorba officinalis*, page 238: *PalDat*
Picture: *Cerinthe minor*, page 238: *PalDat*
Picture: *Pinguicula ehlersiae*, page 238: *PalDat*
Picture: *Polygala chamaebuxus*, page 238: *PalDat*
Picture: *Moltkia petraea*, page 239: *PalDat*
Picture: *Symphytum caucasicum*, page 239: *PalDat*
Picture: *Polygala major*, page 239: *PalDat*
Picture: *Buglossoides arvensis*, page 239: *PalDat*
Picture: *Echinopepon wrightii*, page 239: *PalDat*
Picture: *Utricularia vulgaris*, page 239: Weber, Martina
Picture: *Lathyrus sylvestris*, page 240: Weber, Martina
Picture: *Tussilago farfara*, page 240: Weber, Martina
Picture: *Parthenocissus* sp., page 240: Weber, Martina
Picture: *Centaurea jacea*, page 240: Weber, Martina
Picture: *Cichorium intybus*, page 240: Weber, Martina
Picture: *Lysimachia punctata*, page 240: Weber, Martina
Picture: *Dictamnus albus*, page 241: Weber, Martina
Picture: *Scaevola aemula*, page 241: Weber, Martina
Picture: *Vicia faba*, page 241: Weber, Martina
Picture: *Vitaliana primuliflora*, page 241: *PalDat*
Picture: *Rumex* sp., page 241: Zetter, Reinhard
Picture: *Phacelia campanularia*, page 241: *PalDat*
Picture: *Lythrum hyssopifolia*, page 242: *PalDat*
Picture: *Pardoglossum* sp., page 242: *PalDat*
Picture: *Tetramerium nervosum*, page 242: *PalDat*
Picture: *Cynoglossum officinale*, page 242: *PalDat*
Picture: *Phacelia tanacetifolia*, page 242: *PalDat*
Picture: *Myosotis ramosissima*, page 242: *PalDat*
Picture: *Pseuderanthemum alatum*, page 243: *PalDat*
Picture: *Combretum fruticosum*, page 243: *PalDat*
Picture: *Omphalodes linifolia*, page 243: *PalDat*
Picture: *Medinilla scortechinii*, page 243: *PalDat*
Picture: *Meriania selvaflorensis*, page 243: *PalDat*
Picture: *Phacelia campanularia*, page 243: *PalDat*
Picture: *Alocasia odora*, page 244: Ulrich, Silvia
Picture: *Stylochaeton bogneri*, page 244: Ulrich, Silvia
Picture: *Ambrosina bassii*, page 244: Ulrich, Silvia
Picture: *Synandrospadix vermitoxicus*, page 244: Ulrich, Silvia
Picture: *Spathicarpa sagittifolia*, page 244: Ulrich, Silvia
Picture: *Spathiphyllum* sp., page 244: Ulrich, Silvia
Picture: *Pinellia ternata*, page 245: *PalDat*
Picture: *Populus alba*, page 245: *PalDat*
Picture: *Chlorospatha dodsonii*, page 245: *PalDat*
Picture: *Aglaodorum griffithii*, page 245: *PalDat*
Picture: *Phoebe sheareri*, page 245: *PalDat*
Picture: *Posidonia* sp., page 245: *PalDat*
Picture: *Orchidantha maxillarioides*, page 246: *PalDat*
Picture: *Aristolochia arborea*, page 246: *PalDat*
Picture: *Triglochin maritima*, page 246: *PalDat*
Picture: *Gnetum gnemon*, page 246: *PalDat*

Picture: *Cytinus hypocistis*, page 246: *PalDat*
Picture: *Trillium chloropetalum*, page 246: *PalDat*
Picture: *Cedrus atlantica*, page 247: Weber, Martina
Picture: *Picea* sp., page 247: Ulrich, Silvia
Picture: *Pinus strobus*, page 247: Weber, Martina
Picture: *Pinus cembra*, page 247: Weber, Martina
Picture: *Picea* sp., page 247: Weber, Martina
Picture: *Pinus cembra*, page 247: Weber, Martina
Picture: *Discocleidion rufescens*, page 248: *PalDat*
Picture: *Euphorbia peplus*, page 248: Weber, Martina
Picture: *Fatsia japonica*, page 248: *PalDat*
Picture: *Begonia heracleifolia*, page 248: *PalDat*
Picture: *Lysimachia vulgaris*, page 248: *PalDat*
Picture: *Limnanthes douglasii*, page 248: *PalDat*
Picture: *Merinthopodium neuranthum*, page 249: *PalDat*
Picture: *Merinthopodium neuranthum*, page 249: *PalDat*
Picture: *Butomus umbellatus*, page 249: *PalDat*
Picture: *Blumenbachia hieronymi*, page 249: *PalDat*
Picture: *Omphalodes verna*, page 249: *PalDat*
Picture: *Rhamnus cathartica*, page 249: *PalDat*
Picture: *Salix retusa*, page 250: *PalDat*
Picture: *Anchusa cretica*, page 250: *PalDat*
Picture: *Nematanthus strigillosus*, page 250: *PalDat*
Picture: *Fouquieria columnaris*, page 250: *PalDat*
Picture: *Coris monspeliensis*, page 250: *PalDat*
Picture: *Astragalus tragacantha*, page 250: *PalDat*
Picture: *Dianthus carthusianorum*, page 251: *PalDat*
Picture: *Teucrium pyrenaicum*, page 251: *PalDat*
Picture: *Babiana ecklonii*, page 251: *PalDat*
Picture: *Zea mays*, page 251: *PalDat*
Picture: *Dionaea muscipula*, page 251: *PalDat*
Picture: *Potentilla incana*, page 251: *PalDat*
Picture: *Knautia drymeia*, page 252: *PalDat*
Picture: *Tulipa sylvestris*, page 252: *PalDat*
Picture: *Cucurbita pepo*, page 252: *PalDat*
Picture: *Camellia japonica*, page 252: *PalDat*
Picture: *Agrostemma githago*, page 252: *PalDat*
Picture: *Passiflora citrina*, page 252: *PalDat*
Picture: *Rosa pendulina*, page 253: *PalDat*
Picture: *Passiflora suberosa*, page 253: *PalDat*
Picture: *Agave asperrima*, page 253: *PalDat*
Picture: *Erythronium dens-canis*, page 253: *PalDat*
Picture: *Erythronium dens-canis*, page 253: *PalDat*
Picture: *Potentilla erecta*, page 253: *PalDat*
Picture: *Avena sativa*, page 254: *PalDat*
Picture: *Gladiolus illyricus*, page 254: *PalDat*
Picture: *Poa pratensis*, page 254: *PalDat*
Picture: *Triticum aestivum*, page 254: *PalDat*
Picture: *Plantago lanceolata*, page 254: *PalDat*
Picture: *Poa angustifolia*, page 254: *PalDat*
Picture: *Anemone transsilvanica*, page 255: *PalDat*
Picture: *Opuntia basilaris*, page 255: *PalDat*
Picture: *Ranunculus lanuginosus*, page 255: *PalDat*
Picture: *Portulaca grandiflora*, page 255: *PalDat*
Picture: *Sideritis syriaca*, page 255: *PalDat*
Picture: *Corydalis cava*, page 255: *PalDat*
Picture: *Stigmaphyllon lindenianum*, page 256: *PalDat*
Picture: *Tristellateia australasiae*, page 256: *PalDat*
Picture: *Banisteria muricata*, page 256: *PalDat*
Picture: *Malpighia glabra*, page 256: *PalDat*
Picture: *Malva moschata*, page 256: Weber, Martina
Picture: *Malva alcea*, page 256: Weber, Martina
Picture: *Fumaria officinalis*, page 257: *PalDat*
Picture: *Cucurbita pepo*, page 257: *PalDat*
Picture: *Ribes aureum*, page 257: *PalDat*
Picture: *Costus barbatus*, page 257: *PalDat*
Picture: *Amphitecna macrophylla*, page 257: *PalDat*
Picture: *Opuntia phaeacantha*, page 257: *PalDat*
Picture: *Cobaea scandens*, page 258: Weber, Martina
Picture: *Juglans* sp., page 258: Preusche, Philipp
Picture: *Phaleria capitata*, page 258: *PalDat*
Picture: *Dysphania ambrosioides*, page 258: *PalDat*
Picture: *Atriplex patula*, page 258: *PalDat*
Picture: *Stellaria graminea*, page 258: *PalDat*
Picture: *Cryptomeria* sp., page 259: Zetter, Reinhard
Picture: *Cryptomeria japonica*, page 259: *PalDat*
Picture: *Cryptomeria japonica*, page 259: *PalDat*
Picture: *Metasequoia glyptostroboides*, page 259: *PalDat*
Picture: *Metasequoia glyptostroboides*, page 259: *PalDat*
Picture: *Cunninghamia lanceolata*, page 259: *PalDat*
Picture: *Pachira sessilis*, page 260: Weber, Martina
Picture: *Pachira quinata*, page 260: *PalDat*
Picture: *Centaurea segetum*, page 260: *PalDat*
Picture: *Arbutus unedo*, page 260: *PalDat*
Picture: *Persicaria bistorta*, page 260: *PalDat*
Picture: *Tilia platyphyllos*, page 260: Weber, Martina
Picture: *Euphorbia tithymaloides*, page 261: *PalDat*
Picture: *Justicia brandegeeana*, page 261: *PalDat*
Picture: *Schaueria flavicoma*, page 261: *PalDat*
Picture: *Echinops exaltatus*, page 261: *PalDat*
Picture: *Pachira aquatica*, page 261: *PalDat*
Picture: *Pachira aquatica*, page 261: *PalDat*
Picture: *Sanguisorba cretica*, page 262: *PalDat*
Picture: *Sanguisorba cretica*, page 262: *PalDat*
Picture: *Sarcopoterium spinosum*, page 262: *PalDat*
Picture: *Sarcopoterium spinosum*, page 262: *PalDat*
Picture: *Sanguisorba minor*, page 262: *PalDat*
Picture: *Veratrum nigrum*, page 262: *PalDat*

Picture: *Aechmea allenii*, page 263: *PalDat*
Picture: *Colchicum autumnale*, page 263: *PalDat*
Picture: *Sanchezia nobilis*, page 263: *PalDat*
Picture: *Whitfieldia lateralis*, page 263: *PalDat*
Picture: *Broussonetia papyrifera*, page 263: *PalDat*
Picture: *Quesnelia lateralis*, page 263: *PalDat*
Picture: *Betula humilis*, page 264: Weber, Martina
Picture: *Betula pendula*, page 264: *PalDat*
Picture: *Campanula saxatilis*, page 264: *PalDat*
Picture: *Oenothera fruticosa*, page 264: *PalDat*
Picture: *Cucumis melo*, page 264: *PalDat*
Picture: *Carya* sp., page 264: Zetter, Reinhard
Picture: *Amsonia ciliata*, page 265: *PalDat*
Picture: *Tetrapollinia caerulescens*, page 265: *PalDat*
Picture: *Knautia arvensis*, page 265: *PalDat*
Picture: *Clarkia unguiculata*, page 265: *PalDat*
Picture: *Cordia cylindrostachya*, page 265: *PalDat*
Picture: *Maclura pomifera*, page 265: *PalDat*
Picture: *Phyteuma spicatum*, page 266: Weber, Martina
Picture: *Myriophyllum spicatum*, page 266: *PalDat*
Picture: *Campanula alpina*, page 266: *PalDat*
Picture: *Aechmea fulgens*, page 266: *PalDat*
Picture: *Aechmea tomentosa*, page 266: *PalDat*
Picture: *Asyneuma canescens*, page 266: *PalDat*
Picture: *Alnus glutinosa*, page 267: *PalDat*
Picture: *Legousia speculum-veneris*, page 267: *PalDat*
Picture: *Campanula garganica*, page 267: *PalDat*
Picture: *Irlbachia pendula*, page 267: *PalDat*
Picture: *Elatostema ambiguum*, page 267: *PalDat*
Picture: *Campanula rapunculoides*, page 267: *PalDat*
Picture: *Alnus viridis*, page 268: *PalDat*
Picture: *Pterocarya* sp., page 268: Zetter, Reinhard
Picture: *Drosera kansaiensis*, page 268: *PalDat*
Picture: *Paramoltkia doerfleri*, page 268: *PalDat*
Picture: *Ulmus minor*, page 268: *PalDat*
Picture: *Megaskepasma erythrochlamys*, page 268: *PalDat*
Picture: *Stellaria holostea*, page 269: *PalDat*
Picture: *Cobaea scandens*, page 269: *PalDat*
Picture: *Ipomoea batatas*, page 269: *PalDat*
Picture: *Helianthium bolivianum*, page 269: *PalDat*
Picture: *Calystegia sepium*, page 269: *PalDat*
Picture: *Aechmea azurea*, page 269: *PalDat*
Picture: *Bassia scoparia*, page 270: *PalDat*
Picture: *Bassia scoparia*, page 270: *PalDat*
Picture: *Plantago major*, page 270: *PalDat*
Picture: *Juglans regia*, page 270: *PalDat*
Picture: *Buxus sempervirens*, page 270: *PalDat*
Picture: *Kallstroemia maxima*, page 270: *PalDat*
Picture: *Liquidambar* sp., page 271: Zetter, Reinhard
Picture: *Pavonia multiflora*, page 271: Weber, Martina
Picture: *Arenaria ciliata*, page 271: *PalDat*
Picture: *Lavatera thuringiaca*, page 271: *PalDat*
Picture: *Whitfieldia elongata*, page 271: *PalDat*
Picture: *Alisma lanceolatum*, page 271: *PalDat*
Picture: *Carex remota*, page 272: *PalDat*
Picture: *Cercidiphyllum japonicum*, page 272: *PalDat*
Picture: *Sagittaria sagittifolia*, page 272: *PalDat*
Picture: *Caldesia parnassifolia*, page 272: *PalDat*
Picture: *Schoenoplectus lacustris*, page 272: *PalDat*
Picture: *Scirpus sylvaticus*, page 272: *PalDat*
Picture: *Lythrum salicaria*, page 273: *PalDat*
Picture: *Asperugo procumbens*, page 273: *PalDat*
Picture: *Lumnitzera racemosa*, page 273: *PalDat*
Picture: *Cynoglossum officinale*, page 273: *PalDat*
Picture: *Justicia furcata*, page 273: *PalDat*
Picture: *Pachystachys lutea*, page 273: *PalDat*
Picture: *Zamioculcas zamiifolia*, page 274: Zetter, Reinhard
Picture: *Zamioculcas zamiifolia*, page 274: Zetter, Reinhard
Picture: *Zamioculcas zamiifolia*, page 274: *PalDat*
Picture: *Monstera deliciosa*, page 274: *PalDat*
Picture: *Gonatopus angustus*, page 274: *PalDat*
Picture: *Gonatopus angustus*, page 274: *PalDat*
Picture: *Pedicularis palustris*, page 275: *PalDat*
Picture: *Pedicularis palustris*, page 275: *PalDat*
Picture: *Pedicularis rostratocapitata*, page 275: *PalDat*
Picture: *Iris histrioides*, page 275: *PalDat*
Picture: *Cephalostemon riedelianus*, page 275: *PalDat*
Picture: *Limnanthes douglasii*, page 275: *PalDat*
Picture: *Gonatopus boivinii*, page 276: Ulrich, Silvia
Picture: *Victoria regia*, page 276: *PalDat*
Picture: *Pedicularis gyroflexa*, page 276: *PalDat*
Picture: *Passiflora amethystina*, page 276: *PalDat*
Picture: *Acacia dealbata*, page 276: *PalDat*
Picture: *Acacia dealbata*, page 276: *PalDat*
Picture: *Mimulus* sp., page 277: Weber, Martina
Picture: *Thunbergia alata*, page 277: Weber, Martina
Picture: *Mimulus guttatus*, page 277: *PalDat*
Picture: *Thunbergia alata*, page 277: *PalDat*
Picture: *Claytonia perfoliata*, page 277: *PalDat*
Picture: *Berberis vulgaris*, page 277: *PalDat*
Picture: *Aphyllanthes monspeliensis*, page 278: *PalDat*
Picture: *Berberis amurensis*, page 278: *PalDat*
Picture: *Crocus speciosus*, page 278: *PalDat*
Picture: *Mimulus guttatus*, page 278: *PalDat*
Picture: *Thunbergia laurifolia*, page 278: *PalDat*
Picture: *Bignonia magnifica*, page 278: *PalDat*

Picture: *Dracocephalum austriacum*, page 279: Weber, Martina
Picture: *Salvia glutinosa*, page 279: *PalDat*
Picture: *Plectranthus esculentus*, page 279: Weber, Martina
Picture: *Sanguisorba officinalis*, page 279: Weber, Martina
Picture: *Polygala myrtifolia*, page 279: Weber, Martina
Picture: *Alnus glutinosa*, page 279: Weber, Martina
Picture: *Dracocephalum austriacum*, page 280: *PalDat*
Picture: *Asperula tinctoria*, page 280: *PalDat*
Picture: *Justicia menesii*, page 280: *PalDat*
Picture: *Borago officinalis*, page 280: *PalDat*
Picture: *Polygala myrtifolia*, page 280: *PalDat*
Picture: *Legousia speculum-veneris*, page 280: *PalDat*
Picture: *Iris domestica*, page 281: Weber, Martina
Picture: *Nuphar lutea*, page 281: Weber, Martina
Picture: *Bessera elegans*, page 281: Weber, Martina
Picture: *Freesia* sp., page 281: Weber, Martina
Picture: *Fritillaria pontica*, page 281: Weber, Martina
Picture: *Iris pseudacorus*, page 281: Weber, Martina
Picture: *Lilium martagon*, page 282: *PalDat*
Picture: *Galanthus nivalis*, page 282: *PalDat*
Picture: *Doryanthes palmeri*, page 282: *PalDat*
Picture: *Allium ursinum*, page 282: *PalDat*
Picture: *Cabomba palaeformis*, page 282: *PalDat*
Picture: *Asphodeline lutea*, page 282: *PalDat*
Picture: *Lachenalia aloides*, page 283: Weber, Martina
Picture: *Catopsis floribunda*, page 283: *PalDat*
Picture: *Iris pseudacorus*, page 283: *PalDat*
Picture: *Iris reichenbachii*, page 283: *PalDat*
Picture: *Vriesea neoglutinosa*, page 283: *PalDat*
Picture: *Paradisea liliastrum*, page 283: *PalDat*
Picture: *Nuphar lutea*, page 284: *PalDat*
Picture: *Nuphar lutea*, page 284: Zetter, Reinhard
Picture: *Liriodendron tulipifera*, page 284: *PalDat*
Picture: *Liriodendron tulipifera*, page 284: *PalDat*
Picture: *Anaphyllopsis americana*, page 284: *PalDat*
Picture: *Anaphyllopsis americana*, page 284: *PalDat*
Picture: *Allium sphaerocephalum*, page 285: *PalDat*
Picture: *Asphodelus fistulosus*, page 285: *PalDat*
Picture: *Bessera elegans*, page 285: *PalDat*
Picture: *Ceratozamia kuesteriana*, page 285: *PalDat*
Picture: *Cordyline fruticosa*, page 285: *PalDat*
Picture: *Saururus cernuus*, page 285: *PalDat*
Picture: *Guzmania elvallensis*, page 286: *PalDat*
Picture: *Wachendorfia thyrsiflora*, page 286: *PalDat*
Picture: *Chamaedorea microspadix*, page 286: *PalDat*
Picture: *Tradescantia zebrina*, page 286: *PalDat*
Picture: *Tradescantia zebrina*, page 286: *PalDat*
Picture: *Veratrum album*, page 286: *PalDat*
Picture: *Tofieldia calyculata*, page 287: *PalDat*
Picture: *Xerophyta elegans*, page 287: *PalDat*
Picture: *Uvularia grandiflora*, page 287: *PalDat*
Picture: *Uvularia grandiflora*, page 287: *PalDat*
Picture: *Eichhornia crassipes*, page 287: *PalDat*
Picture: *Crinum x amabile*, page 287: *PalDat*
Picture: Arecaceae, page 288: Grímsson, Friðgeir
Picture: *Dianella intermedia*, page 288: *PalDat*
Picture: *Dianella intermedia*, page 288: *PalDat*
Picture: *Dianella caerulea*, page 288: *PalDat*
Picture: *Dianella tasmanica*, page 288: *PalDat*
Picture: *Dianella tasmanica*, page 288: *PalDat*
Picture: *Hedyosmum scaberrimum*, page 289: Grímsson, Friðgeir
Picture: *Hedyosmum scaberrimum*, page 289: *PalDat*
Picture: *Hedyosmum scaberrimum*, page 289: *PalDat*
Picture: *Hedyosmum scaberrimum*, page 289: *PalDat*
Picture: *Hedyosmum brasiliense*, page 289: Grímsson, Friðgeir
Picture: *Hedyosmum brasiliense*, page 289: *PalDat*
Picture: *Aetanthus macranthus*, page 290: Grímsson, Friðgeir
Picture: *Pseudofumaria lutea*, page 290: *PalDat*
Picture: *Pedicularis verticillata*, page 290: *PalDat*
Picture: *Barringtonia asiatica*, page 290: *PalDat*
Picture: *Berberis nervosa*, page 290: *PalDat*
Picture: *Dodecatheon meadia*, page 290: *PalDat*
Picture: *Callistemon coccineus*, page 291: *PalDat*
Picture: *Cuphea procumbens*, page 291: *PalDat*
Picture: *Cassia pulcherrima*, page 291: *PalDat*
Picture: *Myrtus communis*, page 291: *PalDat*
Picture: *Ardisia crenata*, page 291: *PalDat*
Picture: *Onosma visianii*, page 291: *PalDat*
Picture: Poaceae, page 292: Weber, Martina
Picture: *Bromus erectus*, page 292: *PalDat*
Picture: *Cephalanthera longifolia*, page 292: *PalDat*
Picture: *Luzula luzuloides*, page 292: *PalDat*
Picture: *Sansevieria parva*, page 292: *PalDat*
Picture: *Juniperus communis*, page 292: *PalDat*
Picture: *Heliconia* sp., page 293: *PalDat*
Picture: *Cyrtosperma beccarianum*, page 293: *PalDat*
Picture: *Sparganium erectum*, page 293: *PalDat*
Picture: *Neottia nidus-avis*, page 293: *PalDat*
Picture: *Amborella trichopoda*, page 293: *PalDat*
Picture: *Drimys granadensis*, page 293: *PalDat*

Chapter: "Ornamentation"

Picture: *Peperomia rubella*, page 296: *PalDat*
Picture: *Dracunculus vulgaris*, page 296: *PalDat*
Picture: *Mimosa pudica*, page 296: *PalDat*
Picture: *Poikilacanthus macranthus*, page 296: *PalDat*
Picture: *Cynodon dactylon*, page 296: *PalDat*
Picture: *Justicia brandegeeana*, page 296: *PalDat*
Picture: *Justicia carnea*, page 297: Weber, Martina
Picture: *Justicia carnea*, page 297: *PalDat*
Picture: *Peperomia polybotrya*, page 297: *PalDat*
Picture: *Peperomia polybotrya*, page 297: *PalDat*
Picture: *Megaskepasma erythrochlamys*, page 297: *PalDat*
Picture: *Megaskepasma erythrochlamys*, page 297: *PalDat*
Picture: *Viscum album*, page 298: *PalDat*
Picture: *Viscum album*, page 298: *PalDat*
Picture: *Viscum album*, page 298: *PalDat*
Picture: *Nymphaea alba*, page 298: Weber, Martina
Picture: *Erythrochiton brasiliensis*, page 298: *PalDat*
Picture: *Erythrochiton brasiliensis*, page 298: *PalDat*
Picture: *Ocimum basilicum*, page 299: *PalDat*
Picture: *Ocimum basilicum*, page 299: *PalDat*
Picture: *Ocimum basilicum*, page 299: *PalDat*
Picture: *Salvia hians*, page 299: *PalDat*
Picture: *Salvia argentea*, page 299: *PalDat*
Picture: *Salvia glutinosa*, page 299: *PalDat*
Picture: *Phlox drummondii*, page 300: *PalDat*
Picture: *Phlox paniculata*, page 300: *PalDat*
Picture: *Agastache mexicana*, page 300: *PalDat*
Picture: *Agastache mexicana*, page 300: *PalDat*
Picture: *Vitex trifolia*, page 300: *PalDat*
Picture: *Pachystachys lutea*, page 300: *PalDat*
Picture: *Plectranthus ornatus*, page 301: *PalDat*
Picture: *Hyptis suaveolens*, page 301: *PalDat*
Picture: *Sparmannia africana*, page 301: *PalDat*
Picture: *Melittis melissophyllum*, page 301: *PalDat*
Picture: *Prunella grandiflora*, page 301: *PalDat*
Picture: *Andrographis paniculata*, page 301: *PalDat*
Picture: *Iris planifolia*, page 302: *PalDat*
Picture: *Ilex* sp., page 302: *PalDat*
Picture: *Ilex aquifolium*, page 302: *PalDat*
Picture: *Ilex aquifolium*, page 302: *PalDat*
Picture: *Linum flavum*, page 302: *PalDat*
Picture: *Aratitiyopea lopezii*, page 302: *PalDat*
Picture: *Linum* sp., page 303: *PalDat*
Picture: *Hymenocallis tubiflora*, page 303: *PalDat*
Picture: *Linum capitatum*, page 303: *PalDat*
Picture: *Linum capitatum*, page 303: *PalDat*
Picture: *Geranium robertianum*, page 303: *PalDat*
Picture: *Geranium robertianum*, page 303: *PalDat*
Picture: *Jatropha multifida*, page 304: *PalDat*
Picture: *Plumbago auriculata*, page 304: *PalDat*
Picture: *Hymenocallis littoralis*, page 304: *PalDat*
Picture: *Viburnum lantana*, page 304: *PalDat*
Picture: *Geranium pratense*, page 304: *PalDat*
Picture: *Geranium sibiricum*, page 304: *PalDat*
Picture: *Ibicella lutea*, page 305: *PalDat*
Picture: *Ibicella lutea*, page 305: *PalDat*
Picture: *Proboscidea fragrans*, page 305: *PalDat*
Picture: *Pseudofumaria lutea*, page 305: *PalDat*
Picture: *Iris bucharica*, page 305: *PalDat*
Picture: *Iris bucharica*, page 305: *PalDat*
Picture: *Catalpa bignonioides*, page 306: *PalDat*
Picture: *Iris graeberiana*, page 306: *PalDat*
Picture: *Lophophora williamsii*, page 306: *PalDat*
Picture: *Phyllanthus* sp., page 306: *PalDat*
Picture: *Irlbachia pedunculata*, page 306: *PalDat*
Picture: *Banisteria muricata*, page 306: *PalDat*
Picture: *Croton triqueter*, page 307: *PalDat*
Picture: *Croton triqueter*, page 307: Weber, Martina
Picture: *Jatropha podagrica*, page 307: *PalDat*
Picture: *Daphne laureola*, page 307: *PalDat*
Picture: *Daphne cneorum*, page 307: *PalDat*
Picture: *Thymelaea passerina*, page 307: *PalDat*
Picture: *Garcia nutans*, page 308: Weber, Martina
Picture: *Garcia nutans*, page 308: *PalDat*
Picture: *Daphne tangutica*, page 308: *PalDat*
Picture: *Callitriche stagnalis*, page 308: *PalDat*
Picture: *Croton triqueter*, page 308: *PalDat*
Picture: *Thymelaea passerina*, page 308: *PalDat*
Picture: *Galinsoga ciliata*, page 309: *PalDat*
Picture: *Carduus acanthoides*, page 309: *PalDat*
Picture: *Hibiscus trionum*, page 309: *PalDat*
Picture: *Lavatera thuringiaca*, page 309: *PalDat*
Picture: *Pinellia ternata*, page 309: *PalDat*
Picture: *Pinellia ternata*, page 309: *PalDat*
Picture: *Pavonia multiflora*, page 310: Weber, Martina
Picture: *Abutilon megapotamicum*, page 310: Weber, Martina
Picture: *Calendula officinalis*, page 310: Weber, Martina
Picture: *Drosera scorpioides*, page 310: Weber, Martina
Picture: *Nuphar lutea*, page 310: Weber, Martina
Picture: *Alocasia acuminata*, page 310: Ulrich, Silvia
Picture: *Patrinia gibbosa*, page 311: *PalDat*
Picture: *Hieracium hoppeanum*, page 311: *PalDat*
Picture: *Ambrosia artemisiifolia*, page 311: *PalDat*
Picture: *Aster amellus*, page 311: *PalDat*
Picture: *Portulaca grandiflora*, page 311: *PalDat*
Picture: *Nuphar lutea*, page 311: *PalDat*
Picture: *Petasites albus*, page 312: *PalDat*
Picture: *Antennaria dioica*, page 312: *PalDat*
Picture: *Malva neglecta*, page 312: *PalDat*
Picture: *Tanacetum corymbosum*, page 312: *PalDat*

Picture: *Lonicera fragrantissima*, page 312: *PalDat*
Picture: *Ipomoea purpurea*, page 312: *PalDat*
Picture: *Arenga pinnata*, page 313: *PalDat*
Picture: *Zomicarpa riedeliana*, page 313: *PalDat*
Picture: *Helianthus annuus*, page 313: *PalDat*
Picture: *Knautia drymeia*, page 313: *PalDat*
Picture: *Campanula alpina*, page 313: *PalDat*
Picture: *Ulearum sagittatum*, page 313: *PalDat*
Picture: *Scorzonera austriaca*, page 314: *PalDat*
Picture: *Ipomoea batatas*, page 314: *PalDat*
Picture: *Tetrapollinia caerulescens*, page 314: *PalDat*
Picture: *Gynura scandens*, page 314: *PalDat*
Picture: *Stratiotes aloides*, page 314: *PalDat*
Picture: *Hibiscus schizopetalus*, page 314: *PalDat*
Picture: *Mendoncia albida*, page 315: *PalDat*
Picture: *Erica herbacea*, page 315: *PalDat*
Picture: *Rhododendron hirsutum*, page 315: *PalDat*
Picture: *Ledum palustre*, page 315: *PalDat*
Picture: *Aristolochia manshuriensis*, page 315: *PalDat*
Picture: *Moneses uniflora*, page 315: *PalDat*
Picture: *Pithecellobium dulce*, page 316: *PalDat*
Picture: *Pithecellobium dulce*, page 316: *PalDat*
Picture: *Gaultheria myrsinoides*, page 316: *PalDat*
Picture: *Besleria hirsuta*, page 316: *PalDat*
Picture: *Lagerstroemia indica*, page 316: *PalDat*
Picture: *Lagerstroemia indica*, page 316: *PalDat*
Picture: *Streptocalyx poeppigii*, page 317: *PalDat*
Picture: *Canistrum camacaense*, page 317: *PalDat*
Picture: *Lavandula angustifolia*, page 317: *PalDat*
Picture: *Aechmea araneosa*, page 317: *PalDat*
Picture: *Glossoloma ichthyoderma*, page 317: *PalDat*
Picture: *Glossoloma ichthyoderma*, page 317: *PalDat*
Picture: *Nematanthus strigillosus*, page 318: *PalDat*
Picture: *Cyrtosperma beccarianum*, page 318: *PalDat*
Picture: *Maianthemum stellatum*, page 318: *PalDat*
Picture: *Hohenbergia stellata*, page 318: *PalDat*
Picture: *Aechmea ampla*, page 318: *PalDat*
Picture: *Aechmea ampla*, page 318: *PalDat*
Picture: *Ruellia makoyana*, page 319: *PalDat*
Picture: *Impatiens parviflora*, page 319: *PalDat*
Picture: *Erophila verna*, page 319: *PalDat*
Picture: *Bougainvillea* sp., page 319: *PalDat*
Picture: *Viburnum opulus*, page 319: *PalDat*
Picture: *Thladiantha hookeri*, page 319: *PalDat*
Picture: *Ruellia devosiana*, page 320: *PalDat*
Picture: *Persicaria chinensis*, page 320: *PalDat*
Picture: *Fouquieria macdougalii*, page 320: *PalDat*
Picture: *Ardisia elliptica*, page 320: *PalDat*
Picture: *Schotia brachypetala*, page 320: *PalDat*
Picture: *Passiflora amethystina*, page 320: *PalDat*
Picture: *Dionaea muscipula*, page 321: *PalDat*
Picture: *Cephalopentandra ecirrhosa*, page 321: *PalDat*
Picture: *Stenandrium dulce*, page 321: *PalDat*
Picture: *Asarum europaeum*, page 321: *PalDat*
Picture: *Fatsia japonica*, page 321: *PalDat*
Picture: *Hakea kippistiana*, page 321: *PalDat*
Picture: *Turnera ulmifolia*, page 322: *PalDat*
Picture: *Wachendorfia thyrsiflora*, page 322: *PalDat*
Picture: *Cantua buxifolia*, page 322: *PalDat*
Picture: *Cantua buxifolia*, page 322: *PalDat*
Picture: *Roridula gorgonias*, page 322: *PalDat*
Picture: *Calluna vulgaris*, page 322: *PalDat*
Picture: *Hemiptelia* sp., page 323: Grímsson, Friðgeir
Picture: *Clarkia pulchella*, page 323: *PalDat*
Picture: *Luzula campestris*, page 323: *PalDat*
Picture: *Larix decidua*, page 323: *PalDat*
Picture: *Humulus lupulus*, page 323: *PalDat*
Picture: *Clarkia purpurea*, page 323: *PalDat*
Picture: *Hedera helix*, page 324: *PalDat*
Picture: *Lachenalia aloides*, page 324: *PalDat*
Picture: *Fritillaria meleagris*, page 324: *PalDat*
Picture: *Anthericum ramosum*, page 324: *PalDat*
Picture: *Billbergia seidelii*, page 324: *PalDat*
Picture: *Limodorum abortivum*, page 324: *PalDat*
Picture: *Ornithogalum narbonense*, page 325: *PalDat*
Picture: *Ornithogalum narbonense*, page 325: *PalDat*
Picture: *Bomarea hirsuta*, page 325: *PalDat*
Picture: *Bessera elegans*, page 325: Weber, Martina
Picture: *Chelonanthus alatus*, page 325: *PalDat*
Picture: *Plectranthus esculentus*, page 325: Weber, Martina
Picture: *Acantholimon glumaceum*, page 326: *PalDat*
Picture: *Abeliophyllum distichum*, page 326: *PalDat*
Picture: *Eranthemum wattii*, page 326: *PalDat*
Picture: *Strobilanthes roseus*, page 326: *PalDat*
Picture: *Ruellia* sp., page 326: *PalDat*
Picture: *Thlaspi montanum*, page 326: *PalDat*
Picture: *Armeria pinifolia*, page 327: *PalDat*
Picture: *Armeria pinifolia*, page 327: *PalDat*
Picture: *Impatiens parviflora*, page 327: Weber, Martina
Picture: *Persicaria chinensis*, page 327: *PalDat*
Picture: *Kallstroemia maxima*, page 327: *PalDat*
Picture: *Ruellia tuberosa*, page 327: *PalDat*
Picture: *Leontodon saxatilis*, page 328: *PalDat*
Picture: *Cichorium intybus*, page 328: *PalDat*
Picture: *Opuntia basilaris*, page 328: *PalDat*
Picture: *Pfaffia tuberosa*, page 328: *PalDat*
Picture: *Gazania* sp., page 328: *PalDat*
Picture: *Hieracium hoppeanum*, page 328: *PalDat*
Picture: *Tragopogon orientalis*, page 329: Weber, Martina

Picture: *Prenanthes purpurea*, page 329: Weber, Martina
Picture: *Cichorium intybus*, page 329: Weber, Martina
Picture: *Taraxacum* sp., page 329: Zetter, Reinhard
Picture: *Tragopogon dubius*, page 329: *PalDat*
Picture: *Crepis biennis*, page 329: *PalDat*
Picture: *Cyanthillium cinereum*, page 330: *PalDat*
Picture: *Gomphrena celosioides*, page 330: *PalDat*
Picture: *Opuntia polyacantha*, page 330: *PalDat*
Picture: *Scorzonera aristata*, page 330: *PalDat*
Picture: *Ipomoea caerulea*, page 330: *PalDat*
Picture: *Herniaria alpina*, page 330: *PalDat*
Picture: *Heloniopsis kawanoi*, page 331: *PalDat*
Picture: *Trillium grandiflorum*, page 331: *PalDat*
Picture: *Kickxia spuria*, page 331: *PalDat*
Picture: *Lamium purpureum*, page 331: *PalDat*
Picture: *Aspidistra elatior*, page 331: *PalDat*
Picture: *Callisia fragrans*, page 331: *PalDat*
Picture: *Orobanche lutea*, page 332: *PalDat*
Picture: *Trillium chloropetalum*, page 332: *PalDat*
Picture: *Claytonia perfoliata*, page 332: *PalDat*
Picture: *Petrorhagia prolifera*, page 332: *PalDat*
Picture: *Veronica prostrata*, page 332: *PalDat*
Picture: *Ptelea trifoliolata*, page 332: *PalDat*
Picture: *Podophyllum peltatum*, page 333: *PalDat*
Picture: *Solanum torvum*, page 333: *PalDat*
Picture: *Campanula persicifolia*, page 333: *PalDat*
Picture: *Melampyrum pratense*, page 333: *PalDat*
Picture: *Tilia* sp., page 333: Grímsson, Friðgeir
Picture: *Tilia* sp., page 333: Grímsson, Friðgeir
Picture: *Tradescantia spathacea*, page 334: *PalDat*
Picture: *Aspidistra locii*, page 334: *PalDat*
Picture: *Smilax spinosa*, page 334: *PalDat*
Picture: *Clethra arborea*, page 334: *PalDat*
Picture: *Quercus robur*, page 334: *PalDat*
Picture: *Erica pageana*, page 334: *PalDat*
Picture: *Elaeagnus rhamnoides*, page 335: *PalDat*
Picture: *Polyscias filicifolia*, page 335: *PalDat*
Picture: *Cytisus nigricans*, page 335: *PalDat*
Picture: *Reseda luteola*, page 335: *PalDat*
Picture: *Anemone pratensis*, page 335: *PalDat*
Picture: *Drosera kansaiensis*, page 335: *PalDat*
Picture: *Babiana ecklonii*, page 336: *PalDat*
Picture: *Galium lucidum*, page 336: *PalDat*
Picture: *Scirpoides holoschoenus*, page 336: *PalDat*
Picture: Poaceae, page 336: Grímsson, Friðgeir
Picture: *Callisia fragrans*, page 336: *PalDat*
Picture: *Oenothera fruticosa*, page 336: *PalDat*
Picture: *Dianella tasmanica*, page 337: *PalDat*
Picture: *Veronica longifolia*, page 337: *PalDat*
Picture: *Symphytum caucasicum*, page 337: *PalDat*
Picture: *Hordeum bulbosum*, page 337: *PalDat*
Picture: *Juglans* sp., page 337: Grímsson, Friðgeir
Picture: Amaranthaceae, page 337: Grímsson, Friðgeir
Picture: *Carex filiformis*, page 338: *PalDat*
Picture: Cyperaceae, page 338: Grímsson, Friðgeir
Picture: *Viola calcarata*, page 338:
Picture: *Tilia* sp., page 338: Grímsson, Friðgeir
Picture: *Pterocarya* sp., page 338: Grímsson, Friðgeir
Picture: Rosaceae, page 338: Grímsson, Friðgeir
Picture: *Borago pygmaea*, page 339: *PalDat*
Picture: *Epilobium palustre*, page 339: *PalDat*
Picture: *Rhaphidophora africana*, page 339: *PalDat*
Picture: *Trapa* sp., page 339: Grímsson, Friðgeir
Picture: Caryophyllaceae, page 339: Grímsson, Friðgeir
Picture: Amaranthaceae, page 339: Grímsson, Friðgeir
Picture: *Arceuthobium* sp., page 340: Grímsson, Friðgeir
Picture: Caprifoliaceae, page 340: Grímsson, Friðgeir
Picture: *Drosera binata*, page 340: *PalDat*
Picture: *Alisma* sp., page 340: Grímsson, Friðgeir
Picture: *Heliohebe raoulii*, page 340: *PalDat*
Picture: *Maclura pomifera*, page 340: *PalDat*
Picture: *Trapa* sp., page 341: Grímsson, Friðgeir
Picture: Onagraceae, page 341: Grímsson, Friðgeir
Picture: *Pedicularis verticillata*, page 341: *PalDat*
Picture: *Tetragonia tetragonioides*, page 341: *PalDat*
Picture: *Haloragis erecta*, page 341: *PalDat*
Picture: Haloragaceae, page 341: Grímsson, Friðgeir
Picture: *Colutea arborescens*, page 342: *PalDat*
Picture: *Gonatopus angustus*, page 342: *PalDat*
Picture: *Myosotis arvensis*, page 342: *PalDat*
Picture: *Euphorbia palustris*, page 342: *PalDat*
Picture: *Lysimachia nemorum*, page 342: *PalDat*
Picture: *Cucumis sativus*, page 342: *PalDat*
Picture: *Pulmonaria officinalis*, page 343: *PalDat*
Picture: *Hippocrepis emerus*, page 343: *PalDat*
Picture: *Napoleonaea imperialis*, page 343: *PalDat*
Picture: *Euphorbia helioscopia*, page 343: *PalDat*
Picture: *Rumex acetosa*, page 343: *PalDat*
Picture: *Trifolium montanum*, page 343: *PalDat*
Picture: *Ephedra distachya*, page 344: *PalDat*
Picture: *Ephedra* sp., page 344: Grímsson, Friðgeir
Picture: *Ephedra* sp., page 344: Zetter, Reinhard
Picture: *Hemigraphis primulaefolia*, page 344: *PalDat*
Picture: *Pistia stratiotes*, page 344: *PalDat*
Picture: *Pistia stratiotes*, page 344: *PalDat*
Picture: *Welwitschia mirabilis*, page 345: *PalDat*
Picture: *Amorphophallus lacourii*, page 345: *PalDat*

Picture: *Amorphophallus serrulatus*, page 345: Ulrich, Silvia
Picture: *Brillantaisia owariensis*, page 345: *PalDat*
Picture: *Spathiphyllum minor*, page 345: Ulrich, Silvia
Picture: *Spathiphyllum cannifolium*, page 345: *PalDat*
Picture: *Hedychium gardnerianum*, page 346: *PalDat*
Picture: *Anthyllis vulneraria*, page 346: *PalDat*
Picture: *Lithospermum officinale*, page 346: *PalDat*
Picture: *Vinca minor*, page 346: *PalDat*
Picture: *Orchidantha maxillarioides*, page 346: *PalDat*
Picture: *Sempervivum globiferum*, page 346: *PalDat*
Picture: *Lamium album*, page 347: Weber, Martina
Picture: *Androlepis skinneri*, page 347: *PalDat*
Picture: *Trigonia nivea*, page 347: *PalDat*
Picture: *Maxillaria densa*, page 347: *PalDat*
Picture: *Cheilocostus speciosus*, page 347: *PalDat*
Picture: *Whitfieldia lateritia*, page 347: *PalDat*
Picture: *Cardamine pratensis*, page 348: *PalDat*
Picture: *Luffa cylindrica*, page 348: *PalDat*
Picture: *Physostegia virginiana*, page 348: *PalDat*
Picture: *Aechmea azurea*, page 348: *PalDat*
Picture: *Beloperone guttata*, page 348: *PalDat*
Picture: *Razisea citrina*, page 348: *PalDat*
Picture: *Salix daphnoides*, page 349: *PalDat*
Picture: *Persicaria mitis*, page 349: *PalDat*
Picture: *Paradisea liliastrum*, page 349: *PalDat*
Picture: *Trifolium rubens*, page 349: *PalDat*
Picture: *Ecballium elaterium*, page 349: *PalDat*
Picture: *Ajuga genevensis*, page 349: *PalDat*
Picture: *Ruellia brevifolia*, page 350: *PalDat*
Picture: *Buxus sempervirens*, page 350: *PalDat*
Picture: *Impatiens glandulifera*, page 350: *PalDat*
Picture: *Lupinus polyphyllus*, page 350: *PalDat*
Picture: *Opuntia paraguayensis*, page 350: *PalDat*
Picture: *Theobroma cacao*, page 350: *PalDat*
Picture: *Tropaeolum majus*, page 351: *PalDat*
Picture: *Veratrum album*, page 351: *PalDat*
Picture: *Werauhia tarmaensis*, page 351: *PalDat*
Picture: *Erythronium dens-canis*, page 351: *PalDat*
Picture: *Poncirus trifoliata*, page 351: *PalDat*
Picture: *Melilotus officinalis*, page 351: *PalDat*
Picture: *Vigna speciosa*, page 352: *PalDat*
Picture: *Epipactis helleborine*, page 352: *PalDat*
Picture: *Harpochilus neesianus*, page 352: *PalDat*
Picture: *Polygala major*, page 352: *PalDat*
Picture: *Adenia fruticosa*, page 352: *PalDat*
Picture: *Pisum sativum*, page 352: *PalDat*
Picture: *Thladiantha hookeri*, page 353: *PalDat*
Picture: *Pinguicula alpina*, page 353: *PalDat*
Picture: *Billardiera heterophylla*, page 353: *PalDat*
Picture: *Cephalanthera longifolia*, page 353: *PalDat*
Picture: *Aechmea allenii*, page 353: *PalDat*
Picture: *Lathyrus vernus*, page 353: *PalDat*
Picture: *Jasminum nudiflorum*, page 354: Weber, Martina
Picture: Lamiaceae, page 354: Weber, Martina
Picture: *Iris pseudacorus*, page 354: Weber, Martina
Picture: *Iris domestica*, page 354: Weber, Martina
Picture: *Fraxinus excelsior*, page 354: *PalDat*
Picture: *Persicaria* sp., page 354: Zetter, Reinhard
Picture: *Lilium martagon*, page 355: *PalDat*
Picture: *Lilium candidum*, page 355: *PalDat*
Picture: *Phaleria capitata*, page 355: Weber, Martina
Picture: *Phaleria capitata*, page 355: *PalDat*
Picture: *Fumana procumbens*, page 355: *PalDat*
Picture: *Erdtmanipollis*, page 355: Zetter, Reinhard
Picture: *Anthurium gracile*, page 356: *PalDat*
Picture: *Anthurium gracile*, page 356: *PalDat*
Picture: *Geranium pratense*, page 356: Weber, Martina
Picture: *Geranium reuteri*, page 356: *PalDat*
Picture: *Sarcococca pruniformis*, page 356: *PalDat*
Picture: *Sarcococca pruniformis*, page 356: *PalDat*
Picture: Liliaceae, page 357: Weber, Martina
Picture: *Pachira aquatica*, page 357: *PalDat*
Picture: *Pachysandra terminalis*, page 357: *PalDat*
Picture: *Mercurialis perennis*, page 357: *PalDat*
Picture: *Aponogeton masoalaensis*, page 357: *PalDat*
Picture: *Armeria maritima*, page 357: *PalDat*
Picture: *Securigera varia*, page 358: *PalDat*
Picture: *Zelkova* sp., page 358: Zetter, Reinhard
Picture: *Peucedanum cervaria*, page 358: *PalDat*
Picture: *Fagus* sp., page 358: *PalDat*
Picture: *Sedum acre*, page 358: *PalDat*
Picture: *Circaea lutetiana*, page 358: *PalDat*
Picture: *Carpinus betulus*, page 359: *PalDat*
Picture: *Acer negundo*, page 359: *PalDat*
Picture: *Nymphoides peltata*, page 359: *PalDat*
Picture: *Leucadendron discolor*, page 359: *PalDat*
Picture: *Nicotiana tabacum*, page 359: *PalDat*
Picture: *Myrrhis odorata*, page 359: *PalDat*
Picture: *Fagus* sp., page 360: Zetter, Reinhard
Picture: *Populus alba*, page 360: Weber, Martina
Picture: *Dioon edule*, page 360: Weber, Martina
Picture: *Betula humilis*, page 360: Weber, Martina
Picture: *Ceratozamia mexicana*, page 360: Weber, Martina
Picture: *Sanguisorba officinalis*, page 360: Weber, Martina
Picture: *Acer pseudoplatanus*, page 361: *PalDat*
Picture: *Prunus avium*, page 361: *PalDat*
Picture: *Potentilla inclinata*, page 361: *PalDat*
Picture: *Veronica cinerea*, page 361: *PalDat*

Picture: *Amherstia nobilis*, page 361: *PalDat*
Picture: *Gentiana lutea*, page 361: *PalDat*
Picture: *Saxifraga rotundifolia*, page 362: *PalDat*
Picture: *Aesculus hippocastanum*, page 362: *PalDat*
Picture: *Lycium barbarum*, page 362: *PalDat*
Picture: *Chaenomeles sinensis*, page 362: *PalDat*
Picture: *Rubus caesius*, page 362: *PalDat*
Picture: *Allium flavum*, page 362: *PalDat*
Picture: *Geum reptans*, page 363: *PalDat*
Picture: *Sanguisorba minor*, page 363: *PalDat*
Picture: *Saxifraga taygetea*, page 363: *PalDat*
Picture: *Saxifraga taygetea*, page 363: *PalDat*
Picture: *Begonia heracleifolia*, page 363: *PalDat*
Picture: *Cabomba palaeformis*, page 363: *PalDat*
Picture: *Cuphea llavea*, page 364: *PalDat*
Picture: *Crataegus laevigata*, page 364: *PalDat*
Picture: *Malus sylvestris*, page 364: *PalDat*
Picture: *Neoalsomitra sarcophylla*, page 364: *PalDat*
Picture: *Ruta graveolens*, page 364: *PalDat*
Picture: *Helianthemum nummularium*, page 364: *PalDat*
Picture: *Menyanthes trifoliata*, page 365: *PalDat*
Picture: *Prunus laurocerasus*, page 365: *PalDat*
Picture: *Saxifraga tridactylites*, page 365: *PalDat*
Picture: *Acer pseudoplatanus*, page 365: *PalDat*
Picture: *Amorphophallus interruptus*, page 365: Ulrich, Silvia
Picture: *Amorphophallus serrulatus*, page 365: Ulrich, Silvia
Picture: *Blackstonia perfoliata*, page 366: *PalDat*
Picture: *Erodium cicutarium*, page 366: *PalDat*
Picture: *Gentianella austriaca*, page 366: *PalDat*
Picture: *Gelsemium sempervirens*, page 366: *PalDat*
Picture: *Cotinus coggygria*, page 366: *PalDat*
Picture: *Cotinus coggygria*, page 366: *PalDat*
Picture: *Pelargonium tetragonum*, page 367: *PalDat*
Picture: *Pelargonium carnosum*, page 367: *PalDat*
Picture: *Fouquieria macdougalii*, page 367: *PalDat*
Picture: *Fouquieria macdougalii*, page 367: *PalDat*
Picture: *Ailanthus altissima*, page 367: *PalDat*
Picture: *Cistus clusii*, page 367: *PalDat*
Picture: *Solandra longiflora*, page 368: *PalDat*
Picture: *Pelargonium punctatum*, page 368: *PalDat*
Picture: *Gentiana acaulis*, page 368: *PalDat*
Picture: *Polemonium caeruleum*, page 368: *PalDat*
Picture: *Polemonium caeruleum*, page 368: Weber, Martina
Picture: *Polemonium pauciflorum*, page 368: *PalDat*
Picture: *Brugmansia suaveolens*, page 369: *PalDat*
Picture: *Chlorospatha pubescens*, page 369: *PalDat*
Picture: *Schinus molle*, page 369: *PalDat*
Picture: *Solandra maxima*, page 369: *PalDat*
Picture: *Plagiorhegma dubium*, page 369: *PalDat*
Picture: *Amherstia nobilis*, page 369: *PalDat*
Picture: *Anthurium gracile*, page 370: *PalDat*
Picture: *Anthurium gracile*, page 370: *PalDat*
Picture: *Triglochin maritima*, page 370: *PalDat*
Picture: *Ascarina lucida*, page 370: Grímsson, Friðgeir
Picture: *Lophophora williamsii*, page 370: *PalDat*
Picture: *Phillyrea angustifolia*, page 370: *PalDat*
Picture: *Fraxinus* sp., page 371: Grímsson, Friðgeir
Picture: *Tulipa turkestanica*, page 371: *PalDat*
Picture: *Helleborus foetidus*, page 371: *PalDat*
Picture: *Cercidiphyllum* sp., page 371: Grímsson, Friðgeir
Picture: *Parentucellia latifolia*, page 371: *PalDat*
Picture: *Viburnum utile*, page 371: *PalDat*
Picture: *Linum bienne*, page 372: *PalDat*
Picture: *Saxifraga hostii*, page 372: *PalDat*
Picture: *Betula* sp., page 372: Grímsson, Friðgeir
Picture: *Piper auritum*, page 372: *PalDat*
Picture: *Ulmus* sp., page 372: Grímsson, Friðgeir
Picture: *Tsuga* sp., page 372: Grímsson, Friðgeir
Picture: *Ceratostigma plumbaginoides*, page 373: *PalDat*
Picture: *Ilex* sp., page 373: Grímsson, Friðgeir
Picture: *Akebia quinata*, page 373: *PalDat*
Picture: *Polemonium caeruleum*, page 373: *PalDat*
Picture: *Ouratea hexasperma*, page 373: *PalDat*
Picture: *Coffea arabica*, page 373: *PalDat*
Picture: *Geranium robertianum*, page 374: *PalDat*
Picture: *Geranium robertianum*, page 374: Weber, Martina
Picture: *Atriplex sagittata*, page 374: *PalDat*
Picture: *Pulsatilla grandis*, page 374: *PalDat*
Picture: *Quercus robur*, page 374: *PalDat*
Picture: *Plantago maritima*, page 374: *PalDat*
Picture: *Aristolochia arborea*, page 375: *PalDat*
Picture: *Plantago media*, page 375: *PalDat*
Picture: *Aristolochia tricaudata*, page 375: *PalDat*
Picture: *Calliandra tergemina*, page 375: *PalDat*
Picture: *Corydalis cava*, page 375: *PalDat*
Picture: *Corydalis cava*, page 375: Weber, Martina
Picture: *Teucrium chamaedrys*, page 376: *PalDat*
Picture: *Cyanastrum cordifolium*, page 376: *PalDat*
Picture: *Platycapnos tenuilobus*, page 376: *PalDat*
Picture: *Platycapnos tenuilobus*, page 376: *PalDat*
Picture: *Plantago reniformis*, page 376: *PalDat*
Picture: *Plantago lanceolata*, page 376: *PalDat*
Picture: *Amorphophallus stuhlmannii*, page 377: Ulrich, Silvia
Picture: *Amorphophallus stuhlmannii*, page 377: *PalDat*
Picture: *Sarcocapnos enneaphylla*, page 377: *PalDat*

Picture: *Commelina erecta*, page 377: *PalDat*
Picture: *Stanfieldiella imperforata*, page 377: *PalDat*
Picture: *Callistemon comboynensis*, page 377: *PalDat*

Chapter: "Pollen Wall"

Picture: *Plantago major*, page 380: *PalDat*
Picture: *Salvia verticillata*, page 380: *PalDat*
Picture: *Galeopsis tetrahit*, page 380: *PalDat*
Picture: *Galeopsis tetrahit*, page 380: *PalDat*
Picture: *Salvia nemorosa*, page 380: *PalDat*
Picture: *Salvia nemorosa*, page 380: *PalDat*
Picture: *Corylus avellana*, page 381: *PalDat*
Picture: *Ambrosia artemisiifolia*, page 381: *PalDat*
Picture: *Plantago maritima*, page 381: *PalDat*
Picture: *Adonis aestivalis*, page 381: *PalDat*
Picture: *Quercus robur*, page 381: *PalDat*
Picture: *Apium nodiflorum*, page 381: *PalDat*
Picture: *Ailanthus altissima*, page 382: *PalDat*
Picture: *Ailanthus altissima*, page 382: *PalDat*
Picture: *Fraxinus excelsior*, page 382: *PalDat*
Picture: *Fraxinus excelsior*, page 382: *PalDat*
Picture: *Odontites luteus*, page 382: *PalDat*
Picture: *Odontites luteus*, page 382: *PalDat*
Picture: *Salix fragilis*, page 383: *PalDat*
Picture: *Salix fragilis*, page 383: *PalDat*
Picture: *Alangium* sp., page 383: *PalDat*
Picture: *Lomatogonium carinthiacum*, page 383: *PalDat*
Picture: *Quesnelia lateralis*, page 383: *PalDat*
Picture: *Pachysandra terminalis*, page 383: *PalDat*
Picture: *Rhaphidophora africana*, page 384: *PalDat*
Picture: *Orobanche hederae*, page 384: *PalDat*
Picture: *Sauromatum venosum*, page 384: *PalDat*
Picture: *Globba schomburgkii*, page 384: *PalDat*
Picture: *Sauromatum venosum*, page 384: Weber, Martina
Picture: *Spathicarpa* sp., page 384: Weber, Martina
Picture: *Abies* sp., page 385: Zetter, Reinhard
Picture: *Pinus* sp., page 385: Zetter, Reinhard
Picture: *Pinus* sp., page 385: Zetter, Reinhard
Picture: *Tsuga* sp., page 385: Zetter, Reinhard
Picture: *Gonatopus angustus*, page 385: *PalDat*
Picture: *Zamioculcas zamiifolia*, page 385: *PalDat*
Picture: *Solidago canadensis*, page 386: *PalDat*
Picture: *Cichorium intybus*, page 386: Weber, Martina
Picture: Asteraceae, page 386: Ulrich, Silvia
Picture: *Solidago canadensis*, page 386: *PalDat*
Picture: Asteraceae, page 386: Ulrich, Silvia
Picture: *Chaenorhinum minus*, page 386: *PalDat*
Picture: *Mentha aquatica*, page 387: *PalDat*
Picture: *Cobaea scandens*, page 387: Weber, Martina
Picture: *Mentha aquatica*, page 387: *PalDat*
Picture: Brassicaceae, page 387: Ulrich, Silvia
Picture: *Ligustrum vulgare*, page 387: Weber, Martina
Picture: *Melampyrum pratense*, page 387: *PalDat*
Picture: *Bassia scoparia*, page 388: *PalDat*
Picture: Asteraceae, page 388: Halbritter, Heidemarie
Picture: Asteraceae, page 388: Ulrich, Silvia
Picture: *Tanacetum corymbosum*, page 388: Weber, Martina
Picture: Asteraceae, page 388: Ulrich, Silvia
Picture: *Artemisia vulgaris*, page 388: Weber, Martina
Picture: *Amydrium medium*, page 389: *PalDat*
Picture: *Corylus colurna*, page 389: *PalDat*
Picture: *Viola tricolor*, page 389: *PalDat*
Picture: *Juglans regia*, page 389: *PalDat*
Picture: *Ulmus laevis*, page 389: *PalDat*
Picture: *Plantago lanceolata*, page 389: *PalDat*
Picture: *Dieffenbachia humilis*, page 390: *PalDat*
Picture: *Mahonia aquifolium*, page 390: *PalDat*
Picture: *Berberis vulgaris*, page 390: *PalDat*
Picture: *Iris pumila*, page 390: *PalDat*
Picture: *Argyranthemum* sp., page 391: Frosch-Radivo, Andrea
Picture: *Tanacetum corymbosum*, page 391: Weber, Martina
Picture: *Cyanus segetum*, page 391: *PalDat*
Picture: *Cyanus segetum*, page 392: *PalDat*
Picture: *Centaurea jacea*, page 392: Weber, Martina
Picture: *Agrimonia eupatoria*, page 392: *PalDat*
Picture: *Ageratum* sp., page 392: Weber, Martina
Picture: *Nigella arvensis*, page 392: *PalDat*
Picture: *Artemisia vulgaris*, page 392: Weber, Martina
Picture: *Chaenorhinum minus*, page 393: *PalDat*
Picture: *Plantago maritima*, page 393: *PalDat*
Picture: *Clinopodium vulgare*, page 393: *PalDat*
Picture: *Acinos alpinus*, page 393: *PalDat*
Picture: *Pachypodium succulentum*, page 393: *PalDat*
Picture: *Fraxinus excelsior*, page 393: *PalDat*
Picture: *Odontites luteus*, page 394: *PalDat*
Picture: *Ailanthus altissima*, page 394: *PalDat*
Picture: *Androsace maxima*, page 394: *PalDat*
Picture: *Adonis aestivalis*, page 394: *PalDat*
Picture: *Acer platanoides*, page 394: *PalDat*
Picture: *Chaenorhinum minus*, page 394: *PalDat*
Picture: *Glaucium flavum*, page 395: *PalDat*
Picture: *Odontites vulgaris*, page 395: *PalDat*
Picture: *Ranunculus trichophyllus*, page 395: *PalDat*
Picture: *Delphinium elatum*, page 395: *PalDat*

Picture: *Plantago maritima*, page 395: *PalDat*
Picture: *Stachys officinalis*, page 395: *PalDat*
Picture: *Pistia stratiotes*, page 396: *PalDat*
Picture: *Amorphophallus mossambicensis*, page 396: Ulrich, Silvia
Picture: *Mentha aquatica*, page 396: *PalDat*
Picture: *Spathiphyllum blandum*, page 396: *PalDat*
Picture: *Arophyton buchetii*, page 396: *PalDat*
Picture: *Homalomena wallisii*, page 396: *PalDat*
Picture: *Orobanche hederae*, page 397: *PalDat*
Picture: *Thalictrum flavum*, page 397: *PalDat*
Picture: *Ambrosia artemisiifolia*, page 397: *PalDat*
Picture: *Corylus avellana*, page 398: *PalDat*
Picture: *Corylus avellana*, page 398: *PalDat*
Picture: *Syringa vulgaris*, page 398: *PalDat*
Picture: *Syringa vulgaris*, page 398: *PalDat*
Picture: *Acinos alpinus*, page 398: *PalDat*
Picture: *Acinos alpinus*, page 398: *PalDat*
Picture: *Brassica napus*, page 399: *PalDat*
Picture: *Chenopodium album*, page 399: *PalDat*
Picture: *Trisetum flavescens*, page 399: *PalDat*
Picture: *Cereus* sp., page 399: *PalDat*
Picture: *Avena sativa*, page 399: *PalDat*
Picture: *Ornithogalum nutans*, page 399: *PalDat*
Picture: *Apium nodiflorum*, page 400: *PalDat*
Picture: *Apium nodiflorum*, page 400: *PalDat*
Picture: *Quercus robur*, page 400: *PalDat*
Picture: *Acinos alpinus*, page 401: *PalDat*
Picture: *Acinos alpinus*, page 401: *PalDat*
Picture: *Veronica anagallis-aquatica*, page 401: *PalDat*
Picture: *Anthericum liliago*, page 401: *PalDat*
Picture: *Atriplex tatarica*, page 401: *PalDat*
Picture: *Chenopodium album*, page 401: *PalDat*
Picture: *Aristolochia arborea*, page 402: *PalDat*
Picture: *Aristolochia clematitis*, page 402: *PalDat*
Picture: *Pachypodium succulentum*, page 402: *PalDat*
Picture: *Dieffenbachia humilis*, page 402: *PalDat*
Picture: *Apoballis acuminatissima*, page 402: *PalDat*
Picture: *Apoballis acuminatissima*, page 402: *PalDat*
Picture: *Smyrnium perfoliatum*, page 403: Weber, Martina
Picture: *Smyrnium perfoliatum*, page 403: Weber, Martina
Picture: *Smyrnium perfoliatum*, page 403: Weber, Martina
Picture: *Smyrnium perfoliatum*, page 403: Weber, Martina
Picture: *Nigella arvensis*, page 403: Weber, Martina
Picture: *Nigella arvensis*, page 403: Weber, Martina
Picture: *Ligustrum vulgare*, page 404: Weber, Martina
Picture: *Xanthium saccharatum*, page 404: Weber, Martina
Picture: *Centaurea* sp., page 404: Weber, Martina
Picture: *Pelargonium carnosum*, page 404: Weber, Martina
Picture: *Citrus aurantiifolia*, page 404: Weber, Martina
Picture: *Tilia* sp., page 404: Weber, Martina
Picture: *Chlorospatha kolbii*, page 405: *PalDat*
Picture: *Chlorospatha hannoniae*, page 405: *PalDat*
Picture: *Chlorospatha pubescens*, page 405: *PalDat*
Picture: *Chlorospatha oblongifolia*, page 405: *PalDat*
Picture: *Victoria regia*, page 405: *PalDat*
Picture: *Androlepis skinneri*, page 405: *PalDat*
Picture: *Annona muricata*, page 406: *PalDat*
Picture: *Asimina triloba*, page 406: Weber, Martina
Picture: *Drosera scorpioides*, page 406: Weber, Martina
Picture: *Drosera rotundifolia*, page 406: *PalDat*
Picture: *Calolisianthus pendulus*, page 406: *PalDat*
Picture: *Rhododendron hippophaeoides*, page 406: *PalDat*
Picture: *Alnus glutinosa*, page 407: *PalDat*
Picture: *Alnus* sp., page 407: Zetter, Reinhard
Picture: *Alnus glutinosa*, page 407: *PalDat*
Picture: *Alnus incana*, page 407: *PalDat*
Picture: *Alnus incana*, page 407: *PalDat*
Picture: *Aglaia elliptica*, page 407: *PalDat*
Picture: *Agapetes variegata*, page 408: *PalDat*
Picture: *Agapetes variegata*, page 408: *PalDat*
Picture: *Viola tricolor*, page 408: Zetter, Reinhard
Picture: *Viola tricolor*, page 408: *PalDat*
Picture: *Odontites vulgaris*, page 408: *PalDat*
Picture: *Odontites vulgaris*, page 408: *PalDat*
Picture: *Acokanthera oblongifolia*, page 409: *PalDat*
Picture: *Acokanthera oblongifolia*, page 409: *PalDat*
Picture: *Satyria warszewiczii*, page 409: *PalDat*
Picture: *Satyria warszewiczii*, page 409: *PalDat*
Picture: *Viola calcarata*, page 409: *PalDat*
Picture: *Viola calcarata*, page 409: *PalDat*
Picture: *Axinaea lehmannii*, page 410: *PalDat*
Picture: *Axinaea lehmannii*, page 410: *PalDat*
Picture: *Tulipa linifolia*, page 410: *PalDat*
Picture: *Tulipa linifolia*, page 410: *PalDat*
Picture: *Tulipa kaufmanniana*, page 410: *PalDat*
Picture: *Tulipa kaufmanniana*, page 410: *PalDat*
Picture: *Salvia nemorosa*, page 411: *PalDat*
Picture: *Ambrosia artemisiifolia*, page 411: Ulrich, Silvia
Picture: *Amorphophallus stuhlmannii*, page 411: *PalDat*
Picture: *Salvia glutinosa*, page 411: *PalDat*
Picture: *Helicodiceros muscivorus*, page 411: Ulrich, Silvia

Picture: *Galium* sp., page 411: Weber, Martina
Picture: *Jasminum nudiflorum*, page 412: *PalDat*
Picture: *Nigella arvensis*, page 412: *PalDat*
Picture: *Melampyrum nemorosum*, page 412: *PalDat*
Picture: *Betonica officinalis*, page 412: *PalDat*
Picture: *Anemone ranunculoides*, page 412: *PalDat*
Picture: *Consolida regalis*, page 412: *PalDat*
Picture: *Salvia glutinosa*, page 413: *PalDat*
Picture: *Linaria vulgaris*, page 413: *PalDat*
Picture: *Salix x fragilis*, page 413: *PalDat*
Picture: *Mentha aquatica*, page 413: *PalDat*
Picture: *Syringa vulgaris*, page 413: *PalDat*
Picture: *Scilla bifolia*, page 413: *PalDat*
Picture: *Apium nodiflorum*, page 414: *PalDat*
Picture: *Convolvulus tricolor*, page 414: *PalDat*
Picture: *Veronica spicata*, page 414: *PalDat*
Picture: *Sambucus nigra*, page 414: *PalDat*
Picture: *Atriplex tatarica*, page 414: *PalDat*
Picture: *Chenopodium album*, page 414: *PalDat*
Picture: *Sinapis alba*, page 415: *PalDat*
Picture: *Sinapis alba*, page 415: *PalDat*
Picture: *Brassica nigra*, page 415: *PalDat*
Picture: *Brassica nigra*, page 415: *PalDat*
Picture: *Brassica napus*, page 415: *PalDat*
Picture: *Alliaria petiolata*, page 415: *PalDat*
Picture: *Caladenia latifolia*, page 416: *Svojtka, Matthias*
Picture: *Epipactis muelleri*, page 416: *Svojtka, Matthias*
Picture: *Neottia nidus-avis*, page 416: *Svojtka, Matthias*
Picture: *Spiranthes spiralis*, page 416: *Svojtka, Matthias*
Picture: *Habenaria tridactylites*, page 416: *Svojtka, Matthias*
Picture: *Habenaria tridactylites*, page 416: *Svojtka, Matthias*
Picture: *Dactylorhiza maculata*, page 417: *Svojtka, Matthias*
Picture: *Orchis pallens*, page 417 Matthias Svojtka
Picture: *Neottianthe cucullata*, page 417: *Svojtka, Matthias*
Picture: *Ludisia discolor*, page 417 Matthias Svojtka
Picture: *Brassavola cucullata*, page 417: *Svojtka, Matthias*
Picture: *Cyrtopodium polyphyllum*, page 417: *Svojtka, Matthias*
Picture: *Epilobium angustifolium*, page 418: Weber, Martina
Picture: *Epilobium fleischeri*, page 418: *PalDat*
Picture: *Epilobium parviflorum*, page 418: *PalDat*
Picture: *Epilobium hirsutum*, page 418: Weber, Martina
Picture: *Epilobium hirsutum*, page 418: Weber, Martina
Picture: *Epilobium dodonaei*, page 418: *PalDat*
Picture: *Kalmia latifolia*, page 419: *PalDat*
Picture: *Ledum palustre*, page 419: *PalDat*
Picture: *Oenothera biennis*, page 419: *PalDat*
Picture: *Oenothera biennis*, page 419: *PalDat*
Picture: *Clarkia pulchella*, page 419: *PalDat*
Picture: *Clarkia unguiculata*, page 419: *PalDat*
Picture: *Circaea lutetiana*, page 420: Weber, Martina
Picture: *Circaea lutetiana*, page 420: *PalDat*
Picture: *Fuchsia magellanica*, page 420: *PalDat*
Picture: *Lopezia racemosa*, page 420: Weber, Martina
Picture: *Clarkia purpurea*, page 420: *PalDat*
Picture: *Rhododendron hirsutum*, page 420: *PalDat*
Picture: *Trisetum flavescens*, page 421: *PalDat*
Picture: *Corylus avellana*, page 421: *PalDat*
Picture: *Metasequoia glyptostroboides*, page 421: Weber, Martina
Picture: Cupressaceae, page 421: Grímsson, Friðgeir
Picture: Cupressaceae, page 421: Grímsson, Friðgeir
Picture: *Stellaria graminea*, page 421: *PalDat*
Picture: *Chamaecyparis lawsoniana*, page 422: *PalDat*
Picture: *Chamaecyparis lawsoniana*, page 422: *PalDat*
Picture: *Gladiolus illyricus*, page 422: *PalDat*
Picture: *Quercus robur*, page 422: PalDat
Picture: *Ruspolia seticalyx*, page 422: *PalDat*
Picture: *Ruspolia seticalyx*, page 422: *PalDat*
Picture: *Ephedra foeminea*, page 423: *PalDat*
Picture: *Odontites luteus*, page 423: *PalDat*
Picture: *Nigella arvensis*, page 423: *PalDat*
Picture: *Ficaria verna*, page 423: *PalDat*
Picture: *Tilia platyphyllos*, page 423: *PalDat*
Picture: *Sauromatum venosum*, page 423: *PalDat*
Picture: *Akebia quinata*, page 424: *PalDat*
Picture: *Atriplex sagittata*, page 424: *PalDat*
Picture: *Acacia binervia*, page 424: *PalDat*
Picture: *Justicia brandegeeana*, page 424: *PalDat*
Picture: *Calluna vulgaris*, page 424: *PalDat*
Picture: *Acacia binervia*, page 424: *PalDat*
Picture: *Cyperus longus*, page 425: *PalDat*
Picture: *Justicia brandegeeana*, page 425: *PalDat*
Picture: *Delonix regia*, page 425: *PalDat*
Picture: *Scaevola aemula*, page 425: *PalDat*
Picture: *Orobanche hederae*, page 425: *PalDat*
Picture: *Melampyrum nemorosum*, page 425: *PalDat*
Picture: *Actaea spicata*, page 426: *PalDat*
Picture: *Eranthis hyemalis*, page 426: *PalDat*
Picture: *Urtica dioica*, page 426: *PalDat*
Picture: *Ulmus laevis*, page 426: *PalDat*
Picture: *Quercus robur*, page 426: *PalDat*
Picture: *Plantago major*, page 426: *PalDat*

Chapter: "Pollen Class"

Picture: *Iris bucharica*, page 430: *PalDat*
Picture: *Ibicella lutea*, page 430: *PalDat*
Picture: *Bunias orientalis*, page 430: *PalDat*
Picture: *Corylopsis glabrescens*, page 430: *PalDat*
Picture: *Viola alba*, page 430: *PalDat*
Picture: *Orlaya grandiflora*, page 430: *PalDat*
Picture: *Zeylanidium subulatum*, page 431: *PalDat*
Picture: *Podostemum munnarense*, page 431: *PalDat*
Picture: *Hedychium gardnerianum*, page 431: *PalDat*
Picture: *Sauromatum venosum*, page 431: *PalDat*
Picture: *Prenanthes purpurea*, page 431: *PalDat*
Picture: *Gazania* sp., page 431: *PalDat*
Picture: *Ephedra distachya*, page 432: *PalDat*
Picture: *Welwitschia mirabilis*, page 432: *PalDat*
Picture: *Acacia dealbata*, page 432: *PalDat*
Picture: *Chelonanthus purpurascens*, page 432: *PalDat*
Picture: *Silene flos-cuculi*, page 432: *PalDat*
Picture: *Pachypodium saundersii*, page 432: *PalDat*
Picture: *Abies cephalonica*, page 433: *PalDat*
Picture: *Pinus mugo*, page 433: *PalDat*
Picture: *Berberis thunbergii*, page 433: *PalDat*
Picture: *Thunbergia alata*, page 433: *PalDat*
Picture: *Sandersonia vaurantiaca*, page 433: *PalDat*
Picture: *Gagea villosa*, page 433: *PalDat*
Picture: *Acca sellowiana*, page 434: *PalDat*
Picture: *Primula farinosa*, page 434: *PalDat*
Picture: *Moneses uniflora*, page 434: *PalDat*
Picture: *Neottia ovata*, page 434: *PalDat*
Picture: *Sesleria caerulea*, page 434: *PalDat*
Picture: *Typha laxmannii*, page 434: *PalDat*

Index

H. Halbritter et al., *Illustrated Pollen Terminology*, https://doi.org/10.1007/978-3-319-71365-6

Q

R

www.ingramcontent.com/pod-product-compliance
Lightning Source LLC
Chambersburg PA
CBHW081031060726
47593CB00019B/2411